T0185846

Connected and Autonomous Vehicles in Smart Cities

Connected and Autonomous Vehicles in Smart Cities

Edited by

Hussein T. Mouftah, Melike Erol-Kantarci, and Sameh Sorour

CRC Press is an imprint of the
Taylor & Francis Group, an **informa** business

First edition published 2020
by CRC Press
6000 Broken Sound Parkway NW, Suite 300, Boca Raton, FL 33487-2742

and by CRC Press
2 Park Square, Milton Park, Abingdon, Oxon, OX14 4RN

CRC Press is an imprint of Taylor & Francis Group, LLC

Library of Congress Cataloging-in-Publication Data
Names: Mouftah, Hussein T., editor. | Erol-Kantarci, Melike, editor. |
Sorour, Sameh, editor.
Title: Connected and autonomous vehicles in smart cities / edited by
Hussein T. Mouftah, Melike Erol-Kantarci, Sameh Sorour.
Description: First edition. | Boca Raton, FL : CRC Press/Taylor & Francis
Group, LLC, 2021. | Includes bibliographical references and index.
Identifiers: LCCN 2020037215 (print) | LCCN 2020037216 (ebook) |
ISBN 9780367350345 (hardback) | ISBN 9780429329401 (ebook)
Subjects: LCSH: Vehicle-infrastructure integration. | Automated vehicles. |
Smart cities.
Classification: LCC TE228.3 .C658 2021 (print) | LCC TE228.3 (ebook) |
DDC 388.3/124–dc23
LC record available at https://lccn.loc.gov/2020037215
LC ebook record available at https://lccn.loc.gov/2020037216

ISBN: 978-0-367-35034-5 (hbk)
ISBN: 978-0-429-32940-1 (ebk)

Typeset in Times
by codeMantra

Contents

Preface

Since their introduction in science fiction movies (starting from the 1982 Knight Rider TV series), many people have desired to own a connected, autonomous, and electric vehicle (CAEV) that can be called upon, self-drive, self-park, run on battery, and communicate with other CAEVs for safe and pleasant rides. Today, CAEV technologies are gradually maturing toward becoming the dominant reality of future transportation systems and smart cities infrastructure. CAEV testing by tech giants (e.g., Google and Uber) and vehicle manufacturers (e.g., BMW, Mercedes, Volvo, Audi, and Tesla) have already started in different countries. Some forecasts claim that ten million connected and self-driving vehicles will hit the road by 2025. Yet, many scientists and vehicle manufacturers are still skeptical about these forecasts, especially in regard to the expected level of connectivity and automation in the upcoming phase of CAEVs. In addition, many questions on the integration requirements of CAEVs in future smart cities and their ability to improve traffic efficiency, passenger safety, mobility options, and city's operations are still significantly under-investigated. Moreover, studies on CAEVs' charging technologies and smart (possibly two-way) interaction with the power grid are of extreme importance to their success. Furthermore, all CAEVs' operations and interactions with diverse systems must be protected using advanced security measures to ensure citizens' privacy and prevent dangerous situations that may result; such processes are compromised for malicious purposes. Thus, experts from the vehicular networking, automation, electrification, and security fields are all extensively collaborating to address these specific concerns and reach the highest attainable level of connectivity, autonomy, smart charging, and security in the 2025-promised CAEVs. In parallel, transportation and system engineers are implementing a wide range of studies on how such CAEVs can be of maximum benefit to the future of intelligent transportation and smart cities applications.

This book aspires to present a comprehensive coverage of five fundamental yet intertwined pillars paving the road toward the future of CAEV operation and integration in smart cities. It aims to be a complementary reference for smart city decision-makers, automotive manufacturers, utility operators, smart-mobility service providers, telecom operators, communication engineers, power engineers, vehicle-charging providers, university professors, researchers, and students who would like to learn more about the advances in CAEVs connectivity, autonomy, electrification, security, and integration into smart cities and intelligent transportation systems. The book thus accommodates 18 chapters authored by world-renowned experts, all presenting their views on CAEV technologies and applications to smart cities. These chapters are organized in five parts, each focusing on one of the five pillars of the CAEVs' ecosystem.

Part I: CAEVs and Systems Integration for Smart Cities presents the latest advancements on the integration of CAEVs' as well as their enabling systems and infrastructures into smart cities. It also highlights state-of-the-art models and techniques to enable smart mobility solutions using CAEVs in future smart cities. Part I consists of the following five chapters.

Chapter 1 "Connected and Autonomous Electric Vehicle Charging Infrastructure Integration to Microgrids in Future Smart Cities," authored by Mohammad Sadeghi, Melike Erol-Kantarci, and Hussein T. Mouftah, focuses on the integration of CAEVs' charging infrastructure in future smart cities through the concept of microgrids. The chapter first introduces microgrids then presents two case studies that provide extensive analysis on two integration scenarios of microgrids supporting CAEV charging in smart cities.

Chapter 2 "A Hierarchical Management Framework for Autonomous Electric Mobility-on-Demand Services," authored by Nuzhat Yamin, Syrine Belakaria, Sameh Sorour, and Mohamed Hefeida, suggests a novel two-tier management scheme for fleets of CAEVs operating as a mobility on-demand service. The chapter first models and optimizes local dispatching and charging decisions for CAEVs in each neighborhood of a city. It then determines the needed CAEV resources in each of these neighborhoods and derives an optimal vehicle relocation policy from neighborhoods with excess vehicles to those experiencing shortages.

Chapter 3 "Multifaceted Synthesis of Autonomous Vehicles Emerging Landscape," authored by Hossam Abdelgawad and Kareem Othman, sheds light on the challenges and benefits of CAEVs in addressing multiple verticals in smart cities and illustrates their implication on travel behavior, development of cities, legal liabilities, mobility businesses, among others. It also discusses users' perceptions on deploying CAEVs on city roads and the needed infrastructure requirement to facilitate that deployment.

Chapter 4 "Machine Learning Methodologies for Electric Vehicle Energy Management Strategies: A Comprehensive Survey," authored by John Vardakas, Ioannis Zenginis, and Christos Verikoukis, surveys the latest machine learning-based management approaches for CAEV charging and discharging and identifies their advantages, commonalities, and key alterations. The survey then summarizes the different challenges and implementation limitations of these approaches and sheds light on the importance of addressing them to promote more efficient and cost-effective interactions between CAEVs and the power grid.

Chapter 5 "Dynamic Road Management in the Era of CAV," authored by Mohamed Younis, Sookyoung Lee, Wassila Lalouani, Dayuan Tan, and Sanket Gupte, discusses the challenges of dynamic road and traffic management in current cities and how CAEVs will revolutionize such management. It also highlights the implications of the transition phase from human-driven to CAEV-only vehicles on dynamic road management. The chapter finally presents a novel system for traffic management based on both intervehicle/vehicle-to-sensor communications and active assessments of road conditions.

Part II: Networking for Connected Vehicles consists of five chapters that highlight the state-of-the-art innovations in mobile networking and cloud/edge solutions for both ground and aerial CAEVs.

Chapter 6 "VANET Communication and Mobility Sustainability: Interactions and Mutual Impacts in Vehicular Environment," authored by Ahmed Elbery and Hesham Rakha, investigates the mutual impact between vehicular networking and mobility in highly dynamic environments, exemplified by an eco-routing application in downtown Los Angles as a case study. It also studies and quantifies the effect of intervehicle communication delay and packet drops on transportation impediments (i.e., congestions and gridlocks) and their carbon footprints (fuel consumption and CO_2 emissions).

Chapter 7 "Message Dissemination in Connected Vehicles," authored by Anirudh Paranjothi, Mohammed Atiquzzaman, and Mohammad S. Khan, provides an overview of vehicular cloud and fog computing-based techniques and highlights the significance in improving the message dissemination between CAEVs, especially in highly density areas. A dynamic fog computing-based message dissemination scheme is also proposed in this chapter and is proven to achieve faster and more guaranteed message delivery using a realistic large-scale simulation model built in SUMO.

Chapter 8 "Exploring Cloud Virtualization over Vehicular Networks with Mobility Support," authored by Miguel Luís, Christian Gomes, Susana Sargento, Jordi Ortiz, José Santa, Pedro J. Fernández, Manuel Gil Pérez, Gregorio Martínez Pérez, Sokratis Barmpounakis, Nancy Alonistioti, Jacek Cieślak, and Henryk Gierszal, inspects the applicability of several virtualized functions to vehicular networks with multihoming and mobility support. These virtualized functions are examined in different use cases of vehicular networks and intelligent transportation applications, and are evaluated in a realistic vehicular networking environment with cloud-enabled mobility and multihoming functionalities.

Chapter 9 "Data Offloading Approaches for Vehicle to Everything (V2X) Communications in 5G and Beyond," authored by Muhammed N. Avcil and Mujdat Soyturk, starts by describing the challenges introduced by the vehicular telecommunication traffic resulting from smart cities and intelligent transportation services. It then details the novel technological aspects in 5G vehicle-to-everything communications and studies the expected impacts of vehicular workload offloading approaches on resolving these challenges and further empowering smart cities applications.

Chapter 10 "Connected Unmanned Aerial Vehicles for Flexible Coverage, Data Gathering and Emergency Scenarios," authored by Giacomo Segala, Riccardo Bassoli, and Fabrizio Granelli, analyzes the coverage and performance of networks of aerial CAEVs that are serving different smart city applications, including data gathering and emergency scenarios. It starts by providing a thorough overview of the current aerial CAEV network architectures, then provides a detailed performance evaluation of these different architectures using extensive simulations in realistic application environments.

Part III: Localization and Navigation for Autonomous Vehicles illustrates some of the most recent works on the use of multiple sensors, advanced signal processing, and vehicular communication signaling to localize CAEVs with very high accuracy, thus enabling their automated navigation in dense and high-speed environments. Part III consists of the following three chapters.

Chapter 11 "Localization for Vehicular Ad Hoc Network and Autonomous, Vehicles, Are We Done Yet?," authored by Abdellah Chehri and Hussein T. Mouftah, surveys localization techniques for CAEVs and suggests novel collaborative approaches to improve their accuracy. These approaches are based on data fusion and machine learning techniques that both respect CAEV restrictions and requirements, and combine more than one localization methods to produce accurate position estimates.

Chapter 12 "Automotive Radar Signal Analysis," authored by Hassan Moradi and Ashish Basireddy, presents an in-depth explanation of automotive radar, signal processing, and surrounding perception algorithms, which are all crucial to implement safe and efficient autonomous driving systems of CAEVs. It also details the existing processing techniques for radar signals and borrows new methods from the wireless industry to enhance their performance. It finally quantitatively compares the performance of a conventional radar solution with several proposed innovative methods in the very recent literature.

Chapter 13 "Multi-Sensor Precise Positioning for Autonomous and Connected Vehicles," authored by Mohamed Elsheikh and Aboelmagd Noureldin, proposes the use of multiple modalities to achieve high-precision positioning for CAEVs, thus guaranteeing their safe and efficient navigation, especially in dense and high-speed environments. In addition to the well-known global navigation satellite systems, these modalities include cameras, light detection and ranging (LiDAR) systems, vehicular communication signals, and inertial sensors (e.g., gyroscopes, accelerometers, speedometers).

Part IV: Wireless Charging for CAEVs consists of three chapters that highlight some of the cutting-edge technologies enabling wireless charging for both ground and aerial CAEVs, thus revolutionizing their battery-run operation and attractiveness to road-trippers.

Chapter 14 "Deploying Wireless Charging Systems for Connected and Autonomous Electric Vehicles," authored by Binod Vaidya and Hussein T. Mouftah, sheds light on the needed infrastructure and standardization efforts to enable wireless charging for CAEVs. It also depicts an implementation of a wireless charging system providing effective communications and strategized automated reservation for CAEVs to the supply infrastructure.

Chapter 15 "Dynamic Wireless Charging of Electric Vehicles," authored by Sadegh Vaez-Zadeh, Amir Babaki, and Ali Zakerian, first introduces the fundamentals of wireless power transfer, then illustrates the analysis, design considerations, and optimality requirements for dynamic wireless power transfer to CAEVs. It finally discusses the control system requirements to achieve the maximum efficiency in such power transfer procedures.

Chapter 16 "Wirelessly Powered Unmanned Aerial Vehicles (UAVs) in Smart City," authored by Malek Souilem, Wael Dghais, and Ayman Radwan, extends the study in Chapter 15 to aerial CAEVs by highlighting the latest innovations in the area of in-flight wireless power transfer. Approaches for simultaneous wireless power transfer and data communication to multiple aerial CAEVs, through fixed power lines and mobile power infrastructure, are also discussed. The chapter finally provides a brief survey on the different applications of such wirelessly charged aerial CAEVs in future smart cities.

Part V: Network Security for CAEVs finally sheds light on novel cybersecurity approaches and secure firmware update tools for CAEVs and their supporting services. Part V consists of the following two chapters.

Chapter 17 "Cyber Security Considerations for Automated Electro-Mobility Services in Smart Cities," authored Binod Vaidya and Hussein T. Mouftah, starts by introducing the innovative concepts of automated electromobility and peer-to-peer CAEV sharing services and detailing their challenges and security requirements. It then proposes and validates novel entity/prior-binding authentication and authorization mechanisms using conjugated and token-based approaches.

Chapter 18 "Incentivized and Secure Blockchain-based Firmware Update and Dissemination for Autonomous Vehicles," authored by Mohamed Baza, Joe Baxter, Noureddine Lasla, Mohamed Mahmoud, Mohamed Abdallah and Mohamed Younis, proposes blockchain-based technologies to ensure the authenticity and integrity of distributed firmware updates for the CAEV subsystems. It also establishes both a reputation-based reward system to incentivize CAEVs to distribute firmware updates and an attribute-based encryption scheme to guarantee exclusive download of new updates by authorized CAEVs.

This book contains 18 chapters grouped into five parts to make reading easy and pleasant.Each chapter is authored by widely recognized scholars in their respective fields. This book aims to be a handbook for decision-makers, engineers, researchers, academics, and practitioners, who desire to take active part in CAEV and their integration and applications in smart cities.

Hussein T. Mouftah
Melike Erol-Kantarci
Sameh Sorour

Editors

Hussein T. Mouftah has joined the School of Electrical Engineering and Computer Science (was School of Information Technology and Engineering) of the University of Ottawa in 2002 as a Tier 1 Canada Research chair professor, where he became a *University Distinguished Professor* in 2006. He has been with the ECE Department at Queen's University (1979–2002), where he was prior to his departure a full professor and the Department associate head. He has six years of industrial experience mainly at Bell Northern Research of Ottawa (Nortel Networks). He has served as an editor-in-chief of the IEEE Communications Magazine (1995–1997) and IEEE ComSoc Director of Magazines (1998–1999), Chair of the Awards Committee (2002–2003), Director of Education (2006–2007), and Member of the Board of Governors (1997–1999 and 2006–2007). He has been a Distinguished Speaker of the IEEE Communications Society (2000–2007). He is the author or coauthor of 13 books, 75 book chapters, and more than 1800 technical papers, 16 patents, 6 invention disclosures, and 150 industrial reports. He is the joint holder of 24 Best/Outstanding Paper Awards. He has received numerous prestigious awards, such as the 2017 C.C. Gotlieb Medal in Computer Engineering and Science and the 2016 R.A. Fessenden Medal in Telecommunications Engineering of IEEE Canada, the 2016 Distinguished Technical Achievement Award in Communications Switching and Routing of IEEE Communications Society Communications Switching and Routing Technical Committee, the 2015 IEEE Ottawa Section Outstanding Educator Award, the 2014 Engineering Institute of Canada K. Y. Lo Medal, the 2014 Technical Achievement Award of the IEEE Communications Society Technical Committee on Wireless Ad Hoc and Sensor Networks, the 2007 Royal Society of Canada Thomas W. Eadie Medal, the 2007–2008 University of Ottawa Award for Excellence in Research, the 2008 ORION Leadership Award of Merit, the 2006 IEEE Canada McNaughton Gold Medal, the 2006 EIC Julian Smith Medal, the 2004 IEEE ComSoc Edwin Howard Armstrong Achievement Award, the 2004 George S. Glinski Award for Excellence in Research of the University of Ottawa Faculty of Engineering, the 1989 Engineering Medal for Research and Development of the Association of Professional Engineers of Ontario, and the Ontario Distinguished Researcher Award of the Ontario Innovation Trust. Dr. Mouftah is a fellow of the IEEE (1990), the Canadian Academy of Engineering (2003), the Engineering Institute of Canada (2005), and the Royal Society of Canada RSC Academy of Science (2008).

Melike Erol-Kantarci is a Tier 2 Canada research chair in AI-enabled Next-Generation Wireless Networks and associate professor at the School of Electrical Engineering and Computer Science at the University of Ottawa. She is the founding director of the Networked Systems and Communications Research (NETCORE) laboratory. She is also a courtesy assistant professor at the Department of Electrical and Computer Engineering at Clarkson University, Potsdam, NY, where she was a tenure-track assistant professor prior to joining University of Ottawa. She has over 130 peer-reviewed publications which have been cited many times. She was selected for the 2019 list of "N2Women: Stars in Computer Networking and Communications" along with eight other distinguished scientists. She has received the IEEE Communication Society Best Tutorial Paper Award and the Best Editor Award of the IEEE Multimedia Communications Technical Committee in 2017. She is the coeditor of two books: "Smart Grid: Networking, Data Management, and Business Models" and "Transportation and Power Grid in Smart Cities: Communication Networks and Services" published by CRC Press and Wiley, respectively. She has delivered 7 tutorials and more than 30 invited talks around the globe. She is an editor of the IEEE Transactions on Cognitive Communications and Networking, IEEE Internet of Things Journal, IEEE Communications Letters, IEEE Vehicular Technology Magazine, and IEEE Access. She has acted as the general chair or technical program chair for many international conferences and workshops. She is a senior member of the

IEEE and the past vice-chair for Women in Engineering (WIE) at the IEEE Ottawa Section. She is the Chair for "SIG on Green Smart Grid Communications". She is a steering committee member for the IEEE Sustainable ICT Initiative. Her main research interests are AI-enabled wireless networks, 5G and 6G wireless communications, smart grid, cyber-physical systems, electric vehicles, internet of things, and wireless sensor networks.

Sameh Sorour is an assistant professor at the School of Computing, Queen's University, Canada. He is the founder and director of the Queen's Connected and Autonomous Systems and Technologies (Queen's CASTLE) laboratory. He and his research group have published over 100 peer-reviewed articles in the most prestigious journals and conferences in their fields of research. Several of these publications have been cited more than 100 times. His research achievements were the subject of 1 keynote talk in IEEE LCN workshop, 3 tutorials, and more than 15 invited talks around the globe. Dr. Sorour is a senior member of IEEE and serves as an associate editor for IEEE Communications Letters and Frontiers Journal in Communications and Networks. He is also an area editor in the IEEE Canadian Journal of Electrical and Computer Engineering. In addition, he has acted as a symposium chair, track chair, track co-chair, and technical program chair for several IEEE and IEEE-sponsored international conferences. His research and educational interests lie in the broad areas of advanced networking, computing, learning, and intelligence technologies for connected and autonomous systems. The topics of particular interest include wireless networking, edge computing, edge learning, edge intelligence, AI-enabled 5G/6G networks, internet of things, connected and autonomous vehicles, cyber-physical systems, and intelligent transportation systems.

Contributors

Mohamed Abdallah
College of Science and Engineering
Hamad bin Khalifa University
Doha, Qatar

Hossam Abdelgawad
Faculty of Engineering
Cairo University
Giza, Egypt

Nancy Alonistioti
Department of Informatics and Telecommunications
National and Kapodistrian University of Athens
Athens, Greece

Mohammed Atiquzzaman
School of Computer Science
University of Oklahoma
Norman, Oklahoma

Muhammed Nur Avcil
Department of Computer Engineering
Marmara University
Istanbul, Turkey

Amir Babaki
School of Electrical and Computer Engineering
University of Tehran
Tehran, Iran

Sokratis Barmpounakis
Department of Informatics and Telecommunications
National and Kapodistrian University of Athens
Athens, Greece

Ashish Basireddy
Qualcomm Technologies Inc.
San Diego, California

Riccardo Bassoli
Deutsche Telekom Chair of Communication Networks
TU Dresden
Dresden, Germany

Joe Baxter
School of Engineering and Applied Sciences
Western Kentucky University
Bowling Green, Kentucky

Mohamed Baza
Department of Electrical and Computer Engineering
Tennessee Tech University
Cookeville, Tennessee

Syrine Belakaria
Washington State University
Pullman, Washington

Abdellah Chehri
Department of Applied Sciences
University of Quebec in Chicoutimi (UQAC)
Saguenay, Quebec, Canada

Jacek Cieślak
ITTI
Poznań, Poland

Wael Dghais
Instituto de Telecomunicações
Campus Universitário de Santiago
Aveiro, Portugal
and
Laboratory of Elec. and Microelec.
Université de Monastir
Monastir, Tunisia
and
Institut Supérieur des Sciences Appliquées et de Technologie de Sousse
Sousse, Tunisia

Ahmed Elbery
Virginia Tech
Blacksburg, Virginia

Mohamed Elsheikh
Geomatics Engineering
University of Calgary
Calgary, Alberta, Canada

Pedro J. Fernández
Department of Information and Communications Engineering
University of Murcia
Murcia, Spain

Frank H. P. Fitzek
Deutsche Telekom Chair of Communication Networks
TU Dresden
Dresden, Germany

Henryk Gierszal
Adam Mickiewicz University
Poznań, Poland

Christian Gomes
Instituto de Telecomunicações
Department of Electronics, Telecommunications and Informatics
University of Aveiro
Aveiro, Portugal

Fabrizio Granelli
Department of Information Engineering and Computer Science
University of Trento
Trento, Italy

Sanket Gupte
Department of Computer Science and Electrical Engineering
University of Maryland
Baltimore, Maryland

Mohamed Hefeida
University of Idaho
Moscow, Idaho

Mohammad S. Khan
Department of Computing
East Tennessee State University
Johnson City, Tennessee

Wassila Lalouani
Department of Computer Science and Electrical Engineering
University of Maryland
Baltimore, Maryland

Noureddine Lasla
College of Science and Engineering
Hamad bin Khalifa University
Doha, Qatar

Sookyoung Lee
Department of Computer Science and Engineering
EWHA Womans University
Seoul, Korea

Miguel Luís
Instituto de Telecomunicações and Instituto Superior de Engenharia de Lisboa
Lisboa, Portugal

Mohamed Mahmoud
Department of Electrical and Computer Engineering
Tennessee Tech University
Cookeville, Tennessee

Hassan Moradi
Qualcomm Technologies Inc.
San Diego, California

Aboelmagd Noureldin
Electrical and Computer Engineering
Royal Military College of Canada
Queen's University
Kingston, Ontario, Canada

Jordi Ortiz
Department of Information and Communications Engineering
University of Murcia
Murcia, Spain

Kareem Othman
Civil Engineering Department
University of Toronto
Toronto, Ontario, Canada

Anirudh Paranjothi
School of Computer Science
University of Oklahoma
Norman, Oklahoma

Manuel Gil Pérez
Department of Information and Communications Engineering
University of Murcia
Murcia, Spain

Gregorio Martínez Pérez
Department of Information and Communications Engineering
University of Murcia
Murcia, Spain

Ayman Radwan
Instituto de Telecomunicações
Campus Universitário de Santiago
Aveiro, Portugal

Hesham A. Rakha
Virginia Tech
Blacksburg, Virginia

Mohammad Sadeghi
School of Electrical Engineering and Computer Science
University of Ottawa
Ottawa, Ontario, Canada

José Santa
University of Murcia
Murcia, Spain
and
Technical University of Cartagena
Murcia, Spain

Susana Sargento
Instituto de Telecomunicações
Department of Electronics, Telecommunications and Informatics
University of Aveiro
Aveiro, Portugal

Giacomo Segala
Department of Information Engineering and Computer Science
University of Trento
Trento, Italy

Malek Souilem
Instituto de Telecomunicações
Campus Universitário de Santiago
Aveiro, Portugal
and
École Nationale d'Ingénieurs de Sousse
Université de Sousse
Sousse, Tunisia
and
Laboratory of Elec. and Microelec.
Université de Monastir
Monastir, Tunisia

Mujdat Soyturk
Department of Computer Engineering
Marmara University
Istanbul, Turkey

Dayuan Tan
Department of Computer Science and Electrical Engineering
University of Maryland
Baltimore, Maryland

Sadegh Vaez-Zadeh
School of Electrical and Computer Engineering
University of Tehran
Tehran, Iran

Binod Vaidya
School of Electrical Engineering and Computer Science
University of Ottawa
Ottawa, Ontario, Canada

John S. Vardakas
Iquadrat Informatica
Barcelona, Spain

Christos Verikoukis
Telecommunications Technological Centre of Catalonia (CTTC/CERCA)
Barcelona, Spain

Nuzhat Yamin
University of Idaho
Moscow, Idaho

Mohamed Younis
Department of Computer Science and Electrical Engineering
University of Maryland
College Park, Maryland

Ali Zakerian
School of Electrical and Computer Engineering
University of Tehran
Tehran, Iran

Ioannis Zenginis
Iquadrat Informatica
Barcelona, Spain
Baltimore, Maryland

1 Connected and Autonomous Electric Vehicle Charging Infrastructure Integration to Microgrids in Future Smart Cities

Mohammad Sadeghi, Melike Erol-Kantarci, and Hussein T. Mouftah
University of Ottawa

CONTENTS

1.1 INTRODUCTION

In future smart cities, connected and autonomous electric vehicles (CAEVs) are anticipated to be widely utilized, and consequently, their integration to the power grid will be crucial. The charging demand of CAEVs will impose significant load on the power grid. If CAEVs are not charged in a controlled manner, they may cause failures. The growth in the load can accumulate in peak hours and even result in failures during critical grid conditions. Current distribution transformers are not prepared to tolerate this increase in the load profile. Therefore, CAEVs' charging infrastructure requires innovative techniques to protect the smart grid. The electrical grid infrastructure currently undergoes a vital revolution. The simple centralized-unidirectional system of electric power transmission, distribution, and demand-driven control systems of yesterday are gradually evolving into a massive heterogeneous mix of utility grid and microgrids. Microgrids are power system components that are defined as small-scale electricity distribution systems with loads, storage, generation capacity, and islanding capability. Despite the advantages of microgrids, the integration of CAEVs through this new distributed microgrid system results in further complications in the grid. This chapter will focus on charging infrastructure of CAEVs and their integration to future smart cities through microgrids. The chapter first gives an overview of smart cities, microgrids, renewable energy generation resources, and CAEVs. After that, the effect of integrating CAEVs on the operation and management of the microgrids as well as impacts of shifting to further distributed control systems are investigated. Then, the chapter focuses on a study from the literature which provides analysis of such integration scenarios. Finally, future directions and challenges are presented, and the chapter is concluded.

1.1.1 SMART CITIES

Current cities are going under drastic evolution due to the high integration of technology in the daily life of residents. Therefore, establishing a framework to unify all aspects of daily life in a city as well as digitalizing services and embedding intelligence to their functions are crucial. Smart cities are urban areas that widely employ information and communication technologies (ICT), such as different types of sensors collect data in order to manage resources and improve the quality of life in the city [17,29,39,40]. Major economic and environmental changes are among the motivations for the advancement of smart cities [12]. In particular, global warming and its environmental impacts are among the main motives that have been considered in the design and development of smart cities.

In parallel to the developments in smart cities, the power grid, as an important infrastructure supporting the cities, has been undergoing major changes since mid-2000s. The main driver of the change behind the power grid has been the desire to make power generation less dependent on fossil fuels. This required integration of more renewables that are intermittent and are hence called for innovations in storage technologies. On the other hand, less consumption or demand response became another important area, which contributed to lowered peak hour electricity consumption. All of these changes and many others were possible with the integration of ICT.

Besides the integration of ICT, an emerging concept in smart grids has been microgrids. Although military microgrids existed since several decades [32], their coexistence within commercial distribution system became a feasible idea recently. Microgrids play an important role to bring the energy resources (mostly renewable energy) to where it is needed. Implementing microgrid structure in the smart cities may result in efficient generation of renewable energy, less power loss in the grid, and optimized load regulation which all result in less consumption of fossil fuels and reduced gas emissions [13,23]. In the next section, we provide a detailed overview of the state-of-the-art in microgrids.

1.1.2 MICROGRIDS

There are various definitions of a microgrid in the literature [28,33]. One of the most cited definition for the microgrid is provided by the US Department of Energy as *a group of interconnected loads and distributed energy resources within clearly defined electrical boundaries that acts as a single controllable entity with respect to the grid. A microgrid can connect and disconnect from the grid to enable it to operate in both grid-connected or island mode* [41]. According to the provided definition, a microgrid needs to be (1) distinguishable from the rest of the grid system as an independent unit. (2) It should include resources of energy that helps to not rely on distant resources and sustain as a single unit. (3) It should be able to run effectively in the case of losing connection to the main grid. The general block diagram of a microgrid system is demonstrated in Figure 1.1.

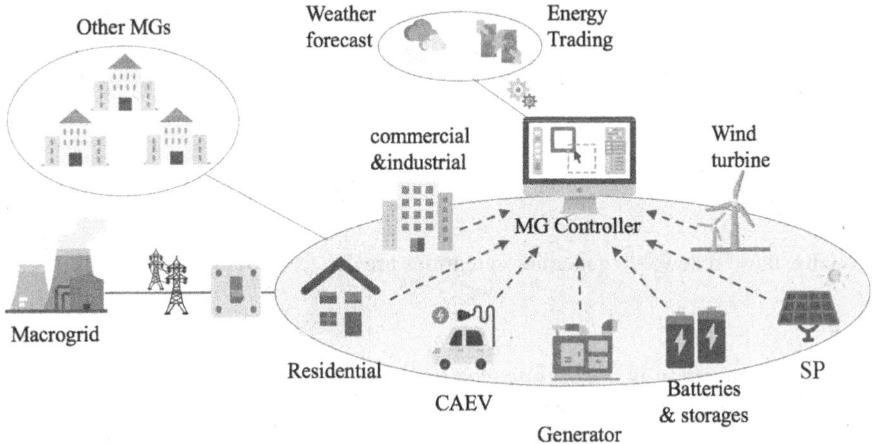

Figure 1.1 Microgrid block diagram.

Several microgrid implementations already exist around the world [27]. Santa Rita Jail is a real implementation example of campus/institutional microgrid [9]. In Ref. [10], the implementation of a military-based microgrid is presented.

Although it is possible to construct a microgrid without renewables, the true potential of opportunities for future microgrids arises from the integration of renewable energy resources. Therefore, renewable energy resources are important components of microgrids. In the next section, we overview the widely used renewable energy generators.

1.1.3 RENEWABLE ENERGY GENERATION RESOURCES

There are no strict criteria in the design of microgrids, and the process of design depends on the specific requirement of the project and economical concerns. There is a broad selection of energy generation resources and storage that can be implemented in the design of microgrids. Diesel engines, microturbines, fuel cells such as solid oxide and alkaline, and renewable generation resources are among generator candidates while sodium-sulfur and lithium-ion batteries, flow batteries such as zinc-bromine and polysulfide bromide batteries, hydrogen from hydrolysis, and kinetic energy storage can be suitable storage choices.

Among generation options, renewable resources gain remarkable attention recently due to the following reasons:

1. low carbon emission
2. less dependence on fossil fuels
3. longer lifespan comparing to conventional resources
4. noise-free in the case of solar panels (SP)
5. low operational cost

Although these sources can be employed effectively in the microgrid, the deployment of these resources has some disadvantages that are summarized in the following:

1. At this moment, the installation cost of renewable generators is relatively higher than conventional resources. As an example, the installation cost of an SP can be up to ten times higher than a diesel generator.
2. The implementation of renewable resources is geographically limited.
3. Renewable resources are lower in terms of energy efficiency comparing to diesel generators.

Despite the aforementioned cons, renewable resources are in the stage of development, and research shows a promising future in the advancement of these resources. Different types of renewable resources can be deployed in a microgrid which can be categorized as the following major types:

→ SP
→ Wind turbines (WT)
→ Mini-hydro

In the following, these renewable generator categories are discussed with more details.

1.1.3.1 SP

The idea of an SP generation is to generate electrical energy from free and limitless solar energy. The efficiency of SPs is impacted by not only environmental parameters such as geographical location, the intensity of solar, temperature, and cloud obstruction but also system parameters such as performance of the employed SP modules, efficiency of converters, inverter and adopted control scheme.

Variation in irradiance and cloud obstruction patterns can crucially impose voltage disturbances which can result in the disconnection of the inverters from the grid and consequently loss of energy. Today's SPs may demonstrate low efficiency in the long term due to cloud coverage and fluctuation in the intensity of solar irradiance. Studies show that SP systems perform five to ten times less efficient comparing to diesel generators [42]. In contrast, SPs have a very long lifetime which sometimes expands to about 20–25 years and the efficiency only drops to 80% after that time.

Numerous microgrid testbeds have employed SPs [2,4,7,20,32]. The Nice Grid project, in Nice, France, is an example of a successful smart SP-based microgrid [7]. Solar power is highly available where the microgrid is located. The microgrid features 2.5 MW of SP generation capacity with an energy storage capacity of 1.5 MW while the interruptible demand is 3.5 MW. One goal of this project is to optimize the distribution grid management, considering the high deployment of SPs. Full functionality in islanded mode which makes microgrid only rely on the SPs is an another objective of this project. Different finance models are also studied to explore the chance of involving the microgrid in the energy market. The project is cofunded by Grid4EU and the French government. In Ref. [7], a microgrid project in Bronsbergen, Netherlands, known as Continuon is discussed. The project includes 109 cottages with SP-covered roofs, a battery energy storage, and a control system that is developed to manage the islanding mode. The block diagram of this microgrid is depicted in Figure 1.2.

1.1.3.2 Wind Turbines

A wind turbine is used to generate electrical energy by converting wind energy into electricity. The wind turbine structure consists of two basic parts: electrical and mechanical parts. The motion energy of wind is captured in the form of rotational energy in the mechanical part, and then in the electrical part, this rotational energy is converted to electrical energy. The wind turbine includes three main components: the tower, the rotor, and the nacelle. The nacelle is equipped with an electrical generator and mechanical power transmission parts. The rotor normally includes more than two blades that extract motion energy from wind. Then, a gearbox is used to transfer the captured rotational energy to the shaft of an electrical generator to generate electric power. Wind turbines can be categorized into two categories: vertical and horizontal axis types. Vertical-type wind turbines are more common for small units and kW ranges up to 100 kW, while larger units are mostly horizontal wind turbines that support the order of MW.

The implementation of wind turbines increases exponentially around the world. More than two-thirds of the installed small wind turbines are only in the United States and China. The main benefit of employing a wind turbine is operating with CO_2 free emission. However, the implementation

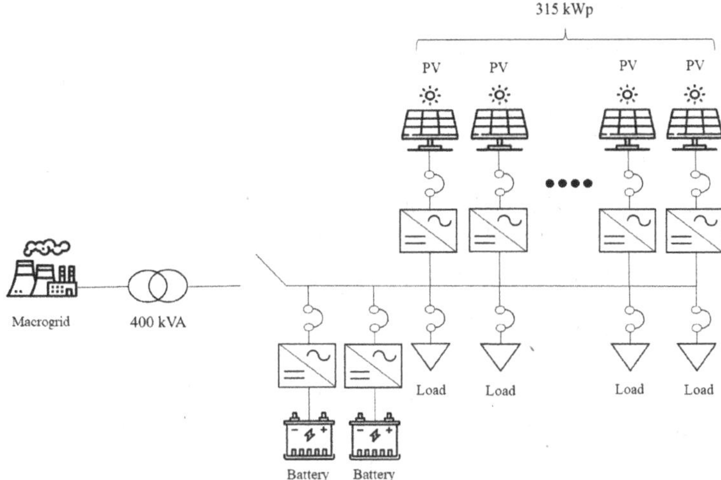

Figure 1.2 Continuon project structure.

cost is still a bottleneck. Inconsistency of wind speed, inability to predict the wind speed accurately, occupying very large areas, esthetics concerns, and bird strikes are among other drawbacks.

Several microgrids around the world adopted wind turbines as a source of energy generation in combination with other sources [1,3,6,32]. The microgrid project Atenea has started in 2013, and its main objective is to support the demanded energy for lighting the facility and testbed types of equipment. This microgrid features 20 kW of wind turbine generation working parallel with 25 kW of SP generation and a 55-kVA diesel generator.

1.1.3.3 Mini-hydro

Hydropower generation technology takes advantage of the kinetic energy of water and converts it to electrical energy. Hydropower can be categorized as storage, run-of-river, pumped storage, and ocean hydropower. Mini-hydro is employed in the microgrid projects equipped with a generator with the capacity up to 10 MW. The potential of mini-hydro is estimated to be 173 GW globally, in which 75 GW of that has been achieved in 2012. China, Brazil, and the United States are the top countries that deploy hydropower. CO_2-free emission is the main advantage of employing this future-generation technology. However, the installation of dams and water reservoirs occupies an area in the order of thousands of kilometers and imposes negative effects on the surrounding environment. Therefore, a comprehensive study on geographical, environmental, and hydrological characteristics is crucial before any construction and installation.

Several microgrid projects included hydropower generation as the renewable resource in their system [5,32,35,38]. The second-largest island in the state of Alaska and the USA, Kodiak island is a community microgrid that utilizes the hydropower since 1980s [5]. Several wind turbine farms and a hydro-turbine are installed on the island. Currently, the Kodiak island microgrid includes 500 kW of hydroelectric, 9 MW of wind generation, and 35 MW of gas/diesel. Additionally, the island is equipped with 2.5 MW/2 MWh of battery storage which helps to guarantee the stability of the microgrid system.

1.1.4 ENERGY TRADING AMONG MICROGRIDS

One interesting aspect of the microgrid technology is the possibility of energy trading among distribution system entities. Energy trading, or in other words, using surplus energy from one microgrid to supply loads at another microgrid, has been a challenging research question for the past few years.

Generally, the problem of energy trading among microgrids can be described as a set of connected microgrids that can exchange energy in a specific region. Each microgrid is equipped with energy generation and storing units (See Figure 1.1). The microgrids are also connected to the macrogrid, and the energy trading can be between one microgrid and the macrogrid or between two or more microgrids. In the literature, it is assumed that for a specific time interval, some microgrids have surplus energy and prefer to sell energy, whereas others suffer from the lack of energy and wish to buy energy. This condition can perfectly be modeled as a game. Therefore, there has been several studies in the literature to investigate the dynamics of energy trading.

In Ref. [24], a game-theoretic approach has been proposed for the distributed energy trading between microgrids. In this study, a set of interconnected microgrids aim to exchange energy with each other and also with the macrogrid. In this market, those microgrids with surplus energy can choose to sell part of their energy and store the rest for the future. Likewise, those microgrids that suffer from shortage of energy or wish to store energy for the future can buy energy. In Ref. [24], two-level continuous-kernel Stackelberg game is employed, in which seller and buyer microgrids are classified as leaders and follower players, respectively. To find the Stackelberg equilibrium of the formulated problem, a backward induction technique is used where the best response of buyers is found in the follower-level game first, and then, the results are plugged into the utility function of each seller to solve the problem at the leader-level game. The numerical evaluation demonstrated that seller can achieve higher utilities as the sellers act as the leaders and have the advantage of choosing their strategies first. The results show that the Nash equilibrium of the Stackelberg game is lower-bounded by half of the optimal cooperative centralized solution.

In Ref. [18], a priority-based energy trading game is proposed in which buyers are prioritized according to the following factors. First, the amount of contribution in the past, in terms of energy, is provided by the buyer to the grid. Second, the amount of energy that is requested by that buyer. This prioritization model is considered in the paper to eliminate the difficulty of energy pricing. An efficient method for energy distribution is proposed in this paper. The aim is to maximize the sum of satisfaction of all the buyers known as social welfare. To do so, an iterative algorithm is used to optimally assign energy to buyers, considering the water filling approach. At the buyer level, the traditional game theory is considered, and the existence of Nash equilibrium is proved. The buyer and seller utilities are selected according to the research in Ref. [34].

In Ref. [45], the authors presented a single-leader, multiple-follower Stackelberg game in which the central power station is considered as a leader (buyer) and followers are the sellers who decide to sell their surplus energy to the central unit. The goal of this paper is to maximize the sum of all followers' utilities while satisfying the minimum cost for the central power station. The approach is decentralized as the main station does not have control over the sellers.

In Ref. [46], the problem of energy trading between microgrids is visited while protecting their private strategies. The system model is adopted from Ref. [24]. In this problem, a market operator is introduced which collects the microgrids' actions and each microgrid chooses its action randomly and individually. To overcome the problem of incomplete information, a new scheme is proposed which combines noncooperative repetitive Stackelberg game and reinforcement learning algorithms. It is shown that learning techniques help each grid to reach the best response achieved by the Stackelberg game. This has been done by connecting the average utility maximization and the best strategy. In the first learning technique, the action set is considered to be finite and the epsilon-optimality is guaranteed. In the second learning technique, the action set is continuous and the numerical results show that in this method by finding the optimal mean, the best response of each player can be achieved.

In Ref. [48], a set of connected microgrids are considered that can transfer energy with each other and the macrogrid. Each microgrid is equipped with the battery to store energy and with local energy generators such as wind turbine and photo voltaic panels. The level of these renewable energy sources is not constant and varies by time and, thereby, should be estimated based on the generation

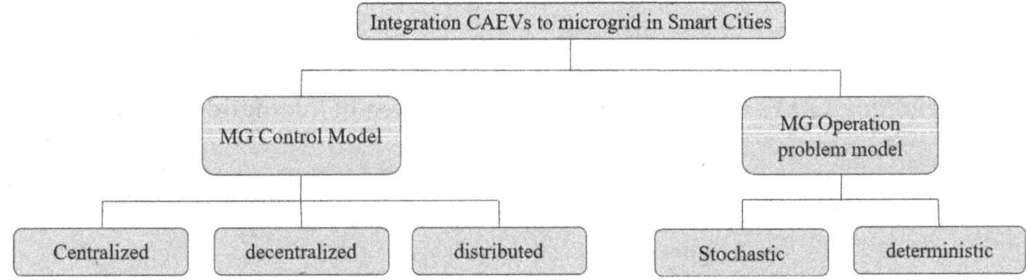

Figure 1.3 Different aspects of CAEVs integration to microgrid in smart cities.

history. Energy trading between microgrid is modeled as noncooperative distributed game, in which each microgrid aims to maximize its own utility function. A hot-booting Q-learning-based approach is implemented to achieve the Nash equilibrium of the dynamic repeated game. As the microgrid interactions in the next state depend on the current battery level, energy generation, and energy transfers, each microgrid can employ Q-learning to learn the best trading strategy. The conventional Q-learning approach starts the learning process with zero information about the game condition. However, in this paper, Hot-booting Q-learning is employed in which the initial values is derived based on the training data achieved during the real experiments and then fed to the Q-learner system to obtain the initial values. The result shows that hot-booting method gains significant efficiency in the convergence time and also increases the overall profit of the players.

In Ref. [47], the author of the work summarized above has improved their work by designing a deep Q-network-based approach. In this paper, the scenario of the problem and the game model is the same as [48]. The block diagram of these two works is depicted in Figure 1.3. Deep Q-network estimates the values of Q-table and, therefore, improves the convergence rate and system performance. Epsilon-greedy algorithm is also considered, which helps the system to avoid staying in the local optimum and let the system search other possibilities. The simulation results show that this approach can be more efficient in terms of players' utility compared to the hot-booting method.

In Ref. [37], the authors presented a coalition framework which employs Bayesian reinforcement learning to address uncertainty in the energy trading process of microgrids. The microgrids are equipped with battery, and with the aid of learning, each microgrid learns the best coalition to join and the best level of battery in each state in the interaction with the environment. The advantage of the proposed method is demonstrated through comparing the given algorithm with the Q-learning and conventional coalitional game theory.

1.2 CAEVs AND THE EFFECT OF INTEGRATING CAEVs TO MICROGRIDS

Plug-in electric vehicles (PEVs) have recently been started to be a part of smart cities as their adoption rate is increasing and the convenience of charging facilities is improving. Automotive manufacturers are also making big advancements toward self-driving or autonomous vehicles which implies that electric and autonomous vehicles will be a part of the future smart cities. Naturally, connectivity will be an essential property of such vehicles for many aspects. In this section, we focus on the integration of CAEVs to the microgrids and their co-existence in the future smart cities.

High level of penetration of CAEVs in the electricity distribution system may affect the operation of microgrids due to their uncertain behaviors. Therefore, considering uncertainty in the problem, design stage is crucial. In the following, different studies on integration of CAEVs to the microgrid and their effects on the operation of microgrids are summarized (Table 1.1).

Table 1.1
Effect of Integrating CAEVs on the Operation Management of Microgrids

References	Objective	Model	Renewable Resources	Storage
H. Kamankesh et al.[19]	Optimizing operation management of microgrids	Stochastic	SP, WT	Battery
S. Bahramara and H. Golpîra [11]	Co-optimizing energy and power management of CAEVs	Stochastic		Battery
A. Kavousi-Fard et al. [22]	Managing the charging demand	Stochastic	SP, WT	
J. Trovão and C. Antunes [43]	Co-optimizing energy and power management of CAEVs	Deterministic		Battery
J. Trovão et al. [44]	Co-optimizing energy and power management of CAEVs	Deterministic		Battery
S. Derakhshandeh et al. [14]	Minimizing charging cost and regulate voltage profile	Deterministic	SP	
I. Zenginis et al. [49]	Optimizing determination of the size of building equipments	Deterministic	SP	Battery

In [22], the effect of CAEVs' charging demand on the optimized operation and the management of microgrids are examined. The authors proposed a novel charging scheme and a smart stochastic framework to manage the charging demand of CAEVs in the residential communities and public charging stations. This framework is evaluated on a microgrid which includes various renewable energy sources such as microturbine, wind turbine, and photo-voltaic fuel cells equipped with battery. The authors concluded that with the cost reduction achieved by the presented framework, higher integration of CAEVs in the microgrids system can be supported.

In [43] and [44], the operation problem of integrating CAEVs to microgrids is visited. The authors presented a scheme for co-optimizing energy and the power management of CAEVs with multiple sources. The management scheme is designed at two levels. Dynamically limiting search space and utilizing a meta-heuristic technique are adopted in these two levels consecutively. The results demonstrate reduction in power loss.

In [14], a dynamic optimal power flow formulation is proposed which satisfies the security of microgrids and industrial constraints while considering CAEVs' energy-related constraints. The proposed scheme is implemented in an industrial microgrid which includes 12 factories with combined heat and power system, SP generation system equipped with SP storage, and various types of CAEVs which are operating in connected and stand-alone modes. The loads and charging rates of the CAEVs are optimized to minimize the charging cost and regulate voltage profile.

In [19], the authors evaluated the impact of CAEVs' charging demand on the optimized operation management of microgrids with different renewable energy sources and battery storage. The authors compared their smart charging strategy with conventional methods through simulation on two test systems, and the results show the superiority of their modified symbiotic organisms' search method in solving the optimal operation management problem. Also, they concluded that although the charging demand of CAEVs result in higher costs for the microgrid, with the proposed smart charging, the cost impact of the charging demand is sufficiently reduced.

In [8], the optimized operation management of smart microgrids and CAEVs' charging infrastructure development are studied. This study shows that although CAEVs are alluring for smart cities and highly accepted due to their green environmental impacts, the standard to connect CAEVs to variable source microgrids has not been fully adopted in the industry.

In [49], the authors study a case in which interconnected cooperative buildings form a microgrid exchanges energy among themselves to enhance their independency to the grid and also reduce power loss. In this setting, the proposed solution aims to optimize the power management as well as capacities of the building equipment. Each building is equipped with SPs, energy storage units, and CAEVs. The load patterns of the buildings, electricity prices and carbon emission taxes, and efficient charging and discharging of the CAEVs are taken into account in the design of the microgrid configuration. The authors proposed a cooperative game theoretical approach, which optimally determines the size of the building equipment and reduces the dependency on the main grid significantly.

In [11], a novel scheme for robust optimization of the microgrid power management is proposed, which encourages buildings in the microgrid to transfer energy among themselves. The uncertain behavior of CAEVs is considered which may impact the microgrid operation. The results show that the proposed solution achieves vital carbon and cost efficiency.

1.3 MICROGRID CONTROL METHODS IN THE PRESENCE OF CAEVs

Integrating and controlling CAEVs in MGs network can be achieved in a centralized (Figure 1.4a), decentralized (Figure 1.4b), or distributed (Figure 1.4c) manner. These three control typologies are further described below.

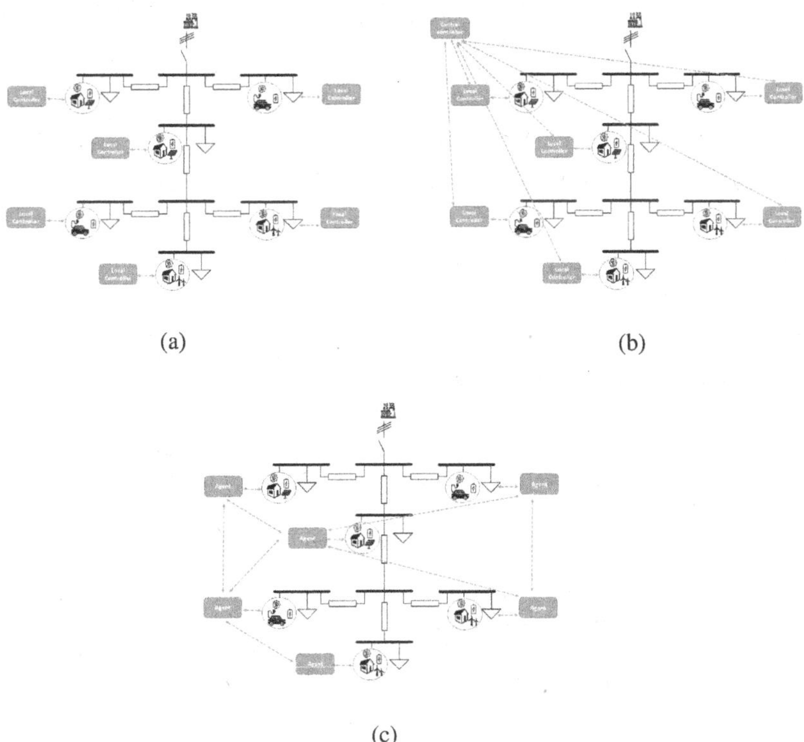

Figure 1.4 Microgrid control typologies. (a) Decentralized, (b) centralized, and (c) distributed.

1.3.1 MICROGRID CENTRALIZED CONTROL

In the centralized approach, all the CAEVs or CAEVs' operators only interact with the central microgrid management. This center is responsible for collecting and analyzing the received data and respond with the appropriate control signal. The central approach results in improved controllability and reduced scalability.

In [21], the authors proposed a central structure to integrate CAEVs in the microgrids' network. A scheme called optimal power set-points calculator is designed, which aims to minimize the variance of active power variance with low and medium voltages substation. This algorithm finds optimized load profiles of CAEVs using evolutionary particle swarm optimization and considering the data collected from the battery, CAEV operators' behavior, and routing pattern.

In [26], a central scheme is presented to balance generation and load and then regularize the voltage in an integrated system. A central control device is implemented which monitors the exchange of active power between the three phases and computes power set points of CAEVs that may connect to microgrid in different phases. This method results in phase balance in the MG.

The central control schemes are well-established and widely implemented in practice for decades. However, the increase in consumers, distributed energy generation and storing, and renewable energy sources make the central control approaches to be less effective. The central approaches are unable to operate and control the future microgrid systems due to the following:

1. The increase in the number of users including the number of CAEVs integrated with microgrids imposes heavy computational load which cannot be handled with the central approach.
2. It is difficult to expand the central control systems, whereas microgrids need to evolve and expand very fast.
3. A single point of failure results in the failure of the whole system. Therefore, these control methods are more suitable for small-scale systems.
4. High level of connectivity is necessary in central approaches, as each agent should be connected to the central unit individually which imposes enormous cost as the number of agents grows.

The aforementioned reasons shifted the system from the central to more distributed structures.

1.3.2 MICROGRID DECENTRALIZED CONTROL

Decentralized and distributed terms are often used interchangeably; however, there is a difference between these two terms. In the decentralized control scheme, each user is either controlled by itself or a higher control level. Most of the decisions are made considering the local measurements, and the number of local connection is limited, which results in lower level of connectivity comparing to the central control approach. These systems are relatively more robust against a single-point failure in the system. If a higher control level or an agent under its control fails, the whole system is still functional. However, due to the low level of connectivity and lack of information exchange between users, global optimization and reliability of the whole network cannot be guaranteed.

1.3.3 MICROGRID DISTRIBUTED CONTROL

Distributed control schemes have the advantages of both centralized and decentralized control while overcoming the difficulties of both of them. In the distributed scheme, neighbor agents can share information among themselves. The sharing of information and the availability of local measurements in the distributed control lead to global optimization and robustness of the whole system. The distributed control has a variety of applications.

In [36], a distributed control scheme for the coordination of CAEVs is proposed to optimize energy sharing and regularization of voltage in a commercial microgrid, which will be able to work as an independent electric unit. The authors designed a CAEV storage controller, which can operate in a distributed or decentralized mode. The controller monitors the reference power of CAEV aggregator. The proposed framework results not only in the efficiency of voltage or frequency regulation and the system robustness but also economically beneficial to CAEV and aggregator operators.

In conclusion, distributed microgrid control approaches are significant revolutions in the microgrid industry, which are very effective tools to coordinate and integrate CAEVs to the microgrid network. Some of the reasons are summarized in the following:

1. Distributed microgrid control is easy to modify, expand, and support scalability.
2. Distributed microgrid control is computationally cost-efficient, as the computational load is distributed among all agents.
3. A single point of failure would not affect the operation of the whole system.
4. Highly adoptable for the expansion and integration of CAEVs to the system.

In the next section, we present a quality-of-service (QoS)-aware charging infrastructure that can be potentially applied to CAEV integration to smart cities.

1.4 QUALITY OF SERVICE IN PLUG-IN ELECTRIC VEHICLE CHARGING INFRASTRUCTURE

When considering integration of CAEVs to microgrids, one of the most important aspects is charging the CAEVs. In [15], a QoS-aware charging mechanism was proposed. According to that scheme, multiple charging requests at the same time can be restricted by the QoS-aware admission control mechanism, considering the energy provisioning while service differentiation is provided to the users. It is assumed that the available energy supply and load on the transformers determine the reserved energy for the charge of CAEVs. The utility grid updates the provisioned energy threshold regularly and transmits it to the energy management system of the distribution substation using wireless communications.

A wireless mesh network including chargers and energy management system is formed using the IEEE 802.11s standard [16]. In this standard, Mesh Point determines the fundamental mesh node. The mesh device that is capable of forwarding and access to client stations is known as Mesh Access Point. The authors assume that this scenario only includes MPs. If a CAEV is plugged-in, a CHARGE-REQ packet is transmitted to the admission control module of the EMS, and then, the SERVICE-STAT message is sent back by the energy management system.

The QoS control parameters β_h and β_l express the amount of power delivered to high-priority and low-priority CAEVs. The admission control relies upon these parameters. Since β_h is chosen to be larger than β_l, high-priority CAEVs have the privilege to be charged faster. Setting β_h and β_l to zero means that charging of CAEVs are prohibited in that period and can help to avoid risking the distribution system by charging. Charging demand can be requested at a later time as long as the CAEV is plugged-in.

Let us assume that CAEVs' charging demand received by the distribution system is following a Poisson process. The mean arrival rate of this Poisson arrival process would be λ. For simplicity and without loss of generality, one can consider that CAEVs' service time is identically and independently distributed exponential distribution with a mean duration of $1/\mu$. In practical systems, CAEVs have various states of charges due to their different charging capacity and driving pattern. Therefore, CAEVs' service time varies in practice. Let x denotes the charging ratio of the battery. Considering M as the maximum permitted CAEVs to be charged, the chance of m CAEVs being charged at a given time can be expressed as the following probability:

$$p(m) = \frac{\left(\frac{(\lambda x/\mu)^m}{m!}\right)}{\sum_{k=0}^{M}\left(\frac{(\lambda x/\mu)^k}{k!}\right)} \quad 0 \leq m \leq M \tag{1.1}$$

$\lambda x/\mu$ denotes the CAEVs and can be shown with ρ_x. The blocking probability can be derived as follows:

$$B = \frac{\left(\frac{(\lambda x/\mu)^m}{m!}\right)}{\sum_{k=0}^{M}\left(\frac{(\lambda x/\mu)^k}{k!}\right)} \quad 0 \leq m \leq M \tag{1.2}$$

In the following, for the sake of simplicity, the equations for the high-priority CAEVs are derived, and derivations for the low-priority CAEVs can be found accordingly. The probability that the power of β_h is delivered to i CAEVs out of m is calculated as follows:

$$p(i|m) = \binom{m}{i} \beta_h^i (1-\beta_h)^{m-i} \tag{1.3}$$

where $0 \leq \beta_h \leq 1$. The total power received by all M CAEVs can be found as follows:

$$D_p = \sum_{i=0}^{M} ip(i) \tag{1.4}$$

where $0 \leq i \leq M$ and $p(i) = \sum_{m=i}^{M} p(i|m)p(m)$.

Plugging $p(i)$ and $z = m-1$ in Eq. (1.4), the deliver power is derived as follows:

$$D_p = (\lambda x/\mu)\beta_h \left\{ \sum_{i=0}^{M}\sum_{z=i}^{M} \frac{\left(\frac{\lambda x/\mu}{z!}\right)^z}{\sum_{k=0}^{M}\left(\frac{(\lambda x/\mu)^k}{k!}\right)} \binom{z}{i} \beta_h^i (1-\beta_h)^{z-i} - \frac{\left(\frac{(\lambda x/\mu)^M}{M!}\right)}{\sum_{k=0}^{M}\left(\frac{(\lambda x/\mu)^k}{k!}\right)} \right\} \tag{1.5}$$

The first term inside the bracket is equal to one, and the second term shows the probability of blocking as calculated in Eq. 1.2. Therefore, the total delivered power to all M CAEVs can be simplified as follows:

$$D_p = \rho_x \beta_h (1-B) \tag{1.6}$$

1.5 PERFORMANCE EVALUATION

In this section, first, the analysis results are presented. The maximum number of CAEVs, M, considered to be equal to 30. The mean arrival rate of the considered Poisson process for the charging demand arrival is set as $\lambda \in \{10, 15, 20\}$ arrivals per hour.

In Figure 1.5, the blocking probability of CAEVs in the proposed mechanism is plotted versus the service time (hour). It is shown that as the rate of charger demand arrival increases, the number of blocked CAEVs with both high and low priority increases as well. Also, as expected, a longer service time results in a higher blocking probability.

In Figure 1.6, the total delivered power to all CAEVs with both high and low priority is presented versus service time, for $\lambda = 20$. It is assumed that all CAEVs consumed all the battery capacities and requested full charge. As it can be seen, as high priority can charge faster, they charge at higher power.

In Figure 1.7, the blocking probability of CAEVs in the proposed mechanism is plotted versus the arrival rate per hour for different values of service time μ. It is shown in the case $\mu = 5$, the probability of blocking starts the rise after $\lambda = 4$ arrivals per hour. In the case $\mu = 10$, the probability of blocking starts the rise after $\lambda = 2$ arrivals per hour, and finally, for the case $\mu = 15$, the probability of blocking starts the rise after $\lambda = 1$. Furthermore, as it has been observed in Figure 1.7, a longer service time results in a higher blocking probability.

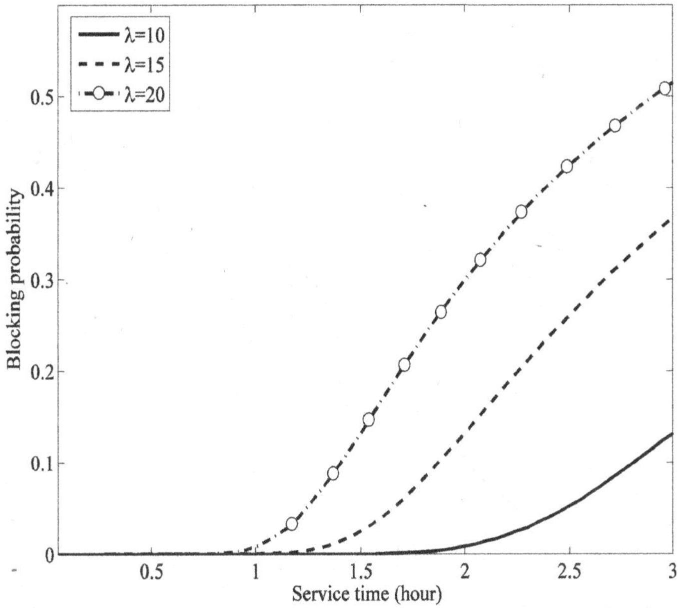

Figure 1.5 Blocking probability of CAEVs versus service time.

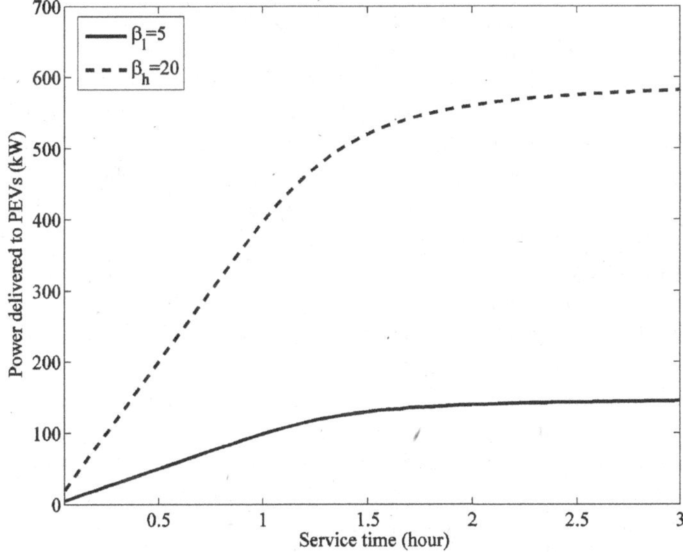

Figure 1.6 Total delivered power to all CAEVs with both high- and low-priority versus service time, for $\lambda = 20$.

1.6 CHALLENGES AND FUTURE DIRECTIONS

The integration of CAEVs to the microgrid communities has evolved significantly over the last decade. However, still major challenges need to be addressed and proper solutions to overcome these challenges are essential. In the following, some of these challenges are outlined:

Load Management: Microgrid communities are designed to transfer energy among themselves to reduce the cost and power loss. However, introducing CAEVs in these communities adds additional

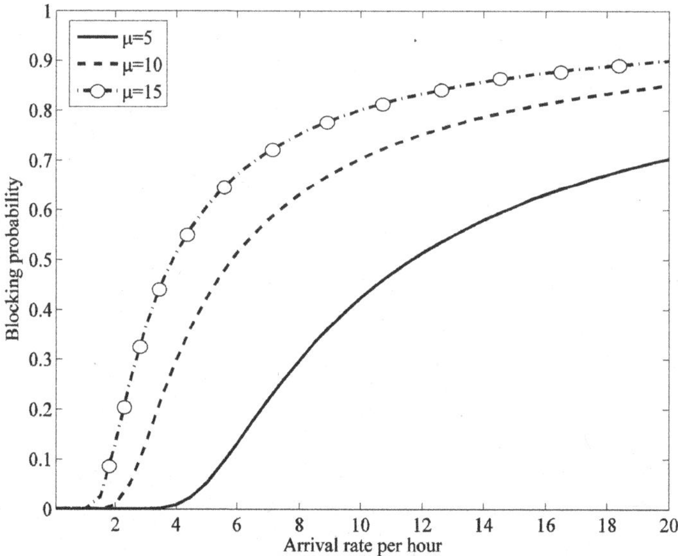

Figure 1.7 Blocking probability versus arrival rate.

load, which affects the stability of the microgrid communities and imposes uncertainty in the energy demand. Therefore, the load management schemes should be reinvestigated to fulfill the challenge of adding CAEVs' loads to the microgrids.

Charging and Discharging: Grid-to-vehicle (G2V) and vehicle-to-grid (V2G) are vastly studied and implemented. However, these technologies are still in the development stages. The two-way power transfer infrastructures are in the design stages, and a wide implementation of these infrastructures relies on the industrial models, standards, and economical justifications.

Distributed Control: As the number of microgrids and CAEVs increases, the complexity of controlling those entities will increase. This will consequently impose heavy computational load that cannot be handled with only one single centralized control unit. Therefore, careful consideration and shifting to distributed control will be essential.

Big Data Analytics: As the number of users with dynamic nature increases in the system, the amount of data that is generated by CAEVs and microgrids increases significantly. Analysis of this huge amount of data is necessary to develop charging policies, design smart-charging schemes, study energy efficiency and capacity of the system, as well as evaluating the financial aspect of energy transactions. Big data analytics is a promising approach to analyze this vast amount of data. However, applying big data on a microgrid system with integrated CAEVs faces various challenges. First, the real-world data are limited. Furthermore, the timeframe for analysis needs to be reduced to seconds to satisfy the microgrid and CAEVs' application requirements. Therefore, fast and effective big data analytic tools are needed to be developed for the microgrid applications [25].

Security and Privacy: As the power system moves toward a more distributed architecture, the urge for sharing more information of each element of the system increases. Although sharing information is crucial for proper data analysis, the privacy of consumers should be taken into account [30,31]. A study on the trade-off between user privacy and data sharing is needed to find the right amount of data sharing that achieves the required data analysis accuracy. Other than that, customers may become more prone to cyberattacks as the widespread of CAEVs increases. Cyberattack protection should be addressed in future studies.

1.7 CONCLUSION

The power grid has been facing major changes in the last two decades in parallel to the developments in smart cities. The interest to make power generation less dependent on fossil fuels attracts the implementation of more renewable energy generators. Another aspect of zero-carbon emission system is adopting CAEVS. The integration of CAEVs through a distributed microgrid system imposes further difficulties. In this chapter, we focused on the integration of CAEVS to the future smart cities through microgrids. First, an overview of smart cities, microgrids, and CAEVs were presented. Then, one study from the literature which provides analysis of such integration scenarios is discussed in detail.

REFERENCES

1. European Commission. 2020 climate & energy package [online]. https://ec.europa.eu/ clima/policies/strategies/2020/.
2. European Research Project cluster. Microgrids and more microgrids projects [online]. http://www.microgrids.eu/default.php.
3. Jofemar Corporation. Factory microgrid description [online]. http://www.factorymicrogrid.com/es/el-proyecto/descripcion-del-proyecto.aspx.
4. Leonardo Energy. The first microgrid in the netherlands [online]. http://www.olino.org/blog/nl/wp-content/uploads/2009/10/the-first-micro-grid-in-the-netherlands-bronsbergen.pdf.
5. Microgrid Media. Microgrid projects map [online]. http://microgridprojects.com/.
6. National Renewable Energy Centre. Atenea microgrid [online]. http://www.cener.com/en/areas/renewable-energy-grid-integration-department/infrastructures-and-technical-resources/microgrid/.
7. Nice Grid. Project architecture and diagram [online]. http://www.nicegrid.fr/en/diagram/.
8. F Ahmad et al. Developments in xevs charging infrastructure and energy management system for smart microgrids including xevs. *Sustainable Cities and Society*, 35:552–564, 2017.
9. E Alegria et al. Certs microgrid demonstration with large-scale energy storage and renewable generation. *IEEE Transactions on Smart Grid*, 5(2):937–943, 2013.
10. WW Anderson. Smart power infrastructure demonstration for energy reliability and security (spiders) final report. Technical report, Naval Facilities Engineering Command Joint Base Pearl Harbor-Hickam United..., 2015.
11. S Bahramara and H Golpîra. Robust optimization of micro-grids operation problem in the presence of electric vehicles. *Sustainable Cities and Society*, 37:388–395, 2018.
12. L Bătăgan. Smart cities and sustainability models. *Informatica Economică*, 15(3):80–87, 2011.
13. VN Coelho et al. A communitarian microgrid storage planning system inside the scope of a smart city. *Applied Energy*, 201:371–381, 2017.
14. S Derakhshandeh et al. Coordination of generation scheduling with PEVs charging in industrial micro-grids. *IEEE Transactions on Power Systems*, 28(3):3451–3461, 2013.
15. M Erol-Kantarci, JH Sarker, and HT Mouftah. Quality of service in plug-in electric vehicle charging infrastructure. *In 2012 IEEE International Electric Vehicle Conference*, Greenville, SC, pp. 1–5, March 2012.
16. GR Hiertz et al. IEEE 802.11 s: The WLAN mesh standard. *IEEE Wireless Communications*, 17(1): 104–111, 2010.
17. A Ibrahim et al. The role of big data in smart city. *International Journal of Information Management*, 36(5):748–758, 2016.
18. A Jadhav and N Patne. Priority based energy scheduling in a smart distributed network with multiple microgrids. *IEEE Transactions on Industrial Informatics*, 13(6):3134–3143, 2017.
19. H Kamankesh et al. Optimal scheduling of renewable micro-grids considering plug-in hybrid electric vehicle charging demand. *Energy*, 100:285–297, 2016.
20. G Kariniotakis, A Dimeas, and F Van Overbeeke. Pilot sites: Success stories and learnt lessons. In *Microgrids*, N. Hatziargyriou (ed.), Piscataway, New Jersey, pp. 206–274, 2013.

21. A Karnama et al. Optimal management of battery charging of electric vehicles: A new microgrid feature. In *2011 IEEE Trondheim PowerTech*, Trondheim, Norway, pp. 1–8, June 2011.
22. A Kavousi-Fard et al. Impact of plug-in hybrid electric vehicles charging demand on the optimal energy management of renewable micro-grids. *Energy*, 78:904–915, 2014.
23. S Khan et al. Artificial intelligence framework for smart city microgrids: State of the art, challenges, and opportunities. In *2018 Third International Conference on Fog and Mobile Edge Computing (FMEC)*, Barcelona, Spain, pp. 283–288, IEEE, 2018.
24. J Lee et al. Distributed energy trading in microgrids: A game-theoretic model and its equilibrium analysis. *IEEE Transactions on Industrial Electronics*, 62(6):3524–3533, 2015.
25. B Li, MC Kisacikoglu, C Liu, N Singh, and M Erol-Kantarci. Big data analytics for electric vehicle integration in green smart cities. *IEEE Communications Magazine*, 55(11):19–25, 2017.
26. JAP Lopes et al. Identification of control and management strategies for LV unbalanced microgrids with plugged-in electric vehicles. *Electric Power Systems Research*, 80(8):898–906, 2010.
27. L Mariam et al. Microgrid: Architecture, policy and future trends. *Renewable and Sustainable Energy Reviews*, 64:477–489, 2016.
28. F Martin-Martínez et al. A literature review of microgrids: A functional layer based classification. *Renewable and Sustainable Energy Reviews*, 62:1133–1153, 2016.
29. H Mouftah et al. *Transportation and Power Grid in Smart Cities: Communication Networks and Services*. Wiley, Hoboken, NJ, 2018.
30. S Mousavian, M Erol-Kantarci, and T Ortmeyer. Cyber attack protection for a resilient electric vehicle infrastructure. In *2015 IEEE Globecom Workshops (GC Wkshps)*, San Diego, CA, pp. 1–6, December 2015.
31. S Mousavian, M Erol-Kantarci, L Wu, and T Ortmeyer. A risk-based optimization model for electric vehicle infrastructure response to cyber attacks. *IEEE Transactions on Smart Grid*, 9(6):6160–6169, November 2018.
32. S Obara and J Morel. *Clean Energy Microgrids*, vol. 1. IET, UK, 2017.
33. D Olivares et al. Trends in microgrid control. *IEEE Transactions on Smart Grid*, 5(4):1905–1919, 2014.
34. S Park et al. Contribution-based energy-trading mechanism in microgrids for future smart grid: A game theoretic approach. *IEEE Transactions on Industrial Electronics*, 63(7):4255–4265, 2016.
35. P Punjad et al. Case study of micro power grid applications in remote rural area of Thailand. In *AORC Technical Meeting*, Tokyo, pp. 1–6, 2014.
36. S Rahman et al. A vehicle-to-microgrid (v2m) framework with optimization-incorporated distributed ev coordination for a commercial neighborhood. *IEEE Transactions on Industrial Informatics*, 16(3):1788–1798, 2019.
37. M Sadeghi and M Erol-Kantarci. Power loss minimization in microgrids using bayesian reinforcement learning with coalition formation. In *2019 IEEE 30th Annual International Symposium on Personal, Indoor and Mobile Radio Communications (PIMRC)*, Istanbul, Turkey, pp. 1–6, September 2019.
38. Highlands and Islands Enterprise. Microgrids a guide to their issues value [online]. https://www.hie.co.uk/media/5957/a-guide-to-microgrids.pdf.
39. T Shelton et al. The 'actually existing smart city'. *Cambridge Journal of Regions, Economy and Society*, 8(1):13–25, 2014.
40. K Su, J Li, and H Fu. Smart city and the applications. In *2011 International Conference on Electronics, Communications and Control (ICECC)*, Ningbo, China, pp. 1028–1031, September 2011.
41. D Ton and M Smith. The US department of energy's microgrid initiative. *The Electricity Journal*, 25(8):84–94, 2012.
42. R Tonkoski, LAC Lopes, and D Turcotte. Active power curtailment of PV inverters in diesel hybrid minigrids. In *2009 IEEE Electrical Power Energy Conference (EPEC)*, Montreal, Canada, pp. 1–6, 2009.
43. J Trovão and C Antunes. A comparative analysis of meta-heuristic methods for power management of a dual energy storage system for electric vehicles. *Energy Conversion and Management*, 95:281–296, 2015.
44. J Trovão et al. A multi-level energy management system for multi-source electric vehicles–an integrated rule-based meta-heuristic approach. *Applied Energy*, 105:304–318, 2013.
45. W Tushar et al. Prioritizing consumers in smart grid: A game theoretic approach. *IEEE Transactions on Smart Grid*, 5(3):1429–1438, 2014.
46. H Wang et al. Reinforcement learning in energy trading game among smart microgrids. *IEEE Transactions on Industrial Electronics*, 63(8):5109–5119, 2016.

47. L Xiao et al. Reinforcement learning-based energy trading for microgrids [online]. https://arxiv.org/abs/1801.06285.
48. X Xiao et al. Energy trading game for microgrids using reinforcement learning. In *Game Theory for Networks*, L Duan et al. (ed.), pp. 131–140. Springer International Publishing, Berlin, Heidelberg, 2017.
49. I Zenginis et al. Cooperation in microgrids through power exchange: An optimal sizing and operation approach. *Applied Energy*, 203:972–981, 2017.

2 A Hierarchical Management Framework for Autonomous Electric Mobility-on-Demand Services

Nuzhat Yamin
University of Idaho

Syrine Belakaria
Washington State University

Sameh Sorour
Queen's University

Mohamed Hefeida
West Virginia University

CONTENTS

2.1 INTRODUCTION

2.1.1 OVERVIEW

The future of private transportation lies in the recently emerged technology – self-driving electric vehicles. With advances in automation of vehicles, along with improvements in electrification and charging, autonomous electric mobility-on-demand (AEMoD) systems are even more favorable. Despite the significant advances in vehicle automation and electrification, the next-decade aspirations for massive deployments of AEMoD services in big cities are still threatened by two major bottlenecks, namely, the communication/computation and charging delays. This chapter discusses the evolution and the role of AEMoD systems, in which self-driving electric vehicles transport passengers in a specified setup, allowing for all localized operational decisions to be made with very low latency by fog controllers located close to the end applications (e.g., each city zone for AEMoD systems). In this chapter, we will present an optimized multiclass charging and dispatching queuing model, with partial charging option for AEMoD vehicles. We will also focus on finding the optimal vehicle-dimensioning for each zone of these models in order to guarantee a bounded response time of its vehicles. A queuing model for multiclass rebalancing and possible in-route charging is developed on top of the system's decentralized fleet management. The stability conditions of this model are first derived. A layered diagram of the whole system has been illustrated to elaborate these operations. The whole scheme is described throughout the chapter with its queuing model and closed-form stability solutions and evaluated within a realistic setting in the city of Seattle to illustrate the merits.

2.1.2 21ST CENTURY URBAN TRANSPORTATION SYSTEM

The population of the world is growing with urbanization. With present trends that emphasize the need for alternative transport types and the need for more personal mobility, travelers' needs are changing. The paradigm of personal urban mobility in recent centuries has been dramatically transformed by private cars, making it possible to move quickly in and out of towns. But urban transport systems today face huge difficulties as personal car ownership demand [1] is on the ups and downs, leading to drastic rises in congestion, demand for parking, enhanced travel times, and carbon footprint [2,3]. More than 7.27 billion individuals live on the earth [4], with over half the population residing in metropolitan areas, according to the latest estimates. By 2050, urban populations will double, tripling already higher than the anticipated amount of cars. This will exacerbate the issues of the moment further. When peak-hour demand is provided for the public transport, it can lead to low effectiveness as vehicles become idle in off-peak hours. Thus, travel and mobility demands are evolving from an emphasis on private automobile ownership to more flexible, public, and private options which incorporate shared-use and multimodal integration. It is vital that these systems are further developed alongside existing services in an integrated fashion to foster a fluid and connected transport system incorporating all modes and people in a seamless fashion, enabling a true complete trip, or point-to-point journey. This obviously requires revolutionary alternatives to maintain personal mobility in the future.

2.1.3 AEMoDs: OPPORTUNITIES AND CHALLENGES

Research on the new mobility alternatives is essential if the increasing mobility requirements, new technologies, and demographics are to be documented, evaluated, and properly planned amid difficult fiscal realities. New business models and the need to choose situational mobility give new possibility in ride sharing, and bike and car sharing. There is also a renewed interest in demand-responsive activities fueled mainly by mobile technology and almost omnipresent smartphones. These fresh developments, along with the traditional transport alternatives, offer true possibilities to create an integrated mobility system that meets the requirements of the varied crossroads. Mobile on-demand (MoD) is poised to contribute to this new ecosystem with connected travelers, infrastructure, innovative operations, and personal mobility needs. These services successfully offered a partial solution to the growing problem of private car ownership [5] by providing one-way vehicle sharing for a monthly subscription fee between dedicated pick-up and drop-off sites, without any concerns regarding vehicle insurance and maintenance costs. Electrifying these MoDs can also decrease the carbon footprint issue gradually. However, the need for additional picking-ups, after dropping-off, and occasionally to fuel and load these MoD vehicles, has considerably hindered the effectiveness of this solution and its impact on the resolution of urban traffic.

Nonetheless, an expected game-changer for the success of these services is the significant advances in vehicle automation and wireless connectivity. Even Tesla's Elon Musk believes that these difficulties may be overcome within the next few years. At Tesla's Autonomous Investor Day in April 2019, Musk announced that he fully expects Tesla to have a million robotaxis on the roads within the next year [6]. With more than 10 million cars on the highways anticipated by 2020 [7], governments and automotive industries' aim to leverage greater wireless connectivity and coordinated optimization on the roads. This is expected to considerably decrease private car ownership by 2025 [9], as individuals' private mobility will further depend on the concept of AEMoD [10]. In brief, AEMoD schemes will allow customers to push a few buttons in an app, so that they get a stand-alone autonomous electric vehicle to transport them door-to-door, without pick-up and drop-off, without dedicated needs for parking, with no carbon emissions [8], no insurance and maintenance costs for vehicles, and additional work and leisure time during trips. AEMoD systems are expected to significantly prevail in all these green, economic, and customer-driven qualities to attract millions of subscribers worldwide and in ensuring on-demand personal urban mobility with flexibility.

In the next decade, AEMoD systems are anticipated to be widely accepted, with demands of tens of thousands per minute in major cities [9,10]. It is evident that the anticipated surge in demand can only be met if AEMoD fleets are properly designed, safe, and effective. The centralized techniques currently in use do not scale well to withstand extensive communication and computational delays. These delays occur during the central collection of customer/vehicle information and optimization of their interactions. The slow charging rate of electric cars is another inconvenience. With the present charging rates and infrastructure, it would not be possible to support the large number of AEMoD vehicles required to meet such demand, resulting in unlimited delays and instability, which will greatly hinder the customers' experience. Recent research works have addressed important problems in AEMoD systems by building different operation models such as a distributed spatially averaged queuing model and a lumped Jackson network model [11]. The system was cast into a closed multiclass BCMP queuing network to solve the routing problem on congested roads [12]. In Ref. [14], a lumped spatial-queuing model was proposed. In Ref. [15], a model-predictive control

(MPC) approach was proposed to optimize the dispatching and scheduling of the vehicles in AMoD systems. In order to address the issue, there were several nonpractical assumptions, and many main factors in these works were not considered in order to simplify the mathematical resolution. In case of massive demands, electrification, and vehicle charging restrictions on system stability, neither of these works proposed any architectural solution for computational latency.

2.1.4 CONTRIBUTIONS

The achievement of AEMoD systems is hampered by two speed limits - communication/computing delays and possible system instability because of the slow charging rate. The first limitation is resolved by exploiting the new and trendy fog-based networking and computing architectures [16]. While long propagation delays remain a key drawback for centralized cloud computing, mobile edge computing (MEC) [17–19] with the proximate access is widely agreed to be a key technology for realizing various applications for next-generation Internet with millisecond-scale reaction time. The privileges brought by this technique will allow handling vehicular networks in need for instantaneous decision-making applications such as autonomous mobility. Consequently, they can also be involved in handling AEMoD system operations in a distributed way. This approach will push the operational decision load close to end customers in each city zone, thus reducing the computational complexity and communication delays. Luckily, this architecture perfectly fits the nature of many AEMoD fleet operations that are mostly local, such as dispatching and charging. Indeed, AEMoD vehicles will be usually directed to pick up customers close to their locations and charge at near-by charging stations. This, thus, makes the fog-based architectures well-suited localized solutions to guarantee low communication and computation latency for such local management operations.

Within this context, the chapter has three contributions.

1. Modeling the AEMoD scheme
2. Mathematics behind ensuring the stability of AEMoD scheme
3. Evaluating the performance of AEMoD scheme

2.2 SMART DISPATCHING AND ROUTING OF AEMoDs

To guarantee the stability and timeliness of future AEMoD systems, given the relatively limited charging resources compared to demand volumes, it is very critical to answer two important operational questions: (1) How to cope with the available charging capabilities of each service zone given the large number of system vehicles? (2) How to smartly manage the dispatching and charging options of different state-of-charge (SoC) vehicles, given the customers' needs and zone resources, in order to minimize the maximum and/or average system response time. System response time is defined as the time elapsed between the instant when an arbitrary customer requests a vehicle and the instant when a vehicle starts moving from its parking or charging spot toward this customer. Motivated by the fact that different customers can be classified in ascending order of their required trip distances (and thus the SoC needed in their allocated vehicles), a multiclass dispatching and charging scheme is modeled for AEMoD vehicles, with options of partial charging for vehicles with nondepleted batteries. Arriving vehicles in each service zone is subdivided into different classes in ascending order of their SoC corresponding to the different customer classes. Different proportions of each class vehicles will then be prompted by the fog controller to either wait (without charging) for dispatching to its corresponding customer class (i.e., customers whose trips will require the SoC range of this class vehicles) with or partially charge to serve the subsequent customer class. Vehicles arriving with depleted batteries will be allowed to either partially or fully charge to serve the first- or last-class customers, respectively. Clearly, the larger the number of classes, the smaller the SoC increase required for a vehicle to move from one class to the next, the smaller the charging time needed to make this transition, the less the burden/requirements on the zone charging resources. On the other hand, given a fixed in-flow rate of vehicles to each city zone, more vehicle/customer classes mean less available in-flow vehicles to each customer class, which may result in longer service delays and even instabilities in their waiting queues. Therefore, determining the optimal number of needed vehicles to stably serve each zone with bounded response time guarantees is a very crucial factor in the operation and key goals of AEMoD systems.

2.2.1 QUEUING MODEL

The system is modeled as a decentralized and multiclass AEMoD framework. This decentralized framework splits cities into adjacent zones. Each service zone is controlled by a fog controller connected to (1) the service request apps of customers in the zone; (2) the AEMoD vehicles; (3) C rapid charging points distributed in the service zone and designed for short-term partial charging; and (4) one spacious rapid charging station designed for long-term full-charging. A central controller is connected to the fog controllers of each of the system's zones as well as the spacious rapid charging stations distributed in the city between these zones, to collect the needed data and perform the required computations for fleet macromanagement. These whole systems can be divided into three layers which will be elaborated in subsequent sections.

1. *Lower Layer*: The lower layer is modeled as a fog-based multiclass charging and dispatching [22] service in each service zone. This approach classifies its incoming vehicles according to their SoC and smartly manages their charging options according to the available charging resources in each zone. This scheme resolves the lack of computational latency in the case of massive demands and relieves limitations on the system's stability due to vehicle charging by allowing vehicles of each class to serve customers from its own class or lower classes with partial charging option for AEMoD vehicles.
2. *Middle Layer*: In order to reduce the longer service delays and instabilities in the waiting queues resulting from the fixed vehicle in-flow rate to the zones, the middle layer will determine the optimal number of needed vehicles (dimensioning) [23] to stably serve each zone with bounded response time guarantees. The decisions on the proportions of each class vehicles to partially/fully charge or directly serve customers are optimized in this layer so as to minimize the total needed vehicles inflow to any given zone.
3. *Upper Layer*: This layer is modeled as a queuing system to perform the joint decisions of multiclass rebalancing and possible in-route charging between zones in order to account for the vehicle deficit (compared to customer demand) in some service zones of lower layer.

2.2.2 SYSTEM PARAMETERS

We will now familiarize the readers about the system parameters and variables needed to briefly elaborate our multiclass AEMoD framework. For a quick review, the list of variables is given in Table 2.1:

AEMoD vehicles enter the service in each zone after dropping off their latest customers in it. Their detection as free vehicles by the zone's controller can be modeled as a Poisson process with rate λ_v. Customers request service from the system according to a Poisson process. Both customers and vehicles are classified into n classes based on an ascending order of their required trip distance and the corresponding SoC to cover this distance, respectively. From the thinning property of Poisson processes, the arrival processes of Class i customers and vehicles, $i \in \{0,\ldots,n\}$, are both independent Poisson processes with rates $\lambda_c^{(i)}$ and $\lambda_v p_i$, where p_i is the probability that the SoC of an arriving vehicle to the system belongs to Class i. Note that p_0 is the probability that a vehicle

Table 2.1
List of System and Decision Parameters

Variables	Definition
λ_v	Total arrival rate of vehicles
p_i	Probability of arrival of a vehicle from Class i
q_0	Probability that a battery-depleted vehicle partially charges
$\overline{q}_0$	Probability that a battery-depleted vehicle fully charges
$q_i, i \neq 0$	Probability that a vehicle in Class i is directly dispatched
$\overline{q}_i, i \neq 0$	Probability that a vehicle in Class i partially charges
μ_c	Service rate of fully charging a battery-depleted vehicle
$\lambda_v^{(i)}$	Arrival rate of vehicles of Class i
$\lambda_c^{(i)}$	Arrival rate of customers served by Class i's vehicles
C	No. of distributed charging points in the service zone of the fog controller

arrives with a depleted battery and is thus not able to serve immediately. Consequently, $\lambda_c^{(0)} = 0$ as no customer will request a vehicle that cannot travel any distance. On the other hand, p_n is also equal to 0, because no vehicle can arrive to the system fully charged as it has just finished a prior trip.

Upon arrival, each vehicle of Class i, $i \in \{1,\ldots,n-1\}$, will park anywhere in the zone until it is called by the fog controller to either (1) serve a customer from Class i with probability q_i or (2) partially charge up to the SoC of Class $i+1$ at any of the C charging points (whenever any of them becomes free), with probability $\overline{q}_i = 1 - q_i$, before parking again in waiting to serve a customer from Class $i+1$. As for Class 0 vehicles that are incapable of serving before charging, they will be directed to either fully charge at the central charging station with probability q_0 or partially charge at one of the C charging points with probability $\overline{q}_0 = 1 - q_0$. In the former and latter cases, the vehicle after charging will wait to serve customers of Class n and 1, respectively.

The full charging time of a vehicle with a depleted battery is exponentially distributed with rate μ_c. Given the uniform SoC quantization among the n vehicle classes, the partial charging time can then be modeled as an exponential random variable with rate $n\mu_c$. Note that the larger rate of the partial charging process is not due to a speed-up in the charging process but rather due to the reduced time of partially charging. The use of exponentially distributed charging times for charging electric vehicles has been widely used in the literature [12,13] to model the randomness in the charging duration of the different battery sizes. The customers belonging to Class i, arriving at rate $\lambda_c^{(i)}$, will be served at a rate of $\lambda_v^{(i)}$, which includes the arrival rate of vehicles that are (1) arrived to the zone with a SoC belonging to Class i and directed to wait to serve Class i customers or (2) arrived to the zone with a SoC belonging to Class $i-1$ and directed to partially charge to be able to serve Class i customers.

The middle-layer operation minimizes the average vehicle inflow rate λ_v to the entire zone [23], given its charging capacity and customer-demand rates, while guaranteeing an average response time limit for each class customers.

In the upper layer, fleet rebalancing [27] occurs to stabilize the whole system. In the conventional MOD literature, "fleet rebalancing" refers to the relocation of excess vehicles in some zones to those suffering from instability (i.e., low vehicle presence compared to customer demands), aiming to adjust the fleet distribution to the spatial variations in customer demands. The central controller makes the decisions in this layer. The collected information by this controller classifies each zone in the system as either a "donor zone"(i.e., having excess vehicles than its customers' needs) or an "acceptor zone" (i.e., having vehicle deficit to meet the demand of its customers). The donor and

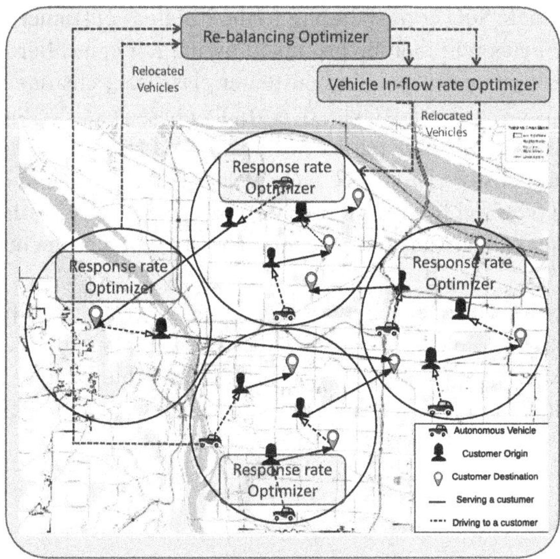

Figure 2.1 Three layers of AEMoD setup.

acceptor zones are designated by the indices $i \in 1,\ldots,n_D$ and $l \in 1,\ldots,n_A$, respectively. Each of these excess vehicles of class j in donor zone i can either:

- Relocate to serve class m in acceptor zone l with a decision probability $P^{(0)}_{di_D,ai_A}$. In this case, the acceptor zone's class m can be only in the range $m \in \{1,\ldots,j-y_{i,l}\}$, where $y_{i,l}$ is a class-egradation term resulting from the loss in the vehicle's SoC while relocating from the donor zone i to acceptor zone l.
- Fully charge in-route at charging station k ($k \in \{1,\ldots,S\}$) before serving class m in the acceptor zone l with a decision probability $P^{(k)}_{ij,kl}$. These vehicles can thus serve any class $m \in \{1,\ldots,n\}$ in the acceptor zone k.

Given the above description and modeling of variables, the entire zone dynamics can thus be modeled by the queuing system depicted in Figure 2.1.

2.3 LOWER LAYER

2.3.1 MULTICLASS CHARGING AND DISPATCHING

A fog-based architecture for AEMoD systems is justified by the fact that many of the AEMoD operations (e.g., dispatching and charging) are localized with very high demand and instantaneous decision-making needs. Indeed, vehicles located in any city zone are the ones that can reach the customers in that zone within a limited time frame. They will also charge in the near-by charging points within the zone. The fog controller in each service zone is responsible for collecting information about customer requests, vehicle inflow to the service zone, their SoC, and the available full-battery charging rates in the service zone. Given the collected information, it can promptly make dispatching and charging decisions for these vehicles in a timely manner. Motivated by the fact that different customers can be classified in an ascending order of their required trip distances (and thus the SoC needed in their allocated vehicles), this section presents a multiclass dispatching and charging scheme for AEMoD vehicles, with options of partial charging for vehicles with nondepleted batteries. Arriving vehicles in each service zone is subdivided into different classes in an ascending order of their SoC corresponding to the different customer classes. Different proportions of each class vehicles will then be prompted by the fog controller to either wait (without charging) for dispatching to its corresponding customer class (i.e., customers whose trips will require the SoC range of this class vehicles) with or partially charge to serve the subsequent customer class. Vehicles arriving with depleted batteries will be allowed to either partially or fully charge to serve the first- or last-class customers, respectively. Clearly, the larger the number of classes, the smaller the SoC increase required for a vehicle to move from one class to the next, the smaller the charging time needed to make this transition, the less the burden/requirements on the zone-charging resources.

Having defined the multiclass dispatching and charging scheme in a city zone, the rest of the section will focus on determining the stability conditions of the system and minimum number of required classes to cope with the charging resources in any arbitrary city zone. The maximum and average response time minimization problems will then be formulated and analytically solved, respectively.

2.3.1.1 System Stability Conditions

We will first deduce the stability conditions of the multiclass dispatching and charging system, using the basic laws of queuing theory. We will then derive an expression for the lower bound on the number n of needed classes that fit the charging capabilities of any arbitrary service zone. Each of the n classes of customers is served by a separate queue of vehicles, with $\lambda_v^{(i)}$ being the arrival rate of the vehicles that are available to serve the customers of the i-th class. Consequently, it is the service rate of the customers i-th arrival queues. We can thus deduce from Figure 2.2 and the system model described above that

$$\begin{aligned} \lambda_v^{(i)} &= \lambda_v\left(p_{i-1}\overline{q}_{i-1} + p_i q_i\right) \qquad && i = 1,\ldots,n-1 \\ \lambda_v^{(n)} &= \lambda_v\left(p_{n-1}\overline{q}_{n-1} + p_0 q_0\right) && \end{aligned} \tag{2.1}$$

Since we know that $\overline{q}_i + q_i = 1$, we substitute $\overline{q}_i$ by $1 - q_i$ in order to have a system with n variables

$$\begin{aligned} \lambda_v^{(i)} &= \lambda_v\left(p_{i-1} - p_{i-1}q_{i-1} + p_i q_i\right) \qquad && i = 1,\ldots,n-1 \\ \lambda_v^{(n)} &= \lambda_v\left(p_{n-1} - p_{n-1}q_{n-1} + p_0 q_0\right) && \end{aligned} \tag{2.2}$$

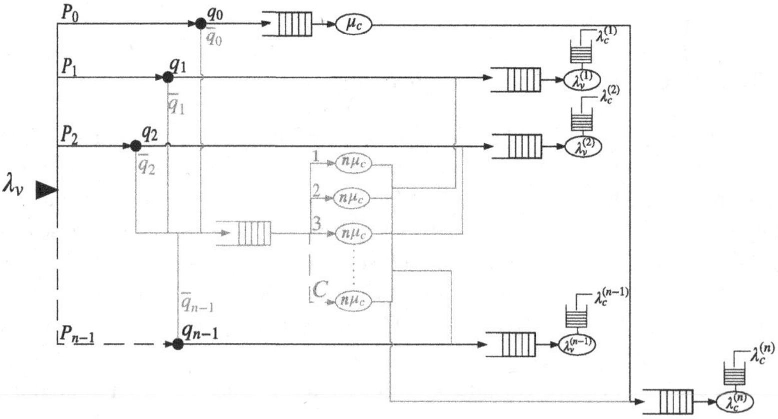

Figure 2.2 Joint dispatching and partially/fully charging model, abstracting an AEMoD system in one service zone.

From the well-known stability condition of an M/M/1 queue [20,21], we have

$$\lambda_v^{(i)} > \lambda_c^{(i)} \qquad i = 1, \ldots, n \tag{2.3}$$

Before reaching the customer service queues, the vehicles will go through a decision step on whether to go to these queues immediately or partially/fully charge. The stability of the charging queues should be guaranteed in order to ensure the global stability of the entire system at the steady state. From the model described in the previous section, and by the well-known stability conditions of M/M/C and M/M/1 queues [20,21], we have the following stability constraints on the C charging points and central charging station queues, respectively,

$$\sum_{i=0}^{n-1} \lambda_v (p_i - p_i q_i) < C(n\mu_c) \tag{2.4}$$
$$\lambda_v p_0 q_0 < \mu_c$$

We will now illustrate the lower bound on the average inflow rate of vehicles for a given service zone, given its rate of customer demands on AEMoD services.

From (2.2) and (2.3), we have

$$\begin{aligned} \lambda_c^{(i)} &< \lambda_v \left(p_{i-1}\overline{q}_{i-1} + p_i q_i\right) & i = 1, \ldots, n-1 \\ \lambda_c^{(n)} &< \lambda_v \left(p_{n-1}\overline{q}_{n-1} + p_0 q_0\right) & i = n \end{aligned} \tag{2.5}$$

The summation of all the inequalities in (2.5) gives a new inequality

$$\sum_{i=1}^{n} \lambda_c^{(i)} < \lambda_v \left[\sum_{i=1}^{n-1} \left(p_{i-1}\overline{q}_{i-1} + p_i q_i\right) + \left(p_{n-1}\overline{q}_{n-1} + p_0 q_0\right)\right] \tag{2.6}$$

$$\sum_{i=1}^{n} \lambda_c^{(i)} < \lambda_v \left[p_0\overline{q}_0 + p_1 q_1 + p_1\overline{q}_1 + \cdots + p_{n-1}\overline{q}_{n-1} + p_0 q_0\right] \tag{2.7}$$

We have $\overline{q}_i + q_i$ so $p_i\overline{q}_i + p_i q_i = p_i$

$$\sum_{i=1}^{n} \lambda_c^{(i)} < \lambda_v \left(p_0 + p_1 + p_2 + \cdots + p_{n-1}\right) \tag{2.8}$$

We have $\sum_{i=0}^{n-1} p_i = 1$ so $\sum_{i=1}^{n} \lambda_c^{(i)} < \lambda_v$

So, for the entire system stability, the inflow rate of vehicles to a given service zone should be strictly more than the total arrival rate of customers belonging to all the classes. In other words,

$$\sum_{i=1}^{n} \lambda_c^{(i)} < \lambda_v \tag{2.9}$$

Furthermore, we will now establish a lower bound on the number of classes n, given the arrival rate of the vehicles λ_v, the full charging rate μ_c, and the number C of partial charging points.

The summation of the inequalities given by (2.4)

$\forall\, i = \{0, \ldots, n\}$ gives the following inequality:

$$\lambda_v \sum_{i=0}^{n-1} p_i - \lambda_v \sum_{i=0}^{n-1} p_i q_i + \lambda_v p_0 q_0 < C(n\mu_c) + \mu_c \tag{2.10}$$

Since $\sum_{i=0}^{n-1} p_i = 1$ (because $p_n = 0$ as described in Section 2.2.2), we get

$$\lambda_v - \lambda_v \sum_{i=1}^{n-1} p_i q_i \tag{2.11}$$

In the worst case, all the vehicles will be directed to partially charge before serving, which means that always $q_i = 0$. Therefore, we get

$$Cn > \frac{\lambda_v}{\mu_c} - 1 \tag{2.12}$$

which can be rearranged to get the number of classes n given by the following inequality to guarantee the stability of the charging queues:

$$n > \frac{\lambda_v}{C\mu_c} - \frac{1}{C} \tag{2.13}$$

2.3.1.2 Maximum Response Time Optimization Problem Formulation

The expected response time of any class is defined as the expected duration between any customers putting a request until a vehicle is dispatched to serve him/her. From the basic M/M/1 queue analysis of the i-th customer class, the expression of this expected response time for the i-th class can be expressed as follows:

$$\frac{1}{\lambda_v^{(i)} - \lambda_c^{(i)}} \qquad i = 1, \ldots, n \tag{2.14}$$

Consequently, the maximum of the expected response times across all n classes of the system can be expressed as follows:

$$\max_{i \in \{1,\ldots,n\}} \left\{ \frac{1}{\lambda_v^{(i)} - \lambda_c^{(i)}} \right\} \tag{2.15}$$

It is obvious that the system's class having the maximum expected response time is the one that have the minimum expected response rate. In other words, we have

$$\arg \max_{i \in \{1,\ldots,n\}} \left\{ \frac{1}{\lambda_v^{(i)} - \lambda_c^{(i)}} \right\} = \arg \min_{i \in \{1,\ldots,n\}} \left\{ \lambda_v^{(i)} - \lambda_c^{(i)} \right\} \tag{2.16}$$

Consequently, minimizing the maximum expected response time across all classes is equivalent to maximizing their minimum expected response rate. Using the epigraph form [24] of the latter problem, we get the following stochastic optimization problem:

$$\max_{q_0, q_1, \ldots, q_{n-1}} \quad R \tag{2.17a}$$

s.t.

$$\lambda_v \left(p_{i-1} - p_{i-1} q_{i-1} + p_i q_i \right) - \lambda_c^{(i)} \geq R, \; i = 1, \ldots, n-1 \tag{2.17b}$$

$$\lambda_v \left(p_{n-1} - p_{n-1} q_{n-1} + p_0 q_0 \right) - \lambda_c^{(n)} \geq R \tag{2.17c}$$

$$\sum_{i=0}^{n-1} \lambda_v \left(p_i - p_i q_i \right) < C \left(n\mu_c \right) \tag{2.17d}$$

$$\lambda_v p_0 q_0 < \mu_c \tag{2.17e}$$

$$\sum_{i=0}^{n-1} p_i = 1,\ 0 \leq p_i \leq 1 \qquad i = 0,\ldots,n-1 \tag{2.17f}$$

$$0 \leq q_i \leq 1 \qquad i = 0,\ldots,n-1 \tag{2.17g}$$

$$R > 0 \tag{2.17h}$$

The n constraints in (2.17b) and (2.17c) represent the epigraph form constraints on the original objective function in the right-hand side of (2.16), after separation [24] and substituting every $\lambda_v^{(i)}$ by its expansion form in (2.2). The constraints in (2.17d) and (2.17e) represent the stability conditions on charging queues. The constraints in (2.17f) and (2.17g) are the axiomatic constraints on probabilities (i.e., values being between 0 and 1, and sum equal to 1). Finally, constraint (2.17h) is a strict positivity constraint on the minimum expected response rate, which also guarantees the stability of the customer queues when combined with (2.17b) and (2.17c). Indeed, if R is strictly positive, this guarantees that that the stability conditions in (2.3) hold with certainty. Clearly, the above problem is a linear program with linear constraints, which can be solved analytically using Lagrangian analysis which is elaborated in the next section.

2.3.1.3 Optimal Dispatching and Charging Decisions

The problem in (2.17) is a convex optimization problem with second-order differentiable objectives and constraint functions that satisfy Slater's condition. Consequently, the KKT conditions provide necessary and sufficient conditions for optimality. Therefore, applying the KKT conditions to the constraints of the problem and the gradient of the Lagrangian function allows us to find the analytical solution of the decisions q_i. The Lagrangian function associated with the optimization problem in (2.17) is given by the following expression:

$$\begin{aligned} L(\mathbf{q},R,\alpha,\beta,\gamma,\omega) &= -R + \sum_{i=1}^{n-1} \alpha_i \left(\lambda_v \left(p_{i-1}q_{i-1} - p_i q_i\right) + R - \lambda_v p_{i-1} + \lambda_c^{(i)}\right) \\ &+ \alpha_n \left(\lambda_v \left(p_{n-1}q_{n-1} - p_0 q_0\right) + R - \lambda_v p_{n-1} + \lambda_c^{(n)}\right) + \beta_0 \left(\sum_{i=0}^{n-1} \lambda_v \left(p_i - p_i q_i\right) - C\left(n\mu_c\right)\right) \\ &+ \beta_1 \left(\lambda_v p_0 q_0 - \mu_c\right) + \sum_{i=0}^{n-1} \gamma_i \left(q_i - 1\right) - \sum_{i=0}^{n-1} \omega_i q_i + \omega_n R \end{aligned} \tag{2.18}$$

where $\mathbf{q}$ is the vector of dispatching decisions (i.e., $\mathbf{q} = [q_0,\ldots,q_{n-1}]$), and where

$\alpha = [\alpha_i]$, such that α_i is the associated Lagrange multiplier to the i-th customer queues inequality.
$\beta = [\beta_i]$, such that β_i is the associated Lagrange multiplier to the i-th charging queues inequality.
$\gamma = [\gamma_i]$, such that γ_i is the associated Lagrange multiplier to the i-th upper bound inequality.
$\omega = [\omega_i]$, such that ω_i is the associated Lagrange multiplier to the i-th lower-bound inequality.

Applying the KKT conditions to the inequalities constraints of (2.17), we get

$$\begin{aligned} &\alpha_i^* \left(\lambda_v \left(p_{i-1}q_{i-1}^* - p_i q_i^*\right) + R^* - \lambda_v p_{i-1} + \lambda_c^{(i)}\right) = 0 \qquad i = 1,\ldots,n-1 \\ &\alpha_n^* \left(\lambda_v \left(p_{n-1}q_{n-1}^* - p_0 q_0^*\right) + R^* - \lambda_v p_{n-1} + \lambda_c^{(n)}\right) = 0 \\ &\beta_0^* \left(\sum_{i=0}^{n-1} \lambda_v \left(p_i - p_i q_i^*\right) - C\left(n\mu_c\right)\right) = 0 \end{aligned} \tag{2.19}$$

$$\begin{aligned}
&\beta_1^* \left(\lambda_v p_0 q_0^* - \mu_c\right) = 0 \\
&\gamma_i^* \left(q_i^* - 1\right) = 0 \qquad i = 0, \ldots, n-1 \\
&\omega_i^* q_i^* = 0 \qquad i = 0, \ldots, n-1 \\
&\omega_n^* R^* = 0
\end{aligned}$$

Likewise, applying the KKT conditions to the Lagrangian function in (2.18), and knowing that the gradient of the Lagrangian function goes to 0 at the optimal solution, we get the following set of equalities:

$$\begin{aligned}
&\lambda_v p_i \left(\alpha_{i+1}^* - \alpha_i^*\right) = \omega_i^* - \gamma_i^* \qquad i = 1, \ldots, n-1 \\
&\lambda_v p_0 \left(\alpha_1^* - \alpha_n^*\right) = \omega_0^* - \gamma_0^* \\
&\sum_{i=1}^{n-1} \alpha_i^* = 1
\end{aligned} \tag{2.20}$$

From Burke's theorem on the stability condition of the queues, the constraints on the charging queues are strict inequalities and the constraints on R should also be strictly larger than 0. Combining the Burke's theorem and the equations on (2.19), we find that $\beta_0^* = \beta_1^* = 0$ and $\omega_n^* = 0$.

Knowing that the gradient of the Lagrangian goes to 0 at the optimal solutions, we get the system of equalities given by (2.20). The fact that $\beta_i^* = 0$ and $\omega_n^* = 0$ explains the absence of β_i^* and ω_n^* in (2.23). The result given by multiplying the first equality in (2.20) by q_i^* and the second equality by q_0^* combined with the last three equalities given by (2.19) gives

$$\begin{aligned}
&\lambda_v p_i \left(\alpha_{i+1}^* - \alpha_i^*\right) q_i^* = -\gamma_i^* \qquad i = 1, \ldots, n-1 \\
&\lambda_v p_0 \left(\alpha_1^* - \alpha_n^*\right) q_0^* = -\gamma_0^* \\
&\sum_{i=1}^{n-1} \alpha_i^* = 1
\end{aligned} \tag{2.21}$$

(2.21) Inserted in the fifth equality in (2.19) gives

$$\begin{aligned}
&\lambda_v p_i \left(\alpha_{i+1}^* - \alpha_i^*\right) \left(q_i^* - 1\right) q_i^* = 0 \qquad i = 1, \ldots, n-1 \\
&\lambda_v p_0 \left(\alpha_1^* - \alpha_n^*\right) \left(q_0^* - 1\right) q_0^* = 0 \\
&\sum_{i=1}^{n-1} \alpha_i^* = 1
\end{aligned} \tag{2.22}$$

From (2.22), we have $0 < q_0^* < 1$ only if $\alpha_1^* = \alpha_n^*$ and $0 < q_i^* < 1$ only if $\alpha_{i+1}^* = \alpha_i^*$ Since $0 \leq q_i^* \leq 1$, then these equalities may not always be true

if $\alpha_1^* > \alpha_n^*$, and we know that $\gamma_0^* \geq 0$ then $\gamma_0^* = 0$ which gives $q_0^* \neq 1$ and $q_0^* = 0$
if $\alpha_{i+1}^* > \alpha_i^*$, and we know that $\gamma_i^* \geq 0$ then $\gamma_i^* = 0$ which gives $q_i^* \neq 1$ and $q_i^* = 0$
if $\alpha_1^* < \alpha_n^*$, then $\gamma_0^* > 0$ (it cannot be 0 because this will contradict with the value of q_i), which implies that $q_0^* = 1$
if $\alpha_{i+1}^* < \alpha_i^*$, then $\gamma_i^* > 0$ (it cannot be 0 because this contradicts with the value of q_i), which implies that $q_i^* = 1$

Otherwise, if $\alpha_1^* = \alpha_n^* \neq 0$ (they cannot be equal to 0 at the same time, which means that $q_0 = 1$, and we know in advance that this cannot be the case here), we have $q_1^* = \frac{p_0 q_0^*}{p_1} - \frac{\lambda_v p_0 - \lambda_c^{(1)} - R^*}{\lambda_v p_1}$ and $q_{n-1}^* = \frac{p_0 q_0^*}{p_{n-1}} - \frac{\lambda_v p_0 - \lambda_c^{(n)} - R^*}{\lambda_v p_{n-1}}$

Finally, if $\alpha_{i+1}^* = \alpha_i^* \neq 0$ (they cannot be equal to 0 at the same time, which means that $q_i = 1$, and we know in advance that this cannot be the case here), we have $q_i^* = \frac{p_{i-1} q_{i-1}^*}{p_i} - \frac{\lambda_v p_{i-1} - \lambda_c^{(i)} - R^*}{\lambda_v p_i}$ and $q_{i+1}^* = \frac{p_i q_i^*}{p_{i+1}} - \frac{\lambda_v p_i - \lambda_c^{(i+1)} - R^*}{\lambda_v p_{i+1}}$

Therefore, the optimal charging/dispatching decisions of the optimization problem in (2.17) can be expressed as follows:

$$
\begin{aligned}
q_0^* &= \begin{cases} 0 & \text{if } \alpha_1^* > \alpha_n^* \\ 1 & \text{if } \alpha_1^* < \alpha_n^* \end{cases} \\
q_i^* &= \begin{cases} 0 & \text{if } \alpha_{i+1}^* > \alpha_i^* \\ 1 & \text{if } \alpha_{i+1}^* < \alpha_i^* \end{cases} \qquad i = 1, \ldots, n-1 \\
\text{if } \alpha_1^* = \alpha_n^* \neq 0 & \begin{cases} q_1^* = \frac{p_0 q_0^*}{p_1} - \frac{\lambda_v p_0 - \lambda_c^{(1)} - R^*}{\lambda_v p_1} \\ q_{n-1}^* = \frac{p_0 q_0^*}{p_{n-1}} - \frac{\lambda_v p_0 - \lambda_c^{(n)} - R^*}{\lambda_v p_{n-1}} \end{cases} \\
\text{if } \alpha_{i+1}^* = \alpha_i^* \neq 0 & \begin{cases} q_i^* = \frac{p_{i-1} q_{i-1}^*}{p_i} - \frac{\lambda_v p_{i-1} - \lambda_c^{(i)} - R^*}{\lambda_v p_i} \\ q_{i+1}^* = \frac{p_i q_i^*}{p_{i+1}} - \frac{\lambda_v p_i - \lambda_c^{(i+1)} - R^*}{\lambda_v p_{i+1}} \end{cases} \qquad i = 1, \ldots, n-1
\end{aligned}
\tag{2.23}
$$

2.3.1.4 Maximum Expected Response Time

Again, since the problem in (2.17) is convex with differentiable objective and constraint functions, then strong duality holds, which implies that the solution to the primal and dual problems is identical. By solving the dual problem, we can express the optimal value of the maximum expected response time as the reciprocal of the minimum expected response rate of the system.

To get the minimum expected response rate R^* of the entire system, we first start by putting the problem on the standard linear programming form as follows:

$$
\begin{aligned}
&\underset{q_0, q_1, \ldots, q_{n-1}}{\text{minimize}} \quad -R \\
&\text{subject to} \\
&\text{Constraint on costumers arrivals queues} \\
&\lambda_v (p_{i-1} q_{i-1} - p_i q_i) + R \leq \lambda_v p_{i-1} - \lambda_c^{(i)}, \; i = 1, \ldots, n-1 \\
&\lambda_v (p_{n-1} q_{n-1} - p_0 q_0) + R \leq \lambda_v p_{n-1} - \lambda_c^{(n)} \\
&\text{Constraint on charging vehicles queues} \\
&-\lambda_v \sum_{i=0}^{n-1} p_i q_i^* < C(n\mu_c) - \lambda_v \\
&\lambda_v p_0 q_0 < \mu_c \\
&\text{Constraint on probabilities and decisions} \\
&q_i \leq 1 \qquad i = 0, \ldots, n-1 \\
&-q_i \leq 0 \qquad i = 0, \ldots, n-1 \\
&-R < 0 \\
&\sum_{i=0}^{n-1} p_i = 1 \qquad 0 < p_i < 1 \qquad i = 0, \ldots, n-1
\end{aligned}
\tag{2.24}
$$

Writing the problem on its matrix form, we get

$$
\begin{aligned}
&\underset{X}{\text{minimize}} \quad \mathbf{c}^T \mathbf{x} \\
&\text{subject to} \quad \mathbf{A}\mathbf{x} \preceq \mathbf{b}
\end{aligned}
\tag{2.25}
$$

where

$$\mathbf{x}_{(n+1\times1)} = \begin{pmatrix} q_0 \\ q_1 \\ \vdots \\ q_{n-1} \\ R \end{pmatrix} \quad \mathbf{c}_{(n+1\times1)} = \begin{pmatrix} 0 \\ 0 \\ 0 \\ \vdots \\ -1 \end{pmatrix} \quad \mathbf{b}_{(3n+4\times1)} = \begin{pmatrix} \lambda_v p_0 - \lambda_c^{(1)} \\ \vdots \\ \lambda_v p_{n-1} - \lambda_c^{(n)} \\ C(n\mu_c) - \lambda_v \\ \mu_c \\ 1 \\ \vdots \\ 1 \\ \infty \\ 0 \\ \vdots \\ 0 \end{pmatrix} \tag{2.26}$$

$$\mathbf{A}_{(3n+4\times n+1)} = \begin{pmatrix} \lambda_v p_0 & -\lambda_v p_1 & 0 & \cdots & 0 & 1 \\ 0 & \lambda_v p_1 & -\lambda_v p_2 & \cdots & 0 & 1 \\ \vdots & \ddots & \ddots & \ddots & \ddots & \vdots \\ 0 & \cdots & 0 & \lambda_v p_{n-2} & -\lambda_v p_{n-1} & 1 \\ -\lambda_v p_0 & 0 & \cdots & \cdots & \lambda_v p_{n-1} & 1 \\ -\lambda_v p_0 & -\lambda_v p_1 & \cdots & \cdots & -\lambda_v p_{n-1} & 0 \\ \lambda_v p_0 & 0 & \cdots & \cdots & \cdots & 0 \\ & & I_{n+1} & & & \\ & & & & & \\ & & -I_{n+1} & & & \end{pmatrix} \tag{2.27}$$

The matrix form of the Lagrangian function can thus be expressed as follows:

Lagrangian:

$$L(\mathbf{x}, \nu) = \mathbf{c}^T\mathbf{x} + \nu^T(\mathbf{A}\mathbf{x} - \mathbf{b}) = -\mathbf{b}^T + \left(\mathbf{A}^T\nu + \mathbf{c}\right)^T\mathbf{x} \tag{2.28}$$

where ν is the vector of the dual variables or Lagrange multipliers vector associated with the problem 2.25. Each element ν_i of ν is the Lagrange multiplier associated with the i-th inequality constraint $\mathbf{a}_i\mathbf{x} - b_i \leq 0$, where $\mathbf{a}_i$ and b_i are the i-th row and and i-th element of matrix $\mathbf{A}$ and vector $\mathbf{b}$, respectively. In fact, ν is the vector that includes all the vectors α, β, γ, ω as follows:

$$\nu^T_{(1\times3n+4)} = \left(\alpha_1 \dots \alpha_n\ \beta_0\ \beta_1\ \gamma_0 \dots \gamma_n\ \omega_0 \dots \omega_n\right) \tag{2.29}$$

We will use this combined notation for ease and clarity of notation.

The Lagrange dual function is expressed as follows:

$$g(\nu) = \inf_{\mathbf{x}} L(\mathbf{x}, \nu) = -\mathbf{b}^T\nu + \inf_{\mathbf{x}} \left(\mathbf{A}^T\nu + \mathbf{c}\right)^T\mathbf{x} \tag{2.30}$$

The solution for this function is easily determined analytically, as a linear function is bounded below only when it is identically zero. Thus, $g(\nu) = -\infty$ except when $\mathbf{A}^T\nu + \mathbf{c} = \mathbf{0}$, where $\mathbf{0}$ is the all-zero vector. Consequently, we have

$$g(\nu) = \begin{cases} -\mathbf{b}^T\nu & \mathbf{A}^T\nu + \mathbf{c} = \mathbf{0} \\ -\infty & \text{ortherwise} \end{cases} \tag{2.31}$$

For each $\nu \succeq \mathbf{0}$ (i.e., $\nu_i \geq 0\ \forall\ i$), the Lagrange dual function gives us a lower bound on the optimal value of the original optimization problem. This leads to a new equivalent optimization problem, which is the dual problem:

$$\begin{aligned} \max_{\nu} \quad & g(\nu) = -\mathbf{b}^T \nu \\ \text{subject to} \quad & \mathbf{A}^T \nu + \mathbf{c} = \mathbf{0} \\ & \nu \succeq \mathbf{0} \end{aligned} \tag{2.32}$$

Applying Slater's theorem for duality qualification, and as strong duality holds for the considered optimization problem, then solving the dual problem gives the exact optimal solution for the primal problem. This is described by the equality:

$$g(\nu^*) = -\mathbf{b}^T \nu^* = \mathbf{c}^T \mathbf{x}^* = -R^* \tag{2.33}$$

By expanding on the values of $\mathbf{b}$ and ν in the above equation, the optimal value of R^* can be expressed as follows:

$$R^* = \sum_{i=1}^{n} \left(\lambda_\nu p_{i-1} - \lambda_c^{(i)}\right) \alpha_i^* + \sum_{i=0}^{n-1} \gamma_i^* \tag{2.34}$$

2.3.1.5 Average Response Time Optimization Problem Formulation

As stated earlier, the expected response time for each of the classes in the system is expressed as in (2.14).

$$\frac{1}{\lambda_\nu^{(i)} - \lambda_c^{(i)}} \quad i = 1, \ldots, n \tag{2.35}$$

Since our system is divided into n classes, the average expected response time across the different classes is expressed as follows:

$$\frac{1}{n} \sum_{i=1}^{n} \frac{1}{\lambda_\nu^{(i)} - \lambda_c^{(i)}} \tag{2.36}$$

Therefore, minimizing the average expected response time across all the classes of the system, while obeying its stability conditions, can be formulated by the following problem.

$$\underset{q_0, q_1, \ldots, q_{n-1}}{\text{minimize}} \ \frac{1}{n} \sum_{i=1}^{n} \frac{1}{\lambda_\nu^{(i)} - \lambda_c^{(i)}} \tag{2.37a}$$

s.t.

$$\lambda_\nu \left(p_{i-1} - p_{i-1} q_{i-1} + p_i q_i\right) - \lambda_c^{(i)} > 0, \ i = 1, \ldots, n-1 \tag{2.37b}$$

$$\lambda_\nu \left(p_{n-1} - p_{n-1} q_{n-1} + p_0 q_0\right) - \lambda_c^{(n)} > 0 \tag{2.37c}$$

$$\sum_{i=0}^{n-1} \lambda_\nu \left(p_i - p_i q_i\right) < C(n\mu_c) \tag{2.37d}$$

$$\lambda_\nu p_0 q_0 < \mu_c \tag{2.37e}$$

$$\sum_{i=0}^{n-1} p_i = 1, \ 0 \leq p_i \leq 1 \qquad i = 0, \ldots, n-1 \tag{2.37f}$$

$$0 \leq q_i \leq 1 \qquad i = 0, \ldots, n-1 \tag{2.37g}$$

The n constraints in (2.37b) and (2.37c) represent the stability constraints in (2.3) and substituting every $\lambda_\nu^{(i)}$ by its expansion form in (2.2). The constraints in (2.37d) and (2.37e) represent the stability conditions on charging queues. The constraints in (2.37f) and (2.37g) are the axiomatic constraints on probabilities (i.e., values being between 0 and 1, and sum equal to 1).

The above constraints are all linear but the objective function is obviously not. After substituting every $\lambda_v^{(i)}$ by its expansion form in (2.2), the function f can be expressed as follows:

$$\begin{aligned} f(q_0, q_1, \ldots, q_{n-1}) &= \frac{1}{n}\sum_{i=1}^{n}\frac{1}{\lambda_v^{(i)} - \lambda_c^{(i)}} \\ &= \frac{1}{n}\sum_{i=1}^{n-1}\frac{1}{\lambda_v(p_{i-1} - p_{i-1}q_{i-1} + p_i q_i) - \lambda_c^{(i)}} \\ &+ \frac{1}{n\left(\lambda_v(p_{n-1} - p_{n-1}q_{n-1} + p_0 q_0) - \lambda_c^{(n)}\right)} \end{aligned} \tag{2.38}$$

We know that a function f is convex, if it is continuous and second-order differentiable, which is the case of our function because it is a sum of continuous and second-order differentiable functions. Moreover, as f is a multivariable function, its Hessian matrix has to be positive semidefinite for the function to be convex. Let H be the Hessian matrix of f such that

$$H_{i,j} = \frac{\partial^2 f}{\partial q_i \partial q_j} \qquad i, j = 0, \ldots, n-1 \tag{2.39}$$

We notice that the Hessian matrix is a symmetric matrix because

$$\frac{\partial^2 f}{\partial q_j \partial q_i} = \frac{\partial^2 f}{\partial q_i \partial q_j} \qquad \forall i, j \tag{2.40}$$

which means that

$$H_{i,j} = H_{j,i} \qquad \forall\, i, j \tag{2.41}$$

The calculation of the terms in the Hessian matrix leads to the following results:

$$\begin{aligned} \frac{\partial^2 f}{\partial^2 q_0} &= \frac{2\lambda_v^2 p_0^2}{\left(\lambda_v(p_0 - p_0 q_0 + p_1 q_1) - \lambda_c^{(1)}\right)^3} + \frac{2\lambda_v^2 p_0^2}{\left(\lambda_v(p_{n-1} - p_{n-1}q_{n-1} + p_0 q_0) - \lambda_c^{(n)}\right)^3} \\ \frac{\partial^2 f}{\partial^2 q_i} &= \frac{2\lambda_v^2 p_i^2}{\left(\lambda_v(p_{i-1} - p_{i-1}q_{i-1} + p_i q_i) - \lambda_c^{(i)}\right)^3} \\ &+ \frac{2\lambda_v^2 p_i^2}{\left(\lambda_v(p_i - p_i q_i + p_{i+1}q_{i+1}) - \lambda_c^{(i+1)}\right)^3}, \qquad i = 1, \ldots, n-2 \\ \frac{\partial^2 f}{\partial^2 q_{n-1}} &= \frac{2\lambda_v^2 p_{n-1}^2}{\left(\lambda_v(p_{n-2} - p_{n-2}q_{n-2} + p_{n-1}q_{n-1}) - \lambda_c^{(n-1)}\right)^3} \\ &+ \frac{2\lambda_v^2 p_{n-1}^2}{\left(\lambda_v(p_{n-1} - p_{n-1}q_{n-1} + p_0 q_0) - \lambda_c^{(n)}\right)^3} \\ \frac{\partial^2 f}{\partial^2 q_{i-1} q_i} &= \frac{-2\lambda_v^2 p_{i-1} p_i}{\left(\lambda_v(p_{i-1} - p_{i-1}q_{i-1} + p_i q_i) - \lambda_c^{(i)}\right)^3} \qquad i = 1, \ldots, n-1 \end{aligned}$$

$$\frac{\partial^2 f}{\partial^2 q_i q_j} = 0 \qquad j > i+1,\ i,j = 0,\ldots,n-1$$

$$\frac{\partial^2 f}{\partial^2 q_{n-1} q_0} = \frac{-2\lambda_v^2 p_0 p_{n-1}}{\left(\lambda_v \left(p_{n-1} - p_{n-1} q_{n-1} + p_0 q_0\right) - \lambda_c^{(n)}\right)^3}. \tag{2.42}$$

Let $x = [x_i]$ be an $n \times 1$ vector such that $0 \leq x_i \leq 1\ \forall\ i$ as our variables q_i on which the function depends are varies that range $q_i \in [0,1]$. Now we know that **H** is a positive semidefinite, if it satisfies the following condition:

$$\mathbf{x}^T \mathbf{H} \mathbf{x} \geq 0 \tag{2.43}$$

$$\mathbf{x}^T \mathbf{H} = \begin{cases} x_0 \frac{\partial^2 f}{\partial^2 q_0} + x_1 \frac{\partial^2 f}{\partial^2 q_0 q_1} + x_{n-1} \frac{\partial^2 f}{\partial^2 q_0 q_{n-1}} & \\ x_{i-1} \frac{\partial^2 f}{\partial^2 q_{i-1} q_i} + x_i \frac{\partial^2 f}{\partial^2 q_i} + x_{i+1} \frac{\partial^2 f}{\partial^2 q_{i+1} q_i} & i = 1,\ldots,n-2 \\ x_0 \frac{\partial^2 f}{\partial^2 q_0 q_{n-1}} + x_{n-2} \frac{\partial^2 f}{\partial^2 q_{n-1} q_{n-2}} + x_{n-1} \frac{\partial^2 f}{\partial^2 q_{n-1}} & \end{cases} \tag{2.44}$$

Multiplying (2.44) by the vector **x** gives

$$\begin{aligned} \mathbf{x}^T \mathbf{H} \mathbf{x} = {} & x_0 \left(x_0 \frac{\partial^2 f}{\partial^2 q_0} + x_1 \frac{\partial^2 f}{\partial^2 q_0 q_1} + x_{n-1} \frac{\partial^2 f}{\partial^2 q_0 q_{n-1}} \right) \\ & + \sum_{i=1}^{n-2} x_i \left(x_{i-1} \frac{\partial^2 f}{\partial^2 q_{i-1} q_i} + x_i \frac{\partial^2 f}{\partial^2 q_i} + x_{i+1} \frac{\partial^2 f}{\partial^2 q_{i+1} q_i} \right) \\ & + x_{n-1} \left(x_0 \frac{\partial^2 f}{\partial^2 q_0 q_{n-1}} + x_{n-2} \frac{\partial^2 f}{\partial^2 q_{n-1} q_{n-2}} + x_{n-1} \frac{\partial^2 f}{\partial^2 q_{n-1}} \right) \end{aligned} \tag{2.45}$$

Simplifying (2.45) gives

$$\mathbf{x}^T \mathbf{H} \mathbf{x} = \sum_{i=0}^{n-1} x_i^2 \frac{\partial^2 f}{\partial^2 q_i} + 2 \sum_{i=1}^{n-1} x_i x_{i-1} \frac{\partial^2 f}{\partial^2 q_{i-1} q_i} + 2 x_0 x_{n-1} \frac{\partial^2 f}{\partial^2 q_{n-1} q_0} \tag{2.46}$$

Substituting (2.42) in (2.46), we get

$$\begin{aligned} \mathbf{x}^T \mathbf{H} \mathbf{x} = {} & \sum_{i=1}^{n-1} \frac{2\lambda_v^2 \left(p_i x_i - p_{i-1} x_{i-1}\right)^2}{\left(\lambda_v \left(p_{i-1} - p_{i-1} q_{i-1} + p_i q_i\right) - \lambda_c^{(i)}\right)^3} \\ & + \frac{2\lambda_v^2 \left(p_0 x_0 - p_{n-1} x_{n-1}\right)^2}{\left(\lambda_v \left(p_{n-1} - p_{n-1} q_{n-1} + p_0 q_0\right) - \lambda_c^{(n)}\right)^3} \end{aligned} \tag{2.47}$$

We can see clearly that $\mathbf{x}^T \mathbf{H} \mathbf{x} \geq 0$ because it is a sum of positive terms. Consequently, **H** is a positive semidefinite, thus making f a convex function. Therefore, the optimization problem we have is convex, which allows us to find an absolute exact solution analytically and numerically.

2.3.1.6 Optimal Dispatching and Charging Decision

As shown above, the problem in (2.37) is a convex optimization problem with second-order differentiable objective functions and constraints that satisfy Slater's condition. Similar to the approach of Section 2.3.1.3, we can apply the KKT conditions to the constraints of the problem and the gradient of the Lagrangian function to find the analytical solution of the decisions q_i. The Lagrangian function associated with the optimization problem in (2.37) is given by the following expression:

$$\begin{aligned}L(\mathbf{q},\alpha,\beta,\gamma,\omega) &= \frac{1}{n}\sum_{i=1}^{n-1}\frac{1}{\lambda_v\left(p_{i-1}-p_{i-1}q_{i-1}+p_iq_i\right)-\lambda_c^{(i)}}\\
&+\frac{1}{n\left(\lambda_v\left(p_{n-1}-p_{n-1}q_{n-1}+p_0q_0\right)-\lambda_c^{(n)}\right)}+\sum_{i=1}^{n-1}\alpha_i\left(\lambda_c^{(i)}-\lambda_v\left(p_{i-1}-p_{i-1}q_{i-1}+p_iq_i\right)\right)\\
&+\alpha_n\left(\lambda_c^{(n)}-\lambda_v\left(p_{n-1}-p_{n-1}q_{n-1}+p_0q_0\right)\right)+\beta_0\left(\sum_{i=0}^{n-1}\lambda_v\left(p_i-p_iq_i\right)-C\left(n\mu_c\right)\right)\\
&+\beta_1\left(\lambda_vp_0q_0-\mu_c\right)+\sum_{i=0}^{n-1}\gamma_i\left(q_i-1\right)-\sum_{i=0}^{n-1}\omega_iq_i\end{aligned} \tag{2.48}$$

where $\mathbf{q}$ is the vector of dispatching decisions (i.e., $\mathbf{q}=[q_0,\ldots,q_{n-1}]$), and where $\alpha=[\alpha_i]$, $\beta=[\beta_i]$, $\gamma=[\gamma_i]$, $\omega=[\omega_i]$ are the vectors of the Lagrange multipliers associated with the inequalities constraints of the problem (2.37) and defined in the same way as explained in Section 2.3.1.3.

By applying the KKT conditions to the inequality constraints of the problem (2.37), we get

$$\begin{aligned}&\alpha_i^*\left(\lambda_c^{(i)}-\lambda_v\left(p_{i-1}-p_{i-1}q_{i-1}^*+p_iq_i^*\right)\right)=0 && i=1,\ldots,n-1\\
&\alpha_n^*\left(\lambda_c^{(n)}-\lambda_v\left(p_{n-1}-p_{n-1}q_{n-1}^*+p_0q_0^*\right)\right)=0\\
&\beta_0^*\left(\sum_{i=0}^{n-1}\lambda_v\left(p_i-p_iq_i^*\right)-C\left(n\mu_c\right)\right)=0\\
&\beta_1^*\left(\lambda_vp_0q_0^*-\mu_c\right)=0\\
&\gamma_i^*\left(q_i^*-1\right)=0 && i=0,\ldots,n-1\\
&\omega_i^*q_i^*=0 && i=0,\ldots,n-1\end{aligned} \tag{2.49}$$

From Burke's theorem on the stability condition of queues, the constraints on the customers' queues and the charging queues are strict inequalities. Combining Burke's theorem with the equations in (2.49), we find that $\beta_0^*=\beta_1^*=0$ and $\alpha_i^*=0\ \forall\ i$. Applying the KKT conditions to the Lagrangian function in (2.48), and knowing that the gradient of the Lagrangian function goes to 0 at the optimal solution, we get the following set of equalities:

$$\begin{aligned}&\frac{\lambda_vp_0}{n}\left(\frac{1}{\left(\lambda_v\left(p_{n-1}-p_{n-1}q_{n-1}^*+p_0q_0^*\right)-\lambda_c^{(n)}\right)^2}-\frac{1}{\left(\lambda_v\left(p_0-p_0q_0^*+p_1q_1^*\right)-\lambda_c^{(1)}\right)^2}\right)\\
&+\gamma_0^*-\omega_0^*=0\\
&\frac{\lambda_vp_i}{n}\left(\frac{1}{\left(\lambda_v\left(p_{i-1}-p_{i-1}q_{i-1}^*+p_iq_i^*\right)-\lambda_c^{(i)}\right)^2}-\frac{1}{\left(\lambda_v\left(p_i-p_iq_i^*+p_{i+1}q_{i+1}^*\right)-\lambda_c^{(i+1)}\right)^2}\right)\\
&+\gamma_i^*-\omega_i^*=0, \qquad i=1,\ldots,n-2\\
&\frac{\lambda_vp_{n-1}}{n}\left(\frac{1}{\left(\lambda_v\left(p_{n-2}-p_{n-2}q_{n-2}^*+p_{n-1}q_{n-1}^*\right)-\lambda_c^{(n-1)}\right)^2}\right.\\
&\left.-\frac{1}{\left(\lambda_v\left(p_{n-1}-p_{n-1}q_{n-1}^*+p_0q_0^*\right)-\lambda_c^{(n)}\right)^2}\right)+\gamma_{n-1}^*-\omega_{n-1}^*=0\end{aligned} \tag{2.50}$$

The fact that $\beta_i^* = 0$ and $\omega_i^* = 0$ explains the absence of β_i^* and ω_i^* in (2.50). Multiplying the first equality in (2.50) by q_0^*, the second by q_i^*, and the third by q_{n-1}^* gives

$$\begin{aligned}
\gamma_0^* &= \frac{\lambda_v p_0}{n} q_0^* \left(\frac{1}{\left(\lambda_v \left(p_0 - p_0 q_0^* + p_1 q_1^*\right) - \lambda_c^{(1)}\right)^2} - \frac{1}{\left(\lambda_v \left(p_{n-1} - p_{n-1} q_{n-1}^* + p_0 q_0^*\right) - \lambda_c^{(n)}\right)^2} \right) \\
\gamma_i^* &= \frac{\lambda_v p_i}{n} qi^* \left(\frac{1}{\left(\lambda_v \left(p_i - p_i q_i^* + p_{i+1} q_{i+1}^*\right) - \lambda_c^{(i+1)}\right)^2} - \frac{1}{\left(\lambda_v \left(p_{i-1} - p_{i-1} q_{i-1}^* + p_i q_i^*\right) - \lambda_c^{(i)}\right)^2} \right) \\
& i = 1, \ldots, n-2 \\
\gamma_{n-1}^* &= \frac{\lambda_v p_{n-1}}{n} q_{n-1}^* \left(\frac{1}{\left(\lambda_v \left(p_{n-1} - p_{n-1} q_{n-1}^* + p_0 q_0^*\right) - \lambda_c^{(n)}\right)^2} \right. \\
& \left. - \frac{1}{\left(\lambda_v \left(p_{n-2} - p_{n-2} q_{n-2}^* + p_{n-1} q_{n-1}^*\right) - \lambda_c^{(n-1)}\right)^2} \right)
\end{aligned} \tag{2.51}$$

Substituting (2.51) in the fifth equality of (2.50) gives

$$\begin{aligned}
& \left(q_0^* - 1\right) q_0^* \left(\frac{1}{\left(\lambda_v \left(p_0 - p_0 q_0^* + p_1 q_1^*\right) - \lambda_c^{(1)}\right)^2} - \frac{1}{\left(\lambda_v \left(p_{n-1} - p_{n-1} q_{n-1}^* + p_0 q_0^*\right) - \lambda_c^{(n)}\right)^2} \right) = 0 \\
& \left(q_i^* - 1\right) qi^* \left(\frac{1}{\left(\lambda_v \left(p_i - p_i q_i^* + p_{i+1} q_{i+1}^*\right) - \lambda_c^{(i+1)}\right)^2} \right. \\
& \left. - \frac{1}{\left(\lambda_v \left(p_{i-1} - p_{i-1} q_{i-1}^* + p_i q_i^*\right) - \lambda_c^{(i)}\right)^2} \right) = 0, \qquad i = 1, \ldots, n-2 \\
& \left(q_{n-1}^* - 1\right) q_{n-1}^* \left(\frac{1}{\left(\lambda_v \left(p_{n-1} - p_{n-1} q_{n-1}^* + p_0 q_0^*\right) - \lambda_c^{(n)}\right)^2} \right. \\
& \left. - \frac{1}{\left(\lambda_v \left(p_{n-2} - p_{n-2} q_{n-2}^* + p_{n-1} q_{n-1}^*\right) - \lambda_c^{(n-1)}\right)^2} \right) = 0
\end{aligned} \tag{2.52}$$

From (2.49), we have $0 < q_i^* < 1 \; \forall \, i$, only if $\gamma_i^* = 0$ and $\omega_i^* = 0$. Since $0 \leq q_i^* \leq 1$, the above result may not always be true.

So, the optimal charging/dispatching decisions of the optimization problem in (2.37) can be expressed as follows:

$$q_i^* = \begin{cases} 0 & \text{if } \omega_i^* \neq 0 \\ 1 & \text{if } \gamma_i^* \neq 1 \end{cases} \qquad i = 0, \ldots, n-1 \tag{2.53}$$

Otherwise, we have:

$$
\begin{aligned}
q_0^* &= \frac{\lambda_v\left(p_0+p_1q_1^*-p_{n-1}+p_{n-1}q_{n-1}^*\right)-\lambda_c^{(1)}+\lambda_c^{(n)}}{2\lambda_v p_0} \\
q_i^* &= \frac{\lambda_v\left(p_i+p_{i+1}q_{i1}^*-p_{i-1}+p_{i-1}q_{i-1}^*\right)-\lambda_c^{(i+1)}+\lambda_c^{(i)}}{2\lambda_v p_i} \qquad i=1,\ldots,n-2 \\
q_{n-1}^* &= \frac{\lambda_v\left(p_{n-1}+p_0q_0^*-p_{n-2}+p_{n-2}q_{n-2}^*\right)-\lambda_c^{(n)}+\lambda_c^{(n-1)}}{2\lambda_v p_{n-1}}
\end{aligned} \tag{2.54}
$$

2.3.1.7 Simulation Results

In this section, we test the merits of the work discussed in Section (2.3.1) using extensive simulations. The metrics used to evaluate these merits are the maximum and average expected response times of different classes. For all the performed simulation figures, the full-charging rate of a vehicle is set to $\mu_c = 0.033$ min^{-1}, and the number of charging points $C = 40$.

For the optimized maximum and average response time solutions, Figures 2.3a and b, respectively, illustrate both the interplay of λ_v and $\sum_{i=1}^{n}\lambda_c^{(i)}$ and the effect of increasing the number of classes n beyond its derived strict lower bound. They depict the maximum and average expected response times for different values of $\sum_{i=1}^{n}\lambda_c^{(i)}$, while fixing λ_v to 15 min^{-1}. For this setting, $n = 12$ is the smallest number of classes that satisfy the stability condition. It is easy to notice that the response times for all values of n increase dramatically when the $\sum_{i=1}^{n}\lambda_c^{(i)}$ approaches λ_v, thus bringing the system closer to its stability limit. As also expected, the figures clearly show that further increasing n beyond its stability, lower bound increases both the maximum and average response times. As explained earlier, this effect occurs when due to the reduced number of available vehicles to each customer class as n grows, given fixed λ_v. We thus firmly conclude that the optimal number of classes is the smallest value satisfying (2.12) as follows:

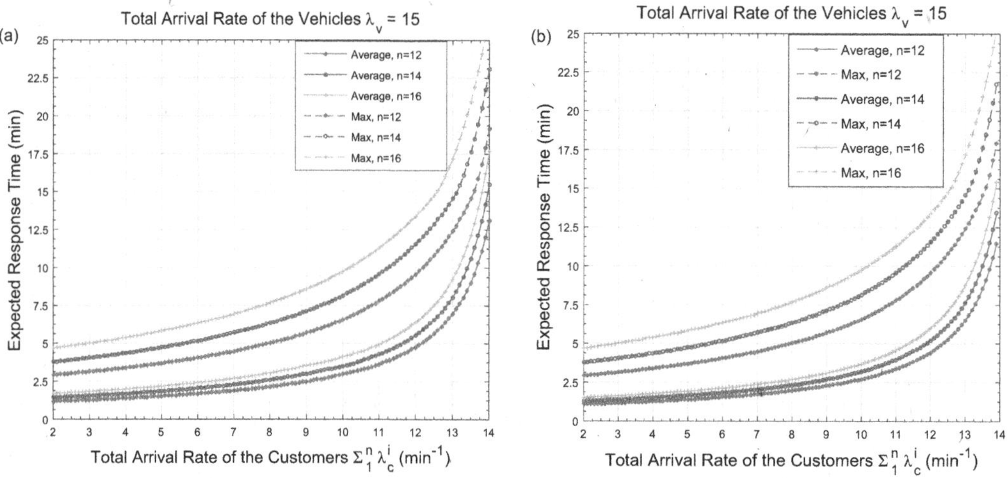

Figure 2.3 (a) Expected response times using the maximum response time optimization solution for different $\sum_{i=1}^{n}\lambda_c^{(i)}$. (b) Expected response times using the average response time optimization solution for different $\sum_{i=1}^{n}\lambda_c^{(i)}$

$$n^* = \begin{cases} \frac{\lambda_v}{C\mu_c} - \frac{1}{C} + 1 & \text{if } \frac{\lambda_v}{C\mu_c} - \frac{1}{C} \text{ is integer} \\ \left\lceil \frac{\lambda_v}{C\mu_c} - \frac{1}{C} \right\rceil & \text{Otherwise} \end{cases} \tag{2.55}$$

For the maximum and average response time optimization solutions, Figures 2.4a and b, respectively, depict the maximum and average expected response time performances for different distributions of the vehicle SoC and customer trip distances, given $\lambda_v = 8$ and thus $n^* = 7$. By decreasing the vehicle SoC distribution, we mean that the probability of an arriving vehicle to have class i SoC is lower than that of it having class $i-1$ SoC $\forall\, i \in \{2,\ldots,n\}$. We can infer from both figures that both the maximum and average response times for Gaussian distributions of trip distances and both Gaussian or decreasing ones for SoCs are the lowest and exhibit the least response time variance. Luckily, these are the most realistic distributions for both variables. This is justified by the fact that vehicles arrive to the system after trips of different distances, which makes their SoC either Gaussian or slightly decreasing. Likewise, customers requiring mid-size distances are usually more than those requiring very small and very long distances.

For the maximum and average response time optimization solutions, Figures 2.5a and b, respectively, compare the maximum and average expected response times performances against $\sum_{i=1}^{n} \lambda_c^{(i)}$, for different decision approaches, namely, our derived optimal decisions in Section 2.3.1.2, always partially charge decisions (i.e., $q_i = 0\ \forall\ i$) and equal split decisions (i.e., $q_i = 0.5\ \forall\ i$); for $\lambda_v = 8$ and thus $n = 7$. The latter two schemes represent nonoptimized policies, in which each vehicle takes its own fixed decision irrespective of the system parameters. The figures clearly show superior maximum and average performances for our derived optimal policies compared to the other two policies, especially as $\sum_{i=1}^{n} \lambda_c^{(i)}$ gets closer to λ_v, which are the most properly engineered scenarios (as large differences between these two quantities resultin very low utilization). Gains of 13.3% and 21.3% in the average and maximum performances, respectively, can be noticed compared to the always charge policy. This demonstrates the importance of our proposed scheme in achieving lower response times and thus better customer satisfaction.

Fig. 2.6 compares the maximum and average expected response time performance given by the maximum and average response time optimization solutions introduced in Section 2.3.1.2 for

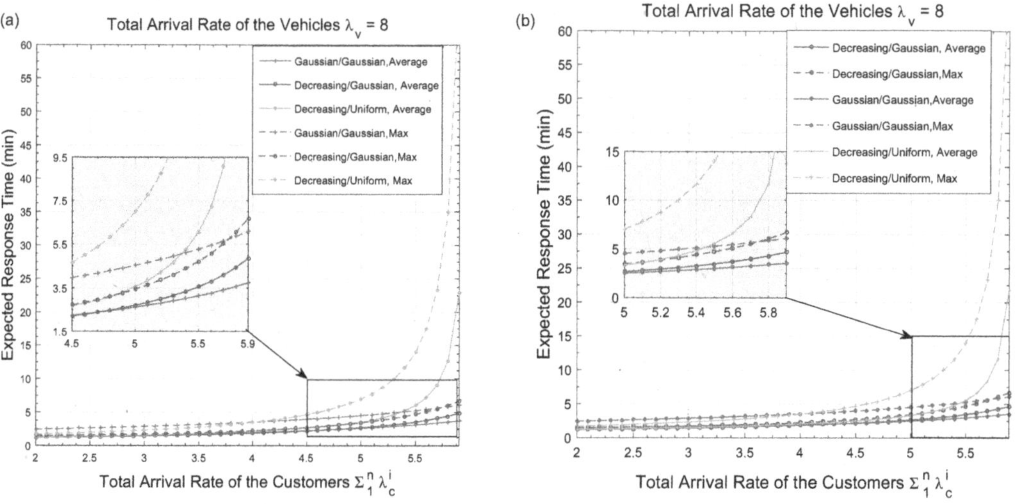

Figure 2.4 (a) Effect of different customer and SoC distributions on the maximum response time optimization solution. (b) Effect of different customer and SoC distributions on the average response time optimization solution.

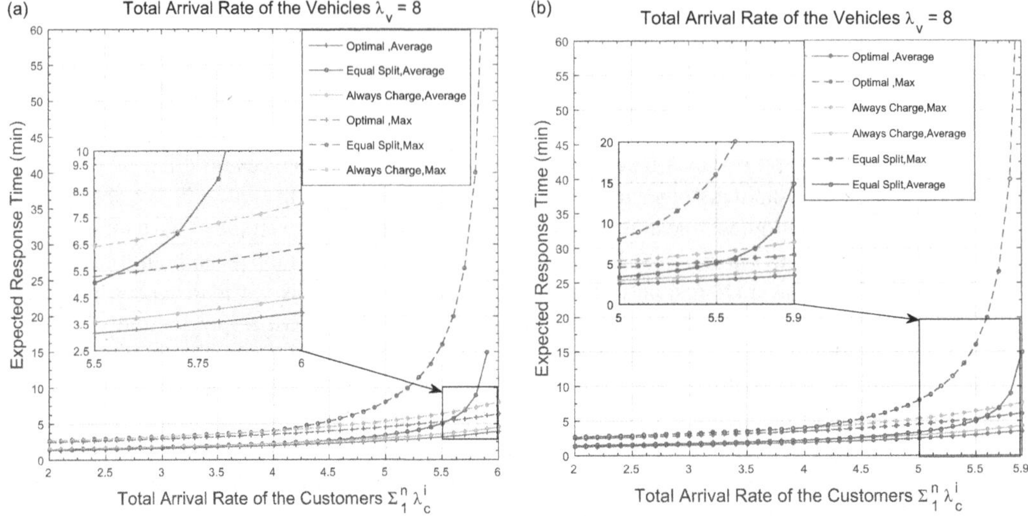

Figure 2.5 (a) Comparison of the maximum response time optimization solution to nonoptimized policies. (b) Comparison of the average response time optimization solution to nonoptimized policies.

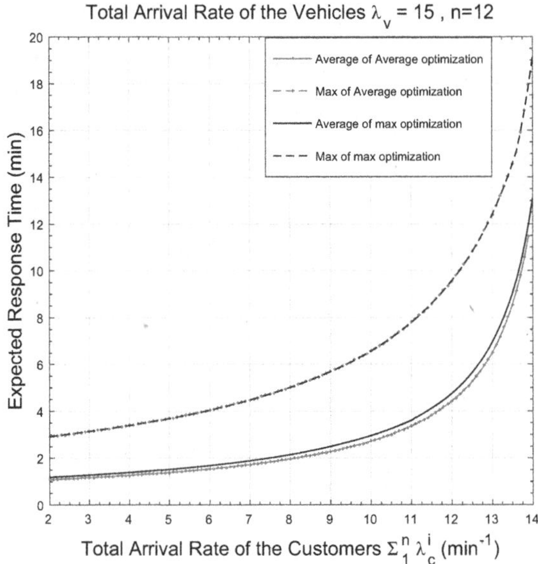

Figure 2.6 Comparison between the maximum minimization and the average minimization of the expected response time.

different values of the $\sum_{i=1}^{n} \lambda_c^{(i)}$ while fixing λ_v to 15 min^{-1} (i.e., $n^* = 12$). We can easily notice that the maximum expected response times achieved by both solutions are the same. On the other hand, the average expected response time given by the average solution is slightly lower than that of the maximum solution. These results suggest that the variance in performance achieved by both solutions is negligible. Consequently, the one that is obtained using less complexity should be used to almost satisfy the minimum value for both metrics. We know from Section 2.3.1.2 that the maximum and average solutions are obtained by solving linear and convex yet nonlinear.

optimizations, respectively. It is well-known that solving the latter requires more computations than the former. For example, when using the interior-point methods, the maximum numbers of iterations for the maximum and average solutions are 10 and 25, respectively. Thus, the use of the maximum solution is recommended in future AEMoD systems due to its lower complexity and its negligible degradation in its average response time performance compared to the average solution.

2.3.2 MULTICLASS DISPATCHING WITH SUBCLASS CHARGING

In this section, we will discuss about a refined model of the previous system, considering the management of one zone (controlled by a fog controller). This scheme mainly differs from the previously studied model in the dispatching (serving) decisions. Despite the valuable results given by deriving the optimal split proportions for this model, vehicles in this model only serve trips fitting their SoC (before or after partially charging), ignoring the fact that they can also serve all customer classes requesting smaller trips (and thus less charge). This is not only impractical but can also result in battery depletion of all vehicles by the end of the service, which can cause instabilities to the entire system. We will address this limitation of the previous model by enabling subclass dispatching; i.e., allowing vehicles of each class to serve customers from its own class as well as all classes, requesting shorter trips. We will then derive the optimal charging and subclasses dispatching proportions to maximize the expected response rate while maintaining stability.

We assume that the variables defined in the previously mentioned model, including the arrival rate of customers in class i $\lambda_c^{(i)}$, the arrival rate of all vehicles λ_v, the classification of vehicles and customers into n classes, and the amount of charging locations C, are the same as mentioned earlier.

Each arriving vehicle in class i , $i \in \{1,\ldots,n-1\}$ will either (1) join vehicles that will serve customers with their current state of charge with probability q_i or (2) charge its battery to the SoC of class $i+1$ with probability $\overline{q}_i = 1-q_i$ and join vehicles that will serve any subclass j with $j \leq i+1$. Vehicles of class 0 will either charge their battery to the fullest with probability q_0 or charge partially with probability $\overline{q}_0 = 1-q_0$.

The novelty in this model is that each vehicle belonging to class i, whether it went through the charging process or not, will serve customers of any subclass $j \leq i$. Considering the above change, each vehicle will be able to serve (1) customers from same class with probability Π_{ii} or (2) customers with a trip distance from any subclass $j\ \forall\ j \leq i$ with probability Π_{ij}.

As widely proposed (e.g., [12,13]), the time needed to fully charge a vehicle's depleted battery is exponentially distributed with rate μ_c. Consequently, the time for the partial charge of a battery from class i to class $i+1$ is represented with an exponential random variable. The rate of this variable is $n\mu_c$. Class i customers arrive at a rate $\lambda_c^{(i)}$. They receive service from vehicles at a rate of $\lambda_{vs}^{(i)}$. This service rate includes the summation of proportions of arrival rates of vehicles that (1) had Class $j\, \forall j \geq i$ SoC when they came to the zone and remained until they could serve customers from Class $i\, \forall i \leq j$ or (2) came to the zone with Class $j-1$ SoC, charged their batteries partially, and remained until they were assigned to serve a subclass $i\, \forall i \leq j$ customer. As shown in Figure 2.7, this system encompasses an M/M/1 queue of each of the n customer classes and for the central charging station. The system also encompasses an M/M/C queue for the C partial charging.

2.3.2.1 Stability Conditions

This section uses queuing theory to derive the following stability conditions. Each class of vehicles with an arrival rate $\lambda_v^{(i)}$ will be characterized by its SoC. Customers will be classified into n classes and served each with a separate queue of vehicles. $\lambda_{vs}^{(i)}$ is the vehicles' inflow rate available to serve customers in class i.

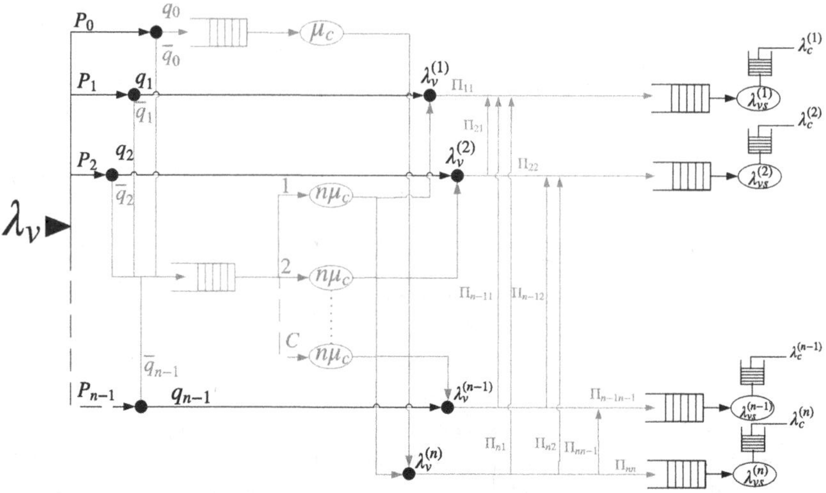

Figure 2.7 Joint charging and dispatching queuing model for multiclass system with subclass service in one city zone.

The expression for λ_v^k can be defined as follows:

$$\lambda_{vs}^{(i)} = \lambda_v \sum_{k=i}^{n-1} (p_{k-1} - p_{k-1}q_{k-1} + p_k q_k)\Pi_{ki} + \lambda_v (p_{n-1} - p_{n-1}q_{n-1} + p_0 q_0)\Pi_{ni},$$

$$i = 1, \ldots, n-1$$

$$\lambda_{vs}^{(n)} = \lambda_v (p_{n-1} - p_{n-1}q_{n-1} + p_0 q_0)\Pi_{nn} \tag{2.56}$$

The stability condition of the customers' queues and charging queues are given by Eqs. (2.73) and (2.74), respectively:

$$\lambda_{vs}^{(i)} > \lambda_c^{(i)}, \ i = 1, \ldots, n \tag{2.57}$$

$$\sum_{i=0}^{n-1} \lambda_v (p_i - p_i q_i) < C(n\mu_c) \tag{2.58}$$

$$\lambda_v p_0 q_0 < \mu_c \tag{2.59}$$

The fog controller places a time limit T on the response time for any class. This improves the response rates and results in an average response time constraint of

$$\frac{1}{\lambda_{vs}^{(i)} - \lambda_c^{(i)}} \leq T \text{ or } \lambda_{vs}^{(i)} - \lambda_c^{(i)} \geq R, \text{with } R = \frac{1}{T} \tag{2.60}$$

2.3.2.2 Subclass Charging and Dispatching Optimization Problem Formulation

Our objective is to minimize the maximum response time of each class. This is determined for each class by averaging the time between a customer request and the assignation of a vehicle to serve him or her. The optimization equation for response time is

$$\max_{i \in \{1,\ldots,n\}} \left\{ \frac{1}{\lambda_{vs}^{(i)} - \lambda_c^{(i)}} \right\} \tag{2.61}$$

From [24], we deduce that we can use the epigraph form to obtain the optimization problem (2.62). The equations in (2.62b) and (2.62c) represent the customer queueing stability conditions. The inequalities (2.62d) and (2.62e) are necessary to maintain stability in the charging queues. Eqs. (2.62f), (2.62g), (2.62h), and (2.62i) represent the fundamental constraints on probabilities. The inequality (2.62j) restricts the response time to positive values. This last constraint comes from a lower bound of the vehicles' arrival rate. This lower bound was derived in the previous work [26].

$$\underset{q_0,\ldots,q_{n-1},\Pi_{11},\ldots,\Pi_{nn}}{\text{maximize}} \quad R \tag{2.62a}$$

s.t

$$\lambda_v \sum_{k=i}^{n-1} (p_{k-1} - p_{k-1}q_{k-1} + p_k q_k)\Pi_{ki} + \lambda_v (p_{n-1} - p_{n-1}q_{n-1} + p_0 q_0)\Pi_{ni} - \lambda_c^{(i)} \geq R \tag{2.62b}$$

$$i = 1,\ldots,n-1$$

$$\lambda_v (p_{n-1} - p_{n-1}q_{n-1} + p_0 q_0)\Pi_{nn} - \lambda_c^{(n)} \geq R \tag{2.62c}$$

$$\sum_{i=0}^{n-1} \lambda_v (p_i - p_i q_i) < C(n\mu_c) \tag{2.62d}$$

$$\lambda_v p_0 q_0 < \mu_c \tag{2.62e}$$

$$\sum_{j=1}^{i} \Pi_{ij} = 1, \; i = 1,\ldots,n \tag{2.62f}$$

$$0 \leq \Pi_{ij} \leq 1, \; i = 1,\ldots,n, \; j = 1,\ldots,i \tag{2.62g}$$

$$0 \leq q_i \leq 1, \; i = 0,\ldots,n-1 \tag{2.62h}$$

$$\sum_{i=0}^{n-1} p_i = 1, \; 0 \leq p_i \leq 1, \; i = 0,\ldots,n-1 \tag{2.62i}$$

$$0 < R \leq \frac{\lambda_v - \sum_{i=1}^{n} \lambda_c^{(i)}}{n} \tag{2.62j}$$

2.3.2.3 Lower Bound Analytical Solutions

The optimization problem in (2.62) is nonconvex and quadratic. The objective and constraint functions are all second-order differentiable. The Lagrange dual optimization and KKT conditions solving for nonconvex problems provide an analytical lower-bound solution for the optimal one. In this section, we solve the problem (2.62) by first deriving an analytical solution, then numerically iterate toward the feasible set. Several methods and algorithms were proposed to increase the accuracy of the lower-bound solutions. For example in Ref. [25], the study proposed the *Suggest-and-Improve algorithm* for nonconvex quadratic problems.

The problem given in (2.62) can be optimized using the following Lagrangian function:

$$L(R,\mathbf{q},\Pi,\alpha,\beta,\gamma,\omega,\mu,\nu,\delta) = -\sum_{i=0}^{n-1} \omega_i q_i - \omega_n (R - \varepsilon_2)$$

$$+ \sum_{i=1}^{n-1} \alpha_i \left[\lambda_c^{(i)} - \lambda_v \sum_{k=i}^{n-1} (p_{k-1} - p_{k-1}q_{k-1} + p_k q_k)\Pi_{ki} - \lambda_v (p_{n-1} - p_{n-1}q_{n-1} + p_0 q_0)\Pi_{ni} + R \right]$$

$$+ \alpha_n \left(\lambda_c^{(n)} - \lambda_v (p_{n-1} - p_{n-1}q_{n-1} + p_0 q_0)\Pi_{ni} + R \right) + \beta_0 \left(\sum_{i=0}^{n-1} \lambda_v (p_i - p_i q_i) - C(n\mu_c) \right)$$

$$+\beta_1(\lambda_v p_0 q_0 - \mu_c) + \sum_{i=0}^{n-1} \gamma_i(q_i - 1) + \gamma_n \left(R - \frac{\lambda_v - \sum_{i=1}^{n} \lambda_c^{(i)}}{n} \right) - R + \sum_{i=1}^{n} \sum_{j=1}^{i} \nu_{ij}(\Pi_{ij} - 1) - \mu_{ij}\Pi_{ij}$$

$$+\sum_{i=1}^{n} \delta_i \left(\sum_{k=1}^{i} \Pi_{ik} - 1 \right) \tag{2.63}$$

where $\mathbf{q} = [q_0, \ldots, q_{n-1}]$ is the vector of charging decisions, $\Pi = [\Pi_{ij}]$ is the vector of dispatching decisions to serve customers, and $\alpha = [\alpha_i]$, $\beta = [\beta_i]$, $\delta = [\delta_i]$, $\gamma = [\gamma_i]$, $\omega = [\omega_i]$, $\mu = [\mu_{ij}]$, and $\nu = [\nu_{ij}]$ are the Lagrange multipliers associated with each constraint.

Three small constants ε_0, ε_1, and ε_2 are added to the inequalities that maintain stability in the charging queues and the positivity condition on the maximum expected waiting time. These positive constants make the inequalities nonstrict, leading to more accurate resolutions. Now, we will find the optimal lower-bound solutions of (2.62).

Applying the KKT conditions to the inequalities constraints of (2.62), we get

$$\begin{aligned}
&\alpha_i^* \left(\lambda_c^{(i)} + R^* - \lambda_v \sum_{k=i}^{n-1} (p_{k-1} - p_{k-1} q_{k-1}^* + p_k q_k^*) \Pi_{ki}^* \right. \\
&\qquad \left. - \lambda_v (p_{n-1} - p_{n-1} q_{n-1}^* + p_0 q_0^*) \Pi_{ni}^* \right) = 0,\ i = 1, \ldots, n-1 \\
&\alpha_n^* (\lambda_c^{(n)} + R^* - \lambda_v (p_{n-1} - p_{n-1} q_{n-1}^* + p_0 q_0^*) \Pi_{nn}^*) = 0. \\
&\beta_0^* \left(\sum_{i=0}^{n-1} \lambda_v (p_i - p_i q_i^*) - C(n\mu_c) \right) = 0. \\
&\beta_1^* (\lambda_v p_0 q_0^* - \mu_c) = 0 \\
&\gamma_i^* (q_i^* - 1) = 0,\ i = 0, \ldots, n-1. \\
&\gamma_n^* \left(R^* - \frac{\lambda_v - \sum_{i=1}^{n} \lambda_c^{(i)}}{n} \right) = 0. \\
&\omega_i^* q_i^* = 0,\ i = 0, \ldots, n-1. \\
&\omega_n^* (R^* - \varepsilon_2) = 0. \\
&\nu_{ij}^* (\Pi_{ij}^* - 1) = 0,\ i = 1, \ldots, n,\ j = 1, \ldots i. \\
&\mu_{ij}^* \Pi_{ij}^* = 0,\ i = 1, \ldots, n,\ j = 1, \ldots i. \\
&\delta_i^* \left(\sum_{j=1}^{i} \Pi_{ij}^* - 1 \right) = 0,\ i = 1, \ldots, n.
\end{aligned} \tag{2.64}$$

Likewise, applying the KKT conditions to the Lagrangian function in (2.63), and knowing that the gradient of the Lagrangian function goes to 0 at the lower-bound solution, we get the following set of equalities:

$$\begin{aligned}
&\frac{\partial L}{\partial q_i} = \lambda_v p_i \left(\sum_{j=1}^{i} \alpha_j^* (\Pi_{i+1j}^* - \Pi_{ij}^*) + \alpha_{i+1}^* \Pi_{i+1i+1}^* - \beta_0^* \right) - \omega_i^* + \gamma_i^* = 0, \quad i = 1, \ldots, n-1. \\
&\frac{\partial L}{\partial q_0} = \lambda_v p_0 \left(\alpha_1^* \Pi_{11}^* - \sum_{j=1}^{n} \alpha_j^* \Pi_{nj}^* - \beta_0^* + \beta_1^* \right) - \omega_0^* + \gamma_0^* = 0 \\
&\frac{\partial L}{\partial \Pi_{ij}} = \alpha_j^* \lambda_v (p_{i-1} q_{i-1}{}^* - p_{i-1} - p_i q_i^*) + \delta_i^* + \nu_{ij}^* - \mu_{ij}^* = 0
\end{aligned} \tag{2.65}$$

$$\frac{\partial L}{\partial \Pi_{nj}} = \alpha_j^* \lambda_v (p_{n-1} q_{n-1}{}^* - p_{n-1} - p_0 q_0^*) + \delta_n^* + \nu_{nj}^* - \mu_{nj}^* = 0$$
$$\frac{\partial L}{\partial R} = -1 + \sum_{i=1}^{n} \alpha_i^* - \omega_n^* + \gamma_n^* = 0.$$

Multiplying each of the partial derivatives in (2.65) by the derivation variable itself combined with the KKT conditions of the variables lower-bound inequalities given by (2.64) gives

$$\begin{aligned}
&\frac{\partial L}{\partial q_i} \times q_i = q_i^* \lambda_v p_i \left(\sum_{j=1}^{i} \alpha_j^* (\Pi_{i+1j}^* - \Pi_{ij}^*) + \alpha_{i+1}^* \Pi_{i+1i+1}^* - \beta_0^* \right) + \gamma_i^* = 0,\ i = 1, \ldots, n-1. \\
&\frac{\partial L}{\partial q_0} \times q_0 = q_0^* \lambda_v p_0 \left(\alpha_1^* \Pi_{11}^* - \sum_{j=1}^{n} \alpha_j^* \Pi_{nj}^* - \beta_0^* + \beta_1^* \right) + \gamma_0^* = 0 \\
&\frac{\partial L}{\partial \Pi_{ij}} \times \Pi_{ij} = \Pi_{ij}^* \left(\alpha_j^* \lambda_v (p_{i-1} q_{i-1}{}^* - p_{i-1} - p_i q_i^*) + \delta_i^* \right) + \nu_{ij}^* = 0 \\
&\frac{\partial L}{\partial \Pi_{nj}} \times \Pi_{nj} = \Pi_{nj}^* \left(\alpha_j^* \lambda_v (p_{n-1} q_{n-1}{}^* - p_{n-1} - p_0 q_0^*) + \delta_n^* \right) + \nu_{nj}^* = 0 \\
&\frac{\partial L}{\partial R} \times R = -R + R \sum_{i=1}^{n} \alpha_i^* - \omega_n^* \varepsilon_2 + \gamma_n^* \left(\frac{\lambda_v - \sum_{i=1}^{n} \lambda_c^{(i)}}{n} \right) = 0
\end{aligned} \tag{2.66}$$

When we inject the result of the first four equations in (2.66) in the KKT conditions on the upper-bound conditions of the variables q_i and Π_{ij}, we find

$$\begin{aligned}
&q_i^* (q_i^* - 1) \left(\sum_{j=1}^{i} \alpha_j^* (\Pi_{i+1j}^* - \Pi_{ij}^*) + \alpha_{i+1}^* \Pi_{i+1i+1}^* - \beta_0^* \right) = 0,\ i = 1, \ldots, n-1. \\
&q_0^* (q_0^* - 1) \left(\alpha_1^* \Pi_{11}^* - \sum_{j=1}^{n} \alpha_j^* \Pi_{nj}^* - \beta_0^* + \beta_1^* \right) = 0 \\
&\Pi_{ij}^* (\Pi_{ij}^* - 1) \left(\alpha_j^* \lambda_v (p_{i-1} q_{i-1}{}^* - p_{i-1} - p_i q_i^*) + \delta_i^* \right) = 0 \\
&\Pi_{nj}^* (\Pi_{nj}^* - 1) \left(\alpha_j^* \lambda_v (p_{n-1} q_{n-1}{}^* - p_{n-1} - p_0 q_0^*) + \delta_n^* \right) = 0
\end{aligned} \tag{2.67}$$

From (2.67), we have:

$0 < q_0^* < 1$ only if $\alpha_1^* \Pi_{11}^* - \sum_{j=1}^{n} \alpha_j^* \Pi_{nj}^* - \beta_0^* + \beta_1^* = 0$
$0 < q_i^* < 1$ only if $\sum_{j=1}^{i} \alpha_j^* (\Pi_{i+1j}^* - \Pi_{ij}^*) + \alpha_{i+1}^* \Pi_{i+1i+1}^* - \beta_0^* = 0$
$0 < \Pi_{ij}^* < 1$ only if $\alpha_j^* \lambda_v (p_{i-1} q_{i-1}{}^* - p_{i-1} - p_i q_i^*) + \delta_i^* = 0$
$0 < \Pi_{nj}^* < 1$ only if $\alpha_j^* \lambda_v (p_{n-1} q_{n-1}{}^* - p_{n-1} - p_0 q_0^*) + \delta_n^* = 0$

Since $0 \le q_i^* \le 1$ and $0 \le \Pi_{ij}^* \le 1$, then these equalities may not always be true

if $\alpha_1^* \Pi_{11}^* - \sum_{j=1}^{n} \alpha_j^* \Pi_{nj}^* - \beta_0^* + \beta_1^* > 0$ and we know that $\gamma_0^* \ge 0$ then $\gamma_0^* = 0$ which gives $q_0^* \ne 1$ and $q_0^* = 0$.
if $\sum_{j=1}^{i} \alpha_j^* (\Pi_{i+1j}^* - \Pi_{ij}^*) + \alpha_{i+1}^* \Pi_{i+1i+1}^* - \beta_0^* > 0$ which gives $q_i^* \ne 1$ and $q_i^* = 0$
if $\alpha_1^* \Pi_{11}^* - \sum_{j=1}^{n} \alpha_j^* \Pi_{nj}^* - \beta_0^* + \beta_1^* < 0$ then $\gamma_0^* > 0$ (it cannot be 0 because this will contradict with the value of q_i), which implies that $q_0^* = 1$.
if $\sum_{j=1}^{i} \alpha_j^* (\Pi_{i+1j}^* - \Pi_{ij}^*) + \alpha_{i+1}^* \Pi_{i+1i+1}^* - \beta_0^* < 0$ then $\gamma_i^* > 0$ (it cannot be 0 because this contradicts with the value of q_i), which implies that $q_i^* = 1$

if $\alpha_j^*\lambda_v(p_{i-1}q_{i-1}{}^* - p_{i-1} - p_i q_i^*) + \delta_i^* > 0$, and we know that $v_{ij}^* \geq 0$, then $v_{ij}^* = 0$ which gives $\Pi_{ij}^* \neq 1$ and $\Pi_{ij}^* = 0$.
if $\alpha_j^*\lambda_v(p_{n-1}q_{n-1}{}^* - p_{n-1} - p_0 q_0^*) + \delta_n^* > 0$ which gives $\Pi_{nj}^* \neq 1$ and $\Pi_{nj}^* = 0$
if $\alpha_j^*\lambda_v(p_{i-1}q_{i-1}{}^* - p_{i-1} - p_i q_i^*) + \delta_i^* < 0$ then $v_{ij}^* > 0$ (it cannot be 0 because this will contradict with the value of Π_{ij}), which implies that $\Pi_{ij} = 1$.
if $\alpha_j^*\lambda_v(p_{n-1}q_{n-1}{}^* - p_{n-1} - p_0 q_0^*) + \delta_n^* < 0$ then $v_{nj}^* > 0$ (it cannot be 0 because this will contradict with the value of Π_{nj}), which implies that $\Pi_{nj} = 1$.

We have also from the KKT conditions given by equation in (2.64) that says either the Lagrangian coefficient is 0 or its associated inequality is an equality:

if $\beta_1^* \neq 0$ we have $q_0^* = \frac{\mu_c}{\lambda_v^* p_0}$
if $\alpha_n^* \neq 0$ we have $\frac{\lambda_c^{(n)} + \lambda_v p_{n-1} q_{n-1}^* \Pi_{nn}^* - \lambda_v p_{n-1} \Pi_{nn}^* + R^*}{\lambda_v p_0 \Pi nn^*}$
if $\alpha_i^* \neq 0$, we have

$$q_i^* = \frac{R^* + \lambda_c^{(i)} + \lambda_v\left[\sum_{k=i+2}^{n-1}(p_{k-1}q_{k-1}^* - p_k q_k^*)\Pi_{ki} + (p_{n-1}q_{n-1}^* - p_0 q_0^*)\Pi_{ni} - \sum_{k=i}^{n} p_{k-1}\Pi_{ki} + p_{i-1}q_{i-1}^*\Pi_{ii}^* - p_{i+1}q_{i+1}\Pi_{i+1i+1}\right]}{\lambda_v p_i(\Pi_{ii} - \Pi_{i+1i+1})}$$

for $i = 1, \ldots, n-1$

Otherwise, by the Lagrangian relaxation:

$q_i^* = \zeta_i(R^*, q^*, \Pi^*, \alpha^*, \beta^*, \gamma^*, \omega^*, \mu^*, v^*, \delta^*)$ for $i = 1, \ldots, n-1$ and $\Pi_{ij}^* = \zeta_{ij}(R^*, q^*, \Pi^*, \alpha^*, \beta^*, \gamma^*, \omega^*, \mu^*, v^*, \delta^*)$ where ζ_i and ζ_{ij} are the solutions that maximize $\inf_{\mathbf{q},\blacksquare} L(\mathbf{q}, \Pi^*, R^*, \alpha^*, \beta^*, \gamma^*, \omega^*, \mu^*, v^*, \delta^*)$

Now in order to find the expression of R^*, we first look at its upper bound-associated condition in (2.64). From there, we can say that if $\omega_n^* \neq 0$ then $R^* = \varepsilon_2$ and if $\gamma_n^* \neq 0$ then $R* = \frac{\lambda_v - \sum_{i=1}^{n}\lambda_c^{(i)}}{n}$

Otherwise, from the last equation in (2.66), if $\omega_n^* = 0$ and $\gamma_n^* = 0$ then

$$\begin{aligned} R^* = &\sum_{i=1}^{n-1}\alpha_i^*\left(\lambda_v\sum_{k=i}^{n-1}(p_{k-1} - p_{k-1}q_{k-1}^* + p_k q_k^*)\Pi_{ki}^* + \lambda_v(p_{n-1} - p_{n-1}q_{n-1}^* + p_0 q_0^*)\Pi_{ni}^* - \lambda_c^{(i)}\right) \\ &+ \alpha_n^*\left(\lambda_v(p_{n-1} - p_{n-1}q_{n-1}^* + p_0 q_0^*)\Pi_{nn}^* - \lambda_c^{(n)}\right) \end{aligned} \tag{2.68}$$

Therefore, the lower-bound solution for the optimization problem in (2.62), obtained from Lagrangian and KKT analysis, can be expressed as follows in (2.69):

$$R^* = \begin{cases} \frac{\lambda_v - \sum_{i=1}^{n}\lambda_c^{(i)}}{n} & \gamma_n^* \neq 0 \\ \varepsilon_2 & \omega_n^* \neq 0 \\ \sum_{i=1}^{n-1}\alpha_i^*(\lambda_v\sum_{k=i}^{n-1}(p_{k-1} - p_{k-1}q_{k-1}^* + p_k q_k^*)\Pi_{ki}^* + \\ \lambda_v(p_{n-1} - p_{n-1}q_{n-1}^* + p_0 q_0^*)\Pi_{ni}^* - \lambda_c^{(i)}) \\ +\alpha_n^*(\lambda_v(p_{n-1} - p_{n-1}q_{n-1}^* + p_0 q_0^*)\Pi_{nn}^* - \lambda_c^{(n)}) & \text{Otherwise} \end{cases}$$

$$q_0^* = \begin{cases} 0 & \alpha_1^*\Pi_{11}^* - \sum_{i=1}^{n}\alpha_i^*\Pi_{ni}^* - \beta_0^* + \beta_1^* > 0 \\ 1 & \alpha_1^*\Pi_{11}^* - \sum_{i=1}^{n}\alpha_i^*\Pi_{ni}^* - \beta_0^* + \beta_1^* < 0 \\ \frac{\lambda_c^{(n)} + \lambda_v p_{n-1}q_{n-1}^*\Pi_{nn}^* - \lambda_v p_{n-1}\Pi_{nn}^* + R^*}{\lambda_v p_0 \Pi nn^*} & \alpha_n^* \neq 0 \\ \frac{\mu_c}{\lambda_v^* p_0} & \beta_1^* \neq 0 \\ \zeta_0(R^*, q^*, \Pi^*, \alpha^*, \beta^*, \gamma^*, \omega^*, \mu^*, v^*, \delta^*) & \text{Otherwise} \end{cases}$$

$$
q_i^* = \begin{cases} 0 & \alpha_{i+1}^* - \alpha_i^* - \beta_0^* > 0 \\ 1 & \alpha_{i+1}^* - \alpha_i^* - \beta_0^* < 0 \qquad\qquad i = 1, \ldots, n-1. \\ \frac{R^* + \lambda_c^{(i)} + \lambda_v \left[\sum_{k=i+2}^{n-1} (p_{k-1} q_{k-1}^* - p_k q_k^*) \Pi_{ki} + (p_{n-1} q_{n-1}^* - p_0 q_0^*) \Pi_{ni} - \sum_{k=i}^{n} p_{k-1} \Pi_{ki} + p_{i-1} q_{i-1}^* \Pi_{ii}^* - p_{i+1} q_{i+1} \Pi_{i+1 i+1} \right]}{\lambda_v p_i (\Pi_{ii} - \Pi_{i+1 i+1})} \\ \alpha_i^* \neq 0 \\ \zeta_i(R^*, q^*, \Pi^*, \alpha^*, \beta^*, \gamma^*, \omega^*, \mu^*, \nu^*, \delta^*) \\ \text{Otherwise} \end{cases}
$$

$$
\Pi_{ij}^* = \begin{cases} 0 & \alpha_j^* \lambda_v (p_{i-1} q_{i-1^*} - p_{i-1} - p_i q_i^*) + \delta_i^* > 0 \\ 1 & \alpha_j^* \lambda_v (p_{i-1} q_{i-1^*} - p_{i-1} - p_i q_i^*) + \delta_i^* < 0 \\ \zeta_{ij}(R^*, q^*, \Pi^*, \alpha^*, \beta^*, \gamma^*, \omega^*, \mu^*, \nu^*, \delta^*) & \text{Otherwise} \end{cases} \tag{2.69}
$$

where solutions ζ_i and ζ_{ij} maximize $\inf_{\mathbf{q},\blacksquare} L(\mathbf{q}, \Pi^*, R^*, \alpha^*, \beta^*, \gamma^*, \omega^*, \mu^*, \nu^*, \delta^*)$

Solving the system using the KKT condition does not allow finding the complete expression for the solution under all possibilities. As shown in Eq. (2.69), we were able to find the exact expression or exact values of the system variables under some conditions but not for all possibilities. We use the added variable ζ_i and ζ_{ij} to refer to the optimal values in the cases in which we did not have hold of the full expression.

2.3.2.4 Simulation Results Analysis

This section includes multiple tests of our enhanced optimized model presented in Section 2.3.2.2 through simulations. The model will be assessed based on the maximum response times among all classes. In the following simulations, $\mu_c = 0.033$ min^{-1} is set as the complete vehicle-charging rate. The number of charging points available is set as $C = 40$.

First, Figure 2.8a describes the performance of the maximum vehicle response time as the customer arrival rate $\sum_{i=1}^{n} \lambda_c^{(i)}$ changes. For this simulation, λ_v = 8 min^{-1} is constant. With these conditions, the smallest amount of classes that satisfies the stability condition in 2.12 is $n = 7$. From queuing theory rules [20,21], the more serving queues a system has, the higher the waiting time

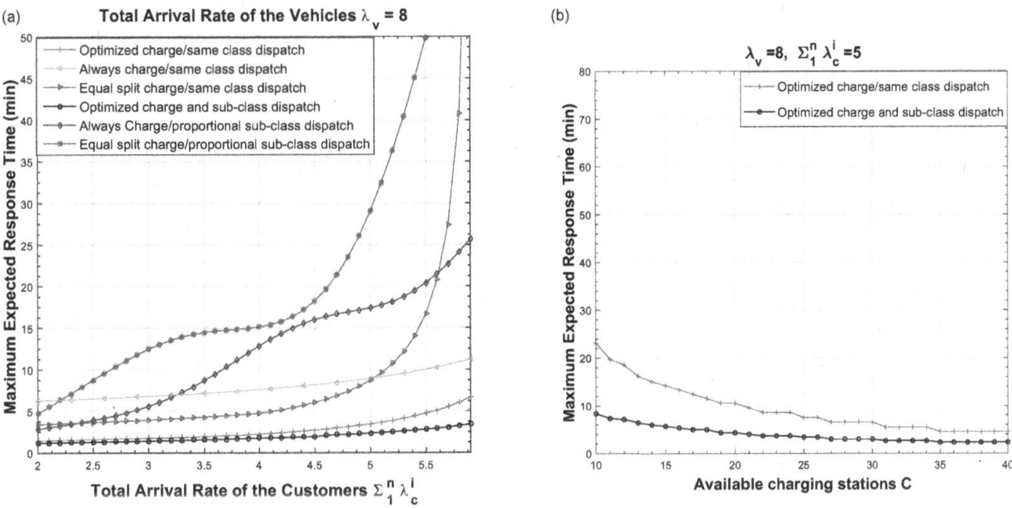

Figure 2.8 (a) Comparison to nonoptimized policies for decreasing SoC distribution. (b) Effect of varying charging points availability.

will be. Moreover, in the previous related work [26], we showed that the number of classes n must remain within its lower bound from 2.12 to avoid damage to the system performance and to avoid an increase in the maximum expected response time.

Figure 2.8a displays maximum expected response times versus $\sum_{i=1}^{n} \lambda_c^{(i)}$ for separate service models, specifically our derived optimal decisions to the following decision sets:

1. Optimized charging decisions (i.e., $q_i \; \forall \; i$) with same-class dispatching (i.e., $\Pi_{ii} = 1 \; \forall \; i$ and $\Pi_{ij} = 0 \; \forall \; i, j \neq i$)
2. Always partially charged decisions (i.e., $q_i = 0 \; \forall \; i$) with same-class dispatching (i.e., $\Pi_{ii} = 1 \; \forall \; i$ and $\Pi_{ij} = 0 \; \forall \; i, j \neq i$)
3. Charging decisions split equally (i.e., $q_i = 0.5 \; \forall \; i$) with same-class dispatching (i.e., $\Pi_{ii} = 1 \; \forall \; i$ and $\Pi_{ij} = 0 \; \forall \; i, j \neq i$)
4. Always partially charged decisions (i.e., $q_i = 0 \; \forall \; i$) with proportional subclasses dispatching decisions (i.e., Π_{ij} proportional to the customers subclasses needs)
5. Equal-split charge decisions (i.e., $q_i = 0 \; \forall \; i$) with proportional subclasses dispatching decisions (i.e., Π_{ij} proportional to the customers' subclasses needs)

These five schemes represent the possible nonoptimized policies, in which all the vehicles are not connected and have predefined decisions without considering the other system parameters. These schemes are possible in case of a nonconnected and optimized system.

Figure 2.8a compares these approaches with a decreasing SoC distribution. The figure clearly shows a superior performance for our optimal proposed scheme contrasted with several other schemes. As $\sum_{i=1}^{n} \lambda_c^{(i)}$ increases and reaches its limit (i.e., λ_v), which are the most properly engineered scenarios, our proposed scheme particularly stands out. Gains of 49.3%, 69.8%, 93.22%, 86.7%, and 94.4% in the performance can be noticed compared to the previously stated policies, respectively. These gains represent the maximum relative gain compared to each of the other approaches over the entire range of the tested $\sum_{i=1}^{n} \lambda_c^{(i)}$.

Figure 2.8b depicts the study of the resilience of our model to critical scenarios including a decrease in the number of charging stations. Charging stations may malfunction naturally due to system failures. Additionally, the fog controller could be deliberately attacked, leading to system error, and the vehicles' inability to use the charging stations. In order to preserve the customer tolerance during critical situations, the fog controller can inform customers on the temporary delay in response times. To study the resilience of our model, we are only comparing the new proposed model to our previously proposed model. The figure shows clearly the advantage brought by the subclass dispatching model. The gain gets higher in critical scenarios and reaches up to 65% with very acceptable maximum expected response time even with very low energy resources. These results emphasize our scheme's ability to provide more efficient and reliable customer service.

2.4 MIDDLE LAYER: OPTIMAL VEHICLE DIMENSIONING

The objective in this layer of operation is to minimize the needed rate of vehicle inflow λ_v to the entire zone with respect to the arrival rate of customers in order to guarantee an average response time limit for customers of every class. We will also shed light on the potential dimensioning and/or response time relaxation solutions for system resilience in extreme cases of very low energy resources and limited actual vehicle inflow.

2.4.1 SYSTEM STABILITY AND RESPONSE TIME LIMIT CONDITIONS

In this section, we first deduce the stability conditions of the proposed system using the basic laws of queuing theory. We will also derive a lower bound on the number of classes n that fits the customer demands, average response time limit, and charging capabilities of any arbitrary service zone.

As shown in Figure 2.2, each of the n customer classes is served by a separate queue of vehicles having a vehicle inflow rate $\lambda_v^{(i)}$. Consequently, $\lambda_v^{(i)}$ represents the service rate of the customer arrival in the ith queue. From the aforementioned vehicle dispatching and charging dynamics in Section 2.2.2, illustrated in Figure 2.2, these service rates can be expressed as follows:

$$\begin{aligned} \lambda_v^{(i)} &= \lambda_v(p_{i-1}\overline{q}_{i-1} + p_i q_i),\ i = 1,\ldots,n-1. \\ \lambda_v^{(n)} &= \lambda_v(p_{n-1}\overline{q}_{n-1} + p_0 q_0) \end{aligned} \tag{2.70}$$

Since $\overline{q}_i + q_i = 1$, $\overline{q}_i$ can be substituted by $1 - q_i$ to get

$$\begin{aligned} \lambda_v^{(i)} &= \lambda_v(p_{i-1} - p_{i-1}q_{i-1} + p_i q_i),\ i = 1,\ldots,n-1 \\ \lambda_v^{(n)} &= \lambda_v(p_{n-1} - p_{n-1}q_{n-1} + p_0 q_0) \end{aligned} \tag{2.71}$$

From the well-known stability condition of an M/M/1 queue, we must have

$$\lambda_v^{(i)} > \lambda_c^{(i)},\ i = 1,\ldots,n \tag{2.72}$$

It is also established from M/M/1 queue analysis that the average response (i.e., service) time for any customer in the i-th class can be expressed as follows:

$$\frac{1}{\lambda_v^{(i)} - \lambda_c^{(i)}} \tag{2.73}$$

To guarantee customers' satisfaction, the fog controller of each zone must impose an average response time limit T for any class. We can thus express this average response time constraint for the customers of the i-th class as follows:

$$\frac{1}{\lambda_v^{(i)} - \lambda_c^{(i)}} \leq T \tag{2.74}$$

which can also be rewritten as:

$$\lambda_v^{(i)} - \lambda_c^{(i)} \geq \frac{1}{T} \tag{2.75}$$

Before reaching the customer service queues, the vehicles will go through a decision step of either going to these queues immediately or partially charging. The stability of the charging queues should be guaranteed in order to ensure the global stability of the entire system at the steady state. From the model described in the previous section, and the well-known stability conditions of M/M/C and M/M/1 queues, we get the following stability constraints on the C charging points and one central charging station queues, respectively:

$$\begin{aligned} &\sum_{i=0}^{n-1} \lambda_v(p_i - p_i q_i) < C(n\mu_c) \\ &\lambda_v p_0 q_0 < \mu_c \end{aligned} \tag{2.76}$$

We will now derive an expression to set a lower bound on the average vehicle in-flow rate to the entire service zone to guarantee both its stability and the average response time limit fulfillment for all its classes, given their demand rates.

From (2.71) and (2.75), we have

$$\begin{aligned} \lambda_c^{(i)} + \frac{1}{T} &\leq \lambda_v(p_{i-1}\overline{q}_{i-1} + p_i q_i),\ i = 1,\ldots,n-1. \\ \lambda_c^{(n)} + \frac{1}{T} &\leq \lambda_v(p_{n-1}\overline{q}_{n-1} + p_0 q_0),\ i = n \end{aligned} \tag{2.77}$$

The summation of all the inequalities in (2.77) gives a new inequality

$$\sum_{i=1}^{n}\lambda_c^{(i)}+\frac{n}{T}\leq\lambda_v\left[\sum_{i=1}^{n-1}(p_{i-1}\overline{q}_{i-1}+p_iq_i)+(p_{n-1}\overline{q}_{n-1}+p_0q_0)\right] \tag{2.78}$$

$$\sum_{i=1}^{n}\lambda_c^{(i)}+\frac{n}{T}\leq\lambda_v\left[p_0\overline{q}_0+p_1q_1+p_1\overline{q}_1+\cdots+p_{n-1}\overline{q}_{n-1}+p_0q_0\right] \tag{2.79}$$

We have $\overline{q}_i+q_i$ so $p_i\overline{q}_i+p_iq_i=p_i$

$$\sum_{i=1}^{n}\lambda_c^{(i)}+\frac{n}{T}\leq\lambda_v(p_0+p_1+p_2+\cdots+p_{n-1}) \tag{2.80}$$

We have $\sum_{i=0}^{n-1}p_i=1$ so $\sum_{i=1}^{n}\lambda_c^{(i)}+\frac{n}{T}\leq\lambda_v$

So, for the entire zone stability, and fulfillment of the average response time limit for all its classes, the average vehicles' inflow rate must be lower-bounded by

$$\lambda_v\geq\sum_{i=1}^{n}\lambda_c^{(i)}+\frac{n}{T} \tag{2.81}$$

Furthermore, we will establish a lower bound on the number of classes n that fits zone's customer demands, average response time limit, and charging capabilities.

The summation of the inequalities given by (2.76) $\forall\, i=\{0,\ldots,n\}$ gives the following inequality:

$$\lambda_v\sum_{i=0}^{n-1}p_i-\lambda_v\sum_{i=0}^{n-1}p_iq_i+\lambda_vp_0q_0<C(n\mu_c)+\mu_c \tag{2.82}$$

Since $\sum_{i=0}^{n-1}p_i=1$ (because $p_n=0$), we get

$$\lambda_v-\lambda_v\sum_{i=1}^{n-1}p_iq_i<\mu_c(Cn+1) \tag{2.83}$$

In the worst case, all the vehicles will be directed to partially charge before serving, which means that always $q_i=0$. Therefore, we get

$$Cn>\frac{\lambda_v}{\mu_c}-1\,, \tag{2.84}$$

which can be rearranged to be

$$n>\frac{\lambda_v}{C\mu_c}-\frac{1}{C} \tag{2.85}$$

From Eqs. (2.85) and (2.81), we have

$$n>\frac{\lambda_v}{C\mu_c}-\frac{1}{C}\geq\frac{\sum_{i=1}^{n}\lambda_c^{(i)}+\frac{n}{T}}{C\mu_c}-\frac{1}{C} \tag{2.86}$$

So, for stabilizing the zone operation given its customer demands, average response time limit, and charging capabilities, the number of classes n in the zone must obey the following inequality:

$$n\geq\frac{\sum_{i=1}^{n}\lambda_c^{(i)}-\mu_c}{C\mu_c-1/T} \tag{2.87}$$

2.4.2 OPTIMAL VEHICLE DIMENSIONING PROBLEM FORMULATION

As previously mentioned, the aim of this section is to minimize the average vehicle inflow rate λ_v to the entire zone, given its charging capacity and customer-demand rates, while guaranteeing an average response time limit for each class customers. Given the described system dynamics in Section 2.2.2 and the derived conditions in Section 2.2.2, the above problem can be formulated as a stochastic optimization problem as follows:

$$\underset{q_0,q_1,\ldots,q_{n-1}}{\text{minimize}}\ \lambda_v \tag{2.88a}$$

s.t

$$\lambda_c^{(i)} - \lambda_v(p_{i-1} - p_{i-1}q_{i-1} + p_iq_i) + \frac{1}{T} \leq 0,\ i = 1,\ldots,n-1 \tag{2.88b}$$

$$\lambda_c^{(n)} - \lambda_v(p_{n-1} - p_{n-1}q_{n-1} + p_0q_0) + \frac{1}{T} \leq 0 \tag{2.88c}$$

Constraints on charging vehicles queues (2.88d)

$$\sum_{i=0}^{n-1} \lambda_v(p_i - p_iq_i) - C(n\mu_c) < 0 \tag{2.88e}$$

$$\lambda_v p_0 q_0 - \mu_c < 0 \tag{2.88f}$$

Constraints on probabilities and decisions (2.88g)

$$\sum_{i=0}^{n-1} p_i = 1,\ 0 \leq p_i \leq 1,\ i = 0,\ldots,n-1 \tag{2.88h}$$

$$0 \leq q_i \leq 1,\ i = 0,\ldots,n-1 \tag{2.88i}$$

$$\lambda_v \geq \sum_{i=1}^{n} \lambda_c^{(i)} + \frac{n}{T} \tag{2.88j}$$

The n constraints in (2.88b) and (2.88c) represent the stability and response time limit conditions of the system introduced in (2.75), after substituting every $\lambda_v^{(i)}$ by its expansion form in (2.71). The constraints in (2.88e) and (2.88f) represent the stability conditions for the charging queues. The constraints in (2.88h) and (2.88i) are the axiomatic constraints on probabilities (i.e., values being between 0 and 1, and sum equal to 1). Finally, constraint (2.88j) is the lower bound on λ_v introduced by (2.81).

The above optimization problem is a quadratic nonconvex problem with second-order differentiable objective and constraint functions. Usually, the solution obtained by using the Lagrangian and KKT analysis for such nonconvex problems provides a lower bound on the actual optimal solution. Consequently, we propose to solve the above problem by first finding the solution derived through Lagrangian and KKT analysis, then, if needed, iteratively tightening this solution to the feasibility set of the original problem.

2.4.2.1 Lower-Bound Solution

The Lagrangian function associated with the optimization problem in (2.88) is given by the following expression:

$$L(\mathbf{q},\lambda_v,\alpha,\beta,\gamma,\omega) = \lambda_v + \alpha_n\left(\lambda_v(p_{n-1}q_{n-1} - p_0q_0 - p_{n-1}) + \lambda_c^{(n)} + \frac{1}{T}\right)$$
$$+ \sum_{i=1}^{n-1} \alpha_i\left(\lambda_v(p_{i-1}q_{i-1} - p_iq_i - p_{i-1}) + \lambda_c^{(i)} + \frac{1}{T}\right)$$

$$+\beta_0\left(\sum_{i=0}^{n-1}\lambda_v(p_i-p_iq_i)-Cn\mu_c+\varepsilon_0\right)+\beta_1(\lambda_v p_0q_0-\mu_c+\varepsilon_1)$$
$$+\sum_{i=0}^{n-1}\gamma_i(q_i-1)-\sum_{i=0}^{n-1}\omega_iq_i-\omega_n\left(\lambda_v-\sum_{i=1}^{n}\lambda_c^{(i)}-\frac{n}{T}\right), \quad (2.89)$$

where $\mathbf{q}$ is the vector of dispatching decisions (i.e., $\mathbf{q}=[q_0,\ldots,q_{n-1}]$), and: $\alpha=[\alpha_i]$, $\beta=[\beta_i]$, $\gamma=[\gamma_i]$, and $\omega=[\omega_i]$ are the Lagrange multipliers associated with each constraint. For more accurate resolutions, two small positive constants ε_0 and ε_1 are added to the stability conditions on the charging queues to make them nonstrict inequalities.

The optimal lower-bound solution of the optimization problem is obtained by solving the equations given by the KKT conditions on the problem equality and inequality constraints. Applying the KKT conditions to the inequalities constraints of (2.87), we get

$$\begin{aligned}
&\alpha_i^*(\lambda_v^*(p_{i-1}q_{i-1}^*-p_iq_i^*-p_{i-1})+\frac{1}{T}+\lambda_c^{(i)})=0, i=1,\ldots,n-1.\\
&\alpha_n^*(\lambda_v^*(p_{n-1}q_{n-1}^*-p_0q_0^*-p_{n-1})+\frac{1}{T}+\lambda_c^{(n)})=0.\\
&\beta_0^*\left(\sum_{i=0}^{n-1}\lambda_v(p_i-p_iq_i^*)-C(n\mu_c)+\varepsilon_0\right)=0.\\
&\beta_1^*(\lambda_v p_0q_0^*-\mu_c+\varepsilon_1)=0\\
&\gamma_i^*(q_i^*-1)=0,\ i=0,\ldots,n-1.\\
&\omega_i^*q_i^*=0,\ i=0,\ldots,n-1.\\
&\omega_n^*\left(\lambda_v^*-\left(\sum_{i=1}^{n}\lambda_c^{(i)}+\frac{n}{T}\right)\right)=0.
\end{aligned} \quad (2.90)$$

Likewise, applying the KKT conditions to the Lagrangian function in (2.88), and knowing that the gradient of the Lagrangian function goes to 0 at the optimal solution, we get the following set of equalities:

$$\begin{aligned}
&\lambda_v^*p_i(\alpha_{i+1}^*-\alpha_i^*-\beta_0^*)=\omega_i^*-\gamma_i^*,\ i=1,\ldots,n-1.\\
&\lambda_v^*p_0(\alpha_1^*-\alpha_n^*-\beta_0^*+\beta_1^*)=\omega_0^*-\gamma_0^*\\
&\sum_{i=1}^{n-1}\alpha_i^*(p_{i-1}q_{i-1}^*-p_iq_i^*-p_{i-1})+\alpha_n^*(p_{n-1}q_{n-1}^*-p_0q_0^*-p_{n-1})\\
&\quad+\beta_0^*\left(\sum_{i=0}^{n-1}(p_i-p_iq_i^*)\right)+\beta_1^*p_0q_0^*-\omega_n^*+1=0
\end{aligned} \quad (2.91)$$

Knowing that the gradient of the Lagrangian goes to 0 at the optimal solutions, we get the system of equalities given by (2.91). Multiplying the first equality in (2.91) by q_i^*, the second equality by q_0^*, and the third equality by λ_v^*combined with the equalities given by (2.90) gives

$$\begin{aligned}
&\lambda_v^*p_iq_i^*(\alpha_{i+1}^*-\alpha_i^*-\beta_0^*)=-\gamma_i^*,\ i=1,\ldots,n-1.\\
&\lambda_v^*p_0q_0^*(\alpha_1^*-\alpha_n^*-\beta_0^*+\beta_1^*)=-\gamma_0^*\\
&\lambda_v^*-\sum_{i=1}^{n}\alpha_i^*(\lambda_c^{(i)}+\frac{1}{T})+\beta_0^*(Cn\mu_c-\varepsilon_0)+\beta_1^*(\mu_c-\varepsilon_1)-\omega_n^*\left(\sum_{i=1}^{n}\lambda_c^{(i)}+\frac{n}{T}\right)=0
\end{aligned} \quad (2.92)$$

Equation (2.92) on substitution in the fifth equality in (2.90) gives

$$\lambda_v^*p_i(\alpha_{i+1}^*-\alpha_i^*-\beta_0^*)(q_i^*-1)q_i^*=0,\ i=1,\ldots,n-1.$$

$$\lambda_v^* p_0(\alpha_1^* - \alpha_n^* - \beta_0^* + \beta_1^*)(q_0^* - 1)q_0^* = 0$$

$$\lambda_v^* = \sum_{i=1}^{n} \alpha_i^* \left(\lambda_c^{(i)} + \frac{1}{T}\right) - \beta_0^*(Cn\mu_c - \varepsilon_0) - \beta_1^*(\mu_c - \varepsilon_1) + \omega_n^* \left(\sum_{i=1}^{n} \lambda_c^{(i)} + \frac{n}{T}\right) \tag{2.93}$$

From (2.93), we have $0 < q_0^* < 1$ only if $\alpha_{i+1}^* - \alpha_i^* - \beta_0^* = 0$. And, $0 < q_i^* < 1$ only if $\alpha_1^* - \alpha_n^* - \beta_0^* + \beta_1^* = 0$. As $0 \leq q_i^* \leq 1$, then these equalities may not always be true

if $\alpha_1^* - \alpha_n^* - \beta_0^* + \beta_1^* > 0$, and we know that $\gamma_0^* \geq 0$, then $\gamma_0^* = 0$ which gives $q_0^* \neq 1$ and $q_0^* = 0$.
if $\alpha_{i+1}^* - \alpha_i^* - \beta_0^* > 0$, and we know that $\gamma_i^* \geq 0$, then $\gamma_i^* = 0$ which gives $q_i^* \neq 1$ and $q_i^* = 0$
if $\alpha_1^* - \alpha_n^* - \beta_0^* + \beta_1^* < 0$, then $\gamma_0^* > 0$ (it cannot be 0 because this will contradict with the value of q_i), which implies that $q_0^* = 1$.
if $\alpha_{i+1}^* - \alpha_i^* - \beta_0^* < 0$, then $\gamma_i^* > 0$ (it cannot be 0 because this contradicts with the value of q_i), which implies that $q_i^* = 1$

We have also from the KKT conditions given by equation in (2.90) that says either the Lagrangian coefficient is 0 or its associated inequality is an equality:

if $\beta_1^* \neq 0$ we have $q_0^* = \frac{\mu_c}{\lambda_v^* p_0}$
if $\alpha_n^* \neq 0$ we have $q_0^* = \frac{p_{n-1}q_{n-1}^* - p_{n-1}}{p_0} + \frac{\lambda_c^{(n)} + \frac{1}{T}}{\lambda_v p_0}$
if $\alpha_i^* \neq 0$, we have $q_i^* = \frac{p_{i-1}q_{i-1}^* - p_{i-1}}{p_i} + \frac{\lambda_c^{(i)} + \frac{1}{T}}{\lambda_v p_i}$
for $i = 1, \ldots, n-1$

Otherwise, by the Lagrangian relaxation:
$q_i^* = \zeta_i(\alpha^*, \beta^*, \gamma^*, \lambda_v^*, q^*)$ for $i = 1, \ldots, n-1$
where $\zeta_i(\alpha^*, \beta^*, \gamma^*, \lambda_v^*, q^*)$ is the solution that maximizes the function $\inf_{\mathbf{q}} L(\mathbf{q}, \alpha^*, \beta^*, \gamma^*, \lambda_v^*)$

Now in order to find the expression of λ_v^*, we first look at the last equation in (2.90). From there, we can say that if $\omega_n^* \neq 0$ then $\lambda_v^* = \sum_{i=1}^{n} \lambda_c^{(i)} + \frac{n}{T}$

Otherwise, from the third equation in (2.93) if $\omega_n^* = 0$ then $\lambda_v^* = \sum_{i=1}^{n} \alpha_i^* (\lambda_c^{(i)} + \frac{1}{T}) - \beta_0^*(Cn\mu_c - \varepsilon_0) - \beta_1^*(\mu_c - \varepsilon_1)$.

So, by summarizing the solution, we get

$$\lambda_v^* = \begin{cases} \sum_{i=1}^{n} \lambda_c^{(i)} + \frac{n}{T} & \omega_n^* \neq 0 \\ \sum_{i=1}^{n} \alpha_i^* (\lambda_c^{(i)} + \frac{1}{T}) - \beta_0^*(Cn\mu_c - \varepsilon_0) - \beta_1^*(\mu_c - \varepsilon_1) & \omega_n^* = 0 \end{cases}$$

$$q_0^* = \begin{cases} 0 & \alpha_1^* - \alpha_n^* - \beta_0^* + \beta_1^* > 0 \\ 1 & \alpha_1^* - \alpha_n^* - \beta_0^* + \beta_1^* < 0 \\ \frac{p_{n-1}q_{n-1}^* - p_{n-1}}{p_0} + \frac{\lambda_c^{(n)} + \frac{1}{T}}{\lambda_v p_0} & \alpha_n^* \neq 0 \\ \frac{\mu_c}{\lambda_v^* p_0} & \beta_1^* \neq 0 \\ \zeta_0(\alpha^*, \beta^*, \gamma^*, \lambda_v^*, q^*) & Otherwise \end{cases}$$

$$q_i^* = \begin{cases} 0 & \alpha_{i+1}^* - \alpha_i^* - \beta_0^* > 0 \\ 1 & \alpha_{i+1}^* - \alpha_i^* - \beta_0^* < 0 \\ \frac{p_{i-1}q_{i-1}^* - p_{i-1}}{p_i} + \frac{\lambda_c^{(i)} + \frac{1}{T}}{\lambda_v p_i} & \alpha_i^* \neq 0 \\ \zeta_i(\alpha^*, \beta^*, \gamma^*, \lambda_v^*, q^*) & Otherwise \end{cases} \tag{2.94}$$

$$i = 1, \ldots, n-1.$$

where $\zeta_i(\alpha^*, \beta^*, \gamma^*, \lambda_v^*, q^*)$ is the solution that maximizes $\inf_{\mathbf{q}} L(\mathbf{q}, \alpha^*, \beta^*, \gamma^*, \lambda_v^*)$

2.4.2.2 Solution Tightening

As stated earlier, the closed-form solution derived in the previous section from analyzing the constraints' KKT conditions does not always match with the optimal solution of the original optimization problem and is sometimes a nonfeasible lower bound on our problem. Unfortunately, there is no method to find the exact closed-from solution of nonconvex optimization. However, starting from the derived lower bound, we can numerically tighten this solution toward the feasibility set of the original problem. We will use *Suggest-and-Improve algorithm* whenever the KKT condition-based solution is not feasible and tightening is required.

2.4.3 SIMULATION RESULTS

In this section, we test both the performance and merits of the proposed dimensioning solution for the considered multiclass AEMoD system. The metric of interest in this study is the optimal vehicle inflow rate to an arbitrary zone of interest. The performance of the proposed dimensioning solution is tested for two possible SoC distributions for inflow vehicles, namely, the decreasing and Gaussian distributions. The former distribution better models the more probable active-vehicle-dominant inflow scenarios; as such, vehicles typically exhibit higher chances of having lower battery charge. The latter distribution models the rarer relocated-vehicle-dominant inflow scenarios; as such, vehicles typically charge for a random period before relocating to the zone of interest. Customers' trip distances are always assumed to follow a Gaussian distribution because customers requiring mid-size distances are usually more than those requiring very small and very long distances. For all the performed simulation studies, the full-charging rate of a vehicle is set to $\mu_c = 0.033\ \text{min}^{-1}$. Moreover, for Figures 2.2, 2.3, and 2.4, the number of charging poles C is set to 40.

The first important finding of this study is that the obtained solutions using the closed-form expressions in (2.94) (i.e., the one derived by applying the KKT conditions) were always feasible solutions to the original problem in (2.88), for the entire broad range of system parameters employed in our simulations. Thus, the derived closed-form solution is in fact the optimal dimensioning solution for a broad range of system settings, and no tightening is needed.

Figure 2.9 shows the trade-off relation between the average response time limit, total customer-demand rate, and the optimal vehicle inflow rate, for both vehicles' SoC distributions. This curve can be used by the fog controller to get a rough estimate (without exact demand information per class nor optimization of the dispatching and charging dynamics) on its required inflow rate (and thus whether it needs extra vehicles or have excess vehicles to relocate) for any given customer-demand rate and desired response time limit.

Figure 2.10 illustrates the effect of increasing the number of classes n beyond its lower bound introduced in (2.87) for both variable total customer-demand rate (while fixing the average response time limits to 5 minutes) and variable average response time limits (while fixing the total customer-demand rate to 5 min^{-1}) in the left and right subfigures, respectively. Both decreasing and Gaussian SoC distributions are considered. In both subfigures, the lower bound on the number of classes varies depending on the values of the average response time and the total customer-demand rate, with maximum values of 14 and 11 for the employed values in the left and right subfigures, respectively. The results in both figures clearly show that increasing n beyond its lower bound increases the required vehicle inflow to the zone. We thus conclude that the optimal number of classes is the smallest integer value satisfying (2.87).

Figure 2.11 compares the performance of our proposed optimal vehicle dimensioning scheme with other nonoptimized approaches (in which vehicles follow a fixed dispatching/charging policy irrespective of the system parameters) for different values of total customer-demand rate (with $T = 5$) and average response time limit (with $\sum_{i=1}^{n} \lambda_c^{(i)} = 5$). The two nonoptimized approaches are the always-charge approach (i.e., $q_i = 0\ \forall\ i$) and the equal-split approach (i.e., $q_i = 0.5\ \forall\ i$). The figure clearly shows the superior performance of our derived optimal policy compared to the two

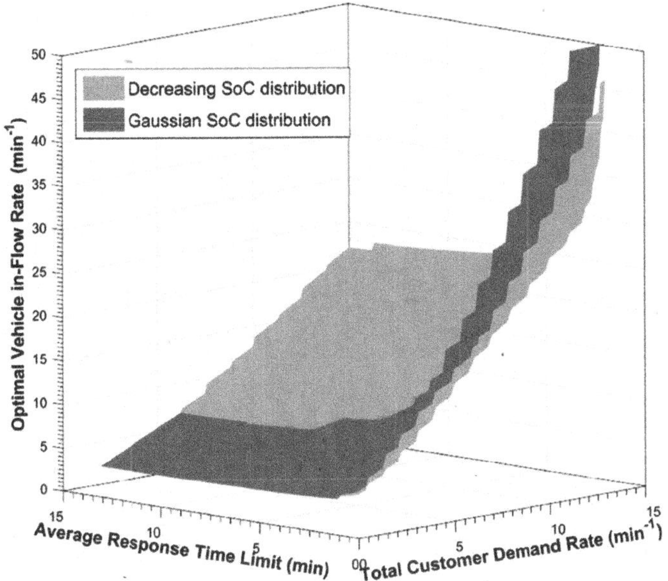

Figure 2.9 Effect of varying the average response time limit and total customer-demand rate.

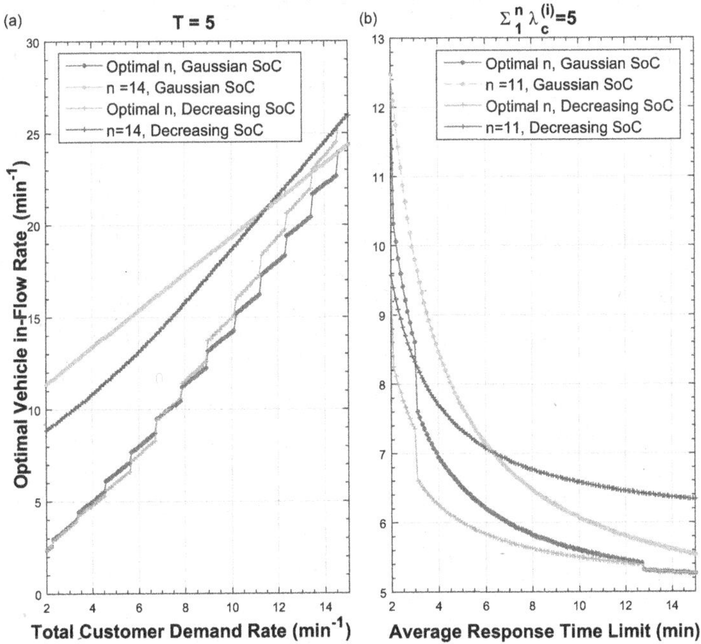

Figure 2.10 Effect of increasing the number of classes: **Left**: Against the total demand rate for an average response time limit of 5 min. **Right**: Against the average response time limit for a total customer demand rate of 5 min^{-1}.

nonoptimized policies, especially for large total customer-demand rates and lower average response time limits. For $\sum_{i=1}^{n} \lambda_c^{(i)} = 10\ \text{min}^{-1}$ in the left subfigure, 36% and 44.4% less vehicle inflow rates are required compared to always-charge and equal-split policies, respectively, for the more typical decreasing SoC distribution. These reductions reach 57.6% and 42.5%, respectively, for $T = 10$ min

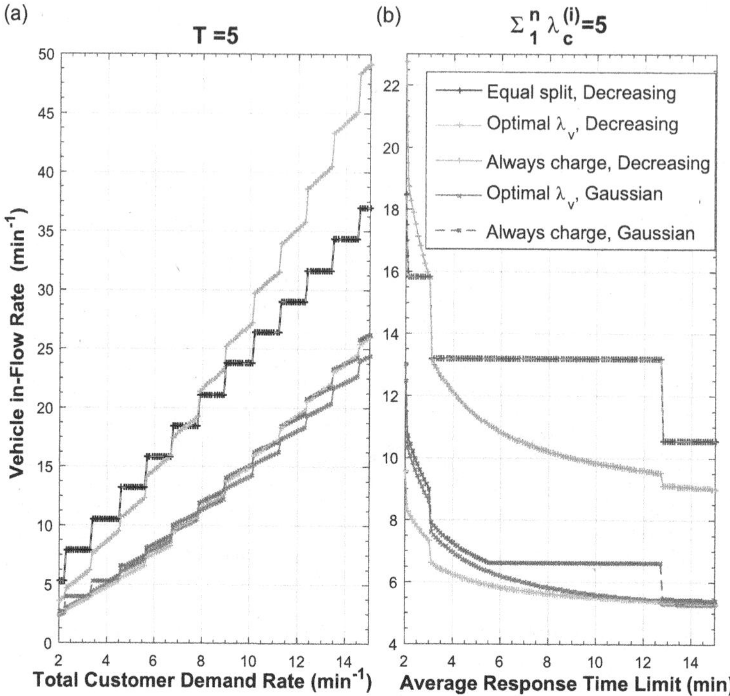

Figure 2.11 Comparison to non-optimized policies **Left**: Against the total demand rate for an average response time limit of 5 min. **Right**: Against the average response time limit for a total customer demand rate of 5 min^{-1}.

in the right subfigure. The always-charge policy is exhibiting less increase in the required vehicle inflow rate when the SoC follows a Gaussian distribution. However, some considerable gains can still be achieved using our proposed optimized approach in this less frequent SoC distribution setting. Noting that these gains can be higher in more critical scenarios, the results demonstrate the importance of our proposed scheme in establishing a better engineered and more stable system with less vehicles.

Finally, we studied the resilience requirements for our considered model in the critical scenarios of sudden reduction in the number of charging sources within the zone. This reduction may occur due to either natural (e.g., typical failures of one or more stations) or intentional (e.g., a malicious attack on the fog controller blocking its access to these sources). The resilience measures that the fog controller can take in these scenarios are both to notify its customers of a transient increase in the vehicles' response times given the available vehicles in the zone and request a higher vehicle in-flow rate to gradually restore its original response time limit.

The left subfigure of Figure 2.12 depicts the maximum response time values of the system for different numbers of available charging poles for a vehicle inflow rate $\lambda_v = 8$ min^{-1} and a total customer-demand rate of 5 min^{-1}. For a Gaussian distribution of vehicles' SoC, the response time increases dramatically when the number of charging poles drops below 20. On the other hand, the degradation in response time was much less severe when the SoC of vehicles follows the decreasing distribution. Luckily, the decreasing SoC distribution is the one that is more probable especially at the time just preceding the failure (where most vehicles arriving to the zone are active vehicles).

As for the recovery from this critical scenario and restoration of the original response time limit, the dimensioning framework presented in Section (2.4.2) can be employed to determine the new optimal value of vehicle inflow rate. The right subfigure in Figure 2.12 depicts the optimal

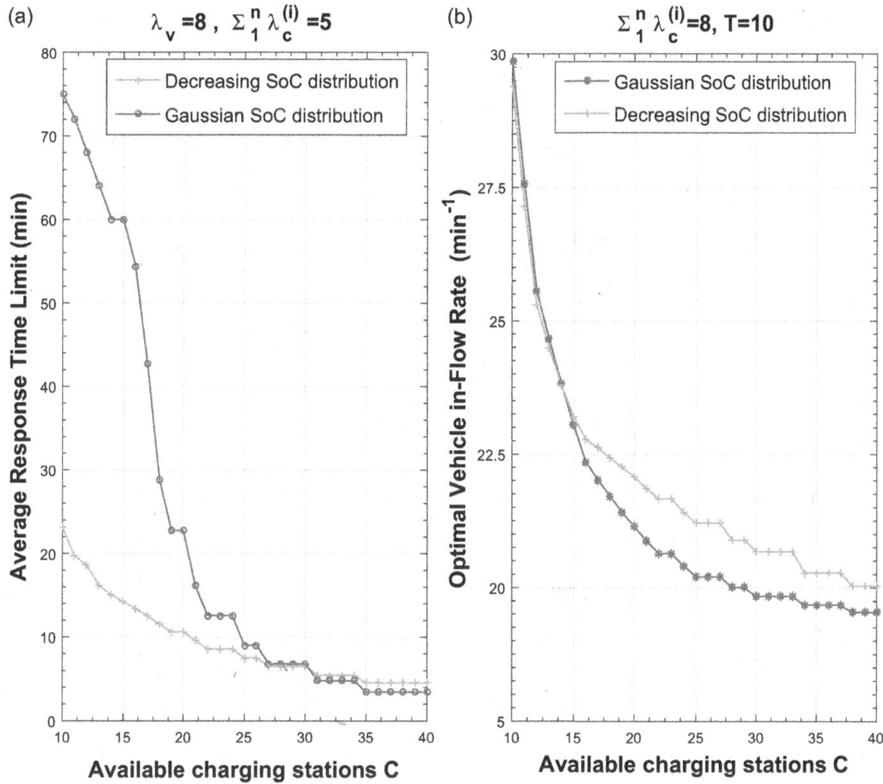

Figure 2.12 Effect of varying the charging point availability **Left**: For a vehicle inflow rate of 8 min^{-1} and a total customer demand rate of 5 min^{-1}. **Right**: For a total customer demand rate of 8 min^{-1} and an average response time limit of 10 min.

vehicles' inflow λ_v^* for different values of available charging poles C. In this simulation, the total customer-demand rate is set to $\sum_{i=1}^{n} \lambda_c^{(i)} = 8\ \text{min}^{-1}$, and the average response time limit is restored back to $T = 10$ min. The figure shows that the Gaussian SoC distribution case, which would be luckily the dominant case in this zone after failure time (due to the domination of relocated vehicles called in by the fog controller to recover from the failure event), exhibits lower need of vehicle inflow rate to restore the system's conventional operation.

2.5 UPPER LAYER: FLEET REBALANCING WITH IN-ROUTE CHARGING

In the conventional MoD literature, "fleet Rebalancing" refers to the relocation of excess vehicles in some zones to those suffering from instability (i.e., low vehicle presence compared to customer demands) is known as fleet rebalancing, aiming to adjust the fleet distribution to the spatial variations in customer demands. Rebalancing of automated mobility on demand systems was first presented as a research problem in Ref. [28]. Optimal rebalancing flows for the vehicles are obtained by solving a linear program. In Ref. [29], the relation to queuing theoretical concepts was established. In Ref. [30], the relation of the rebalancing effort to the underlying distributions of origins and destinations was established, and it was shown that for general distributions, the total minimal rebalancing distance is strictly more than zero. In Ref. [31], the rebalancing problem was solved with an MPC algorithm which performs well but does not scale to large systems. Yet, the interzone management of such approach has not been investigated. Hence, we present the upper layer scheme to solve the fleet rebalancing problem, with possible in-route charging, in decentralized multiclass AEMoD systems.

In this section, we will present the mathematical formulation and simulation results of the upper layer of the scheme, which is the fleet rebalancing layer. The central controller makes the decisions in this layer as mentioned in the queuing model. The collected information by this controller classifies each zone as "donor" or "acceptor' zones.

2.5.1 SYSTEM STABILITY CONDITIONS

In this section, we deduce the stability conditions of our proposed model using the basic laws of queuing theory. We know from the previous section that the excess vehicles of class $i_D \in \{1,\ldots,n\}$ at donor zone $d \in \{1,\ldots,n_D\}$ become available following a Poisson process with rate λ_{di_D} and that each of them takes a decision with probability $P^{(k)}_{di_D,ai_A}$ to go to charge in station $k \in \{1,\ldots,S\}$ (with $k=0$ denoting the case of no in-route charging) and then serve class $i_A \in \{1,\ldots,n\}$ at acceptor zone $a \in \{1,\ldots,n_A\}$. Given the Burke's theorem and the aggregation property of Poisson processes, the total inflow rate of relocated vehicles from all donor zones to class i_A at acceptor zone a is also a Poisson process with a rate denoted by $\lambda_{vs}^{(ai_A)}$, which can be expressed as follows:

$$\lambda_{vs}^{(ai_A)} = \sum_{d=1}^{n_D}\sum_{k=0}^{S}\sum_{i_D=1}^{n} \lambda_{di_D} P^{(k)}_{di_D,ai_A} \quad {}_{\substack{a=1,\ldots,n_A\\ i_A=1,\ldots,n}} \tag{2.95}$$

These vehicles join the local vehicles of class i_A in zone a $\left(\text{arriving at rate } \lambda_v^{(ai_A)}\right)$ in serving the customers of this class $\left(\text{arriving at rate } \lambda_c^{(ai_A)}\right)$. From this M/M/1 queueing system to be stable, we must have [20,21]

$$\lambda_{vs}^{(ai_A)} + \lambda_v^{(ai_A)} > \lambda_c^{(ai_A)} \quad {}_{\substack{a=1,\ldots,n_A\\ i_A=1,\ldots,n}} \tag{2.96}$$

In addition, the excess vehicles from every class of every donor zone may go to charge in station k, each of which having a charging rate μ_c. The following conditions must then hold for the stability of the charging stations' queues:

$$\sum_{d=1}^{n_D}\sum_{a=1}^{n_A}\sum_{i_D=1}^{n}\sum_{i_A=1}^{n} \lambda_{di_D} P^{(k)}_{di_D,ai_A} < \mu_c \quad k=1,\ldots,S \tag{2.97}$$

Consequently, the above stability conditions must constrain any decision-making of excess vehicles when relocating from donor to acceptor zones with or without in-route charging. They will thus be used as constraints in the problem formulation introduced in the next section.

2.5.2 MAXIMUM RESPONSE TIME OPTIMIZATION PROBLEM FORMULATION

We will now introduce our considered formulation of the multiclass rebalancing problem with possible in-route charging, which aims to minimize the maximum expected response time across all acceptor zones' classes while guaranteeing the system's stability.

In queuing theory terms, the aforementioned definition of the response time of the i_A^{th} customer class and a^{th} zone is the expected time spent by the customers in this class and zone's queue. From M/M/1 queue analysis, the expression of this expected response time can thus be expressed as follows:

$$\frac{1}{\lambda_{vs}^{(ai_A)} + \lambda_v^{(ai_A)} - \lambda_c^{(ai_A)}} \quad {}_{\substack{a=1,\ldots,n_A\\ i_A=1,\ldots,n}} \tag{2.98}$$

Consequently, minimizing the maximum of these response times across all acceptor zones and classes can be expressed as follows:

$$\min_{\substack{P^{k}_{di_D,ai_A} \\ \forall\, i,i_D,k,a,i_A}} \left\{ \max_{\forall\, a,i_A} \left\{ \frac{1}{\lambda_{vs}^{(ai_A)} + \lambda_{v}^{(ai_A)} - \lambda_{c}^{(ai_A)}} \right\} \right\} = \max_{\substack{P^{k}_{di_D,ai_A} \\ \forall\, i,i_D,k,a,i_A}} \left\{ \min_{\forall\, a,i_A} \left\{ \lambda_{vs}^{(ai_A)} + \lambda_{v}^{(ai_A)} - \lambda_{c}^{(ai_A)} \right\} \right\} \tag{2.99}$$

Thus, the considered problem can be expressed as maximizing the minimum response rate of all classes of all acceptor zones, given the stability conditions of the system. It is well-known in optimization theory that the above max-min problem can be rewritten in its epigraph form, and thus, the entire problem can now be formulated as follows:

$$\max_{P^{(k)}_{di_D,ai_A}\ \forall d,i_D,k,a,i_A} \quad R \tag{2.100a}$$

s.t.

$$\sum_{d=1}^{n_D} \sum_{k=0}^{S} \sum_{i_D=1}^{n} \lambda_{di_D} P^{(k)}_{di_D,ai_A} + \lambda_{v}^{(ai_A)} - \lambda_{c}^{(ai_A)} \geq R, \ \forall\, a, i_A \tag{2.100b}$$

$$\sum_{d=1}^{n_D} \sum_{a=1}^{n_A} \sum_{i_D=1}^{n} \sum_{i_A=1}^{n} \lambda_{di_D} P^{(k)}_{di_D,ai_A} < \mu_c, \quad \forall\, k \tag{2.100c}$$

$$\sum_{a=1}^{n_A} \sum_{i_A=1}^{n} \sum_{k=0}^{S} P^{(k)}_{di_D,ai_A} = 1, \quad \forall\, d, i_D \tag{2.100d}$$

$$0 \leq P^{(k)}_{di_D,ai_A} \leq 1, \quad \forall\, d, i_D, a, i_A, k \tag{2.100e}$$

$$P^{(0)}_{di_D,ai_A} = 0, \quad \forall\, i_A > (i_D - y_{d,a}) \ \text{ and } \ \forall\, d, i_D, a \tag{2.100f}$$

$$R > 0 \tag{2.100g}$$

$$\begin{aligned} L(\mathbf{P}, R, \alpha, \beta, \gamma, \nu, \mu, \omega, \delta) = &-R + \sum_{a=1}^{n_A} \sum_{i_A=1}^{n} \alpha^{(ai_A)} \left(R + \lambda_{c}^{(ai_A)} - \sum_{d=1}^{n_D} \sum_{k=0}^{S} \sum_{i_D=1}^{n} \lambda_{di_D} P^{(k)}_{di_D,ai_A} - \lambda_{v}^{(ai_A)} \right) \\ &+ \sum_{k=1}^{S} \beta_k \left(\sum_{d=1}^{n_D} \sum_{a=1}^{n_A} \sum_{i_D=1}^{n} \sum_{i_A=1}^{n} \lambda_{di_D} P^{(k)}_{di_D,ai_A} - \mu_c \right) + \sum_{d=1}^{n_D} \sum_{i_D=1}^{n} \gamma_{di_D} \left(\sum_{a=1}^{n_A} \sum_{i_A=1}^{n} \sum_{k=0}^{S} P^{(k)}_{di_D,ai_A} - 1 \right) \\ &- \sum_{d=1}^{n_D} \sum_{a=1}^{n_A} \sum_{k=0}^{S} \sum_{i_D=1}^{n} \sum_{i_A=1}^{n} \mu^{(k)}_{di_D,ai_A} P^{(k)}_{di_D,ai_A} + \sum_{d=1}^{n_D} \sum_{a=1}^{n_A} \sum_{k=0}^{S} \sum_{i_D=1}^{n} \sum_{i_A=1}^{n} \nu^{(k)}_{di_D,ai_A} \left(P^{(k)}_{di_D,ai_A} - 1 \right) - \omega R \\ &+ \sum_{d=1}^{n_D} \sum_{a=1}^{n_A} \sum_{i_D=1}^{n} \sum_{\substack{m> \\ (i_D - y_{d,a})}}^{n} \delta^{(0)}_{di_D,ai_A} P^{(0)}_{di_D,ai_A} \end{aligned} \tag{2.101}$$

The $n_A \cdot n$ constraints in (2.100b) represent the epigraph form constraints of the original objective function in the right-hand side of (2.99), after substituting every $\lambda_{vs}^{(ai_A)}$ by its expansion form in (2.95). The S constraints in (2.100c) represent the stability conditions for the charging queues. The constraints in (2.100d) and (2.100e) are the axiomatic constraints on probabilities, asserting that their values should be between 0 and 1, and their sum should equal to 1. The constraint in (2.100f) assures that the relocating vehicles from class i_D of donor zone d to any acceptor zone a without in-route charging (i.e., $k = 0$) cannot serve classes i_A that are more than $i_D - y_{i,l}$. Lastly, the constraint in (2.100g) is a strict positivity constraint on the minimum expected response rate. It can be easily inferred that this strict positivity constraint, along with the constraints in (2.100b), guarantees the stability of customer queues in acceptor zones expressed in (2.96).

It is evident that the above problem is a linear program with linear constraints, which can be solved analytically using Lagrangian analysis. This analysis is illustrated in the next section.

2.5.3 OPTIMAL REBALANCING AND CHARGING DECISIONS

From the aforementioned description, the problem in (2.100) is a convex optimization problem with second-order differentiable objective and constraint functions, which means that it satisfies Slater's conditions. Consequently, the KKT conditions [24] provide necessary and sufficient conditions for its optimal solution. Therefore, applying the KKT conditions to the constraints of the problem and the gradient of the Lagrangian function allows us to find the optimal analytical solution of the decisions $P^{(k)}_{di_D,ai_A}$. The Lagrangian function associated with the optimization problem in (2.100) is given by the expression in (2.101), where $\mathbf{P} = [P^k_{di_D,ai_A}]\ \forall\ d, i_D, k, a, i_A$, and $\alpha = \left[\alpha^{(ai_A)}\right]$, $\beta = [\beta_k]$, $\gamma = [\gamma_{di_D}]$, $\mu = [\mu^{(k)}_{di_D,ai_A}]$ and $\nu = [\nu^{(k)}_{di_D,ai_A}]$, $\delta = \left[\delta^{(0)}_{di_D,ai_A}\right]$, ω are the Lagrange multiplier vectors associated with the constraints in (2.100b), (2.100c), (2.100d), the lower- and upper-bound constraints in (2.100e), (2.100f), and (2.100g), respectively.

By applying the KKT conditions and Lagrangian analysis, we get the optimal solution of the problem in (2.100).

Applying the KKT [24] conditions to the inequality constraints, we get

$$
\begin{aligned}
&\alpha^{*(ai_A)}\left(R^* + \lambda_c^{(ai_A)} - \sum_{d=1}^{n_D}\sum_{k=0}^{S}\sum_{i_D=1}^{n}\lambda_{di_D}P^{*(k)}_{di_D,ai_A} - \lambda_v^{(ai_A)}\right) = 0, \qquad \forall a, i_A\\
&\beta_k^*\left(\sum_{d=1}^{n_D}\sum_{a=1}^{n_A}\sum_{i_D=1}^{n}\sum_{i_A=1}^{n}\lambda_{di_D}P^{*(k)}_{di_D,ai_A} - \mu_c\right) = 0,\ \forall k = 1,\ldots,S\\
&\gamma^*_{di_D}\left(\sum_{a=1}^{n_A}\sum_{i_A=1}^{n}\sum_{k=0}^{S}P^{*(k)}_{di_D,ai_A} - 1\right) = 0, \qquad \forall d, i_D\\
&\mu^{*(k)}_{di_D,ai_A}P^{*(k)}_{di_D,ai_A} = 0, \qquad \forall d, i_D, k, a, i_A\\
&\nu^{*(k)}_{di_D,ai_A}\left(P^{*(k)}_{di_D,ai_A} - 1\right) = 0, \qquad \forall d, i_D, k, a, i_A\\
&\omega^* R^* = 0,\\
&\delta^{*(0)}_{di_D,ai_A}P^{*(0)}_{di_D,ai_A} = 0, \qquad \forall d, i_D, a, i_A > i_D - y_{d,a}
\end{aligned}
\tag{2.102}
$$

Likewise, applying the KKT conditions to the Lagrangian function in (2.101), knowing that the gradient of the Lagrangian function goes to 0 at the lower-bound solution, and multiplying each of the partial derivatives by the derivation variable itself combined with the KKT conditions of the variables' lower-bound inequalities, we get the following set of equalities:

$$
\begin{aligned}
&\frac{\delta L}{\delta R} \times R = -R + R\sum_{a=1}^{n_A}\sum_{i_A=1}^{n}\alpha^{*(ai_A)} = 0\\
&\frac{\delta L}{\delta P^{(0)}_{di_D,ai_A}} \times P^{(0)}_{di_D,ai_A} = P^{*(0)}_{di_D,ai_A}\left(\gamma^*_{di_D} - \alpha^{*(ai_A)}\lambda_{di_D} + \nu^{*(0)}_{di_D,ai_A}\right) = 0\\
&\frac{\delta L}{\delta P^{(k)}_{di_D,ai_A}} \times P^{(k)}_{di_D,ai_A} = P^{*(k)}_{di_D,ai_A}\left[(\beta_k^* - \alpha^{*(ai_A)})\lambda_{di_D} + \gamma^*_{di_D} + \nu^{*(k)}_{di_D,ai_A}\right] = 0
\end{aligned}
\tag{2.103}
$$

When we inject the result of the equations of (2.103) in the KKT conditions on the upper-bound conditions of the variables $P^{(0)}_{di_D,ai_A}$ and $P^{(k)}_{di_D,ai_A}$, we find

$$\begin{aligned} &P^{*(0)}_{di_D,ai_A}(P^{*(0)}_{di_D,ai_A}-1)(\gamma^*_{di_D}-\alpha^{*(ai_A)}\lambda_{di_D})=0\\ &P^{*(k)}_{di_D,ai_A}(P^{*(k)}_{di_D,ai_A}-1)\left[(\beta^*_k-\alpha^{*(ai_A)})\lambda_{di_D}+\gamma^*_{di_D}\right]=0 \end{aligned} \tag{2.104}$$

From (2.104), we get

$$\begin{aligned} &0<P^{*(0)}_{di_D,ai_A}<1 \text{ only if } (\gamma^*_{di_D}-\alpha^{*(ai_A)}\lambda_{di_D})=0\\ &0<P^{*(k)}_{di_D,ai_A}<1 \text{ only if } (\beta^*_k-\alpha^{*(ai_A)})\lambda_{di_D}+\gamma^*_{di_D}=0 \end{aligned}$$

As $0\le P^{*(k)}_{di_D,ai_A}\le 1$, then these equalities may not always be true. If $(\gamma^*_{di_D}-\alpha^{*(ai_A)}\lambda_{di_D})>0$, and we know that $\delta^{*(0)}_{di_D,ai_A}\ge 0$, then $\delta^{*(0)}_{di_D,ai_A}=0$, which implies that $P^{*(0)}_{di_D,ai_A}\neq 1$ and $P^{*(0)}_{di_D,ai_A}=0$. If $(\gamma^*_{di_D}-\alpha^{*(ai_A)}\lambda_{di_D})<0$, then $\delta^{*(0)}_{di_D,ai_A}>0$ (it cannot be 0 because this contradicts with the value of $P^{*(0)}_{di_D,ai_A}$, which implies that $P^{*(0)}_{di_D,ai_A}=1$.

We also have from the KKT conditions given by equation in (2.102) that either the Lagrangian coefficient is 0 or its associated inequality is an equality. So, if $\alpha^{*(ai_A)}\neq 0$, we have $\forall\, l,m,\ P^{*(0)}_{di_D,ai_A}=\frac{1}{\lambda_{di_D}}\left(R+\lambda_c^{(ai_A)}-\lambda_v^{(ai_A)}-\sum_{d=1}^{n_D}\sum_{i_D=1}^{n}\sum_{k=1}^{S}\lambda_{di_D}P^{*(k)}_{di_D,ai_A}-\sum_{\substack{x=1\\xy\neq di_D}}^{n_D}\sum_{y=1}^{n}\lambda_{xy}P^{*(0)}_{xy,ai_A}\right)$, and $\forall\ l,m, P^{*(k)}_{di_D,ai_A}=\frac{R+\lambda_c^{(ai_A)}-\lambda_v^{(ai_A)}-\sum_{\substack{x=1\\xy\neq di_D}}^{n_D}\sum_{y=1}^{n}\sum_{\substack{z=0\\z\neq k}}^{S}\lambda_{xy}P^{*(z)}_{xy,ai_A}}{\lambda_{di_D}}$.

If $\beta^*_k\neq 0$, we have, $\forall k=1,..,S\ P^{*(k)}_{di_D,ai_A}=\frac{\mu_c-\sum_{\substack{x=1\\xy,vw\neq di_D,ai_A}}^{n_D}\sum_{y=1}^{n_A}\sum_{v=1}^{n}\sum_{w=1}^{n}\lambda_{xy}P^{(k)}_{xy,vw}}{\lambda_{di_D}}$.

Therefore, the optimal joint rebalancing and in-route charging solution of the optimization problem are expressed as in (2.105).

$$\begin{aligned} P^{*(0)}_{di_D,ai_A}&=\begin{cases}0, & \text{if } (\gamma^*_{di_D}-\alpha^{*(ai_A)}\lambda_{di_D})>0\\ 1, & \text{if } (\gamma^*_{di_D}-\alpha^{*(ai_A)}\lambda_{di_D})<0\\ \dfrac{R+\lambda_c^{(ai_A)}-\lambda_v^{(ai_A)}-\sum_{d=1}^{n_D}\sum_{i_D=1}^{n}\sum_{k=1}^{S}\lambda_{di_D}P^{*(k)}_{di_D,ai_A}-\sum_{\substack{x=1\\xy\neq di_D}}^{n_D}\sum_{y=1}^{n}\lambda_{xy}P^{*(0)}_{xy,ai_A}}{\lambda_{di_D}}, & \text{if } \alpha^{*(ai_A)}\neq 0\end{cases}\\ P^{*(k)}_{di_D,ai_A}&=\begin{cases}0, & \text{if } (\beta^*_k-\alpha^{*(ai_A)})\lambda_{di_D}+\gamma^*_{di_D}>0\\ 1, & \text{if } (\beta^*_k-\alpha^{*(ai_A)})\lambda_{di_D}+\gamma^*_{di_D}<0\\ \dfrac{R+\lambda_c^{(ai_A)}-\lambda_v^{(ai_A)}-\sum_{\substack{x=1\\xy\neq di_D}}^{n_D}\sum_{y=1}^{n}\sum_{\substack{z=0\\z\neq k}}^{S}\lambda_{xy}P^{*(z)}_{xy,ai_A}}{\lambda_{di_D}}, & \text{if } \alpha^{*(ai_A)}\neq 0\\ \dfrac{\mu_c-\sum_{\substack{x=1\\xy,vw\neq di_D,ai_A}}^{n_D}\sum_{y=1}^{n_A}\sum_{v=1}^{n}\sum_{w=1}^{n}\lambda_{xy}P^{(k)}_{xy,vw}}{\lambda_{di_D}}, & \text{if } \beta^*_k\neq 0\end{cases} \end{aligned} \tag{2.105}$$

2.5.4 MAXIMUM EXPECTED RESPONSE TIME

Now, in order to find the expression of R^*, we first look at its upper-bound-associated condition in (2.102). From Burke's theorem on the stability condition of the queues, the constraints on R should also be strictly larger than 0. Combining the Burke's theorem and the equations in (2.102), we conclude that $\omega^*=0$. Therefore, from the first equation in (2.103), the minimum expected response rate R^* of the entire system can be expressed as follows:

$$R^* = \sum_{a=1}^{n_A} \sum_{i_A=1}^{n} \alpha^{*(ai_A)} \left(\lambda_c^{(ai_A)} - \lambda_v^{(ai_A)} - \sum_{d=1}^{n_D} \sum_{k=1}^{S} \sum_{i_D=1}^{n} \lambda_{di_D} P_{di_D,ai_A}^{*(k)} \right) \quad (2.106)$$

Finally, the maximum expected response time is the reciprocal of R^*.

2.5.5 SIMULATION RESULTS

We will now examine the hypothetical deployments of AEMoD in a realistic environment mimicking an AEMoD system's decentralized operation in the city of Seattle, using models and procedures from previous sections.

Figure 2.2 shows the city mapped out into 19 zones. High-traffic downtown zones, with typical deficiency in vehicles especially in rush hours, are treated as acceptor zones, are denoted by Roman numbers (I, II, ..., X), and are outlined with red circles on the map. On the other hand, outer uptown zones will typically have excess vehicles after dropping customers. They are thus assumed to be donor zones, denoted with Arabic numbers (1, 2, ..., 9), and are outlined with blue circles on the map. The size (i.e., diameter) of each zone was determined based on average travel time across the zone. Indeed, each zone circumference is decided so that it takes an average of 5–10 minutes to drive to any point within it. Inside each zone, multiclass charging with subclass dispatching (lower layer) is being functioned. While going from one zone to another zone, the vehicle inflow rate dimensioning (middle layer) is performed and rebalancing (upper layer) is performed when vehicles from donor zones are going to acceptor zones to serve customers. In addition, Table 2.2 shows the percentage of loss in vehicles' SoC when traveling between donor zones (rows) to acceptor zones (columns). This percentage loss in SoCs has been calculated by multiplying the average loss in SoC per kilometer (for typical speeds at rush hours) with the distances between zone centers and normalizing the result to the full-battery charge. These percentage SoC losses are then mapped to class degradation terms $y_{i,l}$ by determining the class variation corresponding to this loss. This class variation is calculated based on assuming that close to depletion and fully charged batteries can serve up to classes 1 and n, respectively, and that the variation between these two extremes across all classes is linear. For example, vehicles relocating from zone 9 to zone 1 loose 12.26% of their charge, which corresponds to one-class degradation when $n = 10$, two-class degradation when is $n = 20$, and so on (Figure 2.13).

The following parameters have been set as constants for all the presented simulations: The total (i.e., across all classes) average arrival rate of local vehicles to each acceptor zone has been set to $\lambda_v^{(ai_A)} = 15$ mins^{-1}, the number of in-route charging stations is set to $S = 5$, and the number of

Table 2.2
Loss of SoC (in %)for Interzone Routing of Vehicles

Zone	I	II	III	IV	V	VI	VII	VIII	IX	X
1	1.26	1	1.65	2	2.39	2.39	2.61	3.26	3.32	4.32
2	2.26	1.55	2.16	1.52	3.19	2.71	3.16	2.97	3.9	5.00
3	2.35	1.23	1.71	0.68	2.74	2.61	2.90	3.42	3.45	3.97
4	1.32	2.42	1.32	1.77	1.26	2.48	1.97	2.8	2.32	3.45
5	4.00	4.42	2.87	3.39	2.61	2.03	2.68	2.39	1.97	1.19
6	3.97	4.03	2.90	2.45	2.55	1.90	1.77	1.06	1.55	1.16
7	10.06	10.26	4.10	4.13	3.35	4.19	4.23	3.19	2.29	1.52
8	4.65	4.55	2.97	3.48	2.71	2.97	2.81	2.35	1.58	0.87
9	12.26	26.23	10.35	10.19	10.10	3.68	4.06	2.94	3.77	3.32

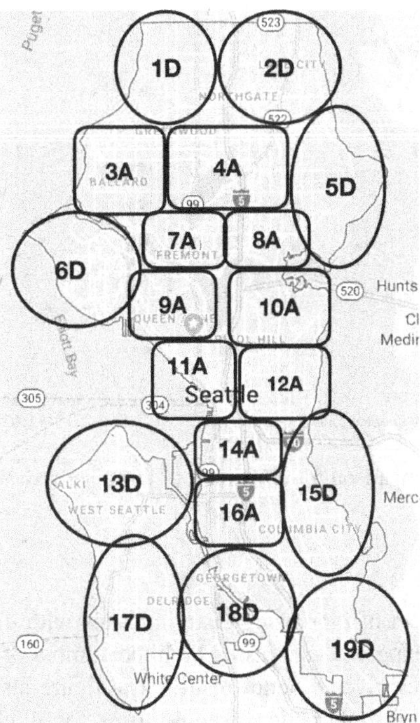

Figure 2.13 Sample system model with 19 fog zones in the city of Seattle.

classes in each zone is set to $n = 10$. Arriving customers' trips to each zone are assumed to be Gaussian-distributed across classes (i.e., more trips corresponding to mid-range classes), whereas both an SoC Gaussian (i.e., more vehicles' SoC corresponding to mid-range classes) and decreasing (i.e., the probability of an arriving vehicle's SoC belonging to class i is higher than that belonging to class $(i+1)\ \forall\ i \in \{1,\ldots,n-1\}$) distributions were tested (with Gaussian being the default if not otherwise stated). The selection of the customers' trips distribution is justified by the fact that customers trips are usually more mid-distance as opposed to very small and very long distances. On the other hand, vehicles' SoC distributions can be either Gaussian or decreasing depending on whether the time allowed for vehicles to partially charge between customer trips is intermediate (corresponding to average-demand levels) or very limited (corresponding to peak-demand levels), respectively.

We compare our multiclass rebalancing scheme with two baseline schemes, namely, the proportional split (i.e., excess donor vehicles are split equally among $k = 0,\ldots,S$ and then relocated to acceptor zones' classes proportional to their customer arrival rates) and the equally split (i.e., excess donor vehicles are split equally among both $k = 0,\ldots,S$ and all acceptor zones' classes) schemes.

Figure 2.14a compares the performance of our proposed solution with the baseline decision approaches for different total customer arrival rates in each acceptor zone. From the figure, it is evident that our proposed scheme gives better performance than the other two schemes for the entire range. Most importantly, our optimized rebalancing scheme exhibits stable performance especially when the rate of customer arrivals considerably exceeds the local vehicle-arrival rate of 15 vehicles/min (i.e., where rebalancing is what maintains the stability of the system). On the contrary, the other two schemes become unstable much faster. Moreover, all three schemes yield lower response times when the charging rate at each station is higher (i.e., when $\mu_c = 0.2\ \text{min}^{-}1$ corresponding to typical 5 mins fast-charging of one Nissan Leaf vehicle).

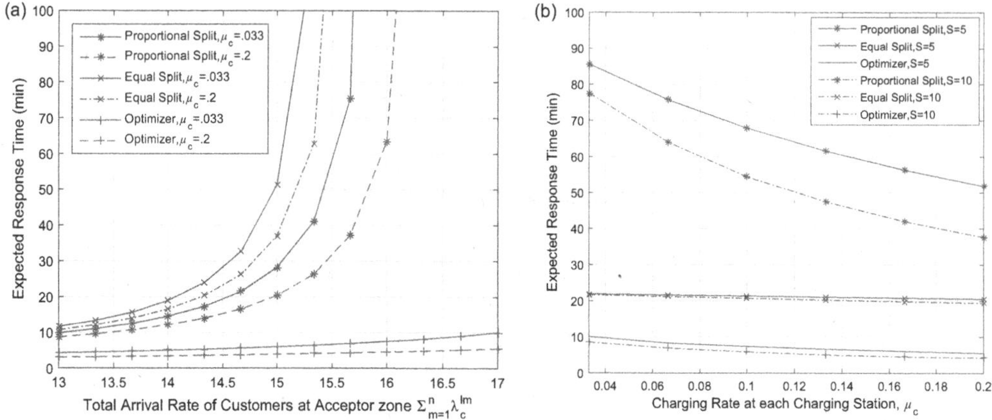

Figure 2.14 (a) Performance comparison for different total customer arrival rates and (b)performance comparison for different charging rates.

Figure 2.14b shows the effect of increasing charging rates with different numbers of charging stations on response time. As expected, increasing both the number of charging stations throughout the city and their charging rates gives a better result. The figure also shows the effectiveness of our proposed scheme in accomplishing better response times for all charging settings. Figure 2.15 shows the effect of different arriving vehicle SoC distributions at both donor/acceptor (using this same order of the legend) zones for different total customer arrival rates per acceptor zone. We can deduce from the figure that the maximum response times for Gaussian distributions of vehicle SoC are lower than that of the decreasing ones. For decreasing SoC distributions, we had to relocate 30% more excess vehicles from donor zones to ensure stability at the acceptor zones.

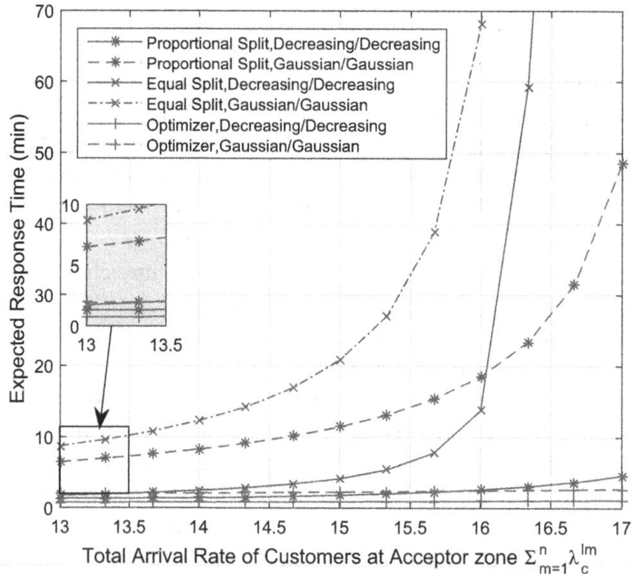

Figure 2.15 Effect of different vehicle-arrival rates distributions on the maximum response time optimization solution and nonoptimized policies.

2.6 CONCLUSION

This chapter overviewed the recent outcomes on the Autonomous Electric Mobility-on-Demand system modeling, control, and assessment technologies. Case studies have shown that it would be much more stable and more convenient to access mobility in an AEMoD system compared to traditional mobility systems based on private vehicle ownership. However, further studies are imperative to develop effective system-wide coordination algorithms for complicated AEMoD systems as part of a multimodal transport network and fully evaluate the financial advantages associated with them. Future study into this subject should follow two primary directions: effective control algorithms for increasingly more realistic models and eventually for actual test beds, and financial analyses for economic assessment.

Despite the great potentials of the marco(i.e., interzone)management in each city zone in simplifying and speeding up the fleet micromanagement, it can further be improved by investigating splitting the zones based on proximity and demand fulfillment (i.e., total service rate of vehicles must be more than the total arrival rate of customers) before running relocations separately. Another important big-picture challenge is making optimized decisions on possible in-route charging of such relocating vehicles and, taking into account, the loss of time due to charging. These are some additional directions open for future research.

REFERENCES

1. W. J. Mitchell, C. E. Borroni-Bird, and L. D. Burns, "*Reinventing the Automobile: Personal Urban Mobility for the 21st Century*". Cambridge, MA: The MIT Press, 2010.
2. D. Schrank, B. Eisele, and T. Lomax, "TTIs 2012 Urban Mobility Report," Texas A&M Transportation Institute, Bryan, TX, 2012.
3. United Nations Environment Programme (UNEP), "The Emissions Gap Report 2013 - UNEP," Technical Report, 2013.
4. The World Bank. World Development Indicators: Motor Vehicles (per 1000 people), November 2014. [online] Available: http://data.worldbank.org/indicator/IS.VEH.NVEH.P3/countries/1W-CN?display=default.
5. A. Santos, N. McGuckin, H. Y. Nakamoto, D. Gray, and S. Liss, "Summary of Travel Trends: 2009 National Household Travel Survey," Technical Report, 2011.
6. C. Nguyen, "Elon Musk says Tesla owners could make up to 30,000 a year turning their cars into 'robotaxis'". *Business Insider*, April 23, 2019. [Online] Available: https://www.businessinsider.com/elon-musk-tesla-robotaxi-app-make-money-repurposing-cars-autonomous-2019-4
7. Digitalist Magazine, "IoT and Smart Cars: Changing the World for the Better," *Digitalist Magazine*, August 30, 2016. [Online] Available: http://www.digitalistmag.com/iot/2016/08/30/iot-smart-connected-cars-willchange-world-04422640
8. X. Tan and A. Leon-Garcia. Autonomous Mobility and Energy Service Management in Future Smart Cities: An Overview, UV 2018, 1–6.
9. "Transportation Outlook: 2025 to 2050," *Navigant Research, Q2'16*, 2016. [Online] Available: http://www.navigantresearch.com/research/transportation-outlook-2025-to-2050.
10. "The Future Is Now: Smart Cars and IoT in Cities," *Forbes*, June 13, 2016. [Online] Available: http://www.forbes.com/sites/pikeresearch/2016/06/13/the-future-is-now-smartcars/63c0a25248c9
11. R. Zhang, K. Spieser, E. Frazzoli, and M. Pavone, "Models, algorithms, and evaluation for autonomous mobility-on-demand systems," *In Proceedings of American Control Conference*, Chicago, IL, 2015.
12. K. Zhang, Y. Mao, S. Leng, Y. Zhang, S. Gjessing, and D. H. K. Tsang, "Platoon-based electric vehicles charging with renewable energy supply: A queuing analytical model," *In Procedings of IEEE International Conference onCommunications (ICC16)*, Washington, DC, USA, 2016.
13. H. Liang, I. Sharma, W. Zhuang, and K. Bhattacharya, "Plug-in electric vehicle charging demand estimation based on queuing network analysis," *IEEE Power and Energy Society General Meeting*, Washington, DC, USA, 2014.
14. R. Zhang and M. Pacone, "Control of robotic mobility-on-demand systems: A queuing-theoretical perspective," *ACM International Journal of Robotics Research*, 35, 186–203, 2016.

15. C. Korkas, S. Baldi, S. Yuan, and E. B. Kosmatopoulos, "An adaptive learning-based approach for nearly-optimal dynamic charging of electric vehicle fleets," *IEEE Transactions on Intelligent Transportation Systems*, 19(7), 2066–2075, 2018.
16. Cisco, "Fog Computing and the Internet of Things: Extend the Cloud to Where the Things Are," *Cisco White Paper*, 2015. [Online] Available: http://www.cisco.com/c/dam/en_us/solutions/trends/iot/docs/computing-overview.pdf
17. P. Mach and Z. Becvar, "Mobile edge computing: A survey on architecture and computation offloading", *IEEE Communication Surveys & Tutorials*, 19(3), 1628–1656, Third quater 2017.
18. Y. Mao, C. You, J. Zhang, K. Huang, and K. B. Letaief, "A survey on mobile edge computing: The communication perspective," *IEEE Communication Surveys & Tutorials*, 19(4), 2322–2358, Fourth quarter 2017.
19. ETSI, "Mobile-edge computing—Introductory technical white paper," White Paper, Sophia Antipolis, France, September 2014.
20. A. Papoulis and S. Pillai, *Probability, Random Variables, and Stochastic Processes*, 4th ed., International Edition, New York: McGraw-Hill, 2002.
21. A. L. Garcia, *Probability, Statistics, and Random Processes for Electrical Engineering*, 3rd ed., Upper Saddle River, NJ: Prentice Hall, 2008.
22. S. Belakaria, M. Ammous, S. Sorour, and A. Abdel-Rahim, "Multi-Class Management with Sub-Class Service for Autonomous Electric Mobility On-Demand Systems", *IEEE Transactions on Vehicular Technology*, vol. 68, no. 7, pp. 7155–7159, July 2019.
23. S. Belakaria, M. Ammous, S. Sorour, and A. Abdel-Rahim, "Optimal vehicle dimensioning for multi-class autonomous electric mobility on-demand systems," *IEEE International Conference on Communications (ICC)*, Kansas City, MO, USA, 2018.
24. S. Boyd and L. Vandenberghe, "*Convex Optimization*", 1st ed. Cambridge: Cambridge University Press, 2015.
25. S. Boyd and J. Park, "*General Heuristics for Nonconvex Quadratically Constrained Quadratic Programming*". Stanford, CA: Stanford University, 2017.
26. S. Belakaria, M. Ammous, S. Sorour, and A. Abdel-Rahim, "Optimal vehicle dimensioning for multi-class autonomous electric mobility on-demand systems," *IEEE International Communication Conference (ICC)*, Kansas City, MI, 2018.
27. N. Yamin, L. Smith, S. Sorour, and A. Abdel-Rahim, "Fleet re-balancing with in-route charging for multi-class autonomous electric MoD systems," *In Proceedings of IEEE International Conference on Communications (ICC)*, Dublin, Ireland, 2020.
28. M. Pavone, S. L. Smith, and E. F. D. Rus, "Robotic load balancing for mobility-on-demand systems," *The International Journal of Robotics Research* 31(7), 839–854, 2012.
29. R. Zhang and M. Pavone," Control of robotic mobility-on-demand systems: A queueing-theoretical perspective," *The International Journal of Robotics Research*, 35(1–3), 186–203, 2016.
30. K. Treleaven, M. Pavone, and E. Frazzoli, "An asymptotically optimal algorithm for pickup and delivery problems," *50th IEEE Conference on Decision and Control and European Control Conference (CDC-ECC)*, 584–590, Orlando, FL, USA, 2011
31. R. Zhang, F. Rossi, and M. Pavone, "Model predictive control of autonomous mobility on-demand systems," *IEEE International Conference on Robotics and Automation (ICRA)*, Stockholm, Sweden, 2016.

3 Multifaceted Synthesis of Autonomous Vehicles' Emerging Landscape

Hossam Abdelgawad
Cairo University

Kareem Othman
University of Toronto

CONTENTS

3.1 INTRODUCTION

The advent of autonomous vehicles (AVs), connected vehicles (CVs), electric vehicles (EVs), not to mention the intertwined synergies among them – connected autonomous electric vehicles, has merely disrupted the transport sector at multiple fronts. The realm of "shared mobility" can be simply looked at as an overall arching mechanism enabling mobility as a service. This chapter will touch upon the above-mentioned four "mobility disruptions" with focus on AVs.

These mobility disruptions are empowered by a myriad of "Technological Enablers", including advances in IPv6, sensors technology, abundance of data, more connected devices and internet of things, artificial intelligence (AI) models, and computational power; all the above has resulted in witnessing new innovative business models.

While all the above is being realized with an ever-fast-growing pace, "Cities" and population continues to grow with a soaring urbanization trend – indicating that close to 60% of the world population lives in urban areas. To better face this urbanization challenge, Cities have embarked determinedly into being Smart (and others endeavor *Intelligent, Magnet, Resilient, Innovative, Sustainable*, and lately *Wise*), with varying degrees of success.

As the AVs' technology continues to emerge, a number of "Key Industry Players" have been at the forefront including automotive original equipment manufacturer (OEMs), Tech Giants such as Google and Apple, Transportation Network Companies, and the private sector enabling a vibrant landscape for startups and new entrances into the mobility ecosystem.

Although the above-mentioned mobility disruptions have significant potential, there is a multitude of "Implications" that shall be considered including demand for transportation service; supply of infrastructure and roadway design requirements; accuracy for mapping, safety, and security; privacy and ethics; users adoption and acceptance; travel behavior; environment and energy; societal impacts; and affordability of travel.

Therefore, Governments are challenged to develop policies (e.g., for electric vehicles), regulations (e.g., for ride-hailing companies and AVs), laws (e.g., data sharing), and champion pilots/testbeds (e.g., AV shuttle pilots in Europe) targeting to beat the forefront in terms of government innovation in the realm of smart cities and beyond.

The above constitutes the "Mind Map" of the chapter as schematically illustrated below, which forms the introductory section of this chapter portraying the big picture with its various pieces of the puzzle.

Current State Mobility **Mind Map**

Mobility Disruptions	Enablers	Cities	Key Players	Implications	Governments
• Automation	• IP6	• Smart	• Automotive	• Demand for Service	• Policies
• Connectivity	• Sensors	• Intelligent	• Tech Giants	• Supply of Infrastructure	• Regulations
• Shared Mobility	• Data	• Magnet	• TNC	• Users Adoption	• Laws
• Electrification	• IoT	• Resilient	• Private Sector	• Travel Behavior	• Pilot / Test Cases
	• AI	• Innovative	• Start Ups	• Environment	• Data Sharing
	• New Business Models	• Sustainable	• New Entrances	• Energy	
		• Wise		• Society	
				• Cost of Travel	

This chapter rather focuses on AVs by addressing the following verticals:

- Benefits of AVs.
- Implication of AVs on Travel Behavior and Development of Cities
- Challenges Associated with AVs
- Infrastructure Requirements and Implications
- User Opinions, Adoption, and Perceptions
- Synopsis on Pilots and Laws and Regulations
- New Value Network and New Business Models

Building on the syntheses and findings from the above intertwined sections, the chapter ends with a framework for incorporating AVs into the realm of smart cities by providing specific directions and recommendations to harness the potential of AVs – and at the same time – address/avoid adverse impacts of AVs.

3.2 BENEFITS

This section discusses the reported benefits of AVs by considering a multitude of dimensions, categorized as follows: *safety, vehicle kilometers traveled (VKT), equity, jobs, system capacity, vehicle ownership, parking, and energy and emissions.*

3.2.1 SAFETY

The safety of AVs is essential to their success in the marketplace (University of Illinois College of Engineering, 2019). Although AVs respond faster than humans and thereof indicate higher safety levels, AVs cannot handle all environmental conditions as the current sensing technology fails to provide reliable information for all travel conditions/situations (Sivak and Schoettle, 2015). While it is argued that recurrent vehicular failures might become obsolete with the technology development, AVs are more complex due to the sensing hardware and the information-processing software. A typical AV has more than 50 processors and accelerators running millions of lines of codes to support the computer vision, planning, and other learning tasks. Additionally, the behavior of AVs is unpredictable in all scenarios unless it is trained in the real world; however, there are endless numbers of scenarios (University of Illinois College of Engineering, 2019). Therefore, the level of development and data to support the safety of autonomy are still at early stages (Lori, 2019). In November 2018, John Krafcik, CEO of Waymo, indicated that AVs will never be able to operate in all possible conditions without human interactions (Stewart, 2019; Tibken, 2018). Another important factor that has contributed to recently reported crashes in AVs is the transition from the conventional mode to autonomous mode. Data collected from 2014 to 2017 by the research team at University of Illinois for 114 AVs that traveled almost 1.1 million VKT showed that the human drivers are 4000 times less likely to have an accident (University of Illinois College of Engineering, 2019). An example of vehicular failure is the fatal accident in 2018 in Arizona when an Uber vehicle detected a pedestrian crossing the street, and the vehicle decided that it does not need to swerve and hit the pedestrian (Walker, 2020).

On the behavioral side, the normal eye contact-feedback-proceed two-way communication between drivers in adjacent cars (or drivers and pedestrians) in conflicting situations is either absent in full AVs' penetration or more confusing in conventional vs AVs' mixed conditions. The perception of drivers/users of advanced degrees of automation indicates a nonpositive view of sense of security and safety (Kyriakidis et al., 2015).

Despite the fact that safety is perceived as one of the key benefits of AVs, it can be stated that (1) the expectation of zero fatalities with self-driving vehicles is not realistic, (2) an AV is not expected to ever perform safely than an experienced driver, and (3) adverse safety outcomes are expected during the transition period when conventional and self-driving vehicles share the same road.

3.2.2 VEHICLE KILOMETERS TRAVELED (VKT), VEHICLE OWNERSHIP, AND CONGESTION

The tandem effect of accepting longer trips/longer travel times in return of adopting AVs would have mixed implications on the perceived benefits related to congestion, travel times, sprawl, and induced demand. It is argued that the benefit of travel time savings in different activities is expected to motivate people to accept longer trips and thereof the likelihood of urban sprawl resulting in more VKT (The Polis Traffic Efficiency and Mobility Working Group, 2018). Despite the fact that the AVs might not directly address congestion, drivers' time will not be considered as an economic loss as it could be spent in a productive activity. Reduced congestion could result in induced demand and thereof attracting more trips.

The combined result of the above positives and negatives is still subject to debate and indecisiveness. For example, in an urban setting, AVs are expected to increase the demand by inducing new users (such as old people, disables, and youngsters) but expected to reduce the household vehicle ownership. The net of both demand increase vs ownership reduction is yet to be decisively positive (Metz, 2018). For example, in the Atlanta Metropolitan Area (Zhang et al., 2018), the implications of AVs on ownership and spatial distribution of unoccupied VKT were investigated combining mixed-integer programming (MIP) methods to optimize VKT and trip assignment to allocate the unoccupied trips on the network. The result indicates that 18.3% reduction in vehicle ownership can be achieved, while VKT is expected to increase (up to 60% in some cases) when compared with the current case. It is concluded that in return for vehicle ownership reduction, a notable amount of unoccupied VKT (at least 13%) will be generated and more vehicles can be reduced if household members start to reschedule daily trips to accommodate AVs.

In Lisbon, simulating two self-driving vehicle concepts (TaxiBot – self-driving cars that can be shared simultaneously by several passengers and AutoVot – pick-up and drop-off single passengers sequentially) indicated that the same mobility demand can be achieved with only 10% of traditional cars, increasing the vehicle utilization from 5% to 60%–75% (International Transport Forum, 2015). This is primarily due to three to five passenger vehicles dominated the self-driving TaxiBot followed by AutoVot. In terms of VKT, it is expected to increase in the range from 6% to 89% for the 50% AV and 100% AV, respectively. The ITF study concluded that for small- and medium-sized cities – if only 50% of vehicles are shared AVs (SAVs), VKT is expected to increase in the range of 30%–90%; this finding holds true regardless of the capacity or availability of an adequate public transit system.

3.2.3 BETTER EQUITY AND POTENTIAL JOB LOSS

The industry is bombarded with dire predictions of jobs lost due to AI, robots, and automation; the relevance of this fear to AVs is no exception. While in one hand, AVs are expected to improve equity for people with limited transportation accessibility (such as old people and disabled); they could also reinforce or accelerate inequity according to the design of the provided system to the public as the service level in terms of speed and quality might be based on the price, i.e., pricing out certain socio-economic stratum (The Polis Traffic Efficiency and Mobility Working Group, 2018). AVs are also expected to impact the transportation sector jobs, with many jobs expected to disappear, and new professions will be induced. It is not clear though how new jobs would be assigned to those who need these jobs and excel at them technically, with a potential for singling out highly skilled/talented professions.

In Washington DC, a study assessed multiple scenarios (considering variants of capacity, value of time, and parking inputs to AVs), resulting in higher level of accessibility (Childress et al., 2015). This was particularly true for remote and rural areas due to the assumption that driving is easier, cheaper, and more enjoyable resulting – on the other hand – in 40% increase in VKT for work trips. In the case when AVs are considered as public utility mode, VKT was reduced by 35% and travelers choose short trips due to high cost with an average trip length reduced by 15% vs the base case. The pricing policy results suggested that sustainable modes such as share-like transit and walking might increase by 140% and 50% and that delays are lower than half of the base case.

Studying the economic effect of AVs from the governmental lens in San Francisco indicates that drivers are likely to be replaced by AVs, which have a significant impact on individuals with lower levels of education and income and, consequently, implications for equity (Clark et al., 2017). On the other hand, many jobs that require a higher level of education cannot be replaced (such as IT and data analytics). Although mobility and accessibility to a transportation mode are often reported as a positive impact of AVs, another equally important side of equity is related to job and skills required. As a result, this disparity by educational attainment has real concerns for equity.

3.2.4 SYSTEM CAPACITY

As AVs run on a system of systems of roads, traffic signals, and intelligent transport systems, there are perceived benefits from utilizing the same system capacity more efficiently and ideally more intelligently via adding the connectivity layer.

Metz (2018) studied the benefits of AVs with a focus on the urban setting, stating that while AVs are expected to increase the lane capacity as a result of platoons and narrower lanes, that capacity potential increase can only be realized with high levels of penetration of AVs and not necessarily in urban areas. On the other hand, investigating the effect of AVs on highways and intersection capacity using macroscopic traffic flow models (Friedrich, 2016) indicates a potential increase in capacity depending on roadway speed limit. At a speed of 80 mph, AVs can increase the highway capacity by a factor of 1.78 and can increase the lane capacity to 3900 vph compared with 2200 vph with 100% AVs' penetration. As the percentage of AVs' penetration decreases to 50%, the capacity might reach 3100 vph. At intersections, AVs would increase the intersection capacity by 40% to 1120 veh/h/lane compared with 800 veh/h/lane for conventional vehicles, and logically, lane capacity increases with the increase in the clearance speed.

In Braunschweig, Germany, the effect of autonomous driving on the traffic management levels with a focus on traffic signal control in mixed traffic operations has been investigated (Wagner, 2016), indicating that at low-demand single intersections the capacity can be doubled. However, if the signal is at capacity, even a minor increase in its capacity can lead to a dramatic improvement. At the city level, AVs can reduce delay dramatically between 5% and 80% with an average of 40%; however, the authors stated that, yet the system becomes faster, it is not necessarily more reliable.

In the Washington DC study by Childress et al. (2015), capacity and VKT was reported indicating that all tested scenarios increase the system capacity (In this study, three scenarios are considered: (1) Technology Changes, But We Don't [i.e., no change in vehicle ownership]; (2) New Technology Drives New Behavior [i.e., no changes in the ownership, but drivers are willing to travel longer trips]; and (3) New Technology Drives New Ownership Models [i.e., similar to scenario 2 but with a complete change in the ownership]) with corresponding increase in VKT. In the first two scenarios, the increase in the capacity offsets the increase in the VKT by allowing the vehicles to travel faster. However, in the third scenario, the reduction in the value of time and parking costs attracted more customers, resulting in high level of demand, so the additional demand offsets the additional capacity gains.

3.2.5 PARKING SPACES

Similar to the paradox of equity gains vs potential job loss, AVs could result in perceived parking benefits the urban settings but could also result in induced demand for mobility and potential loss of governmental revenues.

For example, AVs are expected to reduce the number of off-street parking spaces especially in Central Business District (CBD) areas, which creates room for new development opportunities (Metz, 2018). Theoretically speaking, AVs can park itself without the need for the door space which could enable 20% more free spaces (Catapult Transport Systems, 2017). The result could have a positive impact on search for parking requirements (and positive impact thereof on congestion) and a potential for attracting more trips to development opportunities, especially in dense urban areas.

On-street parking in urban areas – on the other hand – possesses a real challenge for AVs. In 2-way traffic streets, AVs would struggle in such situations not to mention maneuvering narrow streets in the presence of pedestrians or cyclists. In such a case, it might be more plausible to remove on-street parking all together or alter the roadway direction (Catapult Transport Systems, 2017).

In San Francisco, parking represents around 14%–25% of the land use in central urban areas. AVs would reduce the required spaces by almost 90%. However, lower governmental revenues are expected from parking meters, garages, and tickets for illegal parking (Clark et al., 2017).

In the Lisbon study of simulating two self-driving vehicle concepts (TaxiBot, AutoVot), AV could potentially remove the need for on-street parking, indicating that while 80% of the off-street parking can be removed, freeing up an area equivalent to 20% of the city streets area. The potential of economic development and vibrancy to the city is thereof nontrivial.

3.2.6 ENERGY AND EMISSIONS

The above intertwined gains (and, in some cases, losses) of AVs on the mobility sector would have a direct impact on the energy efficiency and emissions. The potential emission gains could result from platooning, smooth driving conditions, and search for parking reduction; also, an anticipated increase in VKT could offset some of these gains.

Platooning of AVs is expected to save 20% of the energy used today (Brown et al., 2014) due to reduction in the aerodynamic drag forces on vehicles (Barth et al., 2014). Fernandes and Nunes (2012) studied the effect of AVs' platooning on communication delays using MATLAB®- and Simulink®-based simulation (with identical vehicle dynamics). The hypothesis is that relaying anticipatory information from the platoon's leader and the followers can positively impact platoon stability. The simulation results indicate that communication delays can be completely canceled out and that AVs can provide a high level of communication and in turn reduction in emissions and energy needed.

Smooth starts and stops can save fuel consumption by 15% (Brown et al., 2014; Barth et al., 2014). On one side, reducing congestion by boosting the system capacity and lowering collision rates (and in turn increasing the speed) reduces emissions (Barth et al., 2014). On the other side, this congestion reduction is expected to provide more opportunities to increase the vehicle speed, resulting in increased energy consumption (Brown et al., 2014). The net effect of both is inconclusive. The same study added that the reduced need for parking search will result in 4% energy savings.

Another futuristic proposition assumes that AVs could allow vehicles to be dramatically light-weighted, as there would be no longer a need for large-framed vehicles for collision-safety purposes. This will reduce the energy consumption dramatically, similar to the fabric and design of the aerospace industry.

In Austin, Texas, simulating the overall impact of AVs on emissions in urban areas indicates that AVs contribute to 11% increase in VKT due to AVs' relocation to cheap parking areas during low demand and that each AV can replace 12 conventional vehicles and 11 parking spaces can be removed for each AV. The net result could be the reduction of CO emissions by 34% and VOC by 49%, portraying that newer AVs' generations might be more environmentally friendly. The study concluded that AVs can reduce energy use and emissions, and quicker vehicle fleet turnover may generate even more benefits, as older and more polluting vehicles are replaced with newer, cleaner ones (Fagnant and Kockelman, 2014).

3.2.7 SYSTEM AND POLICIES

Due to the state of technology development and the need for a transition between conventional driving and automation with its various levels (starting from level 3 and above), certain system requirements and policy implications are still under investigations.

Early levels of AVs require drivers' attention at some situations, indicating that separate lanes might be provided; the benefits of such scenarios are yet to be investigated (with its linkage to equity and societal benefits). Additionally, AVs might unluckily replace public transport partially or completely, meaning a potential for putting public transportation system to an end; however, high-capacity transit is still needed in major corridor, and AVs should support transit and improve access to stops and stations (Litman, 2020; The Polis Traffic Efficiency and Mobility Working Group, 2018; Zmud et al., 2017). Although SAVs reduce the required parking spaces, it increases the need for pick-up and drop-off areas that require better curb management to reduce the risk and conflict (Litman, 2020; International Transportation Forum, 2018). As a result, policy makers are challenged to develop a clear vision and policies for AVs with a structured catalog between the transport authorities, industry, and service providers and not leave it to the market active players alone (The Polis Traffic Efficiency and Mobility Working Group, 2018).

On the other hand, AVs will increase the demand for sharing trips thereof can reduce the number of vehicles in operation and in turn congestion; however, this might attract more trips, and the AVs low cost might shift public transportation passengers to AVs. The dynamics of demand, shared trips, congestion, and public transit ridership are complex and intertwined and are still under investigation (Metz, 2018).

Various public organizations have provided policy guidelines to maximize the benefits of AVs as follows (Litman, 2020; Schlossberg et al., 2018; Shared Mobility Principles for Livable Cities): "promote equity, support public transit, prioritize people over vehicles, support shared vehicles, lead the transition towards clean and renewable energy, in dense areas AVs should only operate in shared fleet".

3.2.8 SUMMARY OF AVs' BENEFITS AND IMPACTS

There is a consensus that VKT are expected to increase; however, the reported increase dramatically differs from one study to another ranging from as low as 4% (Childress et al., 2015) to 89% (International transport forum, 2015).

AVs will reduce the *energy consumption* due to the potential reduction in congestion, the platooning effect, and the smoothing of traffic to avoid the sharp stop and go. However, the anticipated increase in VKT might offset some of the reduction in fuel consumption.

Additionally, AVs will result in *boosting system capacity* due to platooning, allowing for shorter gaps, and more importantly eliminating/reducing the system stochasticity embedded in human behavior. AVs are expected to increase *capacity* by almost two times for the highways and by 40% at intersections. Securing America's Future Energy (SAFE) (2018a) estimated an increase in the capacity by 50%–70% with 50% AVs and 300% increase with 100% AVs. This increase in system capacity might result in *induced demand* and generation of additional trips, especially if urban centers become more viable/attractive with reassignment of parking landuse (see below) to retail and entertainment. The net effect of which is yet to be decided/quantified.

AVs will reduce the need for *parking requirements* dramatically as the off-street parking will be very limited (function of the penetration of AVs) and the on-street parking will be reduced by more than 80% (Clark et al., 2017). Garages will not be required any more, which will free up a space for new developments and increase the land value. This induced land development potential is expected to attract more trips, which in turn can adversely impact VKT and congestion.

The issue of *societal equity* is subject to debate among reported studies. AVs are expected to increase the accessibility for elderly and disabled and offer new opportunities for access to transport modes. However, drivers will be replaced with the new system resulting in loss of jobs. While new jobs are expected to be created, this will only provide new opportunities for individuals with higher educational levels (such as IT, data analytics, and software development). As a result, this disparity by educational attainment has real concerns for equity (Clark et al., 2017; The Polis Traffic Efficiency and Mobility Working Group, 2018; Miller and Heard, 2016; Clements and Kockelman, 2017; Securing America's Future Energy (SAFE), 2018). Moreover, the decrease in demand for personal injury lawyer, doctors, and police officers is expected to hurt career prospects as a result of the increase in safety, especially that vehicle collisions are the most common type of civil trials.

AVs will provide a high level of information that can help *road authorities* in decision-making process; for example, data can be used for dynamic traffic management and visualizing the state of the road. However, governments shall attain to the issues pertained to the shrinkage of certain lines of revenues, including revenue from parking tickets, speed ticketing, drink and drive, civil trials, etc.

A wide range of reported benefits is vividly observed among the literature. Despite the variation of (1) the study context and location, (2) assumptions used, (3) analytical and simulation approaches, there is a clear indication that AVs will result in significant benefits; the economic societal impacts of which are discussed next. Governments are therefore challenged with responding to this rapid and emerging technology with the multifaceted impacts on travel behaviors, system capacity and infrastructure requirements, laws and policies, revenue streams, and demand for jobs.

3.3 ECONOMIC BENEFITS AND SOCIETAL IMPACTS

The transportation industry is a key driving force to many other industries, and the disruption enabled by AVs is expected to impact not only the automotive industry but also electronics and software technology, automobile repairs, health and accidents, land development and parking, digital media and infotainment, insurance, and police and traffic violations.

One of the most comprehensive studies in the US on that respect is the impact of AVs on economy by Clements and Kockelman (2017), synthesizing existing literature and evaluating cost/sales changes across multiple industries. The key findings of the study can be summarized along the following categories:

- *Overall US Economy*: SAVs have a great impact on the economy by an estimated value of $1.2 trillion in total (6.2% of 2017 GDP) or $3800 per person per year as a benefit.

- *Automation Industry*: Hardware represents 90% of the value of current vehicles; however, AVs hardware value is expected to drop to 40% of the value of the vehicle. On the other hand, software constitutes ~10% of vehicle value today; however, AVs' software value may eventually represent almost 40% of the value of the vehicle.
- *Electronics and Software Technology*: Technology companies may have the most gain from the development of AVs with an expected growth from $680 million in 2025 to $15.8 billion in 2040. Software and mapping upgrades will grow from $530 million to $10.6 billion by 2040, meaning that AVs will therefore offer around $26.4 billion revenue in only 15 years.
- *Trucking Industry*: AVs will allow platooning and potentially reducing fuel consumption by 15%. AVs in the trucking industry can save from $100 to $150 billion per year by 2025, with the bulk of these savings due to eliminating drivers' wage. The industry currently employs over 3 million drivers with a shortage of 25,000 drivers. It is therefore expected that AVs will increase the capacity of the logistics companies by avoiding the shortage in the drivers. The role of drivers in the AVs' era might be more technical, which might require training and increase in wage of the drivers but reduce the number of drivers. In summary, AVs would undoubtedly be of massive benefit to trucking industry but could decrease opportunities for the employment of millions of truck drivers.
- *Personal Transportation*: The biggest change as a result of the development of CAVs will likely be transportation modes for short trips offering "on-demand" taxi services with SAVs that would make human-driven taxis obsolete. While the effect on long trips is less clear, it is believed that PT and taxi services will be most disrupted by full automation and SAVs. In total, the changes in PT and taxi services due to the disruptive CAVs' account for $26.5 billion in revenue out of $86 billion (representing 30.8%).
- *Auto Repairs Industry*: As AVs are believed to reduce collision caused by human factor (90%), repair shops are expected to significantly lose a huge toll. In the US, $30 billion were spent on vehicle repairs in 2014. Assuming an extreme case of AVs reducing collision by 90% – with a penetration rate of 100% – the industry will lose significant revenue of $27 billion. On the other hand, the higher vehicle utilization and the expected increase in VKT will require more maintenance, which also means that more exposure on the road results in increasing the probability of collisions. Therefore, the exact impact on the industry is unclear, but collision repair businesses that retain their current model will likely face revenue losses. One might argue that the consumer savings from the reduced repair expenses can be transferred to other goods and services that will deliver greater utility and generate economic activity.
- *Health*: Vehicle collisions alone account for $23 billion in medical expenses (NHTSA 2015), and AVs are expected to cause a reduction of $20.7 billion in these expenses. On the other hand, reduction in the need for supplies and doctors shall be better reallocated to provide better services.
- *Insurance and Legal Profession*: The automobile insurance industry is expected to be disrupted/shrunk by 60% as the liability will be shifted from the individuals to the software and hardware manufactures; automakers and the vehicle's software providers will likely become the main responsible party and will need to purchase insurance for technical failure of the automobiles. Along the same lines, vehicle collisions account for around 35% of all civil trials, indicating a decrease in demand for personal injury lawyers that would hurt their career prospects.
- *Construction and Land Development*: The reduction in the need for parking spaces and garages will open the opportunity for redevelopment of existing garages. Parking lots and garages represent almost one-third of the land area creating dead zones in urban centers – where land is needed the most. A parking space is valued at $6300 and the total value

for all spaces is estimated at $4.5 trillion. Potential reusing of this area in the real-estate industry will potentially increase the land value by 5%.

- *Digital Media and Marketing*: AVs will provide new opportunities for digital media as commuters who usually spend time vigilantly watching the road will demand greater integration of digital media features into their automobiles. The e-commerce industry could receive a large leap from this added free time. On the other hand, a possible loss in demand for radio is expected with the hype for vehicle infotainment and entertainment.
- *Police (Traffic Violations)*: As human errors are expected to diminish, the importance of the cops will decrease. Drunk driving, speeding, and other traffic violations will become less frequent and the size of the police force will therefore decrease or relocate to other services. Although fine values vary from one context to another, CAVs are estimated to account for a $5 billion decrease in government revenue. Some of this loss may be recovered, however, through savings from the decreased need for traffic police.

In the UK, KPMG (2015) showed similar trends to the US when reported on the economic effect of AVs indicating notable social, industrial, and economic benefits, as summarized below:

- *Overall Economic Potential*: The economic and social benefits of the AVs are expected to be 51£ billion (2% of the total GDP in UK in 2019) by 2030 and 121£ billion (4.6% of the total GPD in UK in 2019) by 2040.
- *Mobility*: AVs can save 2,500 lives and prevent 25,000 serious accidents.
- *Jobs*: Contrary to other studies, KPMG foresees that AVs will provide additional 320,000 jobs by 2030 in the UK (25,000 jobs in the automation industry alone). Other related sectors such as telecoms, creative industries, and media will also generate additional jobs as they serve new markets created by AVs. Additionally, AVs are expected to open up new opportunities in fields such as software, vehicle cyber security, and data harvesting and matching.

Securing America's Future Energy (SAFE) (2018b) studied the economic impact of AVs in the US by estimating the consumer benefits using consumer surplus. The analysis showed as follows:

- *Economic Value*: AVs' economic benefits will be almost $800 billion (4.1% of the GDP in 2019) by 2050 from reduction in collisions, value of time, reduction in fuel consumption, and environmental benefits. AVs can result in a social benefit of $58 billion by 2050.
- *Jobs and Accessibility*: AVs will increase the accessibility in the depressed regions. Additionally, AVs will generate new job opportunities replacing the eliminated jobs.
- *Safety and Congestion*: It is expected that AVs will reduce accidents by 94% with an economical value of $118 billion (0.6% of GDP in 2017) and $385 (2% of GDP in 2017) billion due to the quality of life.
- *Congestion and Capacity*: AVs can reduce the distance between vehicles shortening the headway from 2 to 0.5 seconds, resulting in increase in capacity by 50%–70%. AVs will increase the capacity by 50% and reduce congestion with an economic value of $48 billion dollars (0.2% of GDP in 2017) due to fuel and time savings. 100% AVs will increase the capacity by 3.2 times (5500 veh/h/lane). The expected economic value of congestion reduction is $71 billion (0.36% of GDP in 2017). The platooning effect in autonomous trucks will reduce fuel consumption by 20%. If all trucks travel in platoons, the savings will be almost $3.4 billion per year.

Albeit the above expected productivity increasing potential, there will be a significant employment displacement associated with another massive opportunity for new business models to emerge with AVs, which will in turn create new employment (Securing America's Future Energy (SAFE), 2018a).

Another concern about AVs is its affordability for the public. According to the National Automobile Dealers Association, Americans spend on average $30,000 for a new car; the cost of an AV is almost tenfold (Tannert, 2014; Nunes and Hernandez, 2019). Some argue that autonomous taxis can circumvent this cost disparity. As a result, a research in Massachusetts Institute of Technology (MIT) by Nunes and Hernandez (2019) attempted to address this claim and estimated if autonomous taxis can become cost-competitive by owning older vehicles for the city of San Francisco. The results showed that autonomous taxis will cost customers three times more than the conventional affordable vehicles.

On the other hand, an economic analysis by the US Department of Transportation (2017) showed that a fleet of AVs might cause a reduction in the operating costs by 30–50 cents per mile. In TUM, Germany, an economic analysis by Ongel et al. (2019) showed that autonomous taxis can reduce the total operating costs by 60%–75% compared with the conventional taxis.

Although SAVs will be cheaper than human-operated taxis, it offers minimal service quality; consequently, passengers may encounter previous occupants' garbage, stains, and odors, and there will be no driver to help passengers or ensure their safety. SAVs are expected to be like public transportation where passengers share their space with strangers who might be friendly or unpleasant. Moreover, pick-up and drop-off of passengers will cause additional delays, indicating that many travelers might prefer to own their private AV (Litman, 2020). The study concluded that operating costs of AVs will be cheaper than traditional taxis but higher than the traditional transit and traditional vehicles.

The summary of the above multidimensional benefits/impacts is that the economic effects of CAVs will likely extend beyond the simple crash, productivity, and fuel saving into every facet of the economy; therefore, individual businesses must adopt to this disruptive change otherwise soon they suffer from "who moved my cheese?" syndrome. To attempt addressing this question, it is pivotal to shed some light on the implications of AVs on travel behavior and development of cities.

3.4 IMPLICATION OF AVs ON TRAVEL BEHAVIOR AND DEVELOPMENT OF CITIES

The reported benefits and challenges mentioned earlier will result in certain *implications* on driving behavior and travel patterns and can only be achieved if certain infrastructure requirements are available. This section discusses the implication of AVs on the development of cities at large by considering a number of dimensions, including travel behavior, travel cost and affordability, parking strategy and demand, and penetration rate and fleet size.

3.4.1 TRAVEL BEHAVIOR

Embracing AVs implies certain changes to private cars travel, ownership of private cars, driving hours, waiting time for AVs, trip time and distance, and the potential for shared trips.

In Berlin, Germany, the implications of AVs on *travel behavior* were simulated based on the assumptions pertained to fleet size of AVs replacing all private cars, low demand vs overloaded system, and passenger embarkation and disembarkation time (Bischoff and Maciejewski, 2016). The study findings indicate that only a fleet of almost 10,000 AVs can server all passengers in the city with an average waiting time ranging from 0.5 to 2.5 minutes in the peak period; over a period of 7.5 h/day resulting in travel time increase by 17% due to empty trips. On an average, AVs are busy 7.5 h/day and each AV can replace 10–12 conventional vehicles. This implies high utilization of AVs at the expense of minimal waiting time but notable increase in VKT.

Using speculative scenarios approach, the implication of AVs on the *customers' behavior* using *system dynamic modeling* was investigated by Gruel and Stanford (2016). In this study, three scenarios are considered: (1) Technology Changes, But We Don't (i.e., no change in vehicle ownership);

(2) New Technology Drives New Behavior (i.e., no changes in the ownership, but drivers are willing to travel longer trips); and (3) New Technology Drives New Ownership Models (i.e., similar to scenario 2 but with a complete change in the ownership). The study findings indicate that across all scenarios; VKT will increase, which in turn implies higher energy consumption and emissions. The increase in VKT differs significantly between the scenarios. Urban sprawl is likely to increase resulting in public transit suffering low coverage and service standard. It is clear that AVs alone cannot address the current mobility challenges, and thereof ownership levels of AVs must be controlled, attractiveness of traveling by AVs shall be reduced, and public transit systems' attractiveness should increase.

In the greater Munich metropolitan area, the implications of AVs were investigated *using stated preference survey and simulation models* focusing on VKT and average trip duration (Moreno et al., 2018), concluding that 4 AVs could likely replace ten conventional vehicles with the same customer satisfaction as the average waiting time was <5 minutes and 95% of the waiting time is lower than 10 minutes. In all modelled scenarios (with different fleet sizes), AVs would increase the VKT by up to 8% due to empty trips and increasing the fleet size decreases the additional VKT.

In Singapore, the implications of AVs were simulated introducing Autonomous Mobility-on-Demand on a car-restricted zone portraying three levels: (1) long-term (capturing land use, economic activity, and accessibility), (2) midterm (captures agents' activities and travel patterns), and (3) short-term (simulates the microscopic movement of agents) (Azevedo et al., 2016). Passenger waiting time decreases with the increase in the number of the AVs in the fleet until 2500 AVs, and the waiting time remains constant at 5 minutes. The restricted zone and the changes in the transportation system performance on the route choice decisions are significant for the agents driving between specific origin–destination pair. This impact on traffic will inevitably affect the performance of the road network in terms of travel times and congestion levels in the periphery of the CBD.

3.4.2 TRAVEL COST AND AFFORDABILITY

Travel cost is a core element of the basics of utility theory, and the assumption that travelers are rational decision-makers attempting to maximize their utility. Travelers thus assess costs and benefits of their actions/choices which are directly related to the discussion herein and the implications of AVs on trip cost and affordability of travel.

Among the key studies in this subject in the US is the one conducted by Burns et al. (2012) attempting reporting on AVs trip cost in three different cities in the US (Ann Arbor–Michigan, Babcock Ranch–Florida, and Manhattan–New York) using queuing, networking, and simulation models.

- *Ann Arbor* (with a population of 285,000 people) represents a small-medium size city in the US, and it was assumed that AVs would replace all private car trips that are <70 miles. A fleet of 18,000 AVs was able to serve the population with waiting time <1 minutes, resulting in reducing the number of vehicles by more than 80% with an average 75% vehicle utilization. The travel cost dropped thereof from $21 to $2 per day due to reduction in the ownership cost, operating expenses, parking fees, and value of time.
- Babcock Ranch (with a population of 50,000 at build out) represents a small urban area in the US as a future city (under construction). A fleet of 3000–4000 AVs would serve the city efficiently with an average waiting time under 1 minutes, with an average mobility service cost <$3 per day per person or $1 per trip.
- In Manhattan case, it was assumed that AVs would replace the yellow taxicab, and a fleet of 9,000 AV would be able to serve the demand efficiently with an average waiting time <1 minute compared with 5 minutes waiting time for the traditional yellow taxi. The SAVs

model results in significant trip cost reduction to $0.8 per trip (compared with $6.3–$7.8 per trip using the traditional yellow taxi) due to reduction in the ownership cost, operating expenses, and central coordination.

In 2016, Ford showed that it can reduce the cost of SAVs to about $1 per mile which makes it highly competitive to the current taxis that costs $6 per mile. Johnson and Walker (2016) estimated that AVs' cost will be 51 cents per mile in 2025 and 33 cents per mile by 2035, compared with 84 cents per mile for conventional vehicles.

In Germany, an economic analysis by Ongel et al. (2019) showed that autonomous taxis can reduce the total operating costs by 60%–75% compared with the conventional taxis. In UK, the total cost analysis of AVs by Wadud (2017) showed that AVs can reduce the total cost of commercial vehicles by 30% for taxis and 15%–23% for trucks; however, no clear benefits of private use of AVs are shown.

Combining the findings from the above cases, it is clear that a fleet of SAVs has the potential to enhance mobility at a radically lower cost/trip.

3.4.3 PENETRATION RATE AND FLEET SIZE

With the evolution of AV technology levels and users' perceptions of AVs (discussed thereafter in section 5), the penetration rate of AVs vs conventional vehicles and the required minimum fleet size are subject to study and have direct implications on travel behavior, cost, and system performance in terms of congestion, emissions, VKT, parking requirements, etc.

In the Greater Toronto Area (GTA), four scenarios were studied under *different levels of AV penetration* (10%, 50%, 90%) – focused on freeway vs nonfreeway congestion implications (Kloostra and Roorda, 2019). Scenario A1: AV operates on the freeways only; Scenario A2: AV operates on all routes; and Scenarios B1 and B2: same as A1 and A2 but takes into account the additional induced demand due to AVs. While the results show small (1.5%) increases in average trip length at 90% AVs' penetration (without induced demand), significant travel time savings are realized, in the order of 1%–7% at 50% market penetration and 12%–21% at the 90% market penetration. For the induced demand scenarios, all congestion benefits are offset by induced travel, aligned with the "fundamental law of road congestion" stating that VKT increases proportionally with highway capacity and slightly less rapidly for other types of roads.

In the city of Sioux Falls, US – using a simulation platform – the implications of various fleet sizes combined with various demand levels were investigated by reporting on the value of travel time (VOTT) per trip focusing on two types: working vs nonworking trips (Hörl et al., 2016). For a fixed fleet of 1000 AVs, VKT is expected to increase by 60% for the purpose of picking up passengers. During peak periods, waiting time for AVs ranges from 10 to 15 minutes compared with 5 minutes for public transport, and at off-peak, AVs provide a reasonable waiting time of 5 minutes. Sensitivity analysis of various penetration levels indicates that increasing the percentage of AVs reduces the trips completed using other modes, especially for private cars. In terms of public transit, a stable 2%–3% share is reached with the introduction of 2000 AVs, after which AVs are barely used.

In Budapest, Hungary, the implications of different AVs' penetration in urban areas were investigated starting from 0% to 100% AVs with an increment of 20% (Lu et al., 2019). The results showed that the maximum flow (capacity) increases linearly with the increase in AVs' penetration, and 100% AVs can increase flow by 16%–23%. This increase is due to the reduction in the reaction time and shorter headway. Additionally, the increase in density and capacity is slow with low penetration rate up to 40% and then significant increase in the capacity can be achieved. Therefore, high level of penetration is required to realize the benefits of AVs.

In Auckland, New Zealand, the potential loss or gains for different AVs' penetration were investigated for four different scenarios: heavily congested, lightly congested, free flow, and future case taking into account induced demand (three times heavily congested case) (Li and Wagner, 2019). The results showed that higher proportions of AVs can improve the overall performance of roads. For example, under the heavily and lightly congested traffic, 70% AVs can achieve the same travel time of free-flow condition. For the future case, the results showed that 100% AVs can increase the throughput by 88% at the expense of emission increase due to induced demand.

3.4.4 PARKING STRATEGY AND DEMAND

The above discussions and findings on travel behavior, cost and fleet size, and penetration rate would have direct implications on demand for parking and how should cities prepare/adopt for this change.

In Atlanta, US, the implications of *SAVs* were simulated using *Discrete Event Simulation (DES)* with focus on the parking inventory and transportation network by investigating three different parking scenarios: free parking, entrance-based charge, and time-based charge (Zhang and Guhathakurta, 2017). It was found that a fleet of 1000 vehicles could be enough to serve the population and would reduce parking spaces by 4.5%, with one AV removing 20 parking spaces. In terms of parking spaces, AVs parked 20.6, 16.6, and 8.6 times for the three scenarios, respectively (free parking, entrance-based charge, and time-based charge), while the total parking spaces required were 2.42, 2.4, and 1.9/SAV. As expected, VKT is found to be much higher in the case of charged parking; it is 5% higher for the entrance-based charging scenario and 14% higher for the time-based scenario.

Another study in the city of Atlanta, US, by Zhang et al. (2015) investigated the implications of the AVs on *urban parking demand* for a 10-mile × 10-mile *hypothetical city layout* in a grid network of 0.5-mile street segments. A simulation test bed considered only 2% of the population uses AVs, with a trip cost of $0.4 per mile, and maximum two passengers can be served by the same AV based on the customer's preference. It was found that urban parking demand increases with the increase in the fleet of AVs, with an average increase of threefold the number of introduced AVs (so, for example, adding 50 AVs to the system increases the number of parking demand by 150 spaces). And, across the network, 90% of the parking demand for the served population can be removed, as SAVs could replace 14 conventional vehicles. Average waiting time decreases dramatically with the increase in the number of AVs, indicating average waiting time of 7.32 minutes with 550 AVs and 0.12 minutes with 800 AVs. The urban parking distribution is found thereof to be very sensitive to the operations and city strategy.

The above implications tempted two expert workshops conducted in Bremen, Germany (2016) with the first focusing on identifying the parameters influencing the future of mobility and AVs, while the second addressed different future scenarios developed by the experts. The conclusive findings of the experts are indicated as follows:

- AVs will cause high reduction in the number of vehicles due to the high efficiency of the AVs, and SAVs will witness the highest impact resulting in optimizing the vehicle occupancy rate. This will result in a relatively low-cost affordable shared or a public transit mode/trip.
- AVs are expected to change the parking requirements and therefore the urban setting and land-use activities.
- The above combined findings suggest that AVs will likely increase in the VMT and the traffic volume due to the high empty trips.
- The experts stressed on two key questions that require further development and research:
  - How to manage the AVs' evolution to ensure sustainable mobility?
  - With the many looming advantages of AVs and safety, what are the associated regulatory actions that shall be taken before AVs represent a risk?

3.4.5 SUMMARY OF AV IMPLICATIONS

The *required fleet size* to attain certain level of benefits varies dramatically across different countries starting from as low as 1000 AVs in Atlanta, US (Zhang et al., 2015), to more than 100,000 AVs in Berlin (Bischoff and Maciejewski, 2016). This great variance is attributed to different city sizes, transportation network specifics, commuter behaviors, and the operation strategy followed in each study. In most of the studies, each SAV is expected to *replace almost ten conventional* vehicles regardless of the location or the country with a reasonable average waiting time lower than 5 minutes. In other words, it can be stated that AVs will reduce the number of vehicles dramatically with an acceptable level of satisfaction. *AVs utilization* is found to be relatively high and more than 60% in most of the reported studies (Burns et al., 2012; International Transport Forum, 2015) compared with the conventional vehicles in the current case that have an utilization lower than 5%. The above findings shall form a good starting point for simulation-based studies, assumptions, and parameters.

All studies that discussed VKT indicated an expected increase in the VKT due to empty trips. The *level of increase in the VKT* changes dramatically with the change in the number of the AVs in the fleet, ranging from 8% to 90% depending on the fleet size, with an average consensus on *60% VKT increase* in typical scenarios. From the cost perspective, AVs will reduce the *traveling cost* dramatically by more than 80% (Burns et al., 2012), which in turn will increase the attractiveness of AVs in the future. *Parking requirements* can be reduced by more than 80% in the era of AVs (International Transport Forum, 2015; Zhang et al., 2015).

The implication of AVs on travel behavior is vividly allowing people to travel more with much lower waiting time, radically lower costs (as it is used as a shared mode of transportation) and with high level of comfort, which in turn will be attractive to the customers and is expected to *disrupt the travel demand and behavior of the transportation system.* AVs are expected therefore to change the shape and fabric of the cities and especially urban centers with short trips. However, there are obvious negative impacts of AVs including VKT increase, anticipated higher congestion levels, and potentially more emissions. Additionally, the specific use of AVs as a private or shared vehicle will have a great impact on the network operations and requires clear and directive city policies to limit a negative impact on the sustainability of AVs.

It must be noted that traveler's current lack of experience to travel in an AV limits the reliability and validity of stated preference (SP) survey respondents on the actual use of AVs. Additionally, the assumption that the travel demand is unchangeable is not true as AVs will be less expensive (as a shared public mode of transportation) and thereof induced demand is inevitable; the magnitude of which is yet to be succinctly and evidently studied/estimated by the transportation planners. This gap is more important to be addressed now than ever as the anticipated travel demand implications might offset some of the perceived benefits of AVs.

While the authors agree and align on the need for addressing the above questions (refer to section 10 for discussion on a framework for incorporating AVs into the realm of smart cities, and refer to section 8 for synthesis of AVs' regulations across the globe), it is important to pose in the present chapter and commence reflecting on the users of AVs with their options, adoption potential, and perception of benefits/disbenefits.

3.5 USER OPINIONS, ADOPTION, AND PERCEPTIONS

The research and development community have embarked determinedly to assess the potential benefits of AVs and understand their various implications on travel behavior and the development of cities, yet user opinions and perceptions of AVs constitute a significant weight in realizing the potential of any of the above-noted benefits or witnessing some of the mentioned implications. This section endeavors to map users' awareness of the technology, contrast perceptions across countries, dwell into trucks drivers, understand the effect of socioeconomic and demographics characteristics on users' opinions, capture their views on the state of technology, report on perceived benefits and implications, and lastly deduce willingness of users to pay/own AVs.

3.5.1 EFFECT OF AWARENESS/PREVIOUS EXPERIENCE ON PUBLIC OPINION

Users' awareness of the technology and their prior experience of AVs seem to have a notable positive impact on perceptions and acceptance of the technology across a number of cities/countries.

In La Rochelle, France (automated city), a survey was conducted one month after piloting six automated buses (ten passengers capacity) along 1.4 km routes with six stations serving a total of 14,661 people between December 2014 and April 2015 (Piao et al., 2016). Covering 148 respondents and telephone interview with 500 respondents, the overall results imply that 87% of the survey respondents heard about AVs before and the majority were optimistic about the benefits of AVs focusing on energy and emissions reduction (50%). Only 25% of respondents believe that AVs will be safer than conventional vehicles, while 46% believe that it will be as safe as conventional vehicles (CVs) and 39% believe it will be less safe. Security and safety could be an issue for automated buses, with 44% very concerned for night services, compared with 22% for day-time services. The most appealing benefit was "increased mobility for the elderly, disabled, and others" (58% answered "very attractive"). When asked about the modes of transport; the analysis indicates as follows:

- *Buses:* Two-thirds of the respondents would like to take automated buses if both human and automated buses were available because of the cost reduction due to elimination of driver costs.
- *Taxis:* majority of the respondents were intervened with the benefits of auto taxis, with fare reduction being the most attractive benefit with 36% (5/5).
- *SAVs:* One-third of the respondents answered "do not know" due to unfamiliarity with the sharing services and their impact, but releasing a car at the desired place and time is the most attractive benefit with 35% (5/5). Of the AVs' users, 73% prefer to use AVs privately and 27% would use them through services such as car sharing.

In terms of risks associated with AVs, high levels of concern are reported regarding AVs' risks with at least 47% of the respondents were very concerned with any feature. The most notable concern was equipment/system failures with 66% very concerned, followed by legal liability in case of an accident with 56%, and risk of vehicle security with 54%. The study concluded that previous experience with AVs attracts more people toward AVs as 73% people with previous experience would use the AVs compared with 55% of the respondents with no previous experience (Piao et al., 2016).

Aside from the qualitative measures utilized in most user experience surveys, or acceptance measures in automated driving systems (ADSs), Wintersberger et al. (2016) exposed 48 participants to a driving simulator to study the user acceptance riding an AV. Using positive and negative affect schedule (PANAS), affect grid (two-dimensional grid measuring pleasure against sleepiness to arousal), questionnaire, and interviews that were conducted twice, one before and another after the trip, the study was able to analyze participants' emotions. This rather important study and the experimental setup addressed the mental condition and emotional state of the front-seat passengers of an ADS (male or female drivers), compared with human driving, offering the following key findings:

- *Driving Simulator:* Neutral face expressions represent 94%, 4.4% happiness, 0.9% sadness, 0.2% surprise, 0.1% contempt, and 0.1% anger. Participants showed happiness 1.86% (of the pictures) in AVs, while reached 3.99% for female passengers vs 7.6% for male passengers. Sadness was always associated with female drivers showing 1.8%; for AVs and male driver, this percentage was as low as 0.54% and 0.5%, respectively.

- *Affect Grid:* There is no significant difference in the pleasure or the sleeping status. As a result, the participants' self-evaluation of the subjects' current emotional state is not affected by a certain driver.
- *Questionnaire:* No significant difference reported in the safety and security feelings between before and after the experiment, with male drivers feel relatively safer [6/7] when compared with female drivers [5/7]. The possibility to take over the control was requested by 84.5%, which means that the remaining participants can imagine using an AV.

The overall conclusion indicates that participants showed wide acceptance of the AV, and the market is not far from being ready to ADS.

A wide-scale online survey in Australia and New Zealand focused on the attitude toward AVs collecting input from 5,102 Australian and 1,049 New Zealand to evaluate public awareness, perceived benefits, willingness to pay for AVs, and opinions toward AVs (Cunningham et al., 2018). The survey results showed that the majority of people heard about AVs, while <15% have experience with AVs. However, 55.1% of the respondents never heard about a car that can change lanes by itself. Similarly, survey results of 1578 respondents from US, UK, and Australia (501 from US, 527 from UK, and 505 from Australia) indicated that 66% of respondents heard about AVs before (71% in US, 66% in UK, and 61% in Australia), and 56.8% have a positive opinion about AVs vs 13.8% negative opinion (Schoettle and Sivak, 2014).

Comparing the results of a 3-year apart survey conducted in 2014 and later in 2017, Richardson and Davies (2018) analyzed the responses of 199 respondents from the UK to understand the change in public opinion over time. In 2017, the survey showed a relatively negative shift when compared with 2014 results. Even though the public became more aware of AVs (60% of the respondents heard about AVs before in 2014, while 97.5% in 2017), the percentage of respondents with positive opinion dropped by 25% (55% in 2014 vs. 30% in 2017).

3.5.2 COUNTRY/NATIONAL LEVEL INSIGHTS

As the perception of potential AV users is a function of a multitude of dimensions related to country development, economy, advancement in state of technology, research and development funding, public knowledge, education, involvement, income, and local cities piloting/testing, it is important to shed light into comparative analysis among different countries in the body of stated literature.

An international crowdsourcing study conducted three surveys in 112 countries with 8,862 respondents, evaluating different features as follows: (1) "driving behavior", (2) "opinion on ADSs", and (3) examined user acceptance of auditory interfaces in modern cars and fully automated driving (Bazilinskyy et al., 2015). The survey analysis offers the following key findings:

- *Income:* People from high-income countries were more likely to be negative about automated driving than people from low-income countries. 40% of respondents from high GDP countries showed negative opinion, almost 20% from medium GPD countries and almost 13% from low GDP countries. The main concern from high-income countries was related to software glitches.
- *Split:* The public opinion appears to be split, with a significant number of respondents being positive vs negative, with a notable portion does not appear to trust AVs and still prefers to drive manually.

Kyriakidis et al. (2015) conducted 63 questions internet-based surveys with 5000 respondents from 109 countries to investigate the user acceptance, worries, and willingness to pay partially, highly, and fully automated vehicles. The survey results showed as follows:

- *Manual Driving:* Respondents indicated that manual driving is the most enjoyable mode with almost 45% of respondents strongly agree, while fully automated driving would be

the least enjoyable mode with almost 30% strongly agree. 69% of respondents believe that fully automated driving will reach a 50% market share by 2050.
- *Willingness to Pay:* 22% of the respondents are not willing to pay more for a fully automated driving, and only an insignificant 5% are willing to pay more than $30,000 for an AV. People, who drive more, would be willing to pay more for automated vehicles. Countries that register more accidents per inhabitant or per vehicle are more likely to adopt automated driving.
- *Data and Privacy:* The more developed countries were less comfortable with their vehicle-transmitting data with low correlation between the GDP and the average comfortable level.
- *Cyber Security:* Respondents were very concerned about the risks, including software hacking and misuse (average 4/5), legal issues (average 3.8/5), safety (3.8/5), and the least concern is the privacy (3.5/5).

In Slovenia Šinko et al., (2017) created 26 questions online survey with 549 respondents to understand the public general opinion about AVs and the willingness to pay for the technology, followed by a comparison to survey results from India, China, Japan, Australia, UK, and US (Schoettle and Sivak, 2014). The key findings are indicated as follows:

- *Awareness:* 70.3% of respondents heard about AVs before in Slovenia, with senior people had heard about AVs more than younger people. 41% of females evaluated themselves one out of four (never heard about AVs before) compared with 7.3% of males. However, 77.8% of males evaluated themselves three to four out of four compared with 22% of females.
- *Perceptions:* 10.3% of the respondents are very positive toward AVs, 33.5% positive, 38% in the neutral state, 12.4% negative, and 5.8% very negative with an average value of 60% of respondents believed that AVs need more than 10 years to be used, 15% believed that AVs will never be used, 15% stated that conventional vehicles will take the lead vs. AVs because of the high price and risks, and 7.7% stated that AVs will be used within the next 10 years.
- *Country Comparison:* Although Slovenia had higher level of awareness of AVs, the general opinion (measured out of 5) in Slovenia is less positive as the average in China was 4.3, India 4.3, Japan 3.6, US 3.6, UK 3.5, Australia 3.5, and finally Slovenia 3.3, and the results showed that respondents are not willing to pay more for AVs.

3.5.3 SYNOPSIS ON TRUCK DRIVERS

So far, most research regarding truck drivers focuses on safety, distraction, or fatigue of drivers with rare focus on the attitude of drivers toward AVs (Trösterer et al., 2017).

Richardson et al. (2017) created an online and paper-based survey targeting truck drivers (69 respondents) and transport companies (17 respondents) to assess the truck drivers' and fleet managers' opinions about the new autonomous system. The survey analysis showed the following:

- *Awareness:* While 43.6% of the truck drivers do not know about AVs, 90% of the fleet managers already knew about AVs.
- *Jobs:* 25.6% of drivers are in fear of losing their jobs.
- *Benefits:* 51.3% of the truck drivers stated that they would consider AVs as it increases safety and 46.2% because of comfort.
- *Driving Pleasure:* 47% of participants stated that driving pleasure was a key reason behind their choice for this job.
- *Privacy and Liability:* Both the truck drivers and fleet managers are concerned about privacy and liability, with 28.2% of drivers were concerned about the privacy, 51.3% liability,

38.5% safety, and 46.2% about the loss of driving pleasure. On the other side, 40% of managers are concerned about privacy, 60% about liability, 60% about safety, and 40 about the loss of driving pleasure.

Tröster et al. (2017) studied the barriers of semiautonomous trucks for both the truck drivers and companies using semistructured interviews in six companies in Europe. The results showed a contradiction in the opinion of decision-makers and the truck drivers. Companies believe that semiautonomous trucks increase the safety and reduce the stress on the driver, while truck drivers were concerned about the liability in case of accidents and reliability of the autonomous system. Additionally, both the truck drivers and decision-makers were concerned about vehicular failure and required training for drivers. For example, it was found that drivers struggle with the current driving assistant system such as the adaptive cruise control as it fully brakes if a car appears in front, and many drivers stated that they turn it off.

Fröhlich et al. (2018) studied the truck drivers' acceptance of fully autonomous trucks using online and paper questioners that are based on the technology acceptance model with 50 respondents. The results showed that drivers were already exposed to different automation features such as speed control and brake assistance. Drivers were positive toward the new system with an average of 3.2/5. However, drivers were concerned about the safety and reliability of the autonomous system as 2.37/5 trust the system. On average, 3/5 of the respondents believe that the system is not sophisticated. The main result of this study was that drivers are skeptical about automated trucks as only 45% of the respondents accept the system due to the lack of trust and fear of losing their jobs. Additionally, both the driver's age and driving experience have a negative impact on the attitude of drivers.

3.5.4 PERCEIVED BENEFITS AND IMPLICATIONS

Sections 3.3 and 3.5 on benefits and implications of AVs, respectively, focused on a wide range of technical studies ranging from modeling and simulating pilot studies to envisioning hypothetical testing scenarios reporting on a multitude of benefits (safety, congestion, equity, system capacity, parking spaces, energy, and emissions) and implication of AVs' development of cities (travel behavior and travel cost and affordability); this section complements the above with a particular lens on the users' views.

Majority of survey respondents among Australian and New Zealand expressed a concern towards most AVs, with 80.6% of the respondents' concern about having their children in an AV by themselves, and 84.1% concerned about being liable for an accident mistakenly caused by an AV (Cunningham et al., 2018). Data privacy, on the other hand, was the least concern with 67% of the respondents being concerned. In terms of perceived benefits, most respondents did not agree that AVs will reduce the travel time with 34.6% disagree and 25.3% agree, while 73.7% believe that AVs would increase the mobility for people with driving restrictions, 39.1% believe that AVs will improve safety, and 24.5% disagree.

In contrast, the study by Schoettle and Sivak (2014) focusing on US, UK, and Australia indicates that majority of respondents believe that AVs will improve safety (70.4%), improve emergency response (66.9%), lower vehicle emissions (64.3%), reduce energy used (72%), and reduce insurance rates (55.5%). However, respondents did not believe AVs would reduce congestion (51.8%) or reduce the travel time (56.6%). Despite the view that the above results seem to portray a promising era of AVs, respondents are concerned about many issues: system failure (only 3.8% not concerned), liability (only 7.2% not concerned), security (only 9.1% not concerned), data privacy (only 13% not concerned), interaction with pedestrians (only 8.1% not concerned), performance in poor weather (11.1%), and system glitches (5.3%). Moreover, respondents were very concerned with many scenarios: riding with no manual control (54.3%), interaction with commercial vehicles (54.3%), self-driving public transportation (45.9%), and autonomous taxis (42.9%).

In an attempt to understand the public acceptance of 455 respondents in Sydney and Perth, it was found that almost 40% of the respondents are not sure about the positive impact of AVs (increase safety, reduce congestion, lower emissions, reduce travel time, and lower insurance costs) (Greaves et al., 2018). Almost 40% of the respondents have positive opinion about AVs and 20% have negative opinion. Among the top concerns was system hacking (38% very concerned and 30% moderately concerned), 68% were concerned about the safety, 64% were concerned about liability and privacy, and 50% were concerned about losing driving pleasure.

The 3-year apart (2014–2017) study conducted in the UK shows similar concerns and patterns as above, indicating that 18.5% of the respondents were very concerned about trips in AVs, 27% were slightly concerned, 26% in the neutral state, 17% not concerned, and 11.5% not at all concerned. 16.5% of the respondents strongly agree that AVs will increase safety if only AVs in the roads, 25% slightly agree, 21.5% in the neutral state, 25.5% slightly disagree, and 11.5% strongly disagree.

Back in 2013, 107 respondents filled an online survey targeting population with previous experience about AVs in the US (Howard and Dai, 2014); 45% of the respondents believed that increase safety is the most attractive attribute of AVs, 22% believed that the ability for multitasking is the most attractive, and 19% preferred no need for parking. It is clear that with time and advances in AVs levels, users will develop more anxiety about the technology indicating more doubts than trust, especially with the recently reported AVs crashes (Refer to Figure 3.1).

3.5.5 VIEW OF STATE OF TECHNOLOGY

The above perceived views of benefits and implications are clearly functions of AVs level of automation and control. Today, Level 3 is currently at an advanced development state, Level 2 commercialized is underway, and Level 4 looks imminent (Refer to Figure 3.10), yet users have mixed view of the benefits of the technology with tendency toward anxiety and worrisome. This subsection sheds some light on users' perception of the features of autonomy, which is the level of control and comfort with the technology in an attempt to understand whether users are capable of utilizing the technology.

In Australia and New Zealand, respondents showed high level of comfort toward AVs to stay within the lane by itself (57.4%) and control speed (53.4%); however, they were uncomfortable that AVs would follow closely the vehicle at a safe distance (44.7%) or change lanes by itself (42.7%). Majority would feel more comfortable to take the control back from AVs if requested (76.75%), while 73.6% of the respondents would like to drive manually (Cunningham et al., 2018).

In 2017, Abraham et al. evaluated public opinion in the US, and the questions related to features and technology interestingly indicated that 41% of the respondents like most of the features in their current vehicle, 28% are very happy, and 15% like some but do not use most, and only 6% were unhappy. Most of the respondents were found satisfied with the technology integration in their cars today with an average of 8/11. In terms of learning about technology, the most common learning methods are vehicles' manual (63%) and trial and error (59%). However, a small number preferred trial and error (25%), and young adults prefer learning by trial and error method or with the help of a family member.

In 2014, results from US, UK, and Australia showed that 19% of respondents were very concerned about Level 3 AV and 35.5% were moderately concerned with US reporting the highest level of concern. 29.9% of respondents were very concerned about Level 4 AV and 30.5% were moderately concerned with US reporting the highest level of concern (Schoettle and Sivak, 2014).

A year later, the same authors conducted a study focused on understanding the respondents' preference of the different levels of automation (conventional driving, partially automated or fully self-driving), indicating that 96.2% of the respondents preferred to take control of the vehicle when desired (Schoettle and Sivak, 2015). When asked about the preferred method of notification to take control, 59.4% of the respondents preferred a combination of three warnings (sound, vibration, and visual), almost 20% preferred sound and visual and almost 10% preferred sound and vibration. There is no significant difference in the age, nor the gender preference. Interestingly, at the end, people preferred the conventional driving followed by partial AVs, while a small proportion preferred the fully AVs.

3.5.6 EFFECT OF SOCIOECONOMIC AND DEMOGRAPHICS

The above findings on users' perceptions, awareness, view and knowledge of state of technology, national/country levels are presented to provide aggregate views purposely, yet important and insightful they would benefit from shedding some light into the following socioeconomic and demographic characteristics: age groups, gender, and education on the above views and perceptions.

In an online survey in the US, public opinion from 3034 respondents was gleaned to understand the customer satisfaction of their current vehicle, willingness to adopt new technologies, and how they learn about new technologies (Abraham et al., 2017). Younger and middle-aged people were more motivated toward AVs as 40% of the 25–34-year-old participants preferred the maximum level of automation compared with 12.2% for 65–74-years old. The path analysis showed that younger people are willing to pay more for an AV, trust the AV, and comfortable with the idea. On the gender analysis side, males were more comfortable with higher levels of automations, with 53.3% of the males agreed with the idea that AV will take control compared with 40.2% for females.

In another online study targeting 1,578 respondents from US, UK, and Australia, the survey results showed respondents who already heard about AVs were more optimistic about it, with females more concerned about AVs and more pessimistic about the benefits (Schoettle and Sivak, 2014). Younger people were more interested in AVs, with higher education people more aware of AVs' benefits and concerns and generally skeptical.

Similarly, in the study in La Rochelle, younger people of France are more likely to adopt AVs. For respondents aged >65, 56% would consider using automated cars, compared with 62% for people aged between 18 and 34, and 61% for people aged between 35 and 64. Considering the education as a variable, respondents with high education were more positive toward automated cars, with 71% stating that they would consider using automated cars, compared with 52% for people with low education. Comparing males with females, of the male respondents, 64% would consider using automated cars, compared with 55% females.

Another study by Schoettle and Sivak (2015) reinforces a clear gender preference that 19.3% of the males preferred complete AV compared with 12.4% of the females. Females showed higher levels of concern than males, with 40.1% of females were very concerned about fully AVs compared with 30.7% of males. 15.7% of females were concerned about partially AVs compared with 12.2% of the males.

3.5.7 AFFORDABILITY AND WILLINGNESS TO PAY/OWN

A key determining factor in embracing technology – in addition to its added value and benefits – is affordability and willingness of consumers to pay/invest in the technology. This section summarizes the available literature on the subject in an attempt to close the loop on users' perceptions of the AVs.

Among 5,102 Australian and 1,049 New Zealand, majority are not willing to pay for the partially or fully AVs with 65.9% and 56.8%, respectively. Of the respondents willing to pay more for AVs, Australians are willing to pay $14,920 (AUD), while New Zealanders are willing to pay $9286 (NZD) more on average (Cunningham et al., 2018).

One-thousand five-hundred and seventy-eight respondents from US, UK, and Australia indicate that 56.6% of the respondents are not willing to pay more for AVs, while 10% of US respondents are willing to pay at least $5800 and $5130, and $9400 in the UK and Australia (Schoettle and Sivak, 2014).

Comparing a 3-year apart survey in the UK Richardson and Davies (2018) indicated that 16.1% of the respondents were interested in owning an AV in 2017 compared with 18% in 2014, with 38.7% have no interest about AVs in 2017 compared with 36.6% in 2014. Results showed that with the increase in the number of years driving, people became discouraged about AVs.

A recent survey by Deloitte (2019) showed that although OEM continues spending millions and billions of dollars, people are not willing to pay more for AVs. 46% of respondents in Germany are not willing to pay more, 33% in US, and 28% in Japan. In the US, the results showed that 33% of respondents are not willing to pay more, and 42% are willing to pay 500 dollar (Cuneo, 2020).

An international survey by Capgemini Research Institute (2019) with 5,500 respondents from different countries (China, France, Germany, Sweden, US, and UK) showed that 56% of respondents are willing to pay 1%–20% more for an AV, 17% are willing to pay 20%–40% more, and 2% are willing to pay 40% more. For different age groups, it was found that respondents younger than 36 years are willing to pay more than older respondents. Additionally, urban customers are willing to pay more for an AV than respondents who live in small and rural areas.

3.5.8 SUMMARY OF PUBLIC OPINIONS AND PERCEPTION ABOUT AVs

The *level of awareness* indicates that high proportion of respondents are already *aware* about AVs ranging between 64% (Wintersberger et al., 2016) and 97.5% (Richardson and Davies in the UK, 2018); 56.4% of the truck drivers knew about AVs compared with 90% of the managers (Richardson et al., 2017).

The *level of interest* in AVs ranged from 10.3% to 67%. Starting from 2013 to 2016, survey results showed a range from 30% (Kyriakidis et al., 109 countries, 2015) to 67% (Piao et al., 2016) except for the study by Schoettle and Sivak in US (2015); 15.6% showed interest in the fully AVs and 40% in the partially AVs. Suddenly in 2017, the level of interest in AVs declined and ranged from 10.3% (Šinko et al., 2017) to 20% (Panagiotopoulos and Dimitrakopoulos, 2018), and in the survey by Abraham et al. in the US (2017), on an average, 8 of 11 were satisfied with their current vehicles. This decline could be attributed to the widespread news about AVs' fatal accidents and crashes, especially the first AV fatal accident that was reported in July 2016.

The *level of acceptance* of AVs' benefit ranged from 40% (Greaves et al., Australia, 2018) to 72% (Schoettle and Sivak, US, UK and Australia, 2014). *Safety* was perceived as a key benefit for accepting AV technology, in the range from 25% (Piao et al., 2016) to 70.3% (Schoettle and Sivak, US, UK and Australia, 2014) of respondents. This consensus suddenly dropped starting in 2016 as compared with 2013–2015, 45% (Howard and Dai, US, 2014) to 70.4% (Schoettle and Sivak, US, UK and Australia, 2014) believed that AVs will increase safety, and from 2016 to 2018, this percentage dropped in the range of 25% (Piao et al., 2016) to 41.5% (Richardson and Davies, UK, 2018), except for the study by Panagiotopoulos and Dimitrakopoulos (Greece, 2018) with 61%. This decline is similar to the decline in the percentage of interest in AVs.

All the surveys showed *high levels of concerns* about the different risks with notable variation in the reported values:

- *Safety* concerns widely ranged from 10% (Howard and Dai , US, 2013) to 91% (Schoettle and Sivak, US, UK and Australia, 2014) of respondents.
- *Liability* concerns ranged from 25% (Dai and Howard, US, 2013) to 84.1% (Cunningham et al., Australia and New Zealand, 2018) of respondents, and it increased almost linearly with years starting from 2013 to 2018.
- *Privacy* concerns were relatively high and ranged from 64% (Greaves et al., Australia, 2018) to 87% (Schoettle and Sivak, US, UK, and Australia, 2014).
- *Security* concerns were also high and ranged from 54% (Piao et al., 2016) to 91% (Schoettle and Sivak, US, UK, and Australia, 2014).

Gender-specific analysis indicates that males are more interested in AVs than females across several studies. In other words, against each male interested in AVs, 0.75–0.85 female is interested in AVs.

Age group-specific analysis indicates that younger people are more likely to accept and embrace AVs; different surveys considered different age intervals which make the comparison between different surveys somehow challenging, but it can be judiciously concluded that the elderly and senior age groups are more pessimistic toward AVs.

Willingness to pay showed that only a small proportion of respondents are willing to pay for the new technology with 5% willing to pay more than $30,000 in the study by Kyriakidis, Happee, and de Winter (109 countries, 2015) and 10% willing to pay more than $5800, $5130, and $9400 in US, UK, and Australia in the study by Schoettle and Sivak (US, UK, and Australia, 2014). The study by Cunningham et al. (Australia and New Zealand, 2018) in Australia and New Zealand showed that 34.1% are willing to pay $10,077 more. Additionally, recent studies showed the same pattern of the prior studies that a small proportion of respondents are willing to pay more. In US, 75% of respondents are not willing to pay more or willing to pay no more than $500 additional dollars (Deloitte, 2019).

In summary, *levels of concerns* across safety, liability, privacy, and security are almost similar with an average value higher than 55% and high difference between the minimum and the maximum value, especially for safety. Participants showed almost the same *levels of acceptance* with an average of 50% to three main benefits of AVs, specifically, energy reduction, reduced emissions, and increased safety. This interestingly shows a divide in the community in the way safety is being perceived as a benefit vs a potential threat.

It is noteworthy that although *accessibility to elderly and disabled* was among the key drivers/benefits of AVs, little-to-no research focused on this group to understand their acceptance of AVs. Additionally, although elderly was sought to be among the early adaptors of AVs to improve their accessibility, results showed that younger people are more interested in AVs which question the hypothesis that elderly would benefit more from AVs. A wide range of perceptions and statistics are reported across the reviewed studies; such differences in attitudes, opinions, and perceptions could be attributed to a combination of the following reasons: survey implementation time frame/year, country-specific, sampling procedures, and statistical significance and modelling results.

Most of the results are based on respondent's perceptions of AVs, and a notable proportion of respondents did not have previous experience with AVs; this was further depicted by the fact that many studies showed high level of neutral state or "do not know" among respondents. These results might change dramatically in the future as communities and individuals become more aware about the technologies, and therefore, their perceptions and expectations are likely to change, rapidly.

3.6 CHALLENGES ASSOCIATED WITH AVs

Despite the advancements in AVs' technology, numerous challenges still represent obstacles to AVs' progress and wide implementation. This section discusses the following relevant challenges: accuracy, liability and regulations, ethics, and AVs' accidents.

3.6.1 ACCURACY

Sivak and Schoettle (2015) discussed a part of the challenges associated with AVs and the accuracy level required. Specifically, the level of information required of public roads, driveways, and off-road trails will need to be accurately mapped with the location of the streetlights, stop signs, lane marking, and all roadway features. The continuous update of this information is a challenge as some changes are permanent, while others are related to temporary construction at work zones. The accurate identification of roadway features (marking, barriers, curbs, signs, etc.) – across diverse driving environment – is a key challenge.

3.6.2 LIABILITY AND REGULATIONS

One of the key challenges associated with AVs is the required policy changes. One set of legal challenges relates to the rules governing the performance of AVs such as safety rules that are responsible for protecting public from risks and reducing the severity of accidents. Several regulations shall be adjusted to fit AVs' technology such as the rules that require that every car must have a steering wheel, accelerator, or a braking pedal (Texas A&M Transportation Institute, 2017). In fact, the legal requirements are expected to hinder the introduction of AVs more than the technological requirements. The main problem is that most countries consider the car as a *thing* and the driver is a *human*, which means that the liability is weighted on the human driver (Bartolini et al., 2017). For example, 1968 Vienna convention states that drivers are responsible for the control of the vehicle at all times (Martínez-Díaza and Soriguera, 2018), requiring an answer for ... can AVs drive.... One of the proposed solutions is to define AV software as "driver" or changing Vienna convention to fit the new technology. This raises new concerns about liability and ethical responsibility, especially in complex situations. An answer can be programming AVs with different ethical theories and allowing AVs' owner to customize the way they want the vehicle to follow. In this approach, the liability will be similar to the liability nowadays used by the human driver (Bartolini et al., 2017).

Liability is therefore an important challenge that might hinder the introduction of AVs, especially in the initial period when both AVs and conventional vehicles share on the roads. AVs' software is supposed to mimic the human decisions, but some guidance must be programmed to be ethical. The German Ethical Guidelines (Lütge, 2017) gave priority to human life over animals or things. They also stated that AVs must be programmed in a way that reduces the damage (Martínez-Díaza and Soriguera, 2018). Therefore, cities shall discuss liability at three levels: *civil* (damage to a third party), *criminal* (responsibility for death or injury), and *administrative* (driving with the absence of authorized requirements).

In summary, as AVs are still in the initial state, it makes sense that the driver would still be liable; until a future state with a high penetration of AVs, the driver liability might be mitigated. Possibly, a combination between the two extremes could be achieved as the vehicle might alert the driver to take manual control with enough time to react (Bartolini et al., 2017). In addition, AVs will face new challenges related to cybercrime and hacking. The criminal law needs to answer many questions such as (Ilková and Ilka, 2017):

- Who should be held responsible in case an AV is used in a crime (the owner, the manufacturer, or the programmer)?
- As incidents are random events happening in different environments and conditions, will the responsible subject change as a function on the circumstances?

In general, there might be two scenarios: one based on the driver's liability and the second is based on the manufacture's liability. The driver will be liable if the car asked the driver to take manual control of the vehicle at a specific moment and the driver had enough time to take control and take an action to avoid the accident. This approach is expected in this early stage of AVs as there are two modes in AVs (manual mode and automated mode). The second approach is based on the liability of the manufacturer for defective products and software (Bartolini et al., 2017).

Governments, however, are known to follow the principle of blame avoidance due to the safety concerns, which in many cases increase the cost and hinder the technology (Feigenbaum, 2018). Additionally, it must be mentioned that the legal sector is clearly following the development of AVs instead of taking the lead (Bartolini et al., 2017).

3.6.3 ETHICS

Lin (2015) studied the ethical dimension of AVs, which is best depicted through a hypothetical scenario analysis. For example, imagine that an AV had to choose between moving right to hit a young girl or moving left to hit an old person or moving ahead and hit both the girl and the old person. At a first glance, it seems better to move right or left to reduce the loss. It might look less evil to hit the old person as the girl still has her entire life ahead. However, the two choices are unethical in the code of ethics as the Institute of Electrical and Electronics Engineers (IEEE) stated that "to treat fairly all persons and to not engage in acts of discrimination based on race, religion, gender, disability, age, national origin, sexual orientation, gender identity, or gender expression". On the other hand, it might look better to hit the young girl, as the AV might be programmed to protect the owner; in such case, AVs would choose to hit the lighter object.

Another example, if the car has to hit one of two cars it might hit the lighter one to protect the owner. However, if the car was able to identify the model of the apposite vehicle (through vehicle-to-vehicle communication), it looks better to hit the safer vehicle. This approach minimizes the crash impact or called as crash optimization or targeting. However, this approach might decline the sales of some brands known by their high safety levels (such as Volvo) to avoid being targeted for collision. As a result, reasonable ethical principles in the world of AVs might aim to reduce the number of lives loss. It is not only about ethics but also about the public expectations and adaption. The result will not satisfy everyone as expectations and ethics are always challenging each other to all automotive manufacturers.

One of the main challenges in AVs is that unlike the human driving, AVs' decision on how to crash is predefined by a programmer (Goodall, 2014). Gerdes and Thornton (2015) provided an approach that can be used to minimize the collision impact or risks using the cost function. The cost function depends on transferring the ethics into costs, then optimizing the function (different possible outputs) to reach the ideal result. In the cost function, all the objects in the environment are weighted according to their risks, and then the optimal solution is the one that minimizes the risks. However, the previous function might be good for a single vehicle, but it would be suboptimal in complex situations. A possible solution might be evaluating the damage (such as injury, death, and property damage) to all the involved parts then reduce the overall costs (Goodall, 2014). Such a solution, however, requires a lot of information about the surrounding objects and the results of the different choices. To further complicate the matter, the cost function might not be the right approach. Let us imagine a situation where a car has to hit one of two motorcycles: one wearing a helmet and the other do not. In the cost function, the vehicle would choose to hit the one with the helmet to reduce the damage as the chances for survival are higher. Such a behavior will discourage people from wearing helmet to avoid being crashed. Also, vehicles' object classification could be not smart enough to detect whether a moving object has a helmet or not. As a result, it is almost infeasible to address all the information and conditions addressed in a single cost function.

3.6.4 AVs' ACCIDENTS

To date, 9 AV accidents are reported with the first fatal accident dated July 2016 by Tesla in Florida and the most recent in August 2019 – also by Tesla – in Russia. The reported accidents vary spatially and temporarily, with US holding the majority share, and no specific trend over time ranging with highest number occurring in 2018, followed by 2016, and 2019. In terms of technology and automobile manufacturing involved, Tesla valiantly is involved in 7 out of the 9 reported accidents (BBC, 2018; BBC, 2019; Boudetten, 2016; Burke, 2016; Coldewey, 2019; Guardian Staff and Agencies, 2018; Levin, 2018; Levin and Carrie, 2018; Levin and Woolf, 2016; Liping, 2018; Murdock, 2018). In technology advancement cycle, typically front runners bear significant pain and risk on the expenses of leading the industry. This does not negate the fact that Google – technologically – is at comparable level with Tesla, nevertheless Google is involved only in one accident (McFarland, 2016).

Accidents increase the public concerns about AVs and have contributed to discouraging people from adopting AVs. Figure 3.1 summarizes AVs reported accidents worldwide with no attempt to comment on the liability aspect, rather to portray the locations where accidents occurred, provide

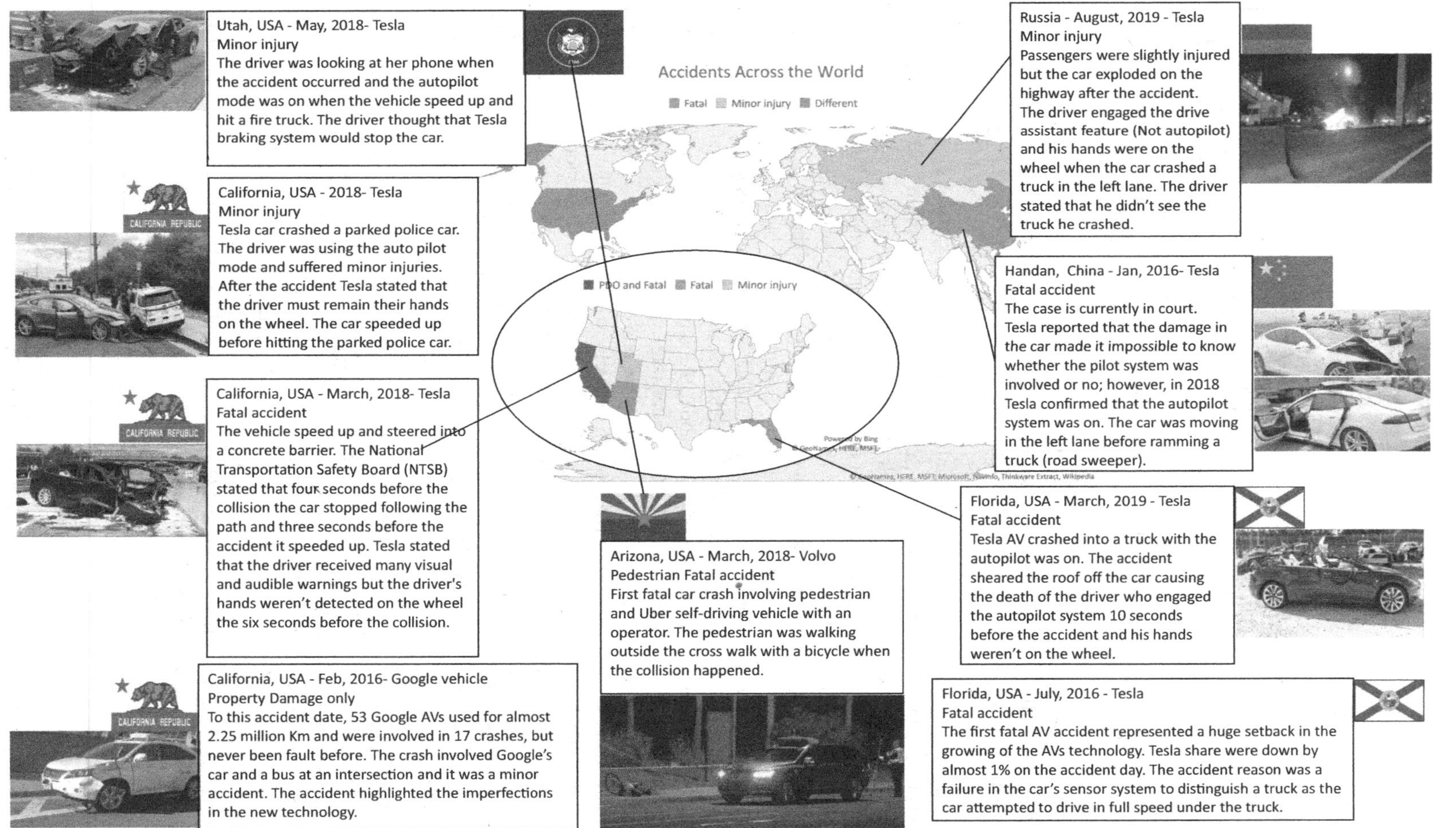

Figure 3.1 Worldwide AVs' reported accidents.

high-level information about involved parties, and the severity of accidents. Despite the exacerbated matter by the majority of reported accidents resulting in fatalities, AVs' opponents rightly argue that – statistically speaking – the millions of VKT by AVs vs the number of accidents indicate that AVs are very safe when compared with human driving.

3.7 INFRASTRUCTURE REQUIREMENTS AND IMPLICATIONS

No matter what is the magnitude of perceived benefits, various implications on travel behavior, or the readiness of Level 3 or 4 or 5, AVs will eventually run on infrastructure related to transport and information and communication technology sectors. This section discusses the infrastructure requirements and implications related to traffic management, road marking and signage, potential feed for safe harbor areas, design of parking for CAVs, fuel and power distribution, impact on bridges, internet and connectivity, and road geometry.

Kockelman et al. (2016) studied the infrastructure requirements for AVs stating that different *sensing technologies require different infrastructure needs.* For example, lane departure requires clear lane marking and signs for sensing these objects. Adaptive cruise control on the other hand and blind-spot monitoring do not require specific infrastructure treatments as they depend on sensing the vehicles around. On the contrary, traffic signal optimization and auto valet parking require extensive vehicle-to-infrastructure (V2I) communication.

3.7.1 TRAFFIC MANAGEMENT

AVs are expected to depend on *accurate mapping* for their entire journey. As cities evolve, roadway maintenance work is expected, resulting in changing the road layout and the locations where the vehicles are expected to navigate. Despite the possibility of using updated maps/Apps, geolocation cones or barriers on the site, or setting a virtual geofence that can be detected by AVs, all these environments possess a significant challenge to AVs (Catapult Transport Systems, 2017). Similarly, *emergency incidents* present a new scenario to AVs as the environment around AV might be challenging due to the stochasticity associated with random incidents (Catapult Transport Systems, 2017).

3.7.2 LANE MARKING AND SIGNAGE

AVs will be primarily guided by *lane marking* and *signs*, requiring clear and consistent marking and signage along the entire journey. Both Tesla and Volvo complained about the poor marking and stated that it is likely to hinder the introduction of AVs (Infrastructure partnerships Australia, 2017; Catapult Transport Systems, 2017). There are many reasons that might cause confusion for AVs such as old marking is invisible, discontinuous marking, white marking in snow condition and marking with poor contrast. Finally, although the road maintenance is well-stipulated, this does not guarantee that it satisfies the AVs' requirements. Similarly, poor road signs can confuse AVs accentuating the need for high level of maintenance for signs. Closely related to the need for high-quality static information are the needs for *standardization* of signs, signals, and marking across multiple jurisdictions and states.

An opportunity of AVs could be harnessing the use of their sensors to accurately report any signs or marking defects to the responsible authorities (Catapult Transport Systems, 2017; Infrastructure Partnerships Australia, 2017; Johnson, 2017).

Additionally, V2I communication can provide a solution for the signs and marking issue. For example, 3M (leadership firm in the US with products related to traffic signage printing and production and marking) provided new smart signs and marking that provide digital information to the vehicle. Advanced Road Markings is a durable and removable magnetic pavement lane marking that can be detected by AVs' sensors even in the most extreme weather case. Smart signs which are compatible with the traditional signs are retroreflective signs that enhance readability, which could enable accurate navigation and faster decision-making for both drivers and automated vehicle systems (3M Science Applied to Life, 2017).

3.7.3 POTENTIAL NEED FOR SAFE HARBOR AREAS

In the era of AVs, it is expected that drivers/passengers will be disengaged from the driving process. However, in the case of malfunctioning or deterioration in the surrounding environment, it is possible that the driver is not ready to take control of the vehicle so AVs will need a safe area to use until the driver takes control. Currently, there are hard shoulders – along highways – for vehicles to stop in case of an emergency; however, many of these shoulders were converted into lanes with emergency refuge area each 2.5 km with a rough length of 100 m. However, there are many concerns about these areas as many accidents took place within proximity of these areas, making these places unsafe for AVs. As a result, safe harbors might be provided along some roads, appropriately designed and with enough length to satisfy the number of vehicles that might need to stop considering the above concerns but also recognizing that AVs might increase the need for such areas. The locations where AVs can stop must be mapped and documented so that AVs can plan to stop if the occupant does not take control (Catapult Transport Systems, 2017).

Safe harbors will be needed in specific areas with extreme conditions or in minimum specific locations in case of AVs' malfunctioning. For example, Arizona is one of the cities that takes the lead in AVs' testing because of the favorable conditions such as good weather and flat land (Greater Phoenix Economical Council). Consequently, the need of harbors will be reduced for the case of AVs' malfunctioning. Generally, the needed harbor areas can be provided by the areas freed up due to the following:

- Removal of on-street parking that can be used as safe harbor during the learning period.
- Reduction in the distance between vehicles that can make the roads narrower, and the remaining space can be used as a safe harbor.

3.7.4 DESIGN OF PARKING FOR CAVs

Connected AVs (CAVs) will be able to search for near-parking spaces (autonomous valet parking), detecting and maneuvering into the space without the human interaction. This provides opportunities for both the users and the infrastructure provider as the user will not have to search for a space to park and this will increase the number of vehicles using the parking area. Theoretically speaking, AVs will park itself without the need for the door space which could enable 20% more free spaces. Furthermore, CAVs can seamlessly and automatically park themselves resulting in maximizing spaces as vehicles could block each other in and let each other out as necessary (Gavanas, 2019). A study by Audi suggested that a parking space can take 2.5 times the conventional vehicles using this method. Figure 3.2 shows the conventional parking layout vs the proposed method (Catapult Transport Systems, 2017).

This opportunity for maximizing the capacity of parking comes with some challenges:

- Exposing vehicles to cyber security threats and the need to safe guard if the vehicle does not respond.
- Need an electronic payment method as there is no occupant in the vehicle (Catapult Transport Systems, 2017).
- Parking garages are often privately owned and thereof do not have a consistent marking or signs system, so AVs are expected to struggle in such environment.
- Drop-off and pick-up areas are required to allow people to call or drop off their cars; these areas may be small at the beginning and expanded with the increase in the use of AVs (Catapult Transport Systems, 2017).

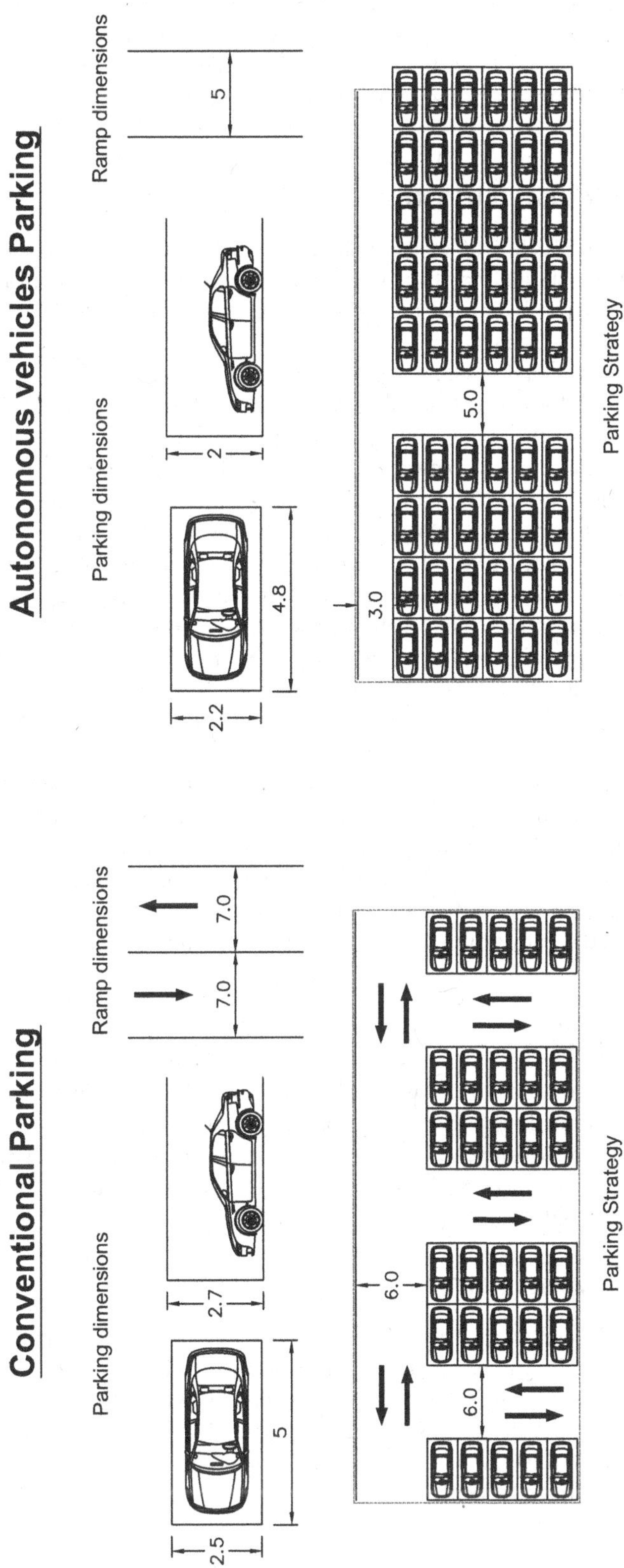

Figure 3.2 Parking Strategy: (a) Conventional parking layout: (b)Autonomous Vehicles parking layout (b). (Adopted from: Catapult transport systems, 2017.)

On-street parking in urban areas possesses a real challenge for AVs. In cases of an AV intends to park in a two-way traffic flow, this parking maneuver is expected to limit the traffic flow and AVs will struggle in such situation, especially if AVs negotiate narrower roads. In such case, it is more convenient to remove the on-street parking or convert the road to be one-way (Catapult Transport Systems, 2017).

3.7.5 FUEL AND POWER DISTRIBUTION

AVs will need to be able to refuel. One of the strategies for AVs' refueling is the use of parking areas as a fuel station. One assumption is that AVs will be electric vehicles where fuel station will be located in the parking location which requires providing the infrastructure with the requirements to be used as a charging station (Johnson, 2017).

Advances in communication and systems controls now pave the way toward intelligent vehicle grid where AVs can receive information from the environment and other vehicles around and feeding it to infrastructure to assist the traffic management (Reissmann, 2019). In the era of Electric AVs (EAVs), there will be coordination between customers and AVs through a cloud system. The customer will send a request to the cloud, and in the same time, EAVs report their availability, then the cloud sends the customer and EAVs the results of the process (Tan and Garcia, 2018). Consequently, AVs can participate in Vehicle to Grid (V2G), due to their autonomous control and advanced vehicular communication technologies. It is possible to arrange an appropriate number of AVs with parking intention in the right location to support V2G services. Therefore, AVs are considered advantageous over ordinary EVs (Lam et al., 2018). For example, Lam et al. (2018) developed an algorithm to coordinate parking in order to serve the V2G using a Coordinated Parking Problem (CPP). Results showed that their algorithm can produce the optimal decisions and provide the required level of coverage.

AVs are still human-dependent for charging processes whether for refuel or recharging. Additionally, the shift to AVs means high vehicle utilization, and the charging requirements might represent a constraint in this case and reduce the level of utilization that can be achieved. The optimum solution for this situation is to allow vehicles to charge while driving. EVs can provide a solution for AVs using inductive pads or wireless charging that provides hands-free EAVs. Wireless charging pads do not depend on any physical connection between the vehicle and the infrastructure (Yamauchi).

3.7.6 IMPACT ON BRIDGES

AVs will enhance the ability of vehicles to move in platoons which is desirable as it saves energy and boasts capacity. Traffic-loading principles in long-span bridges assume a dilution of heavy vehicles by light vans and cars. However, platooning invalidates this assumption by the heavy load it provides over the bridges. It is necessary thereof to study whether the current structural design models shall be revised (Catapult Transport Systems, 2017).

3.7.7 INTERNET AND CONNECTIVITY

AVs mainly depend on GPS and internet for accurate navigation and updated incidences. Google self-driving vehicle, for example, gathers 1 GB of data per second which reflect the significant technological enhancement for short-range communication such as vehicle-to-vehicle communication and long-range communication such as maps and emergency messages (Infrastructure Partnerships Australia, 2017). Continuous, reliable, pervasive internet, and connectivity might not be available in all environments and conditions.

3.7.8 ROAD GEOMETRY

AVs might change the geometric design of roads as roads might become narrower with tighter turning radii. However, it must be noted that such changes will not be convenient for conventional vehicles and will increase the risk of accidents (Johnson, 2017).

In 2018, KPMG studied different countries' *infrastructure* and the *degree to which the infrastructure is ready* to AVs based on six variables: density of electric vehicle charging stations (the number of available stations at specific distance), GSMA Global Connectivity Index for infrastructure (measures the performance of the key enablers of mobile internet adoption), 4G coverage, quality of roads, LPI infrastructure score (ranks countries on six dimensions of trade including customs performance, infrastructure quality, and timeliness of shipments). KPMG's Change Readiness Index to technology and infrastructure indicate that the top three ready countries are Netherlands, Singapore, and Japan – as shown in Figure 3.3. Interestingly, the top countries in this list are not necessarily among the top in AVs advancements in R&D, production, piloting, and testing.

3.8 SYNOPSIS ON PILOTS AND LAWS AND REGULATIONS

In response to the above implications, public perception, and challenges, governments are faced by the need for developing policies (e.g., for electric vehicles and the energy), regulations (e.g., for ride hailing companies, and AVs), laws (e.g., data sharing), and championing pilots/testbeds (e.g., AV

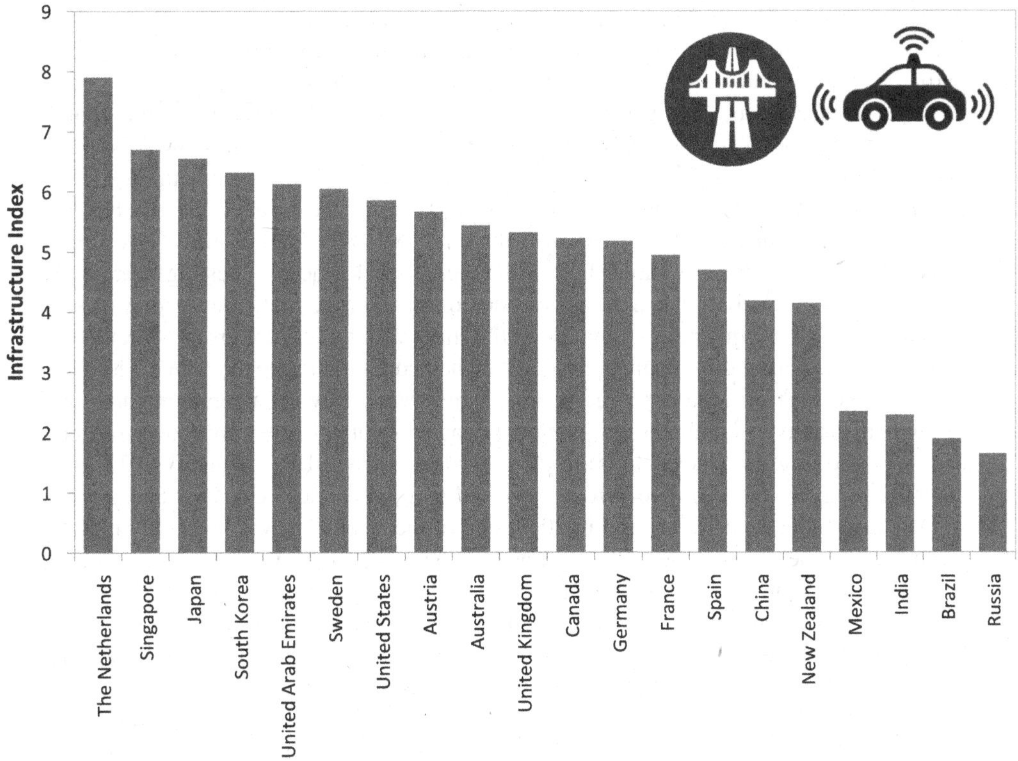

Figure 3.3 Infrastructure score (rank) by country. (Adopted from: Autonomous Vehicle Readiness Index, KPMG International, 2018.)

shuttle pilots in Europe) targeting to beat the forefront in terms of government innovation in the realm of smart cities and beyond. Many pilot studies were conducted in the last few years to understand the real implications of AVs, assess the technology, and partner with service providers. Hereafter, this section provides a synopsis on key pilots, with Figure 3.4 depicting selected relevant pilot studies across the world.

- *Arlington, Texas:* The *main objective* of the pilot is for the city council members to better understand the new technology considering the growing hype of AVs; in particular, increasing public awareness and understanding the level of public acceptance of the new technology. The final goal addresses the city nationally as an innovative testing ground. The pilot study started in 26 August 2017 with more than 1500 trips with the vision to improve the mobility and start a partnership. The city used two low-speed shuttles (with 12 passengers on board) to carry passengers between parking lots and sporting and concrete venues and added pick-up and dropping areas. The passengers were asked to complete a survey by the end of their trip, and they were positively overwhelmed by the new technology. As a result of this successful pilot, the city is planning for a second on-street pilot study (National League of Cities, 2018; Formby, 2017).
- *Boston:* Boston supported the introduction of AVs since 2015, and in 2016, the city was recognized by the World Economic Forum as a focus city for the future of mobility, partnering with two start-ups: nuTonomy and Optimus Ride. The testing embarked into a flexible piloting plan; for example, Optimus Ride was allowed to operate their vehicles in fog and rain for the purpose of learning, and in January 2017, nuTonomy started its first pilot study on the roads of the city park then to the Seaport District and Fort Point neighborhoods by April. The public was not involved in in the pilot until late 2017 when Lyft partnered with nuTonomy to offer ride-sharing services through the Seaport district. As a result of this successful pilot, the city is planning for a second pilot study and nuTonomy is mapping the whole city preparing for the next phase of pilot that will allow AVs to run all over the city (National League of Cities, 2018; Enwemeka, 2018; Marshall, 2018).
- *Arizona, US:* The city represents a unique case in terms of the duration the city has had the AVs and the wide spread of AVs over the city. AVs are running in the city streets since 2016. In 2015, the city showed the willingness to welcome AVs' testing, thereof the arrival of Waymo in 2015. Chandler is considered as one of the first cities testing AVs on public roads, and Waymo put its vehicles on the street with typical drivers as operators for several months during the mapping phase, which helped familiarizing the public with AVs. As AVs have existed in the city streets for over two years, they are a commonplace sight throughout the city's neighborhoods (National League of Cities, 2018). Moreover, General Motors tested its AVs (Chevy Bolts) in the city roads since 2016. In January 2018, Intel started testing its first AVs' fleet which consisted of 100 vehicles. Uber has been testing its AVs in the area since 2016, and in 2018, one of their vehicles was involved in a fatal accident, and the governor temporarily revoked the company's ability to operate in the state and tests are not yet resumed (Apur, 2018; Gitln, 2018).
- *Gothenburg, Sweden:* From May to December 2017, the city embarked into AVs' pilot study for garbage trucks. Volvo group partnered with the Swedish waste and recycling specialist Renova and tested the truck for 6 months that showed that trucks provide high level of efficiency, reduce the congestion, and enhance safety in the working conditions. The pilot results showed 10% reduction in the fuel consumption and that three traditional trucks can be replaced by one autonomous truck, and 50%–70% of the truck drivers are expected to be replaced by 2030 (Apur, 2018; Abbas, 2017).
- *Singapore:* AV pilot studies in Singapore started in 2015 and it is known as one of the world's most AVs friendly countries. The city allowed for tests in a wide range of autonomous vehicles such as autonomous taxis, buses, and shuttles for private and public use.

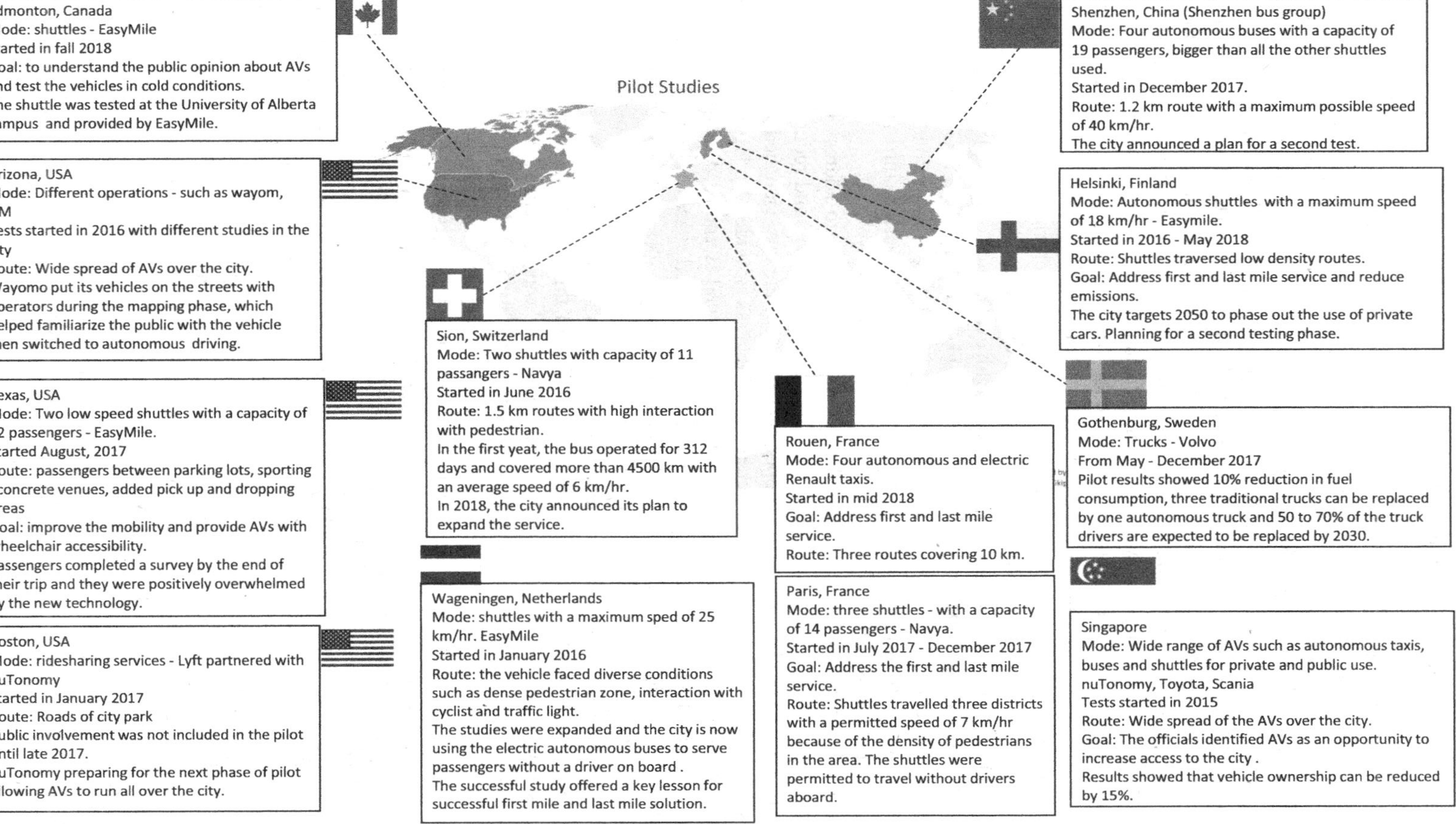

Figure 3.4 Worldwide AVs' pilots.

This environment made the city a fertile environment for AVs' companies such that more than ten companies [e.g., nuTonomy, SMART (Singapore-MIT Alliance for Research and Technology), ST Kinetics & Singapore's Land Transport Authority, Toyota and Scania, Katoen Natie] are testing their vehicles in the city. Singapore therefore works tirelessly to address the technological challenges of AVs, ranging from connected infrastructure to safety regulations. The testing results showed that vehicle ownership can be reduced by 15%. City officials adopted AVs as a mean to increase the access to the city while preparing the city to be in the forefront of international innovation (Apur, 2018; Abdullah, 2017).

- *Wageningen, Netherlands:* A pilot study started in January 2016 so the city is the first one to test autonomous shuttle trips without drivers on a public road. The city used multiple Easymile EZ10-dubbed WePods vehicles with a maximum speed of 25 km/h. The study was expanded and the city is now using the electric autonomous buses to serve passengers without a driver on-board in fixed route between the Wageningen University and Research Center and Wageningen railway station. The successful study offered a key example to other cities with a successful first mile and last mile solution in areas not well-served by public transportation. Moreover, the pilots faced diverse conditions during the testing period such as dense pedestrian zone, interaction with cyclist, and traffic light (Apur, 2018; Boekraad). Netherlands was considered the first ranked country in KPMG Autonomous Vehicle Readiness Index in 2018 and 2019. Four main pillars were considered in the ranking process: policy and legislation, technology and innovation, infrastructure, and public acceptance (KPMG, 2019).
- *Helsinki, Finland:* A pilot study started in 2016 with a vision to address the first and last mile and enhance user experience with public transportation and reduce emissions. The city was one of the early AVs' supporters. In 2016, the city launched a 2-year and 1.2 million SOHJOA AVs based on autonomous shuttles provided by EasyMile with a maximum speed of 18 km/h in suburb north of Helsinki and Hernessaari district, and the shuttles traversed many low-density routes and the study ended successfully in May 2018. The study is considering more complex areas aiming to be a leader in the smart technology and to be carbon neutral by 2035. Moreover, the city targets 2050 to phase out the use of private cars, with the second phase planned to use shuttles supplied by Navya (Apur, 2018).
- *Paris, France:* A pilot study started in July 2017 with three shuttles connecting three districts with a permitted speed of 7 km/h. The study was launched by Ile-de-France Mobilités in partnership with companies Defacto (local partner), Keolis (transport management), and Navya (shuttle technology), with the concept of using AVs as first and list mile solution. The shuttles were permitted to travel without drivers aboard which was the first fully autonomous test in France. The shuttle had a capacity of 14 passengers (11 seats and 4 passengers standing). The study successfully ended in December 2017 (Apur, 2018; Navya, 2017).
- *Rouen, France:* A pilot study started by mid-2018 with 17 drop-off and pick-up zones and a speed limit of 50 km/h in three routes covering 10 km. The study was launched by Rouen Normandy Autonomous Lab Initiative offering four autonomous and electric Renault taxis. The experiment is located in the Technopôle du Madrillet business park and connecting 17 locations within the business park. The main objective is to address the first and last mile problem, and the service is expected to be connected to the public transportation system to provide supplemental on-demand service for the region (Apur, 2018; Rouen, 2018).
- *Shenzhen, China:* A pilot study started in December 2017 using four autonomous buses for 1.2 km route with a maximum speed of 40 km/h. The buses have a capacity of 19 passengers which is higher than all the other shuttles used in any other city. Shenzhen Bus Group (a public transit operator) was the operator of buses. The city announced a plan for

a second test in the proximity of the Southern University of Science and Technology which is 3 km long with ten pre-planned stops (Apur, 2018; Hua, 2017).

- *Sion, Switzerland:* A pilot study started in June 2016 by two shuttles that can serve 11 passengers with 1.5-km routes in the center of Sion with high interaction with pedestrian in the old town area. In the first year only, the bus operated for 312 days and covered more than 4500 km with an average speed of 6 km/h. However, the study temporary stopped in September 2016 as a result of a crash with a truck but with no injury. The algorithms used were reviewed and improved, and in 2018, the city announced its plan to expand the service to Sion's main train station adding additional complexities and obstacles. The town has notably introduced connected traffic lights to the route which communicate with the vehicles as they approach the intersection. The study was launched by Swiss city of Sion and Navya which was responsible for supplying the buses used (Apur, 2018; Welcome to the Future – Smart Shuttle, Sion, Switzerland, 2017).
- *Edmonton, Canada:* In Fall 2018, the city launched the first EAV pilot study in Western Canada with the goal of understanding the public opinion about AVs and tested the vehicles in cold conditions. The shuttle was tested at the University of Alberta, provided by EasyMile. The project is led by the University of Alberta Centre for Smart Transportation and the University of British Columbia (Antoneshyn, 2018; Autonomous Vehicle Pilot Project, 2018).

It is vivid that most pilot studies used small-size shuttle buses, which mean that the focus of the pilot study is to use AVs as a public transportation mode targeting addressing the first and last mile in an attempt to attract ridership to public transportation and reduce vehicle ownership. Moreover, the pilot studies in Singapore started in 2015 and it is known as one of the world's most AVs' friendly countries with results showed reduction in vehicle ownership by 15% by 2018, depicting a great opportunity in a short time. The study in Sweden is the only study focused on testing driverless truck with results showing 10% reduction in fuel consumption, and three traditional trucks can be replaced by one autonomous truck. In other words, the study showed high level of efficiency of AVs' trucking industry.

The above piloting landscape has accentuated the need to either introduce new regulations for testing AVs or revise existing regulations to enable this environment. In an attempt to map out the status of AVs' regulation worldwide, Figure 3.5 is developed to summarize the status of AVs' regulations all over the world, indicating that majority of countries in the global north either permitted AVs' piloting and testing or have specific regulations developed as a function of state/province/city local regulations.

3.9 NEW VALUE NETWORK AND NEW BUSINESS MODELS

Data-driven business models are entering the market because of the abundance of data gathered in the light of big data era, enabled by the multitude of sensors equipped in AVs. The new role of data as a resource triggered massive investment toward data analytics and AI tools, which totally changed the shape of markets (Seiberth and Gründinger, 2018). The automotive industry is among the highest level of data generation creating new business opportunities. Figure 3.6 shows the market leaders and their investment in the automotive industry, portraying three technological giants in China with automotive ambition: Baidu, Didi Chuxing, and Xiaomi (Seiberth and Gründinger, 2018).

Companies will be forced toward these new models to generate revenue from these new business opportunities, creating new *partnerships and consortiums* [e.g., partnership between BMW and Intel, HERE (map provider) and BMW, Audi and Daimler]. The shift in new revenue streams creates a new competitive field that will force the car manufacturers share the profit with the technology

Figure 3.5 Mapping worldwide AVs' regulation status.

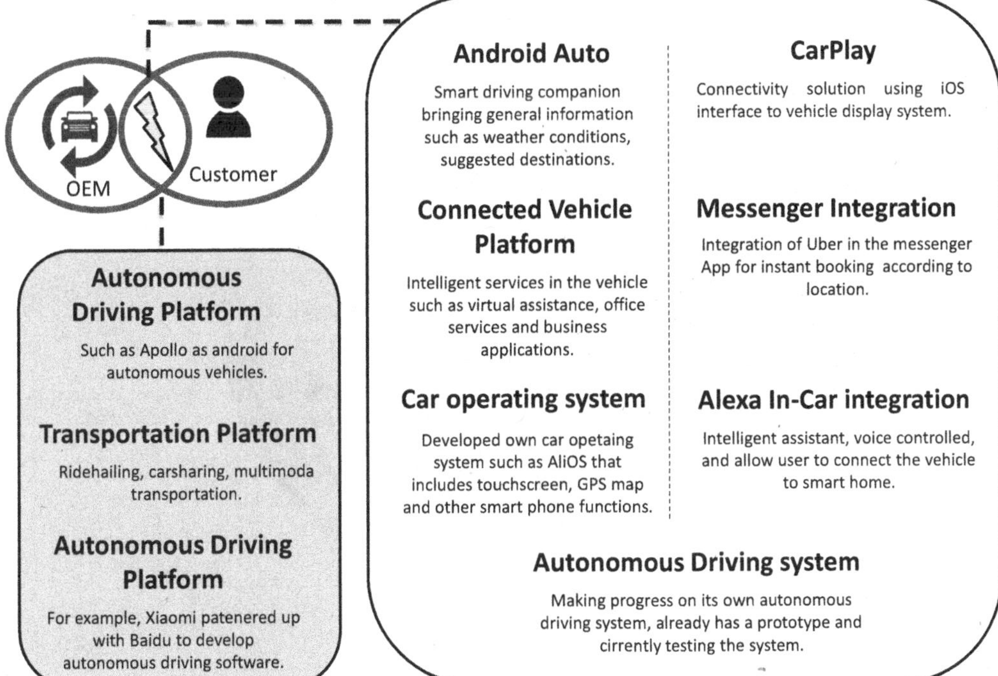

Figure 3.6 Market leaders and their investment in the automation industry. (Adopted from: Seiberth and Gründinger, 2018.)

companies, telecoms, and new entrants. Figure 3.7 shows a new ***value network*** in the era of connected AVs (Seiberth and Gründinger, 2018) harnessing the potential of the new entrants and the IOT platforms who share a sizable amount of the pie.

Recently, three business models emerged for the mobility services (Berrada et al., 2017; Adam and Susan, 2017): (1) business to customer (B2C) models, where the service owner owns a fleet of vehicles and allows people to use it such as car sharing and public sharing; (2) peer-to-peer (P2P) models that allow companies to supervise the transactions among individuals via a virtual platform such as ridesharing; and (3) hire service model, which means a customer hires a driver for a trip such as taxis.

Berrada et al. (2017) discussed three basic business models and their application on AVs, categorizing the models into three main verticals as shown in Figure 3.8.

AVs will provide a new, complex, and diversified mobility that will force the existing players (manufacturers) to compete in multiple fronts and, in many cases, will force them to cooperate with competitors (McKinsey, 2016).

Traditional manufacturers are under pressure to reduce costs and become more efficient, with software competencies becoming the most important factor in the industry. Software is expected to be complex as of the aerospace flight system, resulting in new opportunities for new players who are expected to focus on specific, economic, and attractive market and then expand. Although Google, Tesla, Apple, Baidu, and Uber recently generate significant interest, they only represent the tip of the iceberg (McKinsey, 2016). With such vibrant economic situation, companies need to prepare for disruptive uncertainty and opportunistic partnerships.

The development of AVs besides the service-based economy is expected to create a new mobility era, with a lower vehicle value and the main core will be convenience and customer experience which will drive new business models. Recent examples of this new transition already exist

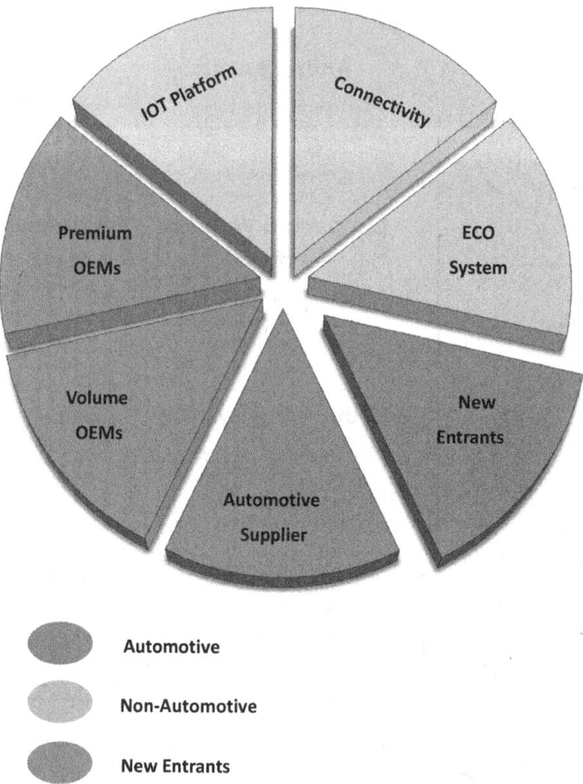

Figure 3.7 New business network in the era of connected AVs. (Adopted from: Seiberth and Gründinger, 2018.)

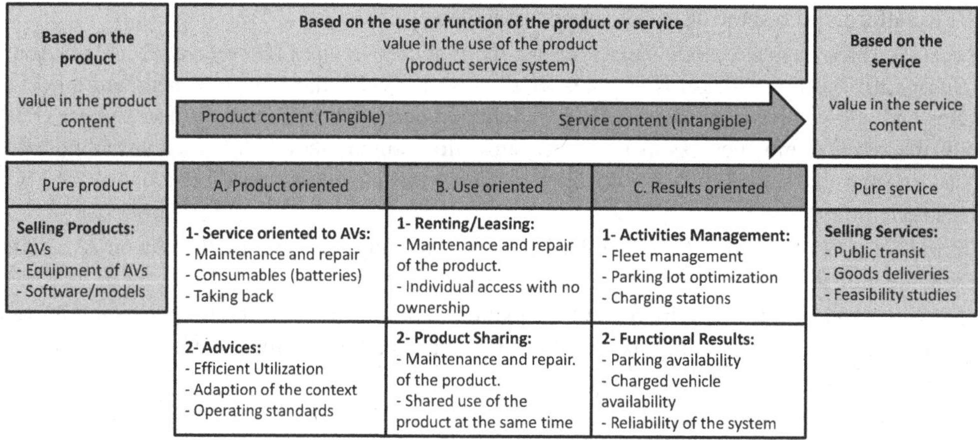

Figure 3.8 Business models applied to the AVs. (Adopted from: Berrada et al., 2017.)

such as Uber. Figure 3.9 shows different evolution cases for AVs and the transformation in the future (Red Chalk Group).

Although OEMs – the technical expertise from the development of safe and efficient vehicles – sales used to be the main source of revenue, the next mobility generation drives the customers away from private ownership. As a result, OEM must adapt and show high level of agility to the emerging

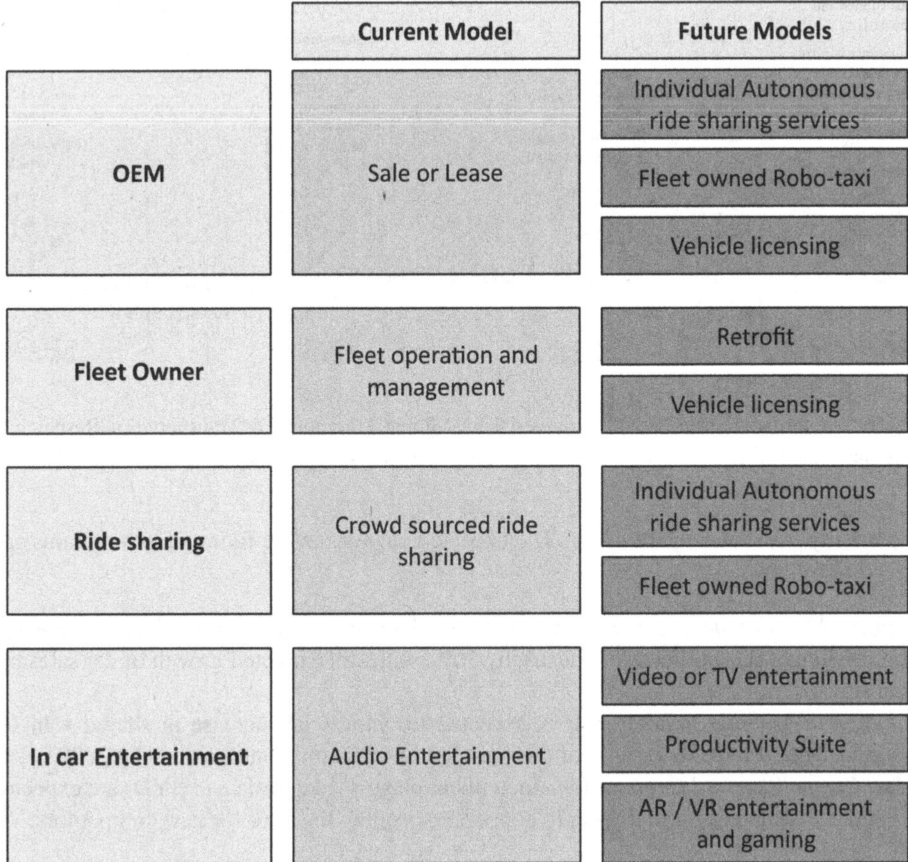

Figure 3.9 Business model transformation in the AVs' era. (Adopted from: RedChalk Group.)

mobility system; for example, many OEMs partnered with shared mobility solutions (e.g., Volvo, Uber, and General Motors with Lyft). Additionally, OEMs could introduce efficient AVs to build robot taxi services and potentially partner with fleet management operators (Red Chalk Group).

AVs will switch drivers to be passive passengers opening the door for memorable trip experience and thereof myriad entertainment features can be offered in AVs' operating system interface. Audio entertainment integration on vehicles currently exists in some vehicles such as Tesla with Spotify, and this integration might be extended in the future to video entertainment such as TV and Netflix.

It is expected that at least four OEMs will skip introducing level 3 AVs, while level 5 AVs are not expected to be introduced before 2025. Figure 3.10 shows the road map to 2030 with obvious partnerships at level 2 AVs, and this partnership is expected to increase with the increase in the level of automation.

In 2018, PWC studied the impact of AVs on the trucking business and the business models in the European Union (EU), indicating that by 2030 trucking logistics costs will decline by 47% due to the high reduction of labor, delivery lead times are expected to incline by 40%, and truck utilization will be as high as 78% compared with 29% in the current state. The above predictions possess a combined challenge to typical trucking companies and calls for expanding their product portfolios by including new powertrains and focusing production on autonomous long-haul trucks to stay competitive (PWC, 2018).

Szmelter and Woźniak (2016) studied the new mobility behavior and their impact on the new business models. They stated that three technologies will transform the automation world in future:

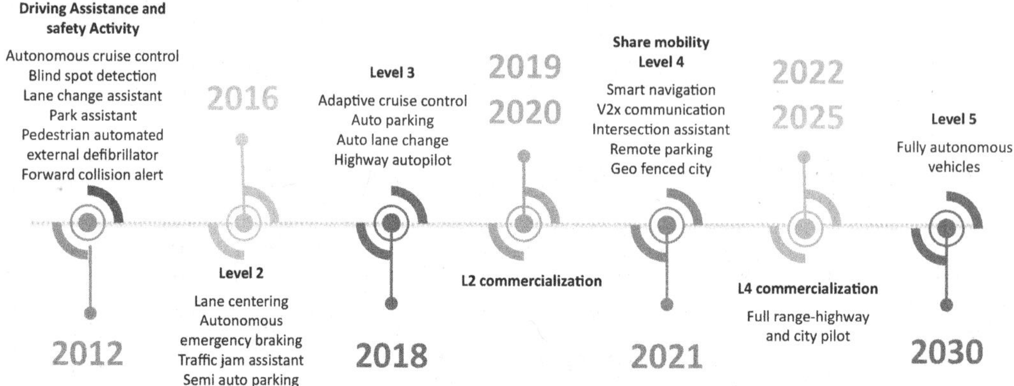

Figure 3.10 AVs' roadmap until 2030. (Adopted from: Global Automotive & Transportation Research Team at Frost & Sullivan, 2018.)

electric mobility, AVs, and solar energy. The expected changes in the future of automations can be categorized as follows:

1. *Shifting Markets and Revenue Pools*: It is expected that new business models will increase the revenue of the automotive industry by 30%, with still expected growth of car sales but at a slow pace at 2%.
2. *Change in Mobility Behavior:* It is expected to witness an increase in shared vehicles estimated at 30% by 2050; 15% of new cars sold will be fully autonomous by 2030.
3. *New Competition and Cooperation:* Incumbent players (like Tesla and BYD) are expected to fiercely compete and find ways to cooperate, opening the door for new competitors.

3.10 FRAMEWORK FOR INCORPORATING AVs INTO THE REALM OF SMART CITIES

This chapter discussed AVs' benefits, implications, challenges, infrastructure requirements, pilot case studies, and finally emerging business models as a result of the above. Although the authors attempted to connect the individual sections throughout the chapter, this is the place to stress on the intertwined relations between the above section and summarize the chapter storyline – as depicted in Figure 3.11 – as follows:

- Reported *benefits* will result in certain *implications* on driving behavior and travel patterns and can only be achieved if certain *infrastructure requirements* are available.
- The general *public* level of *awareness* of the perceptive benefits and the level of *understanding* of AVs' implications require mapping *public opinion* within various groups, ages, backgrounds, etc.
- The current technology state of development and *public opinions* indicate a series of *challenges* that shall be addressed by *piloting/experimenting* and introducing certain laws and regulations for AVs.
- Some of the reported *challenges* are related to *infrastructure requirements*, which in turn can be tested and evaluated using *pilot/case studies*, and equally importantly receive public feedback after trying out piloted AVs on the ground.
- The reported promises and high optimism of AVs *enticed cities and government* to embark on piloting/experimenting with AVs, which triggered the need for laws and regulation especially in response to the challenge of AVs' fatal accidents.

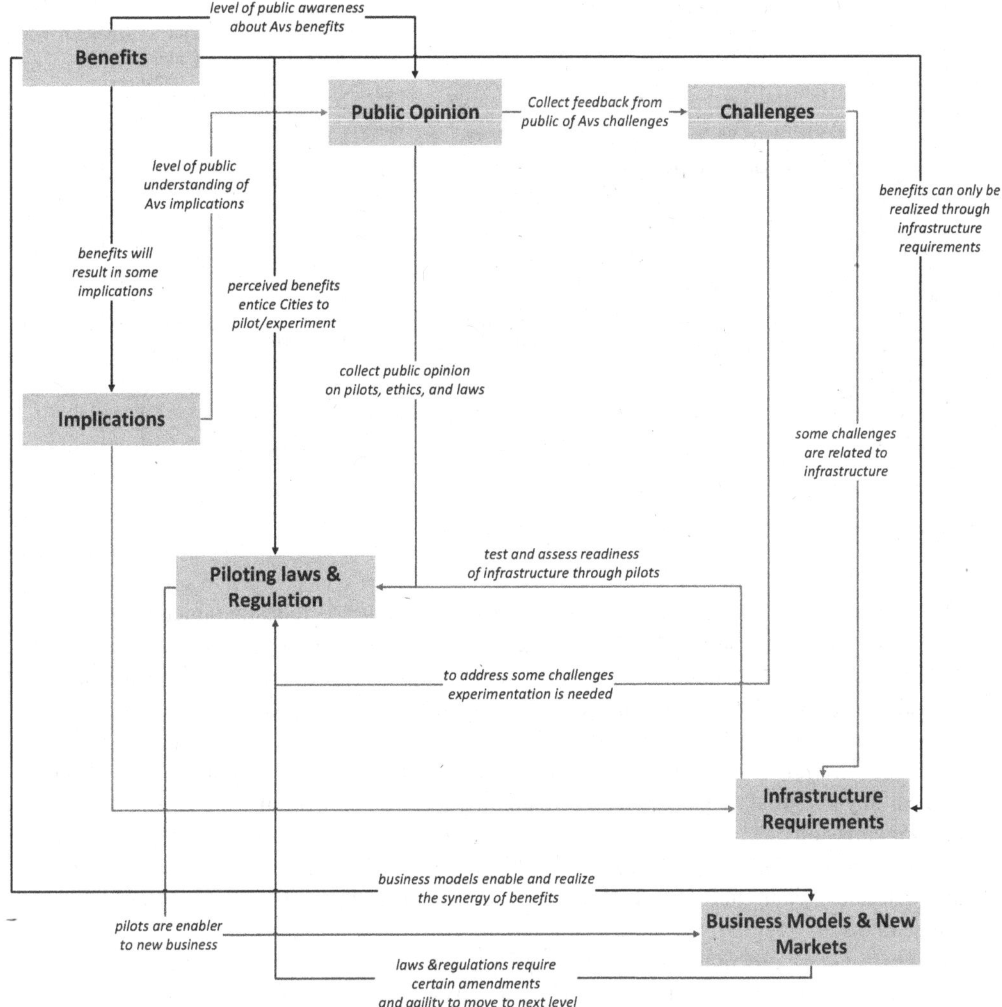

Figure 3.11 Mapping the intertwined relation between AVs' benefits, implications, public opinion, challenges, infrastructure requirements, piloting and regulations, and new business models.

- The new technology and the high interest of forefront cities have enabled a green field for *new business models*, in which pilots are enablers to *new markets*. These new markers and business models have accentuated the need for *revising laws and regulations* and helped harnessing and realizing the potential of the perceived benefits.

Given the above, cities have embarked determinedly into being smart and considering AVs as a key enabler to smart mobility. The following portrays a roadmap for incorporating AVs as part of Smart city initiatives – as depicted in Figure 3.12:

- *PT Integration*: In urban areas, smart cities should focus on integrating AVs into existing public transit services and issuing programs for SAVs.
- *InterCity Full Capacity Gains:* Full capacity gains shall target intercity travel and rural highways.
- *Inclusion Is not an Option:* Cities should ensure involving marginalized low-accessibility population in the discussion of AVs as well as potential job loss vs creation.

Figure 3.12 Framework for incorporating AVs in smart cities.

- *Opportunity for a Shared City*: Cities have an opportunity to redesign/amend the urban setting design to "manage the curb" in the era of shared autonomous vehicles.
- *Rebate and Incentives*: Cities should adopt innovative rebates and incentives programs as a catalyst for early AVs' adoption – with electric vehicle programs, to manage retrofitting existing fleets as well as ownership levels.
- *Electrification and Autonomy*: Cities should embark on electrical autonomous vehicles to address any potential negative environmental impact due to potential increase in VKT and thereof emissions.
- *Connectivity and Autonomy:* Cities should harness the synergetic potential of connectivity and autonomy combined with highest gains expected in the system capacity and safety applications.
- *Economic Gains are Inevitable:* Cities are expected to witness an economic gain in a multitude of ways; the specificity of which is city-dependent considering its current and envision state as well as the level of funding for piloting and agility to new business models and experimentation.
- *Shared and Affordable Mobility*: Albeit the technology hype, cities shall still address the basic need of affordable competitive transport modes, and SAVs are key enablers, additionally address PT first and the last mile challenge.
- *Resource Reallocation and Optimization:* Cities can embark into AVs as a mean for better reallocation of jobs, efficient utilization of resources, and nurturing new talents and skills.
- *Phased-Approach Is a Must*: Cities shall clearly follow a phased-piloting/introduction approach of the new technology with targeted awareness campaigns.
- *No Willingness to Pay More*: Cities should not expect travelers to pay extra for adopting AVs, which can accentuate the need for partnership and integration with PT and shared mobility.
- *Major Cybersecurity and Liability Concerns*: Cities would have to work hard and smart to study and address the very legitimate concerns of cybersecurity, hacking, system failures, and liability. Additionally, cities should adopt an incremental phase – liability approach – while considering the evolution of software development, cyber security, and ethics.
- *Public is at a Divide:* Cities should seize the opportunity that currently the public is at almost 50/50 divide in terms of perceptions of benefits/concerns of AVs. Properly tackling these concerns might enable smart cities to shift gears.

REFERENCES

3M Science Applied to Life. 2017. Helping improve the safety of your road systems today and in the future. https://www.3m.com/.

Abbas, M. 2017. AMCS vehicle technology complements autonomous waste collection truck. AMCS Group. https://www.amcsgroup.nl/nieuws/blog/amcs-vehicle-technology-complements-autonomous-waste-collection-truck/.

Abdullah, Z. 2017. NTU to collaborate with Dutch body on self-driving vehicle research. *The Straits Times*. https://www.straitstimes.com/singapore/transport/ntu-to-collaborate-with-dutch-body-on-self-driving-vehicle-research.

Abraham, H., et al. 2017. Autonomous vehicles and alternatives to driving: Trust, preferences, and effects of age. *Transportation Research Board 96th Annual Meeting*, Washington, D.C.

Adam, S. and Susan, S. 2017. Shared automated vehicles review of business models. The International Transport Forum.

Antoneshyn, A. 2018. Meet ELA, Edmonton's driverless car. *CTV New Edmonton*. https://edmonton.ctvnews.ca/meet-ela-edmonton-s-driverless-car-1.4127402.

Apur. 2018. Impacts and potential benefits of autonomous vehicles.

Autonomous Vehicle Pilot Project. Edmonton. https://www.edmonton.ca/city_government/initiatives_innovation/automated-vehicles.aspx.

Azevedo, L., et al. 2016. Micro simulation of demand and supply of autonomous mobility on demand. *Transportation Research Record Journal of the Transportation Research Board*, 2564, 21–30.

Barth, M., et al. 2014. Vehicle automation and its potential impacts on energy and emissions. In: *Road Vehicle Automation*, Meyer, G. and Beiker, S. (eds). Springer International Publishing: Switzerland, pp. 103–112.

Bartolini, C., et al. 2017. Critical features of autonomous road transport from the perspective of technological regulation and law. *Transportation Research Procedia*, 27, 791–798. doi: 10.1016/j.trpro.2017.12.002.

Bazilinskyy, P., et al. 2015. An international crowdsourcing study into people's statements on fully automated driving. *Procedia Manufacturing*, 3, 2534–2542.

BBC. 2018. Tesla hit parked police car 'while using autopilot'. https://www.bbc.com/news/technology-44300952.

BBC. 2019. Tesla Model 3: Autopilot engaged during fatal crash. https://www.bbc.com/news/technology-48308852.

Berrada, J., et al. 2017. Which business models for autonomous vehicles? *ITS Europe Conference*, Strasbourg.

Bischoff, J. and Maciejewski, M. 2016. Simulation of city-wide replacement of private cars with autonomous taxis in Berlin. *Procedia Computer Science*, 83, 237–244.

Boekraad, M. Provincie Gelderland stopt proef Wepod. https://zelfrijdendeauto.com/provincie-gelderland-stopt-proef-wepod/.

Boudetten, N. 2016. Autopilot cited in death of Chinese Tesla Driver. *The New York Times*. https://www.nytimes.com/2016/09/15/business/fatal-tesla-crash-in-china-involved-autopilot-government-tv-says.html.

Brown, A., et al. 2014. An analysis of possible energy impacts of automated vehicles. In: *Road Vehicle Automation*, Meyer, G. and Beiker, S. (eds). Springer International Publishing: Switzerland, pp. 137–153.

Burke, K. 2016. Lawsuit adds to scrutiny of Tesla's Autopilot. *Automotive News*. https://www.autonews.com/article/20160919/OEM/309199962/lawsuit-adds-to-scrutiny-of-tesla-s-autopilot.

Burns, L., et al. 2012. Transforming personal mobility. *The Earth Institute - Columbia University*.

Capgemini Research Institute. 2019. The autonomous car: A consumer perspective.

Catapult Transport Systems. 2017. Future proofing infrastructure for connected and automated vehicles. Technical Report.

Childress, S., et al. 2015. Using an activity-based model to automated vehicles. *Transportation Research Record*, 2493(1), 99–106.

Clark, B., Larco, N., and Mann, R. 2017. The impacts of autonomous vehicles and e-commerce on local government budgeting and finance. *SSRN Electronic Journal*. doi: 10.2139/ssrn.3009840.

Clements, L. M. and Kockelman, K. M. 2017. Economic effects of automated vehicles. *Transportation Research Record*, 2606(1), 106–114.

Coldewey, D. 2019. Tesla explodes after crash on Russian highway. *Tech Crunch*. https://techcrunch.com/2019/08/11/tesla-explodes-after-crash-on-russian-highway/.

Coren, M. 2019. Investigators found autopilot was engaged in a Tesla crash in Florida. Quartz. https://qz.com/1621235/autopilot-was-engaged-in-a-2019-tesla-model-3-crash-in-florida/.

Cuneo, E. 2020. Auto's future examined in 2020 Deloitte Global Automotive Consumer Study. Autonomous Vehicle Technology. https://www.autonomousvehicletech.com/articles/2228-autos-future-examined-in-2020-deloitte-global-automotive-consumer-study.

Cunningham, M., et al. 2018. A survey of public opinion on automated vehicles in Australia and New Zealand. *28th ARRB International Conference (2018)*, Brisbane, Queensland, Australia.

Deloitte. 2019. Deloitte study: Consumers pump the brakes on autonomous vehicle adoption. https://www2.deloitte.com/global/en/pages/about-deloitte/press-releases/consumers-pump-brakes-on-autonomous-vehicle-adoption.html.

Enwemeka, Z. 2018. NuTonomy can now move its self-driving cars on any road In Boston. wbur. https://www.wbur.org/bostonomix/2018/06/20/nutonomy-self-driving-car-boston-expansion.

Fagnant, D. and Kockelman, K. 2014. The travel and environmental implication of shared autonomous vehicles using agent-based model scenarios. *Transportation Research Part C: Emerging Technologies*, 40, 1–13. doi: 10.1016/j.trc.2013.12.001.

Feigenbaum, B. 2018. Autonomous vehicles: A guide for policy makers. Reason Foundation.

Fernandes, P. and Nunes, U. 2012. Platooning with IVC-enabled autonomous vehicles: Strategies to mitigate communication delays, improve safety and traffic flow. *IEEE Transactions on Intelligent Transportation Systems*, 13(1): 91–106.

Ford. 2016. Investor Day. https://corporate.ford.com/content/dam/corporate/en/investors/investor-events/Press%20Releases/2016/september-2016-ford-investor-deck-for-web.pdf.

Formby, B. 2017. GOP lawmaker wants Texas law to catch up to driverless cars. The Texas Tribune. https://www.texastribune.org/2017/03/09/fort-worth-lawmaker-wants-ensure-texas-ready-driverless-cars/.

Friedrich, B. 2016. The effect of autonomous vehicles on traffic. Autonomous Driving, 317–334.

Fröhlich, et al. 2018. Acceptance factors for future workplaces in highly automated trucks. *Proceedings of the 10th International ACM Conference on Automotive User Interfaces and Interactive Vehicular Applications (AutomotiveUI'18)*, Toronto, Canada.

Gavanas, N. 2019. Autonomous road vehicles: Challenges for urban planning in European Cities. Urban Science, 3(2), 61.

Gerdes, J. C. and Thornton, S. M. 2015. Implementable ethics for autonomous vehicles. In: *Autonomes Fahren: Technische, rechtliche und gesellschaftliche Aspekte*, Maurer, M., Gerdes, J. C., Lenz, B., and Winner, H. (eds). Springer: Berlin Heidelberg, pp. 87–102.

Gitln, J. 2018. Robotaxi permit gets Arizona's OK; Waymo will start service in 2018. Arstechnica. https://arstechnica.com/cars/2018/02/robotaxi-permit-gets-arizonas-ok-waymo-will-start-service-in–2018/.

Global Automotive & Transportation Research Team at Frost & Sullivan. 2018. Global Autonomous Driving Market Outlook, 2018.

Goodall, N. J. 2014. Ethical decision making during automated vehicle crashes. *Transportation Research Record: Journal of the Transportation Research Board*, 2424(1): 58–65.

Greaves, S. P., et al. 2018. ITLS-WP-18-18 autonomous vehicles down under: An empirical investigation of consumer sentiment.

Gruel, W. and Stanford, J. M. 2016. Assessing the long-term effects of autonomous vehicles: A speculative approach. *Transportation Research Procedia*, 13, 18–29.

Guardian Staff and Agencies. 2018. Tesla car that crashed and killed driver was running on autopilot, firm says. *The Guardian*. https://www.theguardian.com/technology/2018/mar/31/tesla-car-crash-autopilot-mountain-view.

Hörl, S., et al. 2016. Simulation of autonomous taxis in a multi-modal traffic scenario with dynamic demand.

Howard, D. F. and Dai, D. 2014. Public Perceptions of Self-Driving Cars: The Case of Berkeley, California.

Hua, C. 2017. Four self-driving buses tested in Shenzhen. *China Daily*. http://www.chinadaily.com.cn/china/2017-12/04/content_35190702.htm.

Ilkova, V. and Ilka, A. 2017. Legal aspects of autonomous vehicles: An overview. *2017 21st International Conference on Process Control (PC)*, Strbske Pleso, Slovakia, 428–433.

Infrastructure Partnerships Australia. 2017. Automated vehicles-do we know which road to take.

International Transport Forum. 2015. Urban mobility system upgrade: How shared self-driving cars could change city traffic.

International Transportation Forum. 2018. The shared-use city: Managing the curb. Organization for Economic Cooperation and Development and the International Transport Forum.

Johnson, C. 2017. Readiness of the road network for connected and autonomous vehicles. The Royal Automobile Club.

Johnson, C. and Walker, J. 2016. Peak Car Ownership Report, The market opportunity of electric automated mobility services. Rocky Mountain Institute. https://rmi.org/insight/peak-car-ownership-report/.

Kloostra, B. and M. J. Roorda. 2019. Fully autonomous vehicles: Analyzing transportation network performance and operating scenarios in the Greater Toronto Area, Canada. Transportation Planning and Technology, 42(2), 99–112.

Kockelman, K., et al. 2016. An assessment of autonomous vehicles: Traffic impacts and infrastructure needs - final report, report, December 2016; Austin, Texas. University of North Texas Libraries, The Portal to Texas History.

KPMG. 2015. Connected and autonomous vehicles: The UK economic opportunity.

KPMG. 2018. Autonomous vehicles readiness index: Assessing countries' openness and preparedness for autonomous vehicles.

KPMG. 2019. Autonomous vehicles readiness index: Assessing countries' preparedness for autonomous vehicles.

Kyriakidis, et al. 2015. Public opinion on automated driving: Results of an international questionnaire among 5000 respondents. *Transportation Research Part F: Traffic Psychology and Behaviour*, 32, 127–140.

Lam, A. Y. S., et al. 2018. Coordinated autonomous vehicle parking for vehicle-to-grid services: Formulation and distributed algorithm. *IEEE Transactions on Smart Grid*, 9(5), 4356–4366.

Levin, S. 2018. Tesla fatal crash: 'Autopilot' mode sped up car before driver killed, report finds. *The Guardian*. https://www.theguardian.com/technology/2018/jun/07/tesla-fatal-crash-silicon-valley-autopilot-mode-report.

Levin, S. and Woolf, N. 2016. Tesla driver killed while using autopilot was watching Harry Potter, witness says. *The Guardian*. https://www.theguardian.com/technology/2016/jul/01/tesla-driver-killed-autopilot-self-driving-car-harry-potter.

Levin, S., and Carrie, J. 2018. Self-driving Uber kills Arizona woman in first fatal crash involving pedestrian. *The Guardian*. https://www.theguardian.com/technology/2018/mar/19/uber-self-driving-car-kills-woman-arizona-tempe.

Li, D. and Wagner, P. 2019. Impacts of gradual automated vehicle penetration on motorway operation: A comprehensive evaluation. *European Transport Research Review*, 11(2019), 1–10.

Lin, P. 2015. Why ethics matters for autonomous cars. In: *Autonomes Fahren: Technische, rechtliche und gesellschaftliche Aspekte*, Maurer, M., Gerdes, J. C., Lenz, B. and Winner, H. (eds). Springer: Berlin Heidelberg, 69–85.

Liping, G. 2018. Tesla confirms 'Autopilot' engaged in fatal crash in China. Ecns.cn. http://www.ecns.cn/2018/02-28/293992.shtml.

Litman, T. 2020. Autonomous vehicle implementation predictions, implications for transport planning. Victoria Transport Policy Institute

Lori, A. 2019. Are self-driving cars safe? Verizon connect. https://www.verizonconnect.com/resources/article/are-self-driving-cars-safe/

Lu, Q., et al. 2019. The impact of autonomous vehicles on urban traffic network capacity: An experimental analysis by microscopic traffic simulation. *The International Journal of Transportation Research, Transportation Letters*. doi: 10.1080/19427867.2019.1662561.

Lütge, C. 2017. The german ethics code for automated and connected driving. *Philosophy & Technology*. doi: 10.1007/s13347-017-0284-0.

Marshall, A. 2018. Massachusetts welcomes self-driving cars: With a couple caveats. Wired. https://www.wired.com/story/massachusetts-boston-self-driving-car-rules-regulations/.

Martínez-Díaz, M and Soriguera, F. 2018. Autonomous vehicles: Theoretical and practical challenges. *Transportation Research Procedia*, 33, 275–282. doi: 10.1016/j.trpro.2018.10.103.

Šinko, S., et al. 2017. Analysis of public opinion on autonomous vehicles. *12th International Conference on Challenges of Europe: Innovative Responses for Resilient Growth and Competitiveness,* Bol, Croatia.

McFarland, M. 2016. Google's self-driving car takes some blame for a crash. *The Sydney Morning Herald*. https://www.smh.com.au/technology/googles-selfdriving-car-takes-some-blame-for-a-crash-20160301-gn71z8.html.

McKinsey & Company. 2016. Automotive revolution- perspective towards 2030. How the convergence of disruptive technology-driven trends could transform the auto industry.

Metz, D. 2018. Developing policy for urban autonomous vehicles: Impact on congestion. *Urban Science* 2(2), 33.

Miller, S. A. and Heard, B. R. 2016. The environmental impact of autonomous vehicles depends on adoption patterns. *Environmental Science & Technology* 50(12), 6119–6121.

Moreno, A. T., et al. 2018. Shared autonomous vehicles effect on vehicle-km traveled and average trip duration. *Journal of Advanced Transportation*, 2018, 1–10.

Murdock, J. 2018. Tesla model S autopilot sped up before crashing into Utah fire truck. *Newsweek*. https://www.newsweek.com/tesla-model-s-crash-car-autopilot-sped-just-utah-firetruck-944251.

National Highway Traffic Safety Administration. 2015. The Economic and Societal Impact of Motor Vehicle Crashes, 2010 (Revised). http://www.nrd.nhtsa.dot.gov/pubs/812013.pdf.

National League of Cities. 2018. Autonomous vehicle-pilots across America. Municipal Action Guid.

Navya be Fluid. 2017. Inauguration of the autonomous shuttles service at la defense in Paris. https://navya.tech/en/inauguration-of-the-autonomous-shuttles-service-at-la-defense-in-paris–2/.

Nunes, A. and Hernandez, K. 2019. The cost of self-driving cars will be the biggest barrier to their adoption. *Harvard Business Review*. https://hbr.org/2019/01/the-cost-of-self-driving-cars-will-be-the-biggest-barrier-to-their-adoption.

Ongel, E., et al. 2019. Economic assessment of autonomous electric microtransit vehicles. *MDPI Sustainability*, 11, 648.

Panagiotopoulos, I. and Dimitrakopoulos, G. 2018. Consumers' perceptions towards autonomous and connected vehicles: A Focus-Group Survey on University Population. *Proceedings of the 6th Humanist Conference*, The Hague, Netherlands, 13–14 June 2018.

Piao, J., et al. 2016. Public views towards implementation of automated vehicles in urban areas. *Transportation Research Procedia*, 14, 2168–2177.

PWC. 2018. The era of digitalizing trucks: Charting your transformation to a new business model.

Red Chalk Group. Transforming mobility: Business models in the age of autonomous vehicles.

Reissmann, M. 2019. The future of autonomous vehicles. Futures platforms https://www.futuresplatform.com/blog/future-autonomous-vehicles.

Richardson, E. and Davies, P. 2018. The changing public's perception of self-driving cars.

Richardson, N., et al. 2017. Assessing Truck Drivers' and Fleet Managers' Opinions towards Highly Automated Driving. Springer International Publishing: Cham.

Rouen. 2018. Final testing before the Rouen Normandy autonomous lab on-demand mobility service opens to the public. Lulop. https://lulop.com/en_EN/post/show/147716/final-testing-before-the-rouen.html.

Schlossberg, M., et al. 2018. Rethinking the street in an era of driverless cars. *Urbanism Next Research*. doi: 10.13140/RG.2.2.29462.04162.

Schoettle, B. and Sivak, M. 2014. A survey of public opinion about autonomous and self-driving vehicles in the U.S., the U.K., and Australia.

Schoettle, B. and Sivak, M. 2015. Motorists' preferences for different levels of vehicle automation.

Securing America's Future Energy (SAFE). 2018a. America's workforce and the self-driving future: Realizing productivity gains and spurring economic growth.

Securing America's Future Energy (SAFE). 2018b. The economic and social value of autonomous vehicles: Implications from past network-scale investments.

Seiberth, G. and Gründinger, W. 2018. Data-driven business models in connected cars, mobility services and beyond, BVDW research, No. 01/18, April 2018, 57 p.

Shared Mobility Principles for Livable Cities. Principles developed by a working group of international NGOs are designed to guide decision-makers and stakeholders toward the best outcomes for all in the transition to new mobility options. www.sharedmobilityprinciples.org.

Sivak, M. and Schoettle, B. 2015. Road safety with self-driving vehicles: general limitations and road sharing with conventional vehicles (University of Michigan, Ann Arbor, Transportation Research Institute, 2015–01).

Stewart, E. 2019. Self-driving cars have to be safer than regular cars. The question is how much. Vox, record. https://www.vox.com/recode/2019/5/17/18564501/self-driving-car-morals-safety-tesla-waymo.

Stoeltje, G. 2017. *Policy Brief: How Does Texas Law Change the Legal Landscape for Automated Vehicles?* Texas A&M Transportation Institute, Transportation Policy Research Center: College Station, TX.

Szmelter, A. and Woźniak, H. 2016. New mobility behaviours and their impact on creation of new business models. *Torun Business Review*, 15, 79–95. doi: 10.19197/tbr.v15i4.59.

Tan, X. and Garcia, A. 2018. Autonomous mobility and energy service management in future smart cities: An overview.

Tannert, C. 2014. Will you ever be able to afford a self-driving car. *Fast Company*. https://www.fastcompany.com/3025722/will-you-ever-be-able-to-afford-a-self-driving-car.

The Polis Traffic Efficiency and Mobility Working Group. 2018. Road vehicle automation and cities and regions.

Tibken, S. 2018. Waymo CEO: Autonomous cars won't ever be able to drive in all conditions. cnet https://www.cnet.com/news/alphabet-google-waymo-ceo-john-krafcik-autonomous-cars-wont-ever-be-able-to-drive-in-all-conditions/.

Tröster, S., et al. 2017. Transport companies, truck drivers, and the notion of semi-autonomous trucks: A contextual examination. *Adjunct Proceedings of the 9th International ACM Conference on Automotive User Interfaces and Interactive Vehicular Applications (AutomotiveUI'17)*, Oldenburg, Germany.

University of Illinois College of Engineering. 2019. Platform for testing of autonomous vehicle safety. *ScienceDaily*, 25 October 2019. www.sciencedaily.com/releases/2019/10/191025170813.htm.

US Department of Transportation. 2017. Autonomous vehicle fleet ownership and operating costs are expected to be half that of traditional vehicles by 2030–2040.

Wadud, Z. 2017. Fully automated vehicles: A cost of ownership analysis to inform early adoption. *Transportation Research Part A: Policy and Practice*, 101, 163–176.

Wagner, P. 2016. Traffic Control and Traffic Management in a Transportation System with Autonomous Vehicles. Springer: Berlin, pp. 301–316.

Walker, A. (2020). Are self-driving cars safe for our cities? https://www.curbed.com/2016/9/21/12991696/driverless-cars-safety-pros-cons.

Welcome to the Future – Smart Shuttle, Sion, Switzerland. 2017. https://growingupwithoutborders.com/europe/switzerland/welcome-to-the-future-smart-shuttle-sion-switzerland/.

Wintersberger, P., et al. 2016. Automated driving system, male, or female driver. *Proceedings of the 8th International Conference on Automotive User Interfaces and Interactive Vehicular Applications – Automotive'UI 16*, Ann Arbor, MI, 51–58.

Yamauchi, M. How will autonomous vehicles charge themselves? Plugless. https://www.pluglesspower.com/learn/solve-last-mile-vehicle-autonomy/.

Zhang, W. and Guhathakurta, S. 2017. Parking spaces in the age of shared autonomous vehicles: How much parking will we need and where? *Transportation Research Record*, 2651(1), 80–91.

Zhang, W., et al. 2015. Exploring the impact of shared autonomous vehicles on urban parking demand: An agent-based simulation approach. *Sustainable Cities and Society*, 19, 34–45.

Zhang, W., et al. 2018. The impact of private autonomous vehicles on vehicle ownership and unoccupied VMT generation. *Transportation Research Part C: Emerging Technologies*, 90, 156–165.

Zmud, J., et al. 2017. Strategies to advance automated and connected vehicles. National Cooperative Highway Research Program; Transportation Research Board; National Academies of Sciences, Engineering, and Medicine.

4 Machine Learning Methodologies for Electric-Vehicle Energy Management Strategies

A Comprehensive Survey

John S. Vardakas and Ioannis Zenginis
Iquadrat Informatica

Christos Verikoukis
Telecommunications Technological Centre of Catalonia (CTTC/CERCA)

CONTENTS

4.1 INTRODUCTION

Automobiles have provided a substantial contribution to the growth of the modern society by providing solutions to the mobility problem in daily rhythms; they are the necessary means of both transportation and personal expression that connects, binds, and integrates in a scattered world [33]. However, automobiles have a huge environmental impact: in the USA, cars and trucks collectively emit nearly 20% of all county's emissions [64]. Electric vehicles (EVs) and hybrid electric vehicles (HEVs) represent a promising technology that is able to achieve significantly lower pollution emissions. In this new era in the transportation section, new research areas have emerged, related to batteries and electric motors (EMs) for EVs/HEVs [32], charging infrastructures [74,90], and energy management systems (EMS).

The design of a well-organized EMS for EVs is highly important for the provision of energy efficiency and low carbon emissions under the same umbrella. The main strategy of an efficient EMS determines the way to optimize the utilization of the energy stored in the vehicle's battery for the case of EVs, or the way to distribute the requested power between the engine and the battery and minimize the fuel consumption, in case of HEVs. To this end, the study of EMS for EVs and HEVs is highly important, where various types of such systems should be identified, and the advantages and limitations of these approaches should be extracted. In this chapter, we present various EMS for EVs that have been presented in the literature. We organize these management systems into three main categories, as illustrated in Figure 4.1. The first category is generally known as rule-based EMS, the second category refers to optimization-based energy management methods, while the third category refers to ML-based energy management strategies.

Rule-based EMS exploits a set of rules that are usually based on human experience and without prior knowledge of the vehicle's route, targeting to determine the operating state of the vehicle's engine and/or its EM. The control decisions of these strategies are only consistent with the current states of the vehicle and its power demand. Rule-based EMSs have gained significant attention from the research and industrial communities, mainly due to low implementation requirements. However, these methods cannot guarantee near-optimal control effects under realistic driving conditions. To this end, optimization-based EMSs are designed to achieve optimum results in terms of energy efficiency and performance of the vehicle. Although optimization-based EMSs show notable performance regarding vehicle's energy economy, their inability to provide real-time control of the vehicle, which requires full knowledge of the vehicle's driving cycle in advance, limits their applicability in real systems, thus they are mainly used for offline simulations and as benchmarks for real-time EMS. Finally, ML-based energy management methods provide real-time control of the vehicle by using either a prediction-based or a learning-based strategy. The former approach considers predicted data and employs the local control solution on a short-term time horizon while achieving near-optimal energy management performance. On the other hand, learning-based approaches consider a large volume of historical and real-time data in order to determine a learning procedure for the vehicle's behavior and for the route characteristics, which will then be used in order to derive the optimal control of the vehicle.

Other surveys on EMSs for EVs can be found in the literature [54,55]. In Ref. [54], the authors present a comprehensive survey on energy management for plug-in hybrid EVs (PHEVs), while they explore synergies of intelligent transportation systems, smart grids, and smart city technologies. The survey presented in Ref. [65] focuses on energy management strategies for hybrid EVs and PHEVs that target to efficiently manage the power flow between fuel and the vehicle's battery. In parallel, energy management strategies for hybrid EVs are also surveyed in Ref. [91] based on bibliometrics by presenting a categorization of the strategies presented in the literature until the year 2015. Machine learning algorithms for energy management have been surveyed in Ref. [55]; however, the main focus of this work is on energy systems.

To position our contribution, we are motivated to present this survey on ML-based EMSs for EVs, which can be used in future research efforts for the development of more sophisticated and effective

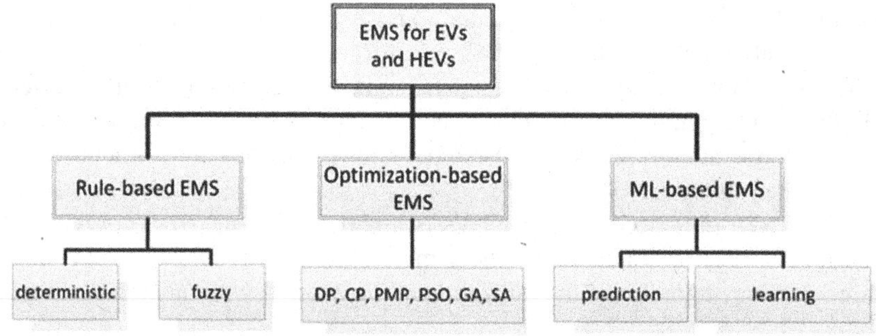

Figure 4.1 Classification of EMS for EVs and HEVs.

EV EMSs. Due to the constant evolution of EV-EMSs, this survey provides a detailed summary of the current status of EMS with a specific emphasis on ML-based approaches. Moreover, our contribution complements the existing surveys by focusing on recent trends on ML algorithms and techniques for EV EMSs, as well as by providing future research directions.

The rest of this work is organized as follows: Section 4.2 presents the first two EMS categories, namely, the rule-based and the optimization-based EMS. In Section 4.3, we tackle an extensive survey on ML-based EMSs that have been proposed in the literature for EVs and HEVs. Finally, Section 4.4 provides concluding remarks of this work together with the lessons learned and future directions. Furthermore, a list of abbreviations used in this survey is listed in Appendix A.

4.2 RULE-BASED AND OPTIMIZATION-BASED EMSs FOR EVs/HEVs

Energy management systems for EVs generally target to satisfy the power requirements of the vehicle, while at the same time preserving its performance. An EMS is a vital component of an EV that is able to manage various energy sources (e.g., engine, battery, or capacitance) with a high flexibility in order to achieve higher performance. Due to the limited amount of energy from those sources, the efficient utilization of this energy is crucial; thus, the main objective of such an EMS is the efficiency maximization of the overall system as well as the minimization of the fuel consumption (for the HEV case). This objective is met by designing efficient EMS strategies, which target to distribute the total power requested between the engine and an energy storage system, in order to optimize the engine work region.

This section provides a discussion on EMS for EVs that belong to two main categories: ruled-based and optimization-based EMS [59]. Rule-based EMSs consider predefined rules based on a set of parameters related to the vehicle (e.g., velocity or engine speed) and to the driver (e.g., torque demand), in order to make real-time control decisions without prior knowledge of the EV's route [53].

4.2.1 RULE-BASED EMS FOR EVs/HEVs

Rule-based EMSs are further categorized into deterministic and fuzzy strategies, based on the definition of the control law rules. Figure 4.2 presents the classification of the rule-based energy management strategies. In deterministic approaches, the energy-management strategies consist of fixed and specific rules and algorithms based on engineering intuition for engine start/stop and battery charging/discharging procedure. Several deterministic approaches have been proposed in the literature, mainly for HEVs: the on/off control strategy targets to maintain the operation of the engine, the EM and the battery within specific and predefined operating ranges, thus achieving nonoptimal efficiency, however with significantly low computational requirements [2]. Extensions of the

Ruled-based EMS for EVs and HEVs		
Deterministic	classical	[12]
	adaptive	[13, 14]
	CD-CS	[15, 16]
	blended	[18–20]
Fuzzy		[22–24]

Figure 4.2 Classification of EMS for EVs and HEVs.

on/off strategy is the power-follower control strategy that considers multiple control modules for the energy management as well as the adaptive rule-based strategy, where a stepwise decision-making process is implemented [3]. Another adaptive rule-based strategy is presented in Ref. [6], targeting to maintain the battery's state-of-charge (SoC) around a predefined set of boundaries. The adaptive strategy of Ref. [6] does not require extensive tuning and detailed knowledge of the operational specifications of the engine; it can be implemented online, while it is able to consider extended travel distances. In addition, the charge-depleting (CD) - charge-sustaining (CS) (or all-electric range) strategies consider that the vehicles use the electrical energy in CD phase and when until its SoC decreases to an allowable minimum value, they switch to the CS mode, where the vehicle is powered by both the engine and the EM, while the battery SoC maintains in the vicinity of the low threshold [31]. The CD-CS approaches have been widely deployed for HEV applications, mainly due to their simplicity, as they mainly consider the vehicle's battery SoC as the indicator in order to follow the predefined rules. However, studies have revealed its suboptimal or nonoptimal fuel-efficiency performance and corresponding electric system efficiency drop during intensive CD mode [95].

In addition, blended strategies are based on the split of the power demand between the internal combustion engine (ICE) and the EM by following specific rules throughout the vehicle's route and can be either engine-dominant or electricity-dominant [24]. For example, a blended rule-based strategy for HEVs has been presented in Ref. [7], which interprets pedal motion of the driver into the necessary propulsion power and, then, determines the appropriate power split between the ICE and the EM based on intuition, operation boundaries, or human expertise. Similarly, the blended strategy of [66] considers the energy demand and the engine's lower output power threshold in order to control the vehicle's turn on and off, as well as its output power range. Another example of deterministic rule-based approach is provided in Ref. [13], in order to determine an optimal EMS for HEV. This solution is implemented online by considering some key threshold values and achieves the reduction of the objective function by 8.28% and the energy consumption by 6.3%, compared to the traditional blended strategy, where the power requirement between the battery and engine is split throughout the driving route.

Fuzzy rule-based EMS strategies have a nonlinear structure and can be considered as a natural extension of basic rules, thus achieving insensitivity to conditions uncertainties, and robustness against the measurement noises and disturbances [19]. In general, fuzzy rules target to increase the operational efficiency of the deterministic rules and fuel economy of the vehicles by using decision-making based on lookup tables in order to compensate for imprecise measurements. Compared to the deterministic approaches, fuzzy-based rules have a higher level of abstraction, and they are more robust and can be easily tuned to incorporate a variety of parameters. For example, the fuzzy-logic approach of [17] takes into account the past and present driving conditions as well as predictions for the length of the vehicle's route, targeting to determine the control modes. Fuzzy rule-based strategies are divided into basic, adaptive, and predictive methods: in the first category, basic strategies perform the straightforward fuzzy-logic steps, where the input data are fuzzified, and a set of human- and expertise-based rules are considered for the determination of the fuzzy output, which is then defuzzified to a set of control signals. Extensions of the basic strategies are the adaptive fuzzy EMS strategies, where the control parameters are adaptive to the current system operating conditions. For example, the method proposed in Ref. [72] considers the trip information of the vehicle and tracks the SoC curve through an adaptive fuzzy logic controller. Finally, predictive fuzzy EMS strategies incorporate predictions of driving conditions that are integrated with the current conditions into a fuzzy controller, thus increasing the performance of the strategy by achieving near-optimal results. Such predictive techniques are presented in Ref. [25] that use the vehicle's velocity-predicted profile and predicted road slope, together with the vehicles current state, in order to achieve the reduction of the engine start and stops, while keeping the battery's SoC into the desired levels, thus increasing the vehicle's energy efficiency.

4.2.2 OPTIMIZATION-BASED EMS FOR EVs/HEVs

In general, the rule-based strategies can be easily and effectively implemented, achieving high robustness and reliability, while they have been widely deployed, mainly due to their low computational load requirements and their effectiveness for real-time supervisory control, especially for low levels of hybridization. However, the main disadvantages of the rule-based methods refer to their limited flexibility that restricts their adaptation to varying driving conditions, their high dependency on the proper design of the considered rules, their inability to provide optimal solutions, as well as their ineffectiveness when they are applied to different vehicle types [70]. Furthermore, rule-based strategies are designed by considering standard driving cycles, that although are able to provide valuable information for the vehicle's performance in terms of power consumption and emissions, they are unable to incorporate real-world conditions with uncertainties as well as to provide optimum results. To this end, optimization-based methods target to regulate the control variables of the system usually through the extraction of as much vehicle-related information as possible and generate an optimum solution (e.g., the minimization of a cost function) within feasible constraints. Optimization methods are generally divided into three main subcategories, namely, offline, prediction-based, and learning-based optimization methods. In this section, offline methods are discussed, while the ML prediction- and learning-based methods are presented in the next section. Figure 4.3 presents the classification of the offline optimization energy management strategies.

The offline optimization methods necessitate the knowledge of the vehicle's route in order to determine the optimal solution (e.g., the minimization of fuel consumption in HEV). In general, offline methods have low complexity, as they are based on historical data for the vehicle; however, they are not able to adapt to real driving conditions for optimum performance. Offline methods include Dynamic Programming (DP) [77], Convex Programming (CP) [58], Pontryagin's Minimum Principle (PMP) [29], Particle Swarm Optimization (PSO) [12], Genetic Algorithm (GA) [28], and Simulated Annealing (SA) algorithm [14].

The optimal solution that is provided by DP methods is generally used as a benchmark in order to explore the potential of other EMS methods. Under a DP approach, the considered problem is discretized over time into a sequence of optimized subproblems, which are then solved by using backward induction, in order to obtain the global optimal solution. The necessity of the backward induction is a limiting factor for DP as it requires the a priori knowledge of the entire duty cycle to optimize fuel economy in HEVs [6]. In any case, DP methods target to derive the optimal solution for the minimization of the fuel consumption during a specific driving cycle, which is given as the vehicle's speed versus the time profile for a specific route. For example, DP was used in Ref. [9] in order to determine the eco-driving cycles with speed limits of the vehicles, which is then used for

Offline optimization-based EMS for EVs and HEVs	
Dynamic Programming	[31–33]
Convex Programming	[34–39]
Pontryagin's Minimum Principle	[40–44]
Particle Swarm Optimization	[14, 20, 45]
Genetic Algorithm	[46–49]
Simulated Annealing	[30, 51–54]

Figure 4.3 Classification of offline optimization-based EMSs.

the determination of the appropriate energy management to reduce fuel consumption. In contrast, the approach in Ref. [1] considers DP in order to determine a joint optimization of both the power-split and the velocity of the vehicle in order to improve the performance of the vehicle. It should be noted that DP methods suffer from the so-called "curse of dimension" ([19]), which prevents its direct adoption in real-world control problems, while the a priori requirement for the knowledge of the driving cycles prevents their application to real-time implementations. An extension of DP, the so-called stochastic DP, is proposed in Ref. [56], in order to alleviate the computational burden of DP especially when they are applied to long routes; however, this approach cannot be considered as a real-time solution, although the control values, which are average results of multiple driving cycles, can be applied in real-time implementations.

On the other hand, CP methods are applied when energy management functions have to handle problems related to nonlinearities and computational burdens, so they are approximated as convex functions. In this way, the system's dynamics are modeled by using convex functions, where the local optima coincide with the global optima, thus guaranteeing the convergence to the global optimum solution. This procedure has been widely used for the solution of EMS optimization problems [22,26,57], while the proof of convexity of the energy management problem for HEVs is provided in [21]. Furthermore, the cooperative EMS of [30] targets to reduce the braking on a hilly terrain of a set of connected HEVs, where CP is applied in order to optimize the velocity, route duration, and battery SoC. It should be pointed out that the CP can only be used when the energy management function and the inequality constraints are convex. However, a number of fixed control parameters, e.g., the engine-on power, cannot be estimated by a convex function [57]; in these cases, randomized heuristic searching algorithm should be applied. The interested reader may refer to [83] for a comparison between CP and DP for HEV-EMS.

PMP methods are analytical methods that are based on the mathematical formulation of the problem, thus they obtain solutions faster compared to other approaches; however, PMPs are offline methods and require the a priori knowledge of the driving cycles [35]. The main idea behind PMP is the fact that it reduces the constrained global optimization problem into a local Hamiltonian minimization problem. PMP has been applied in the EMS proposed in Ref. [23], where the energy management approach is applied to multiple HEVs, given the flexibility in power demand, instead of strictly following the required power from the vehicle level. Similarly, a PMP-based control strategy for HEVs is proposed in Ref. [78], where the optimal control problem based on PMP is converted to a nonlinear programming model, by using the Legendre pseudospectral method. An extension of the PMP, called Approximate PMP, considers a piecewise linear approximation of the engine fuel-rate, in order to obtain a simple convex approximation of the Hamiltonian that is able to achieve a fuel consumption reduction of 6.96% compared to the conventional PMP [29].

Another extension of PMP is the equivalent consumption minimization strategy (ECMS), which is an instantaneous approach derived from PMP, in order to increase the speed of computations. ECMS targets to overcome the main limitation of DP, which is the a priori knowledge of the entire duty cycle, by transforming the global objective function into a local optimization problem and by incorporating a fuel-equivalent cost function in order to convert the vehicle's battery usage into equivalent fuel cost. The main limitation of ECMS, which refers to the fact that the equivalence factor that is applied in order to scale the electrical and fuel costs, depends on different driving tasks and needs constant calibration, led to the development of the adaptive ECMS (A-ECMS) [67], which is discussed in the next section.

Due to the aforementioned restrictions and limitations, more advanced and "intelligent" optimization methods have been used for the derivation of the optimal strategy in EMS for EVs. To this end, PSO methods are considered appropriate for the optimization of the EMS of HEVs that have a degree of noise or irregularity. PSO algorithms belong to a larger class of evolutionary algorithms (EA) and populate particles states, position, and velocity. The particles interact with each other locally, targeting to interchange information, while they are able to store their last best position and group best solution, with the intention to improve the next population. PSO methods require the full a priori knowledge of the entire duty cycle of the vehicle, in order to derive the optimal solution; however, their performance is highly affected by the selection of the constraints both in the initial random and the updated population. By applying PSO in EV EMS problems, the considered particles are potential solutions, which are continuously updated at each iteration based on the current best fuel economy solution for each particle, and the current best global fuel economy solution across all particles [6]. For example, PSO is considered in the EMS strategy proposed in Ref. [89], where it is applied in order to optimize the equivalent factor of a HEV, while an online part of the proposed algorithm provides the optimum control strategy. In order to deal with the main disadvantage of the offline optimization procedures, the strategy was proposed in Ref. [13], where firstly it considers a rule-based approach that contains three operation modes, while then the PSO algorithm is implemented on four threshold values in the considered rule-based strategy.

Genetic algorithms are approaches that also belong to the class of EA and are inspired by the procedure named natural selection, which is a key mechanism of evolution. These algorithms consist of three phases, namely, reproduction, cross-over information, and mutation. In each of the algorithm's iterations, the solution is coded in simulated chromosomes, which the phases encompass randomness in order to guarantee population diversity. Therefore, GA approaches initially assume a set of solutions, which are then evaluated with respect to an objective fitness function, while the best solutions are provisioned so that they can develop, until the set of stopping criteria is fulfilled. This approach is inspired by evolution mechanisms, where the population of solutions is considered as chromosomes, which are able to grow and form the next generation of solutions. An example of the application of GA in EMS for EVs is presented in Ref. [20], where an offline calibration is implemented by using a GA based on the approximation model and a comprehensive evaluation indicator. The results of this approach show a reduction in fuel consumption and exhaust emissions of HEVs. GA algorithms have been combined with other approaches in order to enhance the performance of the strategy. For example, in Ref. [16], a combination of GA, which is used in order to optimize the engine power in a power-split HEV, and quadratic programming (QP) that provides the optimal current of the vehicle's battery, is presented, targeting to optimize the power threshold at which the engine is turned on, thus improving the vehicle's fuel efficiency. In addition, GA has also been utilized in Ref. [51], in order to evaluate the online energy management controller developed for HEVs that consider a driving condition recognition and a GA that is applied to search the optimal values for several control actions offline. Another online approach is presented in Ref. [79], where an easy-to-optimize mathematical representation of EMS using gamma functions is presented, where the coefficients of the proposed strategies are optimized by using GA, thus showing simplicity, computational efficiency, and relatively easy calibration. It should be noted that GA may generate suboptimal solutions and their performance depends on the initial population and tuning parameters, while their main advantages refer to their strong universality, and to the capacity of parallelism detection among different agents, which is particularly beneficial to computing Pareto solutions [33]. Finally, SA algorithms have been widely used in order to overcome the limitations of the CP problems related to the inability to express some control functions as convex, mainly due to their simplicity and higher efficiency, compared to GA. Under SA, a stochastic technique is used in order to search the solution, by taking possible solutions that achieved improvement over the objective function, while the pool of solutions that are suboptimal is also considered [36]. In this way, the algorithm cannot trap in local minima and is able to evolve to the global optimum solution. The EMS optimization problem of [37] is solved by using the Simulated Annealing Particle Swarm Optimization for fuel economy and for six different driving cycles, which are classified by using clustering analysis, while Euclidean proximity was used to recognize the driving cycles. A special case of SA, named Quantum-inspired Simulated Annealing (QSA) [14], is based on the application of the quantum computation concept on the simulated annealing algorithm, in order to solve the mixed-integer nonlinear programming problems. The main advantage of the QSA combination is the utilization of quantum computations, which are based on the uncertainty of quantum states, making them appropriate for solving EMS optimization problems. QSA approaches have been used in order to solve the optimization problems related to optimal placement and sizing of parking lots [61–63], targeting to incorporate the optimal charging management of EVs, by considering their driving patterns and their drivers' behavior with respect to the incentives offered for charging their EVs to the parking lots.

4.3 MACHINE LEARNING-BASED TOOLS FOR EV ENERGY MANAGEMENT

Machine learning is a powerful tool for the implementation of an efficient energy management method for EVs, where the agent (i.e., the decision-maker) is able to "learn" the way to "act" in an optimal way [4]. Under this approach, the agent is able to notice the state of the environment and take an action according to the observed information, while the agent will receive a reward as the outcome of this action. To this end, the agent targets to implement a specific policy that will guarantee the reward, which is based on a mapping from each possible state to an action, by "learning" from its past experience. Figure 4.4 presents the classification of the ML-based energy management strategies.

4.3.1 PREDICTION-BASED EMS FOR EVs

Prediction-based methods target to overcome the main disadvantage of the optimization approaches that are related to the a priori knowledge of the future driving cycles, by providing predictions for the future uncertainty of the driving cycles, in order to enhance the performance of the optimization procedure. This can be achieved by providing an estimation for the vehicle's future velocity over the prediction horizon and its driving habits, which is then combined with the optimal solution of an EMS through the available information (e.g., traffic and preceding vehicles) from an ITS. Through this combination, fuel efficiency is maximized, while the EMS is able to take into account the future driving conditions [69]. Therefore, ML-based prediction methods are able to simplify the extraction of functional dependencies from observations and utilize accurate and efficient predictions in order to achieve real-time implementations of EMS with relatively small computational requirements [5].

Prediction-based methods for EV EMS problems are classified into model-predictive control (MPC) [75] and adaptive ECMS [71]. MPC is a widely used predictive strategy that is able to deal with the multivariable constraint control problem of EMS. An MPC algorithm targets to minimize a series of cost functions, over a specific prediction horizon [81], while they are able to provide power distribution optimization in real time, by considering the upcoming predicted states and by taking

ML-based EMS for EVs and HEVs		
Prediction	MPC	[58, 61–73]
	A-ECMS	[44, 59, 74–78]
Learning	MDP	[80, 81]
	RL	[79, 82–85]
	Q-learning	[21, 60, 86–89]
	D-RL	[90–96]

Figure 4.4 Classification of ML-based EMSs.

advantage of its moving optimization characteristics. Generally, MPC-based EMS strategies show similar performance with corresponding DP approaches under specific working conditions.

MPC has been applied to EMS targeting to provide the battery's SoC planning by considering a predefined route of the vehicle [80]. This approach can be significantly enhanced by considering traffic data, where the optimum battery planning is determined based on real-time traffic data, while MPC can be used in order to provide real-time distribution of the power. Such an approach is presented in Ref. [85], where an adaptive battery planning is developed based on a computationally efficient MPC approach, where predictions are provided by using Markov chains.

It should be noted that the study of the MPC-based EMS showed that it is challenging to acquire the global planning of the battery discharge curve for PHEVs due to the lack of global velocity information. In addition, the MPC's performance is highly affected by the accuracy of the predicted states, a condition that triggers the need for improving the precision of the predicted velocity and the speed of the SoC calculations. To this end, the approach of [15] is based on a hierarchical control framework, where a QP algorithm is used in the upper layers to determine the SoC reference, while in the bottom layer, a velocity-prediction algorithm based on Wavelet Transform technology and Radial Basis Function-Neural Network (NN) is used. The proposed approach shows that it is feasible to provide near real-time SoC planning, as the optimization is limited to 400 ms for a trip.

A number of variations of MPC have been reported in the literature, targeting to improve the efficiency of MPC. These approaches include Stochastic MPC (SMPC), Frozen-Time MPC (FTMPC), and Prescient MPC (PMPC). SMPC considers a Markov model in order to represent the power requested by the driver; therefore, the probability distribution of the requested power can be used in order to minimize the objective function in an expected form. On the other hand, PMPC has the knowledge of the future power request all along the prediction horizon at each time step, while the FTMPC considers that the power request is assumed to be constant over the prediction horizon [96]. A comparison of these MPC variations showed that SMPC outperforms FTMPC in terms of fuel efficiency, while it shows similar performance with PMPC [18]. It is therefore evident that Markov models are able to provide increased performance due to the fact that they can determine probability distributions for the power request or the velocity of the vehicle [40]. Also, hidden Markov models have also been considered for the determination of the future velocity of the vehicle, by considering the current velocity as well as the current torque of the vehicle [47]. The results of the study in Ref. [18] showed that improved fuel economy can be achieved with a hidden Markov chain approach compared to the classical MPC approach. Similarly, the Markov chain-based predictor is used in the dual-loop online intelligent programming of [39] for velocity prediction and energy-flow control. Specifically, a deep fuzzy predictor is developed for predicting driver-oriented velocity prediction, while a finite-state Markov chain is utilized in order to learn transition probabilities between the vehicle's speed and acceleration. This approach demonstrates a significant reduction of the fuel consumption by 9.37% as well as improved computational time compared to the baseline.

Other variations of MPC include linear MPC [8] and Dynamic Programming MPC (DP-MPC) [34,84]. The DP-MPC EMS of [34] is based on a deep neural network speed predictor to predict the future short-term velocity in different prediction horizons, while DP is used to calculate the optimal energy distribution at each MPC control step. In addition, the authors of [84] developed a DP-MPC controller in order to achieve adaptive vehicle velocity control as well as optimal power management in vehicle-following scenarios. A similar approach is presented in Ref. [38]. As reported in Ref. [88], the DP-MPC of [34] is the most accepted, but it requires a long computation time, while the linear MPC can only be used for real-time control due to the fact that the optimization algorithm should be calculated at each control step; however, the simplification of this model leads to increased energy consumption.

In adaptive ECMS, the control parameters are periodically updated based on the current road load of a HEV, so that the fuel consumption is minimized and the battery's SoC is kept within certain boundaries. The effectiveness of the adaptive ECMS is enhanced by considering a modeling method for the route that includes a combination of the vehicle's velocity forecasts and of data provided by Geographical Information Systems (GIS), Global Positioning Systems (GPS), and Intelligent Transportation Systems (ITS) [67]. Therefore, A-EMCSs are based on information that is provided by external sources, thus it requires more computational and memory resources.

The predictive EMS of [10] considers real-time data regarding traffic conditions and traffic light locations in order to provide a strategy that targets to improve the performance of fuel reduction and robust system operation. This is achieved by developing an extension of ECMS by adding a module that generates distance-based predictive speed profiles with varying traffic light-crossing speeds as well as with a module that applies battery SoC feedback control. A similar extension of ECMS is presented in Ref. [27], where a new method based on velocity prediction for evaluating the equivalence factor between fuel cost and electrical cost is presented. This approach considers the fuel and electrical energy use, in order to calculate a probability factor, which is then used to adjust the equivalence factor within a limited range.

The performance of an EMS can be increased by considering route-preview information. Such an approach is provided in Ref. [73], where an adaptive ECMS is considered in order to implement a route-based EMS, which makes full use of the data provided for the vehicle's route and designs the optimal utilization of the battery's energy. Similarly, the adaptive ECMS approach of [92] uses data for the vehicle owner's driving habits and real-time information for the traffic from ITS and GPS, in order to adjust the equivalent factor of the ECMS. In parallel, the role of velocity forecasting is evaluated in Refs. [68,71], where future travel information prediction is applied for obtaining more accurate energy consumption in the full driving cycle, adjusting equivalence factor more reasonably.

It should be noted that both MPC-based and ECMS-based predictive EMS methods are able to significantly improve the performance of HEVs in terms of fuel efficiency by utilizing information regarding traffic, preceding vehicles' velocity and driving habits. As reported in Ref. [44], the effectiveness of adaptive ECMS can be degraded, if the prediction of the future driving cycles is not accurate. In addition, a comparison between MPC and ECMS in terms of fuel efficiency in HEVs showed that the performance of MP-based EMS methods could be considered as a trade-off between ECMS and DP. A study on the performance of DP, MPC, and ECMS-EMS methods showed that their fuel consumptions were 216.39, 228.51, and 242.40 g, respectively [69].

4.3.2 LEARNING-BASED EMS FOR EVs

Learning-based methods target to increase the performance of EMS for EVs with a combination of ML and optimization procedures and have gained significant research interest over the last few years. Learning approaches also overcome the difficulties that prediction-based approaches encounter when significant variations of the predictions exist, by applying learning procedures where both the predictions and the actions are jointly optimized [52]. In this way, the drivers are offered with the flexibility to cover their own needs by utilizing real-time decisions in any context. It should be noted that learning-based methods do not rely on the predefined simple control rules and they do not require the a priori knowledge nor of the driving cycle; the optimal control strategy is derived from the recorded historical driving conditions.

Effective learning-based approaches include Markov Decision Process (MDP), Reinforcement Learning (RL), Deep-RL (D-RL), Q-learning, and NNs.

MDP has been used in the analysis of EMS for vehicles, in order to model the real-world driving, as MDP is a decision-making process under stochastic situations. Specifically, MDP has been considered in order to model the interaction between the EMS and its external environment, which may include both the driver and the vehicle. The latter feature is the main source of uncertainty of the MDP, which directly affects the power demand and the velocity of the vehicle. For example, an MDP approach is used in Ref. [52] to model a discrete time interval charging control process, where, at each time interval, a decision is made on the action to take in order to reach the system's objective. The evaluation of this approach is realized by applying these methods offline by considering the historical data; however, the authors report that their approach can be updated to provide real-time data where the model can be trained incrementally or frequently. In addition, MDP is considered in Ref. [94], as part of the multistep strategies for the model-free predictive control, which also consider RL that is able to optimize the control policy in real time with a maximum prediction length of 65 steps.

RL refers to a set of goal-oriented algorithms that are able to learn how to reach an objective over a series of steps, while they reward for taking the right decisions based on a predefined reward function and they are penalized when they take the wrong ones. RL is a resourceful solution to problems where (1) different actions should be taken along with the changes of system states, while the future state depends on both the current state and the selected action; (2) an expected cumulative return will be optimized, instead of an immediate reward; (3) the agent needs to know only the current state and the reward it receives, while information related to the system input in prior or the detailed system modeling is not needed; (4) the system might be nonstationary to some extent. The last three aforementioned properties differentiate RL from the model-based optimal control, DP, and MPC-based approaches, respectively [45]. In addition, compared to other methods used in EMS for HEVs, such as offline optimization approaches and MPC, the RL approach is able to guarantee real-time and robust performance. An RL-based approach has been presented by Lin et al. [44,45], where a nested learning framework is developed that uses RL in order to obtain the power splits between the EM and the ICE of a HEV, while limiting the battery's SoC within its operational range. In these approaches, the inner loop learning minimizes the fuel usage, while the outer loop minimizes the cost of battery replacement. RL has been combined with Markov models in order to improve the effectiveness of the applied EMS strategy in terms of power demand and velocity predictions [48,50,87].

Q-learning is a well-investigated and effective RL algorithm that has been considered in EMS for HEVs, which targets to learn a policy and then to inform the agent about the action to take, under specific circumstances. For example, a Q-learning has been applied into the EMS of a set of hybrid electric tracked vehicle in the approach presented in Ref. [60], which is demonstrated that it has better performance in terms of fuel economy by 12% compared to the standard binary mode control strategy. The study of the application of Q-learning algorithms in EMS showed that it can give the adequate control orders only if the action value function, which is a judging system that gives action values based on historical data, has been well-trained [81]. In the case of Q-learning, the action value has a form of look-up table matrix, the size of which is determined by the dimensions of state and action variables; therefore, since multidimensional state variables are often used, the computational complexity is highly increased, which makes the application of Q-learning to complex EMS strategies a challenge. To this end, NNs have been considered in order to approximate the Q-matrix in RL. This approach is considered in Ref. [81], where a deep Q-learning-based EMS strategy is presented, based on a deep NN which is employed as the action value function. The comparison of this approach with the corresponding Q-learning method showed that the deep Q-learning approach considers more states, which resulted in a better fuel economy by 5.6% compared to the classical Q-learning approach. Another approach that targets to deal with the limitations of Q-learning in terms of low convergence rate and long computational times is the approach presented in Ref. [19], a new variant of Q-learning, named fast Q-learning is proposed, which demonstrates a faster convergence rate compared to the standard Q-learning, as well as 4.6% improvement in terms of fuel consumption. Similar extensions of Q-learning-based approaches targeting to improve the convergence rate are reported in Refs. [46,49,86]. It should be noted that in these approaches, the curse of dimensionality increases exponentially the size of the look-up table matrix with the incorporation of additional variables, thus limiting the problem to rather small action spaces, in order to deal with the computational complexity issues. To this end, novel solutions that have the capability of handling the high-dimensional state and action space in the actual decision-making process are required.

Over the last years, D-RL has gained a lot of research attention due to the ability of such algorithms to complex problems with high-dimensional sensory input. D-RL methods are combinations of classical RL and NN, where NNs offer the possibilities to expand the application of RL approaches to real-world high-dimensional optimization problems [43]. This combination offers significant advantages for the learning procedure, which comprises an offline deep NN construction phase and an online Q-learning phase, thus allowing the incorporation of further optimization criteria. In addition, D-RL approaches do not require the knowledge of the driving route a priori, while they enable the generalization of the solution to various vehicle-velocity profiles. Such an approach is developed in Ref. [93], where a deep DNN is considered in order to train the offline value functions, while the Q-learning algorithm is applied to determine the online controls, which can be adaptive to different powertrain modeling and driving conditions. Similarly, the D-RL-enabled EMS strategies presented in Ref. [43] consider various driving behaviors, where the authors pointed out the potential adaptability of the agent, which can learn from real driving data, even during the direct usage of the vehicle. Furthermore, the DRL-based method of [11] targets to equilibrate the batteries' SOC of EVs with multiple batteries, which can extend their life span and reduce their frequent maintenance. Other D-RL approaches that deal with the EMS problem can be found in Refs. [41,42,76].

The authors in Ref. [82] highlighted two main drawbacks for the application of RL methods: the first refers to the fact that the agents are optimized in a single cycle, which results in some cases a nonoptimal set of parameters when applied to different driving cycles. This limitation triggers the need for optimizing the EMS strategies on more diversified driving cycles. The second drawback refers to the limited representation of the factors that describe the driving cycles and the environment (e.g., traffic conditions). It is, therefore, crucial to represent the characteristics of the driving cycle in a realistic way. To this end, the approach of [82] incorporates traffic data collected from road sensors that are used by the RL management model, while the NN is trained with a mass of driving data that are generated from traffic simulation.

4.4 CONCLUSIONS AND FUTURE DIRECTIONS

In this chapter, we presented an extensive review of EMS for EVs and HEVs, and we classified them into three categories, namely, rule-based EMS, optimization-based EMS, and ML-based EMS. We reviewed a wide range of EMS strategies of each category and we highlighted the pros and cons of each approach as well as the promising robustness of ML-based algorithms and methods. Based on the above survey, we can focus on the challenges, advantages, and also limitations that arise from the implementation of these EMS methods, which should be addressed in order to make more efficient EMS with real-time response.

Predictive-based approaches significantly increase the performance of an EMS by providing predictions that can be updated in a real-time manner, thus enhancing the overall efficiency of the system by being able to adapt to traffic uncertainties. The major challenge of these approaches is the increase in the prediction accuracy, which can be realized by optimizing the selection of the appropriate prediction horizon, and increase the utilization of data that are provided by external sources (e.g., weather data or traffic data from ITS). On the other hand, another challenge refers to the computational complexity of these approaches, especially in MPC-based methods, which is crucial for the provision of real-time implementations of the EV's EMS.

Learning-based EMSs have significantly improved the effectiveness of the EMS strategies, where the agent is able to learn from real driving data, thus creating an intelligent system that is able to adapt to specific driving routines or any other required characteristics of the vehicle. However, future learning-based EMS strategies should be built up on the current approaches and target to further improve their capabilities, e.g., increased adaptability to the stochastic nature of the driver's behavior, integration of additional vehicle information from external sources, e.g., ITS, classification of the driving conditions (e.g., based on the duration of the trip), and reduction of the computational complexity. Interesting results could be obtained from the implementation of multiagent techniques for connected EV applications, which will consider not only the traffic/road conditions and the drivers' habits but also communication issues such as delays and quality of service. The resolution of these challenges will allow the EMS research and industry to provide better user-experience and high-quality services to the drivers.

ACKNOWLEDGMENT

This work has been partially supported by the "PROGRESSUS" project – an ECSEL Joint Undertaking – (grant 876868).

APPENDIX A

List of abbreviations

A-ECMS	Adaptive Equivalent Consumption Minimization Strategy
AER	All-Electric Range
APMP	Approximate Pontryagin's Minimum Principle
CD	Charge-Depleting
CP	Convex Programming
CS	Charge-Sustaining
DP	Dynamic Programming
DP-MPC	Dynamic Programming Model Predictive Control
D-RL	Deep Reinforcement Learning
EA	Evolutionary Algorithm
ECMS	Equivalent Consumption Minimization Strategy
EM	Electric Motor
EV	Electric Vehicle
FTMPC	Frozen-Time Model Predictive Control
GA	Genetic Algorithm
GIS	Geographical Information System
GPS	Global Positioning System
ICE	Internal Combustion Engine
ITS	Intelligent Transportation System
MDP	Markov Decision Process
ML	Machine-Learning
MPC	Model-Predictive Control
NN	Neural Network
PHEV	Plug-in Hybrid Electric Vehicle
PMP	Pontryagin's Minimum Principle
PMPC	Prescient Model Predictive Control
PSO	Particle Swarm Optimization
QP	Quadratic Programming
QSA	Quantum-inspired Simulated Annealing
RBF	Radial Basis Function
RL	Reinforcement Learning
SA	Simulated Annealing
SA-PSO	Simulated Annealing Particle Swarm Optimization
SMPC	Stochastic Model Predictive Control
SoC	State-of-Charge
WT	Wavelet Transform

REFERENCES

1. Rustem Abdrakhmanov and Lounis Adouane. Efficient ACC with Stop & Go maneuvers for hybrid vehicle with online sub-optimal energy management. In *2017 11th International Workshop on Robot Motion and Control (RoMoCo)*, Wasovo Palace, Poland, pp. 7–14, IEEE, 2017.
2. Ahmed M. Ali, Alhossein Mostafa Sharaf, Hesham Kamel, and Shawky Hegazy. A theo-practical methodology for series hybrid vehicles evaluation and development. Technical Report, SAE Technical Paper, 2017.
3. Ahmed M. Ali and Dirk Söffker. Towards optimal power management of hybrid electric vehicles in real-time: A review on methods, challenges, and state-of-the-art solutions. *Energies*, 11(3):476, 2018.
4. Ethem Alpaydin. *Introduction to Machine Learning*. MIT Press, Cambridge, MA, 2020.
5. Kadir Amasyali and Nora M. El-Gohary. A review of data-driven building energy consumption prediction studies. *Renewable and Sustainable Energy Reviews*, 81:1192–1205, 2018.
6. Rishikesh Mahesh Bagwe, Andy Byerly, Euzeli Cipriano dos Santos, and Zina Ben-Miled. Adaptive rule-based energy management strategy for a parallel HEV. *Energies*, 12(23):4472, 2019.
7. Harpreetsingh Banvait, Sohel Anwar, and Yaobin Chen. A rule-based energy management strategy for plug-in hybrid electric vehicle (PHEV). In *2009 American control conference*, Saint Louis, MO, pp. 3938–3943, IEEE, 2009.
8. Hoseinali Borhan, Ardalan Vahidi, Anthony M. Phillips, Ming L. Kuang, Ilya V. Kolmanovsky, and Stefano Di Cairano. MPC-based energy management of a power-split hybrid electric vehicle. *IEEE Transactions on Control Systems Technology*, 20(3):593–603, 2011.
9. Hippolyte Bouvier, Guillaume Colin, and Yann Chamaillard. Determination and comparison of optimal eco-driving cycles for hybrid electric vehicles. In *2015 European Control Conference (ECC)*, pp. 142–147, IEEE, 2015.
10. Kevin R. Bouwman, Thinh H. Pham, Steven Wilkins, and Theo Hofman. Predictive energy management strategy including traffic flow data for hybrid electric vehicles. *IFAC-Papers Online*, 50(1):10046–10051, 2017.
11. Hicham Chaoui, Hamid Gualous, Loic Boulon, and Sousso Kelouwani. Deep reinforcement learning energy management system for multiple battery based electric vehicles. In *2018 IEEE Vehicle Power and Propulsion Conference (VPPC)*, Chicago, IL, pp. 1–6, IEEE, 2018.
12. Zeyu Chen, Rui Xiong, and Jiayi Cao. Particle swarm optimization-based optimal power management of plug-in hybrid electric vehicles considering uncertain driving conditions. *Energy*, 96:197–208, 2016.
13. Zeyu Chen, Rui Xiong, Kunyu Wang, and Bin Jiao. Optimal energy management strategy of a plug-in hybrid electric vehicle based on a particle swarm optimization algorithm. *Energies*, 8(5):3661–3678, 2015.
14. Zhanghui Chen and Ping Luo. Qisa: Incorporating quantum computation into simulated annealing for optimization problems. In *2011 IEEE Congress of Evolutionary Computation (CEC)*, New Orleans, LA, pp. 2480–2487, IEEE, 2011.
15. Zheng Chen, Ningyuan Guo, Jiangwei Shen, Renxin Xiao, and Peng Dong. A hierarchical energy management strategy for power-split plug-in hybrid electric vehicles considering velocity prediction. *IEEE Access*, 6:33261–33274, 2018.
16. Zheng Chen, Chris Chunting Mi, Rui Xiong, Jun Xu, and Chenwen You. Energy management of a power-split plug-in hybrid electric vehicle based on genetic algorithm and quadratic programming. *Journal of Power Sources*, 248:416–426, 2014.
17. Nicolas Denis, Maxime R. Dubois, and Alain Desrochers. Fuzzy-based blended control for the energy management of a parallel plug-in hybrid electric vehicle. *IET Intelligent Transport Systems*, 9(1):30–37, 2014.
18. Stefano Di Cairano, Daniele Bernardini, Alberto Bemporad, and Ilya V. Kolmanovsky. Stochastic mpc with learning for driver-predictive vehicle control and its application to HEV energy management. *IEEE Transactions on Control Systems Technology*, 22(3):1018–1031, 2013.
19. Guodong Du, Yuan Zou, Xudong Zhang, Zehui Kong, Jinlong Wu, and Dingbo He. Intelligent energy management for hybrid electric tracked vehicles using online reinforcement learning. *Applied Energy*, 251:113388, 2019.
20. Benming Duan, Qingnian Wang, Xiaohua Zeng, Yinsheng Gong, Dafeng Song, and Junnian Wang. Calibration methodology for energy management system of a plug-in hybrid electric vehicle. *Energy Conversion and Management*, 136:240–248, 2017.

21. Sebastian East and Mark Cannon. An admm algorithm for MPC-based energy management in hybrid electric vehicles with nonlinear losses. In *2018 IEEE Conference on Decision and Control (CDC)*, Miami, FL, pp. 2641–2646. IEEE, 2018.
22. Bo Egardt, Nikolce Murgovski, Mitra Pourabdollah, and Lars Johannesson Mardh. Electromobility studies based on convex optimization: Design and control issues regarding vehicle electrification. *IEEE Control Systems Magazine*, 34(2):32–49, 2014.
23. Masood Ghasemi and Xingyong Song. Powertrain energy management for autonomous hybrid electric vehicles with flexible driveline power demand. *IEEE Transactions on Control Systems Technology*, 27(5):2229–2236, 2018.
24. Jeffrey Gonder and Tony Markel. Energy management strategies for plug-in hybrid electric vehicles. Technical Report, SAE Technical Paper, 2007.
25. Ferit Hacioglu, Ilke Altin, Ozgur Aktekin, and Ahmet Sakalli. Predictive rule based optimization techniques for series hybrid electric vehicle. In *2017 25th Mediterranean Conference on Control and Automation (MED)*, Valletta, Malta, pp. 192–197, IEEE, 2017.
26. S. Hadj-Said, G. Colin, A. Ketfi-Cherif, and Y. Chamaillard. Convex optimization for energy management of parallel hybrid electric vehicles. *IFAC-Papers Online*, 49(11):271–276, 2016.
27. Shaojian Han, Fengqi Zhang, and Junqiang Xi. A real-time energy management strategy based on energy prediction for parallel hybrid electric vehicles. *IEEE Access*, 6:70313–70323, 2018.
28. Georges R. Harik, Fernando G. Lobo, and David E. Goldberg. The compact genetic algorithm. *IEEE Transactions on Evolutionary Computation*, 3(4):287–297, 1999.
29. Cong Hou, Minggao Ouyang, Liangfei Xu, and Hewu Wang. Approximate pontryagin's minimum principle applied to the energy management of plug-in hybrid electric vehicles. *Applied Energy*, 115:174–189, 2014.
30. Mattias Hovgard, Oscar Jonsson, Nikolce Murgovski, Martin Sanfridson, and Jonas Fredriksson. Cooperative energy management of electrified vehicles on hilly roads. *Control Engineering Practice*, 73:66–78, 2018.
31. Xiaosong Hu, Nikolce Murgovski, Lars Johannesson, and Bo Egardt. Energy efficiency analysis of a series plug-in hybrid electric bus with different energy management strategies and battery sizes. *Applied Energy*, 111:1001–1009, 2013.
32. Xiaosong Hu, Changfu Zou, Caiping Zhang, and Yang Li. Technological developments in batteries: A survey of principal roles, types, and management needs. *IEEE Power and Energy Magazine*, 15(5):20–31, 2017.
33. Hans Jeekel. *The Car-Dependent Society: A European Perspective*. Routledge, Abingdon, 2016.
34. Guo Jinquan, He Hongwen, Peng Jiankun, and Zhou Nana. A novel MPC-based adaptive energy management strategy in plug-in hybrid electric vehicles. *Energy*, 175:378–392, 2019.
35. Namwook Kim, Aymeric Rousseau, and Daeheung Lee. A jump condition of pmp-based control for phevs. *Journal of Power Sources*, 196(23):10380–10386, 2011.
36. Slawomir Koziel and Xin-She Yang. *Computational Optimization, Methods and Algorithms*, vol. 356. Springer, Berlin, 2011.
37. Zhenzhen Lei, Dong Cheng, Yonggang Liu, Datong Qin, Yi Zhang, and Qingbo Xie. A dynamic control strategy for hybrid electric vehicles based on parameter optimization for multiple driving cycles and driving pattern recognition. *Energies*, 10(1):54, 2017.
38. Zhenzhen Lei, Dongye Sun, Junjun Liu, Daqi Chen, Yonggang Liu, and Zheng Chen. Trip-oriented model predictive energy management strategy for plug-in hybrid electric vehicles. *IEEE Access*, 7: 113771–113785, 2019.
39. Ji Li, Quan Zhou, Yinglong He, Bin Shuai, Ziyang Li, Huw Williams, and Hongming Xu. Dual-loop online intelligent programming for driver-oriented predict energy management of plug-in hybrid electric vehicles. *Applied Energy*, 253:113617, 2019.
40. Tianyu Li, Huiying Liu, and Daolin Ding. Predictive energy management of fuel cell supercapacitor hybrid construction equipment. *Energy*, 149:718–729, 2018.
41. Yuecheng Li, Hongwen He, Amir Khajepour, Hong Wang, and Jiankun Peng. Energy management for a power-split hybrid electric bus via deep reinforcement learning with terrain information. *Applied Energy*, 255:113762, 2019.
42. Yuecheng Li, Hongwen He, Jiankun Peng, and Hong Wang. Deep reinforcement learning-based energy management for a series hybrid electric vehicle enabled by history cumulative trip information. *IEEE Transactions on Vehicular Technology*, 68(8):7416–7430, 2019.

43. Roman Liessner, Christian Schroer, Ansgar Malte Dietermann, and Bernard Bäker. Deep reinforcement learning for advanced energy management of hybrid electric vehicles. In *ICAART (2)*, Funchal, Portugal, pp. 61–72, 2018.
44. Xue Lin, Paul Bogdan, Naehyuck Chang, and Massoud Pedram. Machine learning-based energy management in a hybrid electric vehicle to minimize total operating cost. In *2015 IEEE/ACM International Conference on Computer-Aided Design (ICCAD)*, pp. 627–634, IEEE, 2015.
45. Xue Lin, Yanzhi Wang, Paul Bogdan, Naehyuck Chang, and Massoud Pedram. Reinforcement learning based power management for hybrid electric vehicles. In *2014 IEEE/ACM International Conference on Computer-Aided Design (ICCAD)*, pp. 33–38, IEEE, 2014.
46. Chang Liu and Yi Lu Murphey. Optimal power management based on q-learning and neuro-dynamic programming for plug-in hybrid electric vehicles. *IEEE Transactions on Neural Networks and Learning Systems*, 99:1–33, 2019.
47. Peng Liu, Arda Kurt, and Umit Ozguner. Synthesis of a behavior-guided controller for lead vehicles in automated vehicle convoys. *Mechatronics*, 50:366–376, 2018.
48. Teng Liu and Xiaosong Hu. A bi-level control for energy efficiency improvement of a hybrid tracked vehicle. *IEEE Transactions on Industrial Informatics*, 14(4):1616–1625, 2018.
49. Teng Liu, Xiaosong Hu, Weihao Hu, and Yuan Zou. A heuristic planning reinforcement learning-based energy management for power-split plug-in hybrid electric vehicles. *IEEE Transactions on Industrial Informatics*, 15(12):6436–6445, 2019.
50. Teng Liu, Xiaosong Hu, Shengbo Eben Li, and Dongpu Cao. Reinforcement learning optimized look-ahead energy management of a parallel hybrid electric vehicle. *IEEE/ASME Transactions on Mechatronics*, 22(4):1497–1507, 2017.
51. Teng Liu, Huilong Yu, Hongyan Guo, Yechen Qin, and Yuan Zou. Online energy management for multimode plug-in hybrid electric vehicles. *IEEE Transactions on Industrial Informatics*, 15(7):4352–4361, 2018.
52. Karol Lina López, Christian Gagné, and Marc-André Gardner. Demand-side management using deep learning for smart charging of electric vehicles. *IEEE Transactions on Smart Grid*, 10(3):2683–2691, 2018.
53. Ngoc An Luu, Quoc-Tuan Tran, and Seddik Bacha. Optimal energy management for an Island microgrid by using dynamic programming method. In *2015 IEEE Eindhoven PowerTech*, Eindhoven, The Netherlands, pp. 1–6, IEEE, 2015.
54. Clara Marina Martinez, Xiaosong Hu, Dongpu Cao, Efstathios Velenis, Bo Gao, and Matthias Wellers. Energy management in plug-in hybrid electric vehicles: Recent progress and a connected vehicles perspective. *IEEE Transactions on Vehicular Technology*, 66(6):4534–4549, 2016.
55. Amir Mosavi, Mohsen Salimi, Sina Faizollahzadeh Ardabili, Timon Rabczuk, Shahaboddin Shamshirband, and Annamaria R. Varkonyi-Koczy. State of the art of machine learning models in energy systems, a systematic review. *Energies*, 12(7):1301, 2019.
56. Scott Jason Moura, Hosam K. Fathy, Duncan S. Callaway, and Jeffrey L. Stein. A stochastic optimal control approach for power management in plug-in hybrid electric vehicles. *IEEE Transactions on Control Systems Technology*, 19(3):545–555, 2010.
57. Tobias Nüesch, Philipp Elbert, Michael Flankl, Christopher Onder, and Lino Guzzella. Convex optimization for the energy management of hybrid electric vehicles considering engine start and gearshift costs. *Energies*, 7(2):834–856, 2014.
58. Mitra Pourabdollah, Nikolce Murgovski, Anders Grauers, and Bo Egardt. An iterative dynamic programming/convex optimization procedure for optimal sizing and energy management of PHEVs. *IFAC Proceedings Volumes*, 47(3):6606–6611, 2014.
59. Xuewei Qi, Guoyuan Wu, Kanok Boriboonsomsin, and Matthew J. Barth. Development and evaluation of an evolutionary algorithm-based online energy management system for plug-in hybrid electric vehicles. *IEEE Transactions on Intelligent Transportation Systems*, 18(8):2181–2191, 2016.
60. Xuewei Qi, Guoyuan Wu, Kanok Boriboonsomsin, Matthew J. Barth, and Jeffrey Gonder. Data-driven reinforcement learning–based real-time energy management system for plug-in hybrid electric vehicles. *Transportation Research Record*, 2572(1):1–8, 2016.
61. Mehdi Rahmani-Andebili. Optimal placement and sizing of parking lots for the plug-in electric vehicles considering the technical, social, and geographical aspects. In *Planning and Operation of Plug-In Electric Vehicles, Ed. Mehdi Rahmani-Andebili*, pp. 149–209. Springer, Berlin, 2019.

62. Mehdi Rahmani-Andebili and Haiying Shen. Traffic and grid-based parking lot allocation for pevs considering driver behavioral model. In *2017 International Conference on Computing, Networking and Communications (ICNC)*, Silicon Valley, CA, pp. 599–603, IEEE, 2017.
63. Mehdi Rahmani-Andebili, Haiying Shen, and Mahmud Fotuhi-Firuzabad. Planning and operation of parking lots considering system, traffic, and drivers behavioral model. *IEEE Transactions on Systems, Man, and Cybernetics: Systems*, 49(9):1879–1892, 2018.
64. Razaullah Razaullah, Iftikhar Hussain, and Shahid Maqsood. The impact of supply chain network optimization on total cost and environment. In *2017 International Conference on Industrial Engineering, Management Science and Application*, Seoul, Korea, IEEE, 2017.
65. M.F.M. Sabri, K.A. Danapalasingam, and M.F. Rahmat. A review on hybrid electric vehicles architecture and energy management strategies. *Renewable and Sustainable Energy Reviews*, 53:1433–1442, 2016.
66. Phillip B. Sharer, Aymeric Rousseau, Dominik Karbowski, and Sylvain Pagerit. Plug-in hybrid electric vehicle control strategy: Comparison between EV and charge-depleting options. Technical Report, SAE Technical Paper, 2008.
67. Chao Sun, Hongwen He, and Fengchun Sun. The role of velocity forecasting in adaptive-ecms for hybrid electric vehicles. *Energy Procedia*, 75:1907–1912, 2015.
68. Chao Sun, Hongwen He, and Fengchun Sun. The role of velocity forecasting in adaptive-ECMS for hybrid electric vehicles. *Energy Procedia*, 75:1907–1912, 2015.
69. Chao Sun, Xiaosong Hu, Scott J Moura, and Fengchun Sun. Velocity predictors for predictive energy management in hybrid electric vehicles. *IEEE Transactions on Control Systems Technology*, 23(3): 1197–1204, 2014.
70. Chao Sun, Scott Jason Moura, Xiaosong Hu, J. Karl Hedrick, and Fengchun Sun. Dynamic traffic feedback data enabled energy management in plug-in hybrid electric vehicles. *IEEE Transactions on Control Systems Technology*, 23(3):1075–1086, 2014.
71. Chao Sun, Fengchun Sun, and Hongwen He. Investigating adaptive-ECMS with velocity forecast ability for hybrid electric vehicles. *Applied Energy*, 185:1644–1653, 2017.
72. He Tian, Xu Wang, Ziwang Lu, Yong Huang, and Guangyu Tian. Adaptive fuzzy logic energy management strategy based on reasonable SOC reference curve for online control of plug-in hybrid electric city bus. *IEEE Transactions on Intelligent Transportation Systems*, 19(5):1607–1617, 2017.
73. Mahyar Vajedi, Maryyeh Chehrehsaz, and Nasser L. Azad. Intelligent power management of plug-in hybrid electric vehicles, part I: Real-time optimum SOC trajectory builder. *International Journal of Electric and Hybrid Vehicles*, 6(1):46–67, 2014.
74. John S. Vardakas. Electric vehicles charging management in communication controlled fast charging stations. In *2014 IEEE 19th International Workshop on Computer Aided Modeling and Design of Communication Links and Networks (CAMAD)*, Athens, Greece, pp. 115–119, IEEE, 2014.
75. Hong Wang, Yanjun Huang, Amir Khajepour, Hongwen He, and Dongpu Cao. A novel energy management for hybrid off-road vehicles without future driving cycles as a priori. *Energy*, 133:929–940, 2017.
76. Pengyue Wang, Yan Li, Shashi Shekhar, and William F. Northrop. A deep reinforcement learning framework for energy management of extended range electric delivery vehicles. In *2019 IEEE Intelligent Vehicles Symposium (IV)*, Paris, France, pp. 1837–1842, IEEE, 2019.
77. Ximing Wang, Hongwen He, Fengchun Sun, and Jieli Zhang. Application study on the dynamic programming algorithm for energy management of plug-in hybrid electric vehicles. *Energies*, 8(4):3225–3244, 2015.
78. Hanbing Wei, Yao Chen, and Zhiyuan Peng. Costate estimation of PMP-based control strategy for PHEV using Legendre Pseudospectral method. *Mathematical Problems in Engineering*, 2016:1–9, 2016.
79. Maciej Wieczorek and Miroslaw Lewandowski. A mathematical representation of an energy management strategy for hybrid energy storage system in electric vehicle and real time optimization using a genetic algorithm. *Applied Energy*, 192:222–233, 2017.
80. Guoyuan Wu, Kanok Boriboonsomsin, and Matthew J. Barth. Development and evaluation of an intelligent energy-management strategy for plug-in hybrid electric vehicles. *IEEE Transactions on Intelligent Transportation Systems*, 15(3):1091–1100, 2014.
81. Jingda Wu, Hongwen He, Jiankun Peng, Yuecheng Li, and Zhanjiang Li. Continuous reinforcement learning of energy management with deep q network for a power split hybrid electric bus. *Applied Energy*, 222:799–811, 2018.

82. Yuankai Wu, Huachun Tan, Jiankun Peng, Hailong Zhang, and Hongwen He. Deep reinforcement learning of energy management with continuous control strategy and traffic information for a series-parallel plug-in hybrid electric bus. *Applied Energy*, 247:454–466, 2019.
83. Renxin Xiao, Baoshuai Liu, Jiangwei Shen, Ningyuan Guo, Wensheng Yan, and Zheng Chen. Comparisons of energy management methods for a parallel plug-in hybrid electric vehicle between the convex optimization and dynamic programming. *Applied Sciences*, 8(2):218, 2018.
84. Shaobo Xie, Xiaosong Hu, Teng Liu, Shanwei Qi, Kun Lang, and Huiling Li. Predictive vehicle-following power management for plug-in hybrid electric vehicles. *Energy*, 166:701–714, 2019.
85. Shaobo Xie, Xiaosong Hu, Zongke Xin, and Liang Li. Time-efficient stochastic model predictive energy management for a plug-in hybrid electric bus with an adaptive reference state-of-charge advisory. *IEEE Transactions on Vehicular Technology*, 67(7):5671–5682, 2018.
86. Bin Xu, Dhruvang Rathod, Darui Zhang, Adamu Yebi, Xueyu Zhang, Xiaoya Li, and Zoran Filipi. Parametric study on reinforcement learning optimized energy management strategy for a hybrid electric vehicle. *Applied Energy*, 259:114200, 2020.
87. Yanli Yin, Yan Ran, Liufeng Zhang, Xiaoliang Pan, and Yong Luo. An energy management strategy for a super-mild hybrid electric vehicle based on a known model of reinforcement learning. *Journal of Control Science and Engineering*, 2019:12, 2019.
88. Jingni Yuan and Lin Yang. Predictive energy management strategy for connected 48v hybrid electric vehicles. *Energy*, 187:115952, 2019.
89. Yuping Zeng, Jing Sheng, and Ming Li. Adaptive real-time energy management strategy for plug-in hybrid electric vehicle based on simplified-ecms and a novel driving pattern recognition method. *Mathematical Problems in Engineering*, 2018:12, 2018.
90. Ioannis Zenginis, John Vardakas, Nizar Zorba, and Christos Verikoukis. Performance evaluation of a multi-standard fast charging station for electric vehicles. *IEEE Transactions on Smart Grid*, 9(5): 4480–4489, 2017.
91. Pei Zhang, Fuwu Yan, and Changqing Du. A comprehensive analysis of energy management strategies for hybrid electric vehicles based on bibliometrics. *Renewable and Sustainable Energy Reviews*, 48:88–104, 2015.
92. Yuanjian Zhang, Liang Chu, Zicheng Fu, Nan Xu, Chong Guo, Xuezhao Zhang, Zhouhuan Chen, and Peng Wang. Optimal energy management strategy for parallel plug-in hybrid electric vehicle based on driving behavior analysis and real time traffic information prediction. *Mechatronics*, 46:177–192, 2017.
93. Pu Zhao, Yanzhi Wang, Naehyuck Chang, Qi Zhu, and Xue Lin. A deep reinforcement learning framework for optimizing fuel economy of hybrid electric vehicles. In *2018 23rd Asia and South Pacific Design Automation Conference (ASP-DAC)*, Jeju Island , Korea, pp. 196–202, IEEE, 2018.
94. Quan Zhou, Ji Li, Bin Shuai, Huw Williams, Yinglong He, Ziyang Li, Hongming Xu, and Fuwu Yan. Multi-step reinforcement learning for model-free predictive energy management of an electrified off-highway vehicle. *Applied Energy*, 255:113755, 2019.
95. Yang Zhou, Alexandre Ravey, and Marie-Cécile Péra. A survey on driving prediction techniques for predictive energy management of plug-in hybrid electric vehicles. *Journal of Power Sources*, 412: 480–495, 2019.
96. Yuan Zou, Zehui Kong, Teng Liu, and Dexing Liu. A real-time markov chain driver model for tracked vehicles and its validation: Its adaptability via stochastic dynamic programming. *IEEE Transactions on Vehicular Technology*, 66(5):3571–3582, 2016.

5 Dynamic Road Management in the Era of CAV

Mohamed Younis
University of Maryland

Sookyoung Lee
Ewha Womans University

Wassila Lalouani, Dayuan Tan, and Sanket Gupte
University of Maryland

CONTENTS

5.1 INTRODUCTION

5.1.1 ROAD TRAFFIC PROBLEMS

Vehicular traffic congestion has become a daily problem that most people suffer from, especially in urban areas. With the increasing number of vehicles on the roads, slow traffic and congestion have become an unpleasant expectation during daily commutes. According to the nationwide insurance company, "The average urban commuter is stuck in traffic for 34 hours every year and 1.9 billion gallons of fuel, more than five days' worth of the total daily fuel consumption in the United States were wasted due to road congestion" [1]. Worldwide, an average commuter spent an extra 100 hours a year traveling during the evening rush hour alone in 2014, and the number hits 272 hours in 2018 in the Columbia capital, Bogota, where commuters experience the world's greatest traffic jams [2,3]. Moreover, in the largest urban areas across the United States, commuters consume nearly 7 full working days in extra traffic delay in 2017, which is equivalent to over $1000 in personal costs [4]. This not only impacts productivity but also poses a safety hazard. In addition, traffic congestion causes excessive fuel consumption and high doses of pollution, which adds to the negative economic impact on the nation and potential health risks for citizens. According to the 2019 Urban Mobility Report of the Texas A&M Transportation Institute [2], congestion has been persistently growing and is not restricted to large metropolitan areas as indicated by the statistics in Figure 5.1, and an average auto commuter spends 54 hours in congestion and wastes 21 gallons for fuel due to traffic congestion, which translate to $1010 of congestion cost per auto commuter. In 2017, the overall congestion cost in urban areas is about $166 billion due to the extra 8.8 billion hours trip which requires purchase of an extra 3.3 billion gallons of fuel.

5.1.2 CONVENTIONAL AND EMERGING CONGESTION MITIGATION METHODOLOGIES

The transportation community has come to realize that balancing the traffic load on existing roads is a key for effective mitigation of congestion [5]. However, the steps taken by authorities are constrained either by driver's response, e.g., by fostering ride sharing and encouraging the use of mass transient [6–9], or by employing measures that do not adapt based on real-time conditions, e.g., impose a road use pattern following a static schedule that is based on time of the day and day of the week. For example, many cities throughout the U.S. designate high-occupancy vehicle (HOV) lanes and charge tolls to cross tunnels and bridges and to drive over certain segments of highways. Yet, these measures prove ineffective as the driver's choice to avoid expensive routes is more or less unpredictable, as it is difficult to make an educated decision on trading-off time and cost while being behind the wheel. Other cities, such as Washington DC, change the direction of lanes on certain roads during rush hours, while others (e.g., Minneapolis and Seattle) control the vehicle entry rate to highways through ramp metering during these hours. Again, this static schedule factor in road capacity is the main cause of congestion and does not factor in all incidents, traffic light

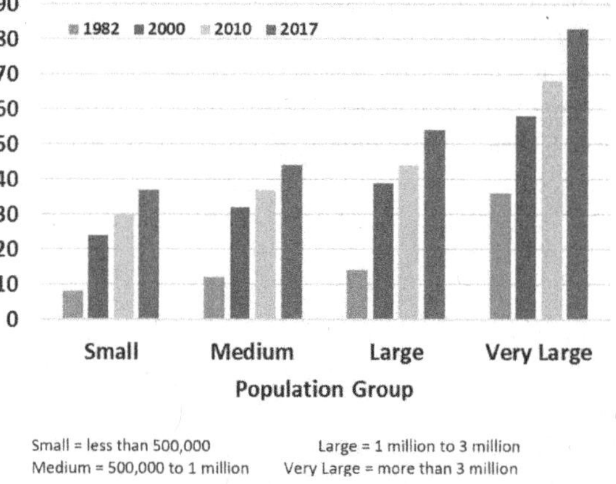

Figure 5.1 With all measures to alleviate traffic congestion, little progress has been made; the effect has stabilized rather than reduced. (Plot is from Ref. [4].)

timing, weather, the effect of these measures on local roads, etc. In addition, conventional means for alerting drivers (e.g., signs and radio updates) lack responsiveness and do not mitigate or prevent traffic congestion.

Given the shortcoming of static schemes, the notion of active (dynamic) traffic management has gained lots of attention in recent years. The key features of the dynamic traffic management (DTM) paradigm are exploiting interaction with the drivers (or vehicles) to predict traffic jams and proactively employ means to avoid them. The examples of DTM-based congestion mitigation schemes include the following [10]:

- *Adjustable Shoulder Use:* converting road shoulder into a lane in response to congestion or accident in order to increase throughput.
- *Varying Speed Limits:* setting the speed limit based on the road condition and vehicle density.
- *Adaptive Ramp Metering:* adjusting the timing of traffic signals at ramp entrances to control vehicle inflow to highways.
- *Dynamic Rerouting*: directing traffic to route alternatives in order to prevent congestion.
- *Adaptive Traffic Signal Timing:* varying the traffic light timing and/or phases to improve throughput and delay at an intersection.

A shared characteristic among these unconventional DTM based schemes is that they involve some form of road reconfiguration, which is a revolutionary view of such major infrastructure.

5.1.3 CONNECTED VEHICLES AND INFRASTRUCTURE

In the early transportation system, a wide variety of traffic monitoring technologies using some sensing technologies such as safety CCTV, traffic video cameras, piezo-electric sensors, inductive loops have been introduced to monitor road conditions and alert motorists through electronic variable-message signs. However, due to the lack of sufficient coverage and high maintenance cost, the transportation systems have evolved by using various types of wireless and mobile technologies such as 2G/3G/4G/LTE/5G, Wi-Fi, GPS, etc. for real-time traffic monitoring [11–13]. Then, intelligent transportation systems (ITSs) have emerged with the advances of information and communication technology, and the prospect of leveraging recent developments in the vehicular ad hoc network (VANET) and wireless sensor network (WSN) areas. Furthermore, it is highly expected in academia and automotive industry that fully autonomous vehicles (AVs) will be fulfilled between 2025 and 2030, and their global dispersion becomes fact between 2030 and 2040 [14–15]. The impact of AVs includes reduced traffic; infrastructure saving; such as parking congestion, increased safety, energy conservation, and pollution reductions; and independent mobility for low-income people [16].

Moreover, the remarkable research results coming from the fields of in-vehicle digital technology, wireless communication, embedded systems, intelligent routing system, sensors and ad-hoc technologies have given rise to the emergence and evolution of connected autonomous vehicles (CAVs). The advent of CAV will lead to a paradigm shift of automobile design from an old-fashioned source of repositioning into a full-scale, smart, and infotainment-rich computing and commuting device. In contrast to human-driven vehicles, CAVs cooperatively share the road that they travel on and can thus be controlled to adaptively handle increased vehicle density and be provided with routes to dynamically optimize the delay for the individual travelers and the vehicular throughput on the road network. Consequently, the arrival of CAV will change the model for how road traffic will be managed and how congestion could be mitigated, and will eventually provide travelers with more safe, accurate, timely decision during a road trip reducing human errors and life-threatening situation on the road [17,18]. In other words, CAV will enable the full realization of the DTM concept.

5.1.4 SCOPE AND ORGANIZATION

This chapter introduces the reader to the notion of DTM in the context of ITS. Particularly, the complications in realizing the full potential of DTM are discussed and how CAV can be instrumental in overcoming these complications. We highlight the various wireless communication technologies for supporting vehicle-to-vehicle (V2V) and vehicle-to-infrastructure (V2I) interaction. We then analyze and enumerate the attributes that can be shared through CAV with the road management infrastructure in order to enable dynamic adaptation of the road configuration as a means for optimizing traffic flow and improving both traveler-centric and system-based performance metrics. Existing CAV-enabled adaptive road reconfiguration techniques are categorized into five groups, namely, autonomous intersection management (AIM), adaptive traffic light control (ATLC), dynamic lane grouping (DLG), dynamic lane reversal (DLR), and dynamic trajectory planning (DTP). We describe each category in detail and compare the various techniques. We further highlight the issues when autonomous and driver-based vehicles coexist on the road and the impact of such a mix on the various road management strategies. Finally, we present our vision for how vehicular traffic will be managed in smart cities and discuss our *I*nternet of *R*adio-equipped *O*n-road and vehicles-carried *A*gile *D*evices (iRoad) project for realizing such a vision. We also report on the results of some of our on-going research and outline future research topics that warrant more investigation.

The chapter is organized as follows. The next section highlights the challenges in implementing DTM. Section 5.3 focuses on how CAV can facilitate DTM and categorizes existing techniques in that regard. In Section 5.4, we highlight the challenge of incorporating DTM in the presence of human-driven vehicles on the road and discuss efforts within our iRoad project for overcoming these challenges. Finally, Section 5.5 concludes the chapter and outlines open-research problems.

5.2 DTM CHALLENGES

As pointed out earlier, DTM opts to respond to incidents and/or congestion in real time. In essence, DTM models a road network as a closed-loop cyber-physical system and ideally provisions the means for autonomous control of such a system. In this section, we highlight the challenges in realizing DTM in practice. The main issues are discussed in the following sections.

5.2.1 DATA COLLECTION

To respond to road incidents and congestion, the status of traffic has to be continually tracked. Traffic condition assessment methodologies can be categorized based on the accuracy of the data collection and on what the road status is needed for. From the vehicle (driver or passenger) perspective, the data are used to detect congestion, estimate arrival time, and decide on the best travel route. On the other hand, a local branch of the department of transportation will be interested in using the data for predicting traffic jams and deadlocks, performing analytics to measure utilization, and assessing criticality of road infrastructure, providing alerts, and diverting vehicles to alternate routes if needed. There are quite a few traffic monitoring systems that gather and disseminate traffic information. These systems can be classified as (1) infrastructure-based systems that are installed and controlled by the authority, and (2) participatory where the data are voluntarily provided by participants or indirectly inferred from other context. Examples of infrastructure-based monitoring systems are on-road sensors, e.g., traffic cameras, loop detectors, laser sensors, and pressure hose. Traffic cameras are the most popular on-road sensors where not only an administrator can get a visual view of the conditions but also computer vision techniques can be applied to recognize and count vehicles in the live video [19,20]. Electromagnetic loops and laser sensors are popular at intersections and are used to determine the traffic signal sequence [21]. Radar is also used at intersections for not only detecting vehicles but also counting them so that the signal timing and sequence are optimally adjusted [22]. Pressure hoses [23] are typically used during field studies to count vehicles and estimate traffic intensity on a road segment during certain duration; generally, they are not durable and not intended for real-time monitoring. Overall, infrastructure-based monitoring systems are expensive, mainly because of the installation cost.

Participatory systems, on the other hand, either (1) exploit the popularity and recent advances in wireless technologies, (2) collect location and contextual data that are voluntarily provided by drivers, or (3) leverage the wealth of vehicle's onboard sensors [24]. For example, Zhang et al. [25] utilize wireless mesh networks to track the movement of specific vehicles; these vehicles are roaming the roads and responding to wireless probes. By localizing the probe responses using mesh relays, the vehicles can be located and their motion pattern and delay can then be correlated to estimate the conditions of the traveled roads. Similarly, routinely traveling vehicles such as buses and taxis are utilized in [26] to report on-road traffic. Prime examples of traffic monitoring systems that rely on voluntarily provided data are Google maps, Waze [27], Inrix [28], and Cellint [29], where the location of mobile individuals is determined through the GPS on their cell phones or portable computing devices while riding their vehicles. By correlating the location information, the motion speed and vehicle density are estimated to infer traffic conditions on road segments [30], and queue lengths are calculated at intersections [31]. Some work even assumes that a traveler will voluntarily provide the primary route and alternatives, which enables predicting vehicle density over time [32]. Meanwhile, systems such as Traffic View [33,34] and SOTIS [35] rely on intervehicle communication in collecting data. Some work has focused on just detecting congestion [36–50]. The detection methodologies vary from a simple monitoring of motion speed [36] to conducting simulation using mobility traces [42,43] or applying fuzzy logic and generic algorithms [37,44–50]. Some approaches try to devise a congestion prediction model as well [51–53]. Finally, multiple modalities have been exploited to improve the fidelity of the traffic assessment. For example, bluetooth Mac scanner is exploited in [54] as an extra modality to boost the accuracy and reliability of the traffic flow measurements made by loop detectors. Other systems, e.g., [38], assume the availability of a high-level traffic report and use the vehicle's local observations for fine-grained assessment, e.g., by checking the travel speed of other vehicles. Figure 5.2 provides a summary of the popular means for collecting traffic data.

5.2.2 ROAD CONFIGURATION

The notion of road parameters' reconfiguration is analogous to adjusting the capacity of and controlling the flow over links in communication networks. By modeling the road infrastructure as a network, one may apply the well-established graph theoretical algorithms to study the performance and predict problems. Particularly, applying network flow analysis techniques will enable estimating

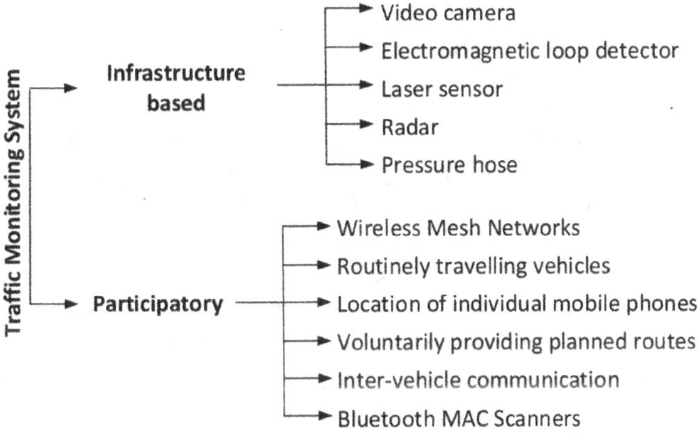

Figure 5.2 Categorization of the data collection methodologies and underlying means.

throughput, identifying bottlenecks, and determining best means for stabilizing the operation and maximizing performance. Road configurability is realized in practice by means such as (1) traffic signal timing, (2) ramp metering, (3) tolls, (4) speed limit, (5) HOV lanes, (6) shoulder, and (7) contraflow lanes. The first four, namely, signal timing, ramp metering, tolls, and speed limit, mainly control the flow to reduce delay and are exploited to prevent congestion in certain travel direction. For example, ramp metering is used to control the inflow to a highway from local roads in order to mitigate slowdown when the vehicles merge after entering the highway and also avoid exceeding the highway capacity. On the other hand, shoulder use and lane reversal boost the capacity of the road, and in case of lane reversal, the increase will be at the expense of reduced capacity in the opposite direction; the goal is to increase vehicle throughput and, consequently, passenger throughput. Meanwhile, the designation of HOV lanes opts to improve passenger throughput only.

Road configuration parameters are currently set based on time of the day and day of the week. Typically, statistics for traffic intensity are used to determine the expected conditions, and consequently, what values are to be assigned to the various parameters in order to optimize contemporary metrics such as vehicular throughput and delay. Safety is also factored in, particularly when it comes to intersection crossing and speed limit settings. The statistics are based on historical data collected during normal circumstances, i.e., in the absence of traffic incidents such as accidents. To realize DTM, road reconfiguration is to be exploited autonomously and in real time based on the traffic status and trend. In other words, the road parameters have to be adaptively adjusted to cope with variations in the traffic patterns. Such an approach will enable effective handling of emerging events that are often experienced sporadically with no predictable patterns. To elaborate, collisions and vehicle breakdown incidents often create traffic jams and may happen at any time. Being able to ease the impact of traffic incidents will be invaluable for both motorists and authorities.

There have been some efforts for supporting optimized road reconfiguration, yet with limited scope. Some approaches exploit dynamic pricing to divert traffic from certain road segments by announcing a toll hike [55–59] and making the speed limit variable to improve flow [60,61]. However, the response is typically slow as the adjustment is centrally controlled and determined by the authority. Some consider road configurability at the planning stage by determining whether HOV lanes should be employed [62]. Adaptive traffic light scheduling is also pursued to deal with vehicle pileup at individual intersections [63–68]. However, the approach does not factor in the impact on other parts of the road. Very few studies, e.g., [69–72], have explored coordination among traffic lights to increase traffic flow; however, none of them considers possible road reconfiguration by changing lane direction. VANET has also been exploited as a means to orchestrate intersection crossing for self-driving cars [73–75]. Finally, the focus of [76] is limited to traffic signal timing. A comprehensive DTM optimization model that factors all means for road reconfiguration is yet to be developed.

5.2.3 COMMUNICATION AND CONTROL

DTM can be implemented in a centralized or a distributed manner. As pointed out above, there are no optimization models that factor all means for road reconfiguration; we further note that centralized control has conventionally been assumed by existing work on DTM. To enable distributed DTM implementation as well as support on-road data collection, means for V2I communication have to be provisioned. Given the vehicle mobility and also to avoid the prohibitive cost of wiring, wireless technologies are considered default for establishing communication links. Basically, connectivity is needed to support interaction among vehicles and between them and road-based data collection and configuration controllers. Communications among the various road units could be wired or wireless; yet wireless links are way less costly to provision for. Interaction among road units could be for coordinated control of the road configuration and for sharing data. In the following, we enumerate popular wireless technologies and analyze their applicability in DTM systems. Table 5.1 summarizes and compares their features.

Table 5.1
Comparison of Different Communication Technologies

Communication Technologies	Range	Frequency	Responding Speed	Need/Support Relay or Not	Support Multiple Vehicles	Applications	Pros	Limits
Long Range WiFi	Multiple kilometers, max 315 km in practical use.	2.4 GHz or 5.8 GHz	Up to 356.33 Mbit/s.	Support	Yes. support MIMO.	1. Bring Internet to remote construction sites or research labs. 2. Connect widespread physical guard posts, e.g. in forest. 3. Transmit real time seismic data in Peru to UCLA.	1. Long range. 2. Support MIMO. 3. Unlicensed spectrum. 4. Smaller, simpler, cheaper antennas (compared to cell or fixed antennas).	1. Line of Sight limitation. 2. Limited to soft obstacles. 3. Lack of commercial service providers.
Cellular Network (LTE)	For optimal performance is <5 km. For reasonable performance is <30 km. For acceptable performance is <100 km.	450 MHz–3700 MHz	Downlink peak rates of 300 Mbit/s, uplink peak rates of 75 Mbit/s. Latency is <70 ms, up to <5 ms.	Need	Yes. Support MIMO.	Cellular Netowrk.	1. Large range. 2. Support MIMO. 3. Support for terminals moving at up to 500 km/h (310 mph). 4. Support for cell sizes from tens of metres radius (femto and picocells) up to 100 km (62 miles) radius macrocells. 5. Support of at least 200 active data clients in every 5 MHz cell.	Needs base stations.

(Continued)

Table 5.1 (*Continued*)
Comparison of Different Communication Technologies

Communication Technologies	Range	Frequency	Responding Speed	Need/Support Relay or Not	Support Multiple Vehicles	Applications	Pros	Limits
Cellular Network (5G)	A few hundred meters.	Below 6 GHz, and 24 GHz–300 GHz	2 Gbit/s. Latency is <30 ms, up to <1 ms.	Need	Yes. Support massive MIMO.	Celluar Netowrk. Internet of Things	1. High data rates. 2. Low latency. 3. Energy saving. 4. Large system capacity and large-scale device connectivity. 5. Support massive MIMO.	1. Need high density base stations. 2. Short range. 3. Poor ability to pass through building walls.
IEEE 802.11p	1 km. Support moving vehicles up to 260 km/h (163 mi/h).	5.85–5.925 GHz	Quicker than legacy WiFi. No need to wait on the association and authentication procedures to complete prior to exchanging data.	No	Yes	Toll collection, vehicle safety services, and commerce transactions via cars.	1. Support data exchange between high- speed vehicles and between the vehicles and the roadside infrastructure. 2. Created specially for Intelligent Transportation Systems (ITS).	No authentication and data confidentiality mechanism.
Dedicated short-range communications (DSRC)	A few hundreds meters, usually <450 m.	5.9 GHz	Max 150 ms latency.	No	Yes	Used in electronic toll collection (ETC) in Europe and Japan.	1. Have multiple channels: one control channel (CCH), six service channels (SCHs). 2. Support vehicles moving in high speed.	Short range.

(Continued)

Table 5.1 (*Continued*)
Comparison of Different Communication Technologies

Communication Technologies	Range	Frequency	Responding Speed	Need/Support Relay or Not	Support Multiple Vehicles	Applications	Pros	Limits
WiFi Direct	<200 m.	2.4 GHz, 5 GHz	<250 Mbps.	No	Yes	1. large files transfer 2. Connection between printers, cameras, scanners, wireless mice and many other common devices.	1. No need require and confirm. 2. No need wireless router (AP). 3. Support diff manufactures. 4. High speed (than bluetooth).	Short range.
Bluetooth	10–300 m.	2.4–2.8 GHz	6–100 ms.	No	Support max 8 connections.	Connection between printers, scanners microphones and so on.	Send small snippets of information even not paired or connected.	Short range. Slow.
LoRa	>10 km. Max 19 km in rural area. Support moving vehicles up to 70 km/h (44 mi/h).	900 MHz	From around 250 bps to 50 kbps.	Support	Yes	Asset tracking. Fire alarms and fire detections. IoT network.	1. Long range and low power consumption. 2. Support both outdoor and indoor environment and other situations. 3. Perform forward error detection and correction.	Proprietary instead of open source.

- *Long Range WiFi:* This technology extends the range of the popular WiFi which does not exceed 100 m in outdoor setups. By employing directional antennas, long-range WiFi achieves a range of multiple kilometers [77]. Other advantages of long-range WiFi include the use of unlicensed spectrum, the incorporation of small and inexpensive antennas, and the availability of reliable and free-licensed software, e.g., DD-WRT [78]. Long-range WiFi is suitable for communication among road configuration units, e.g., between controllers of consecutive traffic signals in order to coordinate timing.
- *Cellular Telecommunication*: For a communication range in excess of 10 km, long-range WiFi is not a viable option. In this case, cellular network is a more appropriate choice that enables the establishment of reliable connections and is supported by well-established service providers. The radius of a cell varies from 1 to 30 km. Yet, the reliance on base stations could constitute an obstacle in low-coverage areas and introduce high latency during heavy network loads [79].
- *IEEE 802.11p*: This IEEE standard is mainly developed to support V2V and V2I communication. The vision is that it serves as the wireless backbone for ITS [80]. The range of the IEEE 802.11p is capped to 1 km [81]. It is also able to support data exchange among fast-moving vehicles.
- *Dedicated Short-Range Communication (DSRC)*: DSRC is based on IEEE 802.11p and is designed to boost on-road safety through the exchange of messages among vehicles. Several alerts are shared among vehicles to avoid collisions, such as forward collision warning, emergency electronic brake lights, blind spot warning (BSW), do not pass warning, intersection collision warning, etc [82]. DSRC also provides a basic safety messaging mechanism, which broadcasts status information of each vehicle, including position, speed, acceleration, and direction, at a frequency of ten times every second over a range of a few 100 m [82]. The safety message data could be leveraged by the road units as well. For example, a traffic signal controller could be augmented with a DSRC transceiver to overhear safety messages in the vicinity and assess the inflow vehicle volume in various directions and the outflow rate. Such assessment can then be used to dynamically set the green time in order to maximize throughput and reduce vehicle waiting time at an intersection.
- *WiFi Direct*: This technology, which is also referred to as WiFi Peer-to-Peer, is used for near-field communication to support data exchange within a range of a few 100 m. It offers the data rate of a typical WiFi, and its support is becoming standard nowadays on smart devices [83]. The distinct feature of WiFi Direct is that it does not need a wireless router and enables devices to establish peer-to-peer links and dynamically form groups. Like DSRC, WiFi Direct can be used to support communication among vehicles and with close-by road units, especially when traffic involves conventional vehicles and the driver's cell phone is used to establish communication links [84].
- *Bluetooth*: Like WiFi Direct, Bluetooth is geared for device-to-device communication. Despite the popularity of Bluetooth, it suffers limitations that diminish its suitability for the realization of DTM systems. Basically, the communication range of Bluetooth is less than around 100 m and supports at most eight connections. It can serve as a secondary means for communication between closely located vehicles.
- *Longe Range (LoRa)*: LoRa is a proprietary communication technology developed by Semtech, for wireless connectivity of Internet of Things (IoT) devices in low-power wide-area networks (LPWANs) [85]. LoRa defines the physical layer protocol while LoRaWAN is developed as the upper layer protocol of the network which is supported by LoRa Alliance with more than 500 members. LoRa can provide a coverage of more than 10 km (max. 19 km in rural area) with very low power consumption [86,87].

5.2.4 TRAFFIC ASSIGNMENT

Traffic assignment refers to how vehicles are routed and plays a profound role in forecasting travel time. In essence, it is a means for controlling the vehicle density to manage traffic and optimize performance. In other words, traffic assignment constitutes the action for closing the traffic control loop. Traffic assignment opts to optimally allocate a set of origin-destination (O-D) pairs to a specific set of paths, i.e., consecutive road segments, according to the criteria set by the system and drivers. The optimization objective could be minimizing the travel distance, maximizing the vehicular throughput, or minimizing fuel consumption. The considered constraints include the infrastructure capacity, safety rules, and traffic regulations. Traffic assignment is generally a very complex optimization as it involves allocating road resources, e.g., lanes and scheduling vehicles' entry. Such optimization problem is NP-hard if done statically, let alone the complexity when conducted in real time where the vehicle arrival rate fluctuates and traffic incidents sporadically take place with no regular pattern. In other words, traffic assignment in DTM has to be formulated as a time-dependent optimization problem.

Figure 5.3 highlights the various classifications of the traffic assignment problem. As indicated in the figure, the traffic assignment optimization can be geared up for system-level optimality criteria where the big picture matters the most. For example, the road capacity utilization could be the main worry, even at the expense of causing inconvenience to some road users. Achieving such optimization objective requires means for controlling traffic flow and vehicle density either through on-road signs/signals, e.g., regulating in-flow rate, changing lane designation, etc., or influencing user selection, e.g., by varying the toll charges. Meanwhile, the objective could be user-centric where the road experience of individuals is targeted. For example, the least arrival time may be the quest of a user regardless of whether the picked route serves as a global optimization metric or not. User centric strategies are the most popular in case of traffic involving human-driven vehicles. When vehicles collectively try to improve the experience of all users, the objective is called dynamic system optimum [88]. The latter is a perfect match for CAV-based scenarios.

The granularity of DTM depends on the underlying traffic flow model. Generally, traffic flow models can be classified into three categories: macroscopic, mesoscopic, and microscopic [89,90]. As the name indicates, microscopic models are fine-grained and factors in vehicle-level behavior, and vehicle-to-vehicle and vehicle-to-road interactions [91]. In case of human-driven vehicles, the driver's response to incidents and traffic conditions is captured as well. For example, traffic flow

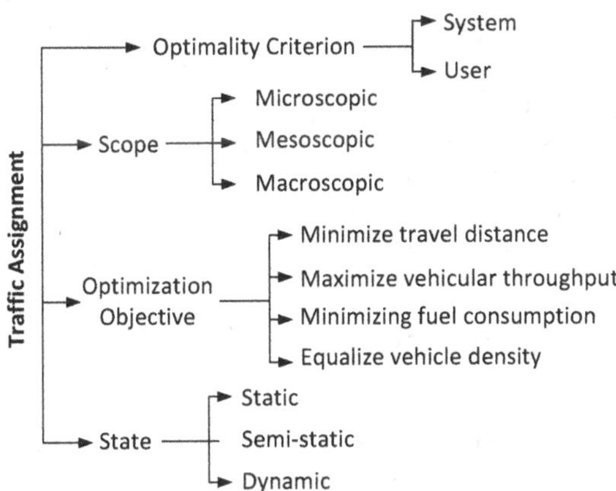

Figure 5.3 General taxonomy of the categorization of the various traffic assignment techniques.

in a construction zone is significantly influenced by drivers, e.g., tailgating, passing speed, etc. Generally, microscopic models involve excessive details and, if adopted, would complicate the traffic assignment process. They could be more suited for CAV given the autonomous control of the involved vehicles. Macroscopic models, on the other hand, are more coarse-grained and categorize traffic in an aggregate term, e.g., average motion speed and average vehicle density [92]. Aghamohammadi and Laval [93] have further classified macroscopic models as continuous and discrete space. The former abstracts the traffic assignment problem to operate on regions, while the latter models the area as road segments (finite number of zones). Continuous-space models are useful when the network is dense such that both the distance between road intersections and their longitude are small compared to the size of the region. Macroscopic models are more popular for managing traffic on highways, major local roads, city-based street grids, etc. [94]. A mesoscopic model falls in-between the microscopic and macroscopic ones where the individual vehicles are considered yet the traffic attributes are captured through a probability distribution function [95].

Whether the traffic assignment is based on user equilibrium or system optimality, popular objectives of the optimization include minimizing the travel time, minimizing the driving distance (shortest path to destination), minimizing fuel consumption, maximizing vehicular throughput, and equalizing the traffic density on the roads [96,97]. The last two objectives are more common for system-level optimization than for individual vehicles. Multiple objectives could also be pursued where a weighting function is employed to reflect the level of importance. Usually, the road network is modeled as a graph with link cost that reflects the optimized attributes. For example, when the travel time is the target of optimization, the average delay for traveling on a road segment will be used as the cost associated with the corresponding link in the graph. The basis for assessing the link cost may be deterministic or stochastic. Deterministic means that it relies on real-time traffic data collected by on-road sensors, e.g., traffic cameras, or through V2V communications. Stochastic means that it includes probability distributions, i.e., mesoscopic models, or historical data sets. Moreover, DTM is an iterative process for a defined traffic flow and a specific period of time. Thus, the link cost is a function of time. The complexity of solving the traffic assignment optimization varies widely based on the objective and link cost functions. For deterministic (scaler) link costs and a linear objective function, a user could apply classical least-cost routing algorithms such as Dijkstra's and Bellman-Ford. When system-level metrics are targeted, the problem is often mapped to multicommodity flow optimization. Such a problem is generally NP-hard; some variants could have polynomial time solutions [98].

Traffic assignment models are classified based on the temporal dimension into three categories: static, semi-static, and dynamic. The difference among them is based on whether the modeling of the flow considers congestion and captures whether there are variations between the in- and outflow for each road segment or zone within the area. The dynamic category reflects instantaneous real-time reaction to incidents and peak rates of vehicle entry to the road network. Therefore, traffic data should be accurate and fresh in order for the complexity of the dynamic strategy to be justified. Dynamic strategies are well-suited for CAV. The semi-static category differs from the dynamic one in the frequency at which situations are assessed and actions are taken; in essence, it strikes a balance between responsiveness and complexity.

5.3 CAV-ENABLED TRAFFIC MANAGEMENT

DTM at intersections has been the main focus in urban areas as junctions are often bottlenecks in road networks. DTM strategies in that context generally follow two methodologies, namely, time management and space management [99]. Work on time management at intersections can be categorized into two groups: (1) traffic light phasing and timing control at signalized intersections, and (2) AIM for controlling semi- or fully AVs at signalized or unsignalized intersections. Meanwhile, space management is generally done through road reconfiguration and is deemed to be instrumental for improving vehicular traffic throughput. The idea is to increase road utilization by (1) DLG which adaptively reassigns turn movements to lanes depending on real-time traffic demands, (2) adaptively reversing contraflow lanes by considering changes in the traffic flow volume, and (3) DTP to factor in coordinated vehicle motion. In the balance of this section, we discuss these techniques and summarize the state-of-the-art.

5.3.1 AUTONOMOUS INTERSECTION MANAGEMENT

Intersection management strives to optimize cycle time, splits, and offsets of traffic light signals (TLSs). In the case of CAVs, vehicle's arrival and request for green light at intersections can be approximately predicted along with its routes. Such prediction is possible as the vehicle may share real-time locations and could be guided by an infrastructure-based controller. Therefore, the problem of the intersection management in CAV is significantly different from the traditional methods. The main focus of the intersection management for self-driving vehicles is on eliminating the potential overlaps of vehicles coming from all conflicting lanes at an intersection and improving passengers' safety and fairness as well as stopping delay, fuel consumption, air quality, and total travel time in comparison to the conventional actuated intersection control using traffic signals and stop signs. In order to achieve the objectives, intervehicle cooperation and/or V2I communication are required for effective intersection operations and management. Controlling and managing CAVs at an intersection can be conducted in a centralized manner, where a single central controller globally decides for all vehicles [100,101], or decentralized where each vehicle determines its own control policy based on the information received from other vehicles on the road, or from a coordinator using V2V or V2I communication [102–104]. For example, the distributed auction-based intersection management approach of [102] employs an automatic bidding system that operates on behalf of the driver. The bidding system is applied at traditional intersections, i.e., stop signs and traffic signals, based on trip characteristics, driver-specified budget, and remaining distance to the destination.

In reality, achieving vehicle's full autonomy is not expected to be instantaneous, and vehicles have been gradually equipped with more and more advanced driver assistance systems (ADAS). Therefore, considering semi-autonomous vehicles' raising issues for AIM systems, in particular, under heavy traffic and difficult driving situations, current ADAS technologies are unable to handle certain dangerous scenario either at intersections or when merging on highways, most notably when dealing with vehicles arriving from the sides. Therefore, a cooperative framework has been proposed for semi-autonomous vehicles to mitigate the risk of collision or deadlocks while remaining compatible with conventional scenarios involving human-driven vehicles [105]. Another semi-AIM allows vehicles with features such as adaptive cruise control to enter an intersection from different directions simultaneously and achieves great reduction in traffic delay at an intersection [106].

In addition, early AIM work has focused on protecting passengers by seeking a safe maneuver for every vehicle approaching an intersection without considering traffic lights and mitigating possible system failure cases that could result from inevitable trajectory overlaps at the intersection [107–109]. Another trajectory-based AIM system optimizes TLS control simultaneously with the autonomous vehicle trajectories based on real-time collected arrival data at detection ranges around the center of the intersection [110]. Deadlocks and starvation (unfairness) are concerns that have also been tackled, where lightweight optimization of trajectories for safe and efficient intersection crossing is proposed [103,111]. Some of the AIM systems focus on minimizing fuel consumption, subject to throughput and safety requirements. The throughput maximization problem with hard safety constraints has been formulated as a decentralized optimal control problem for fuel minimization; an analytical solution has been presented such that vehicles pass an intersection without a full stop, and each vehicle's acceleration and deceleration are optimized [104]. Consequently, transient engine operation is minimized and fuel consumption is saved while also improving travel time.

Another group of AIM considers multiple intersections in a cooperative way [112,113,114]. In [113], a study is reported on how a single AV may cross multiple signalized intersections without stopping in a freeflow mode; an optimal eco-driving algorithm is proposed to generate the acceleration and speed profile by considering multiple intersections jointly rather than dealing with them individually. The multiintersection control is modeled using multiple agents across a network of interconnected intersections. From the multiagent perspective, AVs dynamically modify their scheduled paths based on different navigation policies and in response to minute-by-minute traffic conditions. Therefore, for a large road network, an instance of Braess' paradox may be experienced where opening additional travel options for the vehicles reduces the efficiency of all vehicles in the system. Such paradox is handled in [112]. Meanwhile, Li et al. [114] have developed an intersection automation policy (IAP) for serving requests for green light made by both AV and human-driven vehicles. IAP exploits real-time tracking of vehicle location to predict arrival at intersections along its route where requests for green signals are anticipated. A schedule for green time is then devised based on the phase-time-traffic hypernetwork model, articulated in Figure 5.4, which represents heterogeneous traffic propagation under traffic signal operations. Thus, the signal time and vehicle movements are optimized for all vehicle types.

Like [114], other studies have considered intersection crossing by a mix of AVs and human-driven vehicles [115,116]. Assuming that the vehicle type can be determined, AVs are safely directed through the intersection even if they arrive on a lane that is assigned a red signal. Scheduling platoon crossing an unregulated intersection is also one of the issues addressed in AIM. Generally, the problem is to schedule autonomous platoons through a k-way merge intersection; an intersection crossing involving two-way traffic has been shown as NP-complete [117]. A polynomial-time heuristic has been proposed for planning in which platoons should wait so that others can go through in order to minimize the maximum delay for any vehicle. Recently, the scope of AIM work has been expanded to support the quality of the travel experience from the passenger perspective while still caring for trip safety and efficiency. Dai et al. [118] have proposed an intersection control algorithm that characterizes vehicular kinematic states and smoothness of the vehicle jitter, acceleration, and expected velocity. The algorithm opts to alleviate the vehicle jitter by reducing sudden acceleration and deceleration, and determine the right-of-way of vehicles by striking a balance between the traffic throughput and fairness among vehicles. As a result, the smoothness of vehicular movement has been enhanced, and the travel time of individual vehicles is balanced by controlling the vehicle velocity.

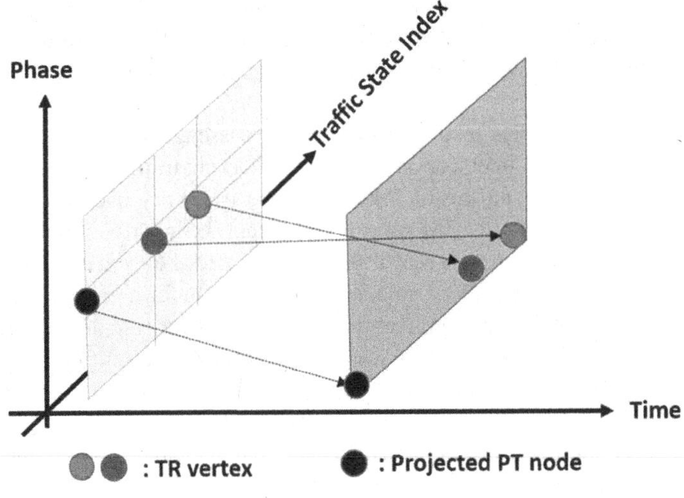

Figure 5.4 Illustration of the phase-time-traffic hypernetwork concept. (Adopted from Figure 7 in [114].)

5.3.2 ADAPTIVE TRAFFIC LIGHT CONTROL

Adjusting the traffic signal timing is the most popular means for DTM [119]. Work that focuses on ATLC covers cases involving human-driven vehicles, AVs, and a mix of both types. Existing schemes can be classified based on the scope and means for collecting real-time traffic data and the objectives of traffic control at an intersection.

Data Collection: Adapting the signal timing in real time requires accurate assessment of the inflow and outflow at the intersection. To measure the arrival rate and queue length, some approaches have placed video cameras, piezo-electric sensors, and inductive loops at the intersection. These measurements are also used to anticipate changes in the traffic pattern and modify signal phases and timing accordingly. These DTM-based approaches have been shown to be quite effective in comparison to that of the pretimed TLSs [120]. In addition, street-mounted sensor nodes have also been used to assess traffic to adjust green lights in order to improve the waiting time, number of stops, and vehicle density [121].

Optimization Objectives: The most common goal of ATLC is to maximize traffic throughput and minimize trip delay by adjusting signal phase and timing at an intersection. Some approaches additionally focus on reducing the total number of stops during the entire travel and thus ameliorating CO_2 emission. On the other hand, the focus of other work is on reducing the time and space complexity for solving the traffic signal control problem while improving the average waiting time [122]. ATLC has been further improved through anticipating traffic fluctuations at an intersection. Such anticipation is by correlating the real-time traffic data at an intersection i_x with neighboring intersections. Such correlation is enabled by data sharing through various infrastructure-to-infrastructure (I2I) technologies. For example, in [120], a WSN has been deployed on the road network to assess upstream and downstream traffic density around i_x to decrease the average waiting time of vehicles crossing an intersection i_x.

Using a broad range of real-time traffic data, a variety of methods to control TLSs have been proposed. For example, rule-based reinforcement learning ATLC is presented in [123], where the traffic lights of neighboring intersections coordinate locally; the work is extended by including an additional hierarchical observer/controller component at the regional level in order to better optimize the ATLC operation [124]. Moreover, multiagent-based algorithms have been applied to traffic light systems [125–131]. For instance, multiple traffic signal controllers, interconnected using IEEE 802.15.4 technology, are employed to dynamically order phases and calculate green time while factoring turns [125]. Figure 5.5 shows all possible turns corresponding to the TLSs at i_C.

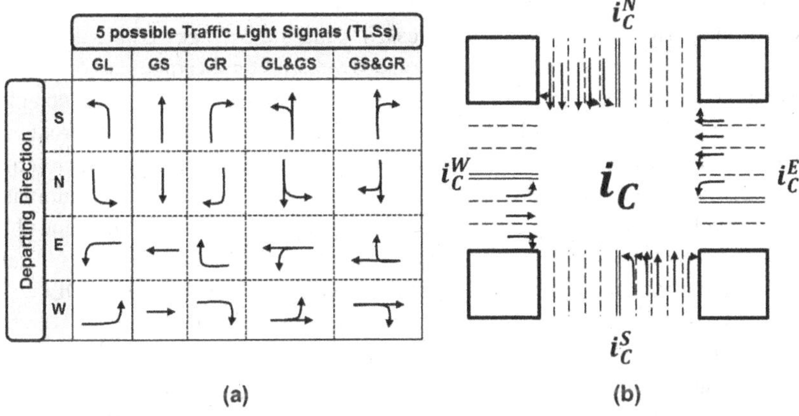

Figure 5.5 (a) A set S of 20 possible TLSs for each lane. (b) An example intersection, where 15 different TLSs?∈ S are used [125].

In addition, a distributed multiagent system has been developed using sensors to monitor traffic volume variations. The system finds the shortest green period during a vehicle trip, so that the experienced waiting time at intersections is minimized. Another group of multiple intersection ATLC algorithms exploit multiagent reinforcement learning algorithms [128–132], where the reactions by local and nearby intersections are considered to adjust the traffic lights timing.

5.3.3 DYNAMIC LANE GROUPING

TLS scheduling generally opts to assign short green duration to less traffic demands; however, low inflow still may occupy unnecessarily large number of lanes while vehicles pile up for making a turn. DLG algorithms opt to overcome such a limitation and increase the utilization of existing road resources by balancing between lane capacity supply and changes in turning demands. DLG algorithms have gained significant attention due to its adaptability to road capacity constraints. The main idea is to relieve traffic congestion and improve throughput and delay at intersections. The basic requirement for an effective DLG strategy is to estimate traffic volume for different movements, which cannot be provided by the inductive loop-detection systems. As discussed in Section 5.2, traffic data can be collected and communicated with various V2V, V2I, and I2I technologies such as road sensors, 802.11p, 802.16, i.e., WiMAX or cellular networks like LTE-V, or 3GPP Cellular-V2X (C-V2X). With the help of these advanced traffic monitoring technologies, one can estimate the vehicle count per turn at intersections.

Some studies have focused on the fundamentals of the DLG concept and formulated the problem mathematically. The popular objective is to maximize lane utilization at an isolated intersection under traffic-demand variation. These studies define a maximum lane flow ratio as the assigned flow is divided by the saturation rate; they strive to minimize changes in such saturation rate among different movements, which could lead to a significant performance degradation at intersections. The effectiveness of DLG has been demonstrated using numerical analysis compared to fixed lane grouping for varying number of lanes, saturated/unsaturated flow, and fixed/adaptive traffic signal timing. The performance is assessed in terms of the average delay at an intersection [133–140]. Some work has pursued DLG by combining two optimization problems, namely, signal timing control and dynamic road space allocation. In such a case, the lane count and the possibility of having to be shared are considered as parameters in the traffic signal timing optimization. The optimal lane group combinations and signal cycles are computed such that the average passing delay at the intersection is minimized [134–138].

Evaluation based on case studies has been conducted by various approaches to validate the performance of DLG. Using microscopic traffic simulation, the benefits of a DLG strategy have been demonstrated and compared to the conventional fixed lane grouping in terms of mobility and sustainability [139]. The performance of DLG is assessed using the average vehicle delay, the number of stops during a trip, the average fuel consumption per vehicle, the average rate of pollutant emissions such as carbon monoxide, hydrocarbons, and oxides of nitrogen and CO_2 per vehicle. Finally, it has been shown that the impact of DLG grows as the traffic volume and the frequency of the turn pattern vary; a DLG strategy is effective in balancing lane flow ratios and reducing intersection crossing delay and, consequently, energy and pollutants emission. Moreover, an automatic screening tool has been developed to identify the intersections for which DLG is advantageous [140]. Four assessment criteria are considered to evaluate traffic supply and demand, namely, (1) safe turning geometry which is a natural and logical prerequisite to qualify an intersection for DLG, (2) volume change to measure traffic fluctuations between time periods, (3) volume-to-lane (V/L) ratio, and (4) volume-to-capacity (V/C) ratio. The V/L and V/C capture the relationship between travel demand and capacity supply at an intersection. Through case studies, V/C has been shown to be the most effective criterion to identify a candidate intersection for DLG with a correct identification rate exceeding 90%; it has also been observed that DLG could reduce the overall intersection crossing delay by $\sim$15% [140].

5.3.4 DYNAMIC LANE REVERSAL

Contraflow lane reversal is used in big cities during rush hours; such scheduling is clearly static and cannot cope with variations of the traffic intensity. DLR would be logistically complicated for traffic involving human-driven vehicles as the flow direction cannot be switched until the lane is empty; often, the police have to be engaged to ensure that. With the emergence of CAVs, DLR is deemed to be a very viable option as AVs can rapidly switch out of the designated lane for flow reversal due to the automatic motion control and the prompt V2I and V2V communications. A number of techniques have been proposed for DLR in CAV [112,141–143], yet turns and collaborative intersection crossing are not addressed. In Ref. [112], DLR in collaboration with AIM has been proposed, where the total traffic volume on a road is monitored every 2 seconds; the road capacity is expanded by reversing the direction of a lane r on the paired road r_{dual} if the traffic demand on r is 1.5 times larger than or equal to that of r_{dual}. Such work has demonstrated how CAV enables efficient utilization of road infrastructure. Duell et al. [141] have developed a DLR for increasing the traffic flow on a congested downtown grid. Although their algorithm factors in dynamically changing traffic condition, it does not consider turns and their conflicts at an intersection. Levin et al. [142] also have focused on mitigating congestion of CAV traffic and formulated a DLR control problem for a single road segment as an integer program. A per-road agent is assumed for managing lanes and communicating with vehicles on the road. Meanwhile, Chu et al. [143] have considered mixed traffic scenarios, yet only CAVs are assumed to travel on reversed lanes. The problem of optimizing schedules and routes on dynamically reversible lanes has been formulated as an integer linear program and evaluated using real-world transportation data.

Some DLR approaches have been specific to particular road layouts and cannot be generalized to other layouts [144–147]. The focus of Li et al. [144] is on a signalized intersection with six lanes and two additional reversible center lanes. Only four scenarios are considered for typical urban morning and evening peak-hours. In Refs. [145,146], a signalized diamond interchange is considered, where the proposed DLR approaches strive to handle the concern of space limitation for different turns in order to reduce oversaturation at the interchange. Krause et al. [145] opt to show the effectiveness of dynamic back-to-back reversible left-turn in collaboration with a TLS controller, while Zhao et al. [146] strive to maximize the reserved capacity of the internal lanes at the intersection considering a fixed set of TLS phases. On the other hand, the focus of [147] is on the applicability of DLR to existing lanes for dynamic left-turn traffic, as articulated in Figure 5.6. With the help of an additional traffic light (presignal) installed at the median opening, exit lanes for left-turn control problem were formulated as a mixed-integer nonlinear program, in which the geometric layout, main signal timing, and presignal timing were integrated and transformed into a series of mixed-integer linear programs. The results have shown significant growth in the intersection capacity and reduction of traffic delay, especially under high left-turn demand.

Adjusting lane direction has been employed as an efficient way to overcome the logistical challenge in handling massive vehicular traffic during evacuation, where people are enabled to safely travel away from a hazardous site [148–156]. Some approaches like [148–150] use preknown incoming traffic volume and road capacities to find the optimal contraflow network configuration that minimizes the evacuation time. Then, lane reversal is usually scheduled once and at the beginning of evacuation. Other studies have expanded the scope of the lane reversal optimization to consider additional factors such as evacuation priority for moving injured people to a hospital or transit priority for low-mobility people in large cities [151,152]. In addition, partial lane reversal is considered in [153–156], where only a subset of the lanes in certain directions are reversed to enable timely handling of evacuees who need urgent care, e.g., elders.

Another prominent application of lane reversal is to manage fluctuated volume of vehicular traffic in a particular direction depending on time or events in urban areas [157–159]. The objective is mainly to overcome the road capacity limitation in order to boost the vehicular throughput and reduce travel time. The focus of the work in this category is on determining the appropriate time

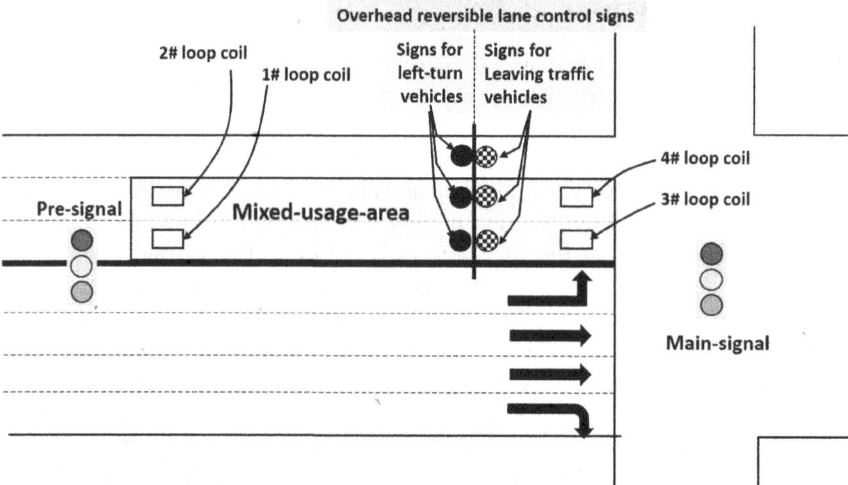

Figure 5.6 Exit lanes for left-turn traffic are controlled with a collaboration of geometric layout, main signal timing, and presignal timing. The location of reversible lane control signs and vehicle detectors is marked. (This figure is adopted from Ref. [147].)

to reverse a lane by monitoring traffic variations in both travel directions on a road segment. For instance, Zhou et al. [157] have developed a self-learning contraflow lane system for controlling tunnel traffic in order to estimate the real-time traffic demand for passing the tunnel and decide when to use contraflow for preventing traffic jams. Meanwhile, Hausknecht et al. [158] have studied the impact of reversing a lane on an unsaturated road on vehicular throughput. They model the maximization of network traffic as a multicommodity flow problem and propose a two-level formulation to calculate the optimal lane reversal configuration. Both approaches do not consider dynamically changing traffic demand and ignore the complication caused by turns at an intersection. On the other hand, Lu et al. [159] pursue a two-tier system for optimizing the reversible lane assignment while considering TLS settings and real-time traffic volume; the higher tier is for optimizing reversible lane assignment based on the total queue length at junctions while the lower tier is for traffic allocation at road segments.

5.3.5 DYNAMIC TRAJECTORY PLANNING

The vehicular traffic assignment for the traditional human-driven vehicles is mainly derived by the stochastic nature of the problem that it is subject to uncertainties related to perception and reaction times of drivers and human-based error. Such uncertainty can be mitigated in CAV. Nonetheless, CAV raises new issues given the fine-grained controllability of AVs. Basically, the spacing between AVs can be significantly reduced, enabling a set of vehicles to travel as a platoon. Thus, forming a platoon, joining and departing of an existing platoon, and setting the appropriate vehicle configuration are unique challenges in the case of CAV. In essence, the vehicles have to collectively determine speed, acceleration, and optimal spacing between them, subject to safety and road condition constraints. Therefore, DTM will not only have to optimally assign traffic but also have to find the optimal vehicle trajectories. Indeed, the path selected or assigned by routing models in DTM will be subject to the multi-CAV motion planning for autonomous or mixed traffic scenarios.

CAV motion planning is categorized in [160,161] into four hierarchical classes: (1) route planning that aims to find the best global route for given O-D pairs and corresponds exactly to the traditional traffic assignment. The individual vehicular route is derived based on traffic statistics (current and anticipated) and does not consider obstacles, road geometry, etc.; (2) path planning which is a bit more fine-grained and considers the constraints of individual road segments that connect the origin and the destination and opts to cope with obstacles and specific flow constraints, e.g., due to construction; (3) maneuver planning which determines the appropriate decision in each step including, for example, "going straight", "going left", etc. It considers further the position and speed of the CAV while taking into account the path that is specified from path planning; (4) trajectory planning which governs the motion of the vehicle and determines the vehicle's transition from one feasible state to another while considering road obstacles and the vehicle's kinematic. In essence, trajectory planning is concerned with the vehicle control.

In the realm of CAV, dynamic traffic assignment constitutes autonomous motion planning to find a feasible route over collision-free path toward destination while taking into account the vehicle dynamics and maneuver capabilities, and respecting the traffic rules and road boundaries [161,162]. The motion planning algorithm would utilize sensor readings and supplement them with data from digital road maps in order to provide a sequence of state transitions for the vehicle controller. In other words, the CAV will iterate over a set of states and actions to implement the plan defined by DTM. The planning problem is quite complicated given the search space and the number of optimization parameters. Many map tessellation solutions have been proposed in the literature in order to tackle such complexity [161]. Search space reduction techniques can be categorized as local or incremental search, depending on the existence of possible prior states. Once the feasible positions are determined, a continuous interaction between path planning and maneuver selection is used in order to ensure respecting road rules and avoiding obstacles. Such interaction is detailed in [161]. It should be noted that such high granularity for motion planning is not warranted for human-driven vehicles due to the presence of a driver. Given the scope of the chapter, we will focus mainly on the individual trajectory optimization, collaborative trajectory optimization and stream-based optimization, and reconfiguration in homogeneous and heterogeneous setup.

Individual Trajectory Optimization: The objective in this category is to minimize the length of the trajectory, its smoothness, and the offset from the central line of the lane, which is used as a reference path, under constraints imposed by the routes [163]. Other metrics include safety [164,165], intersection crossing efficiency [166], and fuel consumption [167]. Multiobjective trajectory optimization has also been considered; Ma et al. [168] promote trajectories that simultaneously optimize travel time and fuel consumption for all vehicles. The optimization constraints usually include the road boundaries, lane restriction, trajectories curvature, speed limit, acceleration rate, and obstacles. The obstacles are generally represented as circles to avoid collision using colliding trajectories detections. Some work also considers maneuvering and traffic rules [169] and generates trajectories that respect the checkpoints determined by path planning, stop signs, traffic lights, turns, lane changes, intersection crossings, turns, and dead-ends. The trajectory efficiency is measured using the distance and time until reaching the next checkpoint and the number of possible collisions with obstacles. The complexity of determining optimal trajectory is very high; some approaches apply machine learning to infer trajectory patterns from human driving [170] and then exploit them for online trajectory generation.

Coordinated Stream of Vehicles: Coordination among a stream of vehicles has been studied by optimizing the trajectory of multiple CAV. Such a problem is very challenging as the vehicles are constrained by car-following models, and each platoon has some characteristics such as acceleration, speed, and safety distance. Two types of optimization have been considered: (1) how to optimize the traffic assignment to avoid frequent switching of traffic lights and the frequency of vehicle stop/start, which will negatively affect the travel time. This category is covered in Section 3.1 of this chapter; (2) how to minimize the traffic flow fluctuation by choosing the appropriate configuration for each vehicle in a platoon or the speed of the leading vehicle in order to improve throughput and other performance metrics. In Ref. [171], the problem is formulated as a mixed integer program and used dynamic programming to solve it. Others, e.g., [168,172], proposed a shooting heuristic that can effectively smooth the trajectory of a stream of vehicles approaching a signalized intersection by detailed control of their acceleration profiles. The shooting heuristic reduces the complexity by representing the trajectory search space as a few segments of analytical quadratic curves. The trajectories are constrained by the vehicle's physical capabilities, safe intervehicle spacing, and traffic signal timing. However, only fixed signal timing and phasing are considered to control vehicle trajectories. On the other hand, some approaches reserve a certain lane for intersection crossing without considering any explicit traffic light [173,174].

Optimization in Heterogeneous Setups: The coexistence of human-driven and AVs limits the ability of CAV to improve the traffic performance using ramp metering, variable speed limits, signal control, etc. This is due to the traffic disturbance caused by drivers which could force cooperative lane changes, or unwarranted variability in intervehicle spacing, and stop-and-go triggered by rubbernecking. To mitigate such disturbance, CAV longitude control has been commonly used, specifically, by maintaining a constant spacing or headway (or time) between successive vehicles. To avoid collisions within a platoon, the CAV controllers have to be designed to ensure string stability. There exist three major approaches for CAV longitudinal control [175]: linear, optimal such as model-predictive control, and artificial intelligence (AI)-based. Linear controllers focus mainly on string stability by determining the appropriate feedback and feed-forward gains to adjust the acceleration. The optimal control approach uses multiobjective formulation that ensures control efficiency (e.g., relative speed to a platoon leader) and comfort (e.g., acceleration) [176]. Motivated by the fact that self-driving can be seen as a data-driven learning, AI-based controllers exploit machine learning techniques instead of parametric rule-based models [177]. Finally, the adoption of electric vehicles is another type of heterogeneity given that they have a greater range of acceleration/deceleration [175,178]. Another source of heterogeneity is user-customized CAVs, where the CAV behavior is influenced by human preference in terms of the desired speed and/or acceleration rates, etc. [175,179] (Table 5.2).

5.4 SMART ROAD VISION AND PRACTICAL ISSUES

As pointed out in the previous section, CAV will revolutionize the transportation industry and will enable the effective management of road traffic to achieve optimal performance. Yet, the reality is that human-driven vehicles will not disappear anytime soon, and numerous practical issues are ought to be considered. In this section, we highlight these issues and present our iRoad vision. The iRoad project, which stands for Internet of Radio-equipped On-road and Vehicles-carried Agile Devices, opts to tackle the challenges in realizing the DTM methodology when a mix of autonomous and human-driven vehicles share the road.

5.4.1 SUPPORT OF HUMAN-DRIVEN VEHICLES

CAV provides three key features that facilitate the implementation of DTM. First, the behavior of an AV is predictable; meaning that if instructed to follow a certain route, or even, trajectory, it indeed does so. Thus, one can estimate the vehicle density on the various road segments with high fidelity. In other words, the data collection is quite easy and accurate when only AVs are on the road. Second, each AV is capable of wireless communication. Such capability will enable instructing a vehicle about route change, sending alerts, and receiving road and vehicle status updates. The third feature is the ability of an AV to precisely and safely maneuver and follow a prescribed trajectory at a fine-grained level. Obviously, these features are not available for human-driven vehicles. Thus, when both autonomous and human-driven vehicles exist on the road, the realization of DTM will be quite challenging and the idealistic view about vehicle compliance would be unrealistic. In other words, human behavior and uncertainty about vehicle navigation and status complicate DTM immensely.

VANETs have been explored as a means to enable DTM for traffic involving contemporary human-driven vehicles [181]. In a VANET, cars are nodes that collaborate in assessing road

Table 5.2
Summary of Traffic Management Algorithms

References	Category	Subcategory	Objectives	Intersection	Junctions	Vehicle Type
[133]	DLG	Optimize space only	Minimize fluctuation of a maximum lane flow ratio and reduce average delay	Signalized	Single or multiple	HV
[134]		Optimize space and time allocations	Minimize average intersection crossing delay		Single	
[135]			Minimize delay			
[136]						
[137]						
[138]						
[139]						
[140]			Reduce delay; increase throughput		Multiple	
[180]			Increase capacity		Single	AV
[148]	DLR	Optimization of lane-based evacuation route	Minimize evacuation time	No signal	Multiple	HV
[149]						
[150]						
[151]						
[152]						
[153]		Optimize partial lane reversal	Reduce evacuation/clear time			
[154]			Prioritize evacuees with urgent care, e.g., elders			
[155]			Reduce evacuation time			
[156]						
[157]		Optimize tunnel lane reversal	Increase accuracy of traffic demand prediction		Tunnel	
[158]		Optimize traffic flow	Increase road network efficiency		Single	
[159]		Optimize traffic assignment	Reduce travel time	Signalized		
[141]		Optimize route selection	Minimize total travel time	No signal		AV
[142]		Optimize traffic assignment	Maximize vehicular flow		Single	
[143]			Optimize travel schedule		Single	
[144]			Optimize layout	Signalized	Arterial roadways	HV
[145]		Focus on left-turn	Reduce delay; increase throughput		Diamond interchanges	
[146]			Expand capacity; reduce congestion			
[147]		Focus on left-turn	Increase capacity; reduce waiting delay		Single	

(Continued)

Table 5.2 (*Continued*)
Summary of Traffic Management Algorithms

References	Category	Subcategory	Objectives	Intersection	Junctions	Vehicle Type
[100]	AIM	Centralized	Minimize total travel time	No signal	Multiple	AV
[101]			Minimize a sum of vehicle exit time		Single	
[102]		Decentralized	Minimize travel time with low budgets	Signalized		
[105]		Decentralized	Increase safety under the high rate of accidents	No signal		Semi-AV
[106]		Trajectory optimization	Decrease traffic delay			Semi-AV
[107]			Improve safety; minimize stops and travel time			AV
[108]			Improve ratio of average trip time to throughput			
[109]			Maximize road capacity			
[110]			Optimize AV trajectories and signal control	Signalized		Mixed
[111]			Improve the computational efficiency			AV
[103]		Decentralized	Improve the computational efficiency	No signal		
[104]			Minimize energy; maximize throughput			
[113]		Multiintersection optimization	Minimize energy consumption and travel time	Signalized	Multiple	
[112]	AIM & DLR		Minimize travel time			
[114]	AIM	Optimize traffic signal timing and vehicle movements	Minimize total delays			Mixed
[115]	AIM & DLG	Hybrid of trajectory and TLS	Minimize delay and maximize throughput		Single	Mixed
[116]	AIM	Trajectory optimization	Balance and maximize traffic flow rate	No signal		Mixed
[117]			Minimize the maximum delay for any vehicle			AV platoon
[118]			Improve fairness, safety and throughput		Single	AV
[120]	ATLC	Optimize green time duration	Minimize average vehicle waiting time	Signalized	Multiple	HV
[121]		Optimize sequence and length of traffic lights	Improve delay and throughput		Single	
[122]		Optimize signal phase and time	Reduce time and space complexity		Single	
[123]		Simulation based	Improve delay and throughput		Multiple	
[124]		Centralized/decentralized	Improve delay reducing fuel consumption			
[125]		Optimize signal phase and green time duration	Improve delay and throughput			
[126]		Optimize signal phase and green time duration	Minimize average vehicle waiting time		Single	
[127]		adjust green time duration	Reduce average vehicle waiting time			
[128]		Optimize signal phase and green time duration	Improve average travel time		Single or multiple	

(*Continued*)

Table 5.2 ***(Continued)***
Summary of Traffic Management Algorithms

References	Category	Subcategory	Objectives	Intersection	Junctions	Vehicle Type
[129]		Optimize signal phase and green time duration	Improve delay and throughput		Multiple	
[130]		Coordinated traffic signal control	Reduce total number of stopped vehicles		Single	
[131]		Reduce computational complexity	Improve delay and throughput		Multiple	
[132]		Optimize signal phase and green time duration	Improve delay and throughput			
[163]	DTP	Optimize individual trajectory	Improves safety and vehicular throughput	No signal	N/A	AV
[164]		Optimize individual trajectory	Increase safety	No signal	N/A	AV
[166]		Multitrajectory optimization	Minimize travel time (highway lanes)	No signal	N/A	AV platoon
[167]		Optimize individual trajectory	Improve fuel efficiency	Signalized	N/A	AV
[168]		Coordinated stream of vehicles	Optimize delay, energy and safety	Signalized	N/A	AV
[169]		Coordinated stream of vehicles	Increase safety	Signalized	N/A	AV
[170]		Optimize individual trajectory	Reduce computation complexity	No signal	N/A	AV
[171]		Coordinated stream of vehicles	Improve safety and throughput	No signal	N/A	AV
[172]		Coordinated stream of vehicles	Improve safety	Signalized	N/A	AV
[173]		Coordinated stream of vehicles	Improve safety and vehicular throughput	N/A	N/A	AV
[174]		Coordinated motion of vehicles	Reduce travel time; increase safety	No signal	N/A	AV
[175]		Handle heterogeneous setups	Analyze safety	No signal	N/A	AV and HV
[176]		Handle heterogeneous setups	Improve safety	No signal	N/A	AV
[177]		Handle heterogeneous setups	Improve safety and fuel consumption	No signal	N/A	AV
[178]		Handle heterogeneous setups	Improve safety	No signal	N/A	AV
[179]		Handle heterogeneous setups	Risk analysis	No signal	N/A	AV

AV, Autonomous vehicle; HV, human-driven vehicles; SLR, static lane reversal.

conditions and sharing safety information [182]. Every participating vehicle also acts as a wireless router to allow nearby cars to connect with each other, creating network topologies that span vast areas [183]. Attempts to deal with congestion using VANETs have been made, mainly by using VANETs to discover and disseminate congestion information [184], providing routing recommendations [185,186] and setting traffic signal timing [25,63–65]. We argue that a VANET that alerts drivers about accidents or traffic bottlenecks would be ineffective in resolving congestion. In addition, routing around bottlenecks and scheduling a traffic light at an intersection ease congestion rather than reduce the probability that it indeed occurs. Moreover, existing VANET-based techniques assume appropriately equipped vehicles, which hinders their applicability since most vehicles on the road today do not have onboard wireless transceivers.

Our iRoad project opts to overcome these limitations and promotes a novel and cost-effective system that can dynamically adapt to traffic conditions and operate autonomously without the need for costly on-road sensors. The iRoad system employs driver-carried smartphones and tablets and leverages the popularity of peer-to-peer (P2P) networking to optimize both road- and driver-centric metrics. The overall methodology can be viewed as managing vehicular traffic using participatory-sensing by the drivers and their vehicles. Unlike existing approaches, e.g., Google maps, Waze [27], Inrix [28], Cellint [29], etc., the goal is to reduce congestion by balancing the load on local roads through (1) predicting increased traffic, (2) accounting for the impact of driver's route choices, and (3) providing route recommendations while factoring in potential changes in the capacity of certain road segments. Every vehicle chooses a route to its destination and generates a set of optional routes. We have developed a P2P system to be used on each vehicle to query other vehicles in its neighborhood on whether they are traveling on any of the edges of its main and optional routes. Based on the data collected from traffic reports, alerts, and vehicles, our system would make a decision for the vehicle to continue on the same route or suggest an alternate route. The combined effect of all vehicles would impact the traffic pattern and allow smooth and faster travel for everyone. In other words, our system opts to achieve user equilibrium. Our preliminary results showed that 20% reduction in travel time could be achieved by such an approach even if only 40% of the drivers follow the route recommendations [32].

Smart road configuration capabilities, e.g., making a lane HOV or switching its direction, are also being exploited as a means for influencing vehicular flow on a particular road. Thus, our iRoad system achieves the DTM goals by providing real-time traffic management and being proactive in preventing traffic congestion [5]. Figure 5.7 shows a functional diagram. Our system has three major players: the vehicles, the VANET, and the smart road configuration controller. The vehicle part constitutes the user, i.e., driver interface in case of human-driven vehicles and the route selection and action planning algorithm. The realization of the vehicle part is as either an app for a smart device or software module that is integrated in the vehicle's dashboard panel. Every vehicle V_i has a unique identifier, e.g., VIN number, and an onboard GPS receiver (smart phone). Neighbors of a node (vehicle) are those it can directly reach through P2P links.

Our system can be implemented using a centralized traffic management controller, where a local server aggregates the data, configures the roads, and provides route recommendations; Figure 5.8 shows the system architecture. Moreover, the system can also be realized in a fully distributed manner where a vehicle (user's smartphone) factors in the data collected in its vicinity in route selection and local road configuration controllers, e.g., traffic light controllers, adjusting their operation to improve traffic flow, and the utilization of the road capacity. All computations for route selection are to be performed on the vehicle itself, e.g., using a smart portable device in case of human-driven vehicles. In such as a case, the driver has to choose a destination and travel route. For AVs, part of this functionality, such as the route selection, could be performed by a road-side unit or a remote traffic management system to optimize some global metrics. While some vehicles may not join the network due to lack of equipment or due to the driver's preference, with the right incentive, the bulk will join a VANET on the road. Incentives could simply better user experience

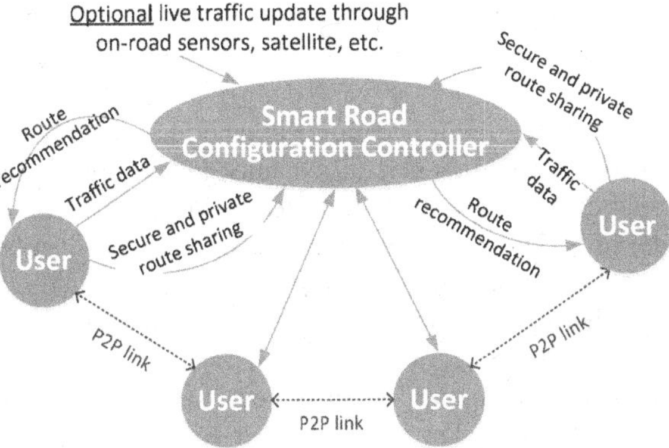

Figure 5.7 A functional diagram of the proposed system. Users may be drivers of contemporary vehicles or AVs. Users securely share their route plans, while factoring in traffic data, collected by participatory sensing and optionally, using on-road sensors and live traffic monitoring facilities when available. The smart road configuration controller (SRCC) may be a regional server or a road-side unit. Route recommendations can be provided by the SRCC or be determined in a distributed manner, i.e., vehicles collaboratively manage their travel paths and implicitly influence road configuration.

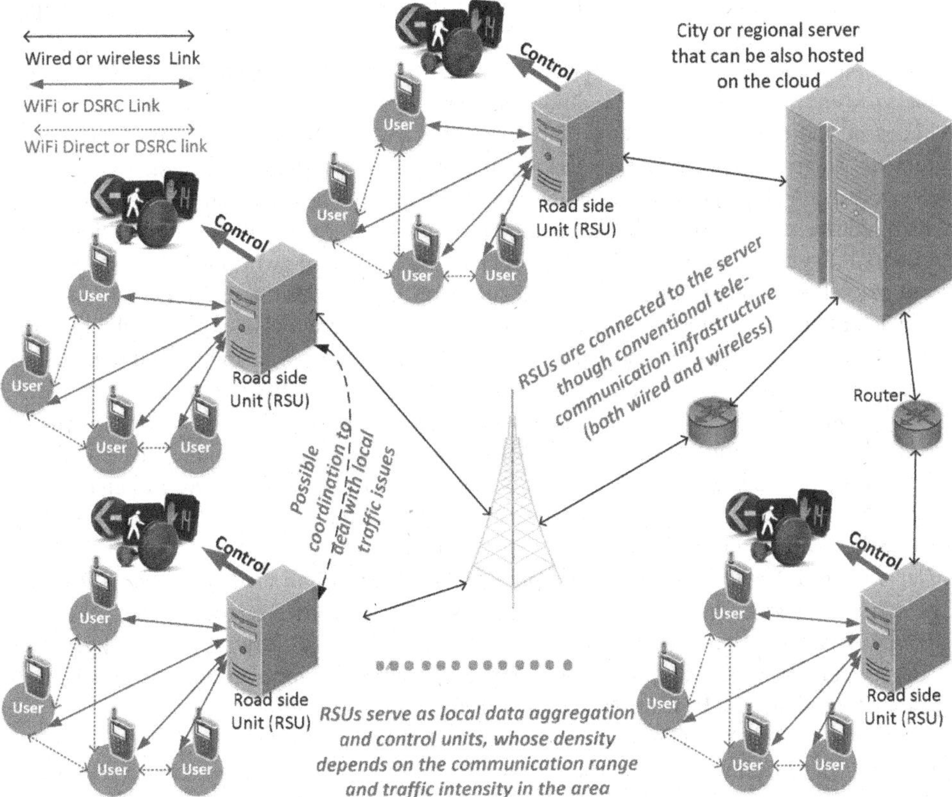

Figure 5.8 Realization of the proposed system, where traffic is autonomously managed locally, e.g., at the level of city downtown area, by road side units, and regionally, e.g., for a large metropolitan or county, by a server possibly while involving authorities for major road configuration decisions. Participatory sensing enables an economic tracking of road conditions and makes drivers part of the traffic management optimizations.

or reduced toll or vehicle registration fees. Although the system is completely autonomous, we expect drivers to decide on staying on the current route or rerouting to suggested paths whenever applicable.

We envision our system as laying the foundation for handling mixed traffic of autonomous and human-driven vehicles where coordination is required among these diverse sets of vehicles and between them, and smart road controllers will be expected and/or mandated. The iRoad system can also facilitate the handling of major evacuation scenarios where traffic can be chaotic. In cases of predicted disasters, such as hurricanes, massive evacuations are often necessary [187]. Our system will enable large-scale sharing of critical information without reliance on the communication infrastructure, which could have been degraded by the disaster, and also enable configuring roads autonomously as needed in order to support organized evacuation. To realize such a system, our research group is tackling the following key tasks. Later, in the section, we summarize some of the results; more can be found in [32,188–193].

1. *Developing Novel Routing Schemes for Preventing Congestion and Slow Traffic:* The key objective is to balance the load on the roads and prevent congestion through intervehicle data sharing while factoring in traffic data that are made available by authorities. Both human-driven and AVs are assumed. The routes planned by other vehicles influence a vehicle's own decision, implying that the system takes advantage of the highly dynamic VANET to make a fine-grained analysis and yield a more optimized travel route. Issues related to uncertainty about the collected data from other vehicles are also being addressed.
2. *Developing a P2P System for Efficient Internetworking of Collocated Devices:* Despite the growing push for VANET technologies, practically, only a small fraction of the vehicles on the road have modems, and AVs are expected to stay a minority in the near future. Therefore, alternative means for intervehicle data sharing is needed. We have developed a P2P system that operates on smart cell phones and tablets to collect and share sensor data among vehicles, and to communicate with road configuration controllers. Specifically, we use Wi-Fi direct to enable internetworking of these devices and extend the protocol stack, and build software libraries for the realization of our system. Our approach not only enables real-time data sharing but also expedites the integration of other prominent technologies like DSRC by providing means for old vehicles to participate.
3. *Developing Adaptive Strategies for Road Configuration to Effectively React to Traffic Problems:* We exploit existing facilities for changing lane specification (e.g., switching traffic direction, designating a lane as HOV, adapting toll assessments, etc.) to enable major traffic flow optimization. We opt to devise optimization models for flow maximization while factoring in possible impacts, e.g., lost toll revenue.
4. *Developing a Protocol to Enable Sharing of Route Information while Protecting Drivers' Location Privacy:* One of the obvious concerns in sharing rides, routes, and other observed/sensed road conditions is how to sustain the privacy of drivers or passengers. In addition, knowing the future location or routes of vehicles can be misused by attackers to conduct physical attacks and robbery [194–196].
5. *Developing Schemes to Guard against Contemporary Security Attacks:* The autonomous nature of the system and the participatory nature of the collected data raise the concern about various attacks launched to disrupt its operation and/or to unjustly gain privileges on the road. An attacker could pretend to be multiple simultaneous vehicles at different locations in order to influence other vehicles' routing decisions. In addition, the attacker can also inject false routes for many nonexisting vehicles to create an illusion of traffic congestion in order to trigger a favorable road reconfiguration or pollute traffic management data [197].

5.4.2 OPTIMIZED ROUTE SELECTION IN MIXED TRAFFIC

Due to the unpredictably varying nature of the vehicular traffic, it is hard to accurately foresee the number of vehicles traveling on the roads in order to avoid congestion. As pointed out above, existing systems opt to mitigate the traffic uncertainty by providing the current status, alerting drivers and providing route recommendations. However, that would not suffice as the response of the individual drivers varies and the rerouting decision made by them may cause congestion somewhere else. Our iRoad system not only provides for an increase in situational awareness but also enables exchanging routing plans among the drivers and, with SRCCs, to improve the vehicle throughput and travel time. Thus, the routes that others take would influence a vehicle's own decision and potentially affect the traffic pattern on certain roads. Such a decision-making process enables even distribution of vehicular traffic as one can predict the condition of alternative routes before changing the travel path. However, these features come at the expense of increased communication and computation overhead. In addition to the intervehicle messaging for data sharing, the routing decision under our model will be more complex than typical. An accurate forecast of traffic intensity is necessary to assure drivers that it would be advantageous to use different routes. Moreover, the effect of potential road reconfiguration has to be considered. While changing the road configuration is an effective means for congestion mitigation, it complicates the routing problem. Basically, the time taken to travel a particular segment will vary due not only to traffic volume but also to changes in road capacities. Such assurance and positive experiences would increase vehicle participation.

To assess the potential impact of sharing travel routes among vehicles and dynamic road configuration, specifically adjusting traffic light timing on performance, we have conducted a preliminary study [32]. The study is based on the following operation model. All vehicles have the same communication range D_R, which is to be carefully set to reduce interference, especially in congested areas. When a vehicle has to retrieve information from others in the vicinity, it broadcasts a message with its location, routes and speed; all vehicles receiving this message are expected to reciprocate by responding with their information. In addition to the route R_C that V_i currently follows, every vehicle has a predefined set of k optional routes $R_O[i], for\ i = \{1,\ldots,k\}$ [185]. Figure 5.9 shows an articulation of a possible configuration. Every interval of time (I_T), a vehicle broadcasts a message that contains its ID and the next "N" edges on route R_C as well as all edges on optional routes.

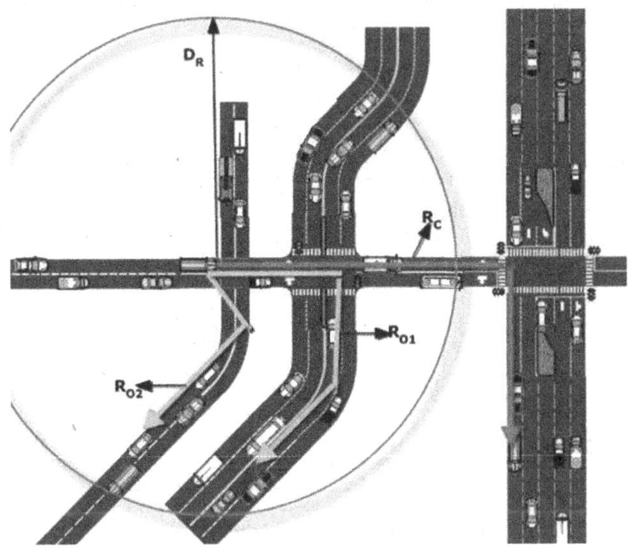

Figure 5.9 Illustration of the assumed system model showing the relevant vehicle parameters.

The value of N is predetermined and is the same for all vehicles. Based on the received updates from neighbors, a node estimates the number of vehicles on the edges of R_C and all $R_O[i]$ and the expected delay based on the vehicle count per unit distance for these edges. Using the delay estimate, the node assesses the travel time by factoring in the waiting time at the traffic lights that will be encountered and decides whether the current route is still the best choice or one of the optional routes would be better. The driver may follow such a recommendation with a probability P_R. On the other hand, an ALTC will use the data to adjust green time. The road with more vehicles would get more green time and vice versa. No inter-ATLC coordination is considered.

We have simulated such an operation model for a 15 × 15 km^2 area in downtown Baltimore. A total of 4000 vehicles are allowed to travel in that area. For each vehicle, the departure time is randomly selected between 0 and 1000, and the origin-destination pair is also chosen randomly within the area. For fair comparison, we use the same set of vehicles with the same departure times and origin-destination pairs as input with and without VANET-enabled autonomous traffic management (VAM). We performed 30 simulation runs, implying 30 different origin-destination pairs, and averaged the results. In the simulation, D_R, I_T, N, and P_R were set to 5 km, 5 units, 5 edges, and 0.85, respectively. We introduced congestion by forcefully stopping two specific vehicles for 30 simulation time units at predefined times when they enter specific edges. Given the goal of the study, we ignored delays due to message queuing and retransmission. We compare the performance of VAM to two other traffic management mechanisms. The first is a centralized routing scheme, similar to that described in [44], which bases the routing decision on the state of the entire set of roads. Obviously, in practice, this approach entails massive state updates. Nonetheless, it would serve as an upper bound for how well the vehicular traffic can be managed. The second baseline approach is called *Autonomos* [198], which alerts vehicles entering a congestion zone to avoid vehicle pile-up. We also compare the performance to the case when no action is taken to mitigate traffic congestion.

Figure 5.10 shows the time until all the simulated vehicles reach their destination. The graph shows that VAM, which assigns routes in real time and relies only on the local state, almost matches the performance of a centralized approach. *Autonomos* shortens the trip time since vehicles receive an alert message when they come close to a congested area, which could be used as a signal to reroute. Yet, *Autonomos* cannot prevent the occurrence of congestion and yields significantly worse results than VAM. The gap between VAM and *Autonomos* grows with the increase in the vehicle population. The plot for the conventional routing shows nonuniformity as compared to the other

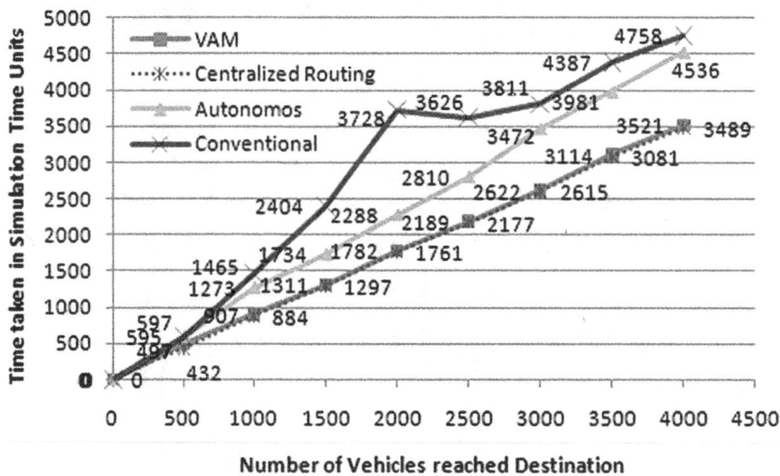

Figure 5.10 Comparing the performance of VAM with contemporary traffic management schemes.

approaches, as it does not deal with congestion. It has a high number of vehicles on the roads in the middle of the simulation, and thus, the delay grows. At the beginning and toward the end of the simulation, the number of vehicles that is still on the roads is fewer and the delay lessens. All the other approaches have provisions for dealing with congestion and hence have a uniformly rising curve. The results have demonstrated a major improvement even with such a scaled down and primitive version of our methodology.

5.5 CONCLUSION AND OPEN RESEARCH PROBLEMS

This chapter has focused on recent advances in mitigating traffic congestion and how the development of CAV has introduced a paradigm shift. The notion of DTM has been explained, and the challenge in applying it has been analyzed. The CAV capabilities are shown to be invaluable and in fact enabler for realizing DTM in practice. Advances made to date have been highlighted, and the existing techniques have been categorized and compared. We have also presented a new vision, namely, *iRoad*, for applying DTM for scenarios involving a mix of autonomous and human-driven vehicles. Sample preliminary results have been discussed as well. The following points out some open issues that are worth further investigation:

- *Supporting Pedestrians Crossing:* As pointed out in Section 5.3, CAV will enable the elimination of traffic signals as vehicles can collaboratively determine the crossing order without causing collisions. With the transition to nonsignalized intersections, other road users like pedestrians and bicyclists will become unprivileged; unless the intersection is empty, these users will not be able to cross safely. According to the 2017 reports of the US National Highway Transportation Safety Authority (NHTSA) [199], about 6000 pedestrians have lost their lives due to road crash with intersection crossing being the most frequent cause. Among the road-related fatalities in urban areas during 2017, pedestrians represent about 25%. Similarly, about 800 pedal cyclists died in 2017 due to road crashes. Relying on obstacle-detection capabilities for AVs would not be sufficient for ensuring safety and effective for sustaining traffic flow as, if unregulated, pedestrians crossing could be disorganized and sporadic. In fact, in crowded areas, e.g., downtown Manhattan in New York City, pedestrian crossing at unsignalized intersection could bring traffic to standstill given the number of locals and visitors typically found on the sidewalks. Thus, developing integrated solutions for road crossing is needed, especially in urban setups.
- *Integrated DTM Optimization Framework:* Road reconfiguration can be a very powerful venue for supporting DTM; as discussed earlier in the chapter, it constitutes a means for closing the control loop. CAV enables a wealth of options such as collaborative intersection crossing and DLR; yet in most publications, these options are being exploited individually rather than in an integrated manner. It is important to study which options leverage and conflict with one another and which options are independent. For example, intersection crossing for CAV is more versatile with additional degree of freedom with more reliable and dynamic road data updates, and is more agile and accurate reaction of vehicles compared to human-driven vehicles. Meanwhile, DLG strategies for CAV provide great flexibility in handling dynamically fluctuated traffic demand. However, applying DLG at nonsignalized intersections is still an unresolved issue, and dynamic motion planning algorithms are needed to deal with the phase sequence constraint [181]. A comprehensive optimization framework is needed.
- *Handling of Mixed Traffic:* CAV enables effective realization of DTM both by providing data and responding to control commands, e.g., to take a system-optimal route. As shown in Section 5.3, CAV makes it easy for exploiting road reconfiguration as a means for mitigating congestion and dealing with traffic incidents. Nonetheless, human-driven

vehicles are still predominant and are expected to coexist on the road with CAV for a long while. Thus, practically CAV-based DTM optimization will have to adapt and prevent the presence of human-driven vehicles from hindering the advantages of CAV. Investigating techniques for DTM in mixed traffic scenarios are necessary. The challenges to be tackled include the fidelity of situation assessment given the diversity of collected data, the effect of driver behavior on trajectory planning for CAV, and scheduling of safe intersection crossing. We envision machine learning techniques to be invaluable in addressing these challenges where CAVs are to self-adapt their operation in various settings, e.g., urban or rural, and varying densities of human-driven vehicles.

- *Vehicle and Infrastructure Security:* In essence, DTM is viable only when accurate data are available. Collecting such data can be via numerous means as discussed in Section 2.2. Some of these means such as electromagnetic loops, laser sensors, pressure hoses, and radars do not identify the vehicles and do not raise privacy concerns. In addition, manipulating these sensors through cyberattacks is not easy, given the physical wiring which mandates the adversary to be present to tamper with sensors. Yet, deploying of these types of sensing is expensive, and the trend is to rely on participatory sensing where the data are collected through V2V or V2I communication. However, such data collection methodology raises privacy and security concerns. Basically, the location and travel route of the participating vehicles should be shared in order for the system to know demand, detect bottlenecks, and anticipate traffic density. Exposing such information about the individual vehicles constitutes invasion of the drivers/passengers privacy and may be exploited by criminals. For example, knowing that someone left home to a remote destination can be exploited for planning a robbery. Moreover, since the DTM system will aggregate the data collected about individual vehicles, Sybil attacks can be launched by a vehicle to gain on-road privilege. A rogue vehicle could claim multiple identities to inflate the density on a certain road and consequently trigger road reconfiguration that better serves its travel path. For example, by sending messages with different identities to a traffic signal controller or a road-side unit, a vehicle could give the false impression that many vehicles are crowding and more green time or lane reversal is warranted. Thus, guarding DTM systems against cyberattacks is very critical. Securing these systems will be very challenging given the resource constraints, heterogeneity, and mobility of the vehicles. The complexity even grows for traffic involving both autonomous and human-driven vehicles. Although some progress has been made [194–197], more efforts are needed to devise comprehensive and holistic security solution.
- *Interaction with Smart City Applications:* Recent technological advances in computation and communication devices have revolutionized how people live and interact. The notion of connected communities and smart cities reflects how societies are being transformed. CAV constitutes an example of what one expects in a smart city. Therefore, the scope of DTM needs to be broadened to cope with emerging applications and services that should be continually accessible and efficiently supported while people are commuting. For example, passengers would expect access to entertainment and social media while riding their vehicles. Thus, route selection and even trajectory planning could be influenced by factors that are not dependent on the road conditions. For example, a vehicle may join a platoon to receive streamed movie or prefers a travel path for which the cell phone coverage is at its best to access live cast of a game. Similarly, vehicles may decide to stay next to each other to enable peer-to-peer wireless battery charging. One may even prefer scenic routes close to landmarks or passing areas with holiday decorations. These unconventional factors will indeed affect DTM; new strategies are needed as societies are getting more and more modernized.

ACKNOWLEDGMENT

Dr. Sookyoung Lee is supported by the National Research Foundation of Korea (NRF) funded by the Korea government (MSIT) (2018R1D1A1B07043671).

REFERENCES

1. http://www.nationwide.com/road-congestion-infographic.jsp.
2. https://corporate.tomtom.com/static-files/69e46ddb-4eac-4014-918f-cd5e7cd0c2c4.
3. https://www.weforum.org/agenda/2019/02/commuters-in-these-cities-spend-more-than-8-days-a-year-stuck-in-traffic/.
4. https://static.tti.tamu.edu/tti.tamu.edu/documents/mobility-report-2019.pdf.
5. U.S. Department of Transportation Federal Highway Administration, "21st century operations using 21st century technologies", Office of Transportation Operations, http://ops.fhwa.dot.gov/aboutus/one_pagers/opstory.html.
6. J. Ding, Y. Zhang, and L. Li, "Accessibility measure of bus transit networks," *IET Intelligent Transportation Systems*, Vol. 2, No. 7, pp. 682–688, 2018.
7. A. Nuzzolo and A. Comi, "Advanced public transport and intelligent transport systems: New modelling challenges," *Transportation Science*, Vol. 12, pp. 674–699, 2016.
8. Y. Dong, S. Wang, L. Li, and Z. Zhang, "An empirical study on travel patterns of internet based ride-sharing," *Transportation Research Part C: Emerging Technologies,* Vol. 86, pp. 1–22, 2018.
9. Y.M. Nie, "How can the taxi industry survive the tide of ride-sourcing? Evidence from Shenzhen, China," *Transportation Research Part C: Emerging Technologies*, Vol. 79, pp. 242–256, 2017.
10. https://ops.fhwa.dot.gov/publications/fhwahop13003/index.htm.
11. T. Zeng, O. Semiari, W. Saad, and M. Bennis, "Joint communication and control for wireless autonomous vehicular platoon systems," *IEEE Transactions on Communications*, Vol. 67, pp.7907–7922, 2019.
12. O. Semiari, W. Saad, M. Bennis, and M. Debbah, "Integrated millimeter wave and sub-6 GHz wireless networks: A roadmap for joint mobile broadband and ultra-reliable low-latency communications," *IEEE Wireless Communications*, Vol. 26, pp. 1–7, 2019.
13. D.J. Fagnant and K. Kockelman, "Preparing a nation for autonomous vehicles: Opportunities, barriers and policy recommendations," *Transportation Research Part A: Policy and Practice*, Vol. 77, pp. 167–181, 2015.
14. W. Bernhart and M. Winterhoff, "Autonomous driving: Disruptive innovation that promises to change the automotive industry as we know it," *Energy Consumption and Autonomous Driving, Proceeding of the 3rd CESA Automotive Electronics Congress*, Paris, Springer, 2016.
15. S. Trommer, E. Fraedrich, V. Kolarova, and B. Lenz, "Exploring user expectations on autonomous driving," *In Proceedings of the Automated Vehicles Symposium*, San Francisco, CA, June 2016.
16. T. Litman, "Autonomous vehicles implementation predictions and implications for transport planning," 2018.
17. S.A. Bagloee, et al., "Autonomous vehicles: Challenges, opportunities, and future implications for transportation policies," *Journal of Modern Transportation*, Vol. 24, No. 4, pp. 284–303, 2016.
18. R. Hussain and S. Zeadally, "Autonomous cars: Research results, issues, and future challenges," *IEEE Communications Surveys and Tutorials,* Vol. 21, No. 2, pp. 1275–1313, 2018.
19. https://new.siemens.com/global/en/products/mobility/road-solutions/traffic-management/on-the-road/smart-detection/video-detection.html.
20. P.F. Alcantarilla, M.A. Sotelo, and L.M. Bergasa, "Automatic daytime road traffic control and monitoring system," *In Proceedings of the 11th IEEE International Conference on Intelligent Transportation Systems*, Beijing, China, October 2008.
21. https://www.lasertech.com/IS-Measure-Traffic-Count.aspx.
22. https://trafficbot.rhythmtraffic.com/in-sync/.
23. https://www.sensourceinc.com/hardware/vehicle-counting-products/.
24. E. Massaro et al., "The car as an ambient sensing platform [point of view]," *Proceedings of the IEEE*, Vol. 105, No. 1, pp. 3–7, 2017.

25. X. Zhang, J. Hong, S.F.Z. Wei, J. Cao, and Y. Ren, "A novel real-time traffic information system based on wireless mesh networks," *In Proceedings of the IEEE Intelligent Transportation Systems Conference*, Seattle, WA, September 2007.
26. F. Calabrese, M. Colonna, P. Lovisolo, D. Parata, and C. Ratti, "Real-time urban monitoring using cell phones: A case study in Rome," *IEEE Transactions on Intelligent Transportation Systems*, Vol. 12, No. 1, pp. 141–151, 2011.
27. https://www.waze.com/.
28. http://inrix.com/.
29. http://www.cellint.com/.
30. Q. Ou, R.L. Bertini, J.W.C. Van Lint, and S.P. Hoogendoorn, "A theoretical framework for traffic speed estimation by fusing low-resolution probe vehicle data," *IEEE Transactions on Intelligent Transport. Systems*, Vol. 12, No. 3, pp. 747–756, 2011.
31. Y. Cheng, X. Qin, J. Jin, and B. Ran, "An exploratory shockwave approach to estimating queue length using probe trajectories," *Journal Intelligent Transportation Systems*, Vol. 16, No. 1, pp. 12–23, 2011.
32. S. Gupte and M. Younis, "Vehicular networking for intelligent and autonomous traffic management," *In Proceedings of the IEEE International Conference on Communications (ICC'12)*, Ottawa, Canada, June 2012.
33. T. Nadeem, S. Dashtinezhad, C. Liao, and L. Iftode, "Trafficview: A scalable traffic monitoring system," *In Proceedings of the IEEE International Conference on Mobile Data Management*, Berkeley, CA, pp. 13–26, 2004.
34. M.D.A. Florides, T. Nadeem, and L. Iftode, "Location-aware services over vehicular ad-hoc networks using car-to-car communication," *IEEE Journal on Selected Areas in Communications*, Vol. 25, pp. 1590–1602, 2007.
35. L. Wischoff, A. Ebner, H. Rohling, M. Lott, and R. Halfmann, "Sotis- a self-organizing traffic information system," *In Proceedings of the 57th IEEE Semiannual Vehicular Technology Conference (VTC-Spring)*, Vol. 4, pp. 2442–2446, April 2003.
36. M. Milojevic and V. Rakocevic, "Distributed road traffic congestion quantification using cooperative vanets," *In Proceedings of 13th Annual Mediterranean in Ad Hoc Networking Workshop (MED-HOC-NET*), Piran, Slovenia, pp. 203–210, June 2014.
37. M.F. Fahmy and D.N. Ranasinghe, "Discovering dynamic vehicular congestion using VANETs," *In Proceedings of the 4th International Conference on Information and Automation for Sustainability (ICIAFS 2008)*, Colombo, Sri Lanka, December 2008.
38. L. Wischhof, A. Ebner, and H. Rohling, "Information dissemination in self-organizing inter-vehicle networks," *IEEE Transactions on Intelligent Transportation Systems*, Vol. 6, No. 1, pp. 90–101, 2005.
39. R. Bauza, J. Gozalvez, and J. Sanchez-Soriano, "Road traffic congestion detection through cooperative vehicle-to-vehicle communications," *In Proceedings of the 35th IEEE Conference on Local Computer Networks (LCN 2010)*, Denver, CO, pp. 606–612, 2010.
40. F. Terroso-Sáenz, M. Valdés-Vela, C. Sotomayor-Martínez, R. Toledo-Moreo, and A.F. Gómez-Skarmeta, "A cooperative approach to traffic congestion detection with complex event processing and vanet," *IEEE Transactions on Intelligent Transportation Systems*, Vol. 13, No. 2, pp. 914–929, 2012.
41. G. Marfia and M. Roccetti, "Vehicular congestion detection and short-term forecasting: A new model with results," *IEEE Transactions on Vehicular Technology*, Vol. 60, No. 7, pp. 2936–2948, 2011.
42. I. Leontiadis, G. Marfia, D. Mack, G. Pau, C. Mascolo, and M. Gerla, "On the effectiveness of an opportunistic traffic management system for vehicular networks," *IEEE Transactions on Intelligent Transportation Systems*, Vol. 12, pp. 1537–1548, 2011.
43. H.R. Varia and S.L. Dhingra, "Dynamic optimal traffic assignment and signal time optimization using genetic algorithms," *Computer - Aided Civil and Infrastructure Engineering*, Vo. 19, No. 4, pp. 260–273, 2004.
44. K. Collins and G. Muntean, "Route-based vehicular traffic management for wireless access in vehicular environments," *In Proceedings of the 68th IEEE Vehicular Technology Conference (VTC-Spring)*, Calgary, BC, Canada, September 2008.
45. G. Araujo, F. De, L P Duarte-Figueiredo, A. Tostes, and A. Loureiro, "A protocol for identification and minimization of traffic congestion in vehicular networks," *In Proceedings of the Brazilian Symposium on Computer Networks and Distributed Systems (SBRC)*, Florianopolis, Brazil, pp. 103–112, May 2014.

46. S.C. Nanayakkara, et al., "Genetic algorithm based route planner for large urban street networks," *In Proceedings of the IEEE Congress on Evolutionary Computation (CEC 2007)*, Singapore, September 2007.
47. A. Ghazy and T. Ozkul, "Design and simulation of an artificially intelligent vanet for solving traffic congestion," *In Proceedings of the IEEE 6th International Symposium on Mechatronics and its Applications (ISMA'09)*, Sharjah, United Arab Emirates, March 2009.
48. Y. Ando, O. Masutani, and S. Honiden., "Performance of pheromone model for predicting traffic congestion," *In Proceedings of the 5th International Joint Conference on Autonomous Agents and Multi-Agent Systems (AAMAS'06)*, Future University-Hakodate, Japan, May 2006.
49. A. Ramazani and H. Vahdat-Nejad, "A new context-aware approach to traffic congestion estimation," *In Proceedings of the* 4th *International Conference on Computer and Knowledge Engineering (ICCKE)*, Washington, DC, pp. 504–508, October 2014.
50. T. Ho and T.H. Heung, "Hierarchical fuzzy logic traffic control at a road junction using genetic algorithms," *In Proceedings of the IEEE World Congress on Computational Intelligence*, Anchorage, AK, November 1998.
51. E. Horvitz, J. Apacible, R. Sarin, and L. Liao, "Prediction, expectation, and surprise: Methods, designs, and study of a deployed traffic forecasting service," *In Proceedings of the 21st Conference on Uncertainty in Artificial Intelligence(UAI 2005)*, Edinburgh, Scotland, July 2005.
52. T. Hunter, R. Herring, P. Abbeel, and A. Bayen, "Path and travel time inference from GPS probe vehicle data," *In Proceedings of the NIPS Workshop on Analyzing Networks and Learning with Graphs*, Whistler, BC, Canada, December 2009.
53. D.B. Work, O.P. Tossavainen, S. Blandin, A.M. Bayen, T. Iwuchukwu, and K. Tracton, "An ensemble Kalman filtering approach to highway traffic estimation using GPS enabled mobile devices," *In Proceedings of the 47th IEEE Conference on Decision and Control (CDC 2008)*, Cancun, Mexico, December 2008.
54. A. Bhaskar, T. Tsubota, and E. Chung, "Urban traffic state estimation: Fusing point and zone based data," *Transportation Research Part C: Emerging Technologies*, Vol. 48, pp. 120–142, 2014.
55. M. Zangui, Y. Yin, and S. Lawphongpanich, "Differentiated congestion pricing of urban transportation networks with vehicle-tracking technologies," *Transportation Research Part C: Emerging Technologies*, Vol. 36, pp. 434–445, 2013.
56. F. Soylemezgiller, M. Kuscu, and D. Kilinc, "A traffic congestion avoidance algorithm with dynamic road pricing for smart cities," *In Proceedings of the 24th IEEE International Symposium on Personal Indoor and Mobile Radio Communications (PIMRC 2013)*, London, UK, September 2013.
57. M. Zangui, Y. Yin, and S. Lawphongpanich, "Sensor location problems in path-differentiated congestion pricing," *Transportation Research Part C: Emerging Technologies*, Vol. 55, pp. 217–230, 2015.
58. B. Zhou, M. Bliemer, H. Yang, and J. He, "A trial-and-error congestion pricing scheme for networks with elastic demand and link capacity constraints," *Transportation Research Part B: Methodological*, Vol. 72, pp. 77–92, 2015.
59. J.A. Laval, H.W. Cho, J.C. Muñoz, and Y. Yin, "Real-time congestion pricing strategies for toll facilities," *Transportation Research Part B: Methodological*, Vol. 71, pp. 19–31, 2015.
60. L. Elefteriadou, S. Washburn, Y. Yin, V. Modi, and C. Letter, "Variable Speed Limit (VSL): Best management practice," Final Report to Florida Department of Transportation, 2012.
61. W. Liu, Y. Yin, and H. Yang, "Effectiveness of variable speed limits considering commuters' long-term response," *Transportation Research Part B: Methodological*, Vol. 81, Part 2, pp. 498–519, 2015.
62. Z. Song, Y. Yin, and S. Lawphongpanich, "Optimal deployment of managed lanes in general networks," *International Journal of Sustainable Transportation*, Vol. 9, No. 6, pp. 431–441, 2015.
63. F. Ahmad, S. Mahmud, G. Khan, and F. Yousaf, "Shortest remaining processing time based schedulers for reduction of traffic congestion," *In Proceedings of the International Conference on Connected Vehicles and Expo (ICCVE)*, Las Vegas, Nevada, December 2013.
64. S. Kwatirayo, J. Almhana, and Z. Liu, "Adaptive traffic light control using VANET: A case study," *In Proceedings of the 9th International Wireless Communications and Mobile Computing Conference (IWCMC 2013)*, Sardinia, Italy, pp. 752–757, July 2013.
65. K. Pandit, D. Ghosal, H. Zhang, and C.-N. Chuah, "Adaptive traffic signal control with vehicular ad hoc networks," *IEEE Transactions on Vehicular Technology*, Vol. 62, No. 4, pp. 1459–1471, 2013.

66. K. Al-Khateeb and J. Johari, "Intelligent dynamic traffic light sequence using RFID," *Journal of Computer Science*, Vol. 4, No. 7, pp. 517–524, 2008.
67. C. Hu and Y. Wang, "A novel intelligent traffic light control scheme," *In Proceedings of 9th International Conference on Grid and Cooperative Computing (GCC)*, pp. 372–376, Nanjing, China, 2010.
68. C. Li and S. Shimamoto, "An open traffic light control model for reducing vehicles' CO_2 emissions based on ETC vehicles," *IEEE Transactions on Vehicular Technology*, Vol. 61, pp. 97–110, 2012.
69. S. Tomforde, et al., "Decentralised progressive signal systems for organic traffic control," *In Proceedings of the 2nd IEEE International Conference on Self-Adaptive and Self-Organizing Systems (SASO'08)*, Venice, Italy, October 2008.
70. V. Hirankitti, J. Krohkaew, and C. Hogger, "A multi-agent approach for intelligent traffic-light control," *In Proceedings of the World Congress on Engineering*, vol. 1., Morgan Kaufmann, London, U.K., 2007.
71. X. Zheng and L. Chu, "Optimal parameter settings for adaptive traffic-actuated signal control," *In Proceedings of* 11th *International IEEE Conference on Intelligent Transportation Systems (ITSC),* Beijing, China, pp. 105–110, 2008.
72. B. Zhou, J. Cao, and H. Wu, "Adaptive traffic light control of multiple intersections in WSN-based ITS," *In Proceedings of the* 73rd *IEEE Vehicular Technology Conference (VTC Spring)*, Yokohama, Japan, May 2011.
73. S. Azimi, G. Bhatia, R. Rajkumar, and P. Mudalige, "Reliable intersection protocols using vehicular networks," *In Proceedings of the ACM/IEEE International Conference on Cyber-Physical Systems (ICCPS 2013)*, Philadelphia, PA, April 2013.
74. R. Hult, G. Campos, P. Falcone, and H. Wymeersch, "An approximate solution to the optimal coordination problem for autonomous vehicles at intersections," *In Proceedings of American Control Conference*, Chicago, IL, pp. 763–768. 2015.
75. R. Azimi, G. Bhatia, R.R. Rajkumar, and P. Mudalige, "Stip: Spatio-temporal intersection protocols for autonomous vehicles," *In Proceedings of the ACM/IEEE International Conference on Cyber-Physical Systems (ICCPS 2014)*, Berlin, Germany, April 2014.
76. https://www.nsf.gov/awardsearch/showAward?AWD_ID=1446813&HistoricalAwards=false.
77. S.D. Assimonis, T. Samaras and V. Fusco, "Analysis of the microstrip-grid array antenna and proposal of a new high-gain, low-complexity and planar long-range WiFi antenna," *IET Microwaves, Antennas and Propagation*, Vol. 12, No. 3, pp. 332–338, 2018.
78. dd-wrt.com, "About DD-WRT". www.dd-wrt.com. Retrieved October 25th, 2019.
79. A. Asadi, Q. Wang, and V. Mancuso, "A survey on device-to-device communication in cellular networks." *IEEE Communications Surveys and Tutorials*, Vol. 16, No. 4, pp. 1801–1819, 2014.
80. S. Eichler, "Performance evaluation of the IEEE 802.11 p WAVE communication standard," *In Proceedings of the 66th IEEE Vehicular Technology Conference (VTC-2007 Fall)*, Baltimore, MD, October 2007.
81. A.M.S. Abdelgader and L. Wu, "The physical layer of the IEEE 802.11 p WAVE communication standard: the specifications and challenges," *In Proceedings of the World Congress on Engineering and Computer Science (WCECS 2014)*, Vol. 2. San Francisco, CA, October 2014.
82. J.B. Kenney "Dedicated short-range communications (DSRC) standards in the United States," *Proceedings of the IEEE*, Vol. 99, No. 7, pp. 1162–1182, 2011.
83. A.A. Shahin and M. Younis, "Alert dissemination protocol using service discovery in Wi-Fi Direct," *In Proceedings of the IEEE International Conference on Communications (ICC 2015)*, London, UK, June 2015.
84. A. Shahin and M. Younis, "Efficient multi-group formation and communication protocol for Wi-Fi direct," *In Proceedings of the 40th Annual IEEE Conference on Local Computer Networks (LCN 2015)*, Clearwater Beach, FL, October 2015.
85. M. Knight and B. Seeber, "Decoding LoRa: Realizing a modern LPWAN with SDR," *Proceedings of the GNU Radio Conference*, Vol. 1. No. 1, September 2016. https://pubs.gnuradio.org/index.php/grcon/article/view/8.
86. J. Sanchez-Gomez, R. Sanchez-Iborra, and A. Skarmeta, "Transmission technologies comparison for IOT communications in smart-cities," *In Proceedings of the IEEE Global Communications Conference (GLOBECOM 2017)*, Singapore, December 2017.
87. R. Sanchez-Iborra, et al., "Performance evaluation of LoRa considering scenario conditions," *Sensors*, Vol. 18, No. 3, p. 772, 2018.

88. C. Wang and Y.-Q. Tang, "The discussion of system optimism and user equilibrium in traffic assignment with the perspective of game theory," *Transportation Research Procedia*, Vol. 25, pp. 2970–2979, 2017.
89. F. Kessels, *Traffic Flow Modelling: Introduction to Traffic Flow Theory through a Genealogy of Models.* Springer, Berlin, 2019.
90. https://ops.fhwa.dot.gov/trafficanalysistools/type_tools.htm.
91. H. Xu, H. Liu, H. Gong, "Modeling the asymmetry in traffic flow (a): Microscopic approach," *Applied Mathematical Modelling*, Vol. 37, No. 22, pp. 9431–9440, 2013.
92. R.M. Velasco and P. Saavedra, "Macroscopic models in traffic flow," *Qualitative Theory of Dynamical Systems*, Vol. 7, No. 1, pp. 237–252, 2008.
93. R. Aghamohammadi and J.A. Laval, "Dynamic traffic assignment using the macroscopic fundamental diagram: A Review of vehicular and pedestrian flow models," *Transportation Research Part B: Methodological*, 2018. doi: 10.1016/j.trb.2018.10.017.
94. A. Spiliopoulou, M. Kontorinaki, M. Papageorgiou, and P. Kopelias, "Macroscopic traffic flow model validation at congested freeway off-ramp areas," *Transportation Research Part C: Emerging Technologies*, Vol. 41, pp. 18–29, 2014.
95. G. Costeseque and A. Duret, "Mesoscopic multiclass traffic flow modeling on multi-lane sections," *In Proceedings of Transportation Research Board* 95th *Annual Meeting*, Washington, DC, January 2016.
96. B.N. Janson, "Dynamic traffic assignment for urban road networks," *Transportation Research Part B: Methodological*, Vol. 25, No. 2–3, pp. 143–161, 1991.
97. Y. Wang, W.Y. Szeto, K. Han, and T.L. Friesz, "Dynamic traffic assignment: A review of the methodological advances for environmentally sustainable road transportation applications," *Transportation Research Part B: Methodological*, Vol. 111, pp. 370–394, 2018.
98. A. Hall, S. Hippler, and M. Skutella, "Multicommodity flows over time: Efficient algorithms and complexity," *Theoretical Computer Science*, Vol. 379, No. 3, pp. 387–404, 2007.
99. W.L. Gisler and N.J. Rowan, "Development of fiberoptic sign displays for dynamic lane assignment," Technical report, Texas Transportation Institute, Austin, Texas, June 1992.
100. H. Mirzaei and T. Givargis, "Fine-grained acceleration control for autonomous intersection management using deep reinforcement learning," *In Proceedings of IEEE SmartWorld, Ubiquitous Intelligence & Computing, Advanced & Trusted Computed, Scalable Computing & Communications, Cloud & Big Data Computing, Internet of People and Smart City Innovation (SmartWorld/SCALCOM/UIC/ATC/CBDCom/IOP/SCI)*, San Francisco, CA, August 2017.
101. J. Wu, A. Abbas-Turki, and A. El Moudni, "Cooperative driving: An ant colony system for autonomous intersection management." *Applied Intelligence*, Vol. 37, No. 2, pp. 207–222, 2012.
102. D. Carlino, S.D. Boyles, and P. Stone, "Auction-based autonomous intersection management." *In Proceedings of the* 16th *International IEEE Conference on Intelligent Transportation Systems (ITSC 2013)*, The Hague, Netherlands, 2013.
103. G.R. de Campos, et al., "Traffic coordination at road intersections: Autonomous decision-making algorithms using model-based heuristics," *IEEE Intelligent Transportation Systems Magazine*, Vol. 9, No. 1, pp. 8–21, 2017.
104. A.A. Malikopoulos, C.G. Cassandras, and Y.J. Zhang, "A decentralized energy-optimal control framework for connected automated vehicles at signal-free intersections," *Automatica*, Vol. 93, pp. 244–256, 2018.
105. F. Altché, X. Qian, and A. de La Fortelle, "An algorithm for supervised driving of cooperative semi-autonomous vehicles (extended)." arXiv preprint arXiv:1706.08046, 2017.
106. T.-C. Au, S. Zhang, and P. Stone, "Autonomous intersection management for semi-autonomous vehicles." In: *Routledge Handbook of Transportation edited by Dusan Teodorovic*. Routledge, Abingdon, pp. 116–132, 2015.
107. J. Lee and B. Park, "Development and evaluation of a cooperative vehicle intersection control algorithm under the connected vehicles environment." *IEEE Transactions on Intelligent Transportation Systems*, Vol. 13, No. 1, pp. 81–90, 2012.
108. Q. Lu and K.-D. Kim, "Intelligent intersection management of autonomous traffic using discrete-time occupancies trajectory." *Journal of Traffic and Logistics Engineering*, Vol. 4, No. (1), pp. 1–6, 2016.
109. J.H. Dahlberg and V. Tuul, "Intelligent Traffic Intersection Management Using Motion Planning for Autonomous Vehicles", 2017.

110. M. Pourmehrab, et al., "Optimizing signalized intersections performance under conventional and automated vehicles traffic." *IEEE Transactions on Intelligent Transportation Systems*, Vol. 21, No. 7, pp. 2864–2873, 2019.
111. Q, Lu and K.-D. Kim, "Autonomous and connected intersection crossing traffic management using discrete-time occupancies trajectory." *Applied Intelligence*, Vol. 49, No. 5, pp. 1621–1635, 2019.
112. M. Hausknecht, T. Au, and P. Stone, "Autonomous intersection management: Multi-intersection optimization," *2011 IEEE/RSJ International Conference on Intelligent Robots and Systems(IROS)*, San Francisco, CA, September 2011.
113. X. Meng and C.G. Cassandras, "A real-time optimal ECO-driving for autonomous vehicles crossing multiple signalized intersections." arXiv preprint arXiv:1901.11423, 2019.
114. P. Li and X. Zhou, "Recasting and optimizing intersection automation as a connected-and-automated-vehicle (CAV) scheduling problem: A sequential branch-and-bound search approach in phase-time-traffic hypernetwork." *Transportation Research Part B: Methodological*, Vol. 105, pp. 479–506, 2017.
115. G. Sharon, S.D. Boyles, and P. Stone, "Intersection management protocol for mixed autonomous and human-operated vehicles." *Transportation Research Part C: Emerging Technologies (Under submission TRC-D-17-00857)*, 2017.
116. C. Wuthishuwong and A.P. Traechtler, "Consensus-based local information coordination for the networked control of the autonomous intersection management." *Complex and Intelligent Systems*, Vol. 3, No. 1, pp. 17–32, 2017.
117. J.J.B. Vial, et al., "Scheduling autonomous vehicle platoons through an unregulated intersection." arXiv preprint arXiv:1609.04512, 2016.
118. P. Dai, et al., "Quality-of-experience-oriented autonomous intersection control in vehicular networks." *IEEE Transactions on Intelligent Transportation Systems*, Vol. 17, No. 7, pp. 1956–1967, 2016.
119. Q. Guo, L. Li, and X. Ban, "Urban traffic signal control with connected and automated vehicles: A survey," *Transportation Research Part C: Emerging Technologies*, Vol. 101, pp. 313–334, 2019.
120. C. Hu and Y. Wang, "A novel intelligent traffic light control scheme," *In Proceedings of the* 9th *International Conference on Grid and Cooperative Computing (GCC)*, pp. 372–376, Nanjing, China, 2010.
121. B. Zhou, J. Cao, X. Zeng, and H. Wu, "Adaptive traffic light control in wireless sensor network-based intelligent transportation system," *In Proceedings of the 72nd IEEE Vehicular Technology Conference*, Ottawa, Canada, 2010.
122. S. Sameh, A. El-Mahdy, and Y. Wada, "A Linear Time and Space Algorithm for Optimal Traffic-Signal Duration at an Intersection," *IEEE Transactions on Intelligent Transportation Systems*, Vol. 16, No.1, pp. 387–395, 2015.
123. H. Prothmann, et al., "Organic traffic light control for urban road networks," *International Journal of Autonomous and Adaptive Communications Systems*, Vol. 2, pp. 203–225, 2009.
124. S. Tomforde, H. Prothmann, J. Branke, J. Hähner, C. Müller-Schloer, and H. Schmeck, "Possibilities and limitations of decentralised traffic control systems," *In Proceedings of the IEEE World Congress on Computational Intelligence*, Barcelona, Spain, pp. 3298–3306, July 2010.
125. S. Lee, M. Younis, A. Murali, and M. Lee, "Dynamic local vehicular flow optimization using real-time traffic conditions at multiple road intersections," *IEEE Access*, Vol. 99, p. 1, 2019.
126. M. Collotta, L.L. Bello, and G. Pau, "A novel approach for dynamic traffic lights management based on Wireless Sensor Networks and multiple fuzzy logic controllers." *Expert Systems with Applications*, Vol. 42, pp. 5403–5415, 2015.
127. M. Elgarej, M. Khalifa, and M. Youssfi, "Traffic lights optimization with distributed ant colony optimization based on multi-agent system," *International Conference on Networked Systems*, Springer International Publishing, Marrakech, Morocco, 2016.
128. B. Abdulhai, S. El-Tantawy, and H. Abdelgawad, "Multiagent reinforcement learning for integrated network of adaptive traffic signal controllers (marlin-atsc): Methodology and large-scale application on downtown Toronto," *IEEE Transactions on Intelligent Transportation Systems*, Vol. 14, No. 3, pp. 1140–1150, 2013.
129. I. Dusparic and V. Cahill, "Autonomic multi-policy optimization in pervasive systems: Overview and evaluation," *ACM Transactions on Autonomous and Adaptive Systems (TAAS)*, Vol. 7, No. 1, p. 11, 2012.
130. A.L.C. Bazzan, D. de Oliveira, and B.C. da Silva, "Learning in groups of traffic signals," *Engineering Applications of Artificial Intelligence*, Vol. 23, No. 4, pp. 560–568, 2010.

131. M. Abdoos, N. Mozayani, and A.L.C. Bazzan, "Holonic multiagent system for traffic signals control," *Engineering Applications of Artificial Intelligence*, Vol. 26, No. 5–6, pp. 1575–1587, 2013.
132. X.-F. Xie, S.F. Smith, and G.J. Barlow, "Schedule-driven coordination for real-time traffic network control," *In 22nd International Conference on Automated Planning and Scheduling (ICAPS)*, pp. 323–331, Atibaia, São Paulo, Brazil, 2012.
133. L. Zhang and G. Wu, "Dynamic lane grouping at isolated intersections: Problem formulation and performance analysis," *Transportation Research Record*, Vol. 2311, No. 1 pp. 152–166, 2012.
134. W.K.M. Alhajyaseen, et al., "The effectiveness of applying dynamic lane assignment at all approaches of signalized intersection," *Case Studies on Transport Policy* Vol. 5, No. 2, pp. 224–232, 2017.
135. K.J. Assi and N.T. Ratrout, "Proposed quick method for applying dynamic lane assignment at signalized intersections," *IATSS Research*, Vol. 42, No. 1, pp. 1–7, 2018.
136. W.K.M. Alhajyaseen, et al., "The integration of dynamic lane grouping technique and signal timing optimization for improving the mobility of isolated intersections," *Arabian Journal for Science and Engineering*, Vol. 42, No. 3, pp. 1013–1024, 2017.
137. X. Li, H. Wang, and J. Chen, "Dynamic lane-use assignment model at signalized intersections under tidal flow," *ICTE 2013: Safety, Speediness, Intelligence, Low-Carbon, Innovation*, pp. 2673–2678, Singapore, Singapore, 2013.
138. Z. Zhong, et al., "An optimization method of dynamic lane assignment at signalized intersection," *IEEE International Conference on Intelligent Computation Technology and Automation (ICICTA)*, Vol. 1, Hunan, China, 2008.
139. G. Wu, et al., "Simulation-based benefit evaluation of dynamic lane grouping strategies at isolated intersections," *2012 15th International IEEE Conference on Intelligent Transportation Systems*, Anchorage, AK, USA, 2012.
140. X. Jiang, R. Jagannathan, and D. Hale, "Dynamic lane grouping at signalized intersections: Selecting the candidates and evaluating performance," *Institute of Transportation Engineers, ITE Journal*, Vol. 85, No. 11, pp. 42–47, 2015.
141. M. Duell, et al., "System optimal dynamic lane reversal for autonomous vehicles," *2015 IEEE 18th International Conference on Intelligent Transportation Systems*, Spain, 2015.
142. M.W. Levin and S.D. Boyles, "A cell transmission model for dynamic lane reversal with autonomous vehicles." *Transportation Research Part C: Emerging Technologies*, Vol. 68, pp. 126–143, 2016.
143. K.F. Chu, A.Y.S. Lam, and V. OK Li, "Dynamic lane reversal routing and scheduling for connected autonomous vehicles." *In Proceedings of the International Smart Cities Conference (ISC2)*, Wuxi, China, September 2017.
144. X. Li, J. Chen, and H. Wang, "Study on Flow Direction Changing Method of Reversible Lanes on Urban Arterial Roadways in China," *Procedia: Social and Behavioral Sciences*, Vol. 96, pp. 807–816, 2013.
145. C. Krause, N. Kronpraset, J. Bared, and W. Zhang, "Operational advantages of dynamic reversible left-lane control of existing signalized diamond interchanges," *Journal of Transportation Engineering*, Vol. 141, No. 5, pp. 73–75, 2014.
146. J. Zhao, Y. Liu, and X. Yang, "Operation of signalized diamond interchanges with frontage roads using dynamic reversible lane control," *Transportation Research Part C: Emerging Technologies*, Vol. 5, pp. 196–209, 2015.
147. J. Zhao, et al., "Increasing the capacity of signalized intersections with dynamic use of exit lanes for left-turn traffic," *Journal of the Transportation Research Board*, Vol. 2355, No. 1, pp. 49–59, 2013.
148. S. Kim, S. Shekhar, and M. Min, "Contraflow transportation network reconfiguration for evacuation route planning," *IEEE Transactions on Knowledge and Data Engineering*, Vol. 20, No. 8, pp. 1115–1129, 2008.
149. C. Xie and M.A. Turnquist, "Lane-based evacuation network optimization: An integrated Lagrangian relaxation and tabu search approach." *Transportation Research Part C: Emerging Technologies*, Vol. 19, No. 1, pp. 40–63, 2011.
150. X.M. Zhang, S. An, and B.L. Xie, "A cell-based regional evacuation model with contra-flow lane deployment," *Advanced Engineering Forum*, Vol. 5, *Trans Tech Publications*, pp. 20-25, 2012.
151. J.W. Wang, et al., "Evacuation planning based on the contraflow technique with consideration of evacuation priorities and traffic setup time." *IEEE Transactions on Intelligent Transportation Systems*, Vol. 14, No. 1, pp. 480–485, 2012.

152. J. Hua, et al., "An integrated contraflow strategy for multimodal evacuation," *Mathematical Problems in Engineering*, Vol. 2014, pp. 1-10, 2014.
153. U. Pyakurel and S. Dempe, "Earliest arrival flow with partial lane reversals for evacuation planning," *International Journal of Operational Research/Nepal (IJORN)*, Vol. 8, No. 1, 27–37, 2019.
154. T.N. Dhamala, U. Pyakurel, and R.C. Dhungana, "Abstract contraflow models and solution procedures for evacuation planning," *Journal of Mathematics Research*, Vol. 10, No. 4, pp. 89–100, 2018.
155. U. Pyakurel, S. Dempe, and T.N. Dhamala, "Efficient algorithms for flow over time evacuation planning problems with lane reversal strategy," TU Bergakademie Freiberg, 2018.
156. U. Pyakurel, H.N. Nath, and T.N. Dhamala, "Partial contraflow with path reversals for evacuation planning," *Annals of Operations Research*, Vol. 283, 591–612, 2019.
157. W.W. Zhou, et al., "An intelligent traffic responsive contraflow lane control system," *Proceedings of VNIS'93-Vehicle Navigation and Information Systems Conference*, Ottawa, Ontario, Canada, Canada, 1993.
158. M. Hausknecht, et al., "Dynamic lane reversal in traffic management," *2011 14th International IEEE Conference on Intelligent Transportation Systems (ITSC)*, Washington, DC, 2011.
159. T. Lu, Z. Yang, D. Ma, and S. Jin, "Bi-level programming model for dynamic reversible lane assignment," *IEEE Access*, Vol. IL6, pp. 71592–71601, 2018.
160. P. Varaiya, "Smart cars on smart roads: Problems of control," *IEEE Transactions on Automatic Control*, Vol. 38, No. 2, pp. 195–207, February 1993.
161. C. Katrakazas, M. Quddus, W.-H. Chen, and L. Deka, "Real-time motion planning methods for autonomous on-road driving: State-of-the-art and future research directions," *Transportation Research Part C: Emerging Technologies*, Vol. 60, pp. 416–442, 2015.
162. S. Zhang, W. Deng, Q. Zhao, H. Sun, and B. Litkouhi, "Dynamic trajectory planning for vehicle autonomous driving," *In Proceedings of the IEEE International Conference on Systems, Man, and Cybernetics*, Manchester, UK, pp. 4161–4166, October 2013.
163. L. Ma, J. Yang, and M. Zhang, "A two-level path planning method for on-road autonomous driving," *In Proceedings of the 2nd International Conference on Intelligent System Design and Engineering Application*, Sanya, Hainan, China, pp. 661–664, January 2012.
164. J. Lee and B. Park, "Development and evaluation of a cooperative vehicle intersection control algorithm under the connected vehicles environment," *IEEE Transactions on Intelligent Transportation Systems*, Vol. 13, No. 1, pp. 81–90, 2012.
165. Z. Li, et al., "Temporal-spatial dimension extension-based intersection control formulation for connected and autonomous vehicle systems", *Transportation Research: Part C Emerging Technologies*, Vol. 104, pp. 234–248, 2019.
166. D. Chen, S. Ahn, M. Chitturi, and D.A. Noyce, "Towards vehicle automation: Roadway 32 capacity formulation for traffic mixed with regular and automated vehicles," *Transportation Research Part B: Methodological*, Vol. 100, pp. 196–221, 2017.
167. H. Jiang, J. Hu, S. An, M. Wang, and B.B. Park, "Eco approaching at an isolated signalized intersection under partially connected and automated vehicles environment," *Transportation Research Part C: Emerging Technologies*, Vol. 79, pp. 290–307, 2017.
168. J. Ma, X. Li, F. Zhou, J. Hu, and B.B. Park, "Parsimonious shooting heuristic for trajectory design of connected automated traffic part II: Computational issues and optimization," *Transportation Research Part B: Methodological*, vol. 95, pp. 421–441, 2017.
169. M. Wang, T. Ganjineh, and R. Rojas, "Action annotated trajectory generation for autonomous maneuvers on structured road networks," *In Proceedings of the* 5th *International Conference on Automation, Robotics and Applications*, Wellington, New Zealand, pp. 67–72, 2011.
170. T. Gu and J.M. Dolan, "Toward human-like motion planning in urban environments," *In Proceedings of the IEEE Intelligent Vehicles Symposium (IV 2014)*, Dearborn, MI, pp. 350–355, June 2014.
171. Y. Wei, et al., "Dynamic programming-based multi-vehicle longitudinal trajectory optimization with simplified car following models," *Transportation Research Part B: Methodological*, Vol. 106, pp. 102–129, 2017.
172. F. Zhou, X. Li, and J. Ma, "Parsimonious shooting heuristic for trajectory design of connected automated traffic part I: Theoretical analysis with generalized time geography," *Transportation Research Part B: Methodological*, Vol. 95, pp. 394–420, 2017.

173. L. Chai, B. Cai, W. ShangGuan, J. Wang, and H. Wang, "Connected and autonomous vehicles coordinating approach at intersection based on space–time slot," *Transportmetrica A: Transport Science*, Vol. 14, No. 10, pp. 929–951, 2018.
174. K. Dresner and P. Stone, "A multi-agent approach to autonomous intersection management," *Journal of Artificial Intelligence Research*, Vol. 31, No. 1, pp. 591–656, 2008.
175. D. Chen, A. Srivastava, S. Ahn, and T. Li, "Traffic dynamics under speed disturbance in mixed traffic with automated and non-automated vehicles," *Transportation Research Part C: Emerging Technologies*, 2019. doi: 10.1016/j.trc.2019.03.017.
176. S. Gong, J. Shen, and L. Du, "Constrained optimization and distributed computation based car following control of a connected and autonomous vehicle platoon," *Transportation Research Part B: Methodological*, Vol. 94, pp. 314–334, 2016.
177. W. Gao, Z.P. Jiang, and K. Ozbay, "Data-driven adaptive optimal control of connected vehicles," *IEEE Transactions on Intelligent Transportation Systems*, Vol. 18, No. 5, pp. 1122–1133, 2017.
178. S. Lefèvre, A. Carvalho, and F. Borrelli, "A learning-based framework for velocity control in autonomous driving. *IEEE Transactions on Automation Science and Engineering*, Vol. 13, No. 1, pp. 32–42, 2016.
179. A. Talebpour, H. Mahmassani, and S. Hamdar, "Multi-regime sequential risk-taking model of car-following behavior: Specification, calibration, and sensitivity analysis," *Transportation Research Record: Journal of the Transportation Research Board*, Vol. 2260, No. 1, pp. 60–66, 2011.
180. W. Weili, J. Zheng, and H.X. Liu., "A capacity maximization scheme for intersection management with automated vehicles," *Transportation Research Procedia*, Vol. 23, pp. 121–136, 2017.
181. T. Kosch, C. Schroth, M. Strassberger, and M. Bechler, *Automotive Inter-Networking*. John Wiley & Sons, Hoboken, NJ, April 9, 2012.
182. H. Hartenstein and K. Laberteaux, *VANET Vehicular Applications and Inter-Networking Technologies.* John Wiley & Sons, Hoboken, NJ, February 2010.
183. F.D. Da Cunha, A. Boukerche, L. Villas, A. Viana, and A.A.F. Loureiro, "Data communication in VANETs: A survey, challenges and applications," Research Report RR-8498, INRIA Saclay, 2014.
184. T. Kitani, T. Shinkawa, N. Shibata, K. Yasumoto, M. Ito, and T. Higashinoz, "Efficient VANET-based traffic information sharing using buses on regular routes," *In Proceedings IEEE Vehicular Technology Conference (VTC Spring)*, pp. 3031–3036, Singapore, Singapore, May 2008.
185. J. Pan, I.S. Popa, K. Zeitouni, and C. Borcea, "Proactive vehicular traffic re-routing for lower travel time," *IEEE Transactions on Vehicular Technology*, Vol. 62, No. 8, pp. 3551–3568, 2013.
186. A. Dua, N. Kumar, and S. Bawa, "A systematic review on routing protocols for Vehicular Ad Hoc Networks", *Vehicular Communications*, Vol. 1, No. 1, pp. 33–52, 2014.
187. U.S. Department of Transportation, "Report to congress on catastrophic hurricane evacuation plan evaluation," June 2006. http://www.fhwa.dot.gov/reports/hurricanevacuation/.
188. A. Shahin and M. Younis, "A framework for P2P networking of smart devices using wi-fi direct," *In Proceedings of the 25th IEEE International Symposium on Personal, Indoor and Mobile Radio Communications (PIMRC* 2014*)*, Washington, DC, September 2014.
189. A. Shahin and M. Younis, "Alert dissemination protocol using service discovery in wi-Fi direct," *In Proceedings of the IEEE International Conference on Communications (ICC 2015)*, London, UK, June 2015.
190. A. Shahin and M. Younis, "Efficient multi-group formation and communication protocol for wi-fi direct," *In Proceedings of the 40th Annual IEEE Conference on Local Computer Networks (LCN 2015)*, Clearwater Beach, FL, October 2015.
191. K. Rabieh, M. Mahmoud, T. Guo, and M. Younis, "Privacy-preserving route reporting scheme for traffic management in VANETs", *In Proceedings of the IEEE International Conference on Communications (ICC 2015)*, London, UK, June 2015.
192. K. Rabieh, M. Mahmoud, T. Guo, and M. Younis, "Cross-layer scheme for detecting large-scale colluding Sybil attack in VANETs", *In Proceedings of the IEEE International Conference on Communications (ICC 2015)*, London, UK, June 2015.
193. S. Olariu, M. Eltoweissy, and M. Younis, "Towards autonomous vehicular clouds," *ICST Transactions on Mobile Communications and Applications*, Vol. 11, No. 7–9, pp. 1–11, 2011.

194. K. Rabieh, M. Mahmoud, and M. Younis, "Privacy-preserving route reporting schemes for traffic management systems," *IEEE Transactions on Vehicular Technology (TVT)*, vol. 66, pp. 2703–2713, March 2017.
195. A. Sherif, K. Rabieh, M. Mahmoud, and X. Liang, "Privacy-preserving ride sharing scheme for autonomous vehicles in big data era," *IEEE Journal on of Things (IoT)*, Vol. 4, pp. 611–618, April 2017.
196. M. Nabil, M. Mahmoud, A. Sherif, A. Alsharif, and M. Abdulla, "Efficient privacy-preserving ride sharing organization for transferable and non-transferable services," *IEEE Transactions on Dependable and Secure Computing (TDSC)*, Vol. 11, No. 7, 1–11, 2019.
197. M. Baza, M. Nabil, N. Bewermeier, K. Fidan, M. Mahmoud, and M. Abdallah, "Detecting sybil attacks using proofs of work and location in VANETs," arXiv:1904.05845, April 2019.
198. A. Wegener, et al., "Designing a decentralized traffic information system: Autonomos," *In Proceedings of the* 16th *ITG/GI - Fachtagung Kommunikation in Verteilten Systemen (KiVS)*, Kassel, Germany, March 2009.
199. US National Highway Transportation Safety Authority, Traffic safety facts, 2017, https://crashstats.nhtsa.dot.gov/Api/Public/ViewPublication/812806.

6 VANET Communication and Mobility Sustainability

Interactions and Mutual Impacts in Vehicular Environment

Ahmed Elbery and Hesham A. Rakha
Virginia Tech

CONTENTS

6.1 INTRODUCTION

Traffic congestion and related air pollution and fuel consumption are critical challenges for smart cities. These problems are attributed to the continuous increase in the cities' population. Specifically, the United Nations (UN) reported in [52] that 54% of the world's population lives in cities, and the forecasted population will be 66% or higher in 2050. Consequently, in a typical smart city, among the most important objectives is the efficient and sustainable transportation, which does not only improve the transportation system mobility but also improves environmental sustainability, i.e., reducing the fuel consumption and pollutant emissions. To achieve these objectives, smart cities rely on integrating information and communication technologies and applying advanced computation methodologies into the transportation system to operate the system more efficiently and to maximize the benefits by better managing the road network resources. This integration builds what is known as Intelligent Transportation Systems (ITSs) [24], which are expected to be the core of future transportation systems.

In ITS, networked sensors, microchips, and communication devices work together to collect, process, and disseminate information about the transportation system. Consequently, enabling the traffic management centre (TMC) to make better-informed decisions to improve the performance of the overall transportation system. For example, the work in [15] on vehicular crowd management depends on a communication system to collect data in real-time and assign routes to vehicles. The correctness and the accuracy of these decisions depend on the accuracy of the data collected at the TMC. Therefore, the communication network is a major component in such feedback ITS applications, as it is responsible for exchanging data between the different sensors/actuators in the network and the TMC.

The vehicular ad hoc network (VANET) [26] was proposed to serve as the communication backbone for ITS applications. Consequently, when studying an ITS application performance, it is imperative to study the impact of VANET communication reliability parameters (packet delay and drop rate) on the mobility and sustainability of the ITS application. Alternatively, the communication system performance is affected by mobility parameters (e.g., speed, traffic demand level, and congestion). For example, navigation applications control vehicle spatial distribution on the road network, consequently altering the vehicle density, which is a major parameter affecting VANET performance. This bidirectional interdependence between mobility and communication creates a mutual impact loop, where mobility affects communication and communication affects mobility. Therefore, understanding such systems, including mobility and communication, is crucial to develop and deploy optimized ITS applications.

Among the ITS techniques, feedback-based eco-routing navigation [19] is a critical application that attempts to minimize the fuel consumption of individual vehicles by routing them along fuel-efficient routes. It utilizes fuel consumption cost information collected in real-time from vehicles currently moving on the network. The eco-routing navigation includes both directions of influence where vehicles' mobility and distribution affect the communication performance, and on the other direction, the communication performance can influence the route assignment, consequently, the mobility and distribution of the vehicles.

Studying and modelling such systems are challenging not only because of the intricate interdependency of the communication and the mobility components but also because of the scale at which the ITS works. Most of the ITS applications operate on a city-level road network that mandates the modelling of tens or hundreds of thousands of concurrent vehicles, which makes such studies more challenging.

To understand how vehicle mobility and vehicular communication affect one another, one needs to understand some basic concepts in both mobility and communication. Therefore, the next section, Section 6.2, introduces some traffic engineering fundamentals, then describes eco-routing navigation. Subsequently, Section 6.3 introduces basic concepts of VANET and its characteristics.

Section 6.4 introduces the model we use to simulate VANET in a city-level road network. Subsequently, how this communication model is integrated with traffic modelling is described. Finally, the case study in the city of Los Angles is presented.

6.2 MOBILITY AND TRAFFIC ENGINEERING FUNDAMENTALS

This section provides an overview of traffic flow fundamentals and the methodologies for modelling transportation systems. It also presents how cars determine the speed and how they follow one another on the roadway, known as car-following models. Finally, this section describes navigation techniques with a focus on eco-routing navigation.

6.2.1 CAR-FOLLOWING BEHAVIOUR

In traffic theory, how cars follow one another and how to determine a car speed at a given time on a given road depend on many parameters. Some parameters reflect the surrounding environment, such as the vehicle headway spacing (h_n) (the distance between a point on the subject car and the same point on the leading car in the same lane; e.g. front to front bumper or rear to rear bumper) and the speed of the leading car (v_{n-1}), as shown in Figure 6.1. Some others represent the road network such as the speed limit and the road capacity. Car-following models are methods that use these parameters to determine how vehicles follow one another on a roadway. So, they characterize the vehicle's longitudinal motion and govern the behaviour of the traffic stream by determining the desired speed at different levels of congestion.

All the car-following models maintain a safety distance between cars following one another. This vehicle spacing is a function of the vehicle speed; the higher the speed, the longer the distance. Thus, the higher the speed, the lower the vehicle density on the road. This relation is called the fundamental diagram (FD) [51], which represents the relationship between speed and density. Figure 6.2 shows the FD calibrated to empirical data.

This FD is crucial for communication in a vehicular environment. The medium access technique in VANET is a contention-based mechanism, where node density is a major parameter. *Thus, this relationship transfers the traffic flow state into communication parameters.*

In Figure 6.2, the right figure, represents the fundamental relationship between the vehicle density and the flow rate on a road segment. This is an important relationship because it identifies

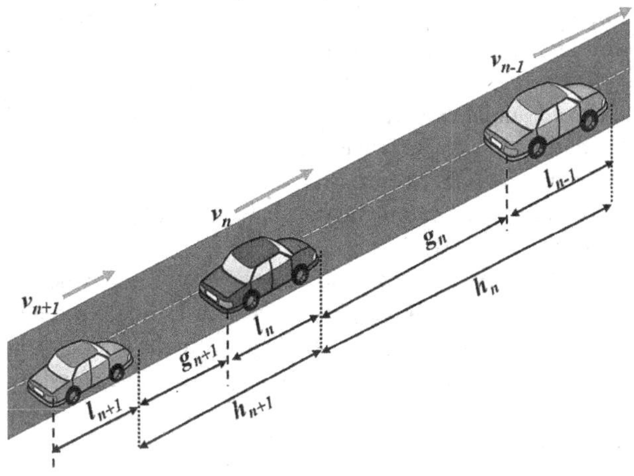

Figure 6.1 Car-following on roads, v_n is a function of h_n, v_{n-1}, and road parameters.

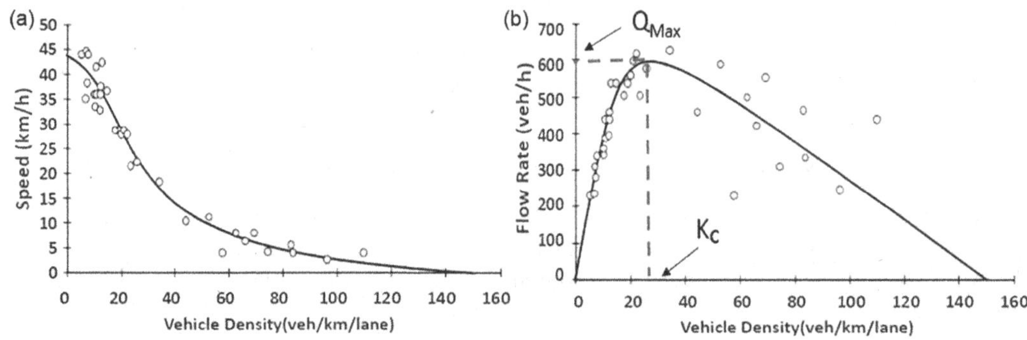

Figure 6.2 The fundamental relationships between vehicle density, speed, and traffic flow [51]. The right figure shows the inverse relationship between vehicle density and speed. The right figure shows the maximum traffic rate Q_{Max} and the jam density K_c of a road segment

two characteristics of the road segment. The first is Q_{Max}, which is the maximum flow that can exit a road segment at any time. The second is K_c which is called the jam density. Having these two parameters for a road segment along with the maximum speed and speed-at-capacity, we can completely characterize the traffic behaviour on this road segment. For example, once the vehicle density on a road segment exceeds K_c (density-at-capacity), the link enters the congested regime, and its throughput starts to decrease as its traffic flow is less than Q_{Max}, as shown in Figure 6.2. So, the objective of any traffic management system is to control network congestion by maintaining the traffic on each road segment less than or equals to K_c.

6.2.2 TRAFFIC MODELLING TECHNIQUES

In traffic theory, many techniques are used to model the transportation system. These techniques can be categorized into three basic types: macroscopic, mesoscopic, and microscopic algorithms.

1. **Macroscopic modelling**

   Macroscopic modelling techniques compute the average number of vehicles traversing each road links based on which the average link parameters can be computed, including speed, travel time, fuel consumption, and emissions. Because of this simplicity, macroscopic models are characterized by high scalability and low computational cost. However, the high scalability and low computational requirements come at the cost of the accuracy, as by using the average link parameters, many transportation phenomena cannot be captured, such as acceleration/deceleration events. These parameters are the predominant factor affecting travel time, delay, fuel consumption, and emissions. Macroscopic systems also cannot model the interactivity between vehicles such as lane changing, gap acceptance techniques, etc. Therefore, macroscopic models lack the needed accuracy.
2. **Mesoscopic modelling**

   Between these two extremes, mesoscopic models are located, where each road link is represented by a queue. These mesoscopic models can estimate a time-varying average for the link parameters based on the number of vehicles in the queue. With the help of the queuing theory, mesoscopic models compute different travel times for different vehicles on the same link at the same time. Therefore, mesoscopic models are more accurate than the macroscopic models. However, because mesoscopic models use average parameters, do not model the trajectories of vehicles on links, and ignore lane-changing behaviour; they are of lower accuracy compared to microscopic models.

3. **Microscopic modelling**

   On the other extreme, microscopic models achieve the highest possible accuracy at the cost of computation and scalability. To achieve the highest accuracy, microscopic models track every individual vehicle on a second-by-second basis. Microscopic models can model and capture all the events that happen in the road network. They also can capture the interactivity between the vehicles on the road as well as between vehicles and control systems such as traffic signals. However, the computation and memory requirements for this level of accuracy limit the scalability and reduce the simulation speed.

Because of the high accuracy of the microscopic modelling, we utilize it in this chapter to model the mobility of vehicles. Therefore, we use the INTEGRATION software [45]. The INTEGRATION software is a microscopic traffic assignment and simulation software that was developed over the past four decades [45,53,54]. It was conceived as an integrated simulation and traffic assignment model and performs traffic simulations by tracking the movement of individual vehicles every decisecond. This allows detailed analyses of lane-changing movements and shock wave propagation. It also permits considerable flexibility in representing spatial and temporal variations in traffic conditions. In addition to estimating stops and delays [10,42,47], the model can also estimate the fuel consumed by individual vehicles, as well as the emissions [4,39]. Finally, the model also estimates the expected number of vehicle crashes using a time-based crash prediction model [6]. The INTEGRATION model was chosen for this study due to its high accuracy in estimating fuel consumption. In addition, the model has not only been validated against standard traffic flow theory [1,10,36,42], but also has been utilized for the evaluation of real-life applications [35,41].

6.2.3 TRAFFIC NAVIGATION

The vehicle route assignment is one of the most important and challenging applications in ITS. Efficient routing can significantly reduce congestion, consequently, save travel time and energy consumption and reduce pollutant emissions. However, this is a challenging task because of the complex relationships between the different cost functions, such as distance, travel time, or fuel consumption. For example, previous studies have shown that standard navigation systems can provide travellers accurate minimum path calculations based on either the shortest distance or the shortest travel time that can achieve some fuel savings [7]. However, previous studies have also shown that in many cases the shortest distance route can produce higher fuel consumption and emission levels due to road grade or higher congestion levels. Similarly, using the shortest travel time routes may also result in higher fuel consumption and emission levels. An example of this case was demonstrated in Ahn and Rakha [3], where significant savings in fuel consumption and emission levels were produced by using the longer travel time and shorter-distance arterial routes.

Thus, most of the navigation systems focus on a single cost function. The most commonly used cost function is the travel time. But, recently, eco-routing [8,19,20,23,34] was proposed to minimize the fuel consumption for individual vehicles. Eco-routing utilizes the fuel consumption cost to route vehicles through the most environmentally friendly routes.

However, even for such a single cost function, there is another challenge. Efficient navigation requires real-time information about the road cost, for which, the best way is to utilize crowdsensing to collect link cost data from drivers through connected vehicles in real-time. For travel time-based navigation, the link cost can be collected from smartphones equipped with a Global Positioning System (GPS), where the smartphone can estimate the travel time it passes on each road link. However, for eco-routing, a GPS-enabled smartphone is not sufficient. It is possible to compute the energy consumption on a smartphone using GPS trajectory data as described in [40]. But, to produce accurate estimation, it needs the vehicle specific information such as vehicle mass, engine size, frontal area, and friction coefficient parameters [55]. So, the most accurate way is to utilize the vehicle's on-board computer and the GPS system to compute the energy/fuel consumption on every road segment that the vehicle passes. The vehicles would then communicate this information to the central navigation system in the TMC. Such a system is called feedback-based eco-routing. In such a system, vehicular communication is an integral component that is responsible for the exchange of information between vehicles and the TMC.

6.2.4 ECO-ROUTING NAVIGATION

Eco-routing navigation techniques were introduced to minimize vehicle fuel consumption and emissions by utilizing the route fuel cost as a metric, based on which, the most environmentally friendly route is selected. Developing and deploying eco-routing navigation techniques are very challenging. One major challenge is the estimation of the route fuel cost. This challenge comes from the fact that the route fuel cost is a function of many parameters, including the route characteristics (i.e., length, maximum speed, grade), vehicle characteristics (e.g., weight, shape, engine, and power), and driving behaviour (vehicle trajectory). It has been demonstrated that it is too difficult to combine all these parameters in a single model, especially because many of these parameters are stochastic and there is a complex interdependency between them.

Therefore, the best way to calculate the route cost is to use a feedback system that collects this data in real-time and fuses it with historical data to estimate the route fuel cost and consequently calculate the best route for the vehicles travelling in real-time [45]. This feedback system is simple and accurately estimates the route cost. But, on the other hand, it requires a communication infrastructure through which this information can be exchanged. In addition to communication, vehicles should be capable of quantifying fuel consumption for each road link. Such a feedback system is illustrated in Figure 6.3, where vehicles use communication to report the fuel cost they experience to the TMC and to request routing information.

Eco-routing was initially introduced in 2006 and was applied to the street network in the city of Lund, Sweden, to select the route with the lowest total fuel consumption and thus the lowest total CO_2 emissions [23]. The streets were divided into 22 classes based on the fuel consumption

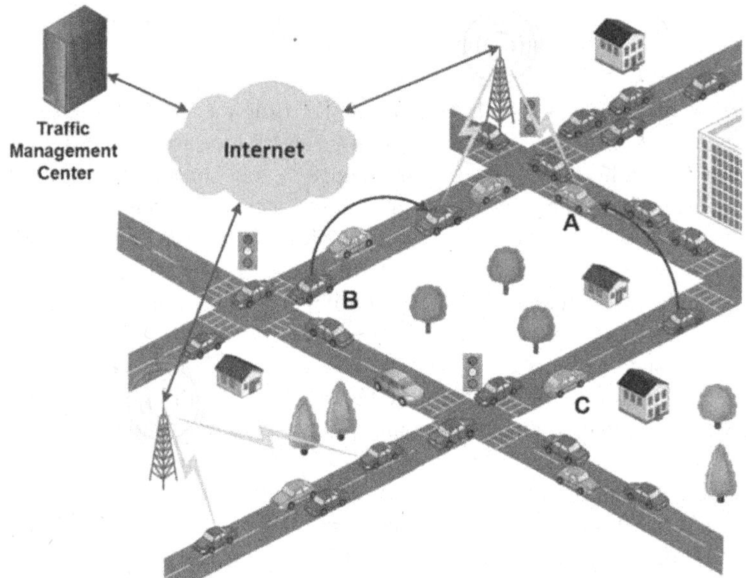

Figure 6.3 Vanet Communication and Navigation in ITS.

factor for peak and non-peak hours, and three vehicle classes were used. This routing technique resulted in 4% average savings in fuel consumption. Ahn and Rakha [2] showed the importance of route selection on the fuel and environment. They demonstrated that the emission and energy-optimized traffic assignment based on speed profiles can reduce CO_2 emissions by 14%–18%, and fuel consumption by 17%–25% over the standard user equilibrium and system optimum assignment. Barth et al. [7] attempted to minimize the vehicle fuel consumption and emission levels by proposing a new set of cost functions that include fuel consumption and the emission levels for the road links. Boriboonsomsin et al. [8] developed an eco-routing navigation system that uses both historical and real-time traffic information to calculate the link fuel consumption levels and then selects the fuel-optimum route. In Ref. [20] we enhanced the eco-routing algorithm developed in Ref. [2] by introducing a new ant-colony based updating technique for eco-routing. In 2017, we developed the first system optimum eco-routing model [21] that uses linear programming and stochastic route assignment to minimize the system-wide fuel consumption. The system optimum eco-routing reduced the fuel consumption by about 36% compared to the user equilibrium model.

However, all these previous efforts did not consider the communication network and its influence on eco-routing system performance. This chapter summarizes our previous work [14,16–18,22] on quantifying the impact of VANET performance on the eco-routing in large-scale smart cities.

6.2.5 ECO-ROUTING CONSIDERING IDEAL COMMUNICATION

As a feedback system, eco-routing assumes that vehicles are connected and equipped with GPS systems. Moreover, vehicles are assumed to be capable of calculating the fuel consumption for each road link they traverse and communicating this information to the TMC. In such a feedback system, the fuel consumption is computed by the vehicle's on-board computer and then communicated to the TMC.

To model this process, microscopic simulation models are used to compute the second-by-second vehicle trajectory, based on which, the fuel consumption rates can be calculated.

One of the commonly used fuel consumption estimation models is the VT-Micro model [38], which was developed as a statistical model from experimentation with numerous polynomial combinations of speed and acceleration levels to construct a dual-regime model, as demonstrated in Eq. 6.1.

$$F(t) = \begin{cases} \exp\left(\sum\limits_{i=1}^{3}\sum\limits_{j=1}^{3} L_{i,j} v^i a^j\right) & \text{if} \quad a \geqslant 0 \\ \exp\left(\sum\limits_{i=1}^{3}\sum\limits_{j=1}^{3} M_{i,j} v^i a^j\right) & \text{if} \quad a < 0 \end{cases} \tag{6.1}$$

where $L_{i,j}$ are model regression coefficients (that should be calibrated for each vehicle type) at speed exponent i and acceleration exponent j, $M_{i,j}$ are model regression coefficients at speed exponent i and acceleration exponent j, v is the instantaneous vehicle speed in (km/h), and a is the instantaneous vehicle acceleration (km/h/s) [38]. Under the ideal communication assumption, the fuel consumption of each road segment is correctly communicated to the TMC. No packet drop or delay is assumed, i.e., all the fuel consumption information is instantaneously delivered to the TMC. This data is used by the TMC to make the vehicle routing decisions that minimize the vehicles' fuel consumption. In this case, the routing decisions are optimized based on complete and accurate cost information.

However, in reality, this information is not completely delivered to the TMC due to packet drops and delays in the communication system. Such incompleteness in the information may result in sub-optimal routing decisions, consequently, resulting in higher fuel consumption.

In order to model such communication impacts, we developed a communication model for the medium access technique in VANET. The next section overviews the VANET communication, and then introduces the model we developed.

6.3 VANET COMMUNICATION

Recent advances in wireless networks have led to a new type of network, the VANET, which is a form of a Mobile Ad Hoc Network (MANET). VANETs are developed as the communication infrastructure of ITS to improve transportation system performance. Consequently, VANET communication is a key enabler for developing new ITS systems that enhance drivers' and passengers' safety and comfort.

By utilizing VANET, ITS attempts to improve safety on the roads and to reduce traffic congestion, waiting times, fuel consumption, and emissions. For example, warning messages sent by vehicles involved in an accident enhance traffic safety by helping approaching drivers to take proper decisions before entering the dangerous crash zone [13,57]. In addition to safety-related applications, VANET can also support other non-safety applications. For instance, information about current transportation conditions facilitates driving by taking new routes in case of congestion, thereby saving time and fuel and reducing emissions [9,33]. Moreover, VANETs are designed to support Quality of Service (QoS) for applications that are time-sensitive, such as audio and video applications. Thus, it enables commuters to have video and/or audio streaming while commuting.

Vehicular networks allow communication-enabled vehicles, roadside units, and other mobile devices to communicate in an ad hoc manner. Therefore, using VANET, moving vehicles can communicate with other vehicles (vehicle-to-vehicle [V2V]), roadside units (RSUs) (vehicle-to-infrastructure [V2I]), or even hand-held mobile devices (vehicle-to-device [V2D]) using a Dedicated Short Range Communication (DSRC) system. DSRC is an enhanced version of the WiFi technology suitable for VANET environments. The DSRC is designed to support communication in vehicular environments, which is characterized by its high mobility that results in rapid topology changes. DSRC has been standardized by IEEE 802.11p [58] in the 1609 family of standards known as Wireless Access in Vehicular Environments (WAVE) [28].

6.3.1 VANET CHARACTERISTICS

Compared to other wireless networks, VANETs are characterized by unique characteristics that distinguish them.

1. *High Mobility*: cars, which are the VANET communication nodes, are characterized by high speeds compared to other network types [25]. This high speed makes VANET a dynamic environment, which brings many challenges to VANETs such as frequent communication link drops and their impact on communication performance.
2. *Predictable and Restricted Mobility Patterns*: in VANET, node movements are restricted by the road network and its properties such as the speed limits and road controls (traffic signals and stop signs). Moreover, node movements are governed by traffic flow theory rules that relate the speed, density, and flow rate to each other. All these rules and constraints make the vehicle trajectories predictable, at least in the short term.
3. *Rapid Topology Changes*: the high speeds of vehicles lead to frequent network topology changes. Consequently, it increases the communication overhead due to higher packet retransmission rates. Additionally, in the case of multi-hop communication, to adapt to frequent topology changes, the routing protocol overhead will be increased to enable faster convergence. This increase in network overhead results in higher background packet traffic demand that significantly affects communication performance.

4. *High Power Availability*: in VANETs, vehicles are equipped with batteries that can be considered as power supplies for all communications and computation tasks. Considering the low communication power required by the vehicle communication devices and the batteries' capacity, these batteries can be considered an infinite power supply for these communication devices because a battery can supply power for weeks without depleting its energy.
5. *Large-Scale Network Size*: VANET networks can cover a city-level road network, which is a very large-scale network. Moreover, high vehicle density makes this network size larger even in the case of small road networks.

6.3.2 VANET CHALLENGES

Due to these unique VANET characteristics, many challenges need to be addressed to improve VANET performance. For example, high mobility in the VANET environment results in shorter communication link lifetimes between moving vehicles, especially for the high relative speed scenarios, such as vehicles moving in opposite directions on highways. Therefore, the MAC mechanism in VANETs is designed to support fast link establishment. Considering the time constraints required by some applications such as safety applications, the MAC mechanism in VANET is designed to support low latency communications to ensure service reliability for safety applications even under high data traffic demand conditions. This feature is enabled in the VANET MAC by adding the QoS features to the MAC mechanism as described earlier.

An example of these challenges is the frequent neighbourhood change due to high mobility. The frequent change must be considered, for example, in the routing algorithms, which should be designed to ensure the quality and continuity of services in such high dynamic communication environments.

One important challenge in VANET is the high vehicle density, which results in high contention in the shared wireless medium. These challenges significantly increase when considering the larger coverage area allowed in VANET compared to other wireless LAN standards. Consequently, the medium access control algorithms in VANET must account for this high vehicle density and must be designed to support the intended service for different applications. The high vehicle density in VANET brings another challenge to routing because the higher the number of communication nodes (vehicles), the higher the data routing update volume and, consequently, the higher the routing overhead.

These challenges must be also considered when designing, developing, and deploying ITS services over VANET communication networks. For instance, time-sensitive ITS applications should be assigned the required quality of Services (QoS). Moreover, ITS service design should consider the network scenarios in which these services are intended to work. For example, efficient navigation services are important in high congested road networks, in which vehicle density is expected to be very high, and here, communication performance is most probably low, including high packet drop rates and longer packet delays. Thus, such navigation services must be designed to work efficiently over this low communication performance.

6.3.3 PHYSICAL SPECIFICATIONS OF VANET

The U.S. Federal Communication Commission assigned a 75-MHz frequency band (5.850–5.925 GHz) to the DSRC vehicle-to-X (V2X) communications. In other countries, counterpart agencies made similar allocations. This spectrum is divided into seven 10-MHz wide channels as shown in Figure 6.4. The middle channel (channel 178) is the control channel (CCH), while channels 172, 174, 176, 180, 182, and 184 are service channels (SCH) that are intended for general purpose data transfer applications.

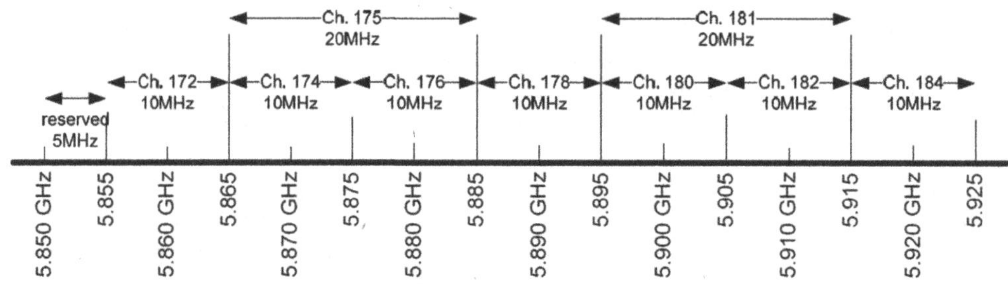

Figure 6.4 DSRC channels in the U.S.

The physical layer of IEEE 802.11p is similar to the IEEE 802.11a standard [29], but IEEE 802.11p uses an Enhanced Distributed Channel Access (EDCA) media access control (MAC) sub-layer protocol that is based on the IEEE 802.11e standard [30], with some modifications to the transmission parameters. IEEE 802.11p supports transmission rates up to 54 Mb/s within ranges up to 1000 m. In IEEE 802.11p, the EDCA provides QoS by supporting four different traffic Access Categories (ACs), namely, Background (BK), Best-effort (BE), Video (VI), and Voice (VO). Each AC has different medium access parameters such as Arbitration Inter-Frame Space (AIFS) and Contention Window (CW) limits.

6.4 MODELLING VANET COMMUNICATION IN LARGE-SCALE NETWORKS

To study the impact of communication on FB-ECO, we utilize the communication model that we developed in [17]. The model has two main components: the Medium Access Control technique (MAC) component and the queuing component. A two-dimensional Markov chain is used to model the MAC, based on the IEEE 802.11p standard [31]. The M/M/1/K queuing model [32] is used to represent the queuing process in the MAC layer of each individual vehicle. The model uses the communication configuration inputs (such as average packet size, average background packet generation rate, communication speed, communication range, and the queue capacity) and the current network condition (such as vehicle density in the sender communication range and the vehicle's connectivity to the RSUs) to compute a set of network performance parameters such as packet drop probability and packet delay.

One important advantage of this communication model over the previous ones is that it considers the MAC layer queue size and its impact on communication performance. For instance, the smaller the queue size, the lower the number of packets that can be queued, and consequently, in the case of high packet traffic rates, many of the packets will be rejected by the queue, which will increase the packet drop ratio. On the other hand, a larger queue size will result in increasing the queuing delay to very long delays. In contrast with the previous models, the model we developed assumes a finite queue size in the MAC layer, which enables the model to consider the queuing process. To build a realistic model, the M/M/1/K queuing model [32] is incorporated into the MAC protocol, so that the back-off technique and the queue interact with each other. Consequently, with the help of this queuing model, we were able to compute both the queuing and processing delay. In addition, the queuing model parameters were used to estimate the throughput and packet drop rate.

The model also supports both saturated and unsaturated data traffic conditions. So, it can be used at different packet generation rates.

The limitation of this model is that it assumes only one Access Category (AC) in the MAC layer, compared to four ACs in the IEEE 802.11p specifications. So, it does not support Quality of Services (QoS), which is supported in the IEEE 802.11p by enabling four ACs. The main purpose of this assumption is to simplify the model, in order to enable modelling of large-scale vehicular systems. This assumption is based on a comparative simulation study we made between a single AC and multiple ACs in VANET. This comparison showed that the performance of a single AC is similar to that of the Best Effort (BE) access category in the full-fledged model.

The detailed description of the model, its assumptions, and derivation are described in details in [17]. However, for the sake of completeness, we will summarize it in the following two sections.

6.4.1 MODELLING THE WAVE MEDIUM ACCESS TECHNIQUE

The basic MAC technique used by VANET is the EDCA, which is similar to the Distributed Channel Access (DCA) with support to quality of service. EDCA supports four traffic access categories. Every access category works as a stand-alone virtual station that has its own queue and its own access parameters.

6.4.1.1 MAC Operation in WAVE

In WAVE, the medium is shared by multiple communication nodes, so, similar to the IEEE802.11e, the IEEE802.11p nodes use a Carrier Sense Medium Access with Collision Avoidance (CSMA/CA) as follows. When an AC has a packet to send, it initializes its back-off counter to a random value within a given range called the contention window (CW). Then, it senses the medium until it becomes idle. If the medium continues to be idle for a specific time period called the Arbitration Inter-Frame Space (AIFS), it counts down its back-off counter. When this counter becomes zero, the AC can send its frame. Within the same station, two ACs can start transmitting at the same time; this situation is known as an internal collision. In this case, the higher priority AC will be granted the transmission, while the lower priority AC will double its CW range and re-initialize the back-off counter and back-off again.

If two stations start sending at the same time, the collision of the two signals will destroy both the frames. So, when a station sends a frame, it has to wait for an acknowledgment (ACK). If an ACK was not received within a specific time period, a collision is assumed, and the station must double its CW range and tries to retransmit the frame.

6.4.1.2 Representing the System in Markov Model and Its Solution

The MAC operation described above is represented by the model shown in Figure 6.5. A queue of size K packets is used to en-queue packets arriving from the upper layer. A two-dimensional Markov chain is utilized to represent the MAC process. State 0 in the Markov chain represents the system-empty state (both the system and the queue are empty). Each of the other states is defined by (i, j), where i and j are the back-off stages and back-off counter value, respectively. The Markov chain has a set of stages. In each stage, the back-off counter is initialized with a random value depending on the stage number (starting with a contention window size of 4, i.e., $w_0 = [0,3]$). Then the back-off counter is decremented with probability p_{idle} (the probability that the medium is idle for a time period equals to AIFS) moving the system from state (i, j) to state $(i, k-1)$. The stage ends when the back-off counter reaches 0 (i.e., state $(i, 0)$), at this point, the node will send the frame and waits for the acknowledgement. If an acknowledgement is not received within a given interval, the system will move to the next stage (stage $i+1$) where the size of the contention window is doubled. The maximum contention window size is 64 (i.e. [0, 63]), after this, the MAC does not increase it when moving to the next stage. This process is repeated for stages from 0 to 15. If the transmission failed in stage 15, the frame will be dropped. Table 6.1 shows the symbols used for this model.

To solve this model, we start from the Markov chain and derive all the state probabilities as a function of $P(0,0)$. The summation of all state probabilities should be equal to 1. We incorporate the queuing parameters into the model.

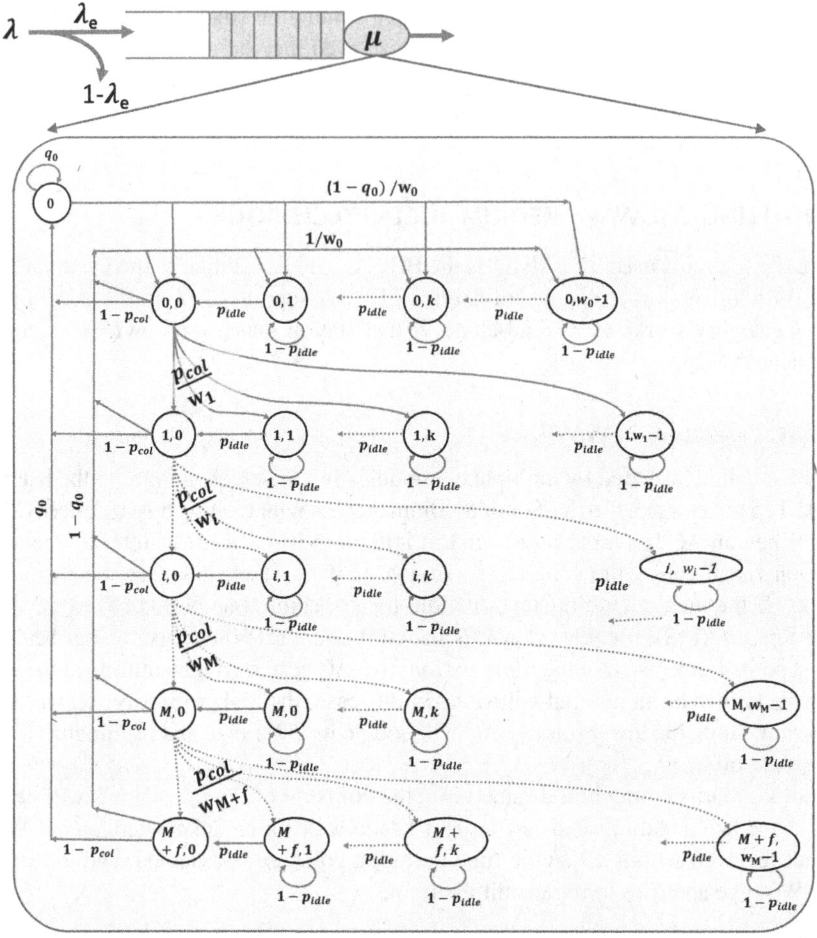

Figure 6.5 Markov chain model for the medium access.

From the Markov chain in Figure 6.5, we can compute the probability that the system is in state $P(0,j)$. $P(0,j)$ can be expressed as:

$$P(0,j)=\frac{w_0-j}{w_0}\frac{1-q_0}{p_{\text{idle}}}\left(P(0)+P(M+f-1,0)+(1-p_{\text{col}})\sum_{i=0}^{M+f-2}P(i,0)\right),\quad j=1,2,...,w_0-1. \tag{6.2}$$

And $P(0,0)$ can be expressed as:

$$P(0,0)=(1-q_0)\left(P(0)+P(M+f-1,0)+(1-p_{\text{col}})\sum_{i=0}^{M+f-2}P(i,0)\right). \tag{6.3}$$

From Equations (6.2) and (6.3), we can derive $P(0,j)$ as:

$$P(0,j)=\frac{w_0-j}{w_0}\frac{1}{p_{\text{idle}}}P(0,0)\qquad j=1,2,...,w_0-1. \tag{6.4}$$

$P(i,0)$ and $P(i,j$ can be calculated, as in Eqs. (6.5) and (6.6)

$$P(i,0)=p_{\text{col}}^{i}\,P(0,0)\quad i=0,1,....,M+f-1, \tag{6.5}$$

Table 6.1
The Model Parameters

Symbol	Description
i	The back-off stage number
j	The back-off counter
M	The maximum number of increases of the CW
f	The maximum number of re-transmissions without increasing the CW
w_i	The CW range for stage i
w_0	The initial value for the maximum CW i
α	The CW increasing factor, where $w_i = w_0\alpha^i$. The typical value is 2.
$p_{\text{idle}_{\text{slot}}}$	The probability that a medium is idle in any time slot
p_{idle}	The probability that the medium is idle
q_0	The probability that the system is empty (no packet in the system)
p_{suc}	The probability that a medium is occupied with a successful transmission
p_{fail}	The probability that a medium is occupied with a failed transmission
p_{tran}	The probability that a station starts transmission in any time slot
p_{col}	The probability that the packet collides
$P(i,j)$	The probability that the system is in state (i,j)
λ	The packet arrival rate
μ	The packet service rate
T_{serv}	The packet service time
N	The number of vehicles in communication range
T_{slot}	The length of the time slot (sec)
T_s	The transmission time of a successful frame transmission
T_f	The transmission time of a failed frame transmission
ρ	The traffic intensity for the queuing model

$$P(i,j) = \frac{w_i - j}{w_i}\frac{p_{\text{col}}^i}{p_{\text{idle}}}P(0,0) \quad i = 1,2,\ldots,M+f-1 \quad \text{and} \quad j = 1,2,\ldots,w_i - 1. \tag{6.6}$$

Finally, we can compute $P(0)$ as :

$$P(0) = \frac{q_0}{1-q_0}P(0,0). \tag{6.7}$$

Since the summation of all the probabilities equals 1, we have:

$$P(0) + \sum_{i=0}^{M+f-1} P(i,0) + \sum_{k=1}^{w_0-1} P(0,k) + \sum_{i=1}^{M+f-1}\sum_{k=1}^{w_{i-1}} P(i,k) = 1. \tag{6.8}$$

Notice that the window exponential factor is α for $i \leq M$; that is, $w_i = w_0\ \alpha^i\ \forall\ i \leq M$, and $w_i = w_0\ \alpha^M\ \forall\ i > M$. By plugging w_i into Eq. (6.8) and using some math, we can calculate $P(0,0)$, as shown in Eq. (6.9).

$$P(0,0) = \left(\frac{q_0}{1-q_0} + \frac{1-{p_{\text{col}}}^{M+f}}{1-p_{\text{col}}} + \frac{w_0 - 1}{2\ p_{\text{idle}}} + \frac{1}{2\ p_{\text{idle}}}\left[(\alpha^{M-1}\ w_0 - 1)\frac{{p_{\text{col}}}^{M-1} - {p_{\text{col}}}^{M+f}}{1-p_{\text{col}}}\right.\right.$$
$$\left.\left. + w_0\frac{\alpha\ p_{\text{col}} - (\alpha\ p_{\text{col}})^{M-1}}{1-\alpha\ p_{\text{col}}} + \frac{p_{\text{col}} - (p_{\text{col}})^{M-1}}{1-p_{\text{col}}}\right]\right)^{-1} \tag{6.9}$$

To solve the model, we need to calculate the values of $p_{\text{col}}, p_{\text{idle}}$, and q_0. To do that, we derive the relationship between these three parameters and the state probabilities.

A collision will happen when two or more stations start transmission in the same time slot. Let the probability that a station starts transmitting at a time slot be p_{trans}, then $p_{\text{trans}} = \sum_{i=0}^{M+f-1} P(i,0)$. When a station sends a packet, the probability that this packet collides is $p_{\text{col}} = 1-(1-p_{\text{tran}})^{N-1}$. Therefore, for the entire system, the medium will be idle at any time slot only if no station is sending: $p_{\text{idle}_{\text{slot}}} = (1-p_{\text{tran}})^N$. The station decides that the medium is idle (p_{idle}) after $AIFS$ idle time slots in a row: $p_{\text{idle}} = {p_{\text{idle}_{\text{slot}}}}^{AIFS}$. For the entire system, the probability that a packet is successfully transmitted without collision is $p_{\text{suc}} = N\ p_{\text{tran}}\ (1-p_{\text{tran}})^{N-1}$.

6.4.2 MAC QUEUING USING THE M/M/1/K MODEL

The only missing part to completely solve the model is finding a relation between q_0 and the state probabilities. To solve for q_0, we use the $M/M/1/K$ model, where we assume the packet inter-arrival time, between packets, is exponentially distributed with an average rate λ. Assuming that the service rate is $\mu = \frac{1}{T_{\text{serv}}}$, where T_{serv} is the average packet processing time. T_{serv} is the summation of the average time the packet stays in each stage. The packet can stay for a time T_w in every state, plus the average frame transmission time T_{trav}, which can be calculated as:

$$T_w = p_{\text{fail}}\ T_f + p_{\text{suc}}\ T_s + \frac{1}{p_{\text{idle}}} T_{\text{slot}}, \tag{6.10}$$

$$T_{\text{trav}} = p_{\text{col}}\ T_f + (1-p_{\text{col}})\ T_s, \tag{6.11}$$

where T_s and T_f are the successful transmission time and the failed transmission time, respectively; and T_s and T_f depend on whether MAC uses the basic or advanced (RTS/CTS) access modes.

The term in T_w is the time required by the station while sensing the medium to ensure it is idle. Consequently, the service time T_{serv} can be calculated as:

$$T_{\text{serv}} = T_{\text{slot}} \sum_{i=0}^{M+f-1} \left(\frac{T_w(w_i-1)}{2} + T_{\text{trav}} \right) {p_{\text{col}}}^{i}. \tag{6.12}$$

Now, we can calculate the traffic intensity $\rho = \frac{\lambda}{\mu}$, and subsequently q_0, as:

$$q_0 = \begin{cases} \frac{1-\rho}{1-\rho^{K+1}} & \text{if } \lambda \neq \mu; \\ \frac{1}{K+1} & \text{if } \lambda = \mu. \end{cases} \tag{6.13}$$

Using these equations, we can solve this model and estimate the total communication drop profitability and delay.

6.4.3 COMMUNICATION MODEL VALIDATION

To validate this communication model, we ran extensive simulations for different vehicular traffic demand levels and different numbers of communication nodes considering V2I communication. We used the OPNET software, which is known currently as the Riverbed modeller [50]. The OPNET modeller is a powerful discrete event simulation tool for specification, simulation, and analysis of data and communication networks. The most important OPNET characteristic is that its results are trusted because its implementations of the standard protocols are tested and validated before publishing. Current versions of OPNET support WAVE as an extension of the IEEE 802.11 implementation. For each simulation scenario, we calculated the average network throughput and the average packet delay. The results show an accurate estimation of both throughput and delay of our model compared to the OPNET simulated results (SIM), as shown in Figures 6.6 and 6.7.

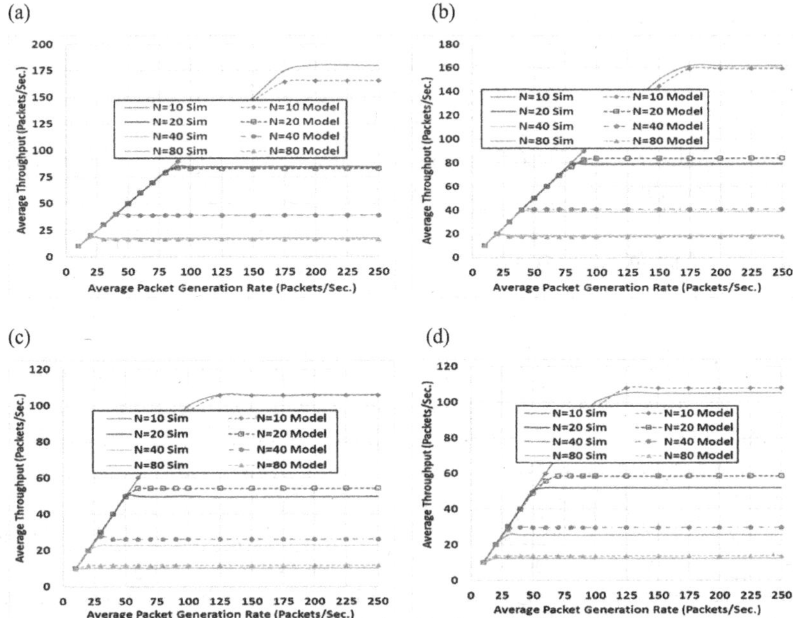

Figure 6.6 Average throughput per vehicle (packets/second) versus packet generation rate. (a) Basic access mode, packet size = 500 bytes, (b) RTS/CTS, packet size = 500 bytes, (c) basic access mode, packet size = 1000 bytes, and (d) RTS/CTS, packet size = 500 bytes.

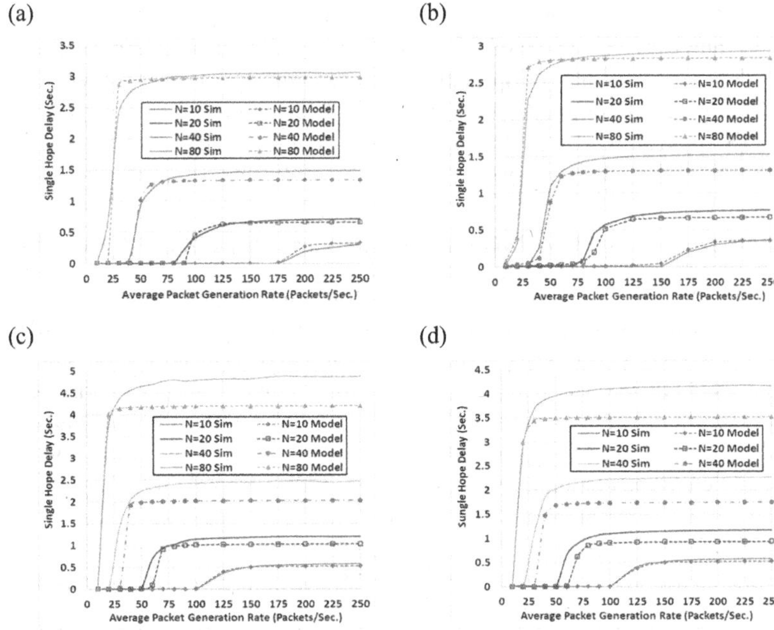

Figure 6.7 Average single hop delay. (a) Basic access mode, packet size = 500 bytes, (b)RTS/CTS access mode, packet size = 500 bytes, (c) basic access mode, packet size = 1000 bytes, and (d) RTS/CTS access mode, packet size = 1000 bytes.

6.5 MODELLING ECO-ROUTING AND VANET

The transportation network is the environment where vehicular communication takes place. Thus, it is essential to integrate the proposed communication model into a scalable transportation simulation software. This section describes how the developed communication model is integrated into the transportation simulator, the INTEGRATION software [45], to capture the inter-dependency of communication and mobility.

6.5.1 THE INTEGRATION SOFTWARE

After validating the communication model, we implemented and incorporated it within the INTEGRATION software [45]. INTEGRATION is a trip-based microscopic traffic assignment, simulation, and optimization model. INTEGRATION is capable of modelling networks with hundreds of thousands of cars. It is characterized by its accuracy, which comes from its microscopic nature, and its small-time granularity; by tracking individual vehicle movements from a vehicle's origin to its final destination at a level of resolution of 1 decisecond. Using this time resolution, INTEGRATION can accurately track vehicles by modelling vehicle car-following based on Rakha-Pasumarthy-Adjerid collision-free car-following model, also known as the RPA model [48]. The RPA model captures vehicle steady-state car-following behaviour using the Van Aerde model [44,56]. In INTEGRATION, movement from one steady state to another is constrained by a vehicle dynamics model described in [39,42]. Vehicle lateral motion is modelled using lane-changing models described in [45]. The model estimates of vehicle delay were validated in [10], while vehicle stop estimation procedures are described and validated in [43]. Vehicle fuel consumption and emissions are modelled using the VT-Micro model [5,46,49]. The INTEGRATION simulation model provides ten basic user equilibrium traffic assignment/routing options [45]. Recently, a system equilibrium routing model was added [21]. One important feature of INTEGRATION is its support for eco-routing traffic assignment, which tries to minimize the fuel consumption and emission levels by assigning vehicles the most environmentally friendly routes.

The resulting framework is capable of modelling large-scale transportation networks while capturing the inter-dependency of the transportation and communication systems. We use this framework to study the impact of communication on the feedback eco-routing application performance as an example ITS application. The following section describes the dynamic eco-routing logic implemented in the INTEGRATION software.

6.5.2 ECO-ROUTING WITH REALISTIC VANET COMMUNICATION MODELLING

In most of the implementations of the transportation applications, a perfect communication network is assumed, which means the drops and delays are assumed to be zero. The INTEGRATION software also makes this assumption. Thus, to build a realistic ITS application simulation framework, we modified the INTEGRATION behaviour to adopt the communication performance parameters and their impact on the ITS applications. The new behaviour is illustrated in Figure 6.8. When a vehicle finishes a road link, instead of directly updating the link cost in the TMC, the vehicle sends this information to the communication module by adding it to the transmission queue. Then, while the vehicle moves, the communication module checks for communication network connectivity. Whenever the vehicle becomes connected to an RSU, the communication module processes the packets in the queue. For each packet, it first calculates its drop probability as described later in Eq. 6.15. If the packet is to be delivered, the communication module calculates its average total delay and inserts it into a time-based ordered queue. So, it will be processed by the updating module in its time of arrival.

To calculate the packet drop probability and total delay, the communication model parameters introduced in Section 6.4 are used.

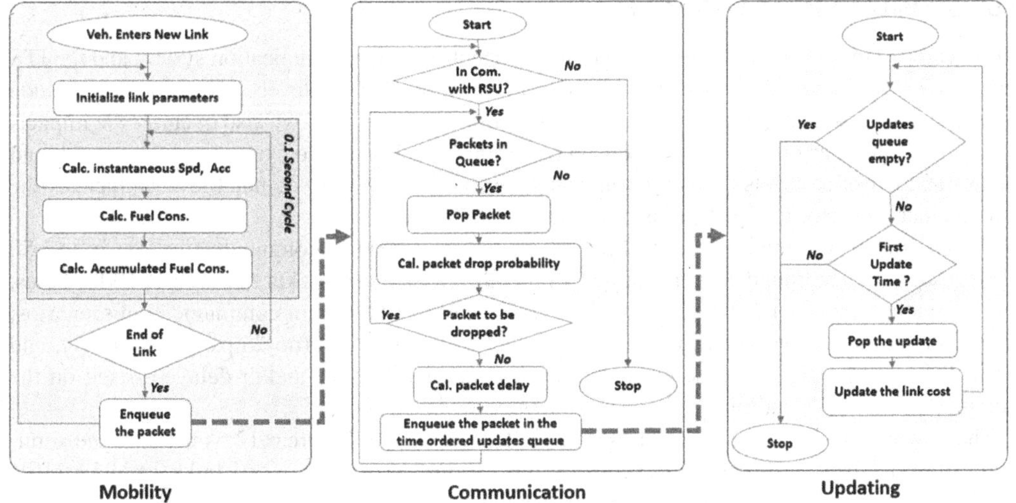

Figure 6.8 Eco-routing with the communication.

With regard to the packet drop probability, the packet will be dropped in two cases. First, if it arrives at the MAC while the queue is full. The second case is when a packet reaches its maximum retransmission attempts. If the queue is full, the packet will be rejected by the queue. We call this the rejection probability P_{rej}. So, P_{rej} is the probability that the queue has K packets. According to the $M/M/1/K$ model [32], P_{rej} can be computed as:

$$P_{\text{rej}} = \begin{cases} \rho^K \frac{1-\rho}{1-\rho^{K+1}} & \lambda \neq \mu; \\ \frac{1}{K+1} & \lambda = \mu. \end{cases} \tag{6.14}$$

For the second case, a packet in the queue might be dropped if it experienced a collision in its last retransmission attempt whose probability is $P(M+f-1,0)\ p_{\text{col}}$. If neither of these cases happened, then the packet will be correctly delivered. Thus, the packet drop probability can be calculated as

$$P_{\text{drop}} = 1-(1-P_{\text{rej}})\Big(1-P(M+f-1,0)\ p_{\text{col}}\Big). \tag{6.15}$$

The total delay of the packet T_{delay} is the summation of both service time T_{serv} and the queuing delay T_q as shown in Eq. 6.16.

$$T_{\text{delay}} = T_{\text{serv}} + T_q, \tag{6.16}$$

where the average service time T_{serv} is calculated by Eq. 6.12, and the average queuing delay T_q can be calculated as:

$$T_q = \frac{1}{\mu - \lambda_{\text{eff}}} \tag{6.17}$$

In Eq. 6.17, λ_{eff} is the effective packet generation rate which is the actual number of packets that enter the queue per unit time. In other words, λ_{eff} is the packets that arrive when the queue is not full. If the Q_n is the probability that the queue has n packets, then λ_{eff} can be calculated as:

$$\lambda_{eff} = \lambda(1-P_{\text{rej}}) = \lambda(1-Q_K) \tag{6.18}$$

Using Eqs. 6.14 and 6.16 through 6.18, the average total packet delay can be calculated.

6.6 SIMULATION AND RESULTS

The developed model is used to study the mutual impacts of the communication system and the ITS feedback-based eco-routing application at different car traffic demand levels. These impacts include how traffic congestion and flow are affected by communication errors. We also quantify the impacts of communication errors on the fuel consumption, the travel time, the average vehicle speed, and the average emission levels. This section also shows how the vehicle demand level influences the communication performance in terms of packet drop rate and delay.

We use the V2I communication paradigm assuming a 1000-m communication range and 50 Packets/second background packet generation rate (λ). The average packet size is set at 1000 bytes, and the queue size is set at 64 Packets. In addition, we use two main communication scenarios; the ideal communication scenario assumes a perfect communication (no drops and no delay), and the realistic communication case where the packets can be dropped and/or delayed based on the surrounding network conditions.

The downtown area in the city of Los Angles (LA), shown in Figure 6.12, is used for the simulation analysis. This road network is about 133 km^2. It has 1625 nodes, 3561 links, and 459 traffic signals. With regards to the vehicular traffic demand, we use a calibrated traffic demand, which is based on the data collected from multiple sources, as described in detail in [12]. This traffic demand represents the morning peak hours in the downtown area of the city of LA, which continue for 3 hours from 7:00 am to 10:00 am. We added 1 hour for traffic pre-loading. So, the demand runs for 4 hours. However, we run the simulation for 30000 seconds to give the vehicles enough time to finish their trips. To study the impact of different traffic origin-destination demand (OD). levels, the calibrated traffic rates are multiplied by scaling factors (ODSFs) ranging from 0.1 through 1.0 at a 0.1 increment which generates ten traffic demand levels. The total number of vehicles that are simulated in each of these scenarios is shown in Table 6.2.

6.6.1 THE IDEAL COMMUNICATION CASE

In this part, we study the system behaviour under ideal communication performance, i.e., no packet drop and no delay.

6.6.1.1 Impact of Penetration Rate on Fuel Consumption

Figure 6.9 also demonstrates that, with feedback enabled (penetration rate > 0) and at low traffic demand levels (ODSF = 0.2 and ODSF = 0.4), the penetration rate does not have a significant impact on fuel consumption. The reason for this is that, at low traffic demand levels, vehicles run almost at the free-flow speed, and there are no significant changes in the network status (such as network congestion and the associated increase in the link costs) that need to be updated at the TMC. In these two demand levels, the network is not congested where the average vehicle density is about 5 veh/km/lane and 9 veh/km/lane (in the cases of ODSF = 0.2 and ODSF = 0.4, respectively), as shown in the network fundamental diagram in Figure 6.11a and 6.11b.

It is also clear in Figure 6.9 that, as the demand level increases, the lack of updates at 0.1 penetration rate results in increasing the fuel consumption, compared to higher penetration rates. This means that, as the demand level increases, the importance of having enough updates becomes higher to reflect the changes in the network status.

We notice also, in Figure 6.9, that at high traffic levels (ODSF = 0.8 and ODSF = 1.0), the system performance sometimes becomes worse when increasing the penetration rate. This is reasoned by the route oscillation effect of the shortest path dynamic routing techniques (bang-bang effect) [37]. What happens is that, if we have full and exact information about the network, then all the vehicles will take the best routes which results in the overloading of these routes (especially at high traffic demand levels), resulting in higher congestion on these routes and higher fuel consumption until the re-routing takes place. To avoid temporal oscillations in route choices, a white noise error

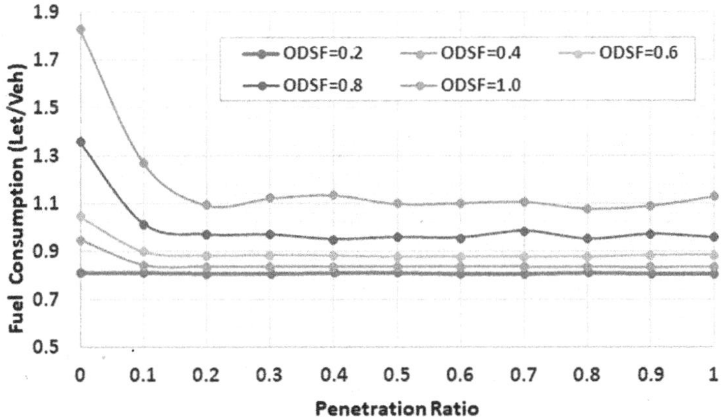

Figure 6.9 Impact of penetration rate on the average fuel consumption at different OD levels.

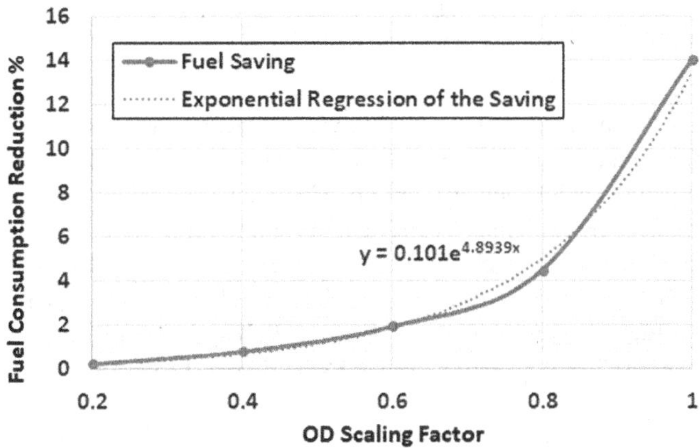

Figure 6.10 The average fuel saving when increasing the penetration rate from 0.1 to 0.2.

function can be introduced into the link cost function. This allows vehicles to select slightly sub-optimum routes, if the costs along alternative routes are very similar and thus distribute the traffic. In our simulation, we introduce random white noise with a coefficient of variation equal to 0.05. Additionally, the low penetration rates implicitly introduce some other white noise to the link cost. Consequently, higher penetration rates sometimes produce worse performance.

Figure 6.10 shows the average fuel saving when increasing the penetration rate just from 0.1 to 0.2 for different traffic demand levels. It demonstrates that this saving is exponentially increasing with the demand level, as shown by the trend line.

From Figures 6.9 and 6.10, we can conclude that a penetration rate of 0.2 results in sub-optimal fuel consumption, which is near to the optimal fuel consumption in all cases. So, we can conclude that a 0.2 penetration rate for the probe vehicles is sufficient.

6.6.1.2 Penetration Rate and Congestion Levels

To understand the results in more depth, we computed the Network Fundamental Diagrams (NFDs) [11,27] for some of these scenarios. Figure 6.11a–6.11d compare these fundamental diagrams for different penetration rates, at different values of the ODFS.

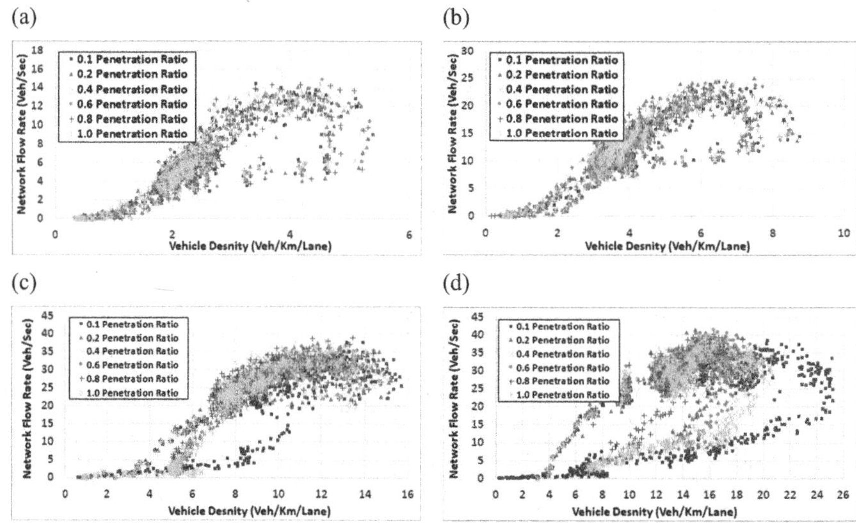

Figure 6.11 The network fundamental diagrams at different traffic rates assuming ideal communication. (a) ODSF = 0.2, (b) ODSF = 0.4, (c) ODSF = 0.8, and (d) ODSF = 1.0.

Figure 6.11a and 6.11b show that, at low traffic demand levels, the impact of the penetration rate is not significant. Meanwhile, Figure 6.11c shows the NFD for ODSF = 0.8, where a penetration rate of 0.1 shows relatively high vehicle density combined with low traffic flow, which reflects temporary network gridlocks, that can be explained by the lack of routing information at a 0.1 penetration rate, which produces inaccurate routing decisions. The figure also shows that, for the higher penetration rates, starting at 0.2, the gridlocks do not exist. These gridlocks at the low penetration rate are reflected in the higher fuel consumption rates in Figure 6.9. This also supports our conclusion that a market penetration rate of 20% is sufficient to deploy the FB-ECO systems and achieve acceptable performance.

Figure 6.11d shows that, at high traffic demand level (full demand rates), the impact of the penetration rate on network congestion becomes more significant. It shows that at a 0.1 penetration ratio, the network congestion becomes higher and the overall network enters the congested regime. It also shows that increasing the penetration ratio results in better system performance. It also shows that the overall network behaves the best at a penetration ratio of 0.6 or 0.8.

6.6.2 REALISTIC COMMUNICATION CASE

This part studies the system performance in the case of realistic communication assuming V2I, where the packets may be dropped or delayed and its impact on the traffic load the connection availability.

6.6.2.1 RSU Allocation

Since we focus on V2I communication, we have to allocate the RSUs in the network. RSU allocation is shown to be critical to the performance of the communication network and consequently the performance of the eco-routing application [19]. The most economical method is to install the RSUs at traffic signals locations to use the already-installed connections and power sources. In the road network of downtown LA, there are 459 traffic signals. Consequently, we need to identify the traffic signal locations to install the RSUs to achieve the best coverage with the minimum cost, which is a min-max coverage problem. To achieve this objective, we use a greedy algorithm shown in Algorithm 1.

Algorithm 1 Select traffic signals to install RSUs

1: **procedure** SELECT TRAFFIC SIGNALS ▷ Select the minimum number of traffic signals to install RSUs in such a way that maximizes the coverage
2: $S \leftarrow \{S_i : i = 1,2....\}$
3: $G \leftarrow \phi$ ▷ The initial solution
4: **while** $S \neq \phi$ **do** ▷ There are uncovered signals
5: $C_i = \{S_j \in S : D_{i,j} < R_{Com}\}$ ▷ Recalculate the coverage
6: Select $S_i \in S$ that maximizes$||C_i||$
7: $G \leftarrow G \cup S_i$
8: $S \leftarrow S \setminus C_i$
9: **return** G ▷ The selected signals

Assuming that the distance between the traffic signals S_i and S_j is $D_{i,j}$, and C_i is the set of traffic signals covered by S_i. In other words, $C_i = \{S_j : D_{i,j} < R_{\text{Com}}\}$, where R_{Com} is the communication range. The algorithm starts with S includes all the traffic signals and empty set G of selected signals. It calculates the coverage for each signal in S. Then, it selects the traffic signal that covers the maximum number of uncovered signals, add it to the selected signals G, and remove it along with all the signals it covers from S. Steps 5–8 are repeated until S becomes empty. This algorithm does not guarantee that the entire network will be covered, but it ensures coverage of the maximum signalized intersections with the minimum cost (minimum number of RSUs). Figure 6.12 shows the coverage map in the cases of 1000-m communication range.

Figure 6.12 The LA downtown area and the coverage map for 1000-m communication ranges.

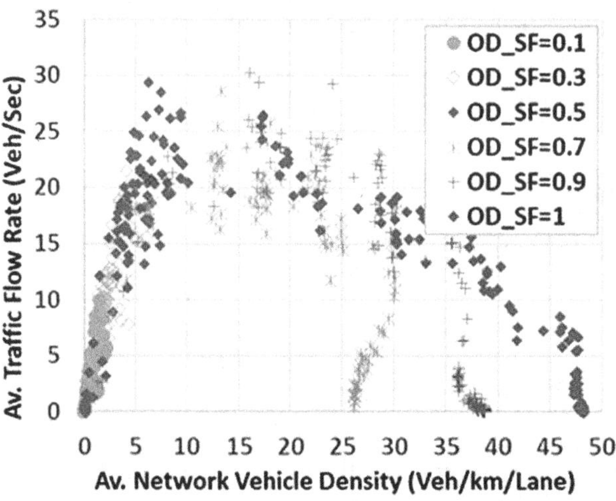

Figure 6.13 The network fundamental diagrams with realistic communication.

6.6.2.2 Communication System Impact on Eco-Routing System Performance

We ran the network at different vehicular traffic demand levels for the realistic communication system. Figure 6.13 shows the NFD in the realistic communication case assuming 100% penetration rate. It shows that the network entered the congestion regime lower traffic demand levels, at an ODSF= 0.7, compared to Figure 6.11. It also shows that there is gridlock in the network that results in a large number of vehicles being unable to exit the network at ODSFs of 0.7 through 1.0 in the realistic communication case.

Table 6.2 shows that in the case of ideal communication only 0.63% and 2.59% did not complete their trips for the two highest traffic demand levels, respectively. It also shows that 0.04% and 1.02% were not able to enter the network because there are no available spots to enter the network. However, in the realistic communication case, at a traffic scale of 0.7, about 16.88% of the vehicles were not able to complete their trips, and 16.45% were not able to enter the network. At the full traffic demand, it shows that 22.41% of the vehicles did not reach their final destination and 36.39%

Table 6.2
Vehicles Count Comparison for Different Traffic Scaling Factors

		Vehicles Started and Finished		% Vehicles Entered but didn't Finish		% Vehicles deferred	
OD SF	**Total No. of Vehicles**	**Ideal**	**Realistic**	**Ideal**	**Realistic**	**Ideal**	**Realistic**
0.1	50,273	100	100	0	0	0	0
0.2	107,047	100	100	0	0	0	0
0.3	164,499	100	100	0	0	0	0
0.4	222,326	100	100	0	0	0	0
0.5	277,973	100	100	0	0	0	0
0.6	338,366	100	100	0	0	0	0
0.7	394,313	100	74.67	0	16.88	0	8.45
0.8	450,670	100	66.87	0	21.05	0	12.08
0.9	507,427	99.33	60.83	0.63	20.01	0.04	19.16
1	563,626	96.39	41.2	2.59	22.41	1.02	36.39

had no chance to enter the network because of the congestion on the network. This higher congestion is accompanied by lower traffic rates exiting the network in the realistic communication case.

From Table 6.2 and Figures 6.11 and 6.13, we can conclude that the realistic communication results in packet drops and delays that lead to incorrect routing decisions that cause the network to be highly congested at a lower traffic demand compared to the ideal communication cases.

6.6.2.3 Quantifying the Realistic Communication Impact on Mobility Sustainability

In the realistic communication scenarios, the incorrect routing and high congestion levels resulting from the packet drops and delays are expected to result in higher fuel consumption levels, longer travel times, and longer delays compared to the ideal communication case. Figure 6.14 compares the mobility parameters in the two cases for the different traffic levels.

Figure 6.14a shows that: (1) at the low traffic demand levels, OSDF = 0.1, 0,2, and 0.3, the average fuel consumption is not affected by the communication system, (2) as the OSDF increases, the average vehicle fuel consumption in the case of realistic communication case becomes significantly higher than the ideal communication case, and (3) at the two highest traffic demand levels, OSDF = 0.9 and OSDF = 1.0, the average vehicle fuel consumption level in the realistic communication case becomes lower. The average travel time and delay have the same behaviour, as shown in Figure 6.14b and c. The emission levels in Figure 6.14d–f have a similar trend.

The third conclusion may appear counter-intuitive at first glance, given that the higher vehicular traffic demand should result in higher vehicle density levels, higher packet drop rates, and longer delays that should lead to incorrect routes, higher congestion levels, and higher fuel consumption levels. These results can be attributed to two folded reasons. First, these high vehicular demand levels produce high congestion levels, as shown in Figure 6.11, consequently, a larger number of vehicles are not able to complete their trips, as shown in Table 6.2. These vehicles are not accounted for in the fuel consumption estimates because the simulation software only includes the statistics of vehicles that complete their trips. Thus, the estimated fuel consumption counts only for 60.83% and 41.2% of the vehicles in the last two scenarios, respectively. The second reason is that most of the trips that finished are short trips because the short trips do not experience similar levels of congestion as do the longer trips. Figure 6.14g shows that the average distance per trip significantly decreased for higher traffic demand levels in the realistic communication scenarios. This shorter distance justifies the lower average vehicle fuel consumption estimates, emissions, and travel times in the highest traffic demand levels.

Figure 6.14h shows the average speed in the network for the different traffic demand levels in both the ideal and the realistic communication cases. It shows that the average speed at the highest traffic level is reduced from 46 km/h to about 33 km/h which is also consistent with the high congestion levels in realistic communication cases.

6.6.2.4 Impact of Realistic Communication and Penetration Ratio

Figure 6.15 compares the impact of the market penetration rate on the system performance in both the ideal communication and realistic communication cases. It demonstrates that, at lower traffic demand rates (ODSF = 0.2 and 0.4), the average vehicle fuel consumption rates in the ideal communication and realistic communication are similar (i.e., the impact of communication during these low traffic demands is not significant). However, as the traffic demand level increases, the average vehicle fuel consumption in the realistic communication case becomes significantly higher than that of the ideal communication. We also notice that, at moderate traffic levels (0.6 and 0.8), the impact of communication is very small at low penetration rates (0.1– 0.3), and increases with the penetration ratio. This behaviour is attributed to the fact that at these moderate levels increasing the penetration ratio, results in higher packet traffic rates, hence higher competition over the wireless medium, consequently, higher drop rates and longer delays.

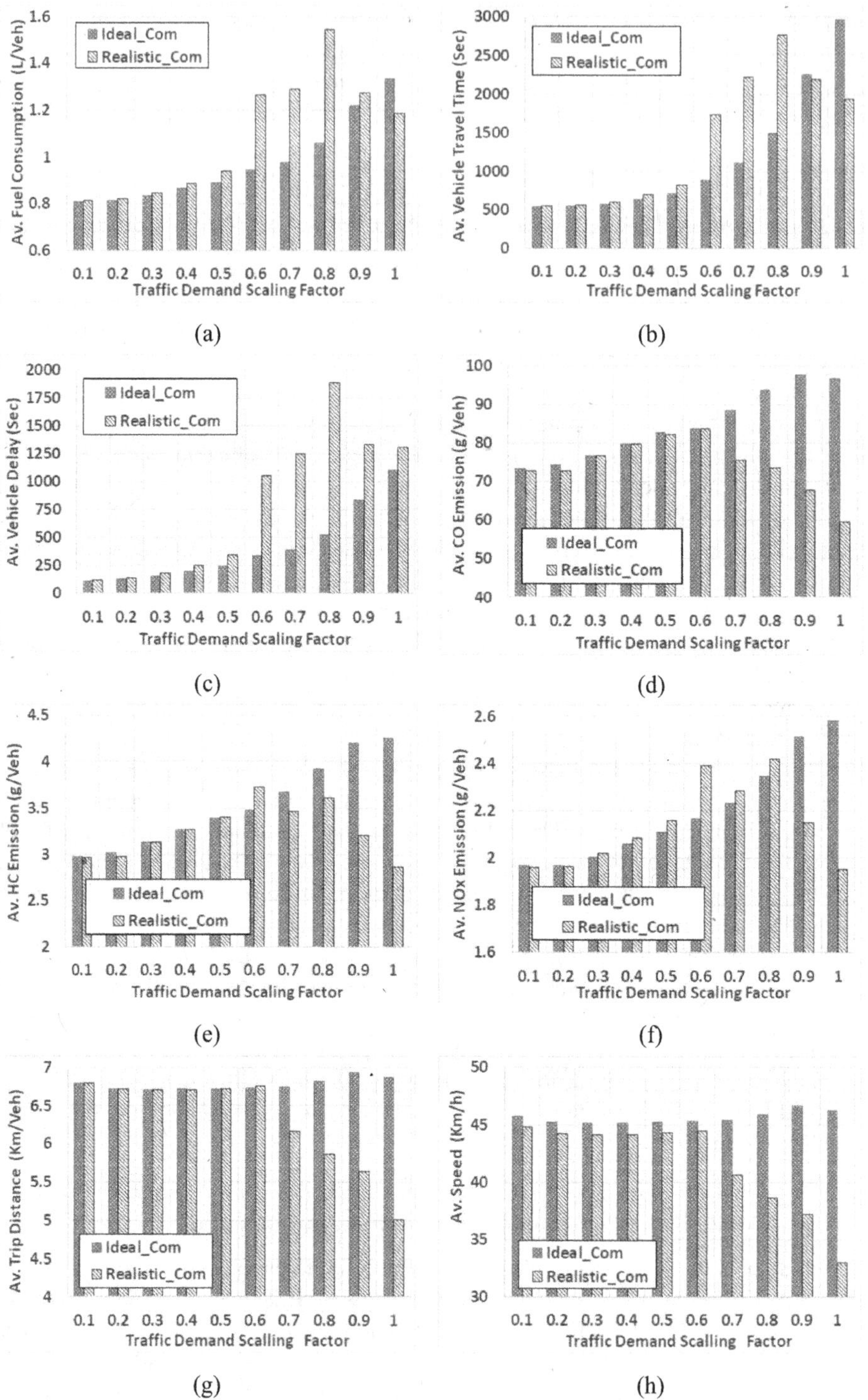

Figure 6.14 The ideal communication vs. the realistic communication outputs. The figures show that considering realistic communication performance (in terms of packet drop rate and delay) can have significant impacts on mobility sustainability parameters.

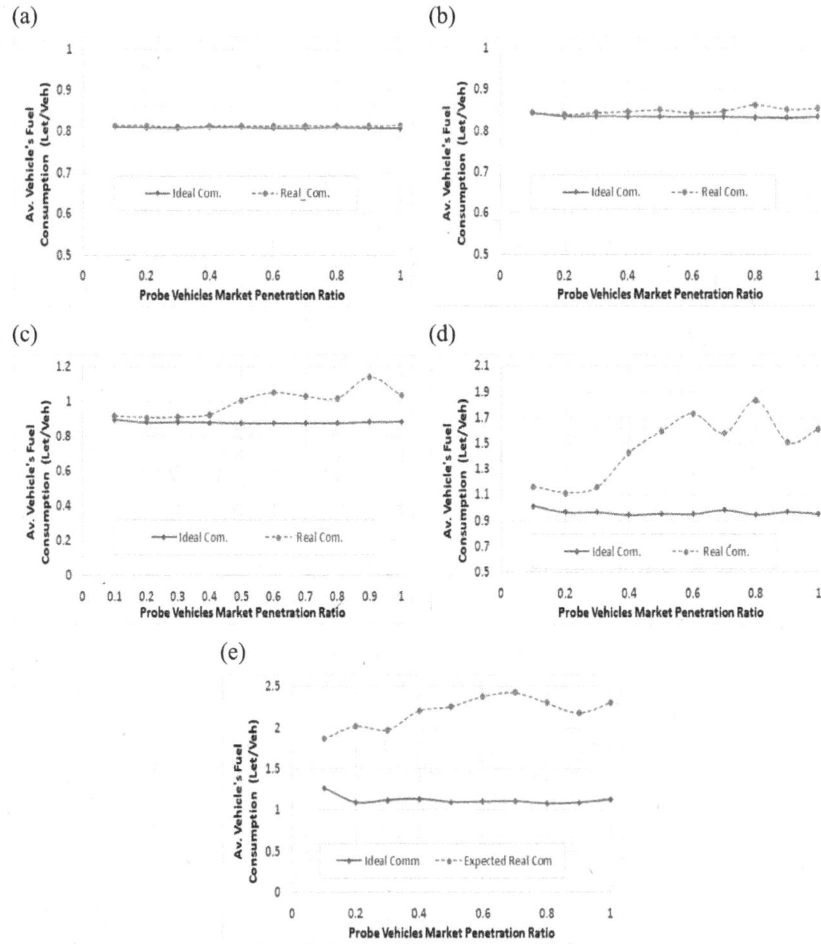

Figure 6.15 The impact of connected vehicles market penetration ratio on the system performance in the realistic communication case. (a) ODSF = 0.2, (b) ODSF = 0.4, (c) ODSF = 0.6, (d) ODSF = 0.8, and (e) ODSF = 1.0.

6.7 CONCLUSION

In this chapter, we introduced some basic concepts of traffic flow theory and vehicular ad hoc network modelling. Subsequently, we introduced a VANET MAC layer modelling framework that captures realistic packet drop probabilities and delays. This model is used to study the impact of vehicular communication on a promising ITS application, namely; the feedback eco-routing traffic assignment. The developed tool is used to study the mutual impact of communication and mobility.

The simulation results on a real-world road network show that in the case of low vehicular traffic demand, the communication performance (in terms of packet drop and delay) does not have a significant impact on the dynamic eco-routing performance. However, as the vehicular traffic demand increases, this impact becomes significant. At a certain congestion level, it can result in routing gridlocks due to incorrect routing decisions. Consequently, it is imperative to consider these mutual interactions of the communication and traffic mobility systems when deploying such systems, especially in highly congested areas.

The simulation results also show that the dynamic eco-routing system can work properly at low market penetration rates of 20% when the traffic demand is low. At high congestion levels, the results show that the eco-routing application works better at low market penetration levels, because of the lower competition over the wireless medium.

REFERENCES

1. Hesham A. Rakha and Brent Crowther. Comparison of greenshields, pipes, and Van Aerde car-following and traffic stream models. *Transportation Research Record*, 1802(Paper No. 02-2143):248–262, 2002.
2. Kyoungho Ahn and Hesham Rakha. Field evaluation of energy and environmental impacts of driver route choice decisions. In *Intelligent Transportation Systems Conference, ITSC 2007*, Seattle, WA, pp. 730–735, IEEE, 2007.
3. Kyoungho Ahn and Hesham Rakha. The effects of route choice decisions on vehicle energy consumption and emissions. *Transportation Research Part D: Transport and Environment*, 13(3):151–167, 2008.
4. Kyoungho Ahn, Hesham Rakha, and Antonio Trani. Microframework for modeling of high-emitting vehicles. *Transportation Research Record: Journal of the Transportation Research Board*, 1880:39–49, 2004.
5. Kyoungho Ahn, Hesham A Rakha, Antonio Trani, and Michel Van Aerde. Estimating vehicle fuel consumption and emissions based on instantaneous speed and acceleration levels. *Journal of Transportation Engineering*, 128(2):182–190, 2002.
6. Alex Avgoustis, Michel Van Aerde, and Hesham A. Rakha. Framework for estimating network-wide safety impacts of intelligent transportation systems. In *Intelligent Transportation Systems Safety and Security Conference*, Miami, FL, 2004.
7. Matthew Barth, Kanok Boriboonsomsin, and Alex Vu. Environmentally-friendly navigation. In *Intelligent Transportation Systems Conference (ITSC) , Seattle, WA 2007*, , pp. 684–689, IEEE, 2007.
8. Kanok Boriboonsomsin, Matthew Barth, Zhu Weihua, and Alex Vu. Eco-routing navigation system based on multisource historical and real-time traffic information. *IEEE Transactions on Intelligent Transportation Systems*, 13(4):1694–1704, 2012.
9. Sasan Dashtinezhad, Tamer Nadeem, Bogdan Dorohonceanu, Cristian Borcea, Porlin Kang, and Liviu Iftode. TrafficView: A driver assistant device for traffic monitoring based on car-to-car communication. In *2004 IEEE 59th Vehicular Technology Conference. VTC 2004-Spring (IEEE Cat. No.04CH37514)*, 5:2946–2950, 2004.
10. Francois Dion, Hesham Rakha, and Youn-Soo Kang. Comparison of delay estimates at under-saturated and over-saturated pre-timed signalized intersections. *Transportation Research Part B: Methodological*, 38(2):99–122, 2004.
11. Jianhe Du, Hesham Rakha, and Vikash V. Gayah. Deriving macroscopic fundamental diagrams from probe data: Issues and proposed solutions. *Transportation Research Part C: Emerging Technologies*, 66:136 –149, 2016; Advanced network traffic management: From dynamic state estimation to traffic control.
12. Jianhe Du, Hesham A. Rakha, Ahmed Elbery, and Matthew Klenk. Microscopic simulation and calibration of a large-scale metropolitan network: Issues and proposed solutions. In *Annual Meeting of the Transportation Research Board*, Washington, DC, 2018.
13. Tamer ElBatt, Siddhartha K. Goel, Gavin Holland, Hariharan Krishnan, and Jayendra Parikh. Cooperative collision warning using dedicated short range wireless communications. In *Proceedings of the 3rd International Workshop on Vehicular Ad Hoc Networks*, VANET '06, New York, ACM, pp. 1–9, 2006,.
14. Ahmed Elbery and Rakha Hesham. City-wide eco-routing navigation considering vehicular communication impacts. *Sensors*, 19(2):290, 2019.
15. Ahmed Elbery ; Hossam S. Hassanein ; Nizar Zorba ; Hesham A. Rakha. Vanet-based smart navigation for vehicle crowds: Fifa world cup 2022 case study. In *2019 IEEE Global Communications Conference (GLOBECOM)*, Taipei, Taiwan, pp. 1–6, December 2019.
16. Ahmed Elbery. Large-scale modeling of smart cities considering the mutual impact of transportation and communication systems. PhD thesis, Virginia Tech University, 2018.

17. Ahmed Elbery and Hesham Rakha. VANET communication impact on a dynamic Eco-routing system performance: Preliminary results. In *2018 IEEE International Conference on Communications Workshops (ICC Workshops)*, Kansas City, MO, pp. 1–6, 2018.
18. Ahmed Elbery, Hesham Rakha, and Mustafa ElNainay. Vehicular communication and mobility sustainability: The mutual impacts in large-scale smart cities. arXiv preprint arXiv:1908.08229, 2019.
19. Ahmed Elbery, Hesham Rakha, Mustafa Elnainay, Wassim Drira, and Fethi Filali. Eco-routing using V2I communication: System evaluation. In *2015 IEEE 18th International Conference on Intelligent Transportation Systems*, Spain, pp. 71–76, IEEE, 2015.
20. Ahmed Elbery, Hesham Rakha, Mustafa Y. ElNainay, Wassim Drira, and Fethi Filali. Eco-routing: An ant colony based approach. In *VEHITS*, Rome, Italy, pp. 31–38, 2016.
21. Ahmed Elbery and Hesham A. Rakha. A novel stochastic linear programming feedback eco-routing traffic assignment system. In *Transportation Research Board 96th Annual Meeting, TRB 2017*, Washington, DC, 2017.
22. Ahmed Elbery, Hesham A. Rakha, and Mustafa ElNainay. Large-scale modeling of VANET and transportation systems. In *International Conference on Traffic and Granular Flow*, Washington DC, pp. 517–526, Springer, 2017.
23. Eva Ericsson, Hanna Larsson, and Karin Brundell-Freij. Optimizing route choice for lowest fuel consumption-potential effects of a new driver support tool. *Transportation Research Part C: Emerging Technologies*, 14(6):369–383, 2006.
24. Stephen Ezell. Explaining international IT application leadership: Intelligent transportation systems. *The Information Technology and Innovation Foundation (ITIF)*, Washington, DC, 1, 2010.
25. Khalid Abdel Hafeez, Lian Zhao, Zaiyi Liao, Bobby Ngok-Wah Ma. Impact of mobility on VANETs' safety applications. In *Global Telecommunications Conference (GLOBECOM 2010)*, Miami, FL, IEEE, pp. 1–5, 2010.
26. Hannes Hartenstein and Kenneth Laberteaux. *VANET: Vehicular Applications and Inter-Networking Technologies*, volume 1. John Wiley & Sons: New York, 2009.
27. Dirk Helbing. Derivation of a fundamental diagram for urban traffic flow. *The European Physical Journal B-Condensed Matter and Complex Systems*, 70(2):229–241, 2009.
28. IEEE Standard 1609.0-2013 "IEEE Guide for Wireless Access in Vehicular Environments (WAVE) - Architecture", March 2014.
29. IEEE Standard 802.11a-1999 "IEEE Standard for Telecommunications and Information Exchange between Systems - LAN/MAN Specific Requirements - Part 11: Wireless Medium Access Control (MAC) and physical layer (PHY) specifications: High Speed Physical Layer in the 5 GHz band", December 1999.
30. IEEE Standard for Information technology–Local and metropolitan area networks–Specific requirements–Part 11: Wireless LAN Medium Access Control (MAC) and Physical Layer (PHY) Specifications - Amendment 8: Medium Access Control (MAC) Quality of Service Enhancements, November 2005.
31. IEEE standard for information technology–local and metropolitan area networks–specific requirements–part 11: Wireless LAN medium access control (MAC) and physical layer (PHY) specifications amendment 6: Wireless access in vehicular environments, July 2010.
32. Leonard Kleinrock. *Queueing Systems, Volume 2: Computer Applications*, volume 66. John Wiley & Sons: New York, 1976.
33. Tamer Nadeem, Sasan Dashtinezhad, Chunyuan Liao, and Liviu Iftode. Traffic view: Traffic data dissemination using car-to-car communication. *ACM SIGMOBILE Mobile Computing and Communications Review*, 8(3):6–19, 2004.
34. Yu Nie and Qianfei Li. An eco-routing model considering microscopic vehicle operating conditions. *Transportation Research Part B: Methodological*, 55(0):154–170, 2013.
35. Hesham A. Rakha, Michel Van Aerde, Loern Bloomberg, and Xinghua Huang. Construction and calibration of a large-scale microsimulation model of the salt lake area. *Transportation Research Record: Journal of the Transportation Research Board*, 1644:93–102, 1998.
36. Hesham Rakha and Brent Crowther. Comparison and calibration of fresim and integration steady-state car-following behavior. *Transportation Research - Part A: Policy and Practice*, 37(1):1–27, 2003.

37. Hesham Rakha, Michel Van Aerde, Ryerson Case, and Alex Ugge. Evaluating the benefits and interactions of route guidance and traffic control strategies using simulation. In *Conference Record of Papers Presented at the First Vehicle Navigation and Information Systems Conference (VNIS '89)*, Toronto, Ontario, pp. 296–303, September 1989.
38. Hesham Rakha, Kyoungho Ahn, Ihab El-Shawarby, and Sebong Jang. Emission model development using in-vehicle on-road emission measurements. In *Annual Meeting of the Transportation Research Board*, Washington, DC, volume 2, 2004.
39. Hesham Rakha, Kyoungho Ahn, and Antonio Trani. Development of VT-micro model for estimating hot stabilized light duty vehicle and truck emissions. *Transportation Research Part D: Transport and Environment*, 9(1):49–74, 2004.
40. Hesham Rakha, Francois Dion, and Heung-Gweon Sin. Using global positioning system data for field evaluation of energy and emission impact of traffic flow improvement projects: Issues and proposed solutions. *Transportation Research Record*, 1768(1):210–223, 2001.
41. Hesham Rakha, Alejandra Medina Flintsch, Kyoungho Ahn, Ihab El-Shawarby, and Mazen Arafeh. Evaluating alternative truck management strategies along interstate 81. *Transportation Research Record: Journal of the Transportation Research Board*, 1925:76–86, 2005.
42. Hesham Rakha, Youn-Soo Kang, and François Dion. Estimating vehicle stops at undersaturated and oversaturated fixed-time signalized intersections. *Transportation Research Record: Journal of the Transportation Research Board*, 1776:128–137, 2001.
43. Hesham Rakha, Ivana Lucic, Sergio Henrique Demarchi, José Reynaldo Setti, and Michel Van Aerde. Vehicle dynamics model for predicting maximum truck acceleration levels. *Journal of Transportation Engineering*, 127(5):418–425, 2001.
44. Hesham A. Rakha. Validation of Van Aerde's simplified steady-state car-following and traffic stream model. *Transportation Letters: The International Journal of Transportation Research*, 1(13):227–244, 2009.
45. Hesham A. Rakha, Kyoungho Ahn, and Kevin Moran. INTEGRATION framework for modeling eco-routing strategies: Logic and preliminary results. *International Journal of Transportation Science and Technology*, 1(3):259–274, 2012.
46. Hesham A. Rakha, Kyoungho Ahn, and Antonio Trani. Development of VT-Micro framework for estimating hot stabilized light duty vehicle and truck emissions. *Transportation Research, Part D: Transport and Environment*, 9(1):49–74, 2004.
47. Hesham A. Rakha, Alejandra Medina, Heung-Gweon Sin, Francois Dion, Michel Van Aerde, and Jenq Page. Traffic signal coordination across jurisdictional boundaries: Field evaluation of efficiency, energy, environmental, and safety impacts. *Transportation Research Record: Journal of the Transportation Research Board*, 1727:42–51, 2000.
48. Hesham A. Rakha, Praveen Pasumarthy, and Slimane Adjerid. A simplified behavioral vehicle longitudinal motion model. *Transportation Letters: The International Journal of Transportation Research*, 1(2):95–110, 2009.
49. Hesham A. Rakha and Yihua Zhang. Integration 2.30 framework for modeling lane-changing behavior in weaving sections. *Transportation Research Record: Journal of the Transportation Research Board*, 1883:140–149, 2004.
50. Technology Riverbed Technology. https://www.riverbed.com/products/steelcentral/opnet.html, Accessed December 2017.
51. Hanif D. Sherali, Jitamitra Desai, and Hesham Rakha. A discrete optimization approach for locating automatic vehicle identification readers for the provision of roadway travel times. *Transportation Research Part B: Methodological*, 40(10):857–871, 2006.
52. United Nations. World's population increasingly urban with more than half living in urban areas. http://www.un.org/en/development/desa/news/population/world-urbanization-prospects-2014.html. Accessed: 2018-01-30.
53. Michel Van Aerde and Sam Yagar. Dynamic integrated freeway/traffic signal networks: Problems and proposed solutions. *Transportation Research Part A: General*, 22(6):435–443, 1988.
54. Michel Van Aerde and Sam Yagar. Dynamic integrated freeway/traffic signal networks: A routing-based modelling approach. *Transportation Research Part A: General*, 22(6):445–453, 1988.

55. Jinghui Wang, Ahmed Elbery, and Hesham A. Rakha. A real-time vehicle-specific eco-routing model for on-board navigation applications capturing transient vehicle behavior. *Transportation Research Part C: Emerging Technologies*, 104:1–21, 2019.
56. Ning Wu and Hesham A. Rakha. Derivation of Van Aerde traffic stream model from tandem-queueing theory. *Transportation Research Record: Journal of the Transportation Research Board*, 2124(18-27), 2009.
57. Qing Xu , Tony Mak, Jeff Ko, and Raja Sengupta. Medium access control protocol design for vehicle-vehicle safety messages. *IEEE Transactions on Vehicular Technology*, 56(2):499–518, March 2007.
58. Jun Zheng and Qiong Wu. Performance modeling and analysis of the IEEE 802.11p EDCA mechanism for VANET. *IEEE Transactions on Vehicular Technology*, 65(4):2673–2687, 2016.

7 Message Dissemination in Connected Vehicles

Anirudh Paranjothi and Mohammed Atiquzzaman
University of Oklahoma

Mohammad S. Khan
East Tennessee State University

CONTENTS

Advances in connected vehicles based on Vehicular Ad hoc Networks (VANETs) in recent years have gained significant attention in Intelligent Transport Systems (ITS) in terms of disseminating messages in an efficient manner. VANET uses Dedicated Short Range Communication (DSRC) for disseminating messages between vehicles and between infrastructures. In general, DSRC uses a dedicated 5.9 GHz band for vehicular communication [1]. DSRC has one control channel responsible for sending critical messages like information on road accidents, traffic jams, roadblocks, and six service channels responsible for sending non-critical messages like personal messages. The DSRC bandwidth is composed of eight channels that consist of six 10 MHz service channels for non-critical communications, one 10 MHz control channel for critical communications, and one 5 MHz reserved channel for future uses. As a result, VANET emerged as a promising solution for ensuring road safety.

In a VANET environment, also known as a connected vehicular environment, information is disseminated among the vehicles through messages. Two types of messages are disseminated among the vehicles: (1) periodic beacon messages and (2) event-driven messages [2,3]. Periodic beacon messages, also known as Basic Safety Messages (BSMs) include information like vehicle's position, speed, direction, acceleration, braking status, etc. Vehicles typically broadcast BSMs to all neighboring vehicles at an interval of 100–300 ms. The objective of broadcasting BSMs is to be aware of the neighboring vehicles. For example, in a high dense vehicular environment like a downtown region, every vehicle receives approximately 1000 BSMs per second. Event-driven messages are generated when a vehicle encounters an event such as accidents, traffic jams, roadworks, etc. [4].

Three types of communication are possible in VANETs: (1) Vehicle to Vehicle (V2V) communication, (2) Vehicle to Infrastructure (V2I) communication, and (3) Infrastructure to Vehicle (I2V) communication [5,6]. V2V and V2I communication depend on DSRC for disseminating messages among the vehicles. V2V communication is used when the vehicles are in the transmission range of each other where vehicles communicate with each other using a multi-hop technique. For example, when a vehicle encounters a dangerous situation such as road accidents, loss of traction, etc., messages are disseminated to the nearby vehicles using a multi-hop technique [7,8]. V2V communication is purely ad hoc in nature since vehicles communicate with each other directly without any infrastructure. V2V is used for communication among vehicles. V2I is used for communicating with fixed infrastructure and long-distance communication. The components of V2I include traffic lights, cameras, etc. V2I converts the available infrastructure into a smart infrastructure by incorporating various algorithms that use data exchange between vehicles and infrastructures. One such real-time example is Audi's "Time to green" feature that enables the vehicle to communicate with a traffic light when the traffic light is red and display the time remaining until it changes to green on the dashboard. I2V communication is most commonly used at the start of each lane of the intersection, where the fixed infrastructure such as traffic signal transmits the information of wait time to the vehicles in its transmission range. Upon receiving the traffic information, the vehicles broadcast it to the tail end vehicles through V2V communication.

Though DSRC based communications are viable, it is still challenging to disseminate messages in a timely manner when vehicles are not in the transmission range of each other. Furthermore, DSRC communication channels are heavily congested when the vehicle density increases on the road [2,9]. To address these limitations, two emerging paradigms: (1) vehicular cloud computing and (2) vehicular fog computing are being adopted to disseminate messages between the vehicles in a connected vehicular environment. The use of cloud computing in VANETs is commonly known as vehicular cloud computing. Cloud computing has many advantages including, (1) ubiquity, (2) high processing power, and (3) location awareness, resulting in the rapid dissemination of messages among the vehicles [10]. The use of fog computing in VANETs is commonly known as vehicular fog computing, where the computations are performed at the proximity of users. Fog computing is also termed as edge computing as the computations are performed at the edge of a network [2,11]. Vehicular fog computing uses fog nodes for the dissemination of messages among vehicles. Any real-world object can be formed as a fog node by acquiring the properties such as (1) network connectivity, (2) computation, and (3) storage.

In this book chapter, we highlight the significance of message dissemination in connected vehicles based on techniques like vehicular cloud computing, and vehicular fog computing. Our objective is to help the readers better understand the fundamentals of connected vehicles and communication techniques while disseminating messages between vehicles and between infrastructures. In a connected vehicular environment, message dissemination techniques can be modeled and performed with the help of simulations as it offers a cost-effective and scalable mechanism to analyze various parameters and scenarios. The most commonly used simulators are Simulation of Urban Mobility (SUMO) and network simulator (ns). To simulate the trace of vehicles movements, SUMO simulator is used [12]. The output of the SUMO simulator is given as input to the ns simulator.

Ns is a discrete event simulator consisting of many modules including, (1) packet loss model, (2) node deployment model [13,14], and (3) node mobility model for dynamic network topologies to perform the simulation. For a better understanding of the simulation, we provide an overview of Hybrid-Vehcloud [15], a vehicular cloud computing-based message dissemination scheme for guaranteed message delivery at high vehicle dense regions like Manhattan where buildings block radio propagation, and Dynamic Fog for Connected Vehicles (DFCV) [16], a dynamic fog computing-based message dissemination scheme for rapid transmission of messages and guaranteed message delivery at high vehicle densities. In Hybrid-Vehcloud and DFCV, simulations were conducted to measure the performance of our message dissemination protocol based on metrics like end-to-end delay of message delivery, probability of message delivery, Packet Loss Ratio (PLR), and average throughput. The contributions of this book chapter are as follows:

1. We provide an extensive overview of message dissemination techniques in connected vehicles including, challenges and various scenarios involved in message dissemination based on two major classifications: (1) vehicular cloud-based message dissemination, and (2) vehicular fog-based message dissemination.
2. We provide two example frameworks: Hybrid-Vehcloud, a vehicular cloud computing-based scheme, and DFCV, a fog computing-based scheme, which ensures low delay and guaranteed message delivery at high vehicle densities in a connected vehicle environment.
3. We provide a detailed analysis of the performance of various message dissemination protocols based on a defined list of performance metrics like end-to-end delay, PLR, etc.

The rest of the book chapter is structured as follows: Overview of existing message dissemination techniques in VANET is illustrated in Section 7.1. The working principle and message dissemination algorithm of Hybrid-Vehcloud are presented in Section 7.2. The working principle and message dissemination algorithm of DFCV are presented in Section 7.3. Performance evaluation and simulation results of message dissemination techniques are illustrated in connected vehicles in Section 7.4. The comparison of vehicular fog computing and vehicular cloud computing is presented in Section 7.5. The conclusion and future directions of the book chapter are presented in Sections 7.6 and 7.7, respectively.

7.1 BACKGROUND WORK

This section provides an overview of various message dissemination techniques in connected vehicles based on two major classifications: (1) dissemination of messages using vehicular cloud computing, and (2) dissemination of messages using vehicular fog computing.

7.1.1 DISSEMINATION OF MESSAGES USING VEHICULAR CLOUD COMPUTING

Vehicular cloud computing is used to handle complex tasks in a connected vehicular environment including, offloading large files, minimize traffic congestion, etc. One such complex task is obstacle shadowing. The radio transmissions are heavily affected by shadowing effects commonly known as obstacle shadowing [17]. Finding a solution for this problem plays an essential role in dense regions like Manhattan and other downtown areas where buildings block radio wave propagation. To overcome the obstacle shadowing problem in a connected vehicle environment, an emerging technique called vehicular cloud computing is heavily being adopted by researchers and industries.

Roman et al. [18] proposed a cloud computing-based message dissemination scheme for the connected vehicular environment. The proposed scheme guarantees the integrity of messages and lower communication overhead compared to the existing message dissemination schemes. However, it has limitations such as high delay and frequent loss of connectivity. Limbasiya et al. [19] proposed a message confirmation protocol for a connected vehicular environment. The protocol helps the Road Side Units (RSU) to verify the messages obtained from other vehicles using vehicular cloud computing techniques. However, the message confirmation protocol suffers from high communication overhead and high maintenance costs. Vasudev et al. [20] illustrated a message dissemination scheme for smart transportation in vehicular cloud computing. The proposed scheme suffers from high PLR as the number of vehicles increases and thus, is not suitable for high vehicle dense regions like the downtown environment. Bi et al. [21] discussed a message dissemination protocol known as Cross-Layer Broadcast Protocol (CLBP) to disseminate messages between vehicles. However, it is not suitable for vehicle-dense obstacle shadowing regions like Manhattan environment.

Syfullah et al. [10] and [22] discussed the RSU-based critical message dissemination scheme to the nearby vehicles with the help of vehicular cloud networks known as Cloud-assisted Message Downlink dissemination Scheme (CMDS) and Cloud-VANET protocols, respectively. CMDS and Cloud-Vanet are not suitable for high vehicle dense regions as they suffers from large transmission delays. Abbasi et al. [23] proposed a vehicular cloud-based routing algorithm for VANETs. One vehicle act as a cloud leader. The cloud leader collects information about vehicles such as position, speed, etc. through beacon messages and transmits the information to the vehicular cloud. Each vehicle has an operating system and hardware, which can provide an optimal route computed by the vehicular cloud. The cloud leader is also responsible for evaluating and monitoring the resources of other vehicles in its transmission range. However, the proposed approach suffers from high routing overhead.

Abdelatif et al. [24] presented a vehicular cloud-based traffic information dissemination approach for VANETs. The mechanism allows the vehicle to avoid highly congested and road work areas by transmitting messages like Traffic Incident Messages (TIM), event-driven messages, etc. to the vehicles present in the communication range of a vehicular cloud. But, the proposed scheme is not suitable for high dense obstacle shadowing regions. Khaliq et al. [25] proposed a vehicular data collection and data analysis scheme based on cloud computing. Once the vehicle encounters an accident, it generates an alert and transmits the information such as accident location, latitude, longitude, facing direction, etc. to the nearby vehicular cloud. Upon receiving the information, the vehicular cloud transmits the accident information to the nearest hospital and requests the ambulance service immediately. The proposed mechanism is not suitable for Manhattan regions, where the buildings block radio wave propagation from vehicles resulting in frequent disconnection of communication.

Sathyanarayanan [26] illustrated sensor-based emergency messages dissemination scheme using vehicular cloud computing. The vehicle collects information such as engine pressure, machine speed, location, etc. from the sensor at a constant time interval and reports the information to nearby gateways with a unique id and password. When the vehicle encounters an emergency scenario like a road accident, the vehicular cloud connection is established, which broadcasts the messages to all nearby vehicles in its communication range. This approach suffers from high transmission delay and routing overhead when the number of vehicles increases in the system. Mistareehi et al. [27] proposed a distributive architecture for the vehicular cloud. The proposed mechanism consists of Regional Cloud (RC) and Vehicular Cloud (VC). RCs collect the information of vehicles through beacon messages and broadcast it to the vehicles in its transmission range. RCs also communicate with CC to provide a wide range of services to vehicles. For example, CC forwards the information of the stolen vehicle to RCs and vehicles in its transmission range. When a vehicle encounters a stolen car on the road, it transmits the location of the stolen car along with the license plate image to the RC. RC transmits the information of a stolen car to the police department. However, this approach suffers from high maintenance costs and communication overhead.

To overcome the limitations of message dissemination schemes [10,18–27], and to provide an efficient solution for obstacle shadowing problems in a connected vehicle environment, a hybrid vehicular cloud computing technique called Hybrid-Vehcloud is emerged. Hybrid-Vehcloud adopts mobile gateways (such as buses) in the vehicular cloud for messages dissemination in obstacle shadowing regions. The detailed explanation of Hybrid-Vehcloud is illustrated in Section 7.2.

7.1.2 DISSEMINATION OF MESSAGES USING VEHICULAR FOG COMPUTING

Fog computing, also known as edge computing is considered a new revolutionary way of thinking in wireless networking. It is an extension of cloud computing where computations are performed at the edge of the network [2]. The use of fog computing in VANET is commonly known as vehicular fog computing. Vehicular fog computing offers unique services including, location awareness, ultra-low frequency, and context information. The fog nodes can be created, deployed, and destroyed faster when compared to other traditional message dissemination techniques.

Cui et al. [28] proposed an edge computing-based scheme for message authentication in a connected vehicular environment. The part of the vehicle acts as an edge computing node and helps RSU in message authentication tasks, thus reducing the overload on RSU. However, this approach suffers from high PLR. Zhong et al. [29] illustrated a message authentication scheme known as a message authentication scheme for multiple mobile devices in intelligent connected vehicles based on edge computing for connected vehicles based on fog computing techniques. In this scheme, vehicles disseminate messages to mobile devices such as laptops, smartphones, etc. for data processing instead of sending them to the cloud, thus reduce the communication overhead and delay. However, this approach has limitations such as high routing overhead and frequent loss of connectivity.

Yaqoob et al. [30] illustrated the fog-assisted message dissemination scheme named Energy-Efficient Message Dissemination (E^2MD) for connected vehicles. In E^2MD, each vehicle updates status such as speed, position, and direction to the fog server. Thus, in case of critical situations, such as road accidents, the fog server informs the vehicles about road congestion and co-ordinates patrols to clear the road. The shortcomings of E^2MD include high maintenance cost and high delay associated with accessing and allocation of resources in the fog server.

Wang et al. [31] and Grewe et al. [32] illustrated the possibility of mobility-based fog computing for broadcasting information in a vehicular environment. However, it creates instability as the load on the fog servers increases. Noorani et al. [33] proposed a fog computing-based geographical model for VANETs. The objective of this approach is to improve data transmission and reduce the latency in inter-vehicular communications. Here, the authors considered RSUs and base stations as fog nodes responsible for transmitting data packets and providing an optimal path for routing resulting in less delay in data transmission. However, this approach suffers from high PLR.

Xiao et al. [34] presented a fog computing-based data dissemination scheme for VANETs. The roadside infrastructures with small coverage areas such as RSUs, Wi-Fi access points are considered as fog nodes, and roadside infrastructures with large coverage areas such as base stations are considered as cloud nodes. The fog nodes transmit the data received from a vehicle to all other vehicles in its transmission range. Furthermore, the vehicles can request data or available services from the fog nodes. Upon receiving the request, the fog nodes upload the request to the cloud nodes, and the requested items will be provided by the cloud nodes with a bitwise exclusive-or strategy. The shortcomings of this approach include high delays associated with accessing available services. Sarkar et al. [35] discussed the usage of fog computing techniques with the internet of things. The focus of this work is to analyze suitability and applicability of fog computing in latency-sensitive applications. Also, the authors performed a comparison of the traditional cloud computing paradigm with fog computing in terms of maintenance cost and latency. From the simulation results, it is clearly depicted that the fog computing outperforms traditional cloud computing at all simulation times.

Youn [36] proposed a vehicular fog-based scheme for transmitting traffic information in a connected vehicular environment. When a vehicle encounters an accident scenario, it transmits the accident information such as location, etc. to nearby fog nodes. Fog nodes transmit the accident information to the cloud and vehicles in its communication range. However, the proposed scheme is not suitable for high dense obstacle shadowing regions. Tang et al. [37] proposed a hierarchical fog computing model for big data analysis in smart cities. The authors analyzed the case study of a smart pipeline system and constructed a working prototype of the fog nodes to demonstrate its implementation. From the results, it is clearly depicted that fog computing architecture has significant advantages over traditional cloud architecture. However, this approach has limitations such as high routing overhead and frequent loss of connectivity.

To address the shortcomings of message dissemination schemes [28–34,36,37], a fog-based layered architecture, called DFCV is proposed for the dissemination of messages. DFCV uses a three-layered architecture consisting of fog computing and cloud computing techniques, thereby ensuring efficient resource utilization, rapid transmission of messages, decreases in delay, and better QoS. The detailed explanation of DFCV is illustrated in Section 7.3.

7.2 HYBRID-VEHCLOUD MESSAGE DISSEMINATION

Hybrid-Vehcloud is a vehicular cloud computing-based scheme which ensures a low delay and guaranteed message delivery at high vehicle densities in a connected vehicle environment. This section provides an overview of the Hybrid-Vehcloud message dissemination technique based on two major classifications: (1) Dissemination of messages using Hybrid-Vehcloud, (2) Hybrid-Vehcloud message dissemination algorithm.

7.2.1 DISSEMINATION OF MESSAGES USING HYBRID-VEHCLOUD

The working of Hybrid-Vehcloud is illustrated with an example. Assume the vehicles represented in Figure 7.1 are in obstacle shadowing regions and need to disseminate messages between each other. Though the vehicles are in the communication range of an RSU, it is not possible to establish communication due to problems caused by shadowing. The solution to this situation is to deploy a vehicular cloud where the buses act as mobile gateways.

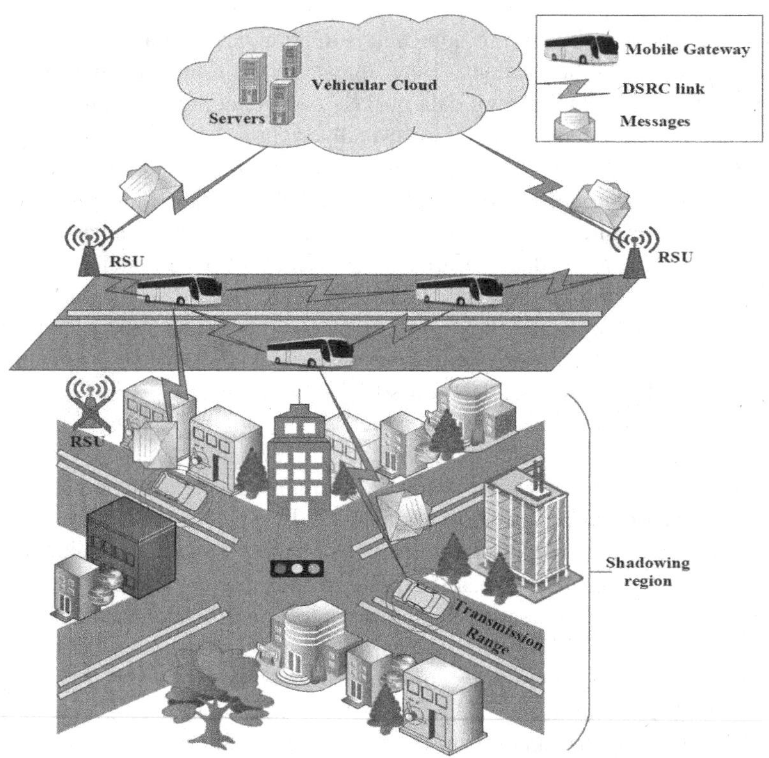

Figure 7.1 Dissemination of messages in connected vehicles using Hybrid-Vehcloud.

The following steps are involved in disseminating messages: First, vehicles collect information like traffic jam, roadblock, etc. and disseminate them to mobile gateways using DSRC protocol. The mobile gateways deploy vehicular cloud based on infrastructure such as RSU and disseminate the message received from vehicles to the cloud. In addition, the mobile gateways transmit their own information like location, access delay, bandwidth, etc. to the cloud. After receiving the information about gateways and input messages, the cloud servers assess the traffic density and determine suitable mobile gateways to disseminate the message. The gateways are selected to maximize the coverage range of vehicles in the targeted area. The messages are then transmitted to the vehicles using DSRC through an appropriate gateway. As mobile gateways are aware of the location of the vehicles probing the situation is not necessary to determine the obstacle shadowing regions. Step-by-step execution of the Hybrid-Vehcloud algorithm is illustrated in Section 7.2.2.

7.2.2 HYBRID-VEHCLOUD MESSAGE DISSEMINATION ALGORITHM

Algorithm 1 Hybrid-Vehcloud (input_msg)

```
scan trans_range (V_x)
calculate n
if n > 0 then
    for (i = 1; i <= n; i++) do
        loc[i] = obstacle_shadowing (i)
        if loc[i] == 1 then
            establish veh_cloud (input_msg)
            print message sent using Hybrid-Vehcloud technique
        end if
    end for
else
    print no nearby vehicles in obstacle shadowing regions
end if
if (V_y == 1) then
    repeat steps 1 to 13
end if
```

The Hybrid-Vehcloud algorithm works as follows: First, the set of neighboring vehicles in the transmission range of a base station associated with a sender is calculated. Then, trans_range (V_x) is used to discover the number of vehicles in the range of the base station. If the number of vehicles is greater than zero, the location of vehicles is determined using the obstacle_shadowing() function. The value 1 represents that the vehicles are in a shadowed region and hence the messages are broadcasted using the vehicular cloud technique. The notations used in Hybrid-Vehcloud are illustrated in Table 7.1.

7.3 DFCV MESSAGE DISSEMINATION

DFCV is a dynamic fog computing-based message dissemination scheme for rapid transmission of messages and guaranteed message delivery at high vehicle densities. This section provides an overview of the DFCV message dissemination technique based on two major classifications: (1) Dissemination of messages using DFCV, (2) DFCV message dissemination algorithm.

Table 7.1
Notations Used in Hybrid-Vehcloud Algorithm

Variables	Purpose
V_x	Vehicle broadcasts the message (sender)
n	Number of vehicles in the transmission range of a sender (i.e., receiver(s))
loc	Either 0 or 1, based on this message the dissemination technique is determined
V_y	New vehicle enters the transmission range of base station associated with a sender

7.3.1 DFCV MESSAGE DISSEMINATION TECHNIQUE

Vehicular fog-based layered architecture, called DFCV, is shown in Figure 7.2. DFCV consists of three layers: (1) Terminal layer, (2) Fog layer, and (3) Cloud layer.

Terminal Layer: This layer is closest to the physical environment and end-user. It consists of various devices like smartphones, vehicles, sensors, etc. As the motive of DFCV approach is to broadcast the messages in a connected vehicular environment, only vehicles are represented in the terminal layer. Moreover, they are responsible for sensing the surrounding environment and transmitting the data to the fog layer for processing and storage.

Fog Layer: The fog layer is located at the edge of a network. It consists of fog nodes, which includes access points, gateways, RSUs, base station, etc. In DFCV, RSUs and base stations play a major role in disseminating the messages. Fog layer can be static at a fixed location or mobile

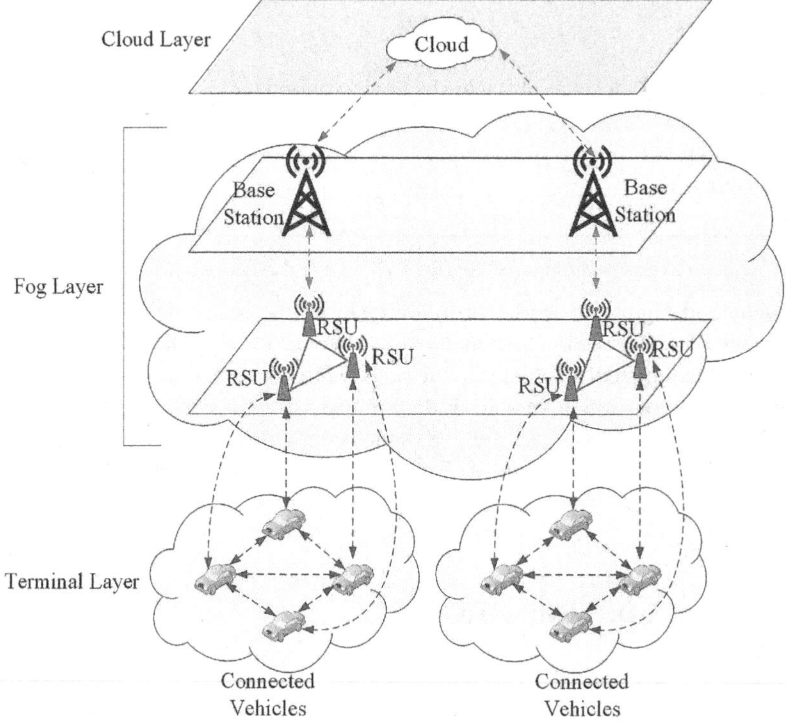

Figure 7.2 Dissemination of messages in connected vehicles using DFCV.

on moving carriers such as in the vehicular environment. Also, they are responsible for processing the information received from the terminal device and temporarily store it or broadcast over the network.

Cloud Layer: The main function of cloud computing in DFCV is to keep track of the resources allocated to each fog node and to manage interaction and interconnection among workloads on a fog layer.

DFCV incorporates all possible scenarios for disseminating the messages including, fog-split and fog-merge. Fog split will occur in two scenarios: Either the capacity of the DFCV is greater than the pre-defined threshold capacity, or the distance between the vehicle increases from the view of the sender, also known as the first observer. Fog merge will occur in two scenarios: Either the capacity of the DFCV is less than the pre-defined threshold capacity, or the distance between the vehicles is less than the minimum distance. Step-by-step execution of DFCV algorithm is illustrated in Section 7.3.2.

7.3.2 DFCV MESSAGE DISSEMINATION ALGORITHM

Algorithm 2 DFCV (input_msg, veh_{send}, veh_{rec})

```
for veh_send ∈ bs_i do
    for v ∈ c do
        calculate distance()
        if (distance > d_min || f_c > th_cap) then
            split (v ∈ c)
            fog_layer = v
            bs_i = send (input_msg)
            veh_rec = bs_i
        else
            merge (v ∈ c)
            fog_layer = v
            bs_i = send (input_msg)
            veh_rec = bs_i
        end if
        print message sent using DFCV technique
    end for
end for
```

DFCV aims to transmit the messages to the neighboring vehicles using fog computing technique. It mainly concentrates on merge and split scenarios as discussed in Section 7.3.1. The split is a primitive operation performed by DFCV using split() function. The steps are as follows: First, the distance between the vehicles is calculated using the distance() function. It is calculated based on the distance from the sender, and then, the capacity of the DFCV is determined using th_cap. The split accomplished when the distance exceeds the minimum distance (d_{min}) or the capacity of the DFCV (f_c) surpasses the threshold capacity. Here, a single fog will split into two parts. After the split, messages are relayed to the base station with the help of the RSU and send() function is used to send the input message to the vehicles in a corresponding base station (bs_i). The notations used in DFCV are illustrated in Table 7.2. Merge is another primitive operation performed by DFCV using merge() function based on the following constraints: The distance is lesser than the minimum distance (d_{min}), or the capacity of the DFCV (f_c) is lesser than the threshold capacity. It combines two or more fog layers under the same base station (bs_i) into a single fog layer. Then, the messages are broadcasted to the neighboring vehicle using send() function.

Table 7.2
Notations Used in DFCV Algorithm

Variables	Purpose
veh_{send}	Set of vehicle(s) that need to transmit messages
veh_{rec}	Intended recipient(s)
th	Threshold capacity of the fog
bs_i	Base station associated with vehsend

DFCV aims to provide rapid transmission of messages and guaranteed message delivery at high vehicle densities. In a connected vehicular environment, many challenges still exist due to the difficulties in the deployment and management of resources. In specific, the current techniques for V2V and V2I communications do not provide guaranteed message delivery resulting in messages being dropped before reaching the destination. It is due to an instability of DSRC, arising from the frequency band used by DSRC, as the number of vehicles increases. Furthermore, the current techniques for message dissemination have limitations such as the efficient utilization of resources, delay constraints due to high mobility and unreliable connectivity, and Quality of Service (QoS). DFCV message dissemination technique addresses the shortcomings of V2V and V2I communications and broadcasts messages to the vehicles using the DFCV message dissemination algorithm. DFCV algorithm can be used in both highway and urban environments as it provides better performance compared to Hybrid-Vehcloud at all vehicle densities. However, DFCV suffers from shadowing effects caused by obstacle shadowing regions.

The objective of the Hybrid-Vehcloud message dissemination technique is to lower the delay and to provide guaranteed message delivery in obstacle shadowing regions at high vehicle densities in a connected vehicular environment. The radio transmissions are heavily affected by shadowing effects caused by obstacles like tall buildings, skyscrapers, etc. To overcome the shadowing effects, the Hybrid-Vehcloud message dissemination algorithm adopts a vehicular cloud for broadcasting the messages in obstacle shadowing regions, where the buses act as mobile gateways. Hybrid-Vehcloud message dissemination algorithm can be used in obstacle regions such as Manhattan and other downtown regions. Hybrid-Vehcloud performs better compared to DFCV in obstacle shadowing regions.

7.4 PERFORMANCE EVALUATION

In a connected vehicular environment, message dissemination techniques including, Hybrid-Vehcloud and DFCV are evaluated using simulations. The most commonly used simulators are SUMO and ns. In this section, we discuss the simulation setup and the most commonly used metrics involved in the simulation of existing message dissemination techniques.

7.4.1 SIMULATION SETUP

Simulations of the message dissemination protocols are performed based on the algorithms. Algorithms illustrate step-by-step execution of the appropriate message dissemination framework. For example, Hybrid-Vehcloud and DFCV perform simulations based on the algorithms discussed in Sections 7.2.2. and 7.3.2. To simulate the trace of vehicle movements, the SUMO simulator is used [38]. The output files of the SUMO simulations are usually in the .xml format, contain information such as vehicle position, trip information, vehicle routes, etc. The output of the SUMO simulator is given as input to the ns simulator. Ns is a discrete event simulator, provides substantial support for simulation of wired and wireless networks [39]. Ns simulator output trace file for every

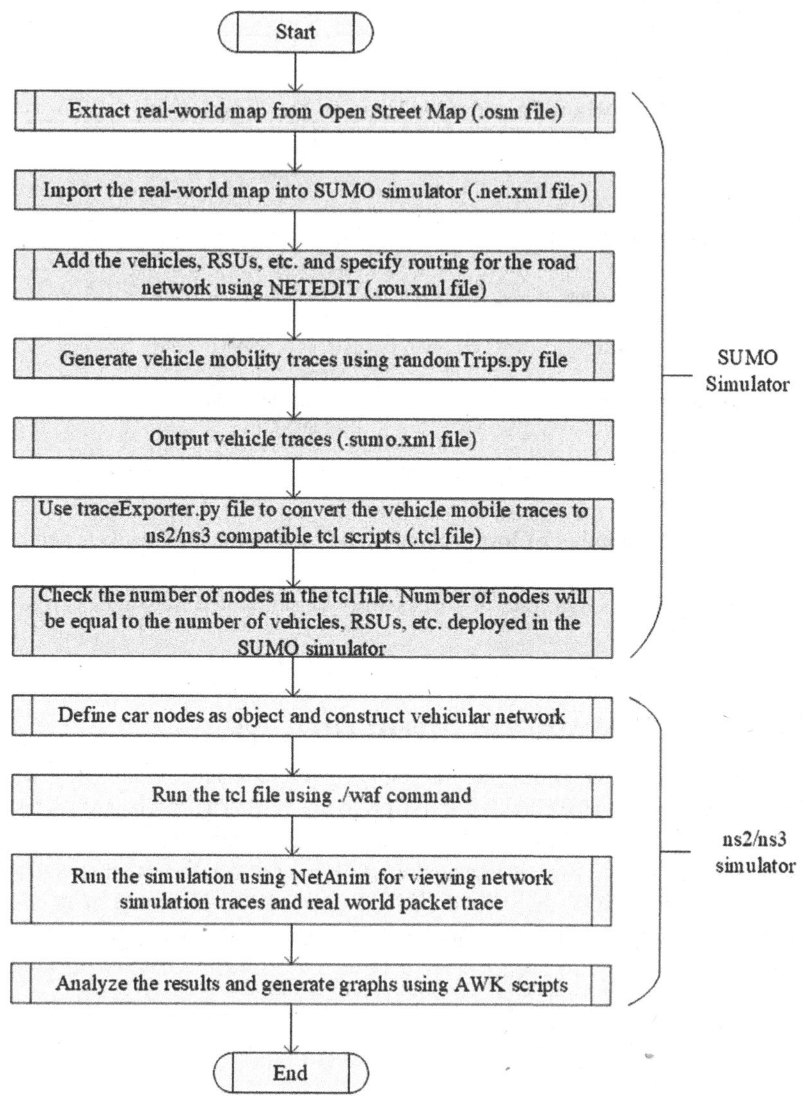

Figure 7.3 Simulation workflow of SUMO and ns2/ns3 simulators.

simulation. From the trace files, simulation data are collected and converted into graphs, represented in Figure 7.3. The simulation files are publicly available for everyone in our GitHub repository [40]. For a better understanding of the simulations, we provide extensive simulation results of Hybrid-Vehcloud and DFCV in Sections 7.4.3. and 7.4.4. The most important parameters used in the simulation are represented in Table 7.3.

7.4.2 PERFORMANCE METRICS

The most important performance metrics in the existing message dissemination protocols are:

1. *End-to-End Delay*: Time is taken for a message to be disseminated across a network from source to destination.
2. *The Probability of Message Delivery*: The probability of the input message delivered to the targeted vehicles.

Table 7.3
Most Important Parameters Used in the Simulation

Parameters	Value
Road length	10 km
Number of vehicles/nodes	50–450
Vehicle speed	30–60 mph
Transmission range	300 m
Message size	256 bytes
Simulator used	ns-2, ns-3, SUMO
Data rate	2 Mbps
Technique used	Vehicular fog, vehicular cloud
Protocol	IEEE802.11p

3. *PLR*: The ratio of the number of lost packets to the total number of packets sent across a network.
4. *Average Throughput*: Average rate of successfully disseminated messages across a communication channel.

7.4.3 PERFORMANCE EVALUATION OF HYBRID-VEHCLOUD

Hybrid-Vehcloud is compared with three previous cloud-based message dissemination schemes: (1) CMDS [22], (2) CLBP [21], and (3) Cloud-VANET protocols [10]. The results are discussed below:

End-to-End Delay: In Hybrid-Vehcloud, knowledge of nearby vehicles significantly reduces the route setup time and propagation time across a network. Hence, it delivers the message much faster when compared to CLBP, CMDS, and Cloud-VANET protocols, represented in Figure 7.4. The end-to-end delay is calculated against the number of vehicles, and it increases as the number of

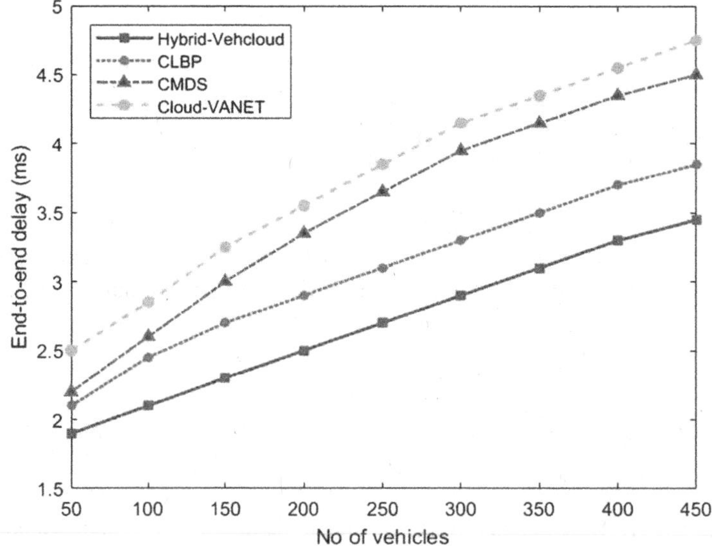

Figure 7.4 Comparison of end-to-end delay of Hybrid-Vehcloud with CLBP, CMDS, and Cloud-VANET approaches.

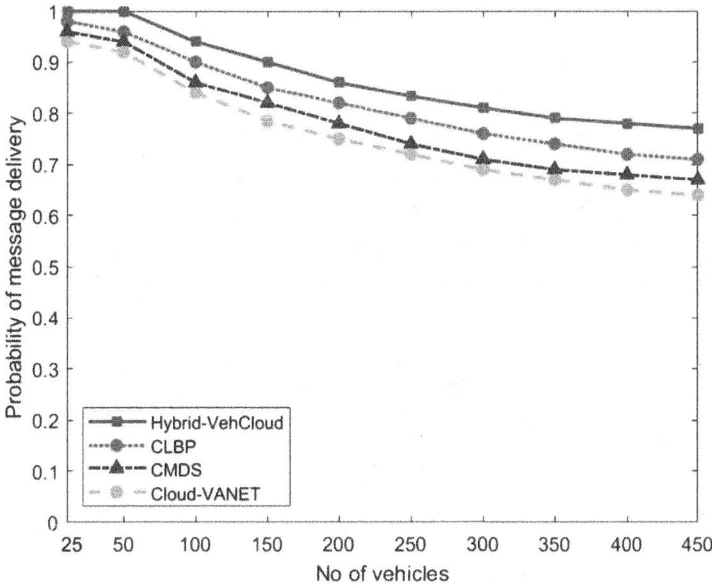

Figure 7.5 Comparison of the probability of message delivery of Hybrid-Vehcloud with CLBP, CMDS, and Cloud-VANET approaches.

users increases in the system due to a large number of messages that need to be delivered within a specific time interval.

Probability of Message Delivery: The probability of message delivery of Hybrid-VehCloud was observed to be higher due to guaranteed message delivery to the vehicles in the obstacle shadowing region, represented in Figure 7.5. The probability of message delivery is calculated against the number of vehicles. For each user, the probability of message delivery is distributed in the range of (0-1). From Figure 7.5, we can observe that the probability of message delivery decreases marginally as the number of users increases due to the increase in load on Hybrid-Vehcloud.

PLR: The PLR of Hybrid-Vehcloud is calculated against the number of vehicles, represented in Figure 7.6. PLR increases marginally as the number of vehicles increases due to the high channel congestion and frequent loss of connectivity. However, PLR of Hybrid-Vehcloud is observed to be lower compared to CLBP, CMDS, and Cloud-Vanet protocols at all vehicle densities.

Average Throughput: In Hybrid-Vehcloud, average throughput is the number of messages disseminated across a communication channel. From Figure 7.7, it can be observed that average throughput increases as the number of vehicle increases in the system, due to a large number of messages disseminated between vehicles. The average throughput of Hybrid-Vehcloud is high when compared to CLBP, CMDS, and Cloud-VANET protocols at all vehicle densities.

7.4.4 PERFORMANCE EVALUATION OF DFCV

DFCV is compared with three previous fog-based message dissemination schemes: (1) Named Data Networking (NDN) with mobility [31], (2) Fog-NDN with mobility [32], and (3) PEer-to-Peer protocol for Allocated REsource (PrEPARE) protocols [41]. The results are discussed below:

End-to-End Delay: In DFCV, as fog nodes are located at the proximity of users, it reduces the time taken for an initial setup across a network from source to destination and disseminate the messages much quicker than existing approaches such as fog-NDN with mobility, NDN with mobility, and PrEPARE protocols. The end-to-end delay is calculated against the number of vehicles and is observed to be lower at all vehicle densities, represented in Figure 7.8.

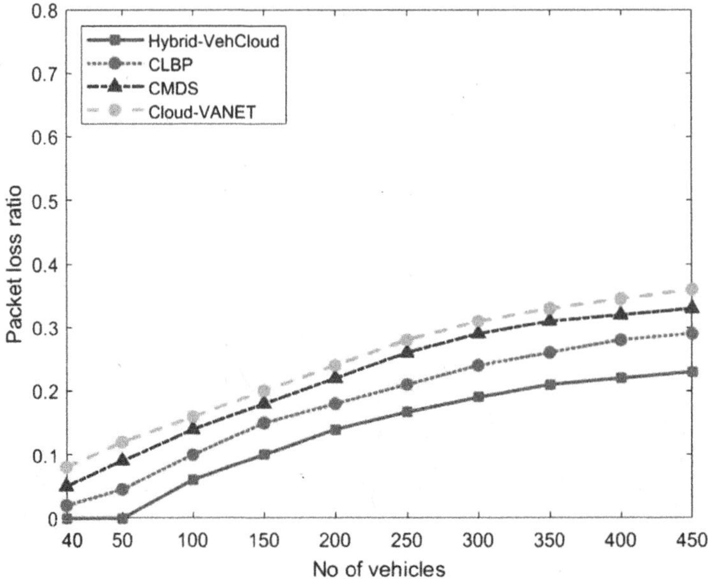

Figure 7.6 Comparison of the packet loss ratio of Hybrid-Vehcloud with CLBP, CMDS, and Cloud-VANET approaches.

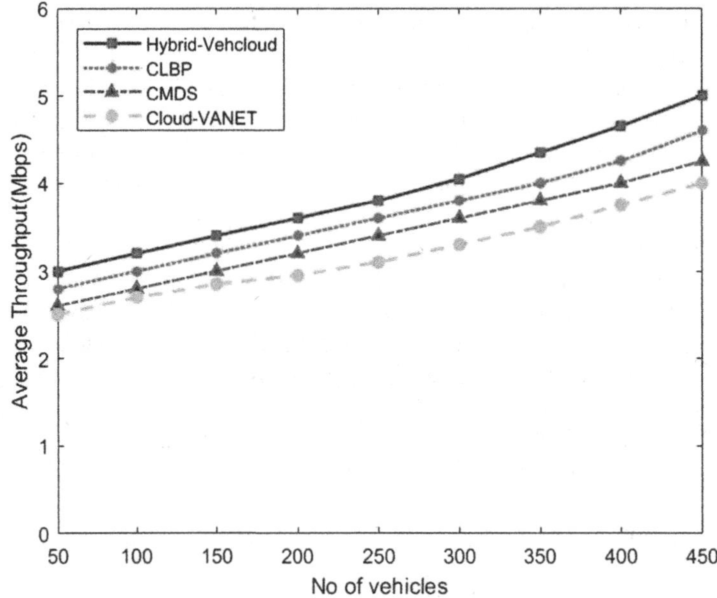

Figure 7.7 Comparison of average throughput of Hybrid-Vehcloud with CLBP, CMDS, and Cloud-VANET approaches.

Probability of Message Delivery: The probability of message of delivery of DFCV was observed to higher like Hybrid-Vehcloud (Section 7.4.3.) as DFCV also provides guaranteed message delivery at all vehicle densities. It is calculated against the number of vehicles, represented in Figure 7.9.

PLR: To observe the ratio of the number of lost packets in a network before reaching the destination, we performed this experiment at a time interval (t) and observed that the PLR of the DFCV

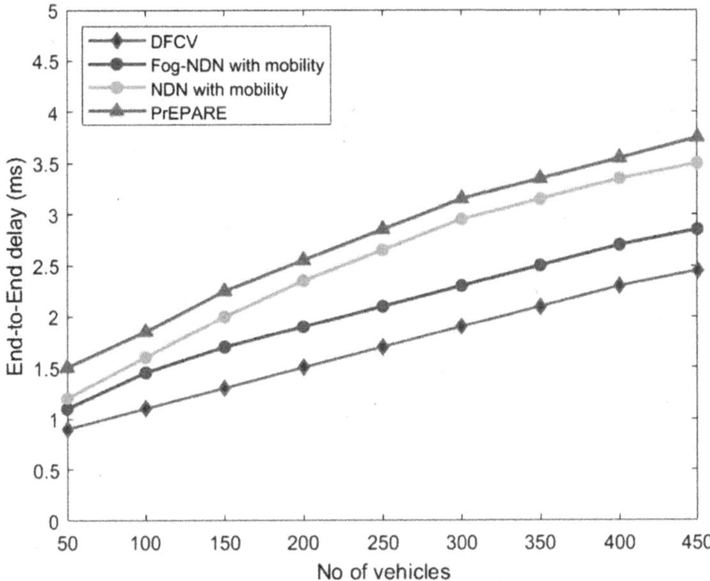

Figure 7.8 Comparison of end-to-end delay of DFCV with NDN with mobility, Fog-NDN with mobility, and PrEPARE approaches.

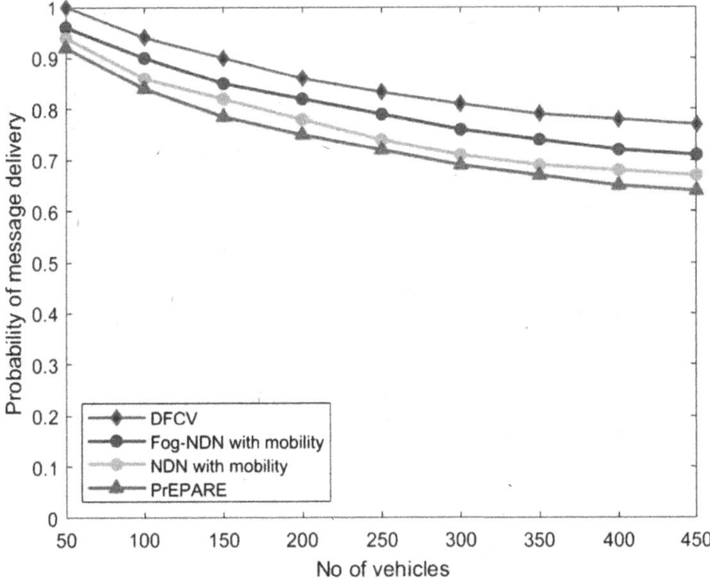

Figure 7.9 Comparison of the probability of message delivery of DFCV with NDN with mobility, Fog-NDN with mobility, and PrEPARE approaches.

approach is lower at high vehicle densities. PLR increases slightly as the number of users increases in the system, as shown in Figure 7.10. It is due to the additional packets generated being more likely to encounter another packet and resulting in a collision.

Average Throughput: The average throughput of DFCV is compared with fog-NDN with mobility, NDN with mobility, and PrEPARE protocols. It is calculated against the number of vehicles,

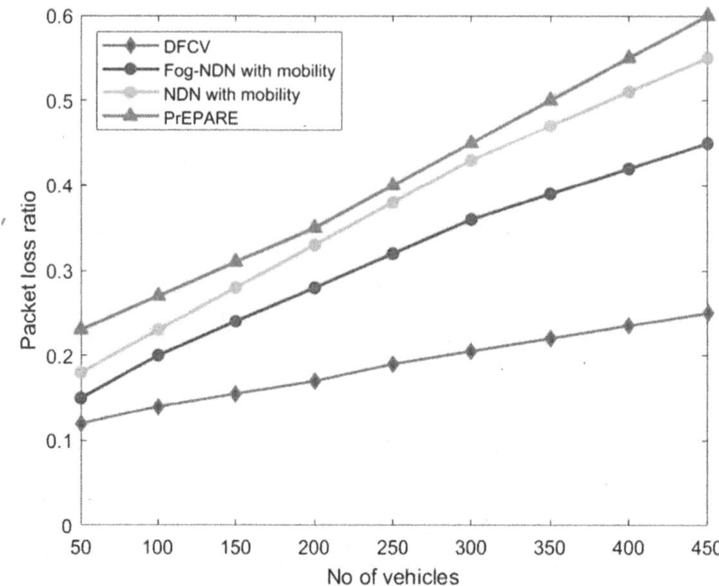

Figure 7.10 Comparison of the packet loss ratio of DFCV with NDN with mobility, Fog-NDN with mobility, and PrEPARE approaches.

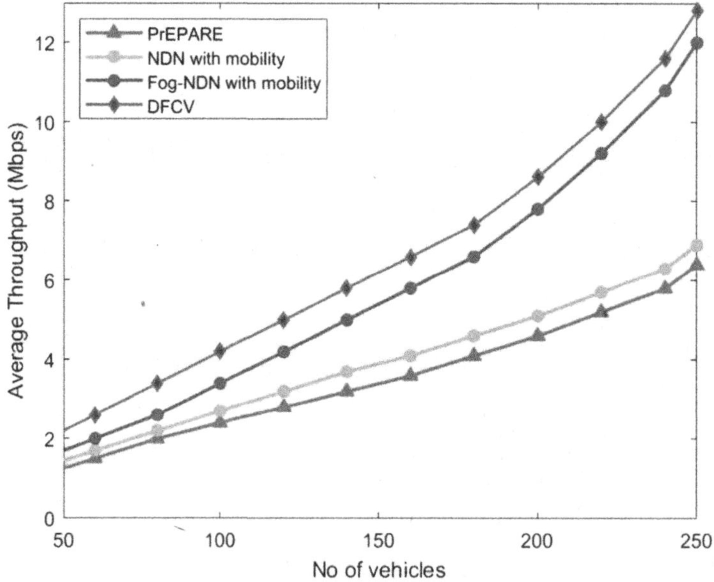

Figure 7.11 Comparison of average throughput of DFCV with NDN with mobility, Fog-NDN with mobility, and PrEPARE approaches.

represented in Figure 7.11. The number of messages disseminated across a network increases as the number of vehicles increases in a system. As DFCV provides guaranteed message delivery at high vehicle densities, the average throughput of DFCV is observed to be higher compared to fog-NDN with mobility, NDN with mobility, and PrEPARE protocols at all vehicle densities.

7.5 COMPARISON OF VEHICULAR FOG COMPUTING AND VEHICULAR CLOUD COMPUTING

This section compares and contrasts between the benefits of the vehicular fog computing against the vehicular cloud computing and vice versa.

7.5.1 ADVANTAGES OF VEHICULAR FOG COMPUTING OVER VEHICULAR CLOUD COMPUTING

1. *Low Latency*: Fog nodes are formed at the proximity of end-users. The processing of data takes place on the edge, much closer to the vehicles, resulting in rapid transmission of messages to other vehicles and fixed infrastructures like RSUs, etc.
2. *Better QoS*: Vehicular fog computing provides better data transmission rates with minimum response time compared to vehicular cloud computing at high vehicle densities. Thus, vehicular fog computing provides better QoS than vehicular cloud computing.
3. *Network Efficiency and Energy Consumption*: Unlike vehicular cloud computing, vehicular fog computing avoids the back and forth transmission between cloud servers. Thus, the bandwidth utilization and energy consumption of vehicular fog computing are much lesser compared to vehicular cloud computing.
4. *Improved Agility of services*: The rapid development of vehicular fog computing allows the users to customize the applications nearer to them instead of sending the changes to vehicular cloud servers and waiting for the response from the cloud servers.
5. *Deployment Cost*: The deployment cost of vehicular fog computing is very less compared to vehicular cloud computing. Any real-world objects that have the properties such as network connectivity, storage, and computing can become a fog node. Thus, the deployment cost of fog is lesser than vehicular cloud computing.

7.5.2 ADVANTAGES OF VEHICULAR CLOUD COMPUTING OVER VEHICULAR FOG COMPUTING

1. *Storage*: The vehicular cloud is centralized, offers wide storage space compared to vehicular fog computing. The users can deploy space constraint applications in the cloud server rather than fog nodes. Fog nodes provide limited storage space to the users, which cannot be used by the applications that require more storage space.
2. *Resource Management*: Vehicular cloud computing manages and dynamically changes the cloud resources based on the departure and arrival of vehicles. Each vehicle negotiates the level of resource sharing directly. Whereas, efficient utilization of resources is being considered as a major research area in vehicular fog computing.
3. *Service*: Vehicular cloud computing is popular in providing services on pay per use to the user based on demand. Vehicular cloud computing offers three types of services such as Network as a Service (NaaS), Storage as a Service (SaaS), and Cooperation as a Service (CaaS). Whereas, vehicular fog computing does not provide services on pay per use to the users.

7.6 CONCLUSION

In this book chapter, we provide an extensive overview of message dissemination techniques in connected vehicles including, challenges and various scenarios involved in message dissemination based on two major classifications: vehicular cloud-based message dissemination, and vehicular fog-based message dissemination. We discussed the working principle of two reliable and efficient message dissemination frameworks: Hybrid-Vehcloud and DFCV to help the readers better understand the fundamentals of connected vehicles, communication techniques, and various scenarios to be considered while disseminating messages. Above all, we have illustrated the performance of various message dissemination protocols including, Hybrid-Vehcloud and DFCV based on performance metrics like end-to-end delay, PLR, average throughput, and probability of message delivery.

7.7 FUTURE DIRECTIONS

In the future, the dissemination of messages in connected vehicles will extend using machine-learning techniques. Machine learning can be implemented among connected vehicles using various techniques including, decision trees, decision tree ensembles, artificial neural networks, and k-nearest neighbors. The use of machine learning in broadcasting messages will monitor the path periodically and transmit the message in an optimized route. This will increase the performance of the system in an efficient manner. Furthermore, integration of VANETs and 5-G will lead to significant enhancements and efficiency in vehicular communications.

With an increase in the number of vehicles, a huge volume of data is being generated. These data should be monitored, analyzed, and managed properly to reduce storage and bandwidth consumption. Advanced data processing and data mining techniques can be applied to handle the large volume of data disseminated from the active participants in a connected vehicular environment. Security and privacy are another major concerns for VANETs. The internet connecting to a large number of vehicles and infrastructures suffers from various network attacks like Denial of Service (DoS) attack, timing attack, Sybil attack, false information attack, black hole attack, tunneling attack, etc. Intrusion Detection System (IDS) can be implemented to detect network attacks caused by the compromised node on the network. The objective of IDS is to protect the network from network-based threats. IDS is located at some special nodes called monitoring nodes. The deployment of the monitoring nodes differs depending on the protocol type and the architecture of the IDS. Furthermore, new secure communication protocols must be investigated under various networking conditions.

REFERENCES

1. A. Paranjothi, U. Tanik, Y. Wang, and M. S. Khan, "Hybrid-Vehfog: A robust approach for reliable dissemination of critical messages in connected vehicles," *Transactions on Emerging Telecommunications Technologies*, vol. 30, pp. 1–11, 2019.
2. A. Ullah, S. Yaqoob, M. Imran, and H. Ning, "Emergency message dissemination schemes based on congestion avoidance in VANET and vehicular fog computing," *IEEE Access*, vol. 7, pp. 1570–1585, 2018.
3. O. S. Oubbati, A. Lakas, P. Lorenz, M. Atiquzzaman, and A. Jamalipour, "Leveraging communicating UAVs for emergency vehicle guidance in urban areas," *IEEE Transactions on Emerging Topics in Computing*, vol. 7, pp. 1–12, 2019.
4. S. S. Karanki and M. S. Khan, "SMMV: Secure multimedia delivery in vehicles using roadside infrastructure," *Vehicular Communications*, vol. 7, pp. 40–50, 2017.
5. A. Paranjothi, M. S. Khan, M. Nijim, and R. Challoo, "MAvanet: Message authentication in vanet using social networks," in *IEEE 7th Annual Ubiquitous Computing, Electronics & Mobile Communication Conference (UEMCON)*, New York, USA, pp. 1–8, 2016.
6. S. Garg, K. Kaur, S. H. Ahmed, A. Bradai, G. Kaddoum, and M. Atiquzzaman, "MobQoS: Mobility-Aware and QoS-driven SDN framework for autonomous vehicles," *IEEE Wireless Communications*, vol. 26, no. 4, pp. 12–20, 2019.
7. A. Paranjothi, M. S. Khan, S. Zeadally, A. Pawar, and D. Hicks, "GSTR: Secure multi-hop message dissemination in connected vehicles using social trust model," *Internet of Things*, vol. 7, no. 1, pp. 1–16, 2019.

8. A. Paranjothi, M. Atiquzzaman, and M. S. Khan, "PMCD: Platoon-merging approach for cooperative driving," *Internet Technology Letters*, vol. 3, no. 1, pp. 1–6, 2020.
9. A. Paranjothi, M. S. Khan, R. Patan, R. M. Parizi, and M. Atiquzzaman, "VANETomo: A congestion identification and control scheme in connected vehicles using network tomography," *Computer Communications*, vol. 151, no. 1, pp. 275–289, 2020.
10. B. Liu, D. Jia, J. Wang, K. Lu, and L. Wu, "Cloud-assisted safety message dissemination in VANET–cellular heterogeneous wireless network," *IEEE Systems Journal*, vol. 11, pp. 128–139, 2015.
11. X. Hou, Y. Li, M. Chen, D. Wu, D. Jin, and S. Chen, "Vehicular fog computing: A viewpoint of vehicles as the infrastructures," *IEEE Transactions on Vehicular Technology*, vol. 65, pp. 3860–3873, 2016.
12. F. Outay, F. Kammoun, F. Kaisser, and M. Atiquzzaman, "Towards safer roads through cooperative hazard awareness and avoidance in connected vehicles," in *31st International Conference on Advanced Information Networking and Applications Workshops (WAINA)*, Taipei, Taiwan, pp. 208–215, 2017.
13. T. M. Behera, S. K. Mohapatra, U. C. Samal, M. S. Khan, M. Daneshmand, and A. H. Gandomi, "Residual energy-based cluster-head selection in WSNs for IoT application," *IEEE Internet of Things Journal*, vol. 6, no. 3, pp. 5132–5139, 2019.
14. T. M. Behera, S. K. Mohapatra, U. C. Samal, M. S. Khan, M. Daneshmand, and A. H. Gandomi, "I-SEP: An improved routing protocol for heterogeneous WSN for IoT based environmental monitoring," *IEEE Internet of Things Journal*, vol. 7, no. 1, pp. 710–717, 2019.
15. A. Paranjothi, M. S. Khan, and M. Atiquzzaman, "Hybrid-vehcloud: An obstacle shadowing approach for vanets in urban environment," in *IEEE 88th Vehicular Technology Conference (VTC-Fall)*, Chicago, USA, pp. 1–5, 2018.
16. A. Paranjothi, M. S. Khan, and M. Atiquzzaman, "DFCV: A novel approach for message dissemination in connected vehicles using dynamic fog," in *16th International Conference on Wired/Wireless Internet Communication (WWIC)*, Boston, USA, pp. 311–322, 2018.
17. S. E. Carpenter, "Obstacle shadowing influences in VANET safety," in *IEEE 22nd International Conference on Network Protocols (ICNP)*, Raleigh, USA, pp. 480–482, 2014.
18. L. Roman and P. Gondim, "Authentication protocol in CTNs for a CWD-WPT charging system in a cloud environment," *Ad Hoc Networks*, vol. 97, pp. 1020–1034, 2020.
19. T. Limbasiya and D. Das, "Secure message confirmation scheme based on batch verification in vehicular cloud computing," *Physical Communication*, vol. 34, pp. 310–320, 2019.
20. H. Vasudev and D. Das, "A lightweight authentication and communication protocol in vehicular cloud computing," in *33rd International Conference on Information Networking (ICOIN)*, Kuala Lumpur, Malaysia, pp. 72–77, 2019.
21. Y. Bi, L. Cai, X. Shen, and H. Zhao, "A cross layer broadcast protocol for multihop emergency message dissemination in inter-vehicle communication," in *IEEE 44th International Conference on Communications (ICC)*, Cape Town, South Africa, pp. 1–5, 2010.
22. M. Syfullah and J. M. Lim, "Data broadcasting on cloud-VANET for IEEE 802.11p and LTE hybrid VANET architectures," in *3rd International Conference on Computational Intelligence & Communication Technology (CICT)*, Ghaziabad, India, pp. 1–6, 2017.
23. M. Abbasi, M. Rafiee, M. R. Khosravi, A. Jolfaei, V. G. Menon, and J. M. Koushyar, "An efficient parallel genetic algorithm solution for vehicle routing problem in cloud implementation of the intelligent transportation systems," *Journal of Cloud Computing*, vol. 9, no. 1, pp. 1–14, 2020.
24. S. Abdelatif, M. Derdour, N. Ghoualmi-Zine, and B. Marzak, "VANET: A novel service for predicting and disseminating vehicle traffic information," *International Journal of Communication Systems*, vol. 33, pp. 1–16, 2020.
25. K. A. Khaliq, O. Chughtai, A. Shahwani, A. Qayyum, and J. Pannek, "Road accidents detection, data collection, and data analysis using V2X communication and edge/cloud computing," *Electronics*, vol. 8, no. 8, pp. 1–28, 2019.
26. P. S. V. SathyaNarayanan, "A sensor enabled secure vehicular communication for emergency message dissemination using cloud services," *Digital Signal Processing*, vol. 85, pp. 10–16, 2019.
27. H. Mistareehi, T. Islam, K. Lim, and D. Manivannan, "A secure and distributed architecture for vehicular cloud," in *4th International Conference on P2P, Parallel, Grid, Cloud and Internet Computing*, Antwerp, Belgium, pp. 127–140, 2019.

28. J. Cui, L. Wei, J. Zhang, Y. Xu, and H. Zhong, "An efficient message-authentication scheme based on edge computing for vehicular ad hoc networks," *IEEE Transactions on Intelligent Transportation Systems*, vol. 20, pp. 1621–1632, 2018.
29. H. Zhong, L. Pan, Q. Zhang, and J. Cui, "A new message authentication scheme for multiple devices in intelligent connected vehicles based on edge computing," *IEEE Access*, vol. 7, pp. 108211–108222, 2019.
30. S. Yaqoob, A. Ullah, M. Akbar, M. Imran, and M. Shoaib, "Congestion avoidance through fog computing in internet of vehicles," *Journal of Ambient Intelligence and Humanized Computing*, vol. 10, no. 5, pp. 1–15, 2019.
31. M. Wang, J. Wu, G. Li, J. Li, Q. Li, and S. Wang, "Toward mobility support for information-centric IoV in smart city using fog computing," in *IEEE 5th International Conference on Smart Energy Grid Engineering (SEGE)*, Oshawa, Canada, pp. 357–361, 2017.
32. D. Grewe, M. Wagner, M. Arumaithurai, I. Psaras, and D. Kutscher, "Information-centric mobile edge computing for connected vehicle environments: Challenges and research directions," in *Proceedings of the Workshop on Mobile Edge Communications*, Los Angeles, USA, pp. 7–12, 2017.
33. N. Noorani and S. A. H. Seno, "SDN-and fog computing-based switchable routing using path stability estimation for vehicular ad hoc networks," *Peer-to-Peer Networking and Applications*, vol. 13, pp. 948–964, 2020.
34. K. Xiao, K. Liu, J. Wang, Y. Yang, L. Feng, J. Cao, and V. Lee, "A fog computing paradigm for efficient information services in VANET," in *IEEE 17th Wireless Communications and Networking Conference (WCNC)*, Marrakesh, Morocco, pp. 1–7, 2019.
35. S. Sarkar, S. Chatterjee, and S. Misra, "Assessment of the suitability of fog computing in the context of internet of things," *IEEE Transactions on Cloud Computing*, vol. 6, no. 1, pp. 46–59, 2015.
36. J. Youn, "Vehicular fog computing based traffic information delivery system to support connected self-driving vehicles in intersection environment," in *3rd International Conference on Advances in Computer Science and Ubiquitous Computing*, Kuala Lumpur, Malaysia, pp. 208–213, 2018.
37. B. Tang, Z. Chen, G. Hefferman, T. Wei, H. He, and Q. Yang, "A hierarchical distributed fog computing architecture for big data analysis in smart cities," in *5th International Conference on BigData & Social Informatics*, Kaohsiung, Taiwan, pp. 1–6, 2015.
38. S. Mallissery, M. Pai, M. Mehbadi, R. Pai, and Y. Wu, "Online and offline communication architecture for vehicular ad-hoc networks using NS3 and SUMO simulators," *Journal of High Speed Networks*, vol. 25, pp. 253–271, 2019.
39. J. A. Sanguesa, S. Salvatella, F. J. Martinez, M. Marquez-Barja, and M. P. Ricardo, "Enhancing the NS-3 simulator by introducing electric vehicles features," in *28th International Conference on Computer Communication and Networks (ICCCN)*, Valencia, Spain, pp. 1–7, 2019.
40. A. Paranjothi, "Hybrid-Vehcloud and DFCV simulation files," February 2020. https://github.com/anirudhparanjothi/Hybrid-Vehcloud-and-DFCV-simulation.
41. R. I. Meneguette and A. Boukerche, "Peer-to-Peer protocol for allocated resources in vehicular cloud based on V2V communication," in *IEEE 15th Wireless Communications and Networking Conference (WCNC)*, San Francisco, USA, pp. 1–6, 2017.

8 Exploring Cloud Virtualization over Vehicular Networks with Mobility Support

Miguel Luís
Instituto de Telecomunicações and Department of Electronics, Telecommunications and Computers Engineering, Instituto Superior de Engenharia de Lisboa

Christian Gomes and Susana Sargento
Instituto de Telecomunicações and Department of Electronics, Telecommunications and Informatics, University of Aveiro

Jordi Ortiz
Department of Information and Communications Engineering, University of Murcia

José Santa
Department of Information and Communications Engineering, University of Murcia and Department of Electronics, Computer Technology and Projects, Technical University of Cartagena

Pedro J. Fernández, Manuel Gil Pérez, and Gregorio Martínez Pérez
Department of Information and Communications Engineering, University of Murcia

Sokratis Barmpounakis and Nancy Alonistioti
Department of Informatics and Telecommunications National and Kapodistrian University of Athens

Jacek Cieślak
ITTI

Henryk Gierszal
Faculty of Physics, Applied Informatics Lab, Adam Mickiewicz University

CONTENTS

8.1 INTRODUCTION

5G vehicular networks provide the possibility to connect cars, people, traffic signs, road sensors and many others, to create smart environments enabling safety services, assisted and autonomous driving and also to provide Internet access to users inside those vehicles, among others. Vehicles with more than one network interface enable the possibility of multihoming solutions where, in order to take advantage of all available resources, vehicles can connect simultaneously to more than one access network. Nevertheless, the dynamic nature of such networks demands mobility solutions to manage the vehicles and user's mobility and to provide constant connectivity.

The next generation of mobile networks, such as 5G, demands rigorous requirements to guarantee a fair balance between latency, reliability and massive broadband. The shared use of resources, allowing the dynamic provision of network and service functions, is foreseen as one of the key enablers for such stringent requirements. Network Functions Virtualization (NFV) presents an innovative solution to support the active and high-performance processing of traffic delivered across mobile networks, including vehicular networks, in a dynamic and on-demand approach [41]. NFV grants more flexibility to network solutions by extending their capabilities and services, due to the ability of deploying and supporting new network services faster and cheaper (according the 5G Infrastructure Public Private Partnership (5G-PPP) program the average service creation time cycle will be reduced from 90 hours to 90 minutes). The key enablers for such evolution are the possibility of decoupling the software from hardware, the consequent flexible network function deployment, and the dynamic scaling of services and functions, granting a more dynamic and granularity control over the network resources according to its needs [29].

Several solutions have been exploring the integration of NFV technologies into the scope of mobile networks. Among the several topics of research, mobility management has been the one attracting more attention, especially when in conjunction with the SDN concept [40]. In vehicular environments, such technologies have been focused on the concept of vehicular Clouds, where the vehicular infrastructure itself is part of the Cloud [20], or on the concept of Infrastructure-as-a-Service in a vehicular network, where the vehicles' resources can be used to provide different types of services [3]. Such technologies have been also used to derive new slicing frameworks to support V2X applications by flexibly orchestrating multi-access and edge-dominated 5G network infrastructures [6], or even to provide flexible services for heterogeneous vehicles [43].

The main contribution of this chapter is to showcase the applicability of softwarization, automation and Cloud virtualization over a real vehicular network with multihoming and mobility support. Several virtualized functions, each one focusing on different Intelligent Transportation System (ITS) subject, were deployed and their performance, with respect to network and application layers, is analyzed and compared to non-virtualized solutions. The evaluation is performed through experimentation using a real vehicular network and supported by a Cloud infrastructure following the NFV reference architecture framework. The list of use cases and its respective virtualized functions include:

- the virtualization of On-Board Unit (OBU) instances in the Cloud to foster 5G vehicular services by extending their capabilities and offloading processing tasks from the same unit to its virtualized representative;
- QoE-aware adaptive streaming by virtualizing context-aware mechanisms that can use no-reference, context-based QoE metric subsets, to allow a single system to cope with many use case scenarios, e.g., to remotely operate automotive cars;
- the safety of Vulnerable Road Users (VRUs) and the intelligence required to effectively predict and avoid vehicular accidents by exploiting a hybrid architecture combining edge and Cloud computing;
- finally, and related to public safety, a virtualized city safety solution to monitor and analyze video streams collected from heterogeneous and distributed sources.

The remainder of this chapter is organized as follows. In the next section we overview the main characteristics of vehicular networks, focusing on mobility management protocols. Section 8.3 presents an introduction to the concept of NFV and its architecture and main components. Section 8.4 describes the real vehicular network implementing the automotive environment used for experimentation and validation, while Section 8.5 details the set of use cases exploring Cloud virtualization over VANETs (Vehicular *Ad hoc* Networks). Section 8.6 identifies key research challenges, and finally, conclusion remarks are summarized in Section 8.7.

8.2 VEHICULAR NETWORKS

The recent advances in the automobile industries and telecommunication technologies brought the focus on ITS, of which Vehicular *ad hoc* Networks (VANETs) gain much more attention. A VANET, a particular branch of mobile *ad-hoc* networks, is characterized by moving vehicles and fixed stations [5]. For a vehicle to be a node of that VANET, it has to be equipped with an On-Board Unit (OBU), responsible for the network connectivity, as well as giving passengers Internet access. On the other hand, Road Side Units (RSUs) deployed alongside the road are connected to the core network, providing Internet access to OBUs. As such, there are two possible types of communications: Vehicle-to-Vehicle (V2V) communications, where the OBUs communicate between themselves, and Vehicle-to-Infrastructure (V2I) communications, where OBUs communicate directly with RSUs, as illustrated in Figure 8.1.

Considering wireless technologies in a vehicular environment, IEEE 802.11p, also known as Wireless Access for Vehicular Environments (WAVE), was the first standard specifically designed to accommodate wireless access in Dedicated Short-Range Communication (DSRC). Its purpose was to work as an enabler for safety applications and improvements in traffic flow in both V2V and V2I environments. It integrates IEEE 802.11a (with transmission rates from 3 to 27 Mb/s, half the

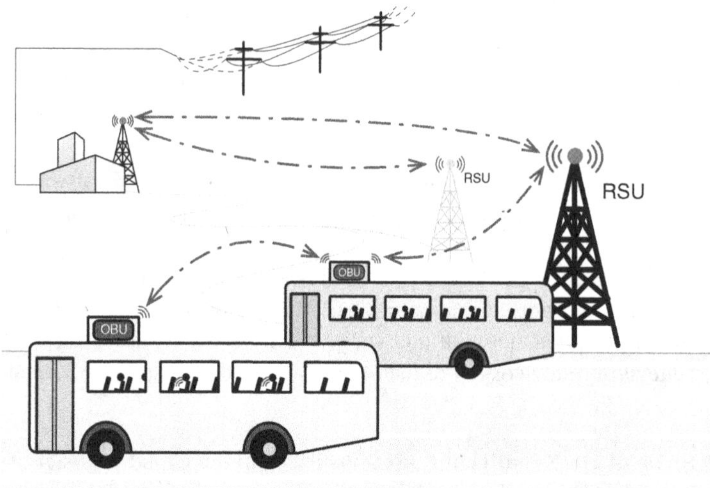

Figure 8.1 Illustrative example of a VANET.

bandwidth assigned in IEEE 802.11a) and IEEE 802.11e (to enhance QoS), making it robust for scenarios with dynamic geographic topology, variable vehicle density and a high number of link failures. Moreover, it does not use any association process, exploring short contact periods.

However, other access technologies can also be used when accounting for specific scenarios. The IEEE 802.11g, n, ac standards are included in nearly every electronic device, and free hotspots are available and scattered throughout some cities. These allow a wider coverage for VANETs and allow for high data rates. However, Wi-Fi presents numerous disadvantages: its range is lower than WAVE, and it performs poorly when used at high speeds or in highly mobile nodes, since the handshake process required for a data exchange needs multiple message exchanges. Nevertheless, such an access technology can complement the use of WAVE in specific scenarios as is the case of traffic lights, gas stations or even in heavy traffic situations.

In the last years, a real alternative for the IEEE 802.11p has been receiving considerable attention. The 3GPP published in Release 14 the C-V2X (Cellular Vehicular-to-Everything) standard (also known as LTE-V or LTE-V2X) for V2V communications [1]. This standard has been designed to support cooperative traffic safety and efficiency in applications and includes two modes of operation depending on if the communication is managed by the infrastructure, or autonomously by the vehicles. The C-V2X can operate in the 5.9 GHz bands as well as in the cellular operators' licensed carrier.

With the availability of multiple wireless access technologies, using a single access technology is a waste of resources and of potential benefits. Thus, under the multihoming concept [30] each OBU is capable of exploring, simultaneously or not, the various access technologies available. The simultaneous connectivity between the OBU and the RSUs, through the available interfaces, is also possible, enabling the choice of the services to be transmitted through each technology. However, several challenges arise and must be taken into account [16], such as packet reordering, network support, interface and application characteristics' estimation, identification of flows and sessions and increased power consumption.

8.2.1 MOBILITY MANAGEMENT

The high mobility and increasing velocity of the vehicles make the mobility management a crucial topic in VANETs. The connections must be seamless and provide continuous access to the users without them noticing any change. The Internet Protocol (IP) was developed assuming that end-nodes are in fixed physical locations, without the need to change their current connection to another network during a data stream transmission. However, this is far from a realistic VANET scenario. Several mobility protocols have been proposed to fulfill the IP gap in the support of mobility.

Mobile Internet Protocol version 4 (MIPv4) [34] is one of the first and well-known mobility protocols. This is more suitable for macro-mobility and is used to provide mobility support when it is used with the Internet Protocol version 4 (IPv4). It introduces the concepts of Home Agent, responsible for managing all mobility and ensures that the mobile node is always reachable, and Foreign Agent, responsible for providing routing services, while the mobile node is outside of its home network. MIPv6 [35] is an improvement to Mobile Internet Protocol (MIP), making use of IPv6 instead of IPv4, thus taking advantage of IPv6 mobility features. Furthermore, with the use of IPv6, some network problems that cannot be handled by IPv4 are also solved, such as triangular routing and shortage of addresses. However, this protocol does not support simultaneous multihoming.

Network Mobility (NEMO) [9] support is an extension of MIPv6 that enables the mobility of an entire network and the users that are connected to it. In this way, an entire network can move to a new one as a whole unit. This extension creates a network mobility management protocol with multihop capabilities but creates scalability problems due to increased tunneling overhead, and it still fails to deliver simultaneous multihoming support. Another approach made to solve the IP mobility challenge is Proxy MIPv6 (PMIPv6) [15], that improves on the bases of MIPv6. These improvements remove the need to have mobility management functionalities in the users' device, decreasing network overhead. However, it fails to fulfill the multihop mobility protocol requirement as well as the simultaneous multihoming support.

Network-Proxy Mobile Internet Protocol version 6 (N-PMIPv6) [27] is a protocol that merges the NEMO and PMIPv6 protocols, creating a protocol similar to PMIPv6 but with support for network mobility. This protocol shows interesting features, allowing the mobility of entire networks through other existing networks, only requiring the end-user to connect to the network provided by his vehicle for Internet access, without the necessity of installing any additional software or added configurations in their devices. Additionally, since the OBUs are the ones responsible for handling the mobility of its connected users, the handover mechanism is assured to be efficient and the number of handovers is considerably reduced to only one, the OBU. However, this approach does not support the specific requirements of vehicular networks, such as the support of broadcast technology, such as IEEE 802.11p, multi-hop through V2V communications, and multihoming. The approach in Ref. [7] has been proposed to address these requirements and is the one used to support mobility and multihoming in the vehicular network in this article.

More recently, the distinct features of Software-Defined Networking (SDN), such as its flexibility, programmability and network abstraction have set the stage for a novel networking paradigm termed as Software-Defined Vehicular Networks (SDVNs) [19]. The convergence of SDN with VANETs is seen as an important direction that can address most of the VANET's current challenges. In particular, the use of SDN's prominent features, such as up-to-date global topology enabled dynamic management of networking resources and efficient networking services to enhance the user experience [23]. These SDN features can meet the advanced demands of VANETs, which include high throughput, high mobility, low communication latency, heterogeneity and scalability, to name a few. In a generic SDVN, the data plane entities (e.g., vehicles) communicate with the control plane entities using the southbound APIs for coordinated and efficient communication. The controllers perform various functions such as routing, information gathering and providing services to end-users based on the instructions and policies received through northbound APIs from the application layer entities. The controller provides an up-to-date network view to the application plane that helps it to manage various services (e.g., security, access control, mobility and QoS) in the network.

8.3 SOFTWARIZATION AND VIRTUALIZATION

With the fifth generation of mobile networks gaining ground, which will include beyond the 5G New Radio and C-V2X technologies for vehicular communications, resource sharing and the use of softwarized networks are becoming increasingly popular [38]. However, new services and applications for the vehicular vertical, targeting important topics such as vehicular safety and multimedia, can also benefit from this technology that provides a dedicated service class of Ultra-Reliable and Low-Latency Communications (uRLLC) supported by high throughput and strong security.

In these important verticals, the concept of NFV comes into play, a technology capable of decoupling software from hardware, enabling flexibility, programmability and extensibility to the network [17]. Through NFV, virtualized functions are available in the Cloud and pushed into the network through the connection to the Cloud. According to the NFV reference architectural framework defined by European Telecommunications Standards Institute (ETSI), the main components of an NFV framework are:

Virtual Network Functions (VNFs): software implementations of network functions previously carried out by dedicated hardware;
Virtual Functions (VxFs): software implementations of a mix of real and virtual network or computing elements, representing network-centric functions and vertical-centric functions;
Network Functions Virtualization Infrastructure (NFVI): hardware and software resources that are required to deploy, execute and manage the VNFs;
Management & Orchestration (MANO): a collection of various functional blocks that, as the name suggests, enables the management and orchestration of NFVI and VxFs. The main blocks within MANO are the Virtualized Infrastructure Manager (VIM) which is responsible for managing the NFVI resources; Virtual Network Function Manager (VNFM) which is responsible for managing the VxFs; and finally the NFV Orchestrator (NFVO) which coordinates NFVI resources from different VIMs.

The MANO system, deployed outside of the vehicular network's scope, is the component in charge of the management and orchestration of the NFVI, as well as the deployment of VxFs. It provides an orchestration service, through the NFVO and a VNFM, as well as a VIM.

As mentioned before the VIM is responsible for the management of the NFVI compute, storage and network resources in a network's domain, such as being responsible for the allocation of NFVI resources to VxFs, keeping a repository with information about which NFVI resources are being used. Several VIM solutions have been proposed, and amongst the most well known, we have OpenStack, OpenVIM and VMware, which are characterized as follows:

- OpenStack[1] is an open-source software solution for cloud computing. It consists of several interrelated services, which are deployed as components and are responsible for managing the computing, networking and storage resources. It also contains other components providing more services, such as a dashboard that enables the management of the OpenStack services. OpenStack is the most prominent open-source VIM solution: it is a very stable and mature project, given the fact that it is used by many renowned companies to support their services.
- OpenVIM[2] is a light implementation of a VIM. It communicates with the NFVI and an OpenFlow controller in order to provide computing and networking capabilities, as well as the ability to deploy the VxFs. The project is currently part of the Open Source

[1] http://www.openstack.org.

[2] https://github.com/nfvlabs/openvim.

MANO (OSM) project, a project that implements an open-source MANO solution aligned with the ETSI's architectural frameworks. It is worth noting that, even though this solution is open source, it is limited in terms of features, and usually it is only used when in conjunction with the OSM project to deploy a quick all-in-one solution used for development.

- VMware is a company that provides cloud computing and virtualization products. One of these products is the VMware vCloud NFV.[3] This product is a commercial implementation of a MANO solution aligned with the ETSI's architectural frameworks. The actual VIM solution provided by the product is the VMware vCloud Director, which is a multi-tenant solution that provides the management of the computing, networking and storage resources. Being a commercial product, this solution provides dedicated end-user support and is very stable.

Regarding the several MANO solutions available in the marketplace, the ones with the highest visibility are Open Source MANO, OpenBaton and RIFT.ware, with the following properties:

- Open Source MANO (OSM)[4] is an ETSI hosted production-quality open-source MANO stack aligned with ETSI NFV architectures. It is an operator-led ETSI community, which includes among its members Bell, Telefónica, Canonical, among others. Its architecture and components include the Network Service Orchestrator, responsible for interacting with the Resource Orchestrator and VNF Configuration and Abstraction components of the OSM solution, in order to support the management of the Network Services (NSs), the management of the MANO repositories, as well as the process of onboarding and configuring NSs and their VNFs; the Resource Orchestrator, responsible for coordinating the compute, storage and network resources that are required for the instantiation of the NS's VNFs, and it provides most of the functions of the NFVO defined by the ETSI MANO framework; the VNF Configuration & Abstraction component, responsible for providing the functions of the VNFM defined by the ETSI MANO framework, and for the configuration of VNFs given in the instructions present in the VNF Descriptors.
- Open Baton[5] is an open-source platform that provides an implementation of the ETSI NFV MANO architecture framework. This platform is composed of several components which provide a number of different services, some of the most important ones are: (1) a driver mechanism that allows different types of VIMs without the need to change anything in the orchestration logic; (2) a generic VNFM able to manage the life cycle of VNFs based on their descriptors; (3) a fault management system which can be used for automatic runtime management of faults. Given the fact that OpenBaton has a driver mechanism that allows for different types of VIMs, it is possible to integrate it with several VIM solutions. It is a project developed by Fraunhofer FOKUS and TU Berlin, and it represents one of the main components of the 5G Berlin initiative.
- RIFT.ware[6] is an NFV MANO solution that is a commercial distribution of the OSM project. It aims to simplify the deployment of multi-vendor NSs, and VNFs in carrier networks and enterprise clouds. It offers everything needed in order to have automated end-to-end service delivery and life cycle management. RIFT.ware is a product offered by Rift.io, a founding member of the OSM project, which has greatly contributed to its progress. All in all, it is a well-established commercial product in the field of MANO.

---

[3]https://www.vmware.com/products/vcloud-director.html.

[4]http://osm.etsi.org.

[5]https://openbaton.github.io/.

[6]https://www.riftio.com/riftware/.

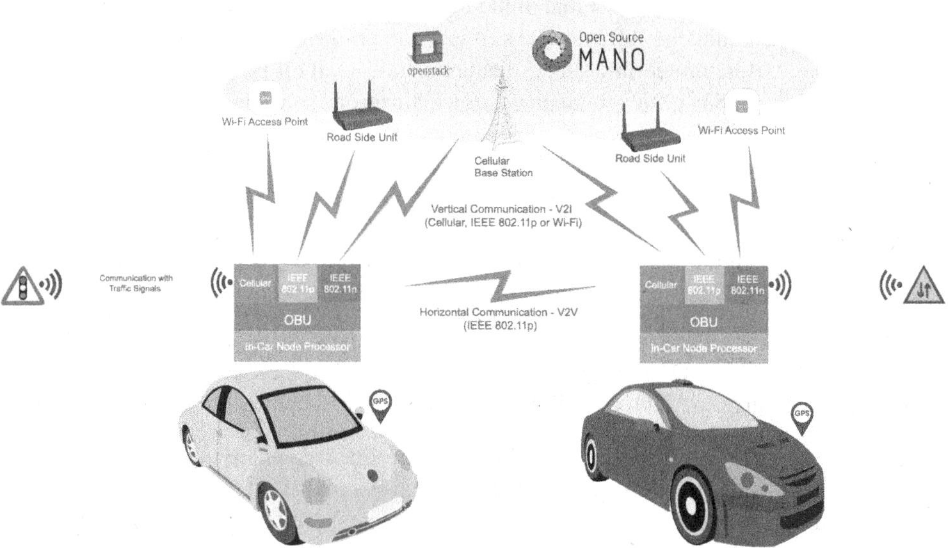

Figure 8.2 IT-Av automotive environment.

8.4 SYSTEM DESIGN

The study presented in this chapter has been performed using a vehicular network with mobility and multihoming support. The mobility approach is based on the N-PMIPv6 architecture presented before, with additional mechanisms developed to support broadcast technologies, multi-hop, transparent handovers and simultaneous multihoming [7]. The main components of such architecture are:

- Local Mobility Anchor (LMA), the home agent for the mobile nodes in the network. It is the topological anchor point for all mobile nodes and it manages their communications;
- RSUs, the mobile access gateways subject to the LMA where clients, on OBUs, can connect to the Internet through them. RSUs behave like points of attachment to OBUs;
- OBUs, acting as mobile routers. They are positioned inside a vehicle and are responsible for creating an access link between the end-user and the mobility network.

The OBUs contain a connection manager responsible for selecting the best connections and technologies for the OBU, from the ones available at the time. The connection manager is also capable of connecting the OBU to multiple RSUs or cellular Base Stations (BSs) using different access technologies. When an OBU is connected to at least two RSUs of different technologies, the process of simultaneous multihoming can happen so that the VANET has the best load balancing for the traffic that flows through the OBU. The main aspect of this architecture is the IPv6 support. IPv6 tunnels are created between the network nodes by the mobility protocol, which are then used for protocol and data communications needed by the nodes. The tunnels between the LMA and the RSU are IPv6/IPv6 tunnels, while the tunnels from the LMA to the OBUs are IPv4/IPv6 tunnels.

This vehicular network, illustrated in Figure 8.2, is deployed in the campus of the University of Aveiro, Portugal, managed by Instituto de Telecomunicações - Aveiro, and it gives support to one of the testbeds available for experimentation under the scope of the European project 5GinFIRE [39]. The automotive testbed, from now on denoted as IT-Av automotive, comprised by the vehicular network and the orchestration services, hosted by 5TONIC laboratory,[7] enables the experimentation of softwarized solutions, by the means of Virtual Functions for the automotive vertical. To enable the

[7]http://www.5tonic.org.

deployment of VxFs, we have explored the use of NFV technologies, following the conventions set by the ETSI, as presented in the previous Section 8.2. In the proposed solution, the MANO system is implemented using open-source software: the Open Source MANO (OSM) for the MANO system and the Cloud computing software OpenStack for the VIM.

8.5 USE CASES

In the following, we present a set of distinct use cases and applications exploring the concept of softwarization and virtualization in the automotive vertical. For each use case we present the concept, the implementation and the performance results, targeting both network and application domains.

8.5.1 SURROGATES

The approach for virtual OBUs and hybrid communications to foster 5G vehicular services, SURROGATES for short, is focused on extending common OBU capabilities by offloading processing tasks from the vehicle side, and virtualizing OBU instances in a Multiple Access/Mobile Edge Computing (MEC) layer implemented in a near cloud. This platform is used to gather data from vehicles and perform pre-processing calculations with the aim of feeding services in the area of mobility and pollution.

There are several proposals in this line in the literature, although they do not reach the flexibility provided by SURROGATES. A MEC layer implemented in drones as road-side units was presented in Ref. [11], but it is statically mapped on physical devices. The improvement implied by MEC for vehicle to VRU communications was evaluated in Ref. [12], using 4G simulation and observing up to 80% of latency reduction as compared with a cloud-based system. In Ref. [24], the MEC paradigm was used to improve the performance of a delay-tolerant network, by providing caching of messages for temporally disconnected vehicles. In Ref. [18], a way to orchestrate computing and caching in vehicular scenarios was proposed between MEC and remote servers. In Ref. [6], a MEC layer was proposed to process information coming from vehicles implementing a vehicular social network, although the potential of NFV and MANO was not exploited to cope with the dynamism of the system. The same lack is found in Ref. [4], where a content delivery network used a MEC-based architecture to speed-up data sharing among vehicles. An interesting feature included in Ref. [33] is the combination of MEC and multiple radio access networks, which is also considered in SURROGATES. Unfortunately, NFV flexibility was not exploited.

8.5.1.1 Concept

In SURROGATES, the OBU communication technologies are exploited to offer hybrid communications as a base enabler to offload processing tasks. The communication channels are used to offload data analytics tasks from OBUs, which should focus on actions of higher priority such as maintaining vehicle connectivity, managing communication flows or applying security measures to data traffic. The approach is to create virtualized images of OBUs in the edge Cloud, where raw data sent from in-vehicle sensors is processed. This idea is illustrated in Figure 8.3.

A base monitoring system, considering sensor data from the vehicle, sends raw information to an OBU virtual instance. Here, pre-processing, aggregation and pattern recognition tasks can be carried out. The processed information is then available for global C-ITS services. The vehicular communication stack includes a set of IPv6 technologies to support network mobility and hybrid communications. Security provision is also born in mind with the establishment of security associations with IKEv2, and the creation of secure data channels with IPsec [13]. Processed or cached information is accessed by vehicular services requiring a global view of the road. This solves the issue of accessing the physical OBU from multiple services and requiring monitoring data when the OBU is temporally offline.

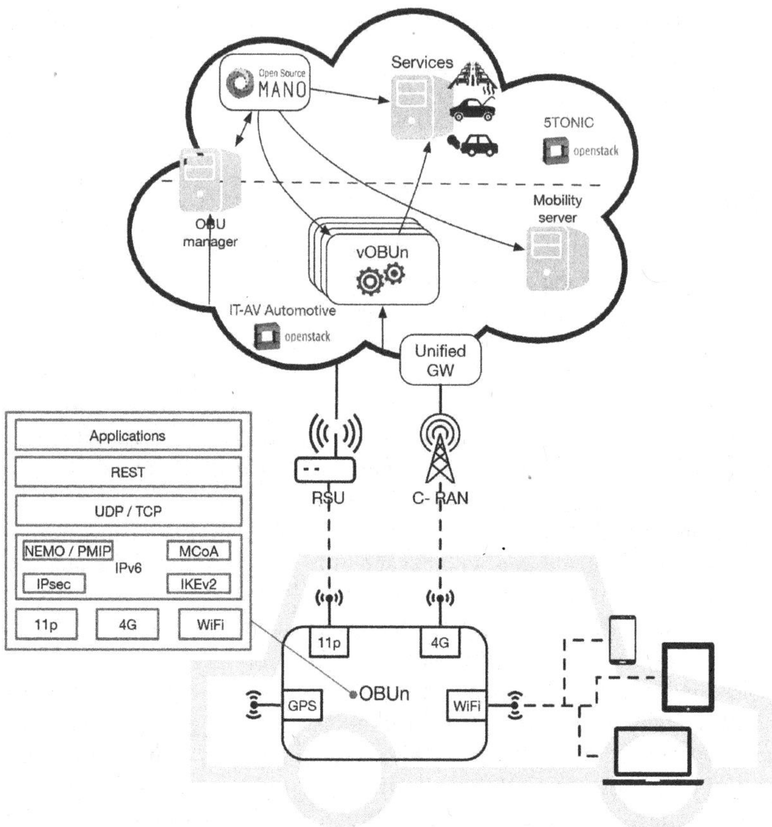

Figure 8.3 Architecture of the SURROGATES use case.

An OBU manager module has been defined to be the initial contact point from physical OBUs wanting to connect to its virtual counterpart. This manager is also a VxF module and communicates with OSM to indicate the need for a virtual OBU instance. OSM is in charge of creating and managing these instances, which could be potentially instantiated on demand. Data to be collected from the OBUs includes navigation information, diagnosis and mechanical monitoring details of the car, and readings from other sensors connected to the OBU.

8.5.1.2 Operation

A modular view of the different virtualized components of SURROGATES is given in Figure 8.4. Three different network services have been deployed in two 5GinFIRE OpenStack domains. Virtual OBUs (vOBUs) were deployed at IT-Av, in order to provide an edge computing layer close to the driving areas of the vehicles. The OBU manager, responsible for vOBU registration was also instantiated at IT-Av. The two remaining VxFs types can be seen onboarded in the 5TONIC testsite. The data analytics module is in charge of periodically collecting data from vOBUs and apply big data algorithms. From the Data Analytics module, the different services get the data collected and processed/aggregated. Such operation, involving the general behavior of the solution, is presented in more detail in Ref. [37].

The operation shown in Figure 8.4, with the arrows connecting the elements of the system, belongs to the registration of the OBUs in the system once the car engine is started. This is necessary in order to assign a vOBU instance to each physical OBU. The process starts with a vOBU solicitation from the OBU, which is processed by the OBU manager. The conceptual design of the solution

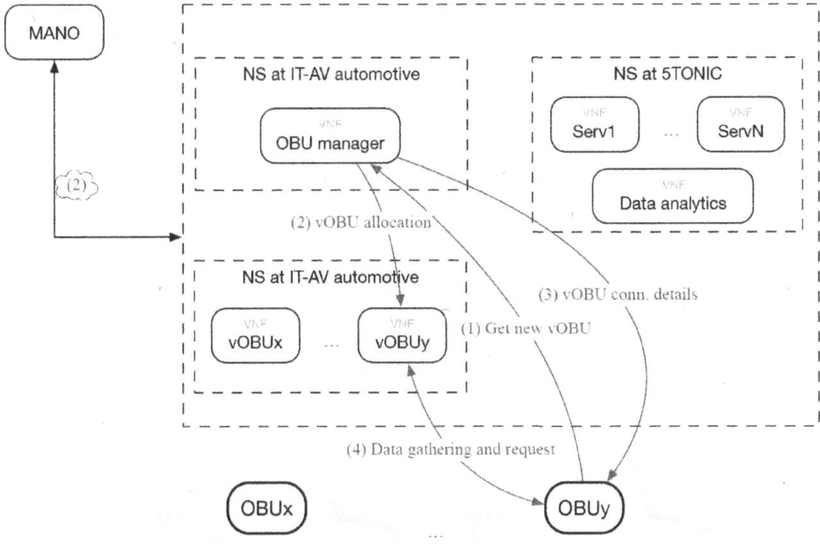

Figure 8.4 Registration of vOBUs in SURROGATES.

considers here communication with the MANO module in order to dynamically instantiate a new vOBU for the physical OBU; however, the OSM platform used in 5GinFIRE at the moment of the project deployment avoids this mechanism. Hence, the OBU manager is in charge now of assigning a vOBU by picking up a free vOBU from a pool of pre-instantiated VxFs. Once the vOBU is allocated, the connection details are given to the physical OBU, which can now start the data gathering process.

The behavior of the system, when data is required from the final services, is illustrated in Figure 8.5. Here it can be seen that there are three different indirection levels where the data requests can be solved, thanks to the multi-layer data processing scheme. In the first one, the data request is processed by the data analytics module, which periodically receives information from the OBUs and processes it in a global manner or, directly, aggregates all data collected. However, if the particular data requested by the service cannot be solved by this module, the data request is then forwarded to the particular vOBU representing a physical OBU. For this to be done, the data analytics module needs to ask the OBU manager about the vOBU impersonating the OBU. Steps 2 and 3 in Figure 8.5 are performed periodically, so it is not needed to perform them at the time the request is made by the service. It is considered that most of the requests could be solved in the first or second indirection levels but, if particular data is required that is not stored in the platform, or if a real-time parameter is needed from the vehicle, the request can be finally forwarded to the physical OBU.

8.5.1.3 Implementation

The monitoring middleware included in the OBUs has been implemented in Python, and it is in charge of periodically asking the OBD-II device for vehicle data. In the current implementation, all available OBD data records supported by the vehicle are collected and then sent using REST communication over TCP to the corresponding vOBU. vOBUs cache the last status values received from the OBUs, acting as the physical one when it can resolve data requests. When OBUs are disconnected due to mobility, vOBU can provide the last read parameters. In the data analytics module, all information collected and resulted from the big data algorithms is saved in a MySQL database. From here, it is possible to provide both current and past values of OBUs.

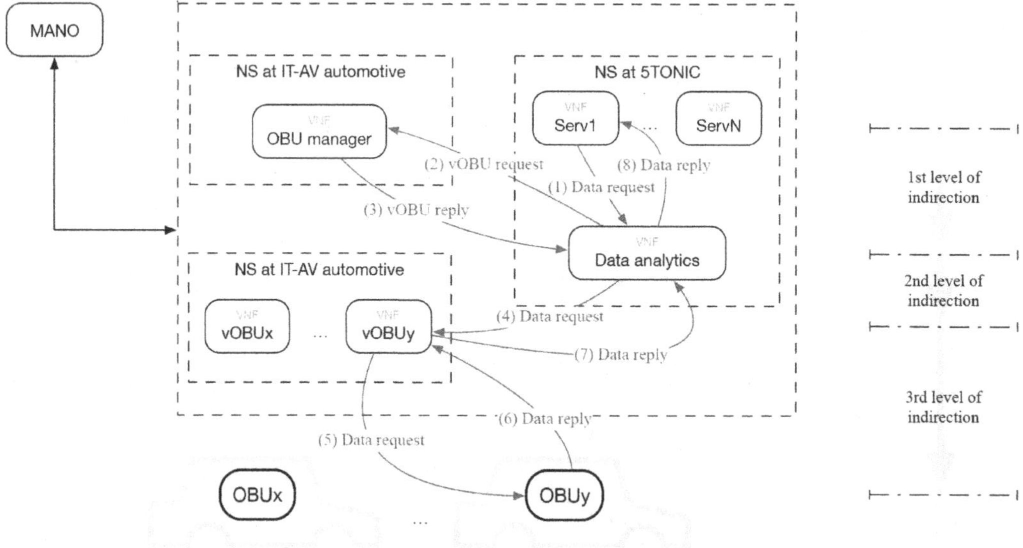

Figure 8.5 Resolution of data requests in SURROGATES.

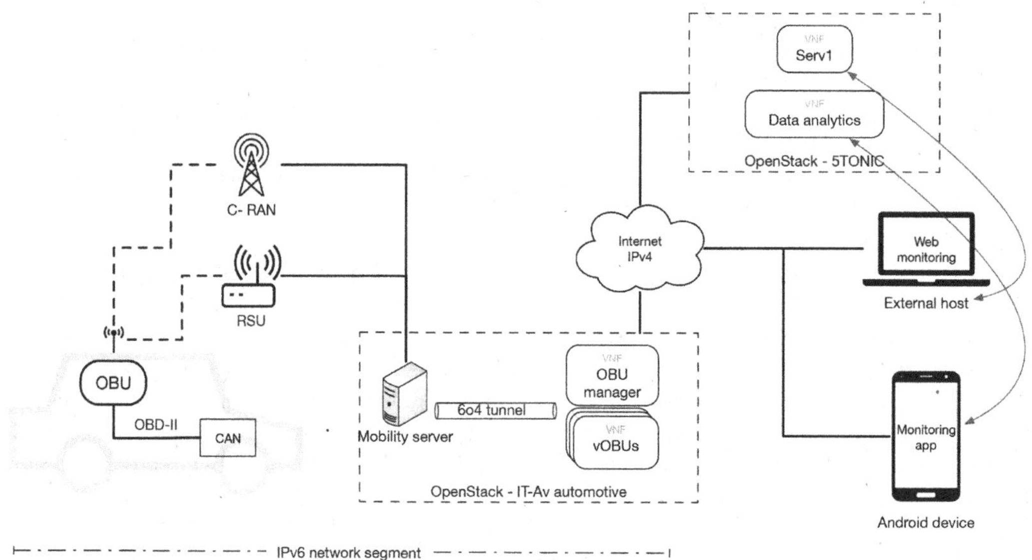

Figure 8.6 Deployment of SURROGATES within the VANET and Cloud ecosystem.

Regarding the final service implemented, in order to enable the validation of the whole platform, Grafana 5.3.1[8] is used in the service VxF. As shown in Figure 8.6, this service can be accessed from a regular terminal through a web browser interface. Additionally, an Android application has been developed to gather data from vehicles directly through the data analytics module from a mobile device. This shows that the platform also allows the federation of data to external services using the developed REST API.

[8]https://grafana.com/.

An important feature of the platform is its IPv6-compliant nature. The solution collects data from vehicles using IPv6 communications over a mobility approach operated with N-PMIPv2. As can be seen in Figure 8.6, vehicles maintain connectivity through IEEE 802.11p and cellular technologies supporting handovers, and the communication between the physical vehicle plane and the edge Cloud virtualization domain is carried out using IPv6. Nevertheless, it is important to note that the combination of the OSM version 2 with OpenStack Pike does not support IPv6 communications in the virtualization domain. Then, a tunneling solution is proposed within the edge Cloud OpenStack domain. Hence, both vOBUs and OBU manager connect to a third VNF that acts as IPv6 vRouter offering tunneled IPv6 over a virtual link descriptor.

8.5.1.4 Performance Evaluation

With the aim of validating both the SURROGATES platform and assess its performance, the solution has been tested under real driving conditions, as illustrated in Figure 8.7. A Fiat 500 from 2012 was mainly used to carry out the tests, equipped with an OBU and taking advantage of IEEE 802.11p connectivity most of the time. Multiple rounds were performed to gather significant results. The OBU is provided with the monitoring software to access the OBD-II interface and report the data collected and is in charge of registering with its virtual counterpart (vOBU) and then continuously send data to it. This data fed the data analytics module and then the Grafana system, allowing end-users to check the past and current status of the vehicle. Additionally, the OBU/vOBU is periodically asked about particular parameters using the Android app, in order to measure the performance of the network in terms of request losses and resolution delay.

The good operation of the solution was checked using both the Grafana web interface and the Android app. Figure 8.8 shows the results obtained in one of the test rounds performed. The plot panels are configurable, but in this view the vehicle speed (upper left), revolutions per minute (upper right), coolant temperature (lower left) and air intake pressure (lower right) are showed. The variability of the speed, driving conditions and the direct relation with the engine revolutions can also be seen in the figure.

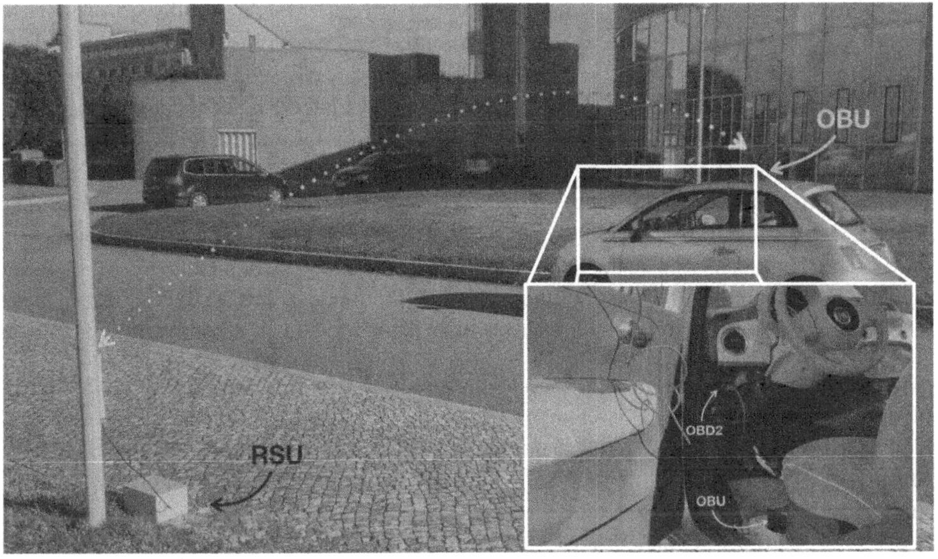

Figure 8.7 Setup of SURROGATES experiment.

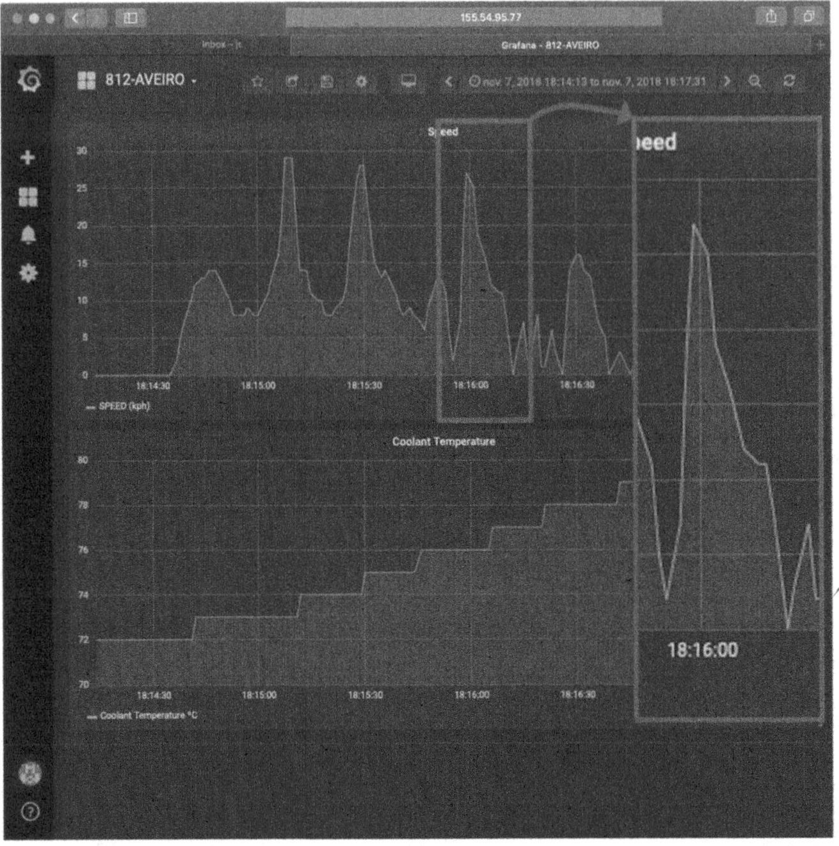

Figure 8.8 Validation test of the SURROGATES solution using Grafana.

A screenshot of the Android app is included in Figure 8.9. Here the vehicle speed is shown in real time, and at the same time a plot is included with the historical values collected. The requests are received by the analytics module, which is configured to forward the request to the associated vOBU. When the vOBU receives the request, it can solve it directly with data previously received, coming from a previous request, or it can directly ask the physical OBU for the engine parameter.

In the next performance tests, we have evaluated the time the system lasts to solve reactive requests from the Android app. To cope with this objective, we have traced all requests and replies through the platform. This way, it is possible to compute the time needed to process each request at each particular module of the system. Given that the vOBU will have the option of solving many requests from different services at the same time, we have emulated this behavior by including a probability parameter in the Android app indicating the likelihood of the vOBU cache hit.

Hence, the behavior of the system when processing the app requests is enumerated as follows:

1. The vOBU receives the data request and it directly replies with the last value collected. This is a cache hit.
2. The vOBU receives the data request and forwards it to the OBU. This is an emulated cache miss. The OBU replies to the request, which is received by the vOBU and forwarded by the platform until reaching the mobile device.
3. The vOBU receives the data request, forwards it to the OBU emulating a cache miss but, unfortunately, the request is lost by the wireless network. In this case the vOBU finally solves the request with a previously collected value.

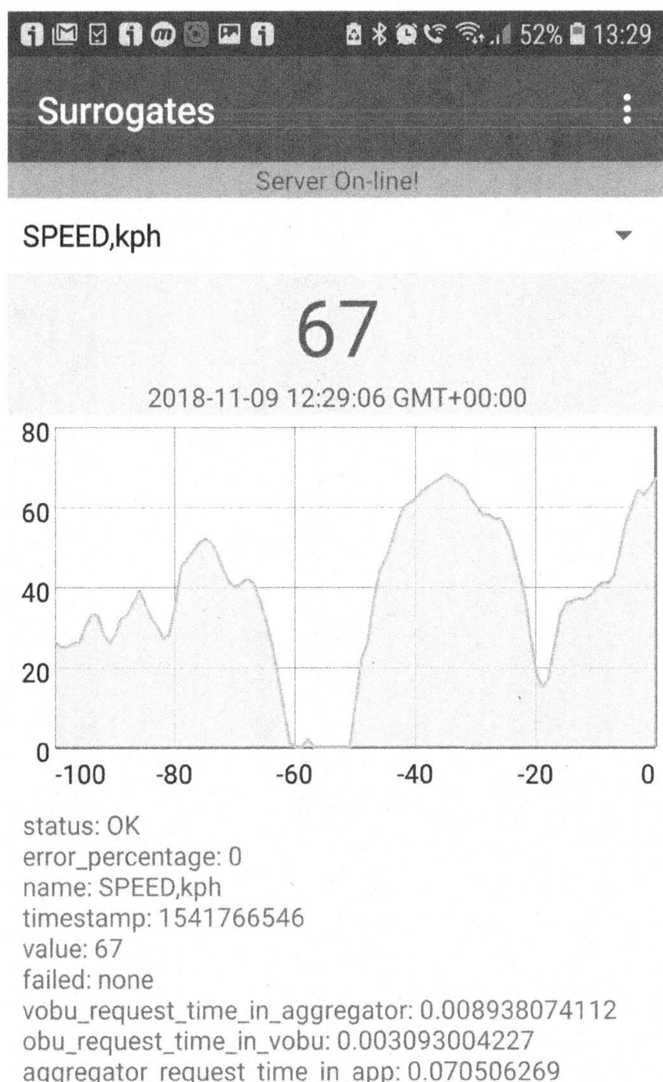

Figure 8.9 Validation test of the SURROGATES solution using the Android app.

The RTT results obtained after a test involving six rounds of experiments are included in Figure 8.10. Here it can be seen how the best absolute results are obtained when the vOBU resolves the data requests, with a medium RTT value of 522 ms. The second-best result is obtained when the physical OBU successfully replies the request with the data parameter. Finally, when the OBU is accessed but the request is lost in the wireless segment, the RTT reaches 650 ms. In these results it is important to see the high impact of the access network, involving most of the processing time by the platform. This is due, not only to the wireless segment of the mobile device, but also to the segment for reaching the 5GinFIRE edge Cloud, where the analytics VxF is placed. It can be seen that this extra overload reaches nearly 500 ms of the overall RTT time of the tests, and it is something out

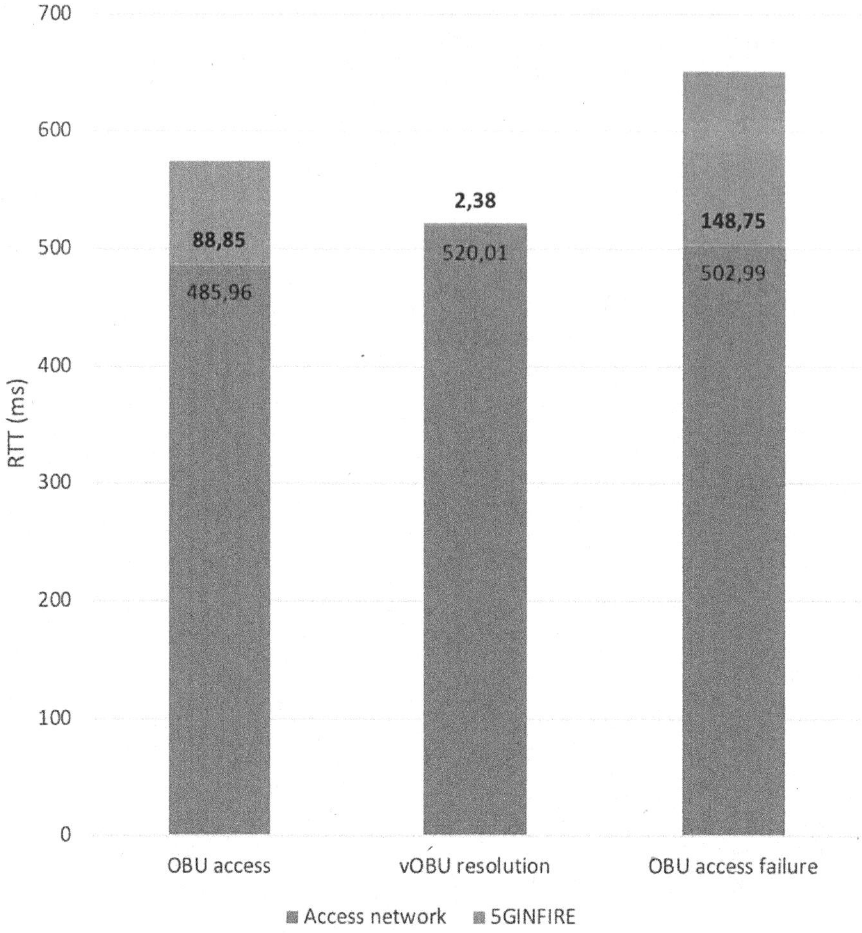

Figure 8.10 RTT delay solving data requests in SURROGATES.

of the scope of the SURROGATES proposal. However, it can be seen clearly the speed up of the SURROGATES proposal when the requests are processed locally by the vOBU. The extra time needed in the worst case of OBU access failure is due to the TCP timeouts involved in the process.

Since the cache hit ratio can be configured in our solution, we have evaluated the operation of the system by varying this parameter. Given that it is expected that many of the requests are solved by the vOBU without requiring access to the physical OBU, we have varied this value from 60% to 100%. As can be seen in Figure 8.11, the improvement is clear, as the success ratio increases. It is important to note that these RTT values come from real tests involving request resolutions of the three previous cases: vOBU resolution by the rate indicated, successful OBU access and unsuccessful OBU access. The sum of the successful and unsuccessful OBU accesses imply the remaining requests until reaching 100%. Nevertheless, the requests lost in the IEEE 802.11p network segment during the tests were quite low, involving only 1.74% of the total requests sent over this channel.

It is important to bear in mind that cache hits will increase as the number of requests is higher, given that the number of services or final users accessing a particular vehicle could be high. In general, it can be said that the whole platform has been validated with a real vehicle monitoring software divided in two different human-machine interfaces, an Android app and a web platform. The performance results indicate speed ups from 80 ms required in a regular (successful) OBU access to 2 ms when using the NFV-powered vOBU.

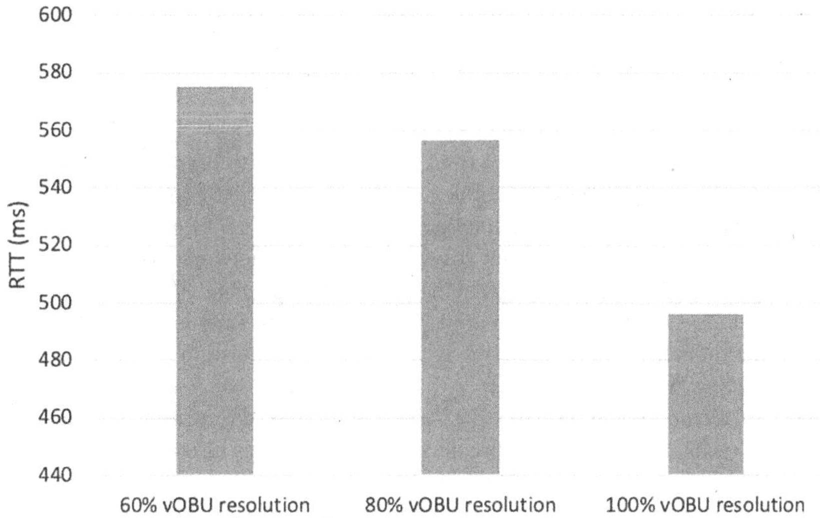

Figure 8.11 RTT delay solving data requests in SURROGATES when varying the vOBU cache hit ratio.

8.5.2 CAVICO

Until recently, *autonomous* vehicles considered supporting tools (e.g. adaptive cruise control, parking assistance, etc.) rather than a vehicle capable of fully performing all actions by itself. However, in the last years, the research toward full autonomy has become very intense; however, even with these sound advancements, there is still a need for a driver to be present inside the vehicle. *Autonomous* cars must perform autonomously; however, there are situations when a car controller becomes malfunctioned, and the driver must take control of the situation. Including aside the legislative requirements, from a business point of view, the need to assign a human driver to such *autonomous* car seems a big waste of resources, as most of the time the driver is passively observing the system state. In the case described above, the human driver – if backed up by the right technology – could oversee the operation of multiple vehicles by being located in a remote office where he or she monitors the number of cars at once.

The CAVICO framework enables the video streaming on the basis of the edge processing in the 5G layer. In such a scenario the video monitoring would play a significant role in order to provide a valuable insight into the car's neighborhood for a remote operator to take the proper action when the *autonomous* car could not (e.g., taking a remote control over a vehicle or sending an instruction on what to do next). Furthermore, to reach Level 4 and 5 autonomy as in the international recommended practice J3016 [36], data transmission between *autonomous* cars will be a need, so the usage of a video controller may help to efficiently manage available resources when considering that transmitted data would include pictures and movie parts as well.

The same problem was tried to be solved for video surveillance systems CCTV (Closed-Circuit TeleVision) in former works done in the Next Generation Multimedia Efficient, Scalable and Robust Delivery (MITSU) project.[9] A robust, efficient and inter-operable video streaming solution was developed to provide seamless and continuous video in wireless links. It enables a convergence of broadcast, telecom and web technologies. The QoE and QoS monitoring allows reaching balanced video encoding and better stream transport. Thanks to it, the content- and QoE-awareness based on video quality assessment methods can be introduced to video streaming systems in order to optimize bandwidth usage and content quality in the function of instant transmission conditions. The solution

[9]http://www.celticplus.eu/project-mitsu

was demonstrated for two use cases: multimedia scenario and security scenario. To create such services an IP Multimedia Sub-System (IMS) can be used in the virtualized environment [14].

8.5.2.1 Concept

Security monitoring of mobile objects (e.g., trucks where payload, motor, fuel or even driver's behavior can be remotely controlled) using mobile networks is not a new concept and there are plenty of solutions, even in the scope of fleet management. However, a rapid growth of technology options, especially the growing popularity and coverage of cellular networks, opens new possibilities for innovative upgrades. There are solutions in the market that leverage 3G+ networks for real-time monitoring of mobile assets, like containers or railway cars using Internet of Things sensors. The acquisition of different metrics in radio networks allows the optimization of a video controller in a lesson-learned adapting process.

Based on ITU-T recommendation P.912 [21], a set of Quality of Experience (QoE) metrics [32] were introduced allowing for an objective prediction of user feelings through automated scripts and without the need for human interaction. Upgraded metrics and software enhancements can lead to the creation of a business solution well-suited for commercial needs in the automotive market. The process of video monitoring also demands Quality of Service (QoS) [10] parameters as well as base station (BS) or access point parameters of the radio interface in order to obtain a more sophisticated solution. All this information can be then used (1) to better adjust transmission in historical-based measures, (2) and to improve the location-awareness mechanism of the controller.

The CAVICO framework was developed as a new VxF dedicated to optimizing video streaming according to the parameters of both network radio interfaces and video streams, as well as use case. Using the CAVICO VxF together with GPS position, image processing and traffic signals, one can create a Network Service (NS) oriented to video monitoring in the automotive domain for different use scenarios. The VxF of the CAVICO video controller is a separate software module running in the Cloud, as illustrated in Figure 8.12. Videos from on-board cameras are streamed via the controller of CAVICO Tools to the user's Monitoring Centre. CAVICO Tools are used to adapt the video according to feedback data processed in the Monitoring Centre in order to optimize the trade-off between perceived video quality (QoE) and instantaneously available transmission parameters in the radio interface (QoS). Other measurements, user's requirements or imposed constraints, such as geo-coded locations, can be also taken into consideration to be more precise in the control loop.

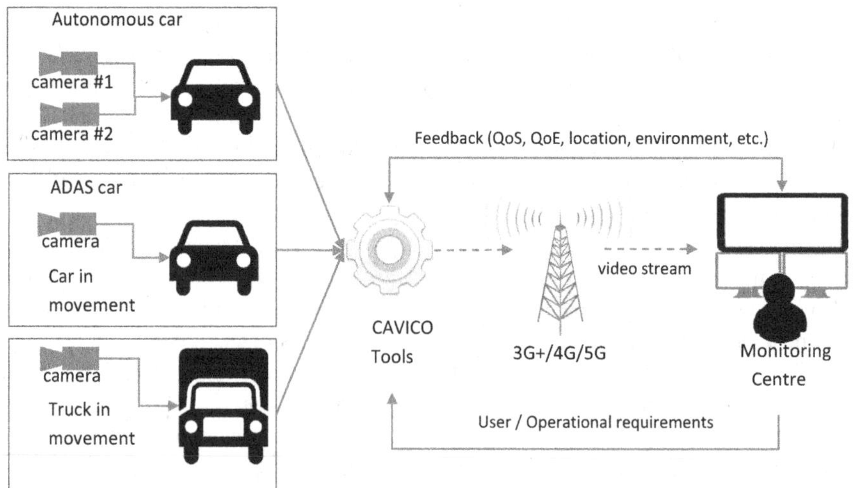

Figure 8.12 Generic CAVICO layout (ADAS – Advanced Driver-Assistance Systems).

8.5.2.2 Operation

The CAVICO system architecture is shown in Figure 8.13. The vehicle is equipped with a camera and a video coder based on Raspberry Pi micro-computer board (the In-car Node Processor) connected to the vehicle's OBU. The vehicular network is responsible for the connection between the OBU and the CAVICO VxF deployed in the IT-Av Cloud. The OBU can use any of the three radio interfaces (IEEE 802.11p/WAVE, IEEE 802.11a,b,g,n/WiFi and cellular 4G links) and is able to switch among them seamlessly to keep the transmission on-going. Additionally, the OBU provides the current geo-location from a GPS receiver. The geographic coordinates are used with OpenStreetMap amenity layers[10] to determine a use case of the driving mode that affects settings of the video coder running in the CAVICO camera handler. Four use cases are distinguished:

1. road, when a vehicle cruises;
2. fuel station, when it is to be refueled;
3. parking, when it goes around parking places;
4. unknown, for other locations.

Depending on the use case, the encoder provides a video stream of different resolutions and frame rates.

The CAVICO video controller uses information about the video QoE [26] and network throughput to compute correction coefficients for resolution (res)

$$C_{\text{res}} = \text{f}\left(\overline{\text{throughput}}, \text{Blockiness, Blockloss}\right), \tag{8.1}$$

and for frame rate (FPS)

$$C_{\text{FPS}} = \text{f}\left(\overline{\text{throughput}}, \text{Temporal Activity}\right), \tag{8.2}$$

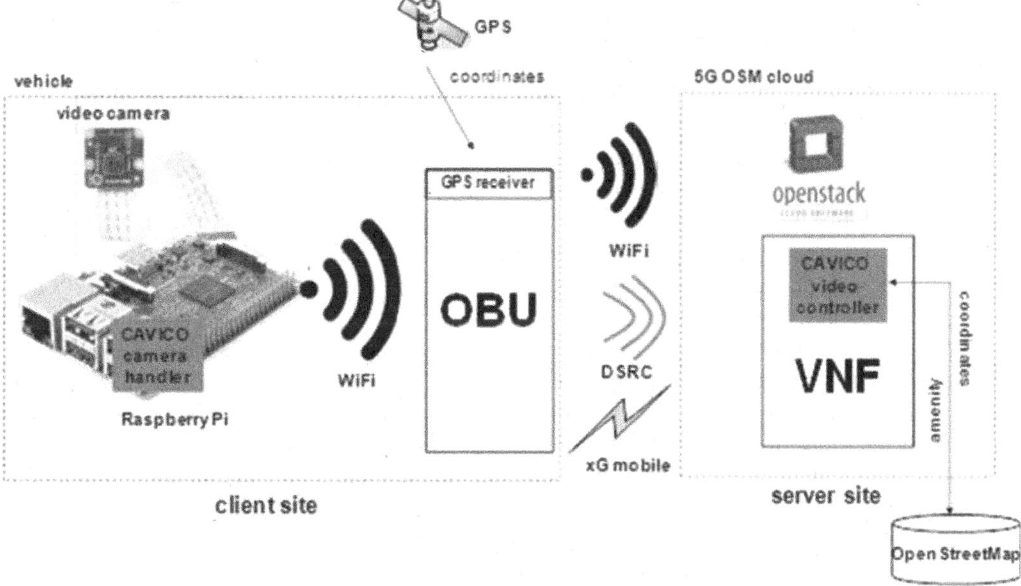

Figure 8.13 Architecture of the CAVICO system.

[10]https://wiki.openstreetmap.org/wiki/Map˙Features#Amenity.

where Temporal Activity represents the dynamic changes of the video, and it is used to adjust the frame rate according to the current mobile traffic and environment limitations. For example, if a device is moving in crowded surroundings, the frame per second (FPS) is increased. If the video is static, the FPS is decreased to reduce the network usage. The Blockiness and Blockloss represent the information about the video distortion, and they are used to reduce the video resolution if the video stream is received with distortions.

The required stream bit rate (bitrate) is calculated as

$$\text{bitrate} = C_{\text{FPS}} \cdot C_{\text{res}} \cdot \text{quantization} \frac{1}{\text{coding}\left(\overline{\text{throughput}}\right)} \tag{8.3}$$

where quantization defines the number of bits used to store a value of one pixel, and the coding function returns a value that is responsible for reducing the connection usage when the average throughput in the network is insufficient.

8.5.2.3 Performance Evaluation

The CAVICO VxF was evaluated in the IT-Av automotive testbed. The client application was installed in the In-Car Node Processor and connected to the OBU, responsible for transmitting the following information to the CAVICO VxF:

1. information about the vehicle position (from a GPS receiver installed in the OBU);
2. transmission channel throughput (measured by the dedicated software);
3. video stream (from the on-board camera).

Positioning data allows the identification of the use case while the channel throughput is used to calculate the video bitrate. The video stream received by the server was analyzed to evaluate a set of QoE parameters using a dedicated software.[11] All these data were used to tune the parameters of the video encoder installed in the vehicle.

The route covered during the tests is represented in Figure 8.14. Several use cases were defined, however, the scenario here presented considered only road and parking use cases. Parking use case occurred during the following periods, between 94.299 and 113.121 s, from 176.861 to 191.831 s, and between 466.932 and 475.424 s. The temporal occurrence of each use case is illustrated in Figure 8.15a, while performance results on network throughput, FPS and horizontal resolution are illustrated in Figure 8.15b–d), respectively.

Every time the vehicle was over a parking place, the GPS information triggered a switch to a lower number of FPS and to a higher video resolution, as illustrated in Figure 8.15c and d. Because the vehicle stayed there for a moment, the throughput increased as the transmission channel became more stable, when compared to a scenario. Once the vehicle left the parking space, the FPS and the video resolution returned to different values, defined by the road use case. We can also notice that in four periods, 51.534, 286.692, 404.302 and 451.812 s, the OBU lost the communication link with the infrastructure, meaning that the system could not recognize the use case. In this case, the system decreased the number of FPS to save the channel capacity and increased the video resolution to provide better video quality in order to compensate from the lower FPS.

This experiment demonstrated, through experimentation in a real case scenario, how several techniques can be merged to provide a new class of services on the basis of partial results. We measured video QoE and network QoS to adjust the video encoder in order to save transmission bandwidth or improve image fidelity depending on the transmission environment and user's expectations. The CAVICO system defined a softwarized architecture, through the use of virtualized functions, to process the video stream and adapt its encoding parameters to the use case, according to the location, the network throughput and the video quality (blockiness, block loss and temporal activity).

---

[11] http://vq.kt.agh.edu.pl/.

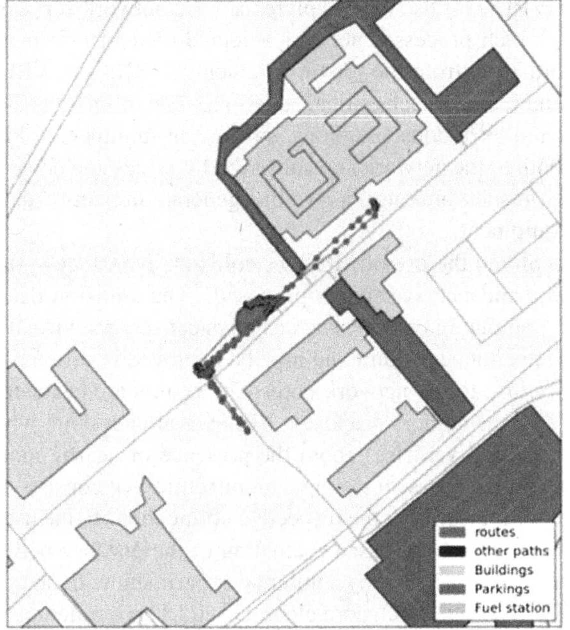

Figure 8.14 Route covered by the OBU during the CAVICO evaluation tests.

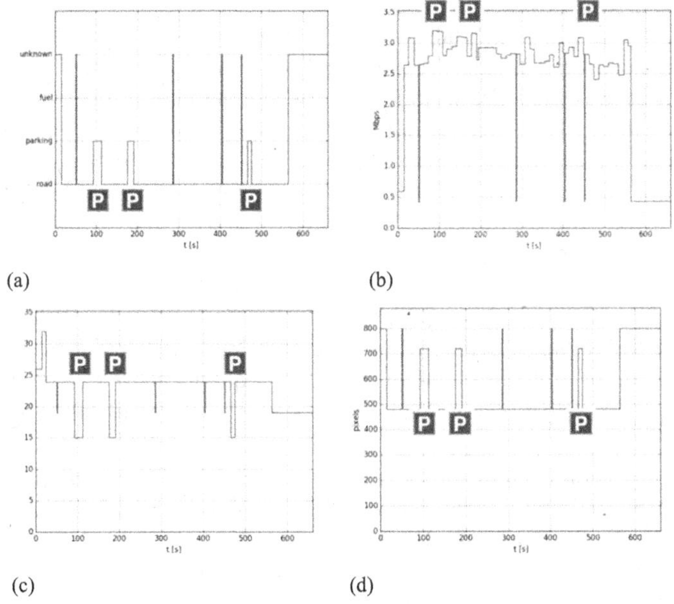

Figure 8.15 Results of the CAVICO experiment: (a) use cases, (b) network throughput, (c) frame rate and (d) horizontal resolution.

8.5.3 VRU-SAFE

Critical V2X applications, such as collision avoidance, VRU safety and hazardous situation detection are identified as key use cases according to 5G Automotive Association (5GAA) [2]. A common characteristic of all aforementioned use cases is the criticality in terms of the temporal aspect. To this end, the VRU-safe use case explores a V2X network service with computing and networking capabilities, which processes network-oriented context information, federated by location and mobility information from the moving elements, OBUs and VRUs, in order to extract meaningful knowledge and identify such critical situations. The ultimate goal is to identify relevant events in cases of predicted hazardous situations, such as an imminent collision, trigger notifications and forward them both to the network and subscribed services and devices (i.e., OBUs) to take the required actions (e.g., instant braking), as well as generate human-readable notifications to be forwarded to the VRU equipment.

Several works have explored the use of MEC for collision avoidance systems. In Ref. [28], an architecture for collision avoidance systems is proposed. The collision detection takes place via Collision Detectors, i.e., computing entities, such as physical servers, virtual machines or containers, that run a collision-detection algorithm and may be deployed in different parts of the network, such as MEC servers, or close to the network core. The results are evaluated through simulation environments. In Ref. [31], the authors propose a VRU warning system, which aims to alert road users (e.g., pedestrians, cyclists, vehicles) about the presence of nearby moving users, in case of hazardous situations. The authors present a simple architecture that consists of a CAM client and a CAM server, responsible for transmitting the respective notifications to the users. The proposed prediction approach is a simple, threshold-based evaluation of the distance between the VRU and the vehicle, while the mechanism is evaluated by a minimal experiment with one scenario that compares Wi-Fi, MEC-enabled LTE and traditional, core cloud-based LTE communication. A 5G/LTE-based protection system for VRUs is described in Ref. [22], which introduces a "context-filter" capable of identifying VRUs in potentially dangerous situations based on several types of contextual information (VRU position, movement direction, accelerations). Particularly, the authors propose the exploitation of smart phone sensor data in order to augment their model. The evaluation of the proposed scheme generates interesting trends in relation to different scenarios, such as different crossing pedestrian curb heights. The improvement implied by MEC for Vehicle-to-Vulnerable Road Users communications was evaluated in Ref. [12] via an extensive simulation-based performance comparison between the conventional and the MEC-assisted network architecture, showcasing considerable latency reduction results in the MEC approach, comparing with centralized cloud-based systems.

8.5.3.1 Concept

The innovation of the VRU-safe concept lies in its architectural approach: VRU-safe service operates in a distributed and dynamic manner exploiting both MEC and VNF-enabled Cloud capabilities, the location/mobility of the involved vehicles/User Equipments (UEs), their connectivity/association per base station (BS)/Access Point (AP), as well as the availability of the MEC/Cloud resources.

Virtualized Cloud resources offer robust performance and scalability, via powerful processing and computing capabilities. On the other hand, MEC capabilities enable the availability of computation resources at the network edge. By processing data locally and accelerating data streams through various techniques, MEC may potentially reduce both the end-to-end latency, as well as the traffic overhead toward the core network. A crucial trade-off, thus, results between the two architectural choices, with the MEC-based processing minimizing the end-to-end transmission delay, and the Cloud-based solution minimizing the processing delays via powerful virtualized, centralized resources.

The high-level architecture of the VRU-Safe concept is depicted in Figure 8.16, which illustrates the two main parts of the deployment, i.e., a) the MEC part, comprising the RSUs and the local VRU-safe instances, and b) the Cloud infrastructure and the virtual VRU-safe components. Light blue arrows illustrate MEC use case and communication, while the orange ones indicate the Cloud operation and triggering of the VRU-Safe VxFs.

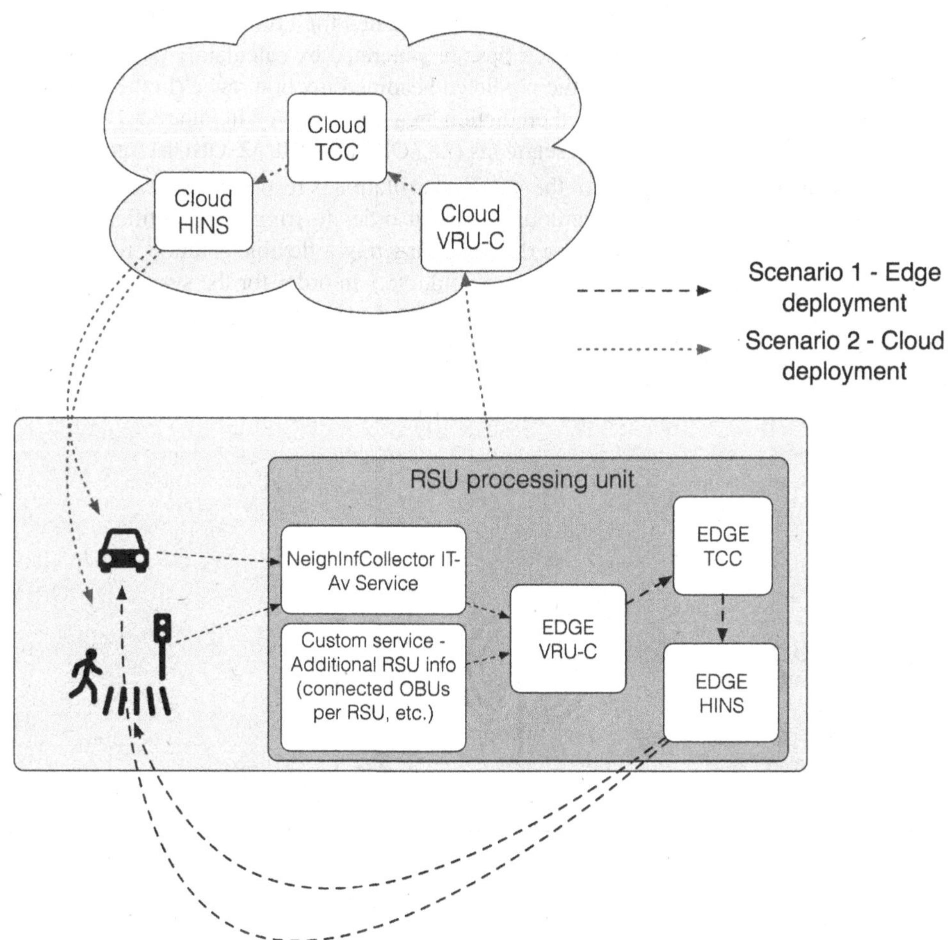

Figure 8.16 VRU-Safe concept.

8.5.3.2 Implementation

VRU-safe is comprised of three key components, each performing one of the main functionalities. The primary components are the Trajectory Computing Component (TCC), the Hazard Identification and Notification Service (HINS) and the VRU-Safe Controller (VRU-C), which is the *brain* of the service, and evaluates the MEC/Cloud computing resources split and selection, i.e., whether and which services need to run locally at the MEC nodes, or be forwarded to the Cloud.

VRU-C module implements the core logic of the service: it receives and pre-processes the context information related to the network environment: the location and mobility characteristics of the involved VRUs and OBUs, their associated radio access points of attachment and the computing and network resources' availability (in terms of utilization load). Based on this information, it decides which TCC and HINS processes will be triggered: a) the MEC computing mode of the system, i.e., the TCC and HINS components running locally, on the road infrastructure as plain software deployed in the RSUs; or b) the virtualized services running on the Cloud-based NFVI in the form of VxFs.

The TCC is the primary computing component of the VRU-safe mechanism that calculates the trajectories of the moving nodes (OBUs and VRUs). The input of the processing module is real-time context information, primarily comprising reported location coordinates, velocity and heading, received from OBUs and VRUs. The trajectories are calculated for a time window of T future time slots, i.e., the prediction window. The trajectories are generated by calculating the future position (in the form of coordinates), along with the predicted heading/direction range (in the form of a set of vectors), resulting in a triangular-shaped prediction area, as illustrated in Figure 8.17, formulated by the predicted coordinates and two line segments (i.e., OBUa1, OBUa2, OBUb1, etc.).

The third and final main component of the VRU-safe solution is responsible for assessing TCC's outputs and identifying potentially hazardous events in order to trigger the notification system. HINS algorithm's implementation has been designed targeting a flexible solution, relying on several parameters (tolerance window, collision threshold, etc.) in order for the system administrator to be able to configure the algorithm with varying triggering events for different use cases, mobility characteristics of the moving nodes (e.g., high/low-speed roads), vehicle/VRU density, etc.. Based on the predicted trajectories that have been generated, the HINS algorithm identifies all potential line segment intersections between a VRU and an OBU. The complete workflow is illustrated in Figure 8.18.

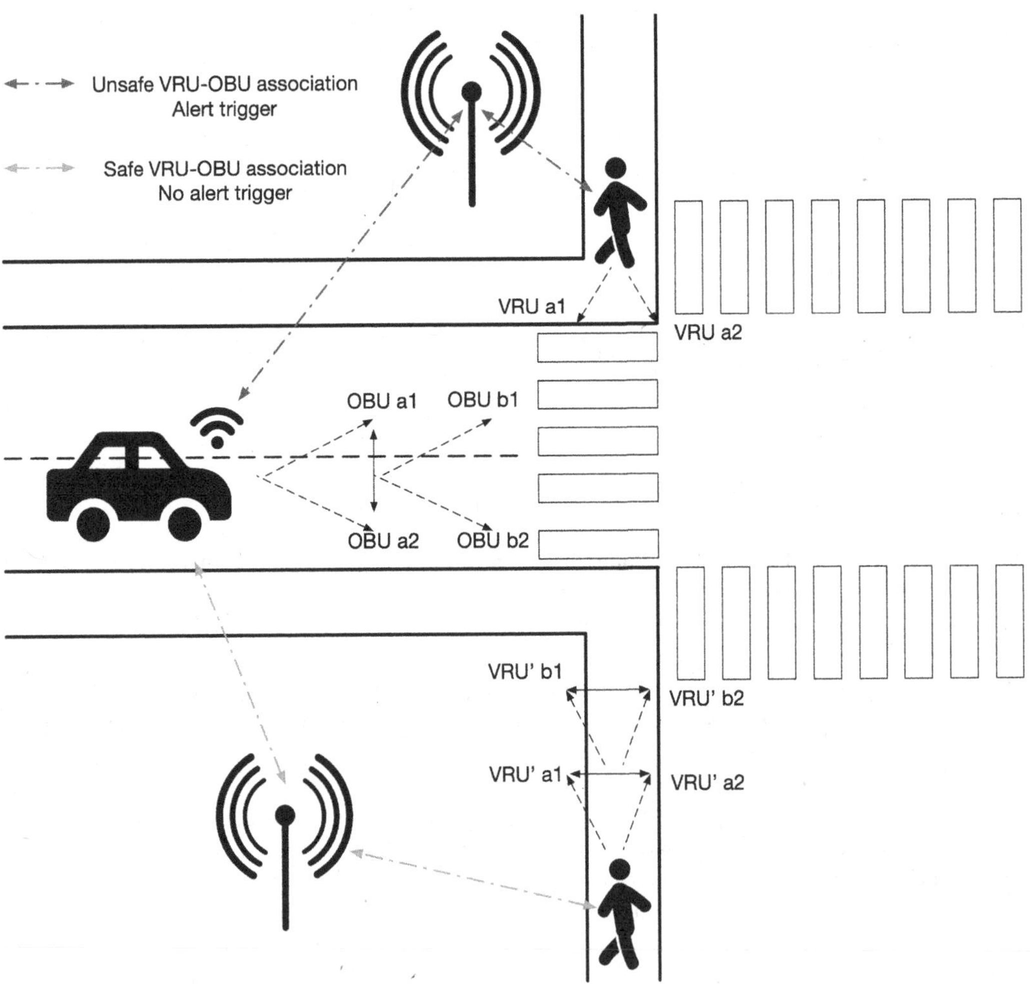

Figure 8.17 Modeling future potential direction and location.

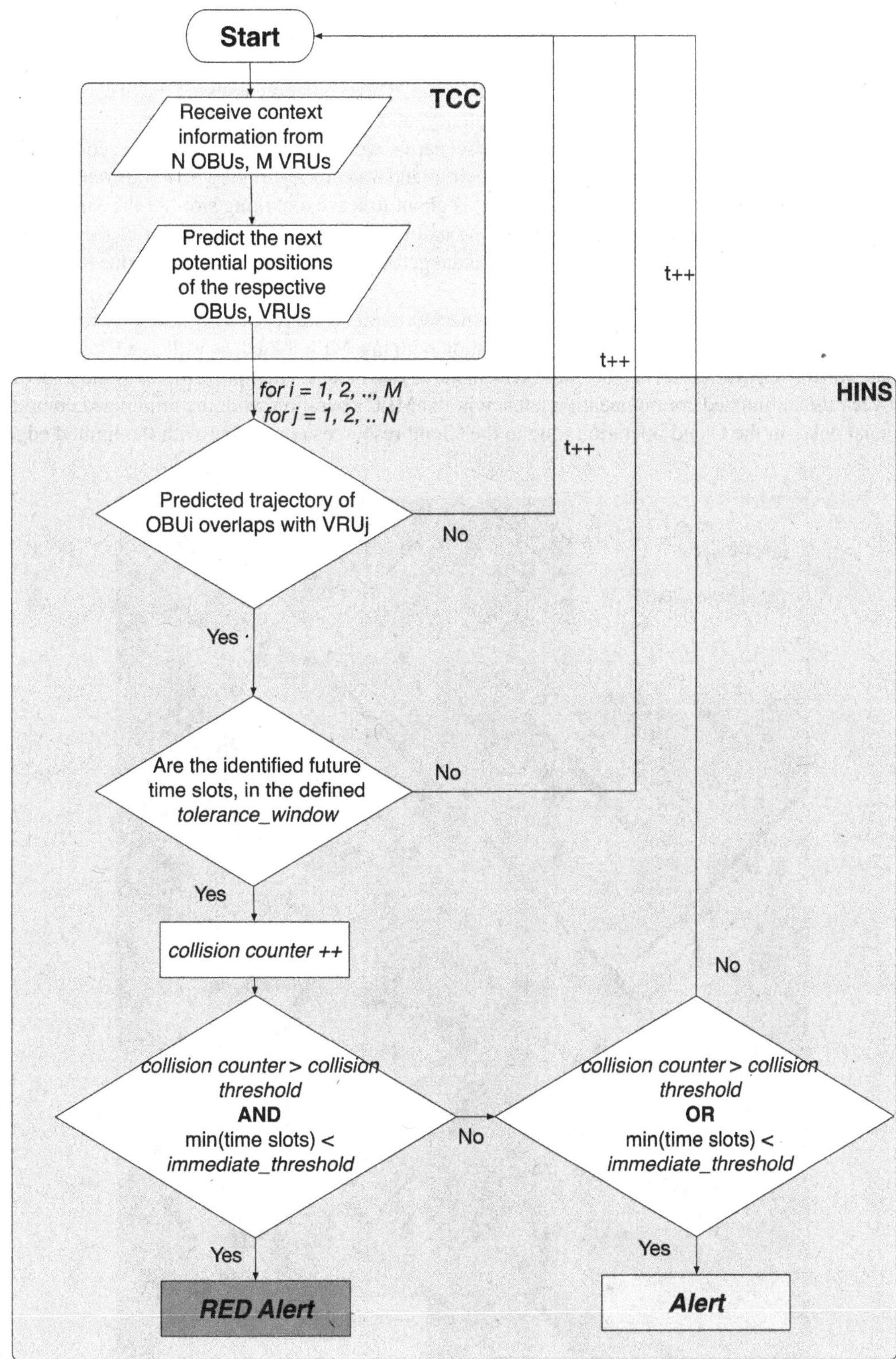

Figure 8.18 VRU-Safe workflow detailing the HINS algorithm.

8.5.3.3 Performance Evaluation

VRU-safe was evaluated in order to assess the overall performance of the proposed hybrid architecture and V2X service, in a real-world network setup. The evaluation assessed the overall performance of the solution in terms of end-to-end latency, and the performance between the MEC and the Cloud operations. Several experimentation scenarios were carried out in order to acquire reliable outcomes, with diverse mobility types, velocities and trajectories. Figure 8.19 illustrates one of those scenarios, where an OBU (marked in red) is about to leave a parking slot. At the same time, a VRU (marked in blue) is approaching from the main road, which passes in front of the parking slot. This scenario was selected as it represents a dangerous situation in actual roads, due to limited visibility because of several obstacles.

In the first use case, VRUs and OBU are connected to the same RSU. The system is capable to process the mobility-related, contextual information both in a MEC-based, as well as a Cloud-based mode; we consider the performance of the system in the two modes, attempting to assess the tradeoff between the minimized communication latency in the MEC operation, with the minimized computational delay in the Cloud operation (due to the Cloud resources comparing with the limited edge

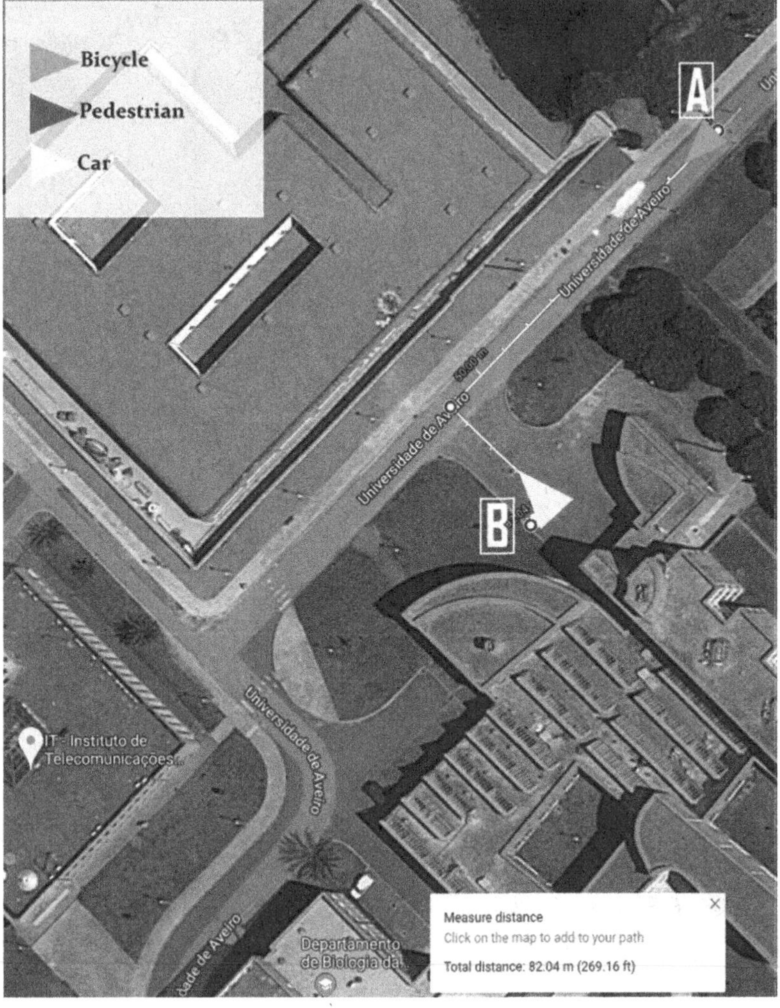

Figure 8.19 Overview of the experimentation scenario.

computing resources). Figure 8.20a and b illustrate the average end-to-end delay in the MEC-based, as well as the Cloud-based operation, respectively, for the different timestamps of the experiment, for which an event was triggered.

The performance between the two schemes (Cloud- and MEC-based) is similar for the majority of the identified events during the scenario execution (with the exceptions of the first event for OBU_1 and the second event for OBU_2). The average end-to-end delay for both systems was monitored at 200–250 ms, with the exception of the two higher performance incidents demonstrating an end-to-end delay of 80–100 ms.

The second scenario involves an identical round of experiments with the same trajectories of VRU and OBU; in this particular set of experiments, however, a second RSU node is also deployed resulting in two RSU nodes serving two mobile elements, one for each RSU. This approach is followed in order to validate the system's performance when no MEC-based operation can take place (as each RSU acquires the mobility information only of its associated mobile element); all the trajectory prediction, as well as hazard identification processes (TCC and HINS modules' operation, respectively), takes place on the Cloud. The TCC and HINS operations, in the form of VxFs, receive contextual information from the moving nodes, via the respective associated RSUs, and forward the respective alert events via the same network paths.

Figure 8.21a and b illustrate the results of the scenario with regard to the end-to-end delay for the Cloud-based operation, for the two OBU-RSU associations. The results illustrate a relatively stable performance for the VxF-based VRU-safe operation; the average delay for OBU is roughly 180 ms, while for the VRU it is slightly lower than 130 ms. Due to the centralized nature of the computing VxFs for mobile nodes, this small difference is explained by a respective difference in the propagation delay of the notifications of around 40–50 ms, caused in the radio part of the end-to-end communication path between each RSU and its associated node.

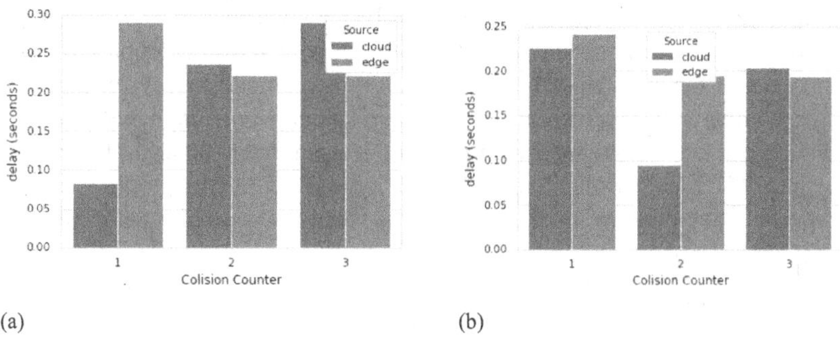

Figure 8.20 Average E2E delay in the hybrid operation: (a) for the OBU, (b) for the VRU.

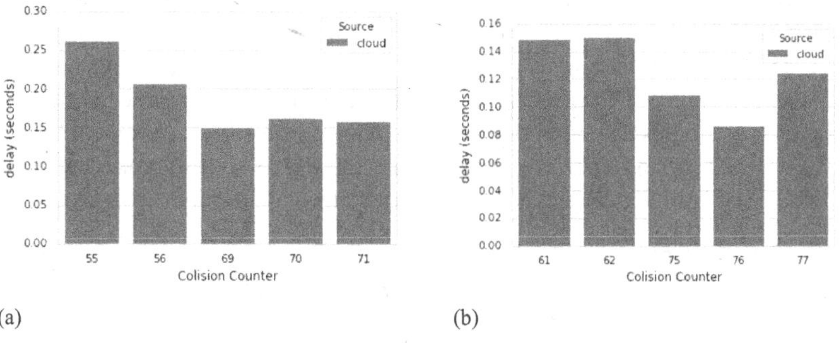

Figure 8.21 Average E2E delay in the Cloud-based operation: (a) for the OBU, (b) for the VRU.

In both scenarios, both operation types always correctly identified the potential hazardous situations and timely forwarded the respective notifications to the VRU and OBU. All in all, the outcomes lead to the conclusion that the Cloud-based system, in most cases, seems to outperform the RSU-based deployment in terms of performance. However, both systems exhibit end-to-end delays (worst case does not exceed 250 ms), which are considered to fully satisfy the requirements of the specific V2X application, due to the fact that the ultimate goal is to timely notify the VRU before the actual potential collision.

The MEC-based solution, in certain scenarios, can potentially provide much lower communication-related end-to-end latency, due to the proximity with the mobile elements. However, limited computational resources can radically influence the RSUs' performance. On the other hand, the Cloud-based operation can provide a much more robust performance, particularly for upscaled scenarios and in cases where multiple RSUs are involved, also aggregating the context/-mobility information from numerous RSUs in one place.

8.5.4 5G-CAGE

Public safety has been flagged as one of the most relevant scenarios for supplying security to citizens in new contexts such as smart cities. Computer vision, big data analytics and Machine Learning (ML) techniques can therefore be used to speed up response times for public safety interventions (e.g., fires and crimes). With 5G, the large volume of data available for public safety applications is expected to increase tremendously, mostly in terms of video and audio streaming, location information and context-awareness. To address these new requirements, the application of intelligent mechanisms to detect certain situations or city objects (e.g., license car plates in which public safety operators are interested for quick reaction) becomes a great challenge targeting big data analytics.

Existing computer vision and intelligent techniques in public safety scenarios are currently applied to operate with TCP/IP protocols, over wired or wireless networks supporting Wi-Fi or 4G. However, intelligent application to public safety with LTE capabilities, for the upcoming 5G technologies, is an open issue currently under investigation whose relevant proposed approaches are being tested in lab environments. High-performance results on intelligent systems can be achieved by feature-based computer vision techniques such as Histogram of Gradients (HoG), as studied in Ref. [25], which can be strengthened with machine learning approaches to better classify computer vision results. Machine learning approaches, or even Deep Learning ones, can also be based on neural network architectures for image-based vehicle analysis, as suggested in Ref. [42].

8.5.4.1 Concept

The 5G-CAGE use case (which stands for 5G-enabled Context and situational Awareness detection with machine learning techniques of city objects in Experimental vertical instances) aims at deploying a virtualized city safety solution, called *City Object Detection* (CODet), as a new NFV Network Service. The 5G-CAGE safety solution is capable of automatically recognizing and detecting certain city elements and objects for a city safety standpoint, specifically license car plates in which the police staff is interested to inquire *when* and *where* such vehicles are (or have been).

The 5G-CAGE use case provides a CODet VxF to detect license car plates which incorporates advanced features for vehicles detection through computer vision and machine learning algorithms. The CODet VxF monitors and analyzes in real time the video sources collected from cameras equipped in vehicles, whose lifecycle can be summarized through the following steps:

1. the CODet VxF is deployed in the Cloud;
2. video streams from selected video sources (e.g., connected cars) are made available in video caches either directly or after video transcoding;
3. an external end-user (e.g. an authorized person of the policy staff) accesses the CODet VxF to configure the detection of a specific license car plate in live video streams;

4. the CODet VxF executes computer vision algorithms on the available video streams to detect the license car plates typed in the previous step;
5. once detected, the CODet VxF sends an alert to the end-user (e.g., the previous policy staff) that attaches a cropped image of the found object – video streams are cropped to only display license car plates due to privacy-preserving issues – as well as other additional information such as the timestamp and location of the detection, capturing device details, etc.

8.5.4.2 Implementation

The implementation and deployment of the 5G-CAGE use case, in which the CODet VxF is embedded, is sketched out below by outlining its framework and its main operational workflow. In the left part of Figure 8.22, the automotive environment testbed in which the CODet VxF was deployed and validated is depicted. As shown in the right part of Figure 8.22, the In-Car Node Processor of the vehicle runs an implemented *TCP Client* and an *HTTP Video Streaming Server*, while the CODet VxF runs the other two implemented elements, a *TCP Server* and a *Flask Web App*. The communications between such components are devoted to supplying the CODet VxF with video streams captured from the In-Car Node Processor (equipped in the vehicle), in addition to geolocation data synchronized with the video streams. The TCP Client gets the GPS coordinates in the In-Car Node Processor and sends them directly to the CODet VxF, along with the capture device URL so that the CODet VxF can access it. The In-Car Node Processor then starts the HTTP Video Streaming Server to forward the video streams. In this way, the VxF gets all the required information (live video streams and GPS coordinates) from the In-Car Node Processor as the vehicle is in motion.

The implementation of the aforementioned components supports multiple vehicles per CODet VxF, each of them running its video streaming server and sending its GPS location. The multi-threaded CODet VxF keeps accessing the video devices of each connected TCP Client, processing

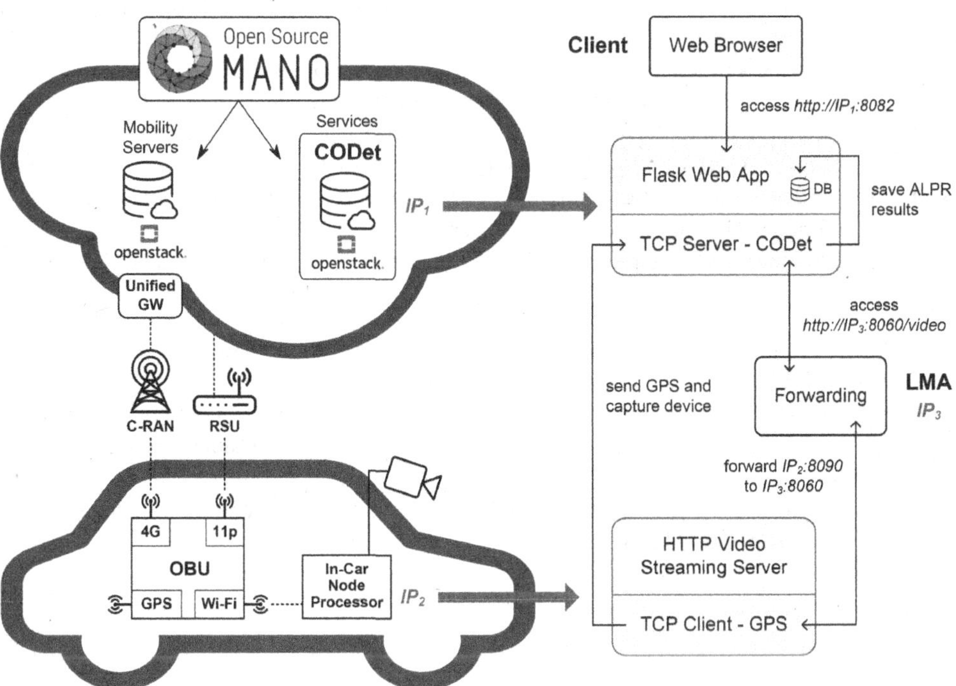

Figure 8.22 Deployment and operational workflow of the 5G-CAGE framework.

Figure 8.23 Visual license car plate results for an end-user interested in its detection.

the gathered video streams of each vehicle with the specific implemented computer vision procedure. Every time a sought license car plate is detected, the CODet VxF saves the event in a local database, while a Flask-based web application (started in the CODet VxF) manages the database and provides the end-user (e.g. policy staff) a way to interact with the 5G-CAGE framework with a simple *Web Browser*.

As an example of operation for an external end-user (e.g., police staff), once the CODet VxF is running, Figure 8.23 depicts the detection of a given license car plate previously registered by the end-user in a Web Browser. The figure shows the place where the license car plate has been detected, indicating its GPS coordinates; the date and time of the precise time of detection; the confidence of the detection result (around 90%) and a cropped image of the license car plate. An embedded web frame is also shown to review the car location on the map.

With respect to the software implementation, the main packages making up the CODet VxF are the following:

the OpenCV library[12] for image processing and computer vision algorithms;

the OpenALPR library[13] for automatic license car plate recognition;

the Python-based Flask library[14] to enable a web application that can be used by external entities or end-users (e.g. police staff) to request the detection of particular license car plates in which they have an interest; and

the GeoPy Python package[15] to access geocoding services and determine the precise location where each detection has taken place.

[12]https://opencv.org.

[13]https://www.openalpr.com.

[14]http://flask.pocoo.org.

[15]https://pypi.org/project/geopy.

It is worth mentioning that, at the core of the automatic license car plate recognition task, the CODet VxF leverages the main OpenALPR pipeline with the following well-known stages:

1. *Detection*: to find possible license car plate regions, using binary classification techniques based on multiscale local binary pattern features; *Binarization*: to isolate individual characters of the license car plate through dynamic thresholding techniques;
2. *Character Analysis*: to test if the overall sizes and locations are consistent with the layout and style of a license car plate, depending on the regulations of each country;
3. *Alignment*: to locate the exact limits of the license car plate using an edge finding procedure; *rectification* to get a fully frontal image of the license car plate;
4. *Character Segmentation*: to find spaces between characters and prepare them to be processed individually; and
5. an *Optical Character Recognition*: process on each individual letter and/or digit, taking into account pattern matching algorithms used by long short-term memory networks.

8.5.4.3 Performance Evaluation

Several experiments were conducted to obtain specific performance results when executing the CODet VxF in order to find out how it behaves from the use case point of view itself in the detection and from the network point of view (additional details in Ref. [8]). To this end, the experiments were executed in two distinct but well-related environments: in local tests by deploying the CODet VxF in a lab scenario and in the IT-Av automotive testbed. The overload that the 5G-CAGE use case execution in a real virtualized environment has, mainly due to delays and network load, can be concluded by comparing its performance results with the ones obtained when it is executed in a local environment.

Figure 8.24 depicts both types of performance results, in which red and blue lines represent the complete processing time of each separate environment, and where the *detection phase* also includes the *pre-processing phase* (pink or green lines) previously executed to achieve the complete processing time. Each point in the graph denotes the processing time for each of the 60 detections, corresponding to the total number of 6 different license car plates identified during the tests. The pre-processing phase corresponds with all the well-known stages introduced above regarding the OpenALPR pipeline, except the OCR interpretation of the detected license car plates. On the other hand, the detection phase (the complete processing time) denotes that the posterior OCR interpretation is the one stage that takes more computational load. It is worth mentioning that peaks in all performance results tend to be caused by input frames eventually containing regions that, though lately correctly discarded, were initially considered as license car plates, making the detection phase longer than usual.

Regarding the local tests in laboratory, the CODet software implementation was executed in a lab setting with a dedicated PC Intel 3.7 GHz CPU with 6 cores and hyperthreading, with a total of 12 real threads running in parallel. The results in processing time are shown in Figure 8.24 with blue and green lines. The total processing time in local tests is 141ms on average, denoted by the blue line. On the other hand, the overall license car plate recognition system of the average confidence is 87.78% in the local tests, which can be considered as acceptable results in detection.

With respect to the 5G-CAGE use case execution with the virtualized platform, whose performance results are shown in Figure 8.24 (red and pink lines), a vehicle with an OBU and a connected camera with a 1280×720 resolution were used in the IT-Av automotive environment. As depicted in Figure 8.24, worse results were obtained than in the local tests, as expected. The reasons for these results are due to the fact that: (1) the CODet VxF deployed in the Cloud was running with much less powerful virtual CPUs, running at 2 GHz, just having only 1 core with no hyperthreading; and (2) network issues in a real environment with low bandwidth and a high volume of traffic. However, the obtained results in terms of performance are still acceptable, with a total processing time around

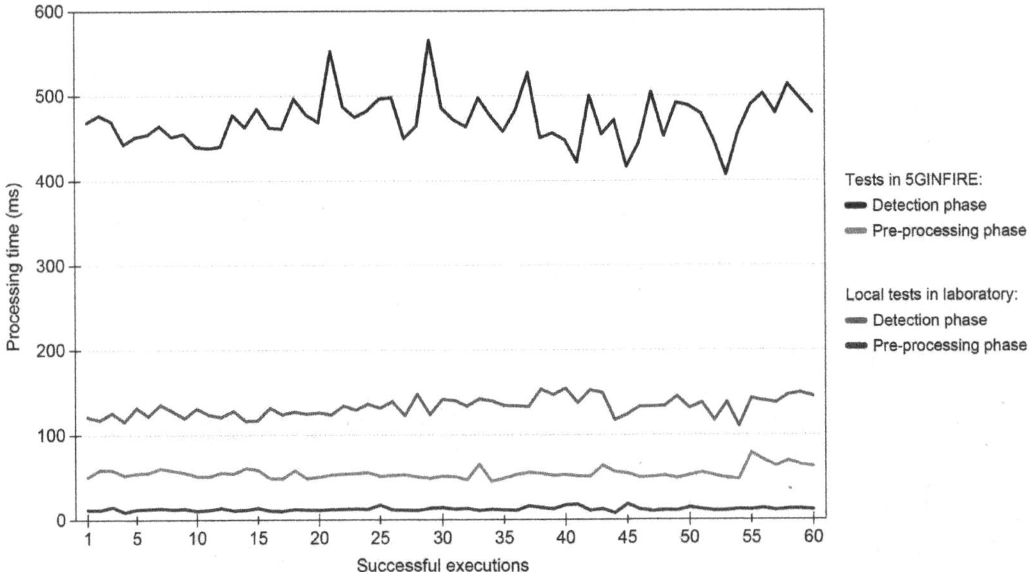

Figure 8.24 Processing time when executing the CODet VxF in the lab and in the IT-Av automotive environment.

of 473 ms and a confidence of 84.11% on average. Results are a bit lower than in the local results, but still consistent with the experiment conditions in a real-time demonstration.

Despite the noticeable difference between the local results and those obtained on a real vehicular environment, it is worth mentioning, as indicated above, that it may be due to the limited resources in the Cloud platform. Because of this, the OCR interpretation, as the last stage in the OpenALPR pipeline that takes more computational load, leads to somewhat worse results in the real environment. In addition to this lack of computational resources, the network overload could also be another drawback for obtaining better and good performance results.

To assess network overload issues, Figure 8.25 illustrates the amount of network traffic required in the 5G-CAGE use case to gather video streams from the vehicle's In-Car Node Processor. The graph displays only the transmission rate of the data plane network interface, using the Speedometer tool in charge of measuring and displaying the rate of data across a network connection. A data rate of 121.2 KB/s on average is shown in Figure 8.25, considering the particular conditions of the real-time experiment.

As shown in Figure 8.25, we can conclude that the volume of network traffic will be an important issue to consider in real automotive scenarios with a higher number of vehicles in motion. However, this traffic congestion could be mitigated decoupling several functions of the CODet implementation and transferring them, when possible, to each vehicle's OBU, in order to reduce the amount of network traffic in the video stream delivery.

8.6 RESEARCH CHALLENGES

Despite the success of the experiments discussed in this chapter, which clearly showed the potential of using virtualized services and applications under the concept of vehicular networks, several research challenges associated with the exploitation of the NFV concept, in particular in mobile and vehicular networks, still remain. Regarding the management and orchestration of virtual functions, current approaches are focused on NFV management without considering the mobility of compute nodes. In the examples discussed in this chapter, although focused on mobile networks, the virtual

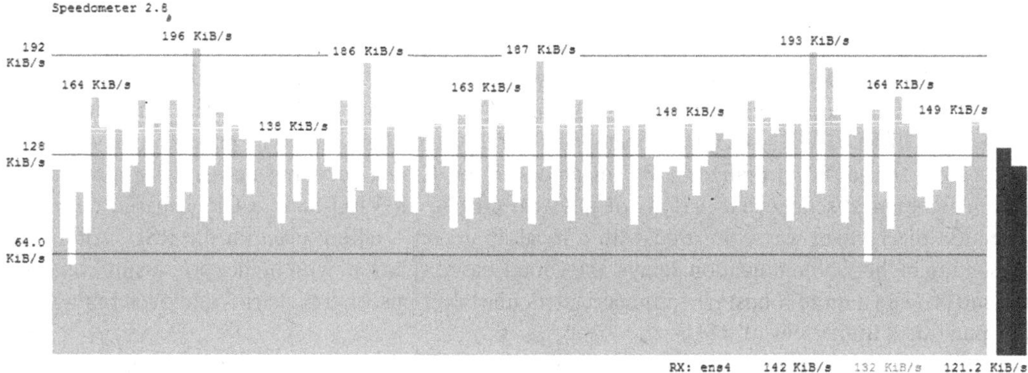

Figure 8.25 Network traffic between the OBU and the RSU when executing the CODet VxF in the IT-Av automotive platform.

services and applications were always deployed on the infrastructure (Cloud or RSUs) with constant connectivity with the NFV orchestrator.

The performance of the NFV is also a challenge. Knowing that a large plethora of devices are able to host virtualized services, most of them may not be able to provide the required processing power to properly accommodate the majority of the network services and applications. Such an issue is aggravated when talking about compute node(s) deployed in vehicles which is, by nature, an energy-constrained element. While hardware acceleration may be used, such a specialization is against the concept of NFV which aims at high flexibility. In such case a trade-off between performance and flexibility must be found.

Another issue regarding the exploration of NFV over mobile networks is the efficient allocation of physical resources. For that, efficient algorithms are required to determine how and where the different services are distributed among the physical resources, which will need to take into account several objectives such as load balancing, eventual recovery mechanisms, energy saving, among others. This task is already challenging in non-mobile environments, and therefore one should expect that the mobility of the physical resources would increase the complexity and operation of the NFV solution.

Other challenging issues such as security, privacy and trust, energy efficiency, resources discovery, Quality-of-Service compliance or even service chain re-composition will also demand special attention for the development and implementation of NFV solutions in vehicular networks where the data volume can change dynamically depending on services used by vehicles, pedestrians and other C-V2X nodes for V2I or V2N (Vehicle-to-Network) communication. Dedicated classes for such services will be needed to provide, e.g., ranging and positing as well as high definition maps and bird's eye view. It is especially challenging if there will be many slices and vertical users deployed in the same network infrastructure. It requires sophisticated and intelligence management of resources as it is done in 5G New Radio (NR) interface.

8.7 CONCLUSIONS

5G vehicular networking is one of the key enabling technologies that will support several mobile applications varying from global Internet services up to active road safety applications. Advances on vehicular radio access technologies bring up better V2V and V2I communication links: more reliable, with higher bandwidths and higher ranges. Such features are essential for the applicability of Cloud virtualization and NFV technologies over vehicular networks where critical virtual functions, deployed somewhere in the non-mobile infrastructure, are explored in a dynamic and on-demand approach.

Supported by a real vehicular network with multihoming and mobility support, along with a Cloud infrastructure with support for virtualization, this chapter presented and evaluated the applicability of several virtual functions targeting the automotive vertical. Four distinct use cases, each one targeting a distinct ITS subject, were tested considering both network and application domains. In SURROGATES the OBU capabilities were extended by offloading processing tasks from the vehicle side up to virtualized OBU instances in the Cloud, speeding up the system response and overcoming disconnected periods of physical OBUs. In VRU-Safe, a Cloud-based collision detection mechanism was compared with a local approach – taking place in the RSU. Although presenting higher communication delays, the Cloud-based solution, with higher processing capability, can provide a more robust performance, particularly for upscaled scenarios and overcoming the computational limitations of RSUs.

In CAVICO, the virtualization of context-aware mechanisms using no-reference, context-based QoE metric subsets, was evaluated by means of a framework capable of optimizing video streaming according to the parameters of both network radio interfaces and video streams, as well as vehicular conditions (road, parking, gas station, etc.). The results showed that, by measuring the video QoE and network QoS, the Cloud-based application is able to adjust the video encoder to save transmission bandwidth or improve image fidelity depending on the transmission environment and user's expectations. Finally, 5G-CAGE experiment has shown that a city object detection mechanism, although executed in the Cloud, is able to identify specific car plates under 500 ms, critical for public safety situations.

The plethora of different use cases that have been implemented and tested through different VxFs in a real vehicular network platform shows the flexibility provided by the 5G-enabled Cloud, and the real potential of a 5G environment in the test and deployment of complex and challenging functionalities, and on the evolution toward 5G softwarized networks in mobile environments to provide services in a secure and resilient way and to ensure safety in V2X applications.

Future directions on virtualization in the scope of vehicular networks will depend on the location of virtualization based on the cloudification paradigm. If the virtualization takes place in the Cloud, the range of coverage is higher, as well as the latency, and therefore critical services should be avoided. In this case, traffic congestion avoidance and environment monitoring are some of the applications to be explored. When in Edge cloud virtualization, for example when using the RSUs as resources of virtualization, the range of coverage is smaller as well as the latency. In this case collision avoidance and autonomous driving can be some of the services to be explored through NFV technologies. Finally, the virtualization in the vehicle gives space for a different type of services such as entertainment content sharing or even mobile social applications. No matter which virtualization topology to be adopted, the flexibility and extensibility provided by the concept of NFV will turn vehicular networks and vehicular communications more agile, efficient and safe.

ACKNOWLEDGMENT

This book chapter was partially supported by the European Commission, under the projects 5GINFIRE under Grant 732497 and by the Spanish Ministry of Science, Innovation and Universities, under the Ramon y Cajal Program (Grant No. RYC-2017-23823). The authors would like to acknowledge Mateusz Rajewski for the technical support.

REFERENCES

1. 3GPP. TS 36.300 E-UTRA and E-UTRAN; Overall description; Stage 2 (v14.8.0, Release 14). Technical Report RTS/TSGR-0236300ve20, ETSI, October 2018.
2. 5G Automotive Association. Toward fully connected vehicles: Edge computing (White Paper), December 2017.

3. M. Abuelela and S. Olariu. Taking VANET to the clouds. In *Proceedings of the 8th International Conference on Advances in Mobile Computing and Multimedia, MoMM '10*, pp. 6–13, New York, ACM, 2010.
4. J. Al-Badarneh, Y. Jararweh, M. Al-Ayyoub, R. Fontes, M. Al-Smadi, and C. Rothenberg. Cooperative mobile edge computing system for vanet-based software-defined content delivery. *Computers and Electrical Engineering*, 71:388–397, 2018.
5. S. Al-Sultan, M.M. Al-Doori, A.H. Al-Bayatti, and H. Zedan. A comprehensive survey on vehicular ad hoc network. *Journal of Network and Computer Applications*, 37:380–392, 2014.
6. C. Campolo, R.D.R. Fontes, A. Molinaro, C.E. Rothenberg, and A. Iera. Slicing on the road: Enabling the automotive vertical through 5G network softwarization. *Sensors*, 18(12):4435, 2018.
7. N. Capela and S. Sargento. An intelligent and optimized multihoming approach in real and heterogeneous environments. *Wireless Networks*, 21(6):1935–1955, 2015.
8. P.E.L. de Teruel, M.G. Perez, F.J. G. Clemente, A.R. Garcia, and G.M. Perez. 5G-CAGE: A context and situational awareness system for city public safety with video processing at a virtualized ecosystem. In *2nd Workshop on Moving Cameras: From Body Cameras to Drones (MCMVS'19)*, Seoul, Korea, 2019.
9. V. Devarapalli, R. Wakikawa, A. Petrescu, and P. Thubert. Network Mobility (NEMO) basic support protocol. RFC 3963, RFC Editor, January 2005.
10. M. El-Hajjar, Y. He, L. Tang, Y. Ren, J. Rodriguez, and S. Mumtaz. Cross-layer resource allocation for multihop V2X communications. *Wireless Communications and Mobile Computing*, 2019:1–16, 2019.
11. H. El-Sayed and M. Chaqfeh. The deployment of mobile edges in vehicular environments. In *2018 International Conference on Information Networking (ICOIN)*, Chiang Mai, Thailand, pp. 322–324, January 2018.
12. M. Emara, M.C. Filippou, and D. Sabella. MEC-assisted end-to-end latency evaluations for C-V2X communications. In *2018 European Conference on Networks and Communications (EuCNC)*, Ljubljana, Slovenia, pp. 1–9, June 2018.
13. P. J. Fernandez, J. Santa, F. Bernal, and A.F. Skarmeta. Securing vehicular IPv6 communications. *IEEE Transactions on Dependable and Secure Computing*, 13(1):46–58, 2016.
14. H. Gierszal, A. Stachowicz, F. Majerowski, B. Kowalczyk, M. Goryński, V. Kassouras, and S. Drakul. Requirements for IMS services and applications over interoperable broadband public protection and disaster relief networks and commercial communication networks. *ACSIS*, 2:933–940, 2014.
15. S. Gundavelli, K. Leung, V. Devarapalli, K. Chowdhury, and B. Patil. Proxy Mobile IPv6. RFC 5213, RFC Editor, August 2008.
16. K. Habak, K.A. Harras, and M. Youssef. Bandwidth aggregation techniques in heterogeneous multihomed devices: A survey. *Computer Networks*, 92:168–188, 2015.
17. B. Han, V. Gopalakrishnan, L. Ji, and S. Lee. Network function virtualization: Challenges and opportunities for innovations. *IEEE Communications Magazine*, 53(2):90–97, February 2015.
18. Y. He, N. Zhao, and H. Yin. Integrated networking, caching, and computing for connected vehicles: A deep reinforcement learning approach. *IEEE Transactions on Vehicular Technology*, 67(1):44–55, 2018.
19. Z. He, J. Cao, and X. Liu. SDVN: Enabling rapid network innovation for heterogeneous vehicular communication. *IEEE Network*, 30(4):10–15, 2016.
20. R. Hussain, J. Son, H. Eun, S. Kim, and H. Oh. Rethinking vehicular communications: Merging VANET with cloud computing. In *4th IEEE International Conference on Cloud Computing Technology and Science Proceedings*, Taipei, pp. 606–609, December 2012.
21. ITU-T. Telecommunication standardization sector international telecommunication union. Subjective video quality assessment methods for recognition tasks. Recommendation P.912, March 2016.
22. A. Jahn, K. David, and S. Engel. 5G / LTE based protection of vulnerable road users: Detection of crossing a curb. In *2015 IEEE 82nd Vehicular Technology Conference (VTC2015-Fall)*, Boston, MA, 2015.
23. D. Kreutz, F.M.V. Ramos, P.E. Verissimo, C.E. Rothenberg, S. Azodolmolky, and S. Uhlig. Software-defined networking: A comprehensive survey. *Proceedings of the IEEE*, 103(1):14–76, 2015.
24. N. Kumar, S. Zeadally, and J.J.P.C. Rodrigues. Vehicular delay-tolerant networks for smart grid data management using mobile edge computing. *IEEE Communications Magazine*, 54(10):60–66, 2016.
25. N. Laopracha and K. Sunat. Comparative study of computational time that HOG-based features used for vehicle detection. In: P. Meesad, S. Sodsee, and H. Unger, editors, *Recent Advances in Information and Communication Technology 2017*, pp. 275–284, Cham, Springer, 2017.

26. M. Leszczuk, M. Hanusiak, M. Farias, E. Wyckens, and G. Heston. Recent developments in visual quality monitoring by key performance indicators. *Multimedia Tools and Applications*, 75:10745–10767, 2014.
27. D. Lopes and S. Sargento. Network mobility for vehicular networks. In *2014 IEEE Symposium on Computers and Communications (ISCC)*, Funchal, Madeira, pp. 1–7, June 2014.
28. M. Malinverno, G. Avino, C.E.E. Casetti, C.F. Chiasserini, F. Malandrino, and S. Scarpina. Edge-based collision avoidance for vehicles and vulnerable users: Benefits for vehicles and vulnerable users an architecture based on MEC. *IEEE Vehicular Technology Magazine*, 15(1):27–35, 2019.
29. R. Mijumbi, J. Serrat, J. Gorricho, N. Bouten, F. De Turck, and R. Boutaba. Network function virtualization: State-of-the-art and research challenges. *IEEE Communications Surveys Tutorials*, 18(1):236–262, Firstquarter 2016.
30. P. Mitharwal, C. Lohr, and A. Gravey. Survey on network interface selection in multihomed mobile networks. In Yvon Kermarrec, editor, *Advances in Communication Networking*, pp. 134–146, Cham, Springer International Publishing, 2014..
31. A. Napolitano, G. Cecchetti, F. Giannone, A.L. Ruscelli, F. Civerchia, K. Kondepu, L. Valcarenghi, and P. Castoldi. Implementation of a MEC-based vulnerable road user warning system. In *2019 AEIT International Conference of Electrical and Electronic Technologies for Automotive, AEIT AUTOMOTIVE*, Turin, Italy, 2019.
32. J. Nawała, M. Leszczuk, M. Zajdel, and R. Baran. Software package for measurement of quality indicators working in no-reference model. *Multimedia Tools and Applications*, 75:10397-10405, 2016.
33. T. Ojanpera, J. Makela, O. Mammela, M. Majanen, and O. Martikainen. Use cases and communications architecture for 5g-enabled road safety services. In *2018 European Conference on Networks and Communications (EuCNC)*, Ljubljana, Slovenia, pp. 335–340, June 2018.
34. C. Perkins. IP Mobility Support for IPv4, Revised. RFC 5944, RFC Editor, November 2010.
35. C. Perkins, D. Johnson, and J. Arkko. Mobility Support in IPv6. RFC 6275, RFC Editor, July 2011.
36. SAE. Surface vehicle recommended practice. Technical Report J3016, Society of Automotive Engineers, June 2018.
37. J. Santa, P.J. Fernández, J. Ortiz, R. Sanchez-Iborra, and A.F. Skarmeta. SURROGATES: Virtual OBUs to foster 5G vehicular services. *Electronics*, 8(2):117, 2019.
38. S.A.A. Shah, E. Ahmed, M. Imran, and S. Zeadally. 5G for vehicular communications. *IEEE Communications Magazine*, 56(1):111–117, 2018.
39. A.P. Silva, C. Tranoris, S. Denazis, S. Sargento, J. Pereira, M. Luis, R. Moreira, F. Silva, I. Vidal, B. Nogales, R. Nejabati, and D. Simeonidou. 5GinFIRE: An end-to-end open5G vertical network function ecosystem. *Ad Hoc Networks*, 93:101895, 2019.
40. S. Wang, X. Zhang, Y. Zhang, L. Wang, J. Yang, and W. Wang. A survey on mobile edge networks: Convergence of computing, caching and communications. *IEEE Access*, 5:6757–6779, 2017.
41. F.Z. Yousaf, M. Bredel, S. Schaller, and F. Schneider. NFV and SDN-key technology enablers for 5G networks. *IEEE Journal on Selected Areas in Communications*, 35(11):2468–2478, 2017.
42. Y. Zhou, H. Nejati, T. Do, N. Cheung, and L. Cheah. Image-based vehicle analysis using deep neural network: A systematic study. In *2016 IEEE International Conference on Digital Signal Processing (DSP)*, Beijing, China, pp. 276–280, 2016.
43. M. Zhu, J. Cao, Z. Cai, Z. He, and M. Xu. Providing flexible services for heterogeneous vehicles: An NFV-based approach. *IEEE Network*, 30(3):64–71, 2016.

9 Data Offloading Approaches for Vehicle-to-Everything (V2X) Communications in 5G and Beyond

Muhammed Nur Avcil and Mujdat Soyturk
Marmara University

CONTENTS

9.1 INTRODUCTION

Intelligent Transportation Systems (ITS) are one of the fundamental bodies of the smart cities comprising resources, assets, activities and transportation networks. ITS aim to provide better services for all consumers and beneficiaries in all aspects including safety, traffic efficiency, cost, sustainability and ecology. Information and Communication Technologies (ICT) enable all these objectives to become real in life providing advanced infrastructure, traffic and mobility management solutions with the use of electronic, wireless and communications technologies. Among these technologies, Vehicle-to-Everything (V2X) technologies constitute the most important and critical component of the communication infrastructure (between the consumer-vehicle-infrastructure-management center) to provide smarter, safer and faster travel in addition to the efficient use of the resources.

For years, vehicle-to-everything (V2X) communications and their applications have attracted a great deal of attention from vehicle manufacturers, service providers and public service authorities due to their great positive contribution to safety, traffic efficiency and comfort driving [5,67]. Expected benefits of V2X applications led to the pilot projects and infrastructure investments by both the public and private organizations, as well as joint initiatives usually in the form of public-private partnerships (PPP) [48]. With the integration of V2X communication in 5G networks, the quantity, the quality and the scope of these applications have increased provisioning more coverage, connectivity and enhanced link quality with the use of 5G connectivity [55].

The 3rd Generation Partnership Project (3GPP) [3] defined and standardized the key 5G-V2X applications [1] and their corresponding specifications [2]. As well as the number of users who want to benefit from these services, the movement patterns of the vehicles affect the service quality of these applications. High and/or continuously changing velocities of vehicles in turn affect their connection to a number of services. In a typical set up with so many fast-moving vehicles, all operations including handover should be scheduled and accomplished precisely to meet the needs and the predefined service requirements. Ultra-low latency and ultra-high reliability requirements introduce challenges in such a dynamic and resource-demanding vehicle topology [52,54]. Meeting both the latency and reliability requirements of the safety-critical applications, and the resource (bandwidth-demanding) requirements of comfort driving applications requires an intelligent and precise planning of all resources, mainly on the access network. Failing to meet these requirements may cause service interrupts and/or degrade the service quality, which eventually disrupts the safety and efficiency of the vehicle traffic and all related services. On the other hand, efficiency during the utilization of resources is a crucial issue at the service provider level since this has direct ramifications on cost and quality [11].

9.2 A BRIEF OVERVIEW OF 5G, V2X COMMUNICATIONS AND APPLICATIONS

Research indicated that one of the most important and comprehensive impacts of 5G technologies on social and economic life is expected to be in the field of Intelligent Transportation Systems (ITS) [14]. 5G technologies' impact on ITS will include not only public and personal transport, but also the passengers, pedestrians and even other autonomous systems that are connected to or benefit from transportation systems. V2X communication which is already incorporated in the 5G and standardized by 3GPP plays a vital role in accessing the network and enable IoT-based connectivity inter-vehicle, and between vehicle and infrastructure. At this stage, it will be useful to explain V2X communication and Connected Vehicles, a concept associated with [61].

Connected Vehicles: The term 'connected vehicle' not only refers to vehicles that connect to the network, but it also covers the services delivered to the vehicle in a continuous and ubiquitous fashion. Connected Vehicle services such as those providing safety, traffic efficiency and comfort driving are generally provided to vehicles by the service providers.

V2X Communication: All of the communication methods required for vehicles to exchange data with other vehicles, infrastructure and environment are called V2X (Vehicle-to-Everything) communication [17]. The communication methods defined with V2X communication (as shown in Figure 9.1) are:

- Vehicle to vehicle communication (V2V)
- Vehicle to infrastructure communication (V2I)
- Vehicle to cellular network communication (V2N)
- Vehicle to portable devices (V2P)

Studies and specifications on V2X communication have been going on for some years, especially intensifying since 2006 and onward. Initially, Dedicated Short Range Communication (DSRC) [29] (a WiFi-based short-distance communication) based standards were defined by IEEE and European Telecommunications Standards Institute (ETSI), in the US and the EU, respectively. V2X communications use a particular set of frequencies in accordance with the industry standard and in compliance with local regulations [56]. US Federal Communications Commission (FCC) allocated 75 MHz of bandwidth in the 5.9 GHz (5850–5925 MHz) spectrum band for DSRC. In the EU, 70 MHz of bandwidth in the 5.9 GHz (5855–5925 MHz) spectrum band is allocated [30]. These

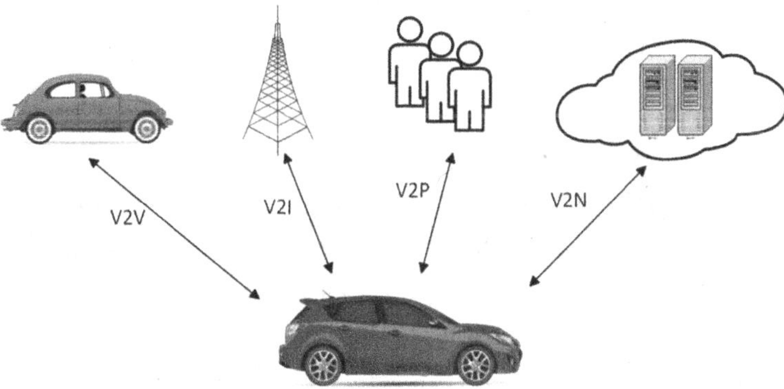

Figure 9.1 Communication modes of V2X.

standards refer to a number of other standards for each layer and various functionalities in the communication protocol stack. US standards include IEEE 802.11p and IEEE 1609 protocol family, while EU standards include ETSI ITS G5 (corresponding to IEEE 802.11p) and a number of related standards.

V2X communications have categorically progressed toward being part of cellular 5G communications as wireless standards evolved. With the increased availability of reliable long-range communications, and hence providing connectivity at everywhere, cellular networks enable the vehicles to stay connected continuously. As part of ongoing standardization efforts to support V2X services in cellular networks, 3GPP defined the standards for V2X communication services in LTE and 5G cellular networks and these definitions are included in Release 14 [23] and Release 15 [25] (Release 16 is in progress).

Both 3GPP [2] and ETSI [22] defined a number of applications and services that use the V2X communications and the infrastructure. These include applications ranging from life sensitive time-critical applications e.g. traffic light violation to traffic efficiency applications e.g. optimal speed advisory and entertainment applications. Each application has specific challenges and therefore specific requirements. According to the demand type and the requirements, these services/applications are generally categorized in two classes: (1) Safety services/applications, (2) non-safety services/applications composed of traffic-efficiency related and comfort driving-related services/applications. As it affects human life/casualties and accidental damages on vehicles, priority has been given to the development of safety applications/services (and subsequently to the traffic efficiency).

V2X communication methods and environment are shown in Figure 9.2. Vehicles exchange information with each other with the use of V2V communication and also access the network infrastructure through the Road Side Units (RSUs) or through the cellular network components e.g. eNodeBs (V2N communication).

The provision of situational awareness is essential in autonomous vehicles and smart city and smart transportation-related applications. For this purpose, one standard method is for vehicles to periodically beacon to the neighboring vehicles and the infrastructure. This approach leads vehicles to enhance their perception of the environment and, therefore, to prevent any possible accident. A second standard method is for the emergency situations that risk the safety of vehicle and/or environment, event-based emergency messages are generated (either by the vehicle or by the infrastructure) and transmitted. These two standard messaging methods provide vehicles the capability to engage and utilize safety-related functions and services (as well as traffic efficiency and comfort driving). Connectivity is essential to situational awareness and emergency messaging. Intermittent connectivity of vehicles, spatio-temporal vehicle traffic characteristics and unavailability of the network infrastructure may lead to inefficient and interrupted V2X services. Therefore, cellular network connectivity is crucial for service availability wherever, whenever.

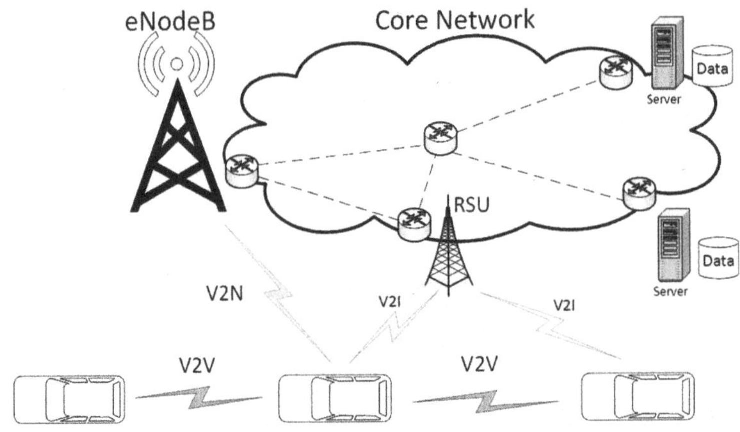

Figure 9.2 Illustration of V2X communication in LTE/5G network.

The market forecasts on autonomous driving and connected cars highlight the importance of connectivity and the service availability for successful penetration in the market. According to [61], 94 million connected cars will be shipped to consumers in 2021 and it is estimated that connected car shipments will account for 82% of the vehicles. In a Statista article [45], the connected car generates per hour 25 GB of sensory data for smart driving and controls, and uses/downloads 1 GB for other online activities such as infotainment. Society of Automobile Engineers (SAE) [18] defines the autonomous driving levels as shown in Figure 9.3. The same driving levels are also referred to in 3GPP standards [1]. Based on IHS Markit forecasts [34], by 2035, 76M vehicles with some level of autonomy (SAE 2-3-4-5) are expected to be shipped, 21M of which are self-driving (SAE 4-5), where the latter is expected to reach 0.6M by 2025. With the increase in the number of autonomous vehicles and connected cars, data requests of vehicles will start to increase more and more day by day.

To meet the service requirements of the autonomous and self-driving cars and their emerging V2X applications, 3GPP defined the specifications in the 5G standards. Table 9.1 lists some of these services and presents their requirements in terms of reliability, data rate and latency. As the degree of automation in the defined services increases, the reliability and latency requirements are

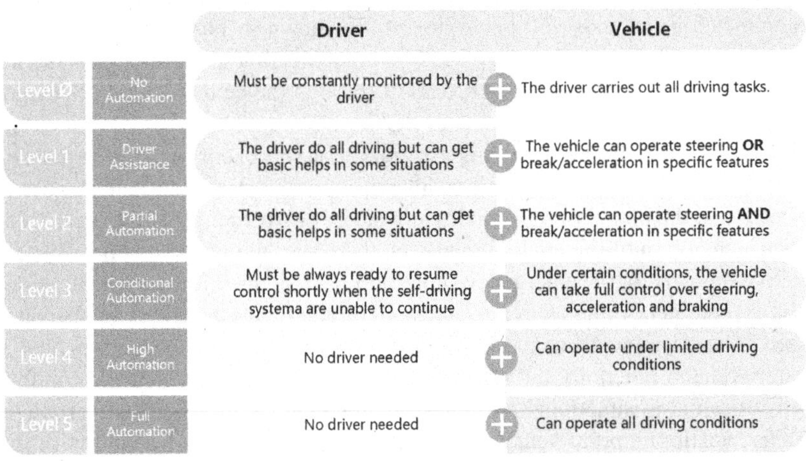

Figure 9.3 SAE automated driving levels [18].

Table 9.1
Performance Requirements of V2X Services for 5G [2]

	Payload (Bytes)	Max Latency (ms)	Reliability (%)
Vehicles Platooning			
Cooperative driving			
- Lowest degree of automation	300–400	25	90
- Highest degree of automation	50–1200	10	99.99
Information sharing between UE supporting V2X application			
- Lower degree of automation	6000	20	-
- Higher degree of automation	-	20	-
Advanced Driving			
Cooperative collision avoidance	2000	10	99.99
Information sharing between UE supporting V2X application			
- Lower degree of automation	6500	100	-
- Higher degree of automation	-	100	-
Emergency trajectory alignment	2000	3	99.999
Cooperative lane change			
- Lower degree of automation	300–400	25	90
- Higher degree of automation	12,000	10	99.99
Extended Sensors			
Sensor information sharing			
- Lower degree of automation	1600	100	99
- Higher degree of automation	-	3	99.999
Video sharing			
- Lower degree of automation	-	50	90
- Higher degree of automation	-	10	99.99
Remote Driving			
Information exchange	-	5	99.999

getting stricter. Smart city applications e.g. remote driving come with the longest requirements lists. It is seen that the defined requirements are very stringent in order to satisfy the required data rate, reliability and latency for emerging V2X services. To meet these requirements, 5G brings new approaches and enhancements on LTE, which are briefly shown in Table 9.2. As the channel bandwidth increases (with variable size feature) in 5G, new modulation schemes are introduced to provide higher reliability and lower latency. Moreover, 5G presents more flexibility in the allocation of the resources compared to the LTE considering the requirements of the application in use. Moreover, millimeter wave (mmWave) support is one key feature in 5G to meet these requirements.

There is critically crucial need on edge computing in the 5G access network to meet the stringent requirements of the delay-sensitive and reliability-sensitive V2X applications e.g. automated driving. Mobile Edge Computing (MEC) allows latency and reliability critical services to be processed and delivered to the vehicles at edge. However, the availability of these services must be guaranteed and connectivity must be provided continuously. There are many factors that may hamper the connectivity and service guarantees, which eventually will affect the availability, efficiency and reliability of the V2X services. These factors are described in Section 9.3 in detail.

Table 9.2
LTE [24] vs 5G [25] Physical Channels and Modulation

Specifications	LTE	5G NR
Full form	Long term evolution	3GPP 5G new radio
Radio frame duration (ms)	10	10
Number of sub-frames in a frame	10	10
Number of slots in a frame	Fixed, 20	Variable, depends on subcarrier spacing
Number of RBs (resource blocks)	100 (maximum for 20 MHz)	100 or more
Subcarrier spacing	Fixed, 15 KHz	Flexible: 15, 30, 60, 120, 240, 480
-For V2X communication	Fixed, 15 KHz	Flexible: 15, 30, 60, 120 KHz
Modulation	QPSK, 16QAM, 64QAM	QPSK, 16QAM, 64QAM, 256QAM
-MCS for V2X Communication	-	QPSK, 16QAM
Carrier bandwidth	1.4/3/5/10/15/20 MHz (for 20 MHz, using carrier aggregation, BW up to 100 MHz can be used)	Variable, (from 100–200 MHz for <6 GHz band, from 100 MHz to 1 GHz for >6 GHz band)
mmWave support	No	Yes

While some of the services e.g. time-critical safety applications are delivered immediately at edge by the access nodes e.g. RSU or eNodeB, non-safety applications usually run on the core network. Data demands such as text message, real-time video streaming, music streaming, web browsing, etc. and the responses to these demands pass through the access network nodes (eNodeB, RSU). Depending on the vehicle density, aggregated large size data requests may cause workload and bottleneck at the access node. Services might be delayed or interrupted [10]. Hence, there is a need for new approaches that avoid bottlenecks and deliver the load among the access network.

9.3 CHALLENGES ON RESOURCE ALLOCATION AND MOTIVATION

It has always been a challenging research problem to optimize the resources in the cellular networks [28,59,65]. Number of the access nodes (base stations) and their placement are major concerns in providing better QoS in cellular networks. Determining the location of the base stations (access nodes) always has been a major concern considering the spatio-temporal data traffic characteristics of mobile users. Generally, base stations are placed considering the criteria which affect QoS of the services as well as the cost of deployment and maintenance [36,37,58]. Because OPEX, CAPEX and maintenance introduce significant cost for the service providers, it is essential as well as an optimization problem to place the minimum number of base stations in the most suitable locations considering the QoS and the cost to the service provider [6,50]. Spatio-temporal characteristics of the mobile users signify this challenging problem due to the high variations in the traffic density along the day and along the week days. Considering the V2X services, we face similar research problems on the placement of the access nodes and the optimization of the resources [11]. We have divided these problems into separate groups as some sections.

Access Node Placement: For a better QoS on V2X services, access nodes must be placed considering the movement patterns of the vehicles and data traffic demands which are usually proportional to the vehicle density [50]. Compared to mobile users in cellular networks, the movement patterns of the vehicles are more predictable in vehicular networks. Vehicles move on the road infrastructure and usually follow a predefined route which is partially common for the neighboring vehicles. With crowd sensing, vehicle mobility patterns and partial routes can be well-estimated [9].

Therefore, vehicular networks present more convenient parameters for the placement of the base stations. On the other hand, the spatio-temporal mobility characteristics and fast mobility present additional challenges. Vehicle density and the speed of the vehicles in the road topology are varying in spatio-temporal manner, but completely different than the mobile users. It is essential to consider the vehicle density for the QoS of the V2X services.

Resource Allocation/Channel Scheduling: Vehicle density on roads is varying in space and in time. Depending on the time of day and day of the week, the density can be at peak while the same space could be sparse in other times, and even, very rare traffic might be observed after midnight [46]. These spatio-temporal characteristics of the vehicles cause the data traffic density to vary in spatio-temporal manner. In sparse areas or times, the access nodes (base stations) can respond to the data requests in a timely manner. On the other hand, in dense vehicle traffic, due to the high data demands of the vehicles in total, the access nodes will serve incoming requests with the available capacity. If the capacity of the base station is sufficient, incoming data requests from the vehicles will be transmitted on time. However, in case of high volume and number of demands in dense traffic, the limited resource will not be sufficient to serve in a timely manner for all demands. There is a need for careful design of the resource allocation and channel scheduling considering the priorities, delay, response time and turnaround time. Vehicles proximity, speed, channel conditions, and other environmental parameters form key design requirements for reliable and better service quality as well as efficient and fair use of the resources [26,35].

MAC Scheduling Algorithms: The base station collects the data requests demanded from the vehicles and schedules these data requests according to some key parameters. The data to be transmitted is obtained from the core network and these data requests are transmitted to the vehicles in the form of resource blocks in accordance with the 3GPP standard defined for LTE and 5G. The MAC scheduling algorithms determine which resource block the data requests will be located in. These scheduling algorithms perform resource block scheduling with a specific method using request information from the vehicles [12].

Fast Mobility and Frequent Topology Changes: Considering the provided services and the mobility characteristics of the mobile terminals in cellular networks, the high speed of the vehicles introduces additional challenges. In general, the speed of mobile terminals in urban areas is 1–3 m/s [33], while the speed of vehicles is 11–17 m/s [21]. In high-speed lanes in the urban areas and in highways, the vehicles' speed is much more faster. High speeds, overtaking, density variation depending on the space and time cause frequent changes in the vehicle topology. Frequent topology changes in the network also affect the resource management of the base stations. Additionally, the frequent and high volume of in-coming and out-going vehicles in the cell will introduce additional signaling and resource management overhead at the base station [40].

Inefficient Resource Utilization and High Latency: Frequent topology changes and high mobility of the vehicles introduce one more problem on the resource allocations. Service demands of the vehicles have to be given in a timely manner. Depending on the amount of data and the speed of the vehicle, the demands might not be met by a single access node, which requires multiple base stations handover the service along the route of the vehicle. For a single type of service of a single vehicle, the channel assignments (based on the resource block assignments in LTE, 4G and 5G) may comprise a number of base stations. For the high-speed vehicles, the connectivity duration with any particular base station will be relatively short. For example, in a high-speed highway with dense vehicle traffic, a single base station will not be able to provide all data demands to all vehicles in its vicinity (cell range). Demands can be satisfied with multiple base stations which are orchestrated to allocate resource in advance considering the mobility characteristics of the vehicles and the available resources. Therefore, there is a need for careful design on the allocation of the resource for such kind of conditions. It is crucially important in two aspects, in the utilization and the efficient use of the resources and in providing the service with QoS requirements defined in the standards. Because of the high mobility of the vehicles, the established communication links with the base stations may fail frequently due to the link impairments (e.g. based on the distance, vehicle speed in addition to the other environmental factors). All these will cause inefficient use of the resources in the network side and additional delay in the user side. Moreover, fairness problem arises contradicting with the efficient and the reliable communication metrics [60,62].

9.4 POSSIBLE SOLUTION APPROACHES

Most of the challenges defined in Section 9.3 are related to (1) the limited capacity of the base stations/core network, (2) varying data traffic demands dependent on the varying vehicle traffic which subjects to spatio-temporal characteristics of the vehicle traffic, (3) vehicle movement characteristics e.g. high speed, fast mobility, intermittent connectivity affecting the frequent handover and QoS metrics and (4) link impairments (dependent on the environment as well as vehicle mobility). While preserving the service quality for all vehicles, utilizing the network resources more efficiently are the major objectives under consideration. To meet these challenges and objectives, some algorithmic approaches without introducing more resource investment are suggested as follows: (1) Understanding the vehicle mobility patterns and spatio-temporal characteristics of the vehicle traffic contributes to design better scheduling algorithms. (2) Moreover, such analysis can aid in load balancing by sharing the load between heavy and light loaded areas/base stations. While using such load balancing approaches, meeting the QoS must be the primary concern in addition to resource utilization. Time-critical (ultra-low latency) and reliability critical (ultra-high reliability) services defined in 3GPP Release 14–16 [23,25] must be met without any disruption. (3) Predicting vehicle mobility would be used to improve the efficiency of the proposed solutions. The summary of the challenges and the possible solution approaches are given in Table 9.3.

One of the most important features of the V2X communications that distinguishes it from the mobile network is that the vehicles move on the road topology which therefore limits the mobility of the vehicles to the defined paths. This feature allows to predict the mobility of the vehicles among the route. Vehicle mobility patterns and spatio-temporal characteristic play a major role in the sense that the demands for all kinds of services can be estimated and precautions can be taken considering the current conditions (e.g. available capacity, time and reliability constraints). Estimating the vehicle routes and mobility aid in estimating the load of the vehicles' data demands,

Table 9.3
Challenges and Possible Solution Approaches

Challenges	Solution Approaches
- Access node placement for better QoS in V2X services/applications - Resource management for fast mobility and frequent topology changes	Considering; - The spatio-temporal movement patterns of the vehicles
- Resource allocation/channel scheduling in highly dense environments *- Efficient MAC scheduling algorithms*	*Considering;* *- The vehicles mobility patterns* *- The priorities, delay, response time and turnaround time of the demands* *- Vehicles proximity to the infrastructure, speed, channel conditions and other environmental parameters*
- Inefficient resource utilization and high latency	Considering; - orchestrated centralized and/or distributed solution comprised of local base stations and allocating resources in advance - The mobility characteristics of the vehicles and the available resources

which eventually aid in balancing the foreseen loads on the communication network. The load at high data demands (heavily loaded) can be shifted to the lightly loaded areas considering the available resource and latency/reliability requirements of the services.

Predicting vehicle mobility also aids in designing efficient MAC scheduling algorithms. By considering the speed of the vehicles, the link quality and the coverage duration in a cell, the vehicle data requests can be met by suitable MAC scheduling methods. This complementary approach helps to meet delay, reliability and throughput constraints. A similar approach that utilizes the vehicles' destination predictions for data offloading has been developed [42].

9.5 WORKLOAD OFFLOADING APPROACHES FOR V2X COMMUNICATIONS IN 5G

In the literature, workload offloading approaches are categorized into two types – data traffic offloading and computation offloading [64]. Data traffic workload offloading process can be done in three different ways; (1) centralized/cloud-based traffic offloading, (2) distributed based traffic offloading with access nodes and (3) selected vehicle-based traffic offloading.

9.5.1 COMPUTATION OFFLOADING

Nodes can upload computing tasks, including programs and data, to the cloud centers to solve over the cellular network. The cloud center fulfills the incoming computing task, and when these tasks are completed, the results are sent back to the nodes over the cellular network. Similarly, cloud centers can transmit their own computational tasks to the nodes using the cellular network. When the nodes have completed these tasks, they send the results back to the cloud center over the cellular network. Transferring tasks to the cloud center or the nodes and gathering the results lead to extra data traffic on the cellular network. Especially when data traffic is heavy, delays occur in the process of downloading the tasks and uploading the results to the cloud or the nodes.

Shi et al. proposed a computation offloading framework COSMOS [53] which is offloading the node computational request to a commercial cloud service provider. Before offloading, COSMOS estimates the offloading benefit of time parameters which are waiting time for connectivity, the transmission time of task and execution time of the task in the cloud. Task offloading occurs only if the offloading benefit value is greater than a threshold value. In the task allocation phase, three heuristic methods are used. The first method is that COSMOS master maintains the offloading requests sent by COSMOS clients and allocates these requests to the COSMOS server. The second method is that COSMOS client controls the workload value of the COSMOS servers and selects one randomly. The third method is that COSMOS master sends a list of the workload value of the COSMOS servers to the COSMOS clients. COSMOS client selects randomly a COSMOS server among that list.

In another study [7], vehicles near the traffic lights are used as computational resource in addition to the centralized cloud center. When vehicles stop at the traffic lights, vehicle information is collected by the traffic lights, and this information is sent to the cloud system. One or more vehicles near the traffic light are clustered with the traffic light and this cluster act as a vehicular cloud (VC) in the centralized cloud (CC). CC sends the computational task to the cluster head (CH) which is traffic light and CH shares this task between cluster members. When the cluster members complete their tasks, they send this information to the CH and the CH transmits the results gathered from vehicles to the CC. In this way, the workload in the cloud center is distributed among the vehicles. The purpose of the proposed architecture is to minimize the processing and network power consumed in a cloud operator's data center.

Authors in Ref. [41] studied the task scheduling problem in vehicular cloud (VC). In VC, vehicles are used as resources, and tasks are assigned to vehicles. If the vehicle leaves the VC, the task assigned to the vehicle must be transferred to another vehicle. This task transfer also adds an extra cost to the VC. For this reason, task scheduling is done by taking into account the vehicle information. The authors proposed algorithms – polynomial time approximation and greedy approximation – for a single task scheduling problem. The goal in this scheme is to ensure that the task is done without any interruptions if the task is assigned to the vehicle.

Authors in Ref. [27] proposed an algorithm related to wireless energy and data offloading using the Markovian Decision Process (MDP) framework. Task execution causes energy consumption. Since battery efficiency is an important parameter in mobile devices, this study focuses on whether the computational tasks will be executed locally on the mobile device or transmitted to other WiFi devices for executing. Local Execution model and WiFi Network and Device Execution Model are formulated based on the task energy consumption. Based on the offloading decision, the problem formulated and it is used in MDP framework.

9.5.2 DATA TRAFFIC WORKLOAD OFFLOADING

In the literature, studies for data traffic workload offloading are categorized as centralized [13,16, 39,43] and distributed [4,32,51,63,66] approaches. These studies generally focus on reducing the workload on the uplink/downlink channels, taking into account the demanded data size and packet delay parameters. In the distributed system, additionally the access node considers the workload densities of neighboring access nodes, and the access node decides which vehicle's data request will be responded. All or some of the requested data will be transferred to the other access node according to the vehicle direction and the requested data size. Studies based on these categorizations are summarized in this section. Finally, challenges and potential solution are summarized in a table (Table 9.4).

9.5.2.1 Centralized/Cloud-Based Workload Offloading

In the centralized system, the amount of workload on the access nodes will be calculated by the cloud/center using the information sent from the access nodes. The centralized system will decide which access node should be used for data transmission considering the access nodes' workload. Genetic algorithm (GA), linear programming (LP) or algorithmic metrics can be efficiently used for this purpose. In the distributed system, access nodes aim to balance the workload themselves locally by exchanging their load information. Each access node calculates its own workload considering data demands and vehicle data (position, speed, etc.), and shares this data with neighboring access nodes. The knowledge of local workload at neighboring access nodes allows the transfer of load from heavy access nodes to light or more appropriate ones.

Sangchul et al. [43] analyzed the Mobility load balancing (MLB) [44] algorithm with vehicular and pedestrian users. Rectangular and circle shape user mobility model is defined and tests are performed on this model. With a centralized approach, the workloads of eNBs are determined

Table 9.4
Offloading Challenges and Possible Solutions for V2X

Challenges	Possible Solutions
The limited capacity of the eNBs	Balancing the workload of the eNBs between neighboring eNBs
High mobility of the vehicles	*Predict the mobility of the vehicles*
Increased data demands because of the spatio-temporal characteristics of V2X	Offloading the data between vehicles or efficient RB allocations considering throughput, delay or fairness
Low link quality	*Efficient eNB placement considering the vehicle data traffic demands*

according to the number of physical resource blocks (PRB) used. Using the workload values of all eNBs, the standard deviation value is calculated. Load transfer occurs between two eNB if the eNB is 70% loaded and the load ratio difference is more than 10%. If workload transfer from eNB to other eNB is performed, the standard deviation is expected to decrease in case of a successful load transfer.

Both in a centralized and distributed system, it is aimed to send the requested data to the vehicle on time. However, in the case of dense data traffic, existing resources may not be able to meet these requests on time. In this case, it is necessary to determine which vehicle's demanded data request will be sent earlier. Some requests may not be sent on time because of the data scheduling strategy or requested data size. In this case, it should be decided which requests will not be sent. Therefore, the available resources need to be used efficiently.

Data demands transmitted to eNB by the vehicles are transmitted to the vehicles by MAC scheduling methods, taking into account the parameters of the vehicles, and placing the demanded data packets in a certain order in the resource blocks as shown in Figure 9.4. In this direction, while the measures such as delay, throughput, fairness, connectivity are taken into consideration, scheduling methods have been developed according to parameters such as SNRI (signal-to-noise ratio improvement) value, vehicle distance from eNB, demanded data size and deadline of the demanded request etc.

In the literature, there are some reviewed studies about the scheduling methods especially for the LTE Systems [12,15,47]. Zain et al. [38] compared the performance of Deficit Round Robin (DRR), Maximum Carrier to Interference Ratio (MAX C/I) and Proportional Fair (PF) MAC scheduling policies in the LTE-Advanced network. In detailed analysis of the performance they evaluated these three algorithms based on some parameters such as throughput, frame delay, packet delay variation (PDV), and they concluded that MAX C/I scheduling method outperforms all the other two scheduling policies.

9.5.2.2 Distributed Based Workload Offloading

One of the major problems in distributed methods is that multiple base stations transfer workloads to the same base station at the same time. In this case, the base station, which has a low workload, can be exposed to excessive workload or even a workload that is over its capacity. Therefore, if different base stations will transfer workloads to the same base station during the load transfer, the amount of load that other base stations want to transfer must be shared between each other. Therefore, in the distributed system, it is important that the base stations should communicate among themselves before workload transfer to other base station [16]. A distributed load balancing algorithm for downloading data from eNB is proposed. First, each node is assigned to an eNB with centralized based, and then eNBs calculate its workload using demanded data of nodes. Then each eNB shares

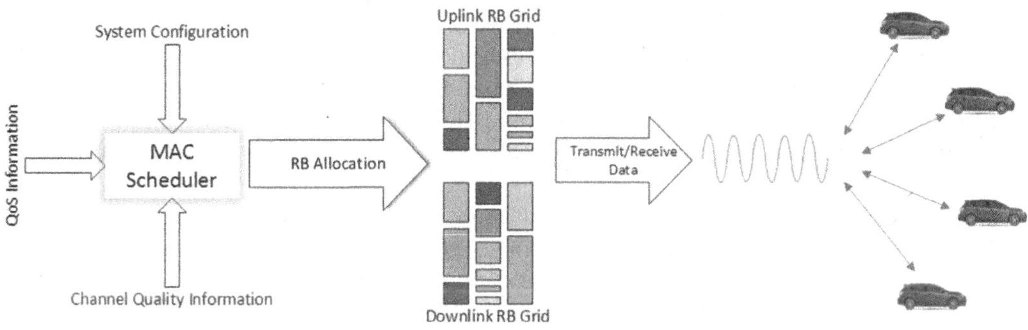

Figure 9.4 MAC scheduler.

the amount of workload with the other eNBs. Since the workload of other base stations is known, the node to which the base station is to be transmitted is determined as distributed. In the proposed algorithm, the logarithmic utility function is used for the utility value of the optimization. Downlink data rate and *SINR* values are used in the algorithm as parameters. Jain's fairness index [31] is used to evaluate the performance of the proposed method.

The authors in Ref. [66] focused on minimizing the cost of downloading content in the hybrid system. The vehicles calculate the cost that occurs when they download the data via WiFi network and cellular network. Authors offer two solutions which are Basic Meet Algorithm (BMA) and Time Slot Algorithm (TSA) to minimize the cost of downloading data. In BMA a cost calculation is made by dividing the road into small road segments, while in TSA cost is calculated by dividing the time into small periods. BMA and TSA algorithms were applied in a simulation environment and performance results were compared with directly downloaded data results from LTE network.

In Ref. [19], authors perform an offloading analysis based on a hybrid network system composed of LTE network and VANET network. The aim of the study is to examine whether the workload of the data requested by the vehicles in the LTE network can be offloaded. The vehicles transmit the requested data over the cellular network. In the cellular network, this is possible in three different ways for transmitting vehicle requests; (1) Transmission of data to the vehicle via RSU, so that the vehicle's request will be transmitted via V2V. (2) Transmitting the data to a relay vehicle and that relay vehicle uploads the data to the destination vehicle. (3) Direct link communication over LTE using the cellular network. This decision-making problem is tried to be solved by using max-min optimization. In this problem, V2V link quality and distance parameters are used. The maximum offloading data is considered as an optimization gain. The authors proposed an extension of this study [20] as VOPP (A VANET Offloading Potential Prediction Model) algorithm.

Newaz Ali et al. proposed a solution for access node load balancing which is cooperative load balancing (CLB) [8]. In rush hour, access node data demands are shared between neighborhood access nodes to equalize the load level between access nodes. Authors proposed an improved approach, modified CLB (MCLB) [42], additionally considering the vehicle's final destination.

9.5.2.3 Selected Vehicle-Based Workload Offloading

In the literature, there are many studies that aim to utilize mobile vehicles as access nodes to share the load among vehicles and access nodes. In these approaches, some vehicles are selected based on some criteria for offloading purposes, which generally are named as seed vehicles. Rather than access nodes' downlink transmissions individually to each vehicle, aggregated data of the vehicles is sent to the seed vehicle which is in a suitable state and is able to share the data with other vehicles with V2V communication. Thus, instead of using the cellular uplink/downlink communication link, the workload on the access node will be reduced by transferring the data using the sidelink of the vehicles. The main purpose of seed vehicle selection is to reduce the workload on the access node and also to cover more vehicles even some vehicles are not in the transmission range of the access node. Vehicles out of the communication range of access nodes can access the network over selected seed vehicles. Seed vehicle selection will be done by considering some parameters such as vehicle speed, vehicle position, demanded data type, demanded data size, request time/deadline, link quality, node degree (number of neighbors), etc. The literature [57] uses the number of neighbors, [39] uses demanded data size and [11] uses the distance between vehicles as the seed vehicle selection criteria.

Stanica et al. [57] proposed three heuristic offloading approaches for floating car data. In this study, each vehicle sends generated sensor data to the selected neighbor vehicle. And that selected vehicle collects all vehicle data to send to the base station (BS). The main parameter of seed vehicle selection for all three algorithms is the number of neighbors in the 1-hop communication range.

In Ref. [39], some vehicles are selected as seed vehicles to download data from the base station and disseminate the data to the other vehicle based on opportunistic communications. Seed vehicles and their neighbors' data demands are sent to the controller over BS. The controller uses these requests and determines content utility value for each vehicle considering vehicle neighbors (9.1). Content utility value for each vehicle is calculated based on the vehicle neighbors' user data interest rate (I_{k,o^*}). Then the highest content utility value vehicle is selected as a seed vehicle. For measuring the performance of the algorithm, it uses the content utility rate (9.2).

$$\text{Utility}(v_i, o^*) = \sum_{v_k \varepsilon \mathcal{L}_{v_i}} I_{k,o^*} \tag{9.1}$$

where O^*: content object, V_i: Vehicle i, L: neighbor list of vehicle i, I_{k,o^*}: interests rate.

$$\text{Content Utility Rate} = \frac{\sum\limits_{o_j \text{ is received by } v_i} I_{i,j}}{\sum I_{i,j}} \tag{9.2}$$

P. Salvo et al. [49] proposed a cluster-based offloading approach. Cluster head (CH) is being selected to transmit the collected messages to the access node. Two parameters are used for selecting a CH, the distance between the neighbor vehicles and whether the message of its own previously transmits by another vehicle. Using these parameters, the CH selection is made for the 1-hop neighborhood. CH collects vehicle data over 1-hop communication sidelink and transmits collected data to the access node over uplink.

9.6 SUMMARY

As one of the main enabling technologies of ITS, V2X communications enable various services and applications for the connected and autonomous vehicles. Real-time data collected from vehicles via V2X communication will provide real-time processing advantage at the edge and ITS management center. These will enrich ITS services and applications with real-time support. V2X communications and services will transform cities into smart cities. On the other hand, with the growth in the number of autonomous vehicles moving around the cities, the enormous size of data will be processed and exchanged among the vehicles and the network. This brings the workload and optimization issues in the network in various aspects. This chapter has provided the challenges and possible solution approaches for workload offloading. A survey on offloading methods is presented to overview the studies for Connected and Autonomous Vehicles in Smart Cities. Analyzing the vehicles' mobility patterns, extracting parameters related to the vehicle traffic and data traffic will provide insights for the optimizations of the resources. This chapter presents an overview of the possible optimization approaches.

REFERENCES

1. 3GPP TR 22.886 V16.2.0. Technical specification group services and system aspects; study on enhancement of 3GPP support for 5G V2X services (release 16). 21-December-2018.
2. 3GPP TS 22.186 V15.4.0. 5G; Service requirements for enhanced V2X scenarios (release 15). 14-June-2019.
3. The 3rd Generation Partnership Project. https://www.3gpp.org/. (Date last accessed 08-January-2020).
4. Fakhar Abbas and Pingzhi Fan. A hybrid low-latency D2D resource allocation scheme based on cellular V2X networks. In *2018 IEEE International Conference on Communications Workshops (ICC Workshops)*, Kansas City, MO, pp. 1–6, IEEE, 2018.
5. Khadige Abboud, Hassan Aboubakr Omar, and Weihua Zhuang. Interworking of DSRC and cellular network technologies for V2X communications: A survey. *IEEE Transactions on Vehicular Technology*, 65(12):9457–9470, 2016.
6. Zeeshan Ahmed, Saba Naz, and Jamil Ahmed. Minimizing transmission delays in vehicular ad hoc networks by optimized placement of road-side unit. *Wireless Networks*, 26(4):2905–2014, 2020.

7. Amal A Alahmadi, Ahmed Q Lawey, Taisir EH El-Gorashi, and Jaafar MH Elmirghani. Distributed processing in vehicular cloud networks. In *2017 8th International Conference on the Network of the Future (NOF)*, London, pp. 22–26, IEEE, 2017.
8. G. G. Md. Nawaz Ali, Edward Chan, and Wenzhong Li. On scheduling data access with cooperative load balancing in vehicular ad hoc networks (VANETs). *The Journal of Supercomputing*, 67(2):438–468, 2014.
9. Noura Aljeri and Azzedine Boukerche. Movement prediction models for vehicular networks: An empirical analysis. *Wireless Networks*, 25(4):1505–1518, 2019.
10. Oluwatosin Ahmed Amodu, Mohamed Othman, Nor Kamariah Noordin, and Idawaty Ahmad. Transmission capacity analysis of relay-assisted D2D cellular networks with M2M coexistence. *Computer Networks*, 164:106887, 2019.
11. Muhammed Nur Avcil and Mujdat Soyturk. Performance evaluation of V2X communications and services in cellular network with a realistic simulation environment. In *2019 1st International Informatics and Software Engineering Conference (UBMYK)*, Ankara, Turkey, pp. 1–6, November 2019.
12. Satheesh Monikandan Balakrishnan, A Sivasubramanian, and SPK Babu. A review of MAC scheduling algorithms in LTE system. *International Journal on Advanced Science, Engineering and Information Technology*, 7(3):1056–1068, 2017.
13. Benjamin Baron, Prométhée Spathis, Hervé Rivano, Marcelo Dias de Amorim, Yannis Viniotis, and Mostafa H Ammar. Centrally controlled mass data offloading using vehicular traffic. *IEEE Transactions on Network and Service Management*, 14(2):401–415, 2017.
14. Federico Boccardi, Robert W Heath, Angel Lozano, Thomas L Marzetta, and Petar Popovski. Five disruptive technology directions for 5G. *IEEE Communications Magazine*, 52(2):74–80, 2014.
15. Francesco Capozzi, Giuseppe Piro, Luigi Alfredo Grieco, Gennaro Boggia, and Pietro Camarda. Downlink packet scheduling in LTE cellular networks: Key design issues and a survey. *IEEE Communications Surveys and Tutorials*, 15(2):678–700, 2012.
16. Diego Castro-Hernandez and Raman Paranjape. A distributed load balancing algorithm for LTE/LTE-A heterogeneous networks. In *2015 IEEE Wireless Communications and Networking Conference Workshops (WCNCW)*, New Orleans, LA, pp. 380–385, March 2015.
17. Shanzhi Chen, Jinling Hu, Yan Shi, Ying Peng, Jiayi Fang, Rui Zhao, and Li Zhao. Vehicle-to-everything (v2x) services supported by LTE-based systems and 5G. *IEEE Communications Standards Magazine*, 1(2):70–76, 2017.
18. SAE On-Road Automated Driving Committee et al. SAE J3016. Taxonomy and definitions for terms related to driving automation systems for on-road motor vehicles. Technical report, SAE International, 2016, (revised June 2018).
19. Ghayet el Mouna Zhioua, Houda Labiod, Nabil Tabbane, and Sami Tabbane. VANET inherent capacity for offloading wireless cellular infrastructure: An analytical study. In *2014 6th International Conference on New Technologies, Mobility and Security (NTMS)*, Dubai, pp. 1–5, March 2014.
20. Ghayet el Mouna Zhioua, Jun Zhang, Houda Labiod, Nabil Tabbane, and Sami Tabbane. VOPP: A VANET offloading potential prediction model. In *2014 IEEE Wireless Communications and Networking Conference (WCNC)*, Istanbul, Turkey, pp. 2408–2413, April 2014.
21. Eva Ericsson. Variability in urban driving patterns. *Transportation Research Part D: Transport and Environment*, 5(5):337–354, 2000.
22. ETSI TR 102 638 V1.1.1: "Intelligent Transport Systems (ITS); Vehicular Communications; Basic Set of Applications; Definitions". 29-June-2009.
23. ETSI TR 121 914 V14.0.0: "Digital cellular telecommunications system (Phase 2+) (GSM);Universal Mobile Telecommunications System (UMTS); LTE; 5G; Release description; Release 14 (3GPP TR 21.914 version 14.0.0 Release 14)". 28-June-2018.
24. ETSI TS 136 211 V13.10.0: "LTE; Evolved Universal Terrestrial Radio Access (E-UTRA); Physical channels and modulation (3GPP TS 36.211 version 13.10.0 Release 13)". 16-July-2018.
25. ETSI TS 138 211 V15.4.0: "5G; NR; Physical channels and modulation (3GPP TS 38.211 version 15.4.0 Release 15)". 2019-04-18.
26. Jin Gao, Muhammad RA Khandaker, Faisal Tariq, Kai-Kit Wong, and Risala T Khan. Deep neural network based resource allocation for V2X communications. In *2019 IEEE 90th Vehicular Technology Conference (VTC2019-Fall)*, Honolulu, Hawaii, pp. 1–5, IEEE, 2019.

27. Jooncherl Ho, Jing Zhang, and Minho Jo. Selective offloading to WiFi devices for 5G mobile users. In *2017 13th International Wireless Communications and Mobile Computing Conference (IWCMC)*, Spain, pp. 1047–1054, IEEE, 2017.
28. Liuwei Huo and Dingde Jiang. Stackelberg game-based energy-efficient resource allocation for 5G cellular networks. *Telecommunication Systems*, 72(3):377–388, 2019.
29. IEEE Standard 1455-1999. IEEE standard for message sets for vehicle/roadside communications. September 19, 1999.
30. ETSI ITS. G5 standard—Final draft ETSI ES 202 663 V1.1.0, Intelligent Transport Systems (ITS); European profile standard for the physical and medium access control layer of Intelligent Transport Systems operating in the 5 GHz frequency band. Technical report, Technical report ETSI, 2011.
31. Rajendra K Jain, Dah-Ming W Chiu, and William R Hawe. A quantitative measure of fairness and discrimination. *Eastern Research Laboratory, Digital Equipment Corporation*, Hudson, MA, 1984.
32. Yang Jinglin, QIN Huabiao, and Xu Ruoqian. Opportunistic mobile data offloading using vehicle movement prediction (DOVP). In *2018 IEEE 14th International Conference on Control and Automation (ICCA)*, Anchorage, AK, pp. 217–222, IEEE, 2018.
33. Robert V Levine and Ara Norenzayan. The pace of life in 31 countries. *Journal of Cross-Cultural Psychology*, 30(2):178–205, 1999.
34. IHS Markit. The connected car, https://ihsmarkit.com/topic/autonomous-connected-car.html. (Date last accessed 14-October-2019).
35. Ahlem Masmoudi, Kais Mnif, and Faouzi Zarai. A survey on radio resource allocation for V2X communication. *Wireless Communications and Mobile Computing*, 2019:12, 2019.
36. Rudolf Mathar and Thomas Niessen. Optimum positioning of base stations for cellular radio networks. *Wireless Networks*, 6(6):421–428, 2000.
37. Rudolf Mathar and Michael Schmeink. Optimal base station positioning and channel assignment for 3G mobile networks by integer programming. *Annals of Operations Research*, 107(1-4):225–236, 2001.
38. Aini Syuhada Md Zain, Mohd Fareq Abd. Malek, Mohamed Elshaikh, Normaliza Omar, and Abadal-Salam T. Hussain. Performance analysis of scheduling policies for VoIP traffic in LTE-Advanced network. In *2015 International Conference on Computer, Communications, and Control Technology (I4CT)*, Sarawak, Malaysia, pp. 16–20, April 2015.
39. Farouk Mezghani, Riadh Dhaou, Michele Nogueira, and André-Luc Beylot. Offloading cellular networks through V2V communications: How to select the seed-vehicles? In *2016 IEEE International Conference on Communications (ICC)*, Kuala, pp. 1–6, May 2016.
40. Mujahid Muhammad and Ghazanfar Ali Safdar. Survey on existing authentication issues for cellular-assisted V2X communication. *Vehicular Communications*, 12:50–65, 2018.
41. Mahmudun Nabi, Robert Benkoczi, Sherin Abdelhamid, and Hossam S Hassanein. Resource assignment in vehicular clouds. In *2017 IEEE International Conference on Communications (ICC)*,Paris, France, pp. 1–6, IEEE, 2017.
42. G. G. Md Nawaz Ali, Md. Abdus Salim Mollah, Syeda Khairunnesa Samantha, and Saifuddin Mahmud. An efficient cooperative load balancing approach in RSU-based Vehicular Ad Hoc Networks (VANETs). In *2014 IEEE International Conference on Control System, Computing and Engineering (ICCSCE 2014)*, Penang, Malaysia, pp. 52–57, November 2014.
43. Sangchul Oh, Hongsoog Kim, and Yeongjin Kim. User mobility impacts to mobility load balancing for self-organizing network over LTE system. In *2018 14th International Conference on Advanced Trends in Radioelecrtronics, Telecommunications and Computer Engineering (TCSET)*, Slavske, pp. 1082–1086, February 2018.
44. Sangchul Oh, Hongsoog Kim, Jeehyeon Na, Yeongjin Kim, and Sungoh Kwon. Mobility load balancing enhancement for self-organizing network over LTE system. In: Olga Galinina, Sergey Balandin, and Yevgeni Koucheryavy, editors, *Internet of Things, Smart Spaces, and Next Generation Networks and Systems*, Springer: Cham, pp. 205–216, 2016.
45. Big Data on Wheels. https://www.statista.com/chart/8018/connected-car-data-generation/. (Date last accessed 14-October-2019).
46. Bahadir K Polat and Mujdat Soyturk. An alternative approach to mobility analysis in vehicular ad hoc networks. In *2016 IEEE Symposium on Computers and Communication (ISCC)*, Messina, Italy, pp. 244–249, IEEE, 2016.

47. S. Radhakrishnan, S. Neduncheliyan, and K.K. Thyagharajan. A review of downlink packet scheduling algorithms for real time traffic in LTE-advanced networks. *Indian Journal of Science and technology*, 9(4):1–5, 2016.
48. 5G Communication Automotive Research and innovation. https://5gcar.eu/. (Date last accessed 24-February-2020).
49. Pierpaolo Salvo, Ion Turcanu, Francesca Cuomo, Andrea Baiocchi, and Izhak Rubin. LTE floating car data application off-loading via VANET driven clustering formation. In *2016 12th Annual Conference on Wireless On-demand Network Systems and Services (WONS)*, Cortina, Italy, pp. 1–8, January 2016.
50. Matheus Ferraroni Sanches, Allan M. de Souza, Wellington Lobato, and Leandro A. Villas. Optimizing infrastructure placement with genetic algorithm: A traffic management use case. In *2019 IEEE Latin-American Conference on Communications (LATINCOM)*, Salvador, Brazil, pp. 1–6, November 2019.
51. Miguel Sepulcre and Javier Gozalvez. Context-aware heterogeneous V2X communications for connected vehicles. *Computer Networks*, 136:13–21, 2018.
52. Vivek Sethi and Narottam Chand. A destination based routing protocol for context based clusters in VANET. *Communications and Network*, 9(3):179–191, 2017.
53. Cong Shi, Karim Habak, Pranesh Pandurangan, Mostafa Ammar, Mayur Naik, and Ellen Zegura. COSMOS: Computation offloading as a service for mobile devices. In *Proceedings of the 15th ACM International Symposium on Mobile Ad Hoc Networking and Computing*, Boston, MA, USA, pp. 287–296, ACM, 2014.
54. Xiao Shu and Xining Li. Link failure rate and speed of nodes in wireless network. In *2007 International Conference on Wireless Communications, Networking and Mobile Computing*, Shanghai, China, pp. 1441–1444, IEEE, 2007.
55. Pranav Kumar Singh, Sunit Kumar Nandi, and Sukumar Nandi. A tutorial survey on vehicular communication state of the art, and future research directions. *Vehicular Communications*, 18:100164, 2019.
56. Mujdat Soyturk, Khaza Newaz Muhammad, Muhammed Nur Avcil, Burak Kantarci, and Jeanna Matthews. Chapter 8 - From vehicular networks to vehicular clouds in smart cities. In: Mohammad S Obaidat and Petros Nicopolitidis, editors, *Smart Cities and Homes*, Morgan Kaufmann, Boston, MA, pp. 149 – 171, 2016.
57. Razvan Stanica, Marco Fiore, and Francesco Malandrino. Offloading floating car data. In *2013 IEEE 14th International Symposium on "A World of Wireless, Mobile and Multimedia Networks" (WoWMoM)*, Madrid, Spain, pp. 1–9, June 2013.
58. Shikha Tayal, P.K. Garg, and Sandip Vijay. Optimization models for selecting base station sites for cellular network planning. In: Jayanta Kumar Ghosh and Irineu da Silva, editors, *Applications of Geomatics in Civil Engineering*. Springer, Berlin, pp. 637–647, 2020.
59. Sahrish Khan Tayyaba and Munam Ali Shah. Resource allocation in SDN based 5G cellular networks. *Peer-to-Peer Networking and Applications*, 12(2):514–538, 2019.
60. Livinus Tuyisenge, Marwane Ayaida, Samir Tohme, and Lissan-Eddine Afilal. Handover mechanisms in Internet of Vehicles (IoV): Survey, trends, challenges, and issues. In *Global Advancements in Connected and Intelligent Mobility: Emerging Research and Opportunities*. IGI Global, Hershey PA, pp. 1–64, 2020.
61. Automotive Industry Trends: IoT Connected Smart Cars & Vehicles. http://www.businessinsider.com/internet-of-things-connected-smart-cars-2016-10. (Date last accessed 14-December-2019).
62. Vladimir Vukadinovic, Krzysztof Bakowski, Patrick Marsch, Ian Dexter Garcia, Hua Xu, Michal Sybis, Pawel Sroka, Krzysztof Wesolowski, David Lister, and Ilaria Thibault. 3GPP C-V2X and IEEE 802.11 p for vehicle-to-vehicle communications in highway platooning scenarios. *Ad Hoc Networks*, 74:17–29, 2018.
63. Shangguang Wang, Tao Lei, Lingyan Zhang, Ching-Hsien Hsu, and Fangchun Yang. Offloading mobile data traffic for QoS-aware service provision in vehicular cyber-physical systems. *Future Generation Computer Systems*, 61:118–127, 2016.
64. Dianlei Xu, Yong Li, Xinlei Chen, Jianbo Li, Pan Hui, Sheng Chen, and Jon Crowcroft. A survey of opportunistic offloading. *IEEE Communications Surveys and Tutorials*, 20(3):2198–2236, 2018.
65. Chia-Hao Yu, Klaus Doppler, Cassio B Ribeiro, and Olav Tirkkonen. Resource sharing optimization for device-to-device communication underlaying cellular networks. *IEEE Transactions on Wireless communications*, 10(8):2752–2763, 2011.

66. Dongdong Yue, Peng Li, Tao Zhang, Junshan Cui, Yu Jin, Yu Liu, and Qin Liu. Cooperative content downloading in hybrid VANETs: 3G/4G or RSUs downloading. In *2016 IEEE International Conference on Smart Cloud (SmartCloud)*, New York, pp. 301–306, IEEE, 2016.
67. Sherali Zeadally, Ray Hunt, Yuh-Shyan Chen, Angela Irwin, and Aamir Hassan. Vehicular ad hoc networks (VANETS): Status, results, and challenges. *Telecommunication Systems*, 50(4):217–241, 2012.

10 Connected Unmanned Aerial Vehicles for Flexible Coverage, Data Gathering and Emergency Scenarios

Giacomo Segala
University of Trento

Riccardo Bassoli
TU Dresden

Fabrizio Granelli
University of Trento

Frank H. P. Fitzek
TU Dresden

CONTENTS

10.1 INTRODUCTION

Unmanned aerial vehicles are increasingly gaining the attention of industry, academia and common enthusiasts for their freedom of movement, reduced costs and flexible usage. Indeed, in recent years, the introduction of low-cost and open microcontrollers has made it possible to experience giant leaps in manoeuvrability, automated piloting and battery lifetime.

In the framework of smart cities and smart regions, availability of drone or other aerial platforms can enable new services, especially where flexible and fast deployment represent the major requirement.

This chapter proposes to describe potential scenarios for application of drones in smart city scenarios. After an introductory description of the drones and their major features, the chapter will focus on the following application scenarios: (1) improved/flexible data coverage, (2) data gathering/acquisition and (3) emergency scenarios.

Connectivity availability represents one of the baseline features of a smart city, since most "smart" services are based on the availability of a data connection with their customers to exchange information in real time. The need to the high availability of such service can represent a big challenge in planning the capacity, due to the presence of temporary events that could drastically modify the typical geographical distribution of users or terminals in an area. Drones or other aerial platforms can enable the deployment of additional on-demand capacity, by bringing cellular or satellite-equipped network nodes in the desired locations and enabling capacity offload and reduce stress on the existing infrastructure. Different networking paradigms exist in this deployment scenario, also supporting multi-hopping and multi-tiered solutions, which will be reviewed in the corresponding section of this chapter.

Another interesting scenario where UAVs can provide several benefits is in gathering or acquiring data on the field by defining proper flying paths in an area. Real-time or delay-tolerant connectivity will enable to use aerial platforms as low cost and effective means to avoid the need for bringing high-cost and long-term networking infrastructure, especially in the case of several sensors spread over a large area or partially moving scenarios.

Finally, a key area in smart cities [1–3] is to effectively manage emergency situations. In emergency situations, the availability of a communication infrastructure is key to enable coordination among the involved emergency personnel as well as to exchange information about the operating unknown environment. Also in this scenario, UAVs could represent a feasible solution, both to bring on-demanding connectivity as well as enabling the deployment of movable sensors to collect data about the environment.

This chapter addresses the analysis of the performance of networks of drones for smart city application scenarios. The chapter is organized as follows: after this introduction, the next section is providing a detailed overview of the existing architectures for building networks of drones. The following section introduces the simulation environment and provides a detailed performance evaluation of the different architectures. Finally, the last section concludes the paper with comments and final remarks.

10.2 ARCHITECTURES FOR NETWORKS OF DRONES

This section discusses the architectures defined in the literature for building a network of drones or Unmanned Aerial Vehicles (UAVs).

Drones, also known as Unmanned Aerial Vehicles (UAVs), represent an emerging technology in the field of telecommunications which can play an important role in this direction. Drones can be deployed easily and rapidly in order to guarantee connectivity to users in unexpected or time-limited missions, such as emergency scenarios due to natural disasters where operators need to work in difficult situations or as a response to a local traffic hotspot. Moreover, drones can be extremely effective as "movable sinks" to gather information from sensors deployed on the ground.

As single-large UAV systems are not particularly efficient to provide coverage above a certain area, and thanks to the proliferation of small-scale and cost-effective drones, multi-UAV systems, typically known as drone networks, are being explored to provide wireless coverage. Thanks to their flexibility and efficiency, networks of drones are expected to play a fundamental role in communication systems of future smart cities.

In the literature [4], four different architectures of drone networks have been suggested. Those architectures will be detailed in the next sections. All of them are appropriate for a typical scenario of future smart cities, where users/citizen require to communicate between each other or towards a server on the Internet in the absence of dedicated infrastructure, while the network of drones flies above the area providing Wi-Fi connectivity. Two different communication patterns can thus be identified: user-to-user and user-to-server.

Without losing generality, we will integrate the network of drones within the existing mobile network infrastructure (if any) by connecting the network of drones to a base station (eNB for 4G LTE, or gNB for 5G), which will play the role of ground control station (GCS). A server will be connected to the Internet through the mobile network infrastructure (Evolved Packet Core). Although GCS typically provides important command and control tasks, in this kind of scenario it simply plays the role of relay, forwarding data packets which pass through it.

Moreover, this situation can also include the scenario of emergency, natural disasters or any situation in which the mobile network infrastructure is not available, since a temporary eNB or gNB can be deployed either by using another movable Aerial Platform (e.g. balloons, bigger UAVs with longer lifetime) or by a satellite node. Figure 10.1 depicts the considered scenario in this chapter.

Any drone network is characterized by UAV-to-UAV communications, where UAVs communicate between each other in order to perform a specific application, and UAV-to-Infrastructure communication where drones communicate with a fixed infrastructure. In most of the cases the infrastructure is represented by the Ground Control Station (GCS), but as described above it could be also a satellite or a high-altitude platform (HAP) such as a balloon.

Figure 10.2 provides a description of the taxonomy of the networks of drones. Four different architectures have been proposed in the literature: Centralized-Mesh network, UAV Ad Hoc network, Multi-Group UAV network and Multi-Layer UAV Ad Hoc network. Centralized-mesh

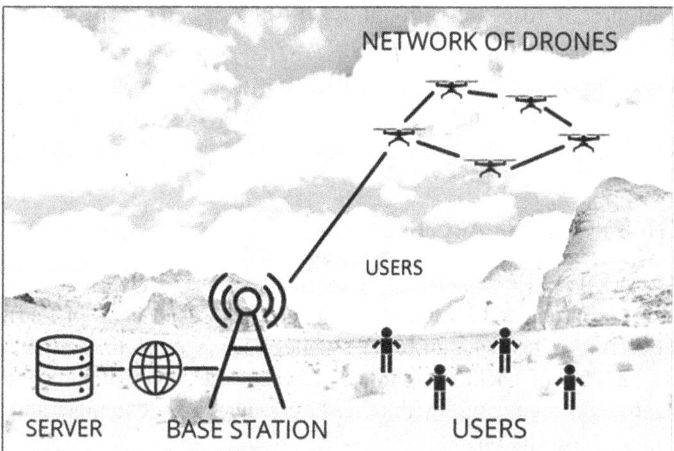

Figure 10.1 A pictorial diagram of the considered scenario: users are connected to the base station via WiFi connectivity provided by drones.

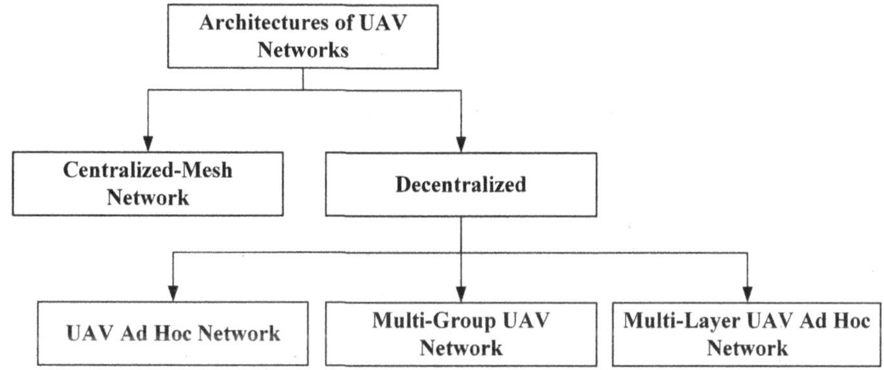

Figure 10.2 Taxonomy of the network of drones architectures.

network represents a variant with respect to the proposed centralized architecture in [4]. UAV Ad Hoc, Multi-Group UAV and Multi-Layer UAV Ad Hoc networks represent decentralized architectures: they contain an ad hoc section and are not fully infrastructure-based. They are categorized as FANETs (Flying Ad Hoc Networks), specific drone networks based on some same concepts of MANETs and currently investigated in the world of research. Each architecture will be described in the next sections.

10.2.1 CENTRALIZED-MESH NETWORK ARCHITECTURE

Unlike the centralized architecture proposed in [1], a hybrid architecture called Centralized-Mesh network has been considered. As in the original architecture, all drones of Centralized-Mesh network are directly attached to the base station, which plays the role of a central node (Figure 10.3).

The Wi-Fi network established between drones and users is a mesh network, so a proper network interface card (NIC) with mesh functionalities following IEEE 802.11s standard is mounted on drones. Users use a mesh NIC as drones, thus unlike the other architectures they are directly involved in forwarding functionalities, collaborating with drones to form the mesh network in order to provide higher reliability within the network.

As a result of new network design, drones and users can communicate directly between each other or passing through other mesh nodes. The communication includes the eNB/gNB only if the destination is not directly reachable through a Wi-Fi mesh network by the sender or when the destination is a server on the Internet.

As users could be connected to more than one drone, the closest one in terms of number of hops is chosen as a gateway, while, if the number of hops is the same, the closest one in terms of Euclidean distance is selected.

10.2.2 UAV Ad Hoc NETWORK ARCHITECTURE

The UAV ad hoc network represents the simplest architecture derived by traditional ad hoc networks (Figure 10.4). No pre-existing infrastructure is expected and all drones of the network mount a specific ad hoc network interface card in order to interconnect them to form an ad hoc network. Only a drone plays the role of backbone/gateway towards the infrastructure, thus it mounts two different radios, one for communicating with other UAVs in an ad hoc manner and the other one for communicating with the base station.

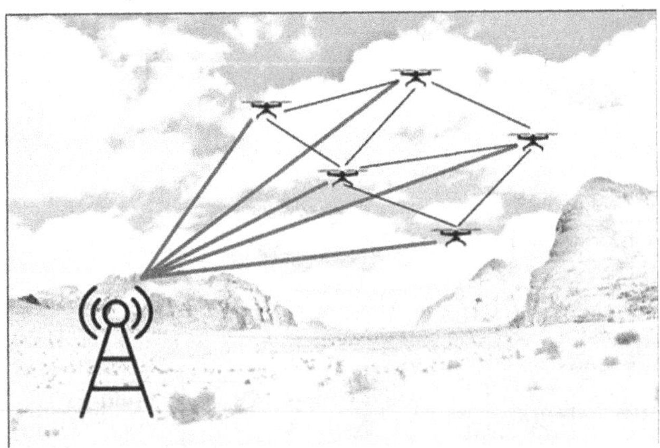

Figure 10.3 Centralized mesh node architecture.

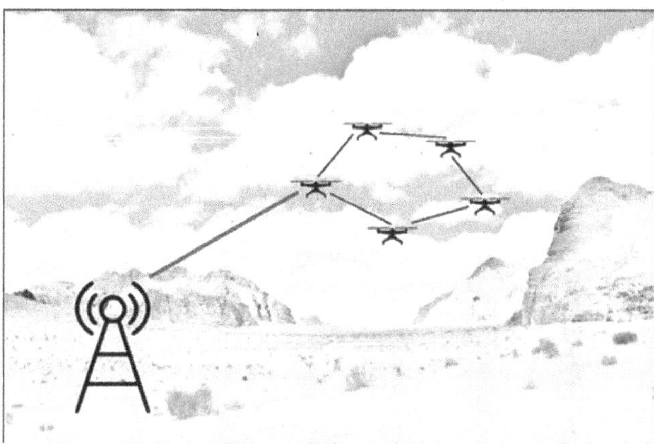

Figure 10.4 UAV ad hoc network architecture.

Moreover, all drones mount a radio for providing Access Point (AP) functionalities in order to guarantee connectivity to users. As a result of this network design, user-to-user communications do not involve the base station but are directly managed by drone network.

10.2.3 MULTI-GROUP UAV NETWORK ARCHITECTURE

Multi-Group UAV network can be considered as a combination of UAV ad hoc network and a purely centralized network. UAV ad hoc networks are attached to the base station, thus enabling more groups of drones to be interconnected within the architecture. The total amount of drones is distributed among a certain number of groups of drones (typically two or four groups) and each single UAV ad hoc network flies above a different subarea, so guaranteeing Wi-Fi connectivity above the entire considered area in a distributed manner. A drone which takes the role of gateway towards infrastructure is chosen within each UAV ad hoc network.

Two different kinds of communications are performed by drones:

Intra-Group Communications (communications between UAVs within the same group), which are performed within the ad hoc network to which UAVs belong to;

Inter-Group Communications (communications between UAVs belonging to different groups), which are performed through the respective backbone UAVs and the GCS.

As a result, intra-group communications are performed if users belong to the same UAV ad hoc network, while inter-group communications are needed if users use the service in different groups. Drones provide APs functionalities in order to guarantee connectivity to the users using different network topologies, such as multi-group (Figure 10.5) and multi-layer (Figure 10.6).

10.2.4 MULTI-LAYER UAV Ad Hoc NETWORK ARCHITECTURE

A multi-layer UAV Ad Hoc network consists of more UAV ad hoc networks interconnected among them in an ad hoc manner and not in a centralized way (as in a Multi-Group UAV network). As in the case of the previous architecture, the Multi-Layer UAV Ad Hoc network provides connectivity in a distributed way. As a result of network design, this architecture is characterized by two layers of UAVs:

The lower layer is given by a UAV ad hoc network which is formed by the UAVs within an individual group;

The upper-layer UAV ad hoc network is composed of the backbone UAVs of all groups.

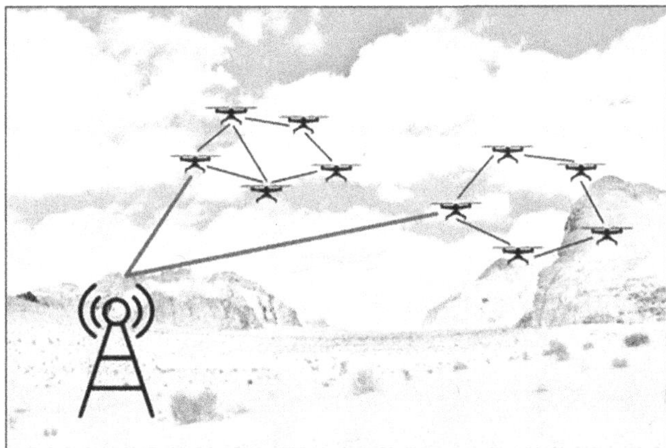

Figure 10.5 Multi-group UAV network architecture.

Figure 10.6 Multi-layer UAV network architecture.

Only one drone among the backbone nodes is attached to the eNB/gNB. As backbone drones are connected in an ad hoc manner among them, they need to be not so far from each other in order to prevent lack of connectivity between groups as far as possible. Accordingly, backbone nodes are expected to be positioned in or around specific geographical locations.

As in the UAV Ad Hoc network, communications between users are directly performed by drones and do not involve the base station. Again, drones provide AP functionalities to users.

10.3 SIMULATION SETUP

All architectures described above have been implemented and tested within the Network Simulator 3 (ns-3), a discrete-event network simulator widely used for research and educational purposes.

As ns-3 does not provide any specific class for modelling objects like drones, they have been modelled as generic nodes, to which different network cards have been installed according to architecture specifics and functionalities.

The main parameters of the WiFi network established between drones and users are listed in Table 10.1

Table 10.1
Parameters of Simulation

Parameter	Value
Standard	IEEE 802.11a (centralized) and IEEE 802.11ac (others)
Rate manager	Constant rate Wi-Fi manager
Transmission mode	OFDM rate 6 Mbps
Pathloss model	Log distance (default) and range propagation
Propagation delay model	Constant speed (default)

A fixed OFDM modulation characterized by a transmission bitrate of 6 Mbps has been used as transmission mode for data and control packets, in order to avoid fluctuations and grant communication ranges of tens of meters.

The network configurations are described in Table 10.2. The number of drones and users have been chosen in order to analyze cases where there are more drones than users and vice versa. In this way, it is possible to evaluate the performance according to the ratio between the number of users and drones, generalizing the scenario which has been investigated. Different values of drones and users have been chosen for Centralized-Mesh architecture. This is due to the high computational load of this kind of architecture as all drones are directly attached to eNB.

As previously reported, User-to-User (UU) and User-to-Server (US) communications are tested in two different scenarios: drones in motion and stationary users or stationary drones and users in motion.

Random Walk 2D mobility model has been applied both in the case of users and drones in case they move. Random Walk represents one of the most used mobility models for studying mobile networks, thus it is reasonable to test it also on drone networks. Moreover, Random Walk 2D implies nodes move not so far from their initial position: as drones take the role of aerial access points and relays, there is no reason they need to be subject to a high mobility.

Considering the need for optimizing the overall coverage area, drones are distributed in a uniform manner: the considered area is divided into several sub-parts and the same number of drones is distributed on each sub-area. Moreover, within each sub-area, positions of drones are chosen randomly, but leaving a minimum distance between a drone and another (40 m). This makes drones distribution more effective and realistic, in order to provide a proper evaluation of the performance of different architectures.

As drones cannot be too close to the ground due to specific regulations concerning public safety but, on the other hand, they cannot be too high in order to offer a good coverage to users; the altitude of drones is chosen randomly between 20 and 60 m, thus drones are located at different heights.

Table 10.2
Parameters of Network Configuration

Variable	Value
Number of drones	4,6,8,10,12 (*centralized*) and 10,15,20,25,30 (*others*)
Number of users	2,4,6,8,10 (*centralized*) and 4,8,12,16,20 (*others*)
Communication types	User-to-User (UU) and user-to-Server (US)
Mobility scenarios	Drones in motion/stationary users and stationary drones/users in motion

Table 10.3
Variables of Simulations

Parameter	Value
Area	350 m^2
Altitude of drones	$20 \leq h \leq 60$ m
Simulation time	130 seconds
Data rate	1 Mbps
Off time	50 seconds
On time	10 seconds
Application protocol	User Datagram Protocol (UDP)
Packet size	512 B

Small drones within a network can work for about 10–15 minutes in the real world. However, running simulations for 10 minutes is unfeasible, thus, an approximation has been necessary. Setting the simulation time equal to 130 seconds is sufficient to evaluate the performance of architectures and make a general comparison among them.

Users send packets with a duty cycle of 10 seconds every minute, while they remain idle for the remaining 50 seconds. Packet size is set to 512 bytes and data rate to 1 Mbps.

The values of the main parameters are summarized in Table 10.3.

10.4 PERFORMANCE EVALUATION AND DISCUSSION

Achieved results are plotted in terms of the ratio between the number of users and drones, making them representative of scenarios characterized by a different area dimension or the number of drones/number of users.

Figure 10.7 represents single transmission performances considering drones in motion, stationary users and user-to-server communication.

Figure 10.8 shows single transmission performances considering drones in motion, stationary users and user-to-user communication.

Figure 10.9 depicts performances considering users in motion, stationary drones and user-to-server communication.

Figure 10.10 depicts single transmission performances considering users in motion, stationary drones and user-to-user communication.

Considering the mobility scenario where drones are in motion while users are stationary at the same locations, performance in terms of Packet Delivery Ratio (PDR) is approximately equivalent among all the architectures of ad hoc nature in user-to-server communications. Only Centralized-Mesh architecture guarantees much higher performance thanks to the direct connection of drones to the base station, as the communication towards the server requires a lower number of hops.

In user-to-user communications, all the architectures provide equivalent values of PDR when there is a good availability of drones while Centralized-Mesh is much more robust in a scenario where there are more users than drones, as users provide forwarding functionalities.

Increasing the number of groups, the performance of Multi-Layer UAV Ad Hoc network gets worse: this is due to the fact that the drones which play the role of backbone/gateway are located in specific positions in order to guarantee the connectivity among groups, so in case of the mobility of drones, groups could be partially affected by lack of connectivity between each other as their backbone nodes move.

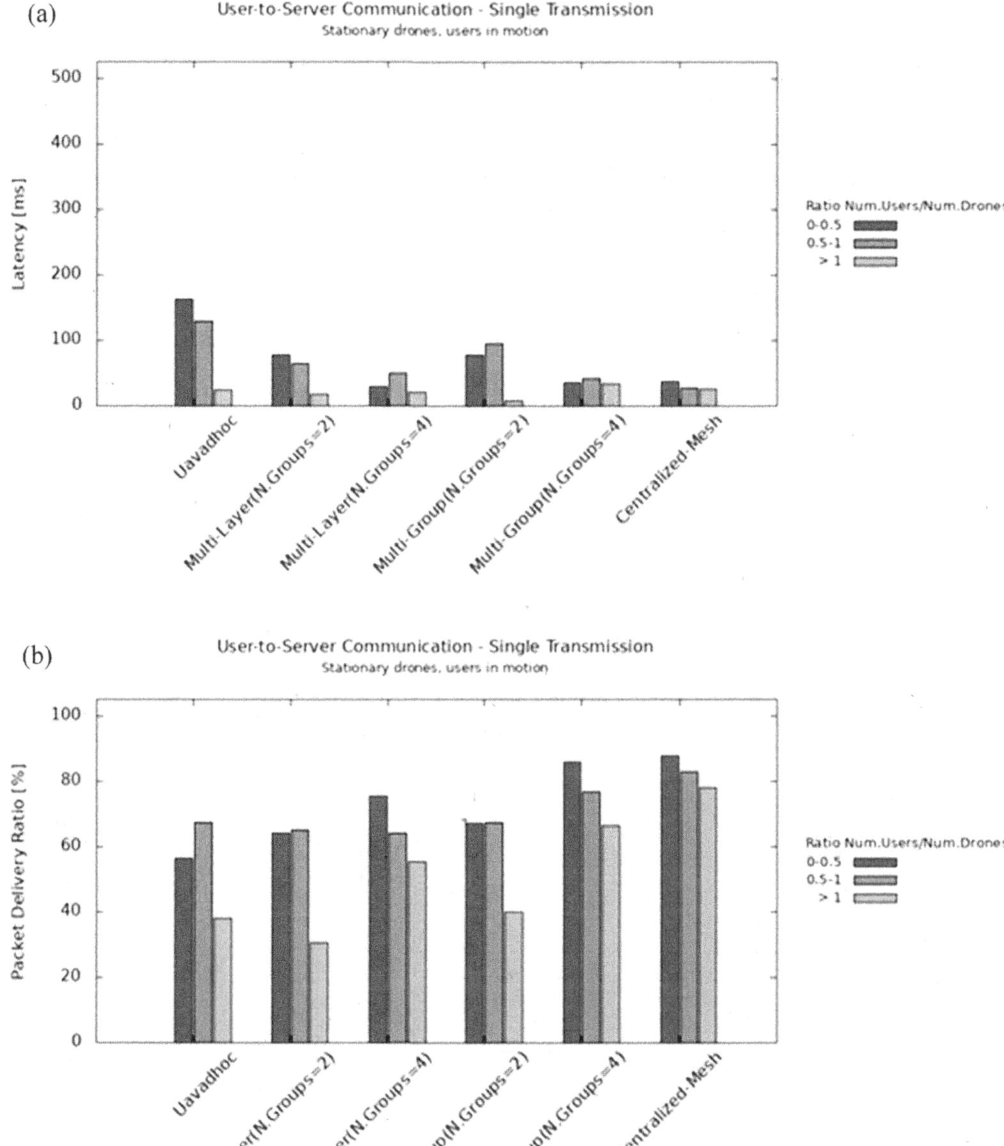

Figure 10.7 Comparison among the UAV network architectures in terms of (a) delay and (b) PDR – user-to-server communication.

Unlike Multi-Layer architecture, Multi-Group network guarantees better performance as the number of groups increases, in particular when the ratio of no. users and no. drones is >1. In both types of communications, latency is very low (under 200 ms) for each architecture.

In the case of stationary drones while users move on the ground, all architectures still guarantee a very low latency in both types of transmissions. With a medium-high availability of drones with respect to users, the packet delivery ratio is approximately constant among all the architectures, especially in user-to-user communications: only Multi-Group (four groups) and Centralized-Mesh

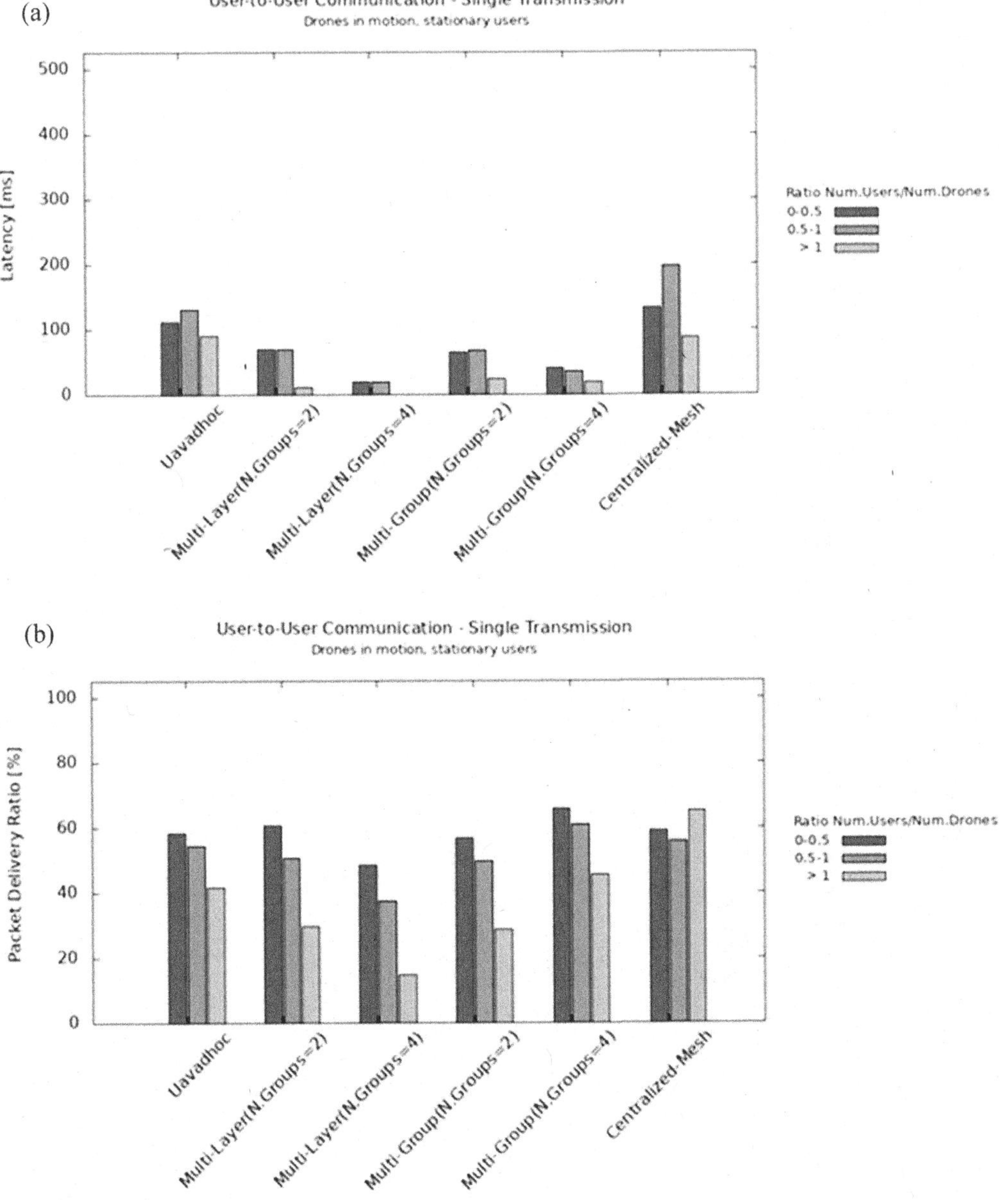

Figure 10.8 Comparison among the UAV network architectures in terms of (a) delay and (b) PDR – user-to-user communication.

architectures provide higher performance regardless of the ratio between no. users and drones is considered. A significant difference in terms of performance is visible between architecture when there are more users than drones in the network: in particular, increasing the number of groups in Multi-Layer and Multi-Group architectures guarantees better performance, especially when users communicate with the server.

Figure 10.11 depicts multiple transmissions performances considering user-to-server communication.

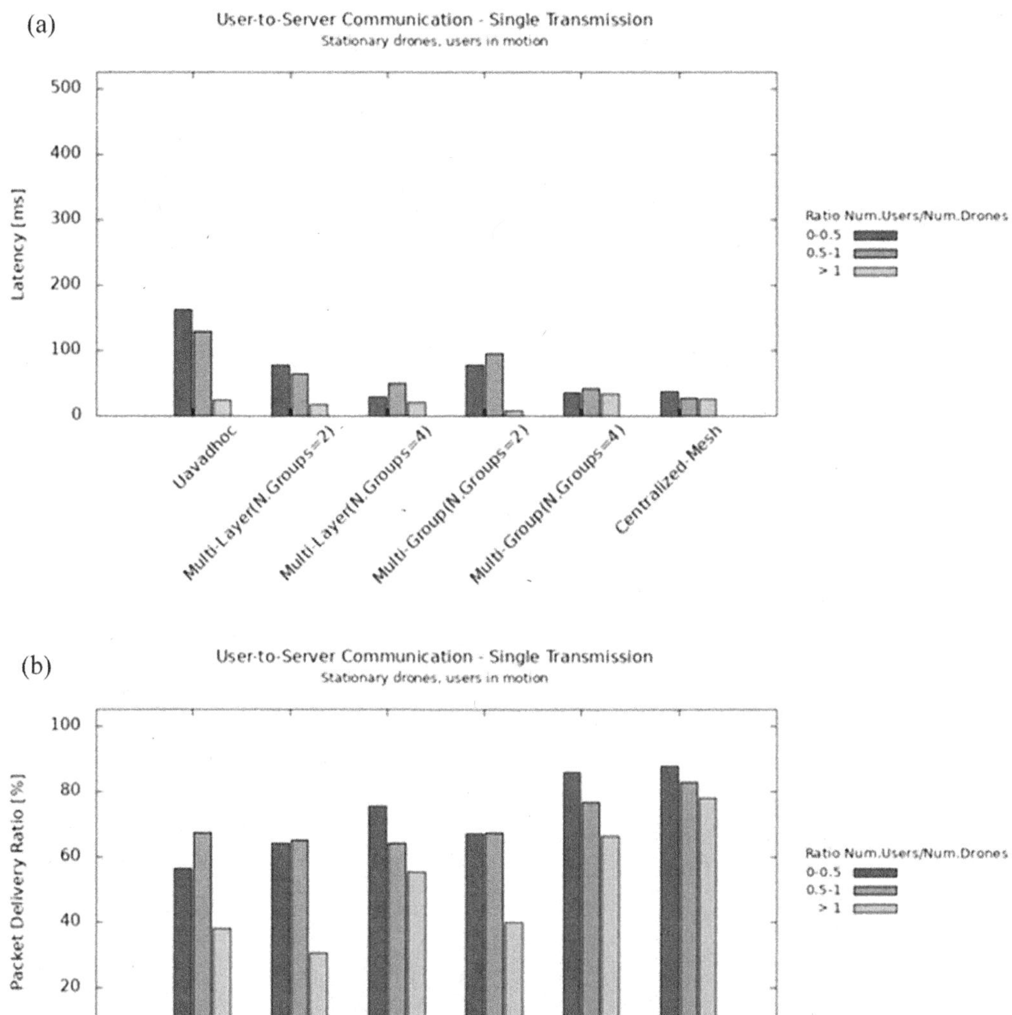

Figure 10.9 Comparison among the UAV network architectures in terms of (a) delay and (b) PDR – user-to-server communication.

Figure 10.12 shows multiple transmissions performances considering user-to-user communication.

By observing the achieved results, it is clear how UAV Ad Hoc Network suffers when more communications occur simultaneously: indeed, it is characterized by higher latency and lower PDR, especially in user-to-server communication, resulting the worst architecture in terms of performance.

In user-to-server communications, Multi-Group (with a no. groups equal to four) and Centralized-Mesh are the most robust architectures, both in terms of latency and PDR, guaranteeing

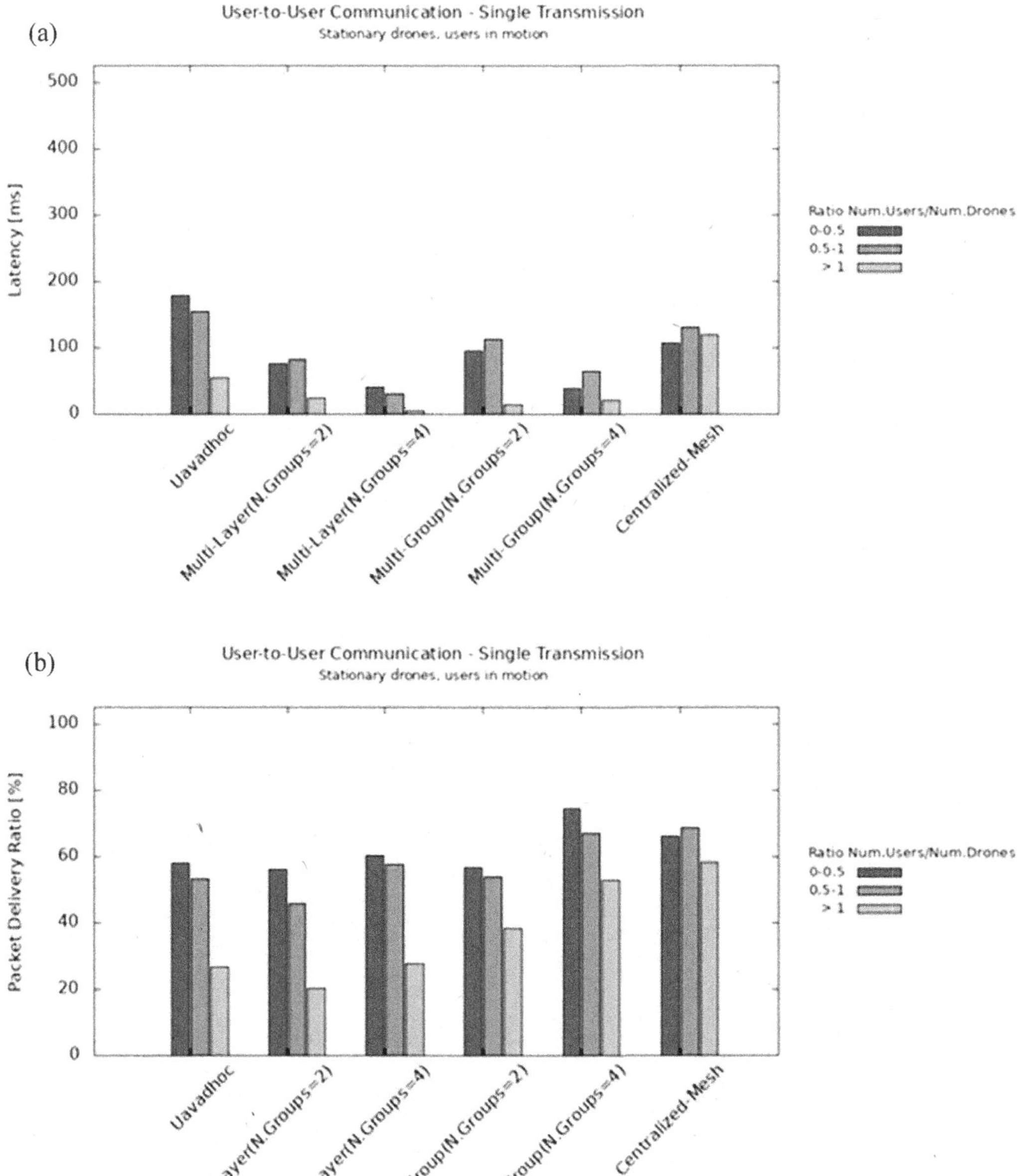

Figure 10.10 Comparison among the UAV network architectures in terms of (a) delay and (b) PDR – user-to-user communication.

very high performance regardless of the ratio between no. users and no. drones. This is mainly due to the fact communications generally need to traverse a smaller number of drones to reach the server as compared to the other architectures. While increasing the number of groups within Multi-Group architecture guarantees lower latency and higher PDR, Multi-Layer does not improve its behaviour both in terms of latency and PDR.

In user-to-user communications, Multi-Group (four groups) and Centralized-Mesh still provide higher values of PDR compared to other architectures, especially when there are more drones than

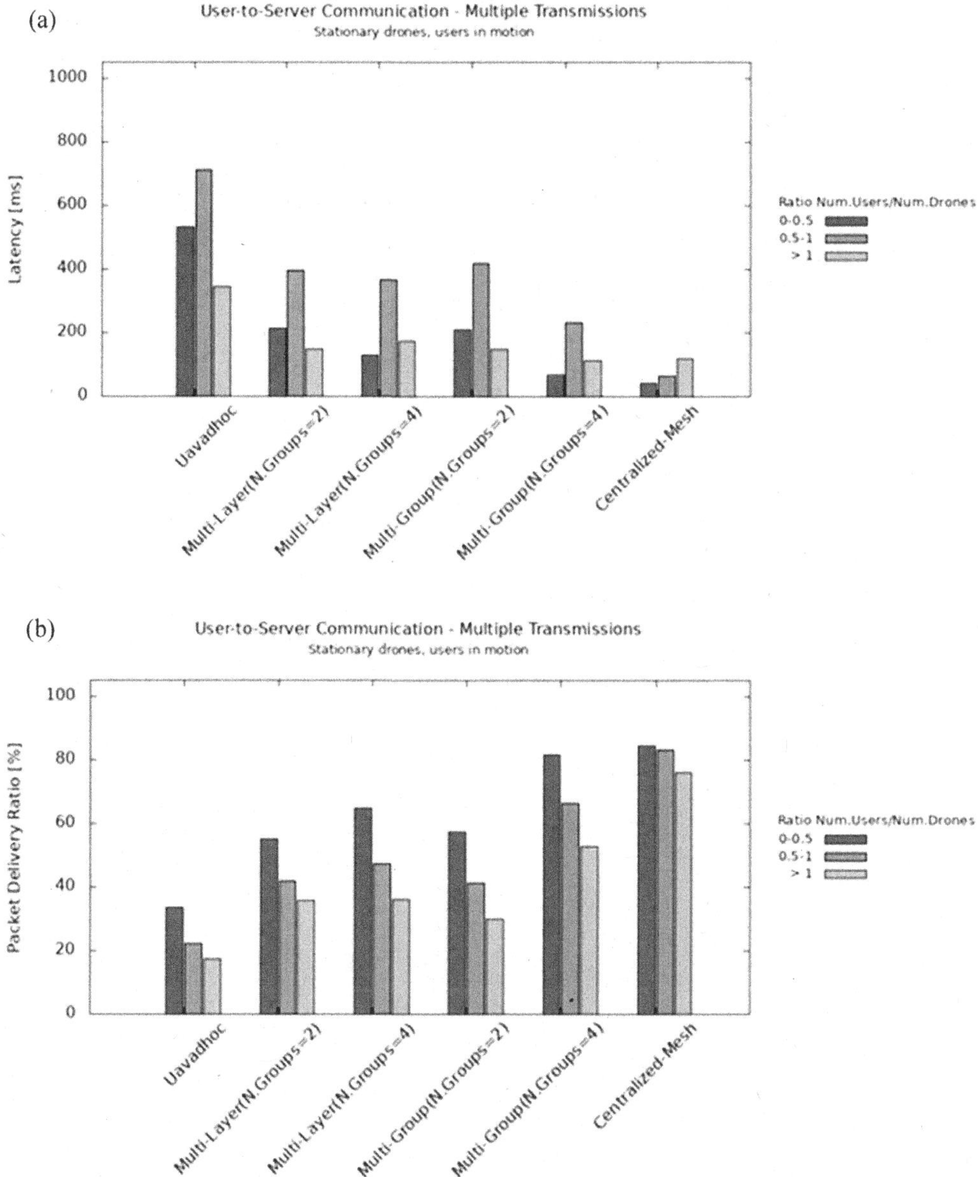

Figure 10.11 Comparison among the UAV network architectures in terms of (a) delay and (b) PDR – user-to-server communications (multiple transmissions).

users within the network, even if Centralized-Mesh suffers a bit from a latency point of view in the same scenario: this can be a consequence of forwarding functionalities of users who need to forward packets while they are involved in communications. In scenarios where there are more users than drones, architectures of ad hoc nature provide equivalent behaviour both in terms of latency and PDR while increasing the number of groups in Multi-Layer and Multi-Group guarantees better performance when there is a greater number of drones than users.

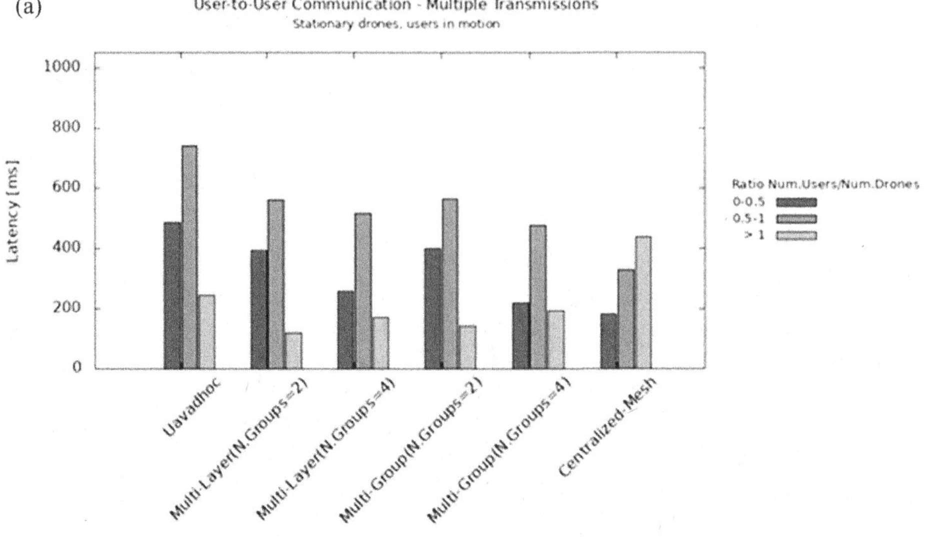

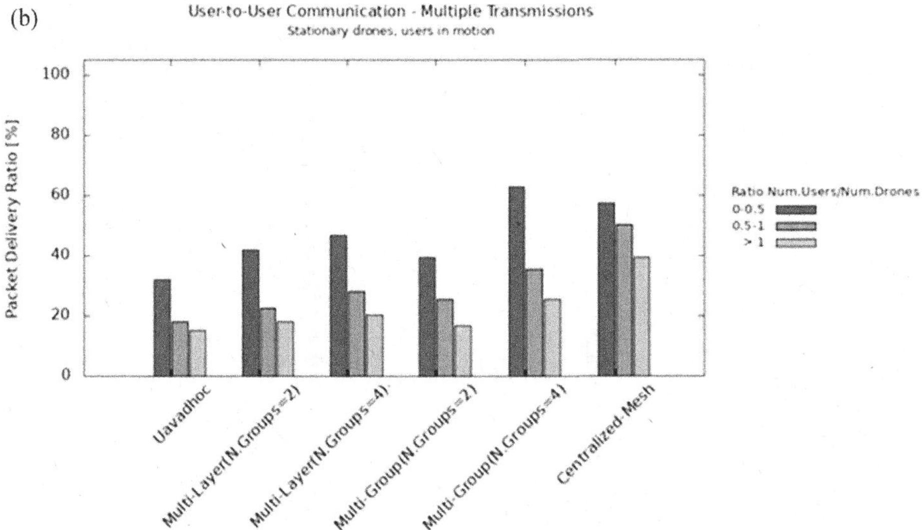

Figure 10.12 Comparison among the UAV network architectures in terms of (a) delay and (b) PDR – user-to-user communications (multiple transmissions).

10.5 CONCLUSION

This chapter discussed different architectures for providing flexible coverage by using UAVs in a smart city/smart region scenario and provides a numerical comparison among the existing solutions.

Based on the achieved results, it is not possible to define which architecture is better than the others. However, Multi-Group UAV (characterized by four groups) and Centralized-Mesh networks have proved to be very robust and provide consistent performance. However, they cannot be considered as the best architectures for all scenarios: the choice of the architecture to use should be done according to the scenario, type of communication and specific requirements of the system such as the availability of drones or the number of expected users within the network.

For example, if a drone network needs to be deployed in search and rescue applications, which typically occur in emergency situations, and multiple transmissions are particularly likely among the users, UAV Ad Hoc network should be avoided – as it is generally affected by a greater latency and a lower PDR than the other architectures: in this case, Multi-Group (four groups) or Centralized-Mesh should be adopted, especially if there is a great availability of drones.

However, if single transmissions are more likely to occur within the network and there is a good availability of drones, UAV Ad Hoc architecture could be used, as it is simpler from an architectural point of view and provides performance comparable with the other architectures, especially in user-to-user communications.

On the other hand, if more communicating users than drones are expected, Centralized-Mesh or ad hoc networks characterized by more groups (Multi-Layer and Multi-Group) should be preferred.

Overall, given the complexity deriving from satisfying all scenarios and the timeliness required to address emergency situations, the possibility to design and deploy programmable and adaptable drone networks could represent an interesting and promising research area.

ACKNOWLEDGEMENT

This work has been partially funded by NATO Science for Peace and Security (SPS) Programme, in the framework of the project SPS G5428 "Dynamic Architecture based on UAVs Monitoring for Border Security and Safety" (DAVOSS).

REFERENCES

1. PiX4D. Drone mapping for smart cities.
2. A. Giyenko and Y. I. Cho. Intelligent UAV in smart cities using IOT. In *2016 16th International Conference on Control, Automation and Systems (ICCAS)*, Gyeongju, Korea, pp. 207–210, October 2016.
3. F. Mohammed, A. Idries, N. Mohamed, J. Al-Jaroodi, and I. Jawhar. Uavs for smart cities: Opportunities and challenges. In *2014 International Conference on Unmanned Aircraft Systems (ICUAS)*, Orlando, FL, pp. 267–273, May 2014.
4. J. Li, Y. Zhou, and L. Lamont. Communication architectures and protocols for networking unmanned aerial vehicles. In *2013 IEEE Globecom Workshops (GC Wkshps)*, Atlanta, GA, pp. 1415–1420, December 9-13, 2013. December 2013.

11 Localization for Vehicular Ad Hoc Network and Autonomous Vehicles, Are We Done Yet?

Abdellah Chehri
University of Quebec in Chicoutimi (UQAC)

Hussein T. Mouftah
University of Ottawa

CONTENTS

11.1 INTRODUCTION

Rarely has there been a technological development that has attracted as much public attention as connected and automated vehicles (CAVS). At the moment, skeptical voices are predominant – at least in public opinion. In the cacophony of promises, fears, and suspicions, there are some arguments in favor of emerging technology [1–4].

During the last decade, there have been considerable developments in automated driving technology. In the future, the vehicles will be able to sense their environment and navigate the surroundings without any human input. Moreover, cars will be able to communicate with other cars, infrastructures, pedestrians, and the cloud [5].

The data exchanged can be used to develop different applications that can enhance road safety, better manage the traffic flow, and provide additional comfort services to the vehicle drivers [6–8]. In self-driving cars, algorithms for controlling lane changes are an essential topic of study. The existing "lane-change" algorithms have one of two drawbacks. Firstly, these algorithms rely on detailed statistical models of the driving environment. Secondly, they can lead to impractical and conservative decisions, such as never changing lanes at all [9]. However, whatever lane-changing strategy being used, an accurate vehicle position estimation is required, particularly under a complex urban environment. Without this information, any of the mentioned applications cannot be achieved [10].

In autonomous driving, it is necessary to know the vehicle position with high precision. Without exact localization, it would be impossible to define the travel plan to the desired destination as well as various driving maneuvers, such as changing lanes and performing turns. Besides, the location will provide other additional useful information, such as taking the next exit or intersection [11].

The transition to fully automated driving requires the exact position of a vehicle to determine whether the autonomous driving mode can be turned on. Most self-localizing vehicles use optical systems to determine their location. They rely on optics to "see" lane markings, road surface maps, and surrounding infrastructure to orient themselves, so that these driverless cars can stay in the lane when road markings are obscured. These optical systems work well in fair weather conditions. Still, it is challenging, maybe even impossible, for them to work when snow covers the markings and surfaces or precipitation obscures points of reference. That would be one of the cases for level 4 systems in which the driver would have to retake full control of the vehicle [12,13]. Designing localization systems for autonomous vehicles involves several challenges due to the following factors:

1. Dynamic topology due to the high and variable speeds of the cars in the network. This causes the connectivity between the cars to change often and rapidly.
2. Objects and other obstacles that exist on roads can interfere with the radio signal and affect their power and strength.

In this chapter, we provide a survey on localization techniques for CAVs, and we open a new way of improving the accuracy of positioning of vehicles through collaboration. For this reason, data fusion techniques and machine learning are needed to combine the localization estimation methods into one model and produce an accurate position estimation. In this survey, all these techniques are discussed in detail as well as the data fusion technique concerning the connected and autonomous vehicles application requirements.

11.2 STANDARD LOCALIZATION SYSTEM IN CAVs

An autonomous vehicle can be defined as "a motor vehicle that uses artificial intelligence, sensors, and global positioning system coordinates to drive itself without the active intervention of a human operator" [14].

Despite the current trend, autonomous vehicles still face enormous challenges in different areas, such as technology maturity, infrastructure investment, consumer acceptance, legislation, and business model [15].

Now in autonomous driving, the challenge is that drivers must trust a technology that they feel is fully delivered. A common approach to self-driving cars is to use detailed high-definition maps that are annotated with precise lane locations, traffic signs, and other metadata that govern the rules of the road [16,17]. These maps are generated offline, which allows the use of complex algorithms that are not necessarily "real-time" to be used by the operating self-driving car [18–20].

The problem of localization in CAVs is subdivided into a few major modules, Figure 11.1 shows one example of such a subdivision [21]. The first module is responsible for perception and gathering information from different surround sensors, satellite navigation, and inertial sensors. The output of this module will typically contain information about both the dynamic environment, such as position and velocity of other road users and the static environment, such as which areas are drivable and which contain static obstacles, signs, lights, etc.

The second module is responsible for real-time vehicle location and mapping. The third module is responsible for planning a trajectory based on information from the other modules. Finally, by using all these modules, the vehicle's control will be executed following the calculated path [22].

11.3 PERCEPTION AND SENSORS

Vehicle perception, as shown in Figure 11.2, can be treated as a process of taking inputs from sensor measurements [23]. These include measurements from perception sensors, such as laser scanners or cameras, which is denoted by Z and frequencies from motion sensors, such as odometry or inertial measurement, which is indicated by U.

The process outputs include the estimated internal vehicle state X, a static map of the surrounding environment M, and a list of moving objects in the vicinity of the vehicle O.

The vehicle state is comprised of variables regarding the vehicle itself, such as speed and its relative position to the map M. The static map of the vehicle environment M contains information about stationary objects as well as their locations on the map. The moving object list O includes information on dynamic objects, their positions, and dynamic states such as velocity and moving direction.

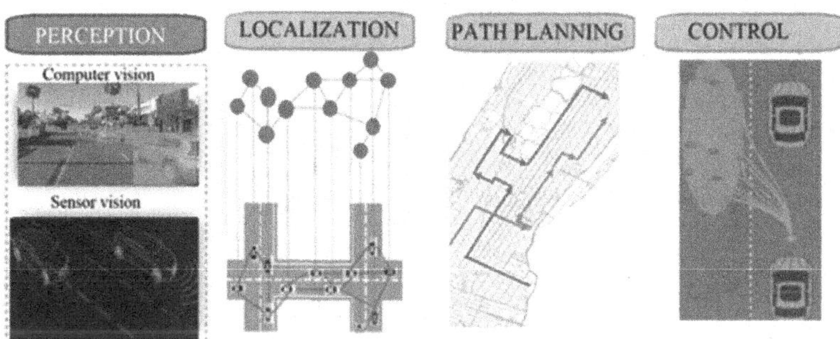

Figure 11.1 Flow of data and processing modules for autonomous driving. Localization here is considered part of the perception module.

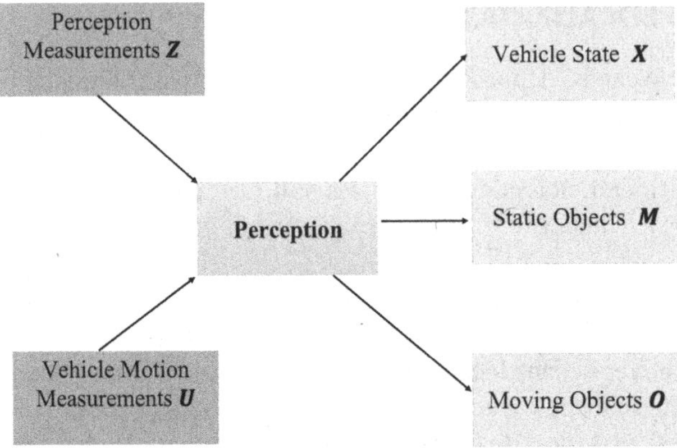

Figure 11.2 The general perception process.

The autonomous vehicle uses a large number of sensors to understand the environment, to locate, and to move [24,25]. For sophisticated driver assistance systems to be possible, these sensors must communicate with each other.

Modern vehicles barely manage without sensors, and the sensors are getting smaller and lighter. With increasing distribution, they will become cheaper, so they can be installed in more and more vehicles. A detailed comparison of these exteroceptive sensors is given in Table 11.1.

11.3.1 ULTRASOUND TECHNOLOGY

The ultrasound technology in vehicles is already mature, and the corresponding sensors are inexpensive and easy to use. Due to the propagation characteristics of sound, ultrasonic sensors work only at short distances and low speeds, for example, in self-parking and blind-spot detection.

11.3.2 FRONT AND REAR RADAR SENSORS

Radar, or radio detection and ranging, uses reflected radio waves to sense surrounding objects. Since radar sensors have been incorporated into vehicle technology, they have become an essential component of driver assistance systems. They can detect stationary and moving objects at short, medium, and long distances and provide important information such as distance, angle, and speed. Unlike optical sensors, such as LiDAR and cameras, radar sensors are significantly less affected by the weather and the lighting conditions. Because of this immunity to interference, they are often

Table 11.1
Exteroceptive Sensors

Sensor	Sensitive to External Factors	Range	Accuracy	Cost
Ultrasound	No	No	Low	Low
Radar sensors	No	High	Medium	Medium
Camera	Yes	Short	Medium	Low
Stereo camera	Yes	Medium (<100 m)	Low	Low
Thermal camera	Yes	Short	Low	Low
Lidar	Yes	Medium (<200 m)	High	High

used in safety-relevant applications, such as adaptive cruise control, collision avoidance systems, and emergency brake assistants [26–28].

Radar systems in short-distance vehicles are currently operated in two frequency bands: K-band at 24 GHz and W-band at 79 GHz. Switching from a lower frequency band to a higher frequency band alleviates interference and offers several other advantages. The higher frequency allows for smaller sensors, better spatial resolution, and larger operating bandwidths. The front radar is an active extrinsic sensor, gathering information about the upcoming obstacles in front of the vehicle. An obstacle detected by the front radar should disable the lane change function in certain situations [29,30].

The rear radar is identical to the front radar. Thus it is also an active extrinsic sensor, but in this case, gathering information about the object behind the vehicle. In case of another car approaching a higher speed than the vehicle's owner, the lane change feature should be blocked.

11.3.3 CAMERA AND VISUAL SENSOR

Cameras are indispensable for object recognition. They provide the car with the information necessary to identify the object perceived at the edge of the road through artificial intelligence such as a pedestrian [31]. Moreover, the high strength of cameras in precise measurement is the angles [32]. For example, it is possible to detect at an early stage whether a vehicle in front will turn. If city traffic requires a wide field of view to capture pedestrians and cyclists, a long range of up to 300 m and a narrow field of view are required on the motorway and highway.

Visual systems continue to evolve toward smaller cameras with higher resolutions and toward stereoscopic cameras. One commonly used technique to recover the 3-D structure is the triangulation method [33]. They are used wherever radar sensors cannot recognize or classify objects. Camera sensors are relatively inexpensive but need a lot of computing power. If the weather and light conditions are favorable, they provide good pictures, but in unfavorable weather and light conditions, the picture quality is often insufficient [34,35].

11.3.4 LiDAR, LASER AND INFRARED SENSORS

The LiDAR (Light Detection and Ranging) is an active extrinsic sensor, emitting a laser beam and obtaining a close-range polar map of the distance versus bearing. The installed sensors have an effective range of 16 m, and they are both fitted on the front corners of the car, sharing the frontal view, while having separate side views and rear-view regions. The disadvantage of LiDAR and infrared sensors is that they are still relatively expensive, which is why both techniques are still relatively little used. LiDAR sensors deliver high-resolution 3-D images, which are essential for autonomous driving [36].

Unfortunately, like cameras, they are quite sensitive to weather conditions. On the other hand, infrared sensors provide good images even at night. However, for a wide application in-vehicle technology, these two techniques must still sink significantly in price [37].

11.3.5 AUTONOMOUS VEHICLE NETWORKS

Vehicular communication systems allow localization that improves safety and maximizes road space utilization. It improves such tasks by enhancing technical applications such as platooning and collision warning.

In Figure 11.3, an example is shown to illustrate the concept of this type of network. As can be seen, two cars (1 and 4) circulate in the same lane, and three cars (5, 3, and 2) circulate in a separate lane but in the same direction as 1 and 4. The five cars can form an autonomous vehicle network as they are close enough to communicate with each other. Every two neighboring cars can have direct

Figure 11.3 Example of an autonomous vehicle network.

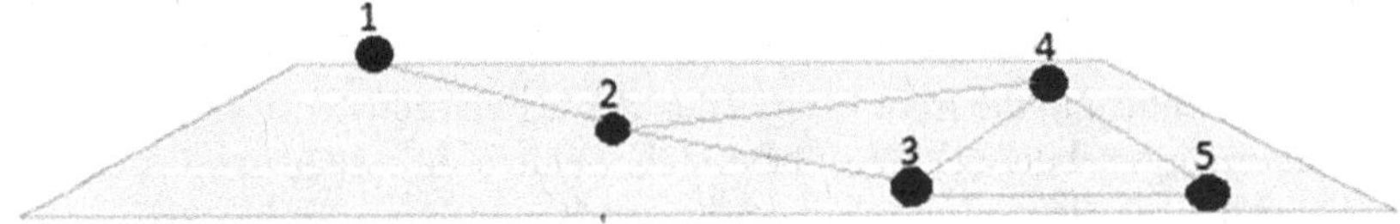

Figure 11.4 Autonomous vehicles network and topology.

communication; based on that, a mesh structure can be made to see more clearly the possible routes to use within the network to send information.

This structure is illustrated in Figure 11.4, where the vehicles are treated as nodes.

Understanding the network performance of the connected vehicle communication, especially in challenging road environments and conditions will be a key to improving performance. One of the crucial characteristics of this network is that the topology can change at any time as the vehicles circulate at different speeds, so some cars can leave the network where some others can join it. This fact enforces that the network needs to have the ability to reconfigure itself to update routes between the nodes, which is one of the challenges of the CAVs network.

11.4 LOCALIZATION SYSTEM DESIGNS FOR AUTONOMOUS DRIVING

Many self-driving car localization approaches rely on the methodology of localization within a prior map, where doing so provides a wealth of knowledge regarding the operating environment of the vehicle, including lanes, traffic signals, and other rules of the road [38].

11.4.1 GLOBAL POSITIONING SYSTEM (GPS)

When talking about the vehicle position, the Global Positioning System (GPS) is the first possibility that comes to mind [39].

Many vehicle manufacturing companies provide the equipment in their cars for localization and tracking distances, vehicle mileage, and speed, etc. The Ford Motor Company developed a telemetric system through GPS that will alert emergency services and locate the vehicle quickly for more assistance [40].

However, some advanced versions of the Global Positioning System exist to solve some of its limitations. For instance, the Precise Positioning Service (PPS); this technique uses the two frequencies to eliminate the atmosphere effect. Another version is called the Differential Global Positioning System (DGPS) which uses two receivers, as shown in Figure 11.5, one of them is always fixed at a known position, and the second one is installed in the moving vehicle [41]. The stationary receiver captures the signal from the satellites and estimates its position based on this information; then, it compares it with its known position to determine the difference. This information is broadcasted to the moving node that used it to compensate for the error. This technique has shown better performance with an accuracy improved up to 3 m.

The disadvantage of this technique is that a fixed receiver should be installed along the road to have good coverage, which means more implementation and maintenance costs. In addition, the

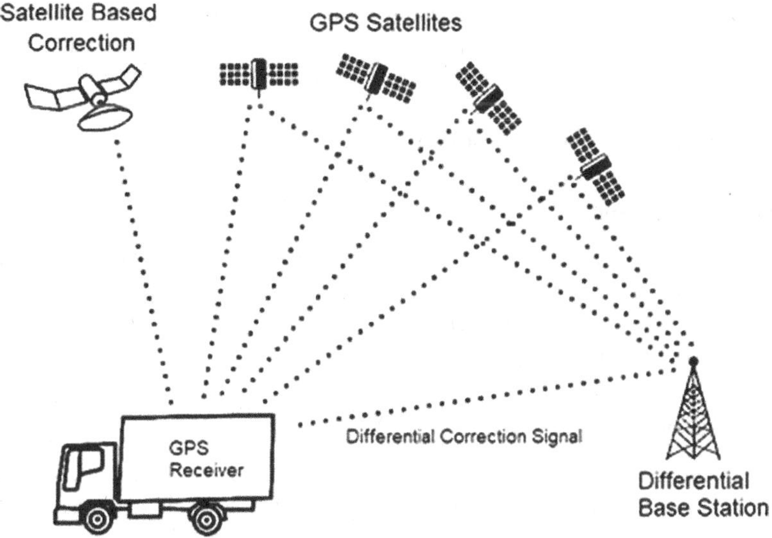

Figure 11.5 Differential Global Positioning System (DGPS).

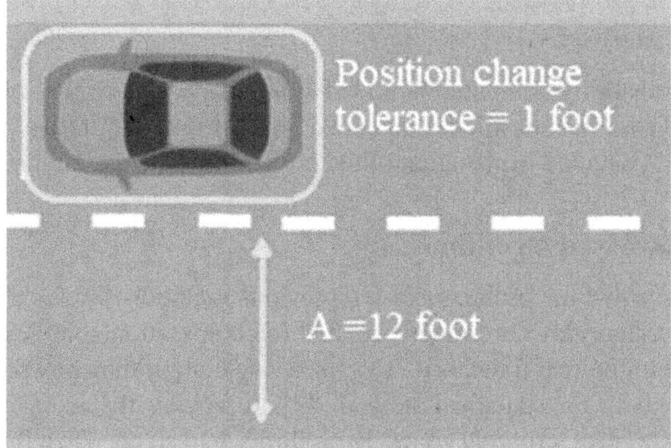

Figure 11.6 Accuracy requirement for autonomous vehicles.

GPS and DGPS show that they cannot keep the same evolution speed as the vehicles. Furthermore, GPS suffers from poor reliability and outages due to many limitations such as signal blockage (i.e., when driving in a tunnel) and multipath (in many densely built-up areas with many high-rise buildings) as well as inadequate accuracy for autonomous vehicles. Typical GPS average accuracy ranges from a few meters to more than 20 m, limiting its implementation to autonomous vehicle applications as decimeter level (30 cm) accuracy is demanded to stay in lane.

A fully autonomous vehicle needs an accurate localization solution paired with the confidence that the localization solution is correct (Figure 11.6). So, the use of the GPS or DGPS in CAVs should be combined with another information source to improve the vehicle's localization.

11.4.2 MAP MATCHING

From its name, the map-matching goal is to align the position of vehicles with an already known map. The motivation behind this technique is to combine the existing positioning with the newly developed geographic information systems that allow access to more accurate mapping data. This method uses different approaches to draw an estimation of the trajectory. Then it compares it with an already known map to adjust any error and generate a more accurate path [42]. This technique is the base of many map-based applications, such as traffic management, driving assistance, etc. [43–46]. Four techniques have been investigated:

11.4.2.1 Incremental Method:

This technique uses measures of consecutive parts of the trajectory of the targeted vehicle and then makes a matching estimation to the road map. These measures are local and deal with angles of the possible curves and the distance between the portions. This technique provides a good trade-off between accuracy and computation speed.

11.4.2.2 Global Method:

Instead of performing a point-by-point matching, as explained in the previous algorithm, this technique uses a curve matching to compare the whole trajectory with the network road map. The global map-matching algorithm relies on two methods to analyze the distance between two curves: *Fréchet distance* and the *weak Fréchet distance* method. These two techniques are discussed in the following paragraphs; also, a detailed study is presented in Ref. [47].

11.4.2.3 Statistical Method:

This algorithm uses probability computation to define the likely path of the vehicle. The algorithm calculates the probability of past positions and then compares it with different paths to conclude the most likely trajectory. A detailed explanation of the probability calculation is provided in Ref. [48].

11.4.2.4 Fuzzy Logic-Based Algorithms:

This technique aims to solve the problems of the previous methods discussed in this section. The disadvantages of those algorithms are mainly related to computation complexity and precision of the estimated distance to the real trajectory. The fuzzy logic algorithm was proposed to overcome the mounting complexity of the statistical method [49]. It predicts the actual characteristics of the traveling road, such as corners, crossing zones, lanes limits, and so on. Then the algorithm finds the position of the vehicle within the trajectory identified in the first part of the problem-solving.

An experiment has been conducted to evaluate the feasibility and the performance of this algorithm. In this experiment, a vehicle was equipped with a GPS and then, using a computer, record its positions at different places on the road [50]. The locations are converted to values representing two-dimensional coordinates.

As mentioned before, the algorithm's goal is to optimize the computation by preforming low memory calculation. The algorithm updates the trajectory positions each time the vehicle leaves the road or changes the direction. This can happen, for instance, when the car goes to a gas station or crosses an intersection.

11.4.3 CELLULAR LOCALIZATION

The cellular localization technique is based on the existing infrastructure used for mobile communication. It uses the signals coming from the base stations to estimate the position of a specific moving device. The area to cover is divided into cellular, and each cellular is equipped with a base station capable of receiving and sending signals within its zone. The moving devices are receiving the signals from any nearby base stations, and only the base station with high strength is taken into consideration [51–53].

Multiple techniques are used, such as ranging methods and the fingerprinting method. In this section, different techniques and algorithms are discussed. In the end, we discuss the use of this method in the context of the CAVs.

11.4.3.1 Ranging Method

11.4.3.1.1 Received Signal Strength Indicator

The main advantage of this technique is the low cost related to distance estimation, as it does not require any complex hardware or computations. The key behind this method is that it uses the attenuation of the traveling signal to compute the distance between the moving car and the transmitting base station. This attenuation is called the path loss, and the following formula describes it:

$$PL(d) = PL(d_0) + 10 \cdot n \cdot \log\left(\frac{d}{d_0}\right) \tag{11.1}$$

Where:

d is the distance between the base station and the moving car.
d_0 is the distance between the base station and a reference receiver close to the base.
PL(d) is the path loss of the received signal with respect to the moving car.
$PL(d)$ is the path loss of the receiving signal with respect to the fixed reference.
n refers to the loss increasing rate.

The issue using this formula is it does not take into consideration the errors that can occur during the signal propagation. These errors can be generated by various problems related to changes in the environment, such as reflections in urban areas, the orientation of the antenna, shadowing, and so on. To reduce the error, another formula can be used:

$$PL(d) = PL(d_0) + 10 \cdot n \cdot \log\left(\frac{d}{d_0}\right) + X_p \tag{11.2}$$

X_p are random variables with zero mean using a Gaussian distribution. The choice of the distribution probability to use is crucial to optimize the error. Some studies show that Weibull distribution can provide better results.

11.4.3.1.2 Time of Arrival method

This method uses the time of sending and receiving a signal to estimate the distance between two nodes. Figure 11.7 shows a scenario where the node A is sending a signal to node B at time t_0. This signal is received by B at time t_1, then B sends back a signal at time t_2 and A receives it at t_3. With this time information and the speed of the signal propagation v for the ultrasonic waves, the distance can be estimated using the following formula:

$$d = \frac{v((t_3 - t_0) - (t_2 - t_1))}{2} \tag{11.3}$$

Where:

v is the propagation speed. Almost all applications that work with this method use ultrasonic waves.
t_0 is the time when A sends the signal to B.
t_1 is the time when B receives the signal from A.
t_2 is the time when B sends back a signal to A.
t_3 is the time when A receives the signal from B.

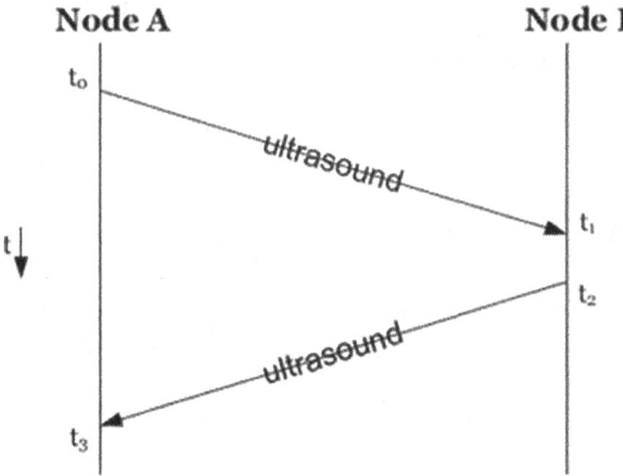

Figure 11.7 Time of arrival method (ToA).

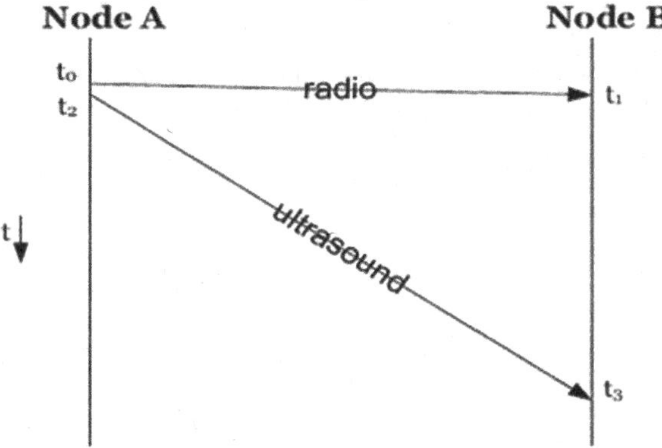

Figure 11.8 Time difference of arrival method.

11.4.3.1.3 Time Difference of Arrival method

The only difference between this method and ToA is that the Time Difference of Arrival method uses two signals instead of only one. One of these two signals should have a slow propagation speed and the second one should have a high propagation speed. Figure 11.8 shows this scenario; the following formula can be used to estimate the distance between A and B:

$$d = ((t_3 - t_0) - (t_2 - t_1))\frac{v_{rf}v_{us}}{v_{rf} - v_{us}} \tag{11.4}$$

Where:

v_{rf} is the propagation speed of the high-speed signal (radio).
v_{us} is the propagation speed of the low-speed signal (ultrasonic).
t_0 is the time when A sends the signal to B.
t_1 is the time when B receives the signal from A.
t_2 is the time when B sends back a signal to A.
t_3 is the time when A receives the signal from B.

11.4.3.1.4 Fingerprinting Method

The main advantage of this technique is the low cost related to the distance estimation process, as it does not require any sophisticated hardware or computations. The concept relies on three phases:

1. Database creation phase.
2. Identification phase.
3. Position estimation phase.

The base station performs the first and the second steps before the targeted moving device is connected to its zone, and the last phase is performed while the vehicle is online. To complete phase one, the base station (BS) starts a survey of its zone by analyzing the received signal strength (RSS) from different transmitters deployed (called access points) within the BS's region. For each received signal, the BS records its position using the RSSI technique. Using statistical estimation, this data, called location fingerprint, is saved in a database for further use. Once a device is online, it sends some measurement samples of its RSS's collected from different nearby access points. The base station uses some estimation algorithms based on the database values to compute the position of the moving vehicle.

11.4.3.2 Limitation of Cellular Localization Accuracy

The cellular localization technique is based on the received signal strength characteristics. Therefore, the precision strongly relies on this measurement, and any change can increase the error significantly. The signal can easily have some attenuation due to the nature of the area of interest, such as in the urban areas where reflection and shadowing are highly probable.

According to some studies [51], the average root mean square error (RMSE) of the localization can reach 250 m in some cases. The localization error could be reduced to 75 m using cooperative positioning techniques for mobile localization in 4G cellular networks [52]. Peral-Rosado et al. provided a feasibility study of the positioning capabilities for future 5G vehicle-to-infrastructure (V2I) networks, considering 5G-like multicarrier signals. Their simulation results show that the position accuracy is above 20 m for 95% of cases [53]. This result is not even acceptable by the inaccurate applications of CAVs.

Therefore, cellular localization is not a suitable technique to use for CAVs application. However, the information collected using this technique can be used to support other technologies to enhance accuracy.

11.4.4 IMAGE AND VIDEO LOCALIZATION TECHNIQUE

Image and video localization technique is a vision-based method that uses the already installed camera in the vehicle, or in some area of interest, to be used for the position estimation process. This technique uses the information gathered by the camera and analyzes it frame-by-frame. An example of the area of interest where this technique can be used is parking lots or tunnels where the cameras are installed and can track the movement of the vehicle [54]. Using this technique, more useful information can be added to other methods and improve the estimation of the vehicle position in CAVs. Some of the information that can be concluded from analyzing the collected data is vehicle density in some areas, incident detection, new construction, and so on.

Li et al. [55] proposed a low-cost localization method utilizing only cameras, where the images obtained from the cameras were down-sampled to a resolution of 800 per 600 pixels to reduce computation time. Kamijo et al. [56] suggested using GPS and inertial motion units (IMUs) for global positioning while using the camera to recognize lane markers for lateral positioning. Using this approach, mean positioning errors of 0.73 m were achieved. Suhr et al. [57] extended the method proposed [56] by the use of cameras for recognition of road markers as well as lane markers to support both lateral and longitudinal positioning.

Some requirements are identified by Yang et al. [58] for these kinds of techniques to be useful:

1. *Robustness*: the control system that uses this technique should be capable of working with the same performance under any weather conditions. For example, during cloudy days, the system should adapt to keep recording a clear image or video about what is happening in the area of interest.
2. *Processing Time*: the control system should be able to track the real-time recording and compute the conclusion as fast as 15 frames per second to output useful data. An example when this condition is required is when a collision should be avoided. So this is an essential requirement.
3. *Adaptability*: some times, the system will experiment with some sudden change in illumination due to clouds, for instance. So it should be able to adapt to any changes to keep recording useful data.

Image and video localization technique has some excellent advantages, such as the availability of such vision-based systems in new cars. This can help to avoid installing new devices and changing the existing infrastructure to detect obstacles or prevent a collision. Also, the related cost is decreasing drastically because of the latest inventions and improvements in the manufacturing process. In addition, some algorithms have been developed to use real-time collected data and make quick decisions and useful conclusions that will support other techniques or systems to estimate the position of a vehicle or obstacle.

However, this technique has also some drawbacks, such as the quick change in illumination. This change is highly probable to happen as the car moves, and it can quickly switch from a sunny zone to a dark area, such as a tunnel. So, the system can easily have errors, as the recording should be in real-time. Another problem is the loss of information. The camera records the scene in 3D. However, the analysis is done in 2D. This can cause some losses, such as the depth of the detected object.

Also, the difference in color and shape of vehicles can cause problems. Finally, the weather changes and the environment will affect the quality of the recording, as shown in Figure 11.9. For instance, in urban areas, the presence of buildings and trees can disturb the background and can lead to significant errors.

In order to analyze the real-time records, many approaches exist; one of them is the sliding window [59]. The concept of this method is based on analyzing the whole image frame by frame. An area of specific size is selected in the targeted image and compared with a known shape of the target, an example of a vehicle. Then, another area is selected and checked, and so on until the whole image is scanned. If no object is detected, the algorithm increases the size of the selected area and repeats the same steps. The problem with this approach is that it takes too much time to scan one image, which is a strict requirement, as explained before, because everything is happening in real-time. To solve this problem, two techniques are discussed in Ref. [59]; the first one relies on using cues to perform the research, and the second one uses cascading classifiers (Figure 11.9).

11.4.5 DEAD RECKONING

Dead reckoning technique is considered to be a GPS backup when the environmental conditions or the road trajectory (such as tunnels, parking lots, etc.) prevent the proper use of the GPS signal. The concept of this technique is based on the use of the last known position in addition to some other information that can be collected from the installed sensors, data, such as speed, acceleration, and orientation. This data can be easily obtained using existing sensors such as gyroscope and odometer [60].

The use of this technique is limited due to its accumulative error; for example, an error of up to 20 m can be reached after 30 seconds of a last known position with a moving speed of about 100 km/h [61].

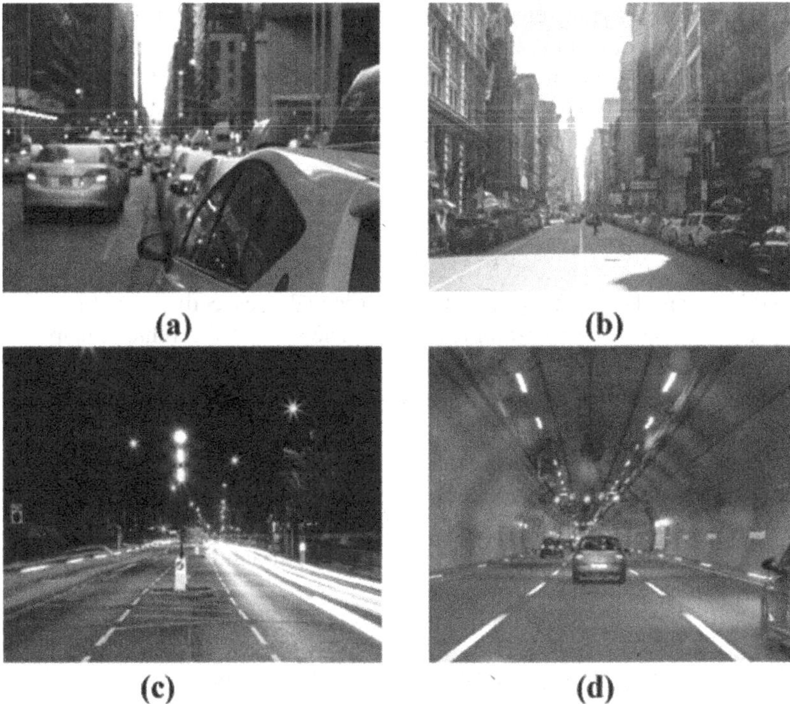

Figure 11.9 Different lighting and image quality. (a) Daylight scene with a higher definition video frame. (b) Daylight scene with low definition video. (c) Night-time scene. (d) In tunnel.

In CAVs the dead reckoning can be used to support other techniques such as map matching to fill the gap that can be caused by the unavailability of the GPS.

11.4.6 DISTRIBUTED AD HOC LOCALIZATION

This technique relies on an important concept used in sensor networks and ad hoc networks. The idea is that the vehicle can estimate its position itself by using information collected from other vehicles connected in the same network. Each vehicle calculates the distance between itself and nearby cars, and then it sends the data to all neighboring nodes.

Some algorithms have been developed to apply this technique. One of them is discussed in Ref. [62], where the studied vehicle which is not equipped with a GPS can estimate its position based on the distance to at least three nearby vehicles equipped with a GPS. However, when this condition is not applied, the algorithm can hardly estimate the position, but it can estimate the direction and the distance from danger or an obstacle such as an accident. Other algorithms can estimate the position using information about the distance from other vehicles by using the received signal strength characteristics.

This method can be useful to support the position estimation process when the GPS information is not available to use like the dead reckoning does. The accuracy depends on the number of nearby vehicles equipped with a GPS.

11.5 SIMULTANEOUS LOCALIZATION AND MAPPING (SLAM)

Knowing the spatial relationship between objects and locations is a crucial requirement for many tasks in autonomous vehicles, such as path planning, target tracking, and control. Such information is often not available before deploying an autonomous vehicle, or previously acquired information is outdated because of changes in the environment [38]. Automatically learning such information is, therefore, a fundamental problem in self-driving, and has received much attention. The problem is generally referred to as simultaneous localization and mapping (SLAM) [63–65].

We consider the vehicle's trajectory as a set of poses (as shown in Figure 11.10). By modeling our noisy odometry and sensory measurements as Gaussian random variables,

$$x_i = f_i\left(x_{i-1}, u\right) + \omega_i \tag{11.5}$$

$$z_k = h_k\left(x_{ik}, M\right) + v_k \tag{11.6}$$

where $f_i(.)$ and $h_k(.)$ are the process and observation models, respectively, and $\omega_i \sim \mathbb{N}(0, \Sigma_i)$ and $va_i \sim \mathbb{N}(0, \Sigma_i)$ are two Gaussian measurement noise. The state of the vehicle, map, control input, and measurement sequences are denoted as $x_{(1:k)} = x_1, x_2, ..., x_k$, $m_{(1:M)} = m_1, m_2, ..., m_k$, $u_{(1:k)} = u_1, u_2, ..., u_k$, and $z_{(1:k)} = z_1, z_2, ..., z_k$, respectively. The goal of SLAM is to find $x_{(1:k)}$ and $m_{(1:M)}$ from $u_{(1:k)}$ and $z_{(1:k)}$ [40].

However, vehicle location and map are both unknown. When the vehicle moves, it accumulates errors in odometry, making it gradually less certain as to where it is. For building an accurate map of the environment, a correct position of the vehicle is needed. Full SLAM refers to the posterior estimation of the entire path. Online SLAM refers to the estimation of the current posterior only [66].

In the probabilistic form, the SLAM problem involves estimating the probability distribution:

$$P\left(x_t, M | z_{0:t} + u_{1:t}\right) \tag{11.7}$$

This probability distribution describes the joint posterior density of the map and vehicle state at time t given the measurements and control inputs up to time t. In general, since data arrives over time, a recursive solution to SLAM is desirable.

Starting with an estimate for the distribution $P(x_t, M|z_{0:t}, u_{1:t})$ at time $(t-1)$, the joint posterior, following a control u_t and measurement z_t, is estimated using a Bayes filtering process.

$$\underbrace{P\left(x_t, M|z_{0:t}, u_{1:t}\right)}_{\text{posterior}\mapsto \text{t}} = \underbrace{P\left(z_t|x_t, M\right)}_{\text{posterior}\mapsto \text{t}} \underbrace{\int P\left(x_t|x_{t-1}, u_1\right) \underbrace{P\left(x_{t-1}, M|z_{0:t-1}, u_{1:t-1}\right)}_{\text{posterior}\mapsto \text{t}-1}}_{\text{predection}} \tag{11.8}$$

For this computation, two probabilities are specified: the sensor measurement model $P(z_t|x_t, M)$ and the vehicle motion model $P(x_t|x_{t-1}, u_1)$. The sensor model describes the probability of making an observation z_t when the vehicle state and a map of the environment are known.

In the literature, solutions to the probabilistic SLAM can be roughly classified according to methods to represent the map M and the underlying technique to estimate the posterior (Eq. 11.8), which involves finding an appropriate representation for the measurement model and motion model [67].

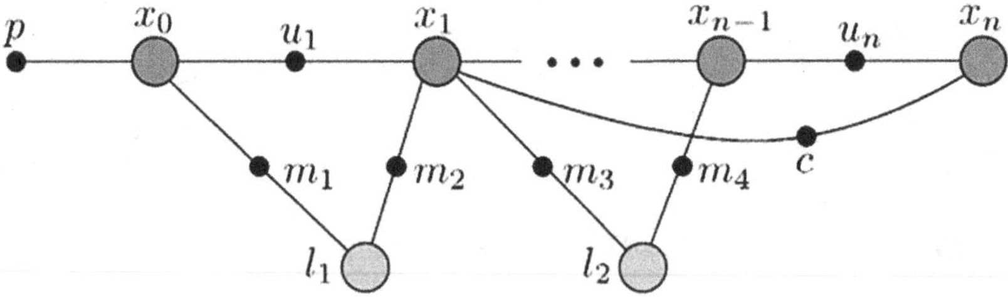

Figure 11.10 Formulation of the SLAM problem, where variable nodes are shown as large circles, and factor nodes (measurements) with small solid circles.

11.5.1 FILTERING SOLUTIONS

The errors ω_i and v_k from Eqs. 11.5 and 11.6 are additive, normally distributed, and independent over time, and the models in themselves are linear. These can be solved by the filtering and smoothing problems optimally, in a mean square error sense, using a Kalman filter [67].

Filtering frameworks approach the SLAM problem from a recursive Bayesian estimation standpoint [68]. With measurements under Gaussian noise, the extended Kalman filter (EKF) is the optimal minimum mean squared error (MMSE) estimator in which system nonlinearities are handled by linearizing the process (Eq. 11.5) and observation models (Eq. 11.6) [69]. Another option, which is used in the appended papers, is the sigma point methods, such as the Unscented Kalman Filter (UKF) [70] and the Cubature Kalman Filter (CKF) [71].

11.5.2 MONTE CARLO SOLUTIONS

The SLAM problem can alternatively be solved using sampling-based approaches in which a finite set of particles can be used to model the posterior distribution (as opposed to using parametric Gaussian densities). Particles are then weighted and resampled as new measurements are processed [67].

11.5.3 SMOOTHING AND MAPPING (LOOP CLOSURE)

Bayesian smoothing of a Markov process can be done optimally in a forward pass, followed by a backward pass. However, for a typical mapping problem, the Markov process does not hold, because when one returns to a previously visited place and recognizes this (loop closure), a direct dependence from that earlier time to now is created. The front-end process of SLAM uses a technique called "loop closure" to recognize previously visited places to make SLAM robust and long term [72]. A pictorial representation of a typical SLAM system is given in Figure 11.11.

11.5.4 iSAM AND GraphSLAM

In the 1990s, SLAM was usually solved using extended Kalman filters [73], where the state vector described the location of landmarks of the map and the most recent robot location. A few years later, the trend shifted toward viewing the SLAM problem as a continuously growing smoothing problem, as seen in iSAM2 [65], MonoSLAM [68], GraphSLAM [74], and iSAM [75]. SLAM-based methods have the potential to replace a priori techniques if their performances can be increased further [76].

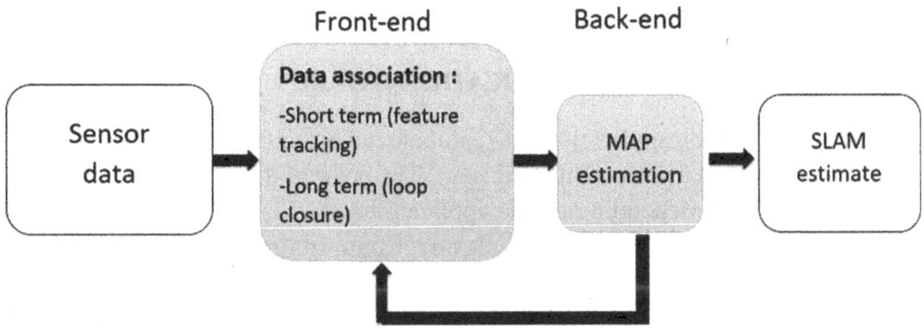

Figure 11.11 Front-end and back-end in a typical SLAM system.

11.6 COOPERATIVE ESTIMATION, FILTERING, AND SENSOR FUSION

Different onboard vehicle sensors have different strengths and weaknesses. The vehicle manufacturers, therefore, combine several different sensor types, which complement each other and thus compensate for each other's boundaries; this is called sensor fusion. Different functional techniques and redundancy through sensor fusion are considered indispensable for accurate and reliable detection of the environment of a vehicle.

Data fusion techniques are those techniques where different localization methods are combined to provide the best estimation possible in terms of cost, accuracy, performance, and complexity. Different available mechanisms can be used depending on the application and its requirements. For instance, filtering techniques such as Kalman filter, and particle filters can be used to improve the accuracy of the localization estimation.

The idea behind the data fusion techniques is to combine different information from different available systems. The data fusion technique will help to estimate the position with the best possible accuracy. Furthermore, it helps to identify additional information such as the lane of the circulation and direction and so on.

Table 11.2 compares the localization techniques mentioned above in terms of sensor configuration, accuracy, and the associated potentials and limitations. Suitability of the methods for autonomous vehicles is based on the robustness and reliability as well as capability for in-lane localization accuracy. Ideally, the accuracy should be below a threshold (usually around 20 cm) at all times. Of course, in straight lines, the longitudinal localization can be less accurate without significant consequences.

11.7 LOCALIZATION SYSTEMS IN USE FOR AUTONOMOUS DRIVING

The intelligent transportation system's main goals are to improve road quality, safety, efficiency, and comfort. The CAVs application plays a significant role in achieving this goal by facilitating the communication between the nodes and providing an efficient transmission in terms of accuracy.

These traditional applications demonstrate some limitations, especially in terms of maintenance, power supply access, coverage, and related implementation cost. The CAVs applications are divided into three main categories:

1. Applications requiring high accurate information: the error tolerance, in this case, is too small. An example of these applications is critical safety applications such as bridge monitoring and high-density intersection monitoring.
2. Applications requiring accurate information: they refer to application transmitting real-time data like position, speed, and so on.
3. Applications requiring inaccurate information: in this case, the error tolerance is higher than other applications. The error, for this kind of application, is concerning the distance estimation of up to 3 m.

11.7.1 APPLICATION FOR ACCURATE LOCATION-AWARE

The objective of the localization application is to estimate the distance between the vehicles by using different information received from neighboring cars. The error of this estimation can be acceptable (between 1 and 5 m) or strict, depending on the application.

In this case, accurate applications, the vehicles exchange information about the free space on the road by using V2V communication. This type of configuration is called a cooperative driving application. In general, there are three types of applications in this group: Platooning Applications, Cooperative Intersection Safety Applications, and Cooperative Adaptive Cruise Control Applications.

Table 11.2
Localization for Autonomous Driving Comparison

Technique	Infrastructure	Accuracy	Advantage	Disadvantages
GPS	Yes	10 m	Low cost	Poor signal availability
DGPS/IMU [77]	Yes	7 m	Low cost	Poor Signal availability
Map matching [78]	-	<1 m	Good accuracy	Requires map updates
Dead reckoning [30]	-	2.5–20 m	Low cost	Low accuracy
Cellular localization [53]	Yes	> 20m	Low cost	Low accuracy
Image/video Localization with lane detection [56]	Yes	1 m	Low Cost	Sensitive to external factors
Distributed Ad hoc localization [62]	-	<20 m	Low cost	Low accuracy
Cooperative positioning localization [52]	Yes	>70 m	Low cost	Low accuracy
Vision-based localization with road marker detection [55]	-	0.5–1.43 m	Low cost	High errors
Camera localization within LiDAR map [79]	-	0.14–0.2 m	Very high accuracy and low cost	Requires map updates
Short rage radar SLAM [28]	-	0.07–0.03 m	Very high accuracy and low cost	Low robustness to dynamic environments
LiDAR SLAM [80,81]	-	0.17–0.03 m	Very high and robust to changes in environment	High cost, high power and processing requirements

11.7.2 COOPERATIVE INTERSECTION SAFETY APPLICATIONS

The Cooperative Intersection Safety applications deal with danger in interstation areas where the vehicles are supposed to perform a safe or blind crossing. The nodes communicate with each other using V2V or V2I communication to exchange information regarding their distance to the crossing zone and their speed. Using this data, cars can cooperate to make a safe crossing.

One of the applications of this technique is the Cooperative Intersection Collision Warning System. The goal of this system is to prevent the collision in the intersection zone by computing an estimation of time to reach the intersection. All vehicles exchange this information and collaborate in order to make a free-collision crossing.

Figure 11.12 illustrates a scenario of this situation where a host vehicle is trying to estimate the time required by the target vehicle to enter the intersection. A new method is discussed in Ref. [82] to estimate the arrival time to the intersection. This method uses the formula from Eqs. (11.1)–(11.4) to compute the time and distance to the intersection for both the host and the target car.

$$\mathrm{TTI}_{\mathrm{host}}(s) = \frac{\text{intersection distance}_{\mathrm{host}}}{\text{host vehicle speed}_{\mathrm{host}}} \tag{11.9}$$

$$\mathrm{TTI}_{n}(s) = \frac{\text{Intersection distance}_{n}}{\text{host vehicle speed}_{n}} \tag{11.10}$$

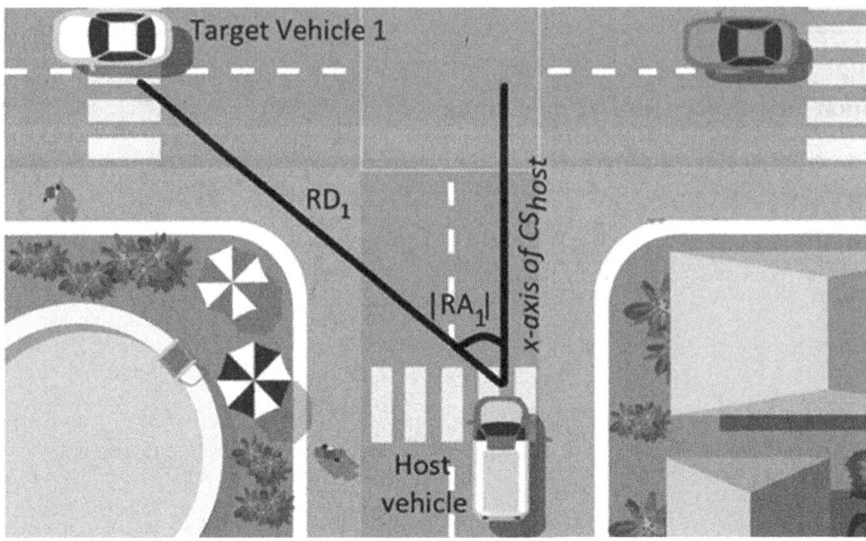

Figure 11.12 Cooperative intersection collision warning system.

$$\text{Intersection distance}_{\text{host}} = \text{RD}_n \cos\left(|RA_n|\right) \tag{11.11}$$

$$\text{Intersection distance}_n = \text{RD}_n \sin\left(|RA_n|\right) \tag{11.12}$$

The collision control system uses this data to estimate the risk, if the TTI is smaller, the risk is higher. To make a rational decision, the system sets a threshold and compares it with the time difference of the host and the target, as shown in formula (11.13).

$$|\text{TTI}_{\text{host}} - \text{TTI}_n| \leq \text{TTI}_{\text{threshold}} \tag{11.13}$$

11.7.3 COOPERATIVE ADAPTIVE CRUISE CONTROL APPLICATION

The cooperative automation of CAVs can introduce benefits to current transportation systems concerning safety, mobility, and environmental sustainability [83]. The Cooperative Adaptive Cruise Control (CACC) is a developed version of the Adaptive Cruise Control (ACC) that uses motion sensors to detect obstacles, like other vehicles. Once a car is detected, the ACC adjusts the speed to keep the same inter-car distance. The driver sets this distance.

However, the CACC deals with the speed itself and not the inter-vehicle distance. In this technique, vehicles use V2V communication to exchange different information such as acceleration to cooperate and keep the same velocity by ensuring a safe distance at any time between cars. Except for the speed adjustment, the driver controls all other functions such as direction.

11.7.4 PLATOONING APPLICATION

The platooning is a technique that has been used to enhance the efficiency in the Automated Highway Systems (AHS). The idea behind this technique is to organize the traffic in order to keep the same distance between a group of vehicles circulating in the same lane (called platoon) and to keep the same velocity for all the nodes. The platooning technique helps to improve the traffic flow by optimizing the free space between cars and following the same distance; also, it decreases the driving danger as it keeps the same variable speed for all vehicles [84]. Moreover, it helps to reduce fuel consumption in the roads, especially for heavy vehicles [85].

A platoon management protocol is discussed in Ref. [86]. The proposed protocol describes some possible maneuvers that can be performed on the road while driving, as shown in Figure 11.13. The main idea of this protocol is based on the coordination layer, where the vehicles exchange different micro-commands to coordinate and make the right decision. The framework of a platoon application is usually composed of a heading vehicle called platoon leader that leads the train of cars, and another car that follows the platoon leader and circulates carefully.

11.7.5 APPLICATION FOR HIGH ACCURATE LOCATION-AWARE

This technique deals with applications that require high accuracy information, such as safety applications. The error tolerance should be small to avoid any misunderstanding of the road environment. Driver assistant systems are one of the main applications that use this technique to provide the drivers with the road conditions (such as curves, speed limit, etc.) and, in some cases, make some automatic changes.

The Cooperative Collision Warning system (CCW) is one of the driver assistant applications. This system is based on vehicle-to-vehicle (V2V) communication to provide the driver with warning messages regarding the road conditions. The CCW of the host vehicle listens to the other car transmissions and collects information about their speed, velocity, position, and direction. Then, it uses this data to generate warning messages and displays to inform the driver about the criticality of the situation. With this technique, the safety can be enhanced as well as the collision anticipation.

For instance, if a platoon header was forced to make an unexpected stop, a collision could happen, especially if the only way to inform the following cars is the header's red brake lights. Typically, the time for a driver to make a decision is between 0.75 and 1.5 seconds. With only 0.75 seconds, the car will continue moving 10.41 m before the driver can hit the break. However, with the CCW system, the platoon header can broadcast instant information to all vehicles, and a faster, automatic decision could be made by the cars to avoid the collision. In Ref. [87], a study was discussed to demonstrate that broadcasting information about the position can significantly reduce the probability of an eventual collision in a platoon.

11.7.6 APPLICATION FOR INACCURATE LOCATION-AWARE

As discussed in the two previous sections, almost all critical applications require accurate information about a car's position to provide services, such as real road flow traffic, routing information, and so on. However, in some cases, such as map localization, the location-aware protocols can accept a certain degree of error (up to 30 m). Even with this fact, the efficiency of this type of algorithm is affected by the error value. Good efficiency means a small error, and vice versa [88].

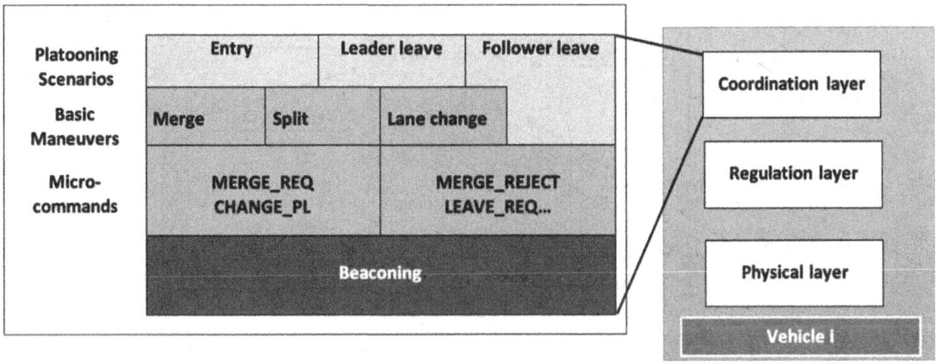

Figure 11.13 Platoon management protocol.

Inaccurate Location Aware applies two main types of protocols: routing algorithms and data dissemination algorithms. The first type uses the information about the localization to find the next nearest-neighbor hopping to the destination; this information will help the protocol to forward the information about a local or received event to the final destination in the case of hop-by-hop transmission. A standard example called Greedy Forwarding is discussed in this section [89]. The main goal of the second type is to inform a neighbor vehicle about certain events or services; the position in this case also plays a crucial role in identifying the destination. Optimized Dissemination of Alarm Messages (ODAM) protocol is discussed as an example of this type of algorithm [90,91].

11.7.7 VETRAC

The main goal of the vehicle tracking system (VETRAC) is to use the WiFi as a communication platform in order to track a moving vehicle with accuracy. The idea is to provide the driver with information that has been gathered from nearby vehicles. This information exchanged is mostly regarding the traffic conditions including any possible obstacle or warning event. The data is displayed on an electronic map in order to make it easy for the driver to get the notification.

The VETRAC uses different servers installed along the road called carries. These available carries around the vehicle are accessible through nearby access points and their job is to connect multiple points and reply to each clients' request. Each available carrier collects information concerning traffic information, such as location, orientation, and so on. Then, the relevant data is exchanged with other vehicles using WiFi.

To illustrate this concept, Figure 11.14 shows the core overview of VETRAC. A detailed architecture is also discussed in Ref. [61]. However, below are the main components of a VETRAC system:

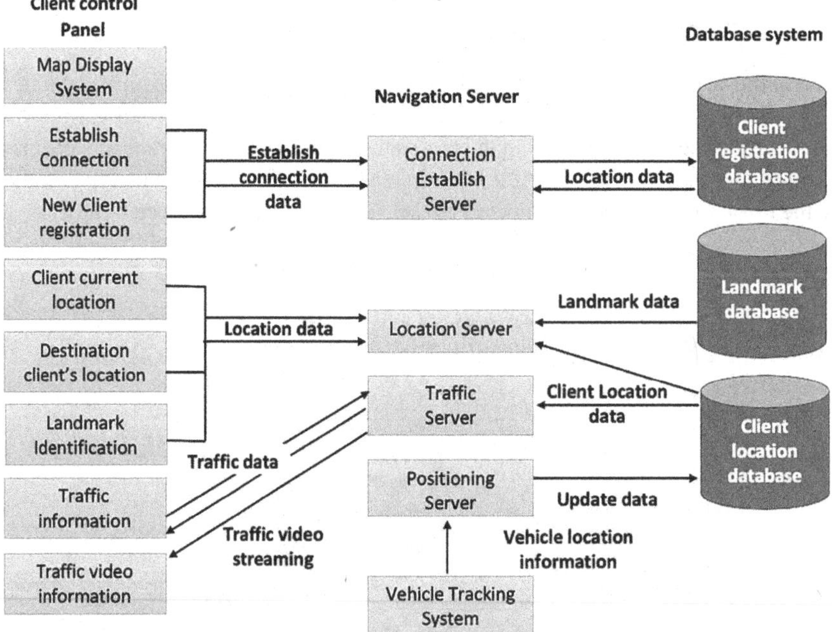

Figure 11.14 VETRAC architecture.

1. *Navigation Server*: also called the carrier, this represents the heart of the vehicles tracking system. It establishes the connection with different vehicles, collects the information regarding the traffic condition, updates the database, and replays to clients' requests.
2. *Vehicle Tracking System*: it allows the whole system to connect to different access points located along the road. Using these access points, the vehicle tracking system tracks the vehicle. Once a vehicle is detected within a specific transmission range of an access point, the system transmits this information to the navigation server.
3. *Map Display*: this is the software that is installed in the vehicle. Using the software, the information will be displayed to the driver.
4. *Client Control Panel*: this is where the user can have access and find different information regarding the traffic conditions.

11.8 CONCLUSION

Cars have been continuously evolving ever since their invention. Autonomous driving is no longer an idea, it's a reality that will change the way we will live. Cars use network services in many ways. Modern telematics finding its way into more and more vehicles, sophisticated driver assistance systems, and online functions (V2X = Vehicle-to-Everything) is almost a reality. Autonomous vehicles are here, and they are here to stay. While their use and acceptance are not yet widespread, that day is coming.

This chapter evaluates the state-of-the-art vehicle localization using onboard sensing systems and their combinations with V2X and V2I systems techniques and provides the potentials and limitations of each method on the implementation of autonomous vehicles.

Sensors are a crucial component to making a vehicle driverless. Cameras, radar, and LiDAR (Light Detection and Ranging) enable an autonomous vehicle to visualize its surroundings, detect objects, and implement interior features such as driver monitoring and customized passenger experiences.

Most localization techniques for CAVs are introduced, such as Geographical Information System (GIS), dead reckoning (DR), cellular and mobile localization, map matching, video, and image processing. The environment in which the solution is to be deployed must be considered as well. Each of the localization techniques has pros and cons and cannot work alone. Therefore, the different positioning systems not only have to work individually but also together. Similar to the aircraft, the systems also need to protect themselves from total failure by a redundant design. It is precisely in these situations that GPS, together with automotive dead reckoning, can take on the role of a fully independent localization system with high precision. Automotive dead reckoning combines satellite navigation data with other data provided by the wheel speed sensors (odometer), gyroscopes (3D), and accelerometers. This system makes accurate positioning possible even if GPS or other global navigation satellite system (GNSS) systems fail. Ultimately, on free highways, it will result in decimated accuracy, while on urban roads with their challenging environments, it will reach an accuracy of <1 m.

It was also shown that, from the performance point of view, the LiDAR techniques show the most significant promise for the localization of autonomous applications. Furthermore, the combination of camera, LiDAR, and radar data with high-definition (HD) maps allows vehicles to position themselves on the map with very high accuracy (within 10 cm) and to detect obstacles in many scenarios. However, this system needed high processing requirements, and its high cost renders it unfeasible from cost-efficiency and commercialization point of view. Therefore, these systems alone are not sufficiently safe, so a driver is still required.

Artificial intelligence (AI) and, especially, deep learning (DL) have gained significant interest in the research field of autonomous vehicles. Some believe that DL will truly allow fully autonomous vehicles, while others believe in the traditional approaches.

REFERENCES

1. Lim, H. S. M., Taeihagh, A. (2018). Autonomous vehicles for smart and sustainable cities: An in-depth exploration of privacy and cybersecurity implications. *Energies*, 11, 1062.
2. Finger, M. Audouin. M. (2019). *The Governance of Smart Transportation Systems: Towards New Organizational Structures for the Development of Shared, Automated, Electric and Integrated Mobility*, 1st ed. Springer International Publishing: Cham, Switzerland.
3. SAE. (2018). Taxonomy and definitions for terms related to driving automation systems for on-road motor vehicles. J3016 Standards; SAE: Warrendale, PA.
4. Lee, U., Jung, J., Jung, S., Shim, D. H. (2018). Development of a self-driving car that can handle the adverse weather. *International Journal of Automotive Technology* 19, 191–197.
5. Pillmann, J., Sliwa, B., Schmutzler, J., Ide, C., Wietfeld, C. (2017). Car-to-cloud communication traffic analysis based on the common vehicle information model. *IEEE Vehicular Technology Conference (VTC-Spring), Workshop on Wireless Access Technologies and Architectures for Internet of Things (IoT) Applications*, Sydney, NSW, 1–5, doi: 10.1109/VTCSpring.2017.8108664.
6. Koopman, P., Wagner, M. (2017). Autonomous vehicle safety: An interdisciplinary challenge. *IEEE Transactions on Intelligent Transportation Systems* 9, 90–96.
7. Aeberhard, M., et al. (2015). Experience results and lessons learned from automated driving on Germany's highways. *IEEE Intelligent Transportation Systems Magazine*, 7(1), 42–57.
8. Duarte, F., Ratti, C. (2018). The impact of autonomous vehicles on cities: A review. *Journal of Urban Technology*, 25, 3–18.
9. Chehri, A., Mouftah, H. T. (2019). Autonomous vehicles in the sustainable cities, the beginning of a green adventure. *Sustainable Cities and Society*, 51, 101751.
10. Kichun, J., et al. (2015). Precise localization of an autonomous car based on probabilistic noise models of road surface marker features using multiple cameras. *IEEE Transactions on Intelligent Transportation Systems*, 16, 3377–3392.
11. Schwarting, W., Alonso-Mora, J., Rus, D. (2018). Planning and decision-making for autonomous vehicles". *Annual Review of Control, Robotics, and Autonomous Systems*, 1, 187–210.
12. SAE J3018. (2015). Guidelines for safe on-road testing of SAE Level 3, 4 and 5 prototype Automated Driving Systems (ADS).
13. Parent, M., et al. (2013). Legal issues and certification of the fully automated vehicles: Best practices and lessons learned. City Mobil 2 Rep.
14. KPMG. (2012). Self-driving cars: The next revolution. Technical report, Center for Automotive Research, Transportation Systems Analysis Group.
15. Lima, P. F. (2018). Optimization-based motion planning and model predictive control for autonomous driving: With experimental evaluation on a heavy-duty construction truck. Ph.D. dissertation, KTH Royal Institute of Technology.
16. Yunpeng, S. (2019). Cooperative platooning and lane changing for connected and automated vehicles: An entropy-based method, Masters Thesis, University at Buffalo.
17. Cho, H., Seo, Y. W., Kumar, B. V. K. V., Rajkumar, R. R. (2014). A multi-sensor fusion system for moving object detection and tracking in urban driving environments. *In Proceedings of the IEEE International Conference on Robotics and Automation*, Hong Kong, 1836–1843.
18. Chen, D., Ahn, S., Chitturi, M., Noyce, D. A. (2017). Towards vehicle automation: Roadway capacity formulation for traffic mixed with regular and automated vehicles. *Transportation Research Part B: Methodological*, 100, 196–221.
19. Hao, J. R., Hu, J., Ma, W. (2017). Swarm intelligence based algorithm for management of autonomous vehicles on arterials.
20. Kato, S., Tsugawa, S. (2001). Cooperative driving of autonomous vehicles based on localization, inter-vehicle communications and vision systems, *JSAE Review*, 22(4), 503–509.
21. Stenborg, E . (2017). Localization for autonomous vehicles. Thesis for the Degree of Licentiate of Engineering, Chalmers University of Technology.
22. Cohen, J. (2018). Self-driving cars and localization. Towards data science.
23. Vu, T.-D. (2009). Vehicle perception: Localization mapping with detection classification and tracking of moving objects. Ph.D Thesis, Institut National Polytechnique de Grenoble.

24. Wang, C.-C., Thorpe, C., Suppe, A. (2003). Ladar-based detection and tracking of moving objects from a ground vehicle at high speeds. *The IEEE Intelligent Vehicles Symposium*, Columbus, OH, USA, 416–421, doi: 10.1109/IVS.2003.1212947.
25. Wang, C.-C., Thorpe, C., Thrun, S. (2003). Online simultaneous localization and mapping with detection and tracking of moving objects: Theory and results from a ground vehicle in crowded urban areas, *Proceedings of IEEE International Conference on Robotics and Automation*, Taipei, Taiwan, 842–849.
26. Greenblatt, N. (2016). Self-driving cars will be ready before our laws are. *IEEE Spectrum*, 53, 46–51.
27. Rasshofer, R. H., Gresser, K. (2005). Automotive radar and lidar systems for next generation driver assistance functions. *Advances in Radio Science*, 3(8), 205–209.
28. Ward, E., Folkesson, J. (2016). Vehicle localization with low cost radar sensors. *2016 IEEE of Intelligent Vehicles Symposium (IV)*, Gothenburg, Sweden, 864–870.
29. Vivet, D., Gérossier, F., Checchin, P., Trassoudaine, L., Chapuis, R. (2013). Mobile ground-based radar sensor for localization and mapping: An evaluation of two approaches. *International Journal of Advanced Robotics Systems (IJARS)*, 10(5), 307–318.
30. Cornick, M., Koechling, J., Stanley, B., Zhang, B. (2016). Localizing ground penetrating RADAR: A step toward robust autonomous ground vehicle localization. *Journal of Field Robotics*, 33(1), 82–102.
31. Qu, X., Soheilian, B., Paparoditis, N. (2015). Vehicle localization using mono-camera and geo-referenced traffic signs. *In Intelligent Vehicles Symposium (IV)*, Seoul, South Korea, 605–610.
32. Arai, S., et al. (2007). Experimental on hierarchical transmission scheme for visible light communication using LED traffic light and high-speed camera. *IEEE 66th Vehicular Technology Conference (VTC-Fall)*, Baltimore, MD, 2174–2178.
33. Ma, Y., Soatto, S., Kosecka, J., Sastry, S. S. (2012). *An Invitation to 3D Vision: From Images to Geometric Models*. Springer: Berlin.
34. Kneip, L., Scaramuzza, D., Siegwart, R. (2011). A novel parametrization of the perspective-three-point problem for a direct computation of absolute camera position and orientation. *Proceedings of IEEE Conference on Computer Vision and Pattern Recognition*, Country, 2969–2976.
35. Lin, D., Fidler, S., Urtasun, R. (2013). Holistic scene understanding for 3D object detection with RGBD cameras, *In 2013 IEEE International Conference on Computer Vision*, Sydney, Australia, 1417–1424.
36. Kuutti, S., Fallah, S., Katsaros, K., Dianati, M., Mccullough, F., Mouzakitis, A. (2018). A survey of the state-of-the-art localization techniques and their potentials for autonomous vehicle applications. *IEEE Internet of Things Journal*, 5(2), 829–846.
37. Kato, S., Takeuchi, E., Ishiguro, Y., Ninomiya, Y., Takeda, K., Hamada, T. (2015). An open approach to autonomous vehicles. *IEEE Micro*, 35(6), 60–68.
38. Woo, A., Fidan, B., Melek, W. W. (2019). Localization for autonomous driving. In *Handbook of Position Location*, (eds S. A. Zekavat and R. M. Buehrer). IEEE Press and John Wiley & Sons, Inc: Hoboken, NJ, 1051–1087.
39. Djuknic, G. M., Richton, R. E. (2011). Geolocation and assisted GPS. *Computer*, 34(2), 123–125.
40. Rohani, M., Gingras, D., Vigneron, V., Gruyer, D. (2015). A new decentralized Bayesian approach for cooperative vehicle localization based on fusion of GPS and VANET based inter-vehicle distance measurement. *IEEE Intelligent Transportation Systems Magazine*, 7(2), 85–95.
41. Tan, H.-S., Huang, J. (2006). DGPS-based vehicle-to-vehicle cooperative collision warning: Engineering feasibility viewpoints. *IEEE Transactions on Intelligent Transportation Systems*, 7(4), 415–428.
42. Quddus, M. A., Ochieng, W. Y., Noland, R. B. (2007). Current map-matching algorithms for transport applications: State-of-the art and future research directions. *Transportation Research Part C: Emerging Technologies*, 15, 312–328.
43. Levinson, J., Montemerlo, M., Thrun, S. (2007). Map-based precision vehicle localization in urban environments. *Robotics: Science and Systems*, Cambridge, MA:MIT Press, 4(3), 121–128.
44. Sharma, K., Jeong, K., Kim, S. (2011). Vision based autonomous vehicle navigation with self-organizing map feature matching technique. *11th International Conference on Control, Automation and Systems*, Gyeonggi-do, 946–949.
45. Brakatsoulas, S., Pfoser, D., Salas, R., Wenk, C. (2005). On map-matching vehicle tracking data. *Proceedings of the 31st International Conference on Very Large Databases*, Trondheim, Norway, 853–864.

46. Lou, Y., Zhang, C., Zheng, Y., Xie, X., Wang, W., Huang, Y. (2009). Map-matching for low-sampling-rate GPS trajectories. *Proceedings of the 17th ACM SIGSPATIAL International Conference on Advances in Geographic Information Systems GIS*, Seattle, WA.
47. Chen, D., Driemel, A., Guibas, L., Wenk, C. (2011). Approximate map matching with respect to the Frechet Distance. *Proceedings of the Thirteenth Workshop on Algorithm Engineering and Experiments, ALENEX 2011*, San Francisco, CA, January 22, 75–83.
48. Pink, O., Hummel, B. (2008). A statistical approach to map matching using road network geometry, topology and vehicular motion constraints. *In the 11th International IEEE Conference on Intelligent Transportation Systems*, Beijing, China.
49. Kim, S., Kim, J.-H. (2001). Adaptive fuzzy-network-based C-measure map-matching algorithm for car navigation system. *IEEE Transactions on Industrial Electronics*, 48(2), 432–441.
50. Jagadeesh, G. R., Srikanthan, T., Zhang, X. D. (2005). A map matching method for gps based real-time vehicle location. *Journal of Navigation*, 57, 429–440.
51. Chen, M., Haehnel, D., Hightower, J., Sohn, T., LaMarca, A., Smith, I., Chmelev, D., Hughes, J., Potter, F. (2006). Practical metropolitan-scale positioning for GSM phones, *In Proceedings of 8th Ubicomp*, Orange County, CA, 225–242.
52. Mayorga, C. L. F., et al. (2007). Cooperative positioning techniques for mobile localization in 4G cellular networks. *IEEE International Conference on Pervasive Services*, Istanbul, 39–44.
53. del Peral-Rosado, J. A., López-Salcedo, J. A., Kim, S., Seco-Granados, G. (2016). Feasibility study of 5G-based localization for assisted driving. *International Conference on Localization and GNSS (ICL-GNSS)*, Barcelona, 1–6.
54. Parra, I., Sotelo, M. A., Llorca, D., Ocaña, M. (2010). Robust visual odometry for vehicle localization in urban environments. *Robotica*, 28(3), 441–452.
55. Li, C., Dai, B., Wu, T. (2013). Vision-based precision vehicle localization in urban environments. *2013 Chinese Automation Congress*, Changsha, 599–604, doi: 10.1109/CAC.2013.6775806.
56. Kamijo, S., Gu, Y., Hsu, L. (2015). Autonomous vehicle technologies: Localization and mapping. *IEICE ESS Fundamentals Review*, 9(2), 131–141.
57. Suhr, J. K., Jang, J., Min, D., Jung, H. G. (2017). Sensor fusion-based low-cost vehicle localization system for complex urban environments. *IEEE Transactions on Intelligent Transportation Systems*, 18(5), 1078–1086.
58. Yang, H., Shao, L., Zheng, F., Wang, L., and Song, Z. (2011). Recent advances and trends in visual tracking: A review. *Neurocomputing*, 74(18), 3823–3831.
59. Ramirez, A. S. (2014). Vehicle localization and tracking from an on-board camera. Masters Thesis, Université Laval.
60. Carlson, C. R., Gerdes, J. C., Powell, J. D. (2004). Error sources when and vehicle dead reckoning with differential wheelspeeds. *Navigation*, 1, 13–27.
61. Parker, R., Valaee, S. (2006). Vehicle localization in vehicular networks. *IEEE 64th Vehicular Technology Conference*, Montreal, Quebec, Canada, 1–5, doi: 10.1109/VTCF.2006.557.
62. Benslimane, A. (2005). Localization in vehicular ad hoc networks. *Systems Communications*, Montreal, Que, Canada, 19–25.
63. Cadena, C., et al. (2016). Past present and future of simultaneous localization and mapping: Toward the robust-perception age. *IEEE Transactions on Robotics*, 32(6), 1309–1332.
64. Bresson, G., Alsayed, Z., Yu, L., Glaser, S.(2017). Simultaneous localization and mapping: A survey of current trends in autonomous driving. *IEEE Transactions on Intelligent Vehicles*, 2(3), 194–220.
65. Kaess, M., Johannsson, H., Roberts, R. , Ila, V., Leonard, J., Dellaert, F. (2011). iSAM2: Incremental smoothing and mapping with fluid relinearization and incremental variable reordering. *IEEE International Conference on Robotics and Automation*, Shanghai, 3281–3288.
66. Thrun, S., Leonard, J. J. (2008). Simultaneous localization and mapping. In: *Springer Handbook of Robotics* (eds B. Siciliano, O. Khatib). Springer-Verlag: Berlin Heidelberg, 871–889.
67. Wolcott, R. W. (2016). Robust localization in 3d prior maps for autonomous driving, Ph.D Thesis, University of Michigan.
68. Davison, A. J., Reid, I. D., Molton, N. D. Stasse, O. (2007). MonoSLAM: Real-time single camera SLAM. *IEEE Transactions on Pattern Analysis and Machine Intelligence*, 26(6), 1052–1067.

69. Julier, S., Uhlmann, J. (1997). A new extension of the Kalman filter to nonlinear systems. *11th International Symposium on Aerospace/Defense Sensing, Simulation and Controls, Multi Sensor Fusion, Tracking and Resource Management II*, Orlando, FL, 182–193.
70. Gustafsson, F., Hendeby, G. (2012). Some relations between extended and unscented Kalman filters. *IEEE Transactions on Signal Processing*, 60(2), 545–555.
71. Arasaratnam, I., Haykin, S. (2009). Cubature Kalman filters. *IEEE Transactions on Automatic Control*, 54(6), 1254–1269.
72. Woo, A., Fidam, B., Mele, W.W. (2019). Localization for autonomous driving. In: *Handbook of Position Location: Theory, Practice, and Advances*, (eds R. Zekavat, R. Buehrer). John Wiley & Sons, Inc: Hoboken, NJ, 1051–1087.
73. Leonard, J. J., Durrant-Whyte, H. F. (1991). Simultaneous map building and localization for an autonomous mobile robot. *Proceedings IROS '91:IEEE/RSJ International Workshop on Intelligent Robots and Systems '91*, Osaka, Japan, 1442–1447, vol. 3, doi: 10.1109/IROS.1991.174711.
74. Thrun, S., Montemerlo, M. (2006). The graph slam algorithm with applications to large-scale mapping of urban structures. *The International Journal of Robotics Research*, 25(5), 403–429.
75. Kaess, M., Ranganathan, A., Dellaert, F. (2008). iSAM: Incremental smoothing and mapping. *IEEE Transactions on Robotics*, 24(6), 1365–1378.
76. Van Brummelen, J., O'Brien, M., Gruyer, D., Najjaran, H. (2018). Autonomous vehicle perception: The technology of today and tomorrow. *Transportation Research Part C: Emerging Technologies*, 89, 384–406.
77. Zhang, F., et al. (2012). A sensor fusion approach for localization with cumulative error elimination. *2012 IEEE International Conference on Multisensor Fusion and Integration for Intelligent Systems (MFI)*, Hamburg, 1–6, doi: 10.1109/MFI.2012.6343009.
78. Milford, M. J., Wyeth, G. F. (2012). SeqSLAM: Visual route-based navigation for sunny summer days and stormy winter nights. *IEEE International Conference on Robotics and Automation*, Saint Paul, MN, 1643–1649.
79. Wolcott, R. W., Eustice, R. M. (2014). Visual localization within LIDAR maps for automated urban driving. *IEEE International Conference on Intelligent Robots and Systems*, Chicago, IL, 176–183.
80. Levinson, J., Thrun, S. (2010). Robust vehicle localization in urban environments using probabilistic maps. *The 2010 IEEE International Conference on Robotics and Automation (ICRA2010)*, Anchorage, Alaska, 4372–4378.
81. Castorena, J., Agarwal, S. (2018). Ground-edge-based LIDAR localization without a reflectivity calibration for autonomous driving. *IEEE Robotics and Automation Letters*, 3(1), 344–351.
82. Cho, H. (2014). Cooperative intersection collision: Warning system based on vehicle-to-vehicle communication. *Contemporary Engineering Sciences*, 7(22), 1147–1154.
83. Wang, Z., et al. (2019). A survey on cooperative longitudinal motion control of multiple connected and automated vehicles. *IEEE Intelligent Transportation Systems Magazine*, 12(1), 4–24, Spring 2020, doi: 10.1109/MITS.2019.2953562.
84. Karbalaieali, S., Osman, O. A., Ishak, S. (2019). A dynamic adaptive algorithm for merging into platoons in connected automated environments. *IEEE Transactions on Intelligent Transportation Systems*,1–12, doi: 10.1109/TITS.2019.2938728.
85. Karlsson, K., Carlsson, J., Larsson, M., Bergenhem, C. (2016). Evaluation of the V2V channel and diversity potential for platooning trucks. *10th European Conference on Antennas and Propagation (EuCAP)*, Davos, 1–5.
86. Amoozadeh, M., Deng, H., Chuah, C. N., Zhang, H. M., Ghosal, D. (2015). Platoon management with cooperative adaptive cruise control enabled by vanet. *Vehicular Communications*, 2(2), 110–123.
87. Tatchikou, R., Biswas, S., Dion, F. (2005). Cooperative vehicle collision avoidance using inter-vehicle packet forwarding. *IEEE Global Telecommunications Conference, GLOBECOM*, St. Louis, MI.
88. Shah, R., Wolisz, A., Rabaey, J. (2005). On the performance of geographical routing in the presence of localization errors ad hoc network applications. *IEEE International Conference on Communications, ICC*, Seoul, 2979–2985, vol. 5, doi: 10.1109/ICC.2005.1494938.
89. Bhoi, S. K., Khilar, P. M. (2015). SIR: A secure and intelligent routing protocol for vehicular ad hoc network. *IET Networks*, 4(3), 185–194.

90. Wang, W., Luo, T. (2016). The minimum delay relay optimization based on nakagami distribution for safety message broadcasting in urban VANET. *IEEE Wireless Communications and Networking Conference*, Doha, 1–6, doi: 10.1109/WCNC.2016.7564751.
91. Allani, S., Yeferny, T., Chbeir, R., Yahia, S. B. (2016). DPMS: A swift data dissemination protocol based on map splitting. *IEEE 40th Annual Computer Software and Applications Conference (COMPSAC)*, Atlanta, GA, 817–822.

12 Automotive Radar Signal Analysis

Hassan Moradi and Ashish Basireddy
Qualcomm Technologies Inc.

CONTENTS

This chapter provides a brief, albeit in-depth, explanation of automotive radar and signal processing. Today's advanced driver assistance system, commonly known as ADAS, and automated driving system, commonly known as ADS, remain at the epicenter of efforts to improve driving safety by leveraging newly developed, efficient surrounding-perception algorithms. Although next generation ADAS is expected to deploy wireless communication features via vehicle-to-vehicle (V2V) and/or vehicle-to-infrastructure (V2I) communication, the use of radar signal processing is inevitable due to its cost-effectiveness. It is widely known that autonomous driving is undergoing a significant technological transformation, warranting its current status as one of the most highly discussed topics among researchers. ADAS novel functionalities are providing the automotive industry with important advancements in both safety and comfort. Such technological advancements are expected to continue and will likely continue to impact our overall transport system. One noteworthy challenge for ADAS/ADS is perceiving the surrounding environment. Typically, various sensors (e.g., camera, radar, and lidar) have been employed to mitigate this challenge. Solutions are based on cost and reliability under varying conditions. The focus of this chapter is radar signal processing. Both a brief description of radar's significance in the automotive industry and a more detailed discussion of existing signal processing techniques are detailed in several sections. Radar signals are poised for enhancement by newly adapted methods borrowed from the wireless industry. This chapter highlights a comparative study, wherein the performance of a conventional radar solution is compared to an innovative method. In doing so, this chapter serves as a brief reference for automotive radar signal analysis.

12.1 AUTOMOTIVE RADAR

Radar has been extensively examined as a solution among industry, military, and academic researchers for solving issues related to speed and range estimations. Most vehicle detection techniques, including ADAS, utilize frequency modulated continuous wave (FMCW) radar. This analog waveform has been used in a number of practical radar environments, ranging from automotive to military to human detection [1]. FMCW is a simple, robust, and efficient solution already used in many practical radar environments. However, in spite of its exceptional properties, FMCW suffers from several drawbacks, including ghost targets problem [2] and high processing load. One important challenge is the technology's ability to tweak analog waveform processing for optimizing reliable, multi-target detection in the presence of interference. The idea of deploying digital waveforms, investigated in [3–6], is coming from the aforementioned drawbacks.

Over the years, radar technology has facilitated speed estimation and object tracking, with a primary focus on military applications. Hence, exploitation of radar signal processing in autonomous driving is highly anticipated. The underlying principle of radar focuses on continuous wave transmission and its reflection for speed and target range estimates. FMCW is designed to work in millimeter wave range (e.g., 24 GHz and 76–81 GHz) and provides multi-target detection at no additional cost for receiver and/or transmitter. Due to its simplicity and cost efficiency, FMCW and its various applications, including ADAS, have been operational in the radar industry for several decades. As previously mentioned, one notable drawback of FMCW is the problem of ghost target detection in multi target scenarios (i.e., traditional triangular FMCW-based radar has been characterized by introducing ghost targets that cause a false alarm at a rate that increases simultaneously with the increased number of targets). There is another type of FMCW in the form of sawtooth waveform, which eliminates the ghost target problem at the cost of high memory usage and processing load. New technologies are needed to mitigate such issues. One possible solution among many options is leveraging generated waveforms based on orthogonal frequency division multiplexing (OFDM) signals. The advantages and disadvantages of this new methodology are discussed below. Additional details about FMCW-based radar systems are available in the list of references at the end of this chapter.

12.1.1 ASSISTED DRIVER SENSORS

Technically, a vehicle must be equipped with various sensors to facilitate autonomous driving. Cameras, radars, and lidars are predominantly used for this purpose. An illustration of such driving-assisted sensors is shown in Figure 12.1. Some current-generation vehicles utilize up to eight cameras and a computer vision algorithm for processing. Careful camera placement aids in detecting objects at varying angles and ranges.

Range resolution can present a problem when detecting objects that are in close proximity (e.g., parking assist and rear moving activities) where low resolution—in the range of one inch—is desired. Since the resolution of electromagnetic radar is directly proportional to the speed of light, traditional FMCW radars are likely unable to provide necessary accuracy. Instead, ultrasonic waves that travel at a speed much lower than electromagnetic waves can be utilized. Hence, ultrasonic sensors in autonomous vehicles (AVs) are often utilized to provide non-microwave signals for close proximity detection applications. In such cases, normally up to 12 ultrasonic sensors are utilized in current vehicles to provide functional vision for detecting various types of objects, e.g., human, wall, or box located in a garage.

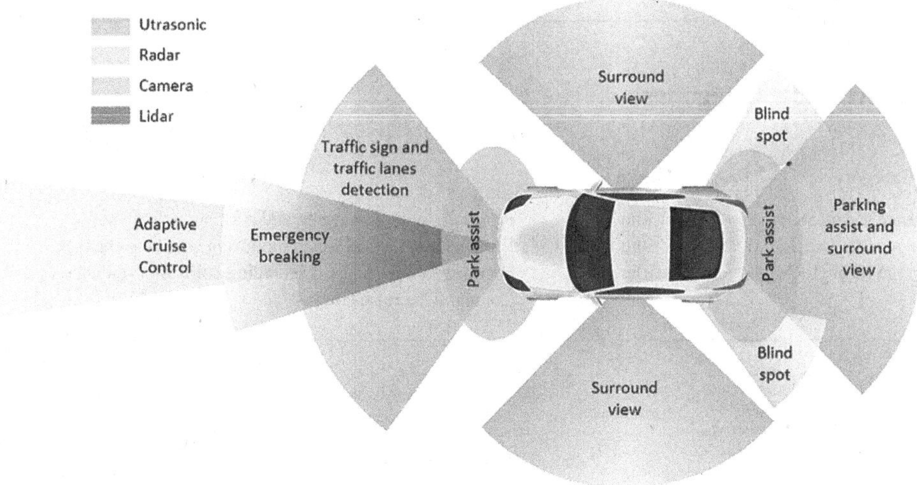

Figure 12.1 Distribution of sensors in an ADAS capable vehicle

Automotive sensors are configured to deliver independent measurements and facilitate a precise understanding of the surrounding environment, including location, motion, and identification. Each sensor has advantages and disadvantages related to metrics for range and velocity resolution, robustness to noise and weather conditions, as well as comparisons for size, weight, and price. Even though a camera sensor can deliver high-range resolution, has small form factor, and is highly accurate in detecting road signs and providing object recognition, it is ineffective under severe atmospheric conditions. Camera-based image processing techniques for vehicle detection also suffer from shortcomings during dark, night-time conditions. Alternatively, lidars provide a highly accurate and reliable perception of the surrounding environment, as well as acceptable range estimates, albeit at a higher cost. Although ultrasonic transceivers are inefficient for detecting objects in long ranges, radar based on this technology provides valuable information in many challenging scenarios and at a lower price. But radars are unable to guarantee object identification. For example, while a radar can detect the location and speed of another vehicle, it is unable to solely recognize the exact edges of a vehicle or to distinguish between a road sign and a stationary vehicle. Knowing these difficulties, one can only imagine associated issues triggered in an AVdriving situation.

To achieve the highest efficiency for radar in AV driving, information from multiple sensors must be merged and/or combined, i.e., a methodology referred to as Perception with an algorithm named Sensor Fusion [7–9]. Perception is, in fact, adept at understanding the environment by leveraging various types of sensors. Perception not only helps with obstacle detection, it also engages in vehicle localization [10], motion tracking, and path planning [11–13]. The former utilizes positioning information (e.g. GPS) to more precisely identify vehicle position, and the latter creates a trajectory for the vehicle to react to surrounding objects and develop an autonomous driving plan. Readers should note that these topics are out of the scope of this chapter.

Existing autonomous and automated vehicles are typically equipped with up to six radars categorized according to their applications rather than range covered:

- Short-range radar (SRR) to detect proximity objects up to 30m
- Mid-range radar (MRR) to watch surrounding up to 50m
- Long-range radar (LRR), forward-looking radar to scan object up to 250m

Automotive radar classifications are listed in Table 12.1, and use for each category is described in Figure 12.2. SRR radars are mainly applicable for parking assist, although ultrasonic sensors appear

Table 12.1
Automotive Radar Categories

Radar Category	Range	Usage
Short-range radar (SRR)	30m	Proximity objects warning and parking assist features
Mid-range radar (MRR)	60m	Blind spot detection, lane monitoring, and lane change
Long-range radar (LRR)	250m	Forward-looking sensors including collision avoidance warning and adaptive cruise control.

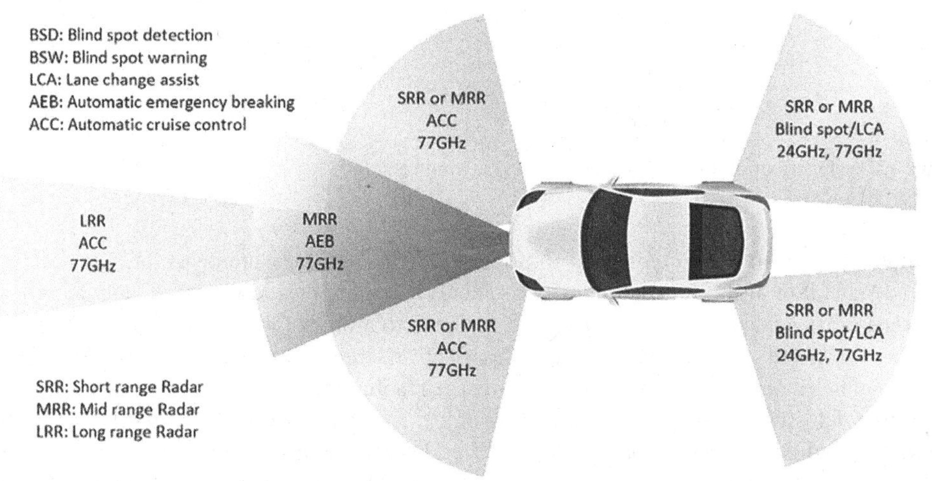

Figure 12.2 Distribution of radars in automotive

to be more efficient due to low range resolution. It can be also used for blind spot monitoring/detection/warning (BSM/BSD/BSW). MRR is the most popular radar currently on roadways and used for BSM/BSD/BSW, lane change assist (LCA), automatic emergency breaking (AEB), parking assist, and automatic cruise control (ACC). LRR radar is quite different from the other types and is used for ACC only. Its signal processing does not deliver high resolution measurements, however, at a long range of 100m, for example, a low resolution is acceptable. There will likely be no safety concerns if a target at this range is detected, even with an acceptable error of a few meters. Another unique property of LRR is that as coverage increases, loss and attenuation of the received signal also increase; hence, a reasonable excuse to keep LRR signal frequency as low as possible (e.g., if a 77 GHz band is used, frequency is limited to a range of 76-77 GHz). Note that the information provided in Figure 12.2 and Table 12.1 are configuration examples of an ADAS/ADS system and vehicles may not follow exact configuration [14]. With such control in accuracy, radar system signal processing resources can be exploited to cover longer ranges and improve coverage. This is an important factor in automotive radar design. Given the trade-offs between coverage and accuracy, measurements from LRR, SRR, and MRR can be combined to guarantee both range/speed resolution and detection accuracy.

International Organization for Standardization (ISO) is an entity on regulating the performance and procedures of driving sensors. [15] is one of the examples on performance requirements and test procedures for lane change assist. In the United States, ADAS and ADS specifications are regulated by National Highway Traffic Safety Administration (NHTSA). As a reference to the performance of blind spot, [16] summarizes findings of a study of BSM by original vehicle manufacturers on standard production vehicles.

12.1.2 SIGNIFICANCE

Radar signals based on electromagnetic waves provide the most efficient method for measuring moving object velocity in our surroundings. Occasionally, camera images accumulated over time are used in automotive industry for this purpose. Such techniques have proven inefficient in harsh weather conditions; however, while radar has been used seamlessly in all weather conditions, basically for line of sight object identification.

With radar, a target is modeled as a point scatterer and characterized by radar cross section (RCS), indicating that the target is assumed to be acting like a point. A typical radar sensor is small in size and uses less processing energy when compared with image analysis for range and velocity estimation. The current role of radar in the automotive industry is solely based on target detection, and, as such, the information from radar suffices for many assisted driving features and applications. However, such information is not adequate for total self-driving capabilities of AVs. Since radar provides independent measurements, and act and a point scatterer, they pave the way for efficient data fusion from multiple sensors for decision making and achieving a robust and precise self-driving system.

In spectrum analysis, millimeter wave (mm wave) refers to electromagnetic waves with frequency above 24GHz, where wavelength is in the millimeter range [17]. Signal wavelength in this frequency range will be less than 12.5 *mm*. In mobile communications, 5G technology serves as a first effort for using mm wave for cellular communications. One important parameter in characterizing high frequency signals is sensitivity of phase-to-propagation distance. One can see that based on the following well known equation, phase and distance are related and that at a small wavelength, a small change in the propagation distance of ΔR will cause a remarkable change in phase:

$$\Delta\phi = \frac{2\pi\Delta R}{\lambda} \quad \text{(one way transmission)} \tag{12.1}$$

where λ denotes wavelength. For a two way transmission, the phase shift will be given by:

$$\Delta\phi = \frac{4\pi\Delta R}{\lambda} \quad \text{(two way transmission, i.e. radar)} \tag{12.2}$$

Additionally, time delay τ for a certain phase shift $\Delta\phi$ in a one way transmission is given by:

$$\tau = \frac{\Delta\phi\ \lambda\ f_c}{2\pi\ c} \tag{12.3}$$

where c is the speed of light and f_c is the wave frequency. The previous equations will be used in range and speed estimation of FMCW-based radars, specifically for a sawtooth waveform in section 12.2.3. Since radar is based on reflected signals, round trip time (RTT) is more applicable in such analyses and is defined as twice of time delay. We denote this as τ_d in our analysis:

$$\tau_d = 2\,\tau \tag{12.4}$$

The RTT delay τ_D is typically an extremely small fraction of a radar signal's total duty cycle. For instance, given a radar with 250 meter range and 50 μs duty cycle, maximum RTT value is only 1.67 μs, which is small relative to duty cycle. In practice, RTT associated with a target is much smaller than such a value. For a target located at a range of 2 m, RTT will be equal to 13 nm. In this way, a practical pulse radar will be either unable or inefficient for detecting such small RTT values. In the next section, three typical radar signal waveforms will be presented.

12.2 WAVEFORMS IN AUTOMOTIVE RADARS

As mentioned earlier, AV relies on multiple sensors to perceive the surrounding environment. Among all safety devices, radar is one of those playing an important role. Interestingly, there is currently a strong push for an efficient AV radar system while pulsed waveform radars prove inefficient due to the short range of target location. The following section reviews FMCW radar fundamentals, which are built on continuous frequency modulation and are well adopted for serving as the most popular existing radar technology in the automotive industry.

Continuous frequency modulated signals can have multiple representations. Mathematical expression depends on variation in frequency with modulated signal time. Typically, frequency versus time can vary as sawtooth, triangular, sinusoidal, or trapezoid shapes. These have similar functionalities and advantages with only minor differences. Among all types, triangular and sawtooth are from up chirp and down chirp frequencies advantage. In this section, an analysis of rectangular, trapezoidal and sawtooth waves is presented, and design details of each waveform are discussed. Additional details about FMCW functionality and implementation can be found in [18] and [19]. This section furthermore demonstrates how digital modulation in the form of OFDM can be used in automotive radars. Advantages and disadvantages of each waveform are presented below.

12.2.1 TRIANGULAR FMCW WAVEFORM

An FMCW is based on a "chirp" signal, which is a sinusoidal wave whose frequency varies linearly with time. Let f_c denote start frequency of the chirp. If wave frequency increases to a value of $f_c + B$, B is known as bandwidth of the chirp. The block diagram of an FMCW transceiver is shown in Figure 12.3. The plot is a simplified block diagram and does not provide details of the low pass filtering (LPF) (i.e., decimation filter) nor of the signal processing unit. At the transmitter side, a voltage-controlled oscillator (VCO), power amplifier (PA), and power divider are deployed. The receiver includes a low noise amplifier (LNA), a mixer, and an LPF [20]. Mixer component output provides the product of respective input signals, which after decimation LPF will result in an intermediate frequency (IF) signal composed of information related to beat frequencies:

As the term "waveform" indicates, FMCW includes periodic up chirp and down chirp frequencies. The model of the waveform ensures the plot of frequency vs. time, as depicted in Figure 12.4, wherein frequency waveforms of both transmitted and received signal in a FMCW system are plotted.

Mathematically, such a triangle continuous-frequency modulated signal can be represented as below:

$$x(t) = \sum_{n=1}^{N_c-1} A_t \cos\left(2\pi\left(f_c\, t + 0.5\alpha\, t^2 - \alpha T_c\, n\, t\right)\right) \tag{12.5}$$

where A_t is the amplitude of the transmitted signal; T_c is the length of one period, which is basically the length of one chirp (aka chirp repetition period and sweep time); f_c is the center carrier or

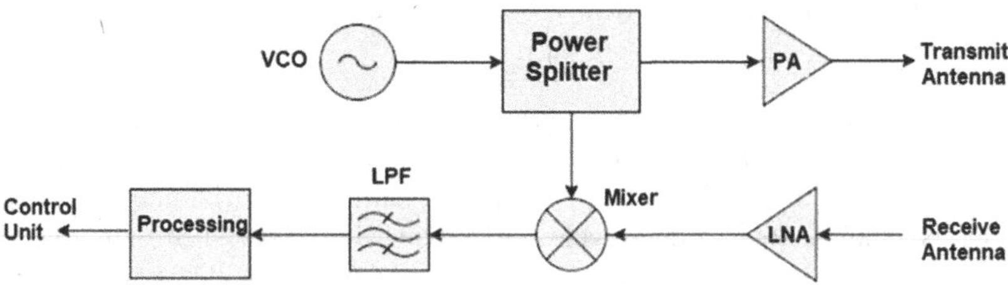

Figure 12.3 FMCW transceiver block diagram

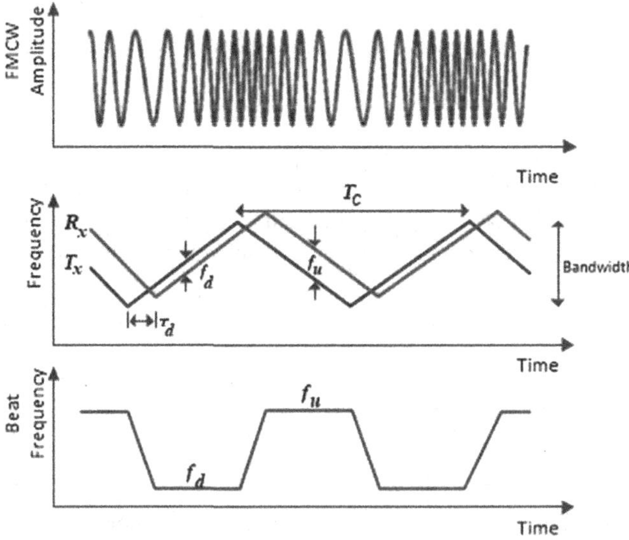

Figure 12.4 Frequency vs. time of transmitted and received signals of an FMCW triangle waveform.

operating frequency; α is the sweep rate (aka chirp rate); and N_c is the total number of chirps. Chirp rate is represented in Hz per second and can be considered as the rate of frequency changes. This phenomenon is represented as:

$$\alpha = B/T_f \tag{12.6}$$

Considering the above transmitted signal, received signal is given by:

$$y(t) = \sum_{n=1}^{N_C-1} \{A_r \cos\left(2\pi\left((f_c + f_D)(t - \tau_d) + 0.5\alpha(t - \tau_d)^2 - \alpha T_c\, n\, (t - \tau_d)\right)\right)\} \tag{12.7}$$

where A_r is the amplitude of the received signal; τ_d is the time delay due to range (i.e., RTT); and f_D is the Doppler frequency. Notably, the bandwidth shift between transmitted and received signals is equal to the Doppler shift. Since frequency resolution is inversely proportional to the time span of data collected, the size of Fast Fourier Transform (FFT) is related to the minimum Doppler frequency, which can be detected by the FMCW system. In other words, speed resolution specifies the data length required for processing. Analogously, range resolution is related to the inverse of chirp bandwidth. These metrics will be discussed later in this chapter.

Traditional FMCW transmitter is designed using a VCO, whose input is based on frequency vs. time requirement. One of the noted advantages of this algorithm is that it does not require high sampling rates for detecting close proximity targets—a major challenge for automotive radar. At the receiver, range and velocity are estimated by solving two equations [21]. System frequency components are impacted by two parameters: linear frequency shift and Doppler frequency shift. The former is given by:

$$f_b = \frac{2R\,B}{T_c\,c} = \frac{2\alpha R}{c} \tag{12.8}$$

which is the frequency shift due to linear frequency changes of the chirp. In (12.8), R is the range of target w.r.t. to radar. Doppler component shift is impacted by target mobility and is equal to:

$$f_D = \frac{2v}{\lambda} \tag{12.9}$$

where v is the relative speed of the target. In (12.9), the factor of 2 emerges due to the round trip property of the radar signal. Beat frequencies for down chirp and up chirp can be calculated as:

$$\begin{aligned} f_d &= f_b - f_D \\ f_u &= f_b + f_D \end{aligned} \tag{12.10}$$

respectively, where f_d and f_u are beat frequencies between transmitted and received signals, as depicted in Figure 12.4. Note that beat frequencies are, in fact, components of intermediate frequency (IF) signal. By combining the equations above, range and velocity of a target can be computed as follows:

$$R = \frac{c\,T_c}{4B}(f_u + f_d) \tag{12.11}$$

$$v = \frac{c}{4f_c}(f_u - f_d) \tag{12.12}$$

Despite the advantage of low complexity in analog domain, continuous frequency modulation suffers from some weaknesses. The equations listed above describe the characterization of a single target in a radar range. Under multiple target scenarios (and even though it is relatively easy to expand these equations), solving them results in multiple range and velocity (i.e., R and v) possibilities. The non-existent pair results in a ghost target, and their number is dependent on the number of actual targets. Mathematically for N_t targets, there will be multiples copies of the reflected signal received at radar location, resulting in $2N_t$ beat frequencies. Using binomial coefficient, a total of $2N_t$ beat frequencies corresponds to

$$N_{FMCW} = 2N_t^2 - N_t \tag{12.13}$$

possible targets, of which $2N_t^2 - 2N_t$ are not real targets. Such targets can be classified further as ambiguous and unambiguous targets. Ambiguous targets can be excluded from the list if targets are outside the desired interval of range and velocity estimations. Generally, since beat frequencies are captured in the analog domain, signal processing won't find corresponding pairs. In fact, continuous wave modulation introduces a high probability of false alarm. Although there are modifications to such a traditional FMCW-based radar (e.g., [2]), the modifications have mainly shown to be either inefficient or involve extensive processing load.

Another issue with continuous wave modulation is the lack of interferer resistance. As the number of vehicles equipped with continuous wave rapidly increases on roadways, interference from signals generated by adjacent radars will greatly impact desired parameter estimates. A major concern about moving vehicle interference is that it can cause incorrect perception of surroundings, leading to false behavior planning of desired vehicle. This indeed raises safety concerns for services (e.g., ranging from emergency breaking systems to autonomous driving) built on FMCW.

12.2.1.1 Triangular FMCW Simulation

This section presents simulation details of the triangle waveform signals employed in continuous wave radar transmission. Such signals serve as a reference for future comparisons and are also known as FMCW radars. Notably, there are no additional improvements or revisions to the legacy continuous wave. Table 12.2 lists parameters employed in the continuous wave simulation.

For simulations reported in this chapter, each target is represented by a two-value format in parentheses:

$$Target\ m : (R_m, v_m) \tag{12.14}$$

where R_m is the range and v_m is the speed of target m.

Table 12.2
FMCW Parameters

Parameter	Name	Value
T_c	Chirp period	500μs
B	Bandwidth	300MHz
α	Sweep rate	0.6 MHz/μs
f_c	Carrier frequency	77GHz

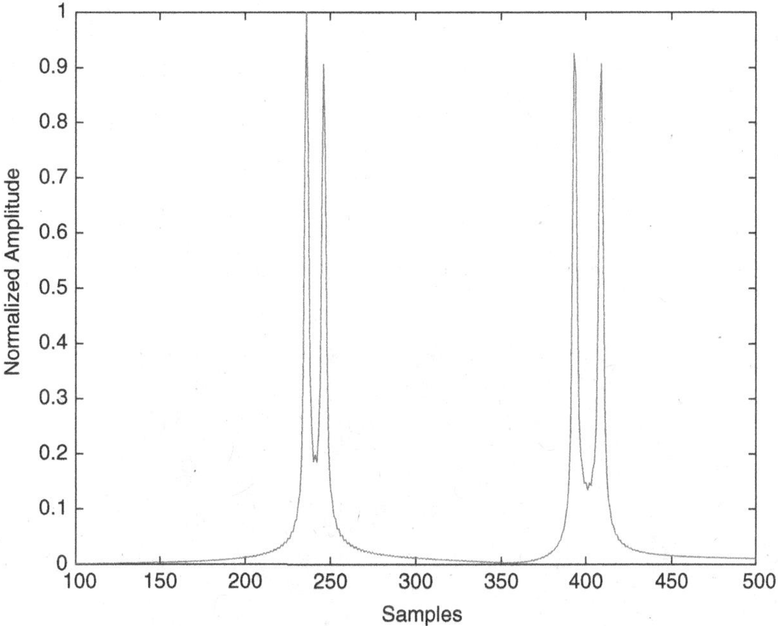

Figure 12.5 FMCW frequency beats for a -target scenario with targets located distant from one another, at (30,40) and (50,60)

As discussed in section 12.2, FMCW performance is adversely impacted by ghost targets. Figure 12.5 shows beat frequencies, and Figure 12.6 and Figure 12.7 demonstrate 3D and 2D representations of locations for a two-target scenario, respectively, albeit relatively distant from each other. Locations are situated at positions (30,40) and (50,60), where the first number denotes velocity and the second refers to range. Four beat frequencies and signal processing are required to populate Eqs. (12.11) and (12.12). Note that according to Eq. (12.13), total number of detectable targets will equal six, four of which are ghost targets. In this case, all ghost targets represent ambiguous targets w.r.t speed estimations, which can be intuitively deleted from the list. The algorithm will accurately find a matching pair for each target, as depicted in Figure 12.5.

In a different scenario, one target location is updated, wherein targets are located at closer proximity. Consider targets located at (30,40) and (33,60). When using FMCW, Figure 12.8 demonstrates that proper beat frequency target detection is impossible without a hypothesis. Consequently, the algorithm is unable to find the matching pair for each target, and all targets are affected by ambiguity, as shown in Figure 12.9. In this case, six targets are detected, even though it was already

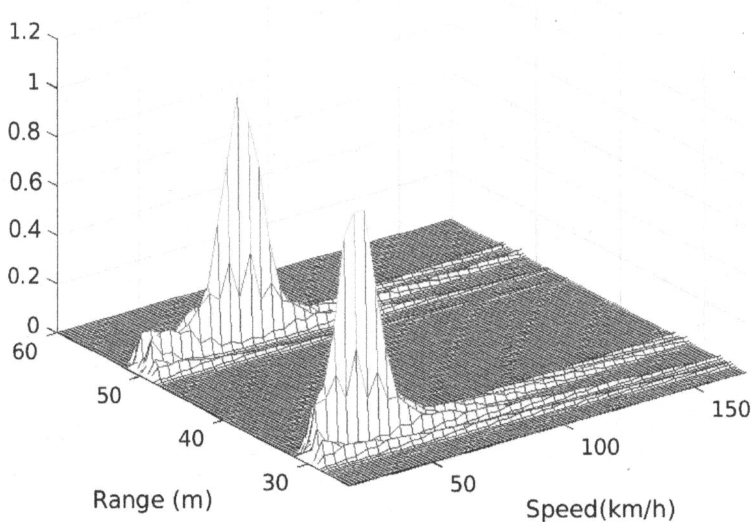

Figure 12.6 3D representation of target detection in the scenario in Figure 12.5.

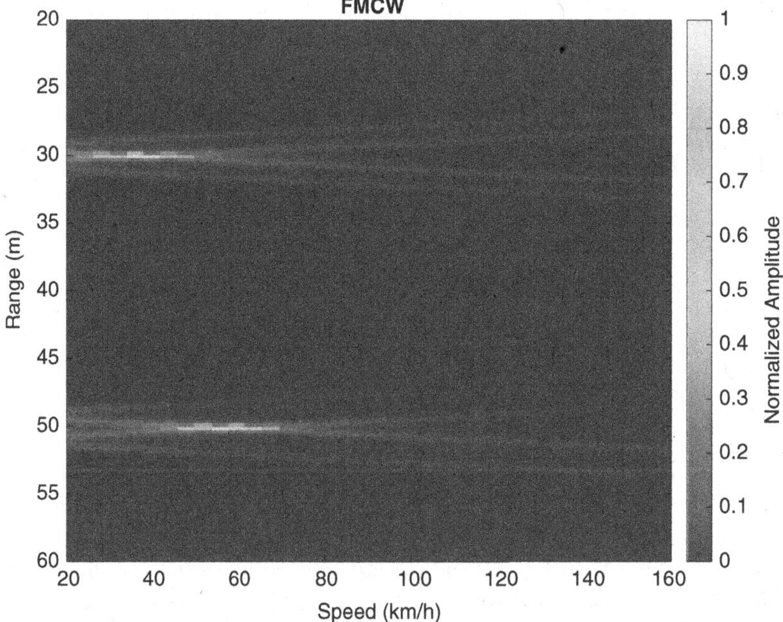

Figure 12.7 2D representation of the detection of the scenario in Figure 12.5. No target ambiguity is observed.

determined that only two are real. Under such a scenario, FMCW demonstrates degraded performance and some other solution might be needed for additional signal processing.

Interference impact on target detection performance is one metric in literature and is of great significance in this area [22–24]. There is no standard method for defining interference in radar systems, especially for continuous wave signal. In this chapter, interference relating to radar signals

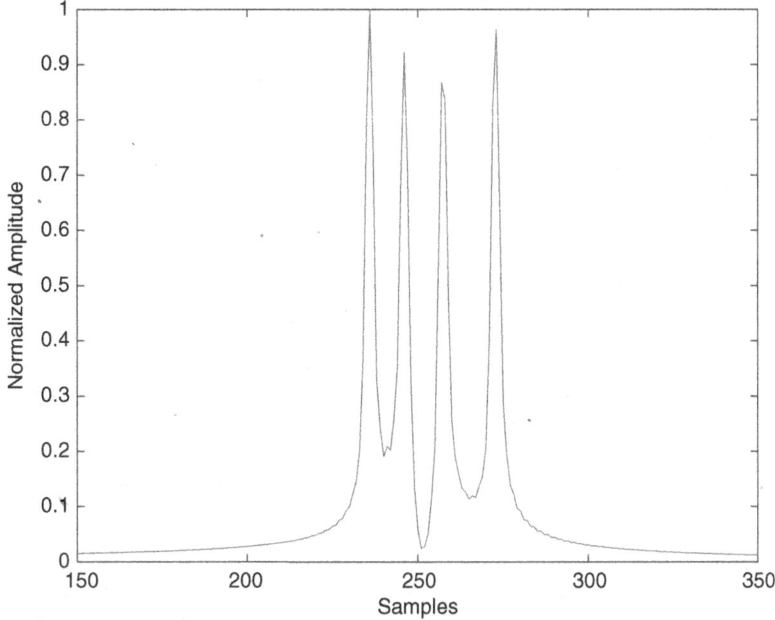

Figure 12.8 FMCW, frequency beats for a 2-target scenario with close targets to each other, (30,40) and (33,60)

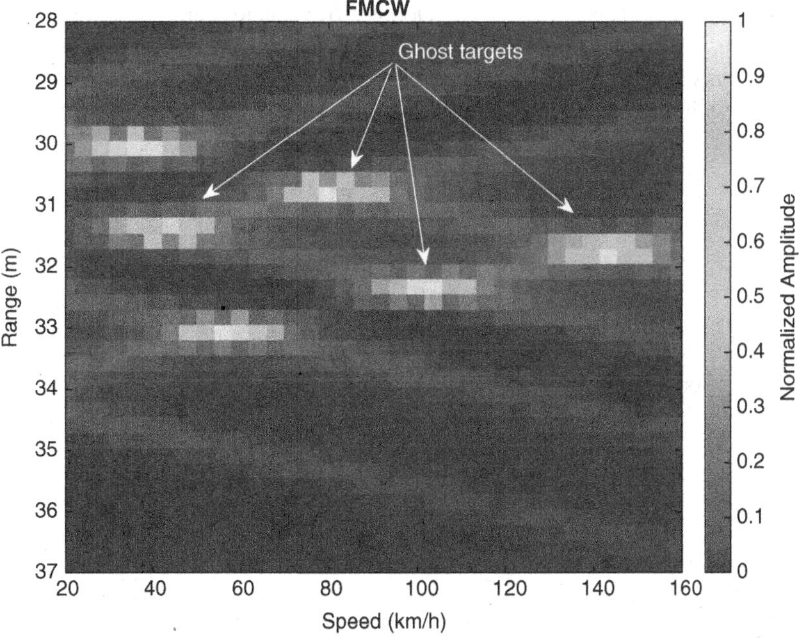

Figure 12.9 FMCW target detection in the 2-target scenario in Figure 12.8. Target ambiguity occurred.

is generated by other vehicles. Under such scenarios, the interfering vehicle (i.e., interferer) may be in various locations. Yet, two main cases might be most interesting: scenario 1, in which the interferer is moving in the same direction as vehicle under test (VUT), and scenario 2, in which the interferer is driving in opposite direction of the vehicle under test. Scenarios 1 and 2 are depicted in Figures 12.10 and 12.11. For simplicity, it is assumed—without the loss of generality—that targets are located at ample distance to avoid ghost targets and maintain an analysis focused on target detection in the presence of interference. Consider targets located at (30,40) and (50,60). Corresponding results are shown in Figures 12.12 and 12.13. For interference scenario 1 depicted in Figure 12.12, FMCW properly detected targets although an additional ghost target was incorrectly detected. For interference scenario 2 depicted in Figure 12.13, FMCW was unable to correctly detect any target.

12.2.2 TRAPEZOIDAL FMCW WAVEFORM

The ghost target problem appeared in triangular FMCW radars has been addressed through several efforts, and potential solutions have been proposed. One solution relies on trapezoidal waveforms rather than a typical periodic triangle. Such FMCW-based waveforms can be fully trapezoidal [25] or in the form of hybrid triangular-trapezoidal waveform [2], as shown in Figure 12.14. The idea is based on the premise that having knowledge about Doppler frequencies will distinguish real targets from ghost counterparts. This implies that when using a trapezoid FMCW wave, a new beat frequency will appear, which is identical to the Doppler frequency of f_D for each target. Thus, a portion of the signal with a flat frequency will aid in recognizing unambiguous Doppler frequencies (i.e., velocities). In the other words, unambiguous velocities are detected during the flat portion of the trapezoid. Given that frequencies remain constant for a period of time, the processing unit can highlight the mismatched velocities in the R-V plane. See Figure 12.9, which demonstrates that knowledge on Doppler frequencies will help for identifying ghost targets.

This technique requires exact estimation of Doppler frequencies by recognizing from up chirp and down chirp beat frequencies. Otherwise, Doppler frequencies cause additional ambiguity for target detection (i.e., if Doppler frequencies are falsely considered as typical up/down frequencies). This incident not only fails to mitigate ambiguity, it also creates new ghost targets. If a trapezoidal FMCW technique is implemented in practice, data must be tailored in the time domain to include

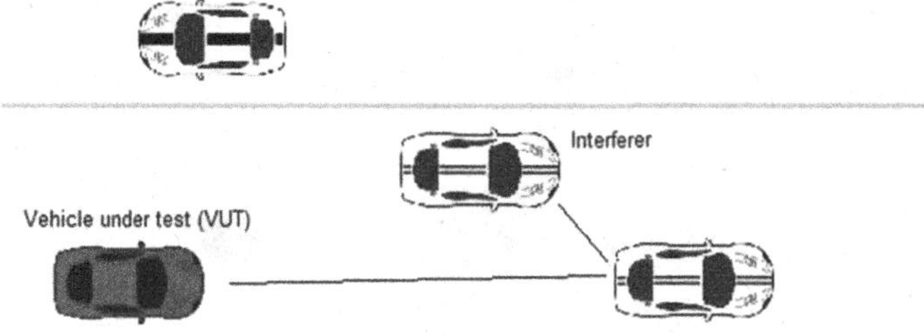

Figure 12.10 Interference scenario 1: Reflection of signal from a different radar

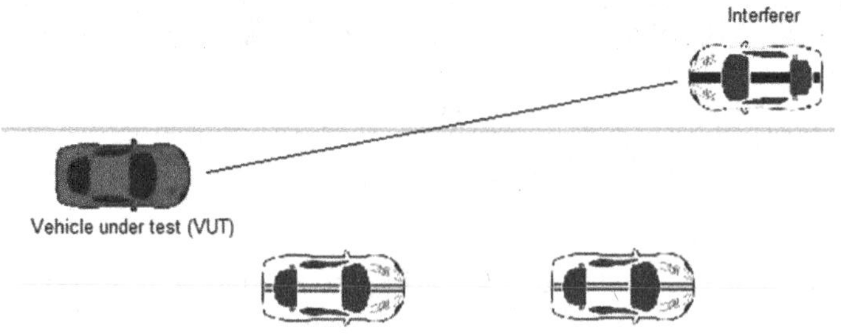

Figure 12.11 Interference scenario 2: Direct signal from a different radar in oncoming traffic

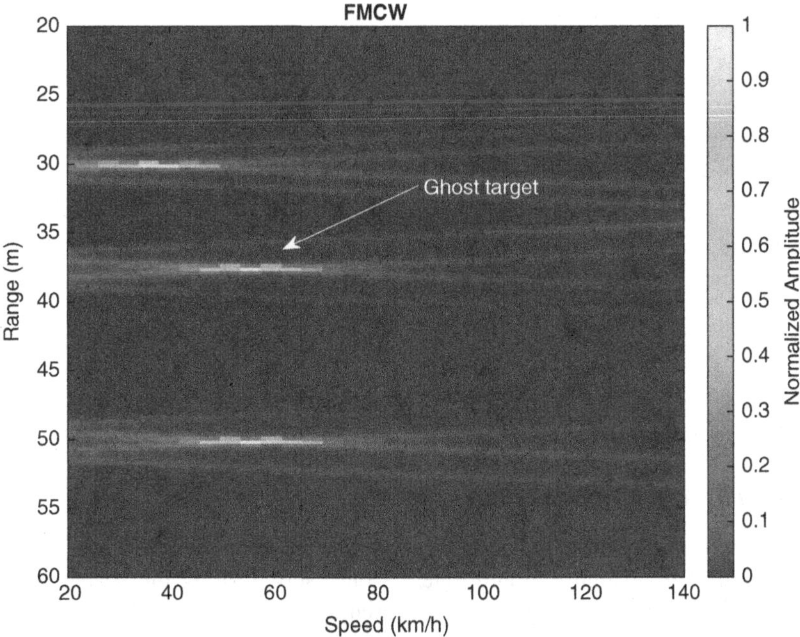

Figure 12.12 Target detection for interference scenario 1. FMCW is able to detect the targets but with addition of one ghost target..

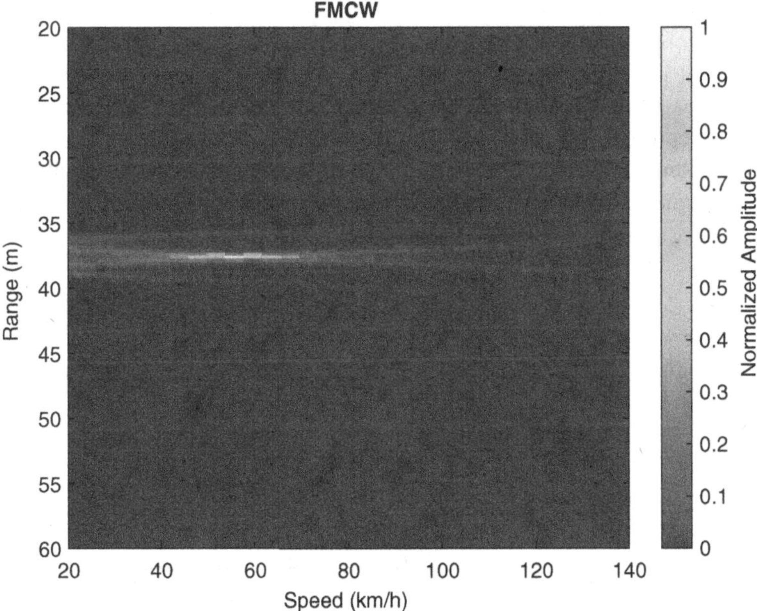

Figure 12.13 Target detection for interference scenario 2. FMCW is not able to correctly detect the targets.

samples during the flat timeslot. FFT processing should then be applied on tailored data to remove up/down beat frequencies. This part of processing is in addition to necessary regular processing for up/down beat frequencies, meaning that trapezoid signal analysis requires additional processing load.

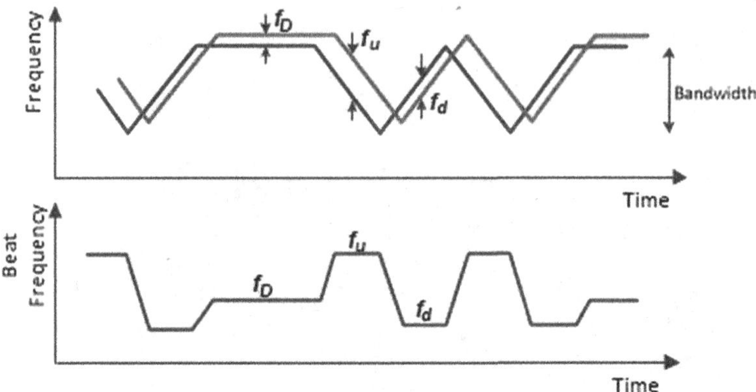

Figure 12.14 Trapezoidal FMCW waveform

12.2.3 SAWTOOTH FMCW WAVEFORM

Although radars built on digital modulation (e.g., OFDM, as we will see shortly) mitigate target ambiguity, they suffer from high sampling rates. Conversely, FMCW diminishes high sampling rate requirements, although target ambiguity is present. In fact, the unambiguity in pairing beat frequencies of the triangle waveform causes uncertainty in target estimation, indicating that the combination of an up chirp and down chirp pair might not be an efficient solution. In the case of sawtooth waveform, only the up chirp section of the signal is present. Thus, signal processing techniques from triangle FMCW radar system related to velocity and range detection in Eqs. (12.11) and (12.12) are not applicable here.

The configuration of a sawtooth FMCW waveform is depicted in Figure 12.15. The frequency of transmitted signal increases linearly as a function of time during a chirp repetition period or sweep time (T_c). Clearly, the sweep time of a sawtooth wave is half that of triangle FMCW. Waveform bandwidth is represented as the frequency span between the minimum and maximum signal frequencies. Parameters f_D, τ, and f_{beat} are specified by location and target velocity.

A sawtooth continuous-frequency modulated signal can be mathematically represented as:

$$x(t) = \sum_{n=1}^{N_c-1} A_t \cos(2\pi f_c(t-(n-1)T_c)) \tag{12.15}$$

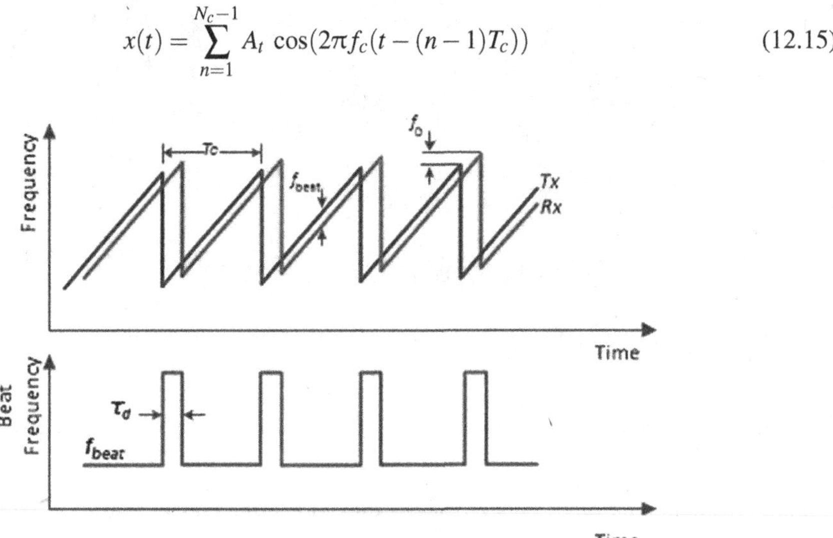

Figure 12.15 Sawtooth FMCW

where T_c is the length of one period, which is basically the length of one chirp (i.e., sweep time); f_c is the center carrier or operating frequency; and N_c is total number of chirps. Similar to triangle waveform, a combination of bandwidth B and sweep time T_c can introduce the parameter known as slope or α, as defined in Eq. (12.6). Although the equation for sweep rate/slope of the chirp remains same, its valve is twice that when compared with triangle waveform, as sweep time is half that for the triangle waveform. Considering the above equation as transmitted signal, the received signal is given by:

$$y(t) = \sum_{n=1}^{N_c-1} \{A_r \cos(2\pi(f_c + f_D)(t - \tau_d - (n-1)T_c)\} \tag{12.16}$$

12.2.3.1 Range, Velocity and Angle Estimation

Since only one single beat frequency is associated with sawtooth waveform, Eqs. (12.11) and (12.12) cannot be exploited from triangle waveform. However, it is still possible to estimate information about the target. Figure 12.15 illustrates that there are several parameters that must be clarified:

τ_D: The horizontal shift in Rx chirp w.r.t .Tx chirp and denotes round trip time, defined in Eq. (12.4)—in fact the time span between the start points of Tx and Rx chirps. The value of τ_D is directly dependent on only target location (i.e., range); target velocity has no impact on this value.

f_D: Vertical shift in Rx chirp w.r.t Tx chirp denotes Doppler frequency based on the target velocity. Target distance has no impact on this parameter.

f_{beat}: Beat frequency in the IF signal contains information relating to range and velocity of a target.

B: Chirp bandwidth.

When taking the above parameters into account, Figure 12.15 shows that the IF signal has a constant frequency tone:

$$f_{beat} = \frac{2\alpha R}{c} + f_D \tag{12.17}$$

as described by Eq. (12.8). The equation represented by Eq. (12.17) indicates that the range estimation requires knowledge on Doppler frequency, which is proportional to the target velocity. Even though the portion of Doppler frequency can be neglected if the target velocity is small compared to the linear frequency change of the chirp, it needs to be considered for a more accurate estimation. In practice, range estimation can be calculated by applying the approximation in Eq. (12.17):

$$R \approx \frac{f_{beat}\, c}{2\alpha} \tag{12.18}$$

Once the calculation of velocity is completed, range estimation can be adjusted by considering Doppler frequency into the beat frequency in (12.17) [26].

There is another alternative equation for range estimation and is given by:

$$R = \frac{c\, \tau_D}{2} \tag{12.19}$$

It is understandable to note in Eq. (12.19) that measuring τ_D involves practical complexity issues potentially due to the small values of RTT in automotive applications.

The estimation of target velocity for sawtooth waveform can be simply realized by the phase difference measured across two consecutive chirps. Due to target displacement, a phase shift proportional to the amount of displacement is observed in the received signal. When measuring the phase, one may want to look at the level of phase where the peak in FFT occurs, as shown in Figure 12.16. Displacement is given by $R_c = v\,T_c$, where v is the velocity of the target, and T_c is the time span between chirps at transmitter or sweep time. Using Eq. (12.2), velocity is calculated as follows:

$$v = \frac{\Delta\phi\,\lambda}{4\pi\,T_c} \tag{12.20}$$

In practice, measuring beat frequency peaks as well as the corresponding phase shifts requires an FFT known as Range-FFT to be calculated individually on each chirp [27]. The FFT applied on IF signal corresponding to each chirp will have peaks in the same location, yet differing phases, as shown in Figure 12.16. Although in this case only two chirps suffice for velocity estimation, a number of N_c equi-spaced chirps are necessary for detecting the velocity of multiple targets at the same range. This process is referred to as a Doppler-FFT, where N_c is the number of chirps building a chirp frame. The combination of Range-FFT and Doppler-FFT form a 2D-FFT so called Range-Velocity plane (R-V plane) which is shown in Figure 12.17. The row dimension refers to the FFT over duration of a chirp and specifies range while the column dimension refers to the FFT over chirps and specifies the velocity.

Note that the range estimation is approximated because Doppler frequency is neglected from the beat frequency. After being shown approximated range and velocity estimation of a sawtooth waveform through equations as well as R-V plane, adjustments can be applied to range. Luckily,

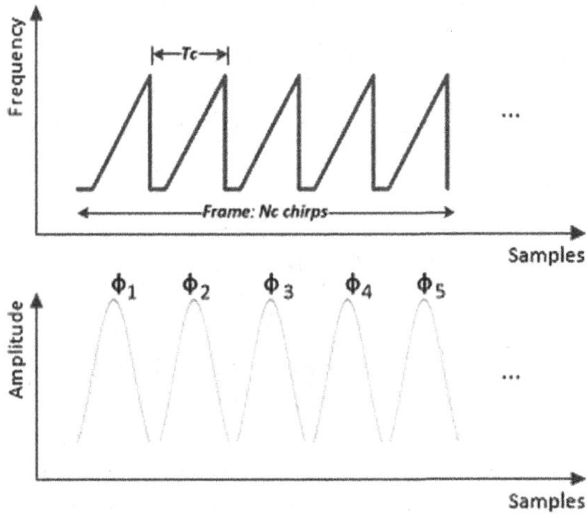

Figure 12.16 Phase shift due to target displacement

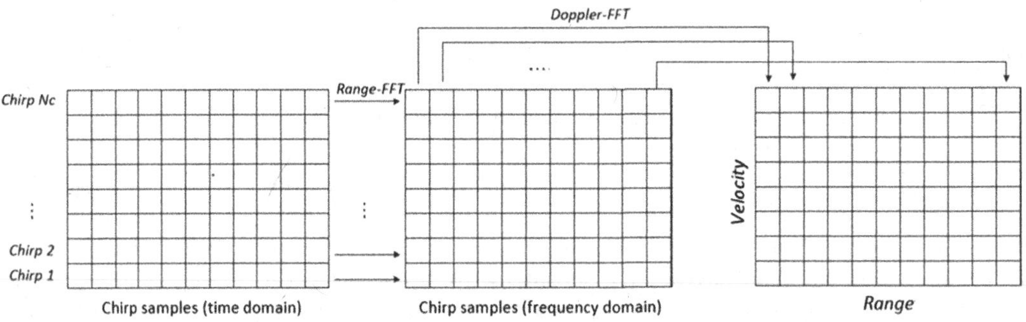

Figure 12.17 Building Range-Velocity plane in FMCW using 2D-FFT

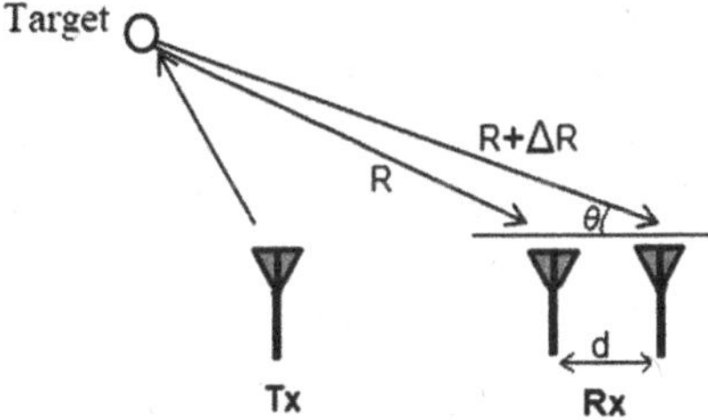

Figure 12.18 Receive Diversity in Radar

the velocity estimate in (12.20) is independent of range approximations. Thus, by combining Eqs. (12.9) and (12.20), the exact equation for range estimation can be given by:

$$R = \frac{c}{2\alpha}\left(f_{beat} - \frac{2v}{\lambda}\right) \tag{12.21}$$

which depends on readily measurable parameters.

To measure angular information about target location (i.e., angle of arrival (AoA)), the idea of receive diversity (RxD) and phase shift is applied. Recall that a small change in location of a target results in a peak phase change in range-FFT. Figure 12.18 shows that if the radar is equipped with two receivers, the received reflected signals will have a phase difference proportional to the distance between antennas and angle of arrival. From Eq. (12.1),

$$\Delta\phi = \frac{2\pi d}{\lambda}\sin(\theta) \tag{12.22}$$

where θ is the angle of target w.r.t. the receiving aperture. The angular information of target is given by

$$\theta = \sin^{-1}\left(\frac{\lambda\,\Delta\phi}{2\pi d}\right) \tag{12.23}$$

12.2.4 FMCW SYSTEM RESOLUTION AND PERFORMANCE

[] FMCW is basically designed to estimate object range, velocity, and angle relative to radar location. It is important to depict FMCW radar performance defined by these parameters. Typical configuration values of an FMCW radar for carrier frequency f_c is 77 GHz with a maximum bandwidth B of 4 GHz , resulting in maximum radar frequency of 81 GHz. Chirp rate or chirp slope α will be equal to 100 MHz/μs.

Range and velocity resolution is an important parameter in evaluating radar performance. The significance increases for special case automotive radars, where targets are located in close proximity and safety is a concern. This section focuses on sawtooth waveform performance evaluation. Concepts are also applicable for triangular waveforms.

12.2.4.1 Range Resolution

Before range resolution analysis, let's first explore the maximum range supported by a radar. The only constraint on maximum range is that beat frequency must be less than the sampling rate (i.e., $f_{beat} < f_s$). Common sense suggests that the highest supported range is dependent on maximum sampling rate supported by receiver side ADC:

$$R_{max} = \frac{c}{2\alpha}\left(f_s - \frac{2v}{\lambda}\right) \tag{12.24}$$

For static targets, $v = 0$ and (12.24) can be simplified.

Range resolution is defined as the minimum separation of two targets that can be unambiguously detected. For FMCW radar to detect two different targets, Eq. (12.21) indicates that two corresponding frequency beats must be detected. Although frequency beat location is highly dependent on target velocity, one can simply assume that targets are static. From (12.21), range resolution can be computed as:

$$R_{res} = \frac{c\,\Delta f}{2\alpha} \tag{12.25}$$

In frequency domain, two varying frequencies can be resolved, given that frequency difference is greater than the value of frequency resolution, which is a known parameter in Fourier analysis and is equal to the inverse of observation period, which in this case, is equal to chirp duration:

$$\Delta f = \frac{1}{T_c} \tag{12.26}$$

By substituting in (12.25), it turns out that range resolution depends only on chirp bandwidth B for static targets:

$$R_{res} = \frac{c}{2B} \quad \text{(static target)} \tag{12.27}$$

Higher chirp bandwidth results in lower range resolution. Note that a lower value for range resolution provided by (12.27) delivers improved radar performance.

Notably, a moving target is more realistic for automotive applications. Range resolution associated with a moving target will be impacted by target velocity and is given by:

$$R_{res} = \frac{c}{2}\left(\frac{1}{B} - \frac{2v}{\alpha\,\lambda}\right) \quad \text{(moving target)} \tag{12.28}$$

12.2.4.2 Velocity Resolution

It is known that target velocity is measured using phase shifts in the reflected wave and that the measurement is unambiguous only if $|\phi| < \pi$. Therefore, maximum target velocity without ambiguity that can be detected by radar is:

$$v_{max} = \frac{\pi\,\lambda}{4\pi\,T_c} = \frac{\lambda}{4\,T_c} \tag{12.29}$$

Eq. (12.29) indicates that v_{max} is inversely proportional to T_c, which denotes chirp repetition time in Figure 12.16

Two varying frequencies can be resolved, given that frequency difference is greater than frequency resolution. In this case, beat frequencies of two targets at different velocities can be determined, as Doppler-FFT has two separate peaks. Thus, to determine velocity resolution, Doppler-FFT must be considered. Since Doppler-FFT has a size of N_c, which is the number of chirps, and as frequency resolution requires phases separated at least $2\pi/N_c$, from (12.20) and by replacing $T_{fr} = T_c\,N_c$,

$$v_{res} = \frac{2\pi\,\lambda}{4\pi N_c\,T_c} = \frac{\lambda}{2\,T_{fr}} \tag{12.30}$$

Note that velocity resolution is inversely proportional to frame time, T_{fr}.

12.2.4.3 Angle Resolution

From 12.23, permitted phase range for avoiding angular ambiguity is $-\pi < \theta < \pi$, which indicates 180° field of view. Maximum angle will be:

$$\theta_{max} = \sin^{-1}\left(\frac{\lambda}{2d}\right) \tag{12.31}$$

$\lambda/2$ spacing results in maximum field of view (i.e., 180°).

Angle resolution is defined as the minimum angle of two targets that appear as separate peaks. Using angle-FFT, angle resolution will be the minimum separation in FFT peaks. The size of angle FFT is equal to the number of receive antennas, N_r. A minimum of two receivers (i.e., $N_r = 2$) is needed to exploit angle estimation. Angle resolution can be computed as [27]:

$$\theta_{res} = \frac{\lambda}{N_r\, d\, \cos(\theta)} \tag{12.32}$$

Note that angle resolution is dependent on θ, meaning that targets located on sides experience a worsening resolution than targets located in front. $\lambda/2$ spacing and $\theta = 0$ result in $\theta_{res} = 2/N_r$ in radian.

12.2.5 OFDM WAVEFORM

This section reviews unique radar waveforms based on OFDM that are categorized as a digital waveform, w.r.t the analog property of FMCW. The review includes system model, transmitted and received signal characteristics, and signal construction mathematical background. This section also discusses the impact of various parameters pertaining to signal construction and receiver implementation (e.g., orthogonality, sampling rate , and subcarrier spacing). The latter part of the review delves into the ease of detecting multiple targets and provides details about added advantages of OFDM waveform in this regard [3–6].

12.2.5.1 System Model

The utilized system model is depicted in Figure 12.19, where an OFDM signal is generated based on a known sequence at the transmitter. Serial-to-parallel block converts input sequence into a block of sequences used as OFDM subcarriers. Like a typical OFDM transmitter, an inverse FFT (IFFT) block is necessary for manipulating the raw data sequence onto orthogonal carriers.

A continuous time domain OFDM signal is given by:

$$x_c(t) = \sum_{k=0}^{N-1} c_k e^{j2\pi(f_c+f_k)t} \tag{12.33}$$

where k denotes the subcarrier index; N is the number of subcarriers; f_c is the carrier frequency; and f_k is the subcarrier frequency, given by $f_k = kf_\Delta$, where f_Δ is subcarrier spacing and c_k is the modulated symbol. The corresponding baseband signal can be expressed as:

$$x(t) = \sum_{k=0}^{N-1} c_k e^{j2\pi f_k t} \tag{12.34}$$

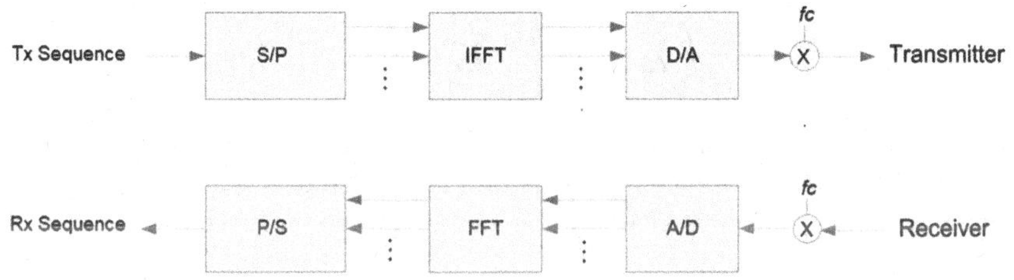

Figure 12.19 OFDM Transceiver block diagram

Without loss of generality, the received signal from a single target reflection is given by:

$$r(t) = \sum_{k=0}^{N-1} \alpha c_k e^{j2\pi(f_k - f_D)(t-\tau_d)} + n(t) \tag{12.35}$$

where α is the attenuation due to path loss, scattering and radar cross section (RCS) of the vehicle (assuming a point scatterer); τ_d is the RTT; f_D is the Doppler frequency; and $n(t)$ is additive noise plus interference. Note that Eq. (12.35) represents the received signal due to a single target. Under a multi-target scenario, received signal is given by:

$$r(t) = \sum_{m=1}^{N_t} \sum_{k=0}^{N-1} \alpha_m c_k e^{j2\pi(f_k - f_{Dm})(t-\tau_{dm})} + n(t) \tag{12.36}$$

where N_t is the number of targets; τ_{d_m} and f_{D_m} are the RTT and Doppler frequency associated with target m, respectively.

OFDM waveform hinges on the appropriate choice of c_k to achieve desired results. Details about using a form of pseudo-noise binary sequences are provided below. Such sequences are generated using shift registers, known as m-sequence. Although m-sequences are not the only option, they are nonzero and periodic with an episode at most $N = 2^{n-1}$, where n is the number of shift registers [28]. Based on this idea, c_0, c_1,.... c_{N-1}, in Eq. (12.33) corresponds to a modulated symbol for each bit of the respective m-sequence. Since these sequences are near to the desired autocorrelation and cross correlation properties, they can be used to introduce vehicle orthogonality. Further details on m-sequence and its assignment to OFDM will be presented in section 12.2.5.3.

12.2.5.2 Range and Velocity Estimation

The impact of range and target velocity on received signal is mathematically represented in Eqs. (12.35) and (12.36). Since c_k symbols are based on a m-sequence generator, target detection requires signal processing to manipulate them from a received signal under a single target or multi-target scenario. Given that targets are located at different speeds or distances, the complex received signal in the form of OFDM includes all copies of the transmitted signal that are received at different time and frequency offsets. Due to the excellent correlation properties of modulated data, straightforward signal processing involving correlation facilitates the extraction of time delay τ_d and frequency offset f_D for each target. When the m-sequence is used as input data to subcarriers, the modulated data has excellent correlation properties in frequency domain. However, as shown in the following equations, such m-sequences can be regarded as having desired correlation properties in time domain, as well.

Let $\left\{c_k^{(1)}\right\}_0^{N-1}$ and $\left\{c_k^{(2)}\right\}_0^{N-1}$ be two different m-sequences of the same length and distributed in frequency domain on an OFDM signaling basis. Subsequently, cross correlation is given by Eq. (12.44). However, Eq. 12.35 indicates that c_k's are, in fact, frequency domain components of the transmitted signal. Corresponding time domain cross correlation is given by:

$$R_x(\tau) = \int_t \sum_{k=0}^{N-1} c_k^{(1)} e^{j2\pi f_k t} \sum_{m=0}^{N-1} c_m^{(2)} e^{-j2\pi f_m (t-\tau)} dt \tag{12.37}$$

where f_m and f_k are corresponding subcarrier frequencies for each sequence. Alternatively, the terms in the summations denote the Fourier transform associated to $c^{(1)}$ and $c^{(2)}$, respectively. The corresponding time domain signal is given by convolution of two sequences:

$$\sum_{k=0}^{N-1} c_k^{(1)} e^{j2\pi k \Delta f t} \sum_{m=0}^{N-1} c_m^{(2)} e^{-j2\pi m \Delta f (t-\tau)} = \mathscr{F}\left\{c_{am}^{(1)} * c_{bm}^{(2)}\right\} \tag{12.38}$$

where $a = -\triangle ft$ and $b = \triangle f(t-\tau)$. However, the convolution in the above equation is equal to:

$$\sum_{n=0}^{N-1} c_{an}^{(1)} c_{b(m-n)}^{(2)} \tag{12.39}$$

Depending m value, the above summation will be equal to N at the m's equal to $m = (a+b)n/b$ or $m = \tau n/(\tau - t)$, given that $\tau/(\tau - t)$ is an integer for any given value of t. However, this is valid only if $\tau = 0$, meaning the integral in (12.37) will have a peak at $\tau = 0$ only, which is given by Eq. (12.44). Subsequently, cross correlation in time is conferred by:

$$R_x(\tau) = \begin{cases} N, & \tau = 0 \\ -1, & \text{otherwise} \end{cases} \tag{12.40}$$

which describes the desired correlation properties of m-sequence-based OFDM signal in time domain. The cross correlation between transmitted and received signals is presented as:

$$R_{cc}(\tau) = \sum_{n=0}^{N-1} r(nT_s)x^*(nT_s - \tau) \tag{12.41}$$

where $T_s = 1/f_s$ is the sampling interval; τ is time shift; and n is sample index in time domain. Eq. (12.41) denotes that cross correlation is carried out in time domain, even though the m-sequence is spread over frequency domain. By substituting r and x from Eqs. (12.34) and (12.36), the resultant equation will be:

$$\begin{aligned} R_{cc}(\tau) = & \sum_{m=1}^{N_t} \sum_{n=0}^{N-1} \sum_{k=0}^{N-1} \alpha c_n c_k e^{-j2\pi f_k (nT_s - \tau)} e^{j2\pi (f_k - f_{Dm})(nT_s - t_{dm})} \\ & + \sum_{n=0}^{N-1} \sum_{k=0}^{N-1} c_k n(nT_s) e^{-j2\pi f_k (nT_s - \tau)} \end{aligned} \tag{12.42}$$

or, simply, the cross correlation for any target m is given by:

$$\begin{aligned} R_{cc}^{(m)}(\tau) = & \sum_{n=0}^{N-1} \sum_{k=0}^{N-1} \Big\{ \alpha c_n c_k e^{j2\pi f_k (\tau - t_{dm})} e^{-j2\pi f_{Dm} (nT_s - t_{dm})} \\ & + c_k n(nT_s) \, e^{-j2\pi f_k (nT_s - \tau)} \Big\} \end{aligned} \tag{12.43}$$

The second part in (12.43) refers to the effective noise portion of the received signal. Note that $R_{cc}(\tau)$ value is highly dependent on timing parameters in the equation. Due to orthogonality of c_k and c_n, the summation has a high value when $n = k$. Also, at this point $nT_s = t_{d_m}$ and $\tau = t_{d_m}$. Cross correlation value becomes negligible at other points. Since the peak occurs at only a single pair of t_{d_m} and f_{D_m} of any given target, there will theoretically be no ambiguity of target detection with such a waveform. Therefore, the impact of interference is expected to be negligible when compared with legacy continuous wave signal. Simulation results and details are discussed later in this chapter.

12.2.5.3 Waveform Construction

As previously discussed, OFDM subcarriers can be modulated using a special form of pseudo-noise binary sequences to facilitate orthogonality among vehicles. One option is using m-sequences by assigning them to users (i.e., vehicles). Typically, binary sequences are generated using shift registers known as linear-feedback shift register (LFSR). For example, as depicted in Figure 12.20, such binary sequences are generated using a set of shift registers with connections based on a generator

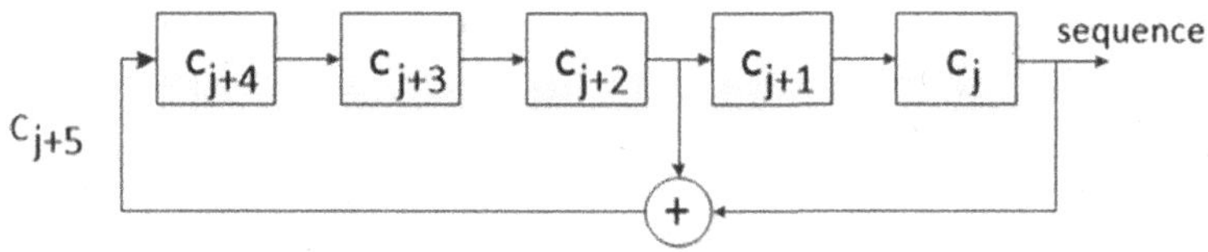

Figure 12.20 An m-sequence generator based on LFSR with $n = 5$

polynomial (i.e., $x^5 + x^2 + 1$). These sequences are nonzero and periodic with an interval at most $2^n - 1$, where n is the number of shift registers [28]. For example, the LFSR for $n = 5$ is shown in Figure 12.20.

Serving as a special case, a sequence with period $2^n - 1$ is denoted as maximum length or, briefly, m-sequence. An interesting property of an m-sequence is that any cyclic shift results in another m-sequence. Furthermore, the correlation between such different sequences is always low. This property serves as a motivation to differentiate multiple m-sequences (e.g., multiple users or vehicles), which is the basis of the work reported in this chapter.

Figure 12.21 illustrates the correlation function of an m-sequence and its cyclic shifted versions. m-sequence used in the plot is obtained using LFSR, shown in Figure 12.20. The cross-correlation function is given by:

$$C(k) = \sum_{m=0}^{N-1} c_1(m)c_1(m-k) = \begin{cases} 2^n - 1, & k = 0 \\ -1, & \text{otherwise} \end{cases} \tag{12.44}$$

Consider $S = \{s_0, s_1, s_2...., s_{N-1}\}$ as a m-sequence with period N. The transmitted signal in this case is given by replacing c_k with s_k in Eq. (12.33):

$$x(t) = \sum_{k=0}^{N-1} s_k e^{j2\pi f_k t} \tag{12.45}$$

Let's now describe how m-sequence can be employed in the ADAS system. According to the earlier described property of m-sequence, a given m-sequence will be orthogonal to any cyclic shifted copy of itself. Hence, when an m-sequence is assigned to a user (i.e., vehicle), the total number of users supported while holding orthogonality will be equal to time-frame N. If N is not large enough, the probability of two or more users having the same m-sequence will be high. On the other hand, since the radar application manages time- and frequency-delayed signals, interference might occur

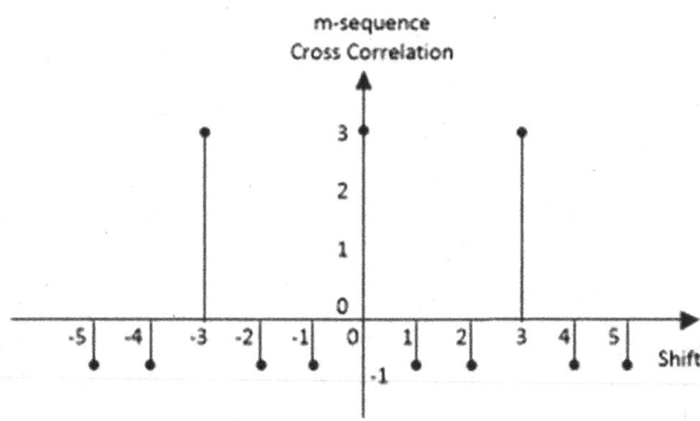

Figure 12.21 Cross correlation of an m-sequence with $n = 2$

between users [29]. An example would be a case wherein the Doppler frequency shift becomes equal to or more than subcarrier spacing. To avoid such issues, a scrambled m-sequence would be used. This procedure is similar to a cell search algorithm in 3gpp 4G technology [30].

Three binary sequences, namely S, C and Z—each of length N, are combined to generate the final sequence. Primary sequences s_0 and s_1 are generated as different cyclic shifts of m-sequence S, which is obtained using a primitive polynomial x^5+x^2+1 given by:

$$s_0(k) = S((k+m_0)\,\mathrm{mod}(N)) \tag{12.46}$$

and

$$s_1(k) = S((k+m_1)\,\mathrm{mod}(N)) \tag{12.47}$$

where m_0 and m_1 are derived from the physical vehicle identity code (VIC). VIC is a unique identification number assigned to each vehicle and could well correspond to vehicle VIN information. m-sequence C is based on primitive polynomial x^5+x^3+1, and two subsequences c_0 and c_1 are created with the same length as primary sequences s_0 and s_1:

$$c_0(k) = C((k+VIC)\,\mathrm{mod}(N)) \tag{12.48}$$

and

$$c_1(k) = C((k+VIC+3)\,\mathrm{mod}(N)) \tag{12.49}$$

Finally, another sequence Z is used to generate final sequence for transmission by each user, based on primitive polynomial $x^5+x^4+x^2+x+1$. The sequence pair associated with Z are calculated as:

$$z_0(k) = Z(k+(m_0\,\mathrm{mod}(8))\,\mathrm{mod}(N)) \tag{12.50}$$

and

$$z_1(k) = Z(k+(m_1\,\mathrm{mod}(8))\,\mathrm{mod}(N)) \tag{12.51}$$

The final combined sequence c_k is constructed using different scrambling sequences when mapped to even and odd resource elements:

$$c_k = \begin{cases} s_0(k)\,c_0(k), & k\,\text{Even} \\ s_1(k)\,c_1(k)\,z_1(k), & k\,\text{Odd} \end{cases} \tag{12.52}$$

12.2.5.4 OFDM Simulation

OFDM radar waveform performance is presented in this section. Like FMCW, waveform performance based on scrambled m-sequence will be evaluated under similar scenarios. Table 12.3 lists parameters used in continuous wave simulation. Targets are originally located at (30,40) and (33,60). In this case, FMCW was facing target ambiguity impacted by ghost targets. Results are demonstrated in Figures 12.22 and 12.23.

By observing OFDM waveform performance plots, superior performance is evident when compared against continuous modulation wave. There are no ghost targets and peak distinctness results in greater range and speed estimation accuracy. Interference impact on target detection performance is also evaluated for OFDM-based radar. Similar interference scenarios were followed for continuous modulation wave, which are depicted in Figures 12.10 and 12.11. It was also assumed that the interferer was occupying the entire OFDM bandwidth throughout the time of measurement. An additional target, when compared with FMCW simulation, is added to more thoroughly evaluate OFDM performance. Targets were located at (30,40), (33,60) and (70,20). Corresponding results are shown in Figures 12.24–12.26. For interference scenario 1 shown in Figure 12.10, the waveform based on m-sequence could properly detect targets without noise and/or interference issues. For interference scenario 2 shown in Figure 12.11, the waveform was still able to detect target location and speed, although the signal appears noisier than the case without interference. These differences are significant, as legacy FMCW demonstrates poor performance under these scenarios.

Table 12.3
Parameter Setting for *m*-Sequence Based OFDM Waveform

Parameter	Name	Value
N	m-sequence period	63
T_s	Symbol duration	0.33 μs
f_c	Carrier frequemcy	77GHz

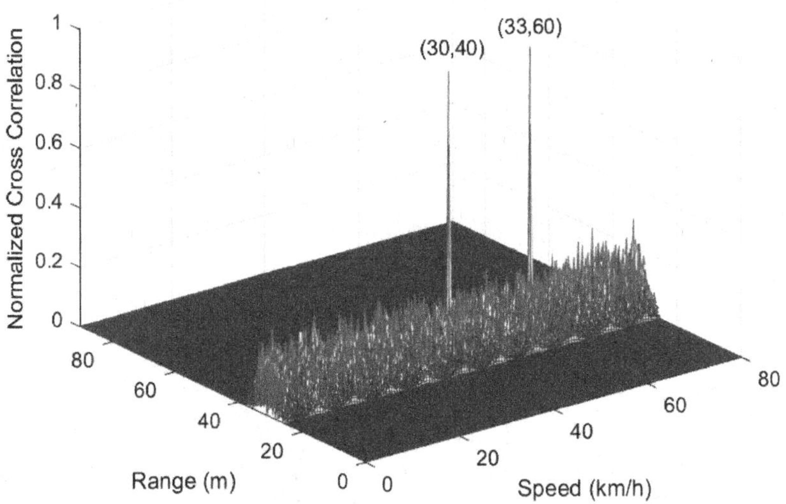

Figure 12.22 3D demonstration of peaks in a two-target scenario. Targets located at (30,40) and (33,60).

12.3 MIMO RADAR

One advantage of mm-wave is improving radar performance when using multiple input multiple output (MIMO) configuration. This technique incorporates both transmit and receive diversity. Due to millimeter wavelength range, aperture sizes are relatively small, making spatial multiplexing and diversity both practical and effective for automotive applications. MIMO has been widely employed in various advanced wireless communication systems. Likewise, in the field of radar signaling, a $N_{Tx} \times N_{Rx}$ configuration MIMO radar involves multiple T_x and R_x antennas, where N_{Tx} is the number of transmit antennas and N_{Rx} is the number of receive antennas.

In section 12.2.3.1, which featured angle estimation, multiple receivers were shown to facilitate angular measurement of a target. The analysis was based on single transmit antenna and multiple receive antenna. For a radar equipped with one Tx and two Rx, a minimum angular resolution was computed. Doubling the number of receivers to four will improve angular resolution by half. This configuration is composed of a single transmitter and multiple receivers (SIMO). Another approach for improving angle resolution by half, is doubling the number of transmit antenna in the form of MIMO. Various antenna configurations can be leveraged to realize the same antenna array configuration. Presumably, the 2x2 configuration can be characterized as a virtual array of a 1x4

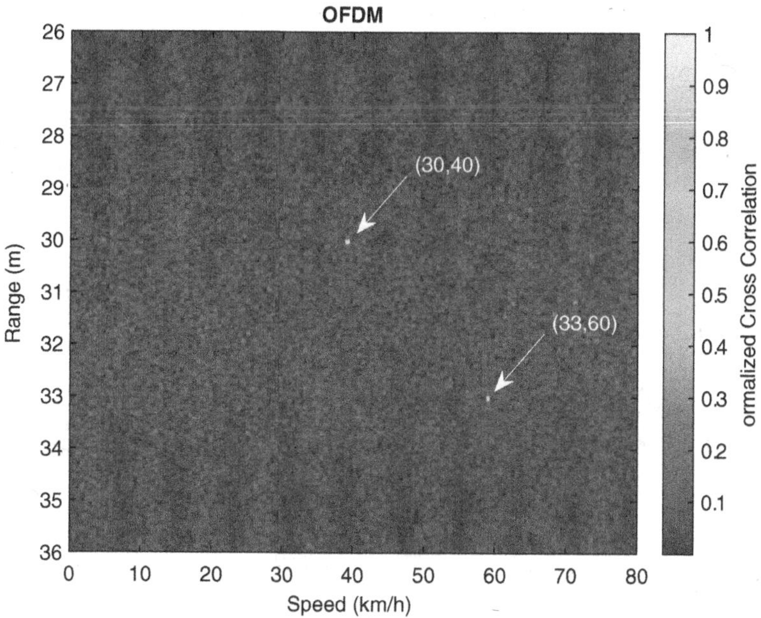

Figure 12.23 Target detection for the two-target scenario in Figure 12.22. Targets located at (30,40) and (33,60).

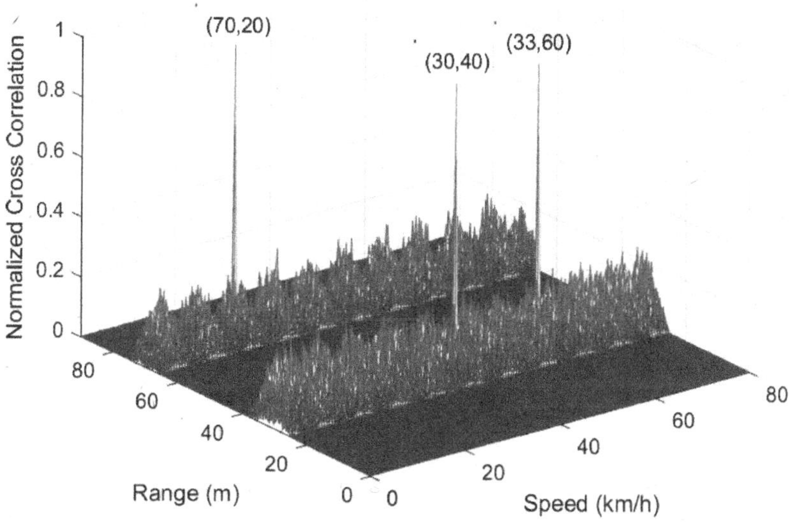

Figure 12.24 A 3D demonstration of peaks in a three-target scenario. Targets located at (30,40), (33,60) and (70, 20).

configuration [31]. Similarly, a 1x8 configuration will deliver the same angle resolution as a 2x4 configuration. Notably, the latter configuration has three fewer antenna than the former, thereby reducing hardware cost and complexity.

Phase analysis under transmit diversity involves additional processing. The received signal at each receive antenna observes an additional phase shift due to spatial distance between transmit

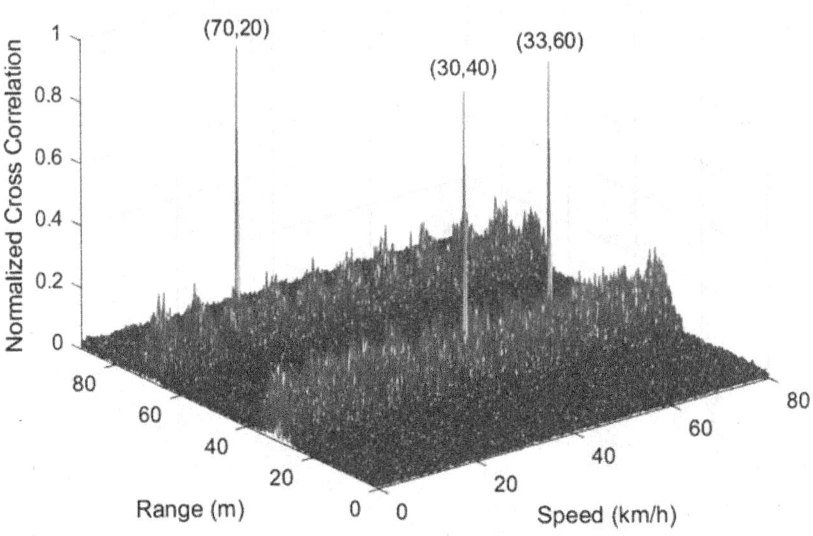

Figure 12.25 Target detection for interference scenario 1. The OFDM waveform correctly detected the targets. Targets are located at (30,40), (33,60) and (70, 20).

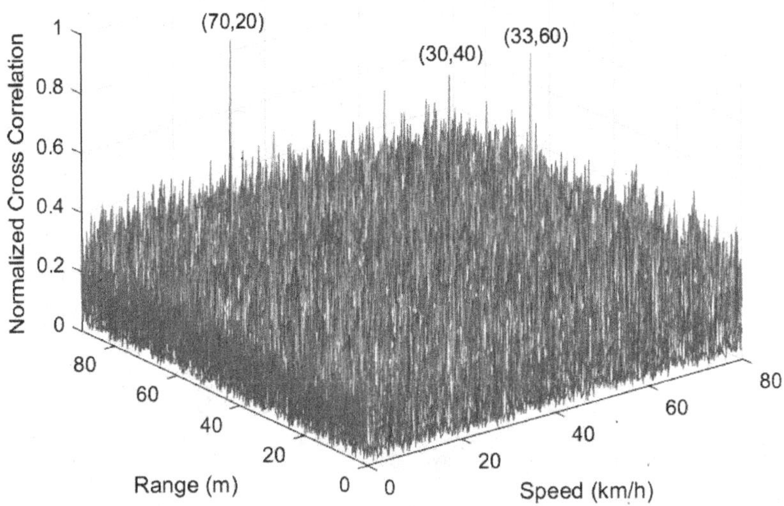

Figure 12.26 Target detection for interference scenario 2. The OFDM waveform remains able to correctly detect targets. Targets are located at (30,40), (33,60) and (70, 20).

antennas. Since it is easier to compute phase processing for SIMO, virtual array synthesis is used to characterize the phase among antennas.

Depending on the positions of the Tx and Rx antennas, the angular information can be related to horizontal or elevation directions. The concept of *Radar Imaging* emerges from MIMO radar and involves two dimensional angular information, where receiving and transmitting apertures

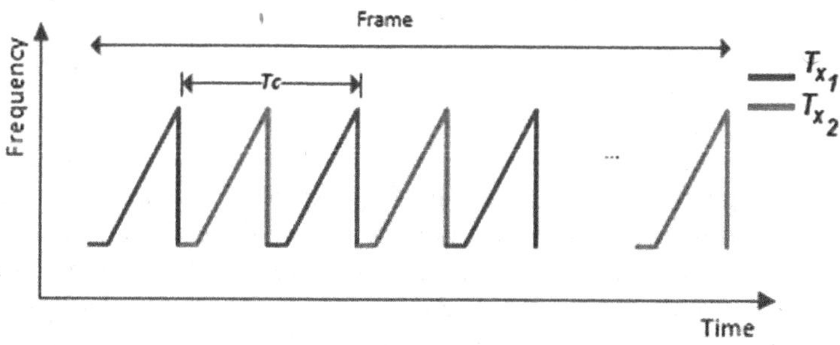

Figure 12.27 TDM-MIMO

are installed in a way to manipulate the phase information in both horizontal plan (H-plane) and elevation plan (E-plane).

Although employing transmit diversity will decrease hardware cost, one constraint must be met: receive antennas must separate corresponding signals from each transit antenna. In the other words, MIMO transmitter requires a mechanism to deploy orthogonality in transmitted signals.

Time Division Multiplexing (TDM) is one solution for MIMO radar, wherein the transmission interval is time-shared between transmitters. Hence, TDM-MIMO requires time synchronization between transmitter and receiver. While this method is simple to implement, it is inefficient, as chirp period and frame time will be longer than a single transmit configuration, and radar performance, including velocity resolution, is impacted (See Eqs. (12.28) and (12.29)). TDM-MIMO is visually depicted for two transmitters in Figure 12.27.

Code Division Multiplexing (CDM) serves as yet another solution for orthogonality requirements of MIMO radar, wherein orthogonality among transmitted signals is realized by encoding. In CDM-MIMO, each transmitter is assigned a code that is orthogonal among other transmitters. The codes are binary sequences of 1 and -1, thus, only the chirp phase—not frequency—will be impacted by the code. For this reason, CDM-MIMO is often referred to as binary phase modulation (BPM) [27], [31] and [32]. In its simplest configuration, wherein the transmitter has only two antennas (i.e., $N_{Tx} = 2$), orthogonal sequences will be $\{1,1\}$ and $\{1,-1\}$ for T_{x1} and T_{x2}, respectively. Assume x_1 and x_2 represent the original transmitter signals before encoding. Without loss of generality and for the sake of simplicity, let's neglect the effect of transmission loss. The combined signal received at a given receiver for two consecutive chirps will be:

$$\begin{aligned} x_c(i) &= x_1 + x_2 \\ x_c(i+1) &= x_1 - x_2 \end{aligned} \tag{12.53}$$

where i is the chirp index. Furthermore, each receiver can extract transmitted signals by combining two executive chirps:

$$\begin{aligned} x_1 &= (x_c(i) + x_c(i+1))/2 \\ x_2 &= (x_c(i) - x_c(i+1))/2 \end{aligned} \tag{12.54}$$

CDM-MIMO fully exploits chirp duration since, unlike TDM, the receiver can completely decode chirps while both transmitters are continuously active. Notably, since the average of consecutive chirps are considered as represented in (12.54), the overall utilization will be same as TDM. The realization of CDM-MIMO depends on the availability of binary orthogonal codes. For a greater number of Tx antennas, a longer code is necessary. Walsh codes and Hadamard codes serve as examples of orthogonal binary sequences.

SUMMARY

Radars are considered a primary tool as driver's assistant and for autonomous driving. Radar signal processing—with its resistance to severe weather conditions—is poised to provide added safety for drivers in an ADAS based environment. Detecting typical on-road targets, obstacles, or objects in close proximity using digitally designed radar can be an expensive venture. Instead, analog radars based on FMCW waveforms are recommended, as these systems can estimate range and velocity in a multi-target scenario. Additionally, utilizing multiples antenna adds angle estimation functionality.

This chapter outlined introducing radars for practical use in automobiles. Triangular, trapezoidal, and sawtooth forms of FMCW were briefly discussed. An OFDM-based solution was presented as a digital form of automotive radar that relies on orthogonal codes or *m*-sequences. Range and frequency resolution was shown to be superior to traditional FMCW with no target ambiguity. It is important to note that the OFDM radar technology suffers from high sampling rate requirements.

The aim of this chapter was to provide a basic understanding of and references to automotive radar signal analysis. For interested readers, a sample code of triangular FMCW signal processing in Matlab is attached to this chapter.

TRIANGULAR FMCW MATLAB SIMULATION CODE

```
% This Matlab code simulates a triangular FMCW radar.
clc;
close all;
clear all;

fc = 77e9; % carrier frequency
c = 3e8;
lambda = c/fc;
car_dist = [30 50]; % add more targets here, e.g. [30 50 20]
car_speed = [40 60]; % add more targets here, e.g. [40 60 10]
Nt=length(car_dist);
range_res = 1;

len=3000;
T=0.03333e-6; %us. sybmol duration
dlen=150; % samples over 1 symbol
Ts=T/dlen; %sampling period in sec
N=2048;
range_max = round(c*Ts*(len-dlen)/2);
v_max = c*N/7e9;

% Triangle FMCW simulation
deltaF = c/range_res; %sweep freq
D=1;% number of triangles (cycles)
Tf=5*1e-4; %length of one period
alph=deltaF/Tf; %sweep rate
R=car_dist;%initial distance of the target
td=2*R/c; %initial delay of returned signal in time
fd=fc*car_speed/c;
fs=3*fc;
td_n=round(td*fs); %initial delay in samples

t = 0:1/fs:(D*Tf)-1/fs;
L=length(t);
L1=L/D;

n=0; %triangle index
nT=round(Tf*fs); %length of one period in samples
a=zeros(1,length(t)); %transmitted signal
b=zeros(1,length(t)); %received signal
```

```
r_t=zeros(Nt,length(t));

h = randn(1,length(t));
awgn = randn(1,length(t));

ft=fc*ones(1,length(t));
ftr=fc*ones(1,length(t));
ft1=fc*ones(1,length(t));
f1=ft1(1);
f2=ft1(1)+deltaF;
r_t(1:Nt,1)=R+norm(car_speed)*1/fs; % range of the target in terms of its velocity
    and initial range
ta(1:Nt,1)=2*r_t(1:Nt,1)/c; % delay for received signal

for i=1:Nt
    r_t(i,:)=R(i)+norm(car_speed(i))*t;
end

ta=2*r_t/c; % RRT for received signal

for d=1:D
    ft1(n*nT+1:(n+0.5)*nT)=f1+2*alph*(t(n*nT+1:(n+0.5)*nT)-n*Tf); %transmitted
        signal
    ft1((n+0.5)*nT+1:(n+1)*nT)=f2-2*alph*(t((n+0.5)*nT+1:(n+1)*nT)-n*Tf-Tf/2);
    n=n+1;
end

%transmitted signal
a=real(exp(-j*2*pi*(ft1.*t)));

for m=1:Nt
    fr1(m,:)=circshift(ft1,td_n(m))+fd(m);
end

%received signal
b=real(exp(-j*2*pi*fr1(1,:).*(t-ta(1,:))))+real(exp(-j*2*pi*fr1(2,:).
    *(t-ta(2,:))));% + add the interference here

figure(1);
plot(ft1,'r');

hold on
plot(fr1(1,:),'b');
hold on
plot(fr1(2,:),'g');
legend('Tx','Target␣1','Target␣2');
xlabel('Time'); ylabel('Frequency');

mixed1=(a.*b); %IF signal (output of mixer)

L2=length(mixed1);
m1=reshape(mixed1,L2/D,D); %generating matrix
[My,Ny]=size(m1');
win=hamming(Ny);
m2=conj(m1).*(win*ones(1,My)); %taking conjugate and applying window for sidelobe
    reduction (in time domain)

ii=1:length(m2(:,1));
M2=(fft(m2,L2)); %first FFT for range information

figure(2)
plot(abs(fft(a)), 'r');
hold on
plot(abs(fft(b)), 'b');
```

```
legend('r','b')
hold on
plot(abs(fft(mixed1)), 'g');
legend('Tx','Rx','IF')
hold on
xlabel('Samples'); ylabel('FFT');

% LPF to pass beat frequencies only
ftl=[ones(D,round(L2/(fs/fc))) zeros(D,L2-2*round(L2/(fs/fc)))
    ones(D,round(L2/(fs/fc)))];
M2=M2.*ftl';

figure(3)
M2_f=(abs(M2(:,1)));
win1 = round(0.000005*L1);
M3_f=M2_f(1:1+2*win1);
plot(1:1+2*win1, abs(M3_f));
xlabel('Samples'); ylabel('Magnitude');
title('Beat␣Frequencies');


pd=zeros(2*win1,4*win1);
for k = 1:2*win1
    for m = k+6:1:2*win1
        pd(m-k,k+m)=M3_f(k)*M3_f(m);
        if (m-k)>8
        else
        end
    end
end

L3=50;
pd2=zeros(L3,4*win1);
pd2=pd(1:L3,:);
sr=round(1/(c/(4*deltaF)));
L4=round(length(pd2(1,:))/sr);

for m=1:L3
    k=0;
    for i = 1:L4
        pd3(m,i)=mean(pd2(m,k+1:k+sr));
        k=k+sr;
    end
end

pd4=pd3(1:round(Tf*v_max*4*fc/c),1:round(4*deltaF*range_max/c));
figure (4)
mesh(0:4:(L3-1)*4,0:1/4:(L4-1)/4, abs(pd3)');grid on
xlabel('Speed(km/h)'); ylabel('Range␣(m)');

figure();
imagesc(0:2*v_max, 0:range_max, abs(pd4)');
colorbar
xlabel('Speed(km/h)'); ylabel('Range␣(m)');
title('FMCW');
```

REFERENCES

1. C. Will, P. Vaishnav, A. Chakraborty and A. Santra, *Human Target Detection, Tracking, and Classification Using 24-GHz FMCW Radar*, IEEE Sensors Journal, Vol. 19, Issue 17, pp. 7283–7299, Sep. 2019.
2. Z. Duan, Y. Wu, M. Li, W. Wang, Y. Liu,and S. Yang, *A novel FMCW waveform for multi-target detection and the corresponding algorithm*, 5th International Symposium on Electromagnetic Compatibility (EMC-Beijing), Beijing, pp.1–4, 2017.

3. P. van Genderen, *A communication waveform for radar*, 2010 8th International Conference on Communications, Bucharest, 2010, pp. 289–292.
4. X. Li, L. Tang and X. Zhang, *Range Estimation of CE-OFDM for Radar-communication Integration*, 2018 IEEE International Conference on Communication Systems (ICCS), Chengdu, China, 2018, pp. 131–135.
5. Ajmal, Maria and Khan, Junaid Ali, *Analysis of Radar Performance in CE-OFDM Signal*, Science International, Vol. 26, Issue 1, pp. 105–111, 2014.
6. Yoke Leen Sit, Christian Sturm, Lars Reichardt, Thomas Zwick, Werner Wiesbeck, *The OFDM Joint Radar-Communication System: An Overview*, The Third International Conference on Advances in Satellite and Space Communications (SPACOMM 2011), pp. 69–74, 2011.
7. Ricardo Omar Chavez-Garcia and Olivier Aycard, *Multiple Sensor Fusion and Classification for Moving Object Detection and Tracking*, IEEE Transactions on Intelligent Transportation Systems, Volume: 17 , Issue: 2 , 2016.
8. Jelena Kocić, Nenad Jovičić, and Vujo Drndarević, *Sensors and Sensor Fusion in Autonomous Vehicles*, 2018 26th Telecommunications Forum (TELFOR), Belgrade, pp. 420–425, 2018.
9. Lawrence A. Klein *Sensor and Data Fusion for Intelligent Transportation Systems*, SPIE Press, 2019.
10. R. Karlsson and F. Gustafsson, *The Future of Automotive Localization Algorithms: Available, reliable, and scalable localization*, IEEE Signal Processing Magazine, vol. 34, no. 2, pp. 60–69, 2017.
11. *Introduction to Sensor Fusion and Autonomous Drive Vehicle Algorithms. Part 9: Autonomous Path Planning*, NXP Online Course: https://www.nxp.com/design/training-events:TRAINING-EVENTS, 2019.
12. Claudine Badue, Rânik Guidolini, Raphael Vivacqua Carneiro, Pedro Azevedo, Vinicius Brito Cardoso, Avelino Forechi, Luan Jesus, Rodrigo Berriel, Thiago Paixão, Filipe Mutz, Lucas Veronese, Thiago Oliveira-Santos, Alberto Ferreira De Souza, *Self-Driving Cars: A Survey*, Cornell University Press, pp. 1–19, 2019.
13. M. Rapp. M. Barjenbrucha, M. Hahnb, J. Dickmannb, and K. Dietmayera, *Probabilistic ego-motion estimation using multiple automotive radar sensors*, Elsevier Journal of Robotics and Autonomous Systems, pp. 136–146, 2017.
14. Choi Jeongdan, et. al, *Key technologies in autonomous driving, radar and lidar*, Kama Web Journal, Vol 350, 2018.
15. ISO 17387:2008, *Intelligent transport systems — Lane change decision aid systems (LCDAS) — Performance requirements and test procedures*, 2018.
16. Forkenbrock, G., Hoover, R. L., Gerdus, E., Van Buskirk, T. R., and Heitz, M, *Blind Spot Monitoring in Light Vehicles — System Performance*, Washington, DC: National Highway Traffic Safety Administration, 2014.
17. Cesar Iovescu and Sandeep Rao, *The fundamentals of millimeter wave sensors*, Texas Instrument, May 2017.
18. M. Jankiraman *FMCW Radar Design*, Artech House, Incorporated, 2018.
19. J. Jr. Lin, Y. Li, W. Hsu, and T. Lee, *Design of an FMCW radar baseband signal processing system for automotive application*, Springer Plus Journal, pp. 1–16, 2016.
20. Suleyman Suleymanov, *Design and implementation of an FMCW Radar signal processing module for automotive applications*, Master's Thesis, University of Twente, 2016.
21. B. R. Mahafza, *Radar Signal Analysis and Processing Using MATLAB* 1st Edition, CRC Press, 2009.
22. G. Kim, J. Mun, J. Lee, *A peer-to-peer interference analysis for automotive chirp sequence radars*, IEEE Transactions on Vehicular Technology, vol. 67, no. 9, pp. 8110–8117, Sep. 2018.
23. R. Singh,D, Saluja, S. Kumar, *Power Controlled Adaptive Range Radar for Self Driving Vehicles*, , IEEE 89th Vehicular Technology Conference (VTC2019-Spring), 28 April-1 May 2019.
24. S. Neemat, O. Krasnov, A. Yarovoy, *An interference mitigation technique for FMCW radar using beat-frequencies interpolation in the STFT domain*, IEEE Trans. Microw. Theory Techn., vol. 67, no. 3, pp. 1207–1220, Mar. 2019.
25. Changsheng Yang, Hailong Tang and Hang An, *Beat-frequency matching for multi-target based on improved trapezoid wave with FMCW Radar*, IEEE International Conference on Signal Processing, Communications and Computing (ICSPCC), Xi'an, pp. 1–4, 2011.
26. Pasi Koivumaki, *Triangular and Ramp Waveforms in Target Detection with a Frequency Modulated Continuous Wave Radar*, Master thesis, Aalto University, 2017
27. Sandeep Rao, *Introduction to mmwave Sensing: FMCW Radars*, Texas Instrument, July 2017.

28. D. V. Sarwate and M. B. Pursley, *Crosscorrelation properties of pseudorandom and related sequences*, Proceedings of the IEEE, vol. 68, no. 5, pp. 593–619, May 1980.
29. X. Hu, Y. Li, M. Lu, Y. Wang, X. Yang, *A Multi-Carrier-Frequency Random-Transmission Chirp Sequence for TDM MIMO Automotive Radar*, IEEE Transactions on Vehicular Technology, vol. 68, no. 4, pp. 3672–3685, April 2019.
30. *3GPP 36.211 Evolved Universal Terrestrial Radio Access (E-UTRA); Physical channels and modulation*, 2019.
31. Sandeep Rao, *MIMO Radar*, Texas Instrument, April 2018.
32. Xinrui Qin, Xiaoqi Yang and Weidong Jiang, *Amplitude-Fluctuation Separation Based on BPM Waveform Reconstruction in MIMO Systems*, Journal of Physics: Conference Series, Volume 1284, conference 1, 2019.

13 Multisensor Precise Positioning for Automated and Connected Vehicles

Mohamed Elsheikh
University of Calgary

Aboelmagd Noureldin
Royal Military College of Canada and Queen's University

CONTENTS

13.1 POSITIONING OF AUTOMATED VEHICLES

New technologies and services are being developed every day to support the evolution of automated vehicles (AVs). The ultimate goal is to provide safety features and driving automation systems to enable driverless operation resulting in more efficient transportation. Recently, terms like "self-driving" and "autonomous" cars have been used by the industry and the public to describe different phases of vehicle automation. This creates the need to agree on a certain terminology that can contribute to and regulate the automation process.

In 2012, the Germany Federal Highway Research Institute (BASt) defined five degrees of vehicle automation, namely *driver only*, *driver assistance*, *partial automation*, *high automation*, and *full automation* [1]. In 2013, the U.S. Department of Transportation's National Highway Traffic Safety Administration (NHTSA) defined similar five levels of automated driving. In 2014, the Society of Automotive Engineers (SAE) International defined six levels of driving automation in its J3016 standard, which was recently updated in June 2018 [2]. Later in 2016, NHTSA and the U.S. Department of Transportation updated their policy to adopt the SAE six levels of vehicle automation [3,4]. These six levels were given numbers and names as follows: *no automation* (level 0), *driver assistance* (level 1), *partial automation* (level 2), *conditional automation* (level 3), *high automation* (level 4), and *full automation* (level 5).

The SAE International standard [2] describes the roles of the driver and the driving automation system in the performance of the dynamic driving task (DDT) in each automation level. It is important to understand that an AV may have multiple features, each associated with a particular level of driving automation. The DDT is divided into the *sustained lateral and longitudinal vehicle motion control* and *object and event detection and response* (OEDR). In levels 0 to 2, the driver is responsible for completing the DDT and DDT fallback even if she/he was partially supported by the driving automation system (level 1 and level 2). On the other hand, the driving automation system performs the entire DDT in levels 3, 4, and 5, and is called the Automated Driving System (ADS). However, in level 3, the driver is called "DDT fallback-ready user" which means she/he is responsible for the DDT fallback. The DDT fallback-ready user should be ready to intervene whenever a failure occurs in the DDT or when the driving automation system is about to leave its operational design domain (ODD). The ADS performs the DDT and DDT fallback in levels 4 and 5, and any human in the car is considered as a passenger. The difference between levels 4 and 5 is that in level 5, the ODD is unlimited, and hence, the full automation is achieved.

The positioning capability of the vehicle navigation system is crucial for the driver or the ADS, or both to efficiently complete their jobs. Nevertheless, the positioning requirements depend on the applied level of automation and its supported features. Furthermore, the required positioning accuracy and the characteristics of each driving level are important factors in the choice of which navigation technology should be used. Table 13.1 describes the six levels of automation defined by SAE, the roles of the driver, and the automation system in each level, example features, and the expected required positioning accuracy for each level.

From the above discussion, the higher the automation level, the more robust the positioning solution should be. Fully automated vehicles require submeter-/centimeter-level accuracy, which should be available all the time and everywhere. This explains why multisensor fusion is essential in AV/connected vehicles (CVs), and why the research is still ongoing on how to effectively integrate the different navigation sensors to increase the accuracy and integrity of the navigation solution. Figure 13.1 summarizes the main requirements for robust positioning.

13.2 DIFFERENT SENSORS FOR AV/CV POSITIONING

Multisensor fusion becomes an essential part of AV/CVs due to the strict safety requirements that have to be satisfied in these vehicles before they can go on the road. Various sensors are currently employed in automated vehicles not just for positioning but also in warning systems. A reliable navigation system relies on the information from both positioning and warning systems, and can even integrate both to provide a robust solution with high integrity. Figure 13.2 gives examples of some of the sensors that can be used for the positioning of AV/CVs. In this section, each sensor is described in terms of its principle of operation, positioning range, advantages, and limitations.

Table 13.1
SAE J3016 Automation Levels and Their Characteristics

	Level 0 No Automation	Level 1 Driver Assistance	Level 2 Partial Assistance	Level 3 Conditional Automation	Level 4 High Automation	Level 5 Full Automation
Lateral and Longitudinal Motion Control	Driver	Driver and System	System	System	System	System
OEDR	Driver	Driver	Driver	System	System	System
DDL fallback	Driver	Driver	Driver	Driver	System	System
OOD	n/a	Limited	Limited	Limited	Limited	Unlimited
Example features	Automatic braking system and blind spot warning	Lane centering or adaptive cruise control	Lane centering and adaptive cruise control	Traffic jam chauffeur	Local driverless taxi and valet parking	Same as level 4 but everywhere
Positioning accuracy	Meter	Submeter	Submeter	Submeter centimeter	Submeter centimeter	Centimeter

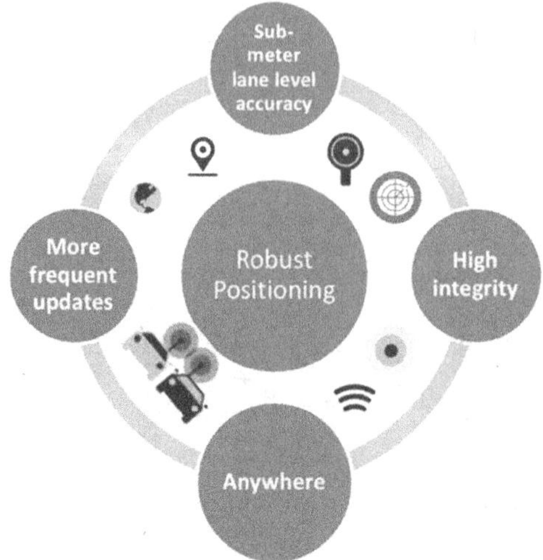

Figure 13.1 Illustration of the main requirements for robust positioning.

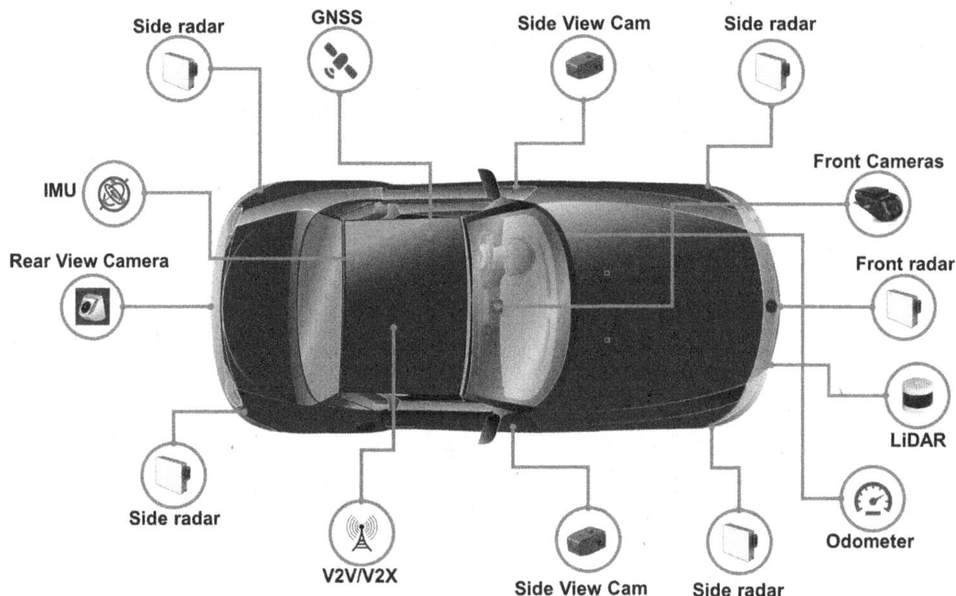

Figure 13.2 Illustration of some of the sensors used for AV/CVs positioning.

13.2.1 GLOBAL NAVIGATION SATELLITE SYSTEMS

When it comes to precise positioning, the Global Positioning System (GPS) is the first thing to come in mind. The GPS is the first fully operational global navigation satellite system (GNSS) that provides all-weather worldwide coverage. Later, the Russian Global Navigation Satellite System (GLONASS) has also achieved its full operational capability. Other GNSS systems are still on the way to provide global coverage such as the European Galileo and the Chinese BeiDou systems.

When GPS was initially designed, it utilized the code measurements with accuracy limited to few meters. The designers did not see the potential for higher accuracy at that time [5]. The concept of using GPS carrier phase measurements to obtain centimeter-level accuracy was first introduced by

Counselman [6,7]. Carrier phase measurements are more precise than the corresponding code measurements because their wavelengths are in the centimeter level, e.g., the wavelength of the GPS L1 frequency is approximately 19 cm. Despite the high precision of the GNSS carrier phase, it suffers from the problem of integer ambiguity, which is the unknown number of full carrier cycles between the receiver and the satellite during the initial signal acquisition. The required high precision will not be obtained unless this ambiguity is resolved [5].

Carrier phase-based differential GPS (DGPS) positioning was the first technique to obtain centimeter-level accuracy from GPS [7]. DGPS, or generally DGNSS, is a method to improve the performance of the GNSS receiver with the aid of information from one or more reference stations at known locations [8]. The main idea behind DGNSS is that if the baseline length, i.e., the distance between the rover receiver and the reference station, is short enough (e.g., <10 km), then some errors such as the atmospheric errors, and the satellite orbit and clock errors are spatially correlated and can be canceled by the differencing techniques. The real-time kinematics (RTK) is a DGNSS technique that can provide a real-time positioning solution with centimeter-level accuracy. Nevertheless, the cost of the reference stations and the baseline length limitations are significant drawbacks of DGNSS.

The precise point positioning (PPP), introduced in 1997 [9], has eliminated the need for local reference stations. PPP provides the absolute position of the receiver based only on the GNSS measurements of the rover receiver and globally distributed precise correction products. Because PPP must deal with all the error sources, which are canceled in DGNSS, its most significant challenge is the convergence time required to resolve the integer ambiguity and reach centimeter-level accuracy [10]. Nevertheless, the capability of PPP to provide a global solution using a single receiver made it very promising and attracted many researchers to develop methods to overcome its challenges [11]. More details about PPP can be found in Section 13.4.1. The aforementioned GNSS precise positioning techniques are summarized in Figure 13.3.

GNSS is a core element in navigation systems of AV/CVs in all levels of autonomy. Despite the precise accuracy that can be obtained from GNSS, this accuracy requires continuous visibility of GNSS satellites and open-sky conditions that are not available in some applications. For example, in land vehicle navigation, there are many challenging environments for GNSS such as urban canyons, tunnels, and overpass bridges. In such scenarios, the GNSS solution will deteriorate or become unavailable, and it must be integrated with other systems to maintain the continuity of the navigation solution.

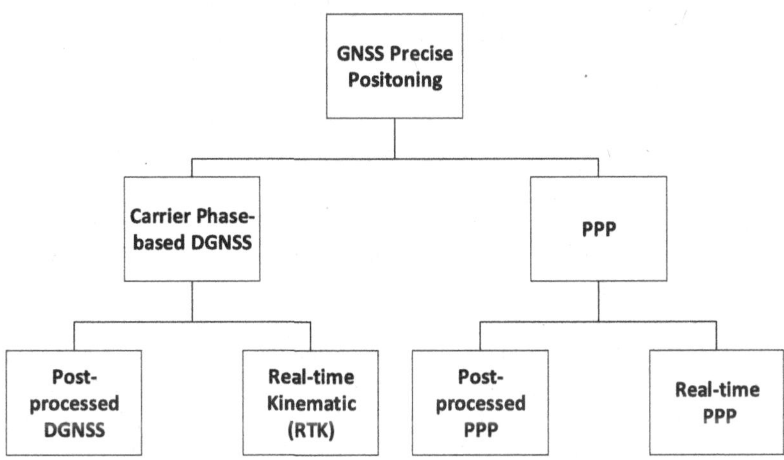

Figure 13.3 GNSS precise positioning techniques.

13.2.2 INERTIAL NAVIGATION SYSTEMS

An inertial navigation system (INS) is a navigation system that includes a navigation processor and an inertial measurement unit (IMU). The IMU consists typically of accelerometers to measure specific forces (acceleration) and gyroscopes to measure angular rotation rates. The INS is an example of dead-reckoning navigation systems. It is autonomous, which means it does not need an external reference and only needs the knowledge of the initial position, speed, and heading information. Moreover, it gives a high output rate of at least 50 Hz compared to the low GNSS data rate which is typically around 10 Hz [12]. Despite the significant advantage of being a standalone and weather- and environment-independent system, the INS has a short-term accuracy. In the long term, the INS solution drifts dramatically away from the correct navigation parameters due to the inherent sensor errors. The mathematical integration of the IMU measurements to obtain the velocity and position causes that any small bias in any of the sensors will quickly grow larger with time [13].

The navigation solution of the INS is calculated through a mechanization process, which is the use of the IMU measurements to calculate the position, velocity, and attitude information. A full IMU system consists of three orthogonal accelerometers and three orthogonal gyroscopes to be able to measure the accelerations and rotations in all directions in a three-dimensional (3D) space. Because the full IMU system has six degrees of freedom, it can represent motions of different dynamics including land and air applications. High-quality INS solution requires high-end sensors, which are usually expensive and bulky. Recently, microelectromechanical system (MEMS) sensors have been used extensively in navigation applications due to their low cost and small size. Nevertheless, the use of MEMS sensors is a trade-off between the cost and the quality of the solution. The fusion of MEMS-based INS with other systems helps to mitigate the effects of their solution drift.

In the land vehicle navigation industry, another approach to reduce the system cost is to use a lower number of sensors. This reduction is based on the land vehicle characteristics that constraint the vehicle motion in almost a horizontal plane [14]. The minimum configuration of IMU sensors for a reduced INS is one vertical gyro and two horizontal accelerometers [15]; nevertheless, a wheel speed odometer may be used to replace the accelerometers in 2D models [16]. This configuration has been studied by many researchers in the past years [15–17]. Some of these studies have presented a simplified 2D solution ignoring the vehicle off-plane motion [16], whereas others have tried to model the pitch and roll angles as a stochastic process [15,17]. A reduced inertial sensor system (RISS) was introduced as a 2D model in Ref. [18] using one vertical gyro and a speed odometer. Later, the measurements from the accelerometers were also incorporated in the RISS model to estimate the off-plane vehicle motion, and it was called 3D RISS [19].

The fact that the INS can be used in all environments made it an essential component of vehicular navigation systems.

13.2.3 ODOMETERS

Odometry is the calculation of the change in position and the speed of an object using motion sensors. The distance traveled and the speed of a land vehicle can be determined by measuring the rotation of its wheels. An *odometer* has been traditionally fitted to the transmission shaft of the vehicle to measure its speed; nevertheless, most of the new vehicles have one on each wheel known as the *wheel speed sensor*, which is used for the antilock braking system [12]. Knowing the forward speed of the vehicle can enhance the navigation solution by either using it to reduce the inertial system model [16,19] or as a measurement update. The vehicle forward speed can be accessed in most of the vehicles through its on-board diagnostics interface. However, the speed obtained from the on-board diagnostics interface is usually provided at a low output rate and with a low resolution which causes quantization errors.

Having access to the raw wheel measurements can result in more precise speed measurements. The wheel measurements are typically represented as pulse counts, with each pulse corresponding to a certain traveled distance through a scale factor in units of (m/pulse). The distance traveled between two epochs is calculated as the difference between the number of pulses at each epoch multiplied by the scale factor. The forward speed can then be calculated by dividing the distance traveled by the time duration between the two epochs. Furthermore, by differencing the left and right wheel measurements, the yaw rate of the vehicle may be calculated through a technique known as *differential odometry* [20].

Odometry is a dead-reckoning technique; therefore, it is also prone to error accumulation over time. Errors in the wheel scale factor can affect both the speed and the yaw rate calculation; therefore, measurements from other sensors are needed to calibrate the scale factor of each utilized wheel. Conversely, the quantization errors are found to have a negligible effect on the speed and the yaw rate estimation in the long term [20]. Land vehicle odometry is also affected by the slipping and skidding of the wheels and the road surface variations, especially in the yaw rate calculation [21].

13.2.4 PERCEPTION SYSTEMS

A vital requirement for the safety of AV/CVs is the self-awareness of the surrounding environment. Various perception systems, which are based on environmental feature-matching techniques, can be used with the AV/CVs such as cameras, light detection and ranging (LiDAR), and radars.

Cameras and vision systems are used to detect features around the vehicle such as road and lane edges. Cameras also are useful in obstacle detection and avoidance. The term *visual odometry* refers to vision-based positioning where the camera frames can be processed and utilized for pose estimation [22,23]. Because visual odometry integrates small incremental motions over time, it is bound to drift due to the variations observed in the scene and, hence, is better to be integrated with other navigation systems to compensate for this drift. Cameras perform well in textured environments with visible features and good illumination; nevertheless, visual odometry performance degrades in low-visibility and low-textured environments.

LiDAR, on the other hand, can operate in degraded-vision environments and can provide accurate measurements of range information with respect to the surrounding objects. LiDAR is also used in mapping applications. However, the problems of using LiDAR in positioning applications are its high cost and that processing LiDAR data is computationally demanding. LiDAR shares the limitations of cameras in challenging weather conditions such as rain, snow, and fog [24]. Leading car manufacturers and pioneers in the automated vehicle industry, such as Tesla and Nissan, have decided to exclude LiDAR from their ADSs and move toward vision and radar [25].

The automotive radar is an all-weather system that has been used for a long time in land vehicles since the early levels of automation in the adaptive cruise control feature. Radars can be used as an alternative to LiDAR to detect objects and to provide range estimation. Nevertheless, most present radar-based odometry and localization methods lack the target submeter-level accuracy required by AV/CVs and cannot be used alone for precise positioning [26,27].

The above discussion shows that perception systems have their own limitations; however, their benefits when combined with GNSS and dead-reckoning systems can improve the accuracy and robustness of the positioning solution of AV/CVs, especially in higher levels of automation, namely, levels 3, 4, and 5.

13.2.5 SENSORS UTILIZATION UNDER DIFFERENT DRIVING CONDITIONS

The characteristics of each navigation sensor impose limitations on the driving conditions where it can be used. Table 13.2 provides a comparison between the different navigation sensors under some driving conditions and environments.

Table 13.2
Utilization of Navigation Sensors under Different Driving Conditions and Environments (✓ Work, × Does not Work, ? Limited Operation)

Sensor	Underground Parking	Tunnels	Downtown	Fog	Rain	Snow and Slippery Roads
GNSS	×	×	?	✓	✓	✓
Vision	✓	✓	✓	×	×	?
LiDAR	✓	✓	✓	?	?	?
Odometer	✓	✓	✓	✓	✓	?
Radar	✓	✓	✓	✓	✓	✓
INS	✓	✓	✓	✓	✓	✓

GNSS positioning is independent of the weather conditions; however, it requires line-of-sight visibility of the GNSS satellites. Therefore, GNSS positioning does not work in covered areas. Another factor that limits GNSS positioning is the multipath effect that is very likely to occur in downtown areas due to high-rise buildings.

Vision systems work efficiently indoors and outdoors if there is enough illumination. Nevertheless, in vision-degraded environments caused by darkness or weather conditions like rain or fog, positioning based on vision systems may not be reliable. Furthermore, the performance of vision systems outdoors might be degraded when snow covers the road signs. On the other hand, LiDARs can work in poor-illuminated environments, but it is still affected by the wet conditions that affect the reflectivity of the surrounding objects [24].

Odometers work in different environments and weather conditions; nonetheless, they depend on the rotation of the vehicle wheels and, hence, affected by slippery road conditions and rood irregularities.

Although radars can generally work in all conditions, they are limited by the number of useful features that can be detected in each frame. The radar localization algorithm can fail if there is insufficient number of features. Inertial sensors are the only navigation sensors that can work everywhere and under any operational conditions. Still, they require updates from other sensors to limit their error drift.

13.3 MULTISENSOR FUSION FOR POSITIONING

Multisensor fusion in navigation applications is the process of integrating measurements from different sensors to provide a more robust and continuous navigation solution than the one that can be obtained by each sensor separately. As can be seen in Section 13.2, each navigation sensor has its own advantages and limitations, and it cannot be relied upon solely for AC/CVs positioning. Fortunately, the characteristics of these sensors are complementary, and hence, fusing their measurements will result in enhanced positioning capability. The first part of this section briefly reviews the main sensor fusion techniques and filters, whereas the second part summarizes the major sensor fusion architectures in positioning applications.

13.3.1 MULTISENSOR FUSION FILTERS

Most of the sensor data fusion techniques are based on the science of statistics and signal estimation [28]. These techniques are based on the Bayes' rule, which provides information about the posterior probability of a certain estimate of the system state x given an observation z using the formula

$$p(x|z) = \frac{p(z|x)p(x)}{p(z)} \tag{13.1}$$

where $p(x)$ is the a priori probability density function (PDF) of x, $p(z)$ is the PDF of the measurement z, and $p(z|x)$ is the conditional probability of z given x. The two examples of the Bayesian filters that are widely used in the navigation field are the Kalman filter and the Particle filter.

13.3.1.1 The Kalman Filter

The Kalman filter [29] is a recursive Bayesian filter that estimates the system states with the aid of some measurements from the system observables. The Kalman filter combines the state prediction from the system model and the updates from the measurements to generate the output solution. This combination depends on the covariance matrices of the prediction and measurement models which represent the uncertainty about each one of them. The Kalman filter outputs are the estimated system states accompanied by the statistical information about the covariance of these estimates. The advantage of Kalman filter, compared to other techniques such as least squares, is the utilization of the information about the deterministic and statistical properties of the system states in addition to the independent measurement updates.

The limitation of the traditional Kalman filter is that it is only optimal for linear systems with Gaussian noise distributions. These two assumptions of linearity and Gaussian PDF are usually not met in the navigation systems. Therefore, other extensions have been developed to address these limitations such as the *extended Kalman filter* (EKF) and the *unscented Kalman filter* [12]. The rest of this section summarizes the basic implementation of the EKF.

The EKF works on the errors of the system states rather than their absolute values. These errors are assumed to be much smaller than the states itself, allowing linear system and measurement models to be applied. The limitations of the EKF are the assumption of a Gaussian PDF and the effect of ignoring the higher-order error terms during the linearization process [30]. Despite these limitations, the EKF and its variants are still major components in the integrated navigation systems [31–33]. In general, the Kalman filter works well in data fusion problems where the system model can be described by a continuous parametric state such as the position, velocity, and attitude of land vehicles [28].

13.3.1.1.1 EKF Prediction

The system dynamic model of the EKF can be described by the first-order differential equation

$$\dot{\mathbf{x}}(t) = f(\mathbf{x}(t)) + \mathbf{w}(t) \tag{13.2}$$

where $\mathbf{x}(t)$ is the vector of system states, $f(.)$ is the nonlinear mapping function of the system states that represents the state transition, and $\mathbf{w}(t)$ is the noise vector which represents the uncertainty in the state evolution model.

In the EKF, working on the errors of the state vector allows the linear Kalman filtering to be applied based on the assumption that these errors are small compared to the state vector itself, and hence, Eq. (13.2) can be modified as follows:

$$\delta\dot{\mathbf{x}}(t) = F(t)\delta\mathbf{x}(t) + \mathbf{w}(t) \tag{13.3}$$

where $F(t)$ is the system dynamic coefficients matrix. The error vector $\delta\mathbf{x}(t)$ represents the difference between the true state value $\mathbf{x}(t)$ and its estimated value $\hat{\mathbf{x}}(t)$, i.e., $\delta\mathbf{x}(t) = \mathbf{x}(t) - \hat{\mathbf{x}}(t)$.

In the discrete–time implementation of the EKF, the prediction of the state vector errors $\delta\mathbf{x}_{\mathbf{k}}^{-}$ and its a priori covariance P_k^{-} at epoch k are achieved through the state transition matrix ϕ_{k-1}, which can be approximated as [34]

$$\phi_{k-1} \approx I + F_{k-1}\Delta t \tag{13.4}$$

where I is the identity matrix, Δt is the time interval between the epochs k and $k-1$, and F_{k-1} is calculated from the posterior state estimate of the last epoch $\hat{\mathbf{x}}_{k-1}^{+}$ using

$$F_{k-1} = \left. \frac{\partial f(\mathbf{x}(t))}{\partial \mathbf{x}} \right|_{\mathbf{x}=\hat{\mathbf{x}}_{k-1}^{+}} \tag{13.5}$$

The EKF predication is performed as follows:

$$\delta \mathbf{x}_{\mathbf{k}}^{-} = \phi_{k-1} \delta \mathbf{x}_{\mathbf{k-1}}^{+} \tag{13.6}$$

$$P_k^{-} = \phi_{k-1} P_{k-1}^{+} \phi_{k-1}^{T} + Q_{k-1} \tag{13.7}$$

where δx_{k-1}^{+} and P_{k-1}^{+} are the posterior state error vector and its covariance matrix from the previous epoch, respectively. The matrix Q represents the covariance of the system noise $\mathbf{w}(t)$.

13.3.1.1.2 EKF Update

The nonlinear measurement model of the EKF is described by

$$\mathbf{z}(t) = h(\mathbf{x}(t)) + \eta(t) \tag{13.8}$$

where $\mathbf{z}(t)$ is the measurement vector, $h(.)$ is a nonlinear mapping function from the system states to the observations domain, and $\eta(t)$ denotes the measurement noise.

Similar to the prediction step, the assumption of small state errors can be used to linearize Eq. (13.8) to become

$$\delta \mathbf{z}(t) = H \delta \mathbf{x}(t) + \eta(t) \tag{13.9}$$

where H is the measurement design matrix which is calculated at epoch k using

$$H_k = \left. \frac{\partial h(\mathbf{x}(t))}{\partial \mathbf{x}} \right|_{\mathbf{x}=\hat{\mathbf{x}}_{k}^{-}} . \tag{13.10}$$

The term $\delta \mathbf{z}(t)$ can be calculated at epoch k using

$$\delta \mathbf{z}_{\mathbf{k}} = \mathbf{z}_{\mathbf{k}} - h(\hat{\mathbf{x}}_{\mathbf{k}}^{-}). \tag{13.11}$$

The difference between the calculated $\delta \mathbf{z}_{\mathbf{k}}$ in Eq. (13.11) and the value $H_k \delta \mathbf{x}_{\mathbf{k}}^{-}$ from the linearization process in Eq. (13.9) is called the *innovation sequence*. The update of the error state vector is performed by multiplying this innovation sequence by the Kalman gain K_k and, then, add the result to the predicted value from Eq. (13.6). The Kalman gain is also used to obtain the posterior covariance matrix P_k^{+} as follows:

$$K_k = P_k^{-} H_k^{T} \{H_k P_k^{-} H_k^{T} + R_k\}^{-1} \tag{13.12}$$

$$\delta \mathbf{x}_{\mathbf{k}}^{+} = \delta \mathbf{x}_{\mathbf{k}}^{-} + K_k(\delta \mathbf{z}_k - H_k \delta \mathbf{x}_{\mathbf{k}}^{-}) \tag{13.13}$$

$$P_k^{+} = (I - K_k H_k) P_k^{-} \tag{13.14}$$

where R is the measurement noise covariance.

13.3.1.1.3 Closed-Loop EKF

In many navigation systems, such as INS, some parameters, such as the sensor biases and scale factors, are better to be calibrated online for better performance and to keep the estimated trajectory close to the true trajectory. This can be achieved by adopting a closed-loop configuration. The closed-loop approach is implemented by feeding back the estimated error states and the new estimated navigation solution to the system model, which means that the error states will be reset to zero after each epoch. Thus, the new estimated errors will be kept small to minimize the effect of the linearization process and ignoring the higher order terms. In the closed-loop EKF, Eq. (13.6) will reduce to $\delta \mathbf{x}_{\mathbf{k}}^{-} = 0$. Nevertheless, care must be taken to the stability of the closed-loop EKF by careful tuning of the filter parameters in addition to the accurate initialization.

13.3.1.2 The Particle Filter

Although the EKF provides a solution for nonlinear estimation problems, the PDF of the states is still approximated by a Gaussian distribution, and hence, only the mean and covariance of the states are propagated. Some of the system models in the navigation applications, such as pattern-matching systems, cannot be fitted with a Gaussian PDF [12]. The challenge for nonlinear non-Gaussian problems is that the required PDF has no closed-form expression. In Ref. [35], the non-Gaussian PDF was approximated numerically through recursive and random sampling of the state space. This approach was the basis for what is currently known as *particle filters*.

Particle filters are recursive implementations of the Monte Carlo method for signal estimation. They approximate the optimal solution to the physical system model rather than applying an optimum filter to an approximate system model (as in the Kalman filter) [36]. Particle filters propagate the posterior PDF of the system states over time in the form of a set of random samples of the state space, known as *particles*. Each particle x^i is assigned a normalized weight w^i such that $\sum_i w^i = 1$. The PDF can now be approximated as follows:

$$p(x_k|z_k) \approx \sum_{i=1}^{N_p} w_k^i \delta(x_k - x_k^i) \tag{13.15}$$

where N_p is the number of utilized particles.

Equation (13.15) indicates that an accurate representation of the actual PDF requires a sufficiently large number of particles. Therefore, the primary challenge for employing particle filters is the computational cost, especially in real-time applications. However, due to the continuous improvements to the computational power of the embedded systems, particle filters are currently used in real time in fields such as computer vision, map matching, tracking, and robotics [37,38].

More details about the different types of particle filters and their implementations can be found in Ref. [37,39,40].

13.3.2 MULTISENSOR FUSION ARCHITECTURES

The fusion filters described in Section 13.3.1 provide an algorithmic way of fusing the data from multiple sensors. On the other hand, the architecture of a multisensor system defines the organization of the different system components and how the information from different sensors is handled. GNSS/INS integration is the core of most navigation systems. Other systems can be added according to the cost, accuracy, environment, and degree of robustness requirements [12]. This section starts with a description of the main integration structures for GNSS/INS integration followed by a brief discussion of more generalized architectures for adding more sensors.

13.3.2.1 GNSS/INS Integration Architectures

There are several ways in the literature to classify the GNSS/INS integration approaches; however, they can be generally categorized into three basic architectures: loosely coupled, tightly coupled, and ultratightly coupled integration [41].

13.3.2.1.1 Loosely Coupled Integration

In the loosely coupled integration, the INS and GNSS operate independently and provide completely separate navigation solutions, and then, both solutions are combined through a fusion filter. Figure 13.4 shows the block diagram of the loosely coupled GNSS/INS integration, where the option of using the closed-loop configuration, as described in Section 13.3.1.1, is represented by the dotted feedback line. The loosely coupled approach is simple, and the separate solutions prevent a fault in one of them from affecting the other. However, this approach is based on the final GNSS positioning solution and, hence, at least four visible GNSS satellites. Because the GNSS solution itself is usually obtained through a navigation filter, the loosely coupled integration suffers from the correlated measurements of the cascaded GNSS and integration filters [41].

13.3.2.1.2 Tightly Coupled Integration

In the tightly coupled approach, shown in Figure 13.5, the integration is performed on the GNSS measurement level. This approach can significantly improve the performance by using one filter, especially when less than four satellites are visible, or the satellite geometry is degraded [42]. The drawback of this mode is that there is no inherent GNSS stand-alone solution; nevertheless, an independent GNSS solution may be maintained in parallel if required. Also, the GNSS tracking loops do not benefit from the INS information, and this has been the motivation for the third approach, the ultratightly coupled integration.

13.3.2.1.3 Ultratightly Coupled Integration

The term *ultratight*, or sometimes called *deep integration*, reflects how the output of the integration filter is not just used to correct the INS output as in the previous types, but it is fed back as an aiding information for the GNSS receiver tracking loops to help in the signal acquisition and tracking. The ultratight architecture is actually a vector-based GNSS receiver with the addition of the IMU information as an input to the navigation filter, which becomes the integration filter [43]. In this case, the GNSS tracking loops operate only on the errors in the aiding information and in the receiver clock instead of the absolute motion dynamics. Therefore, most of the received signal dynamics are removed, which in turn helps to reduce the tracking loop bandwidth [42]. With a narrower bandwidth, the noise and jamming rejection of the receiver increase, allowing the acquisition and tracking of weaker signals.

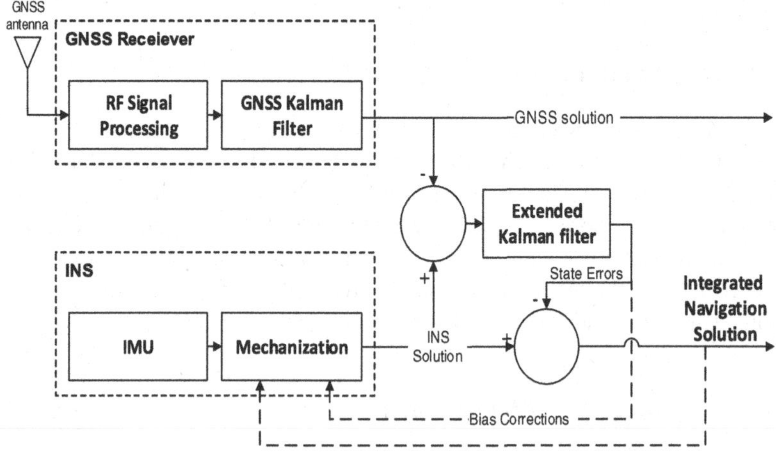

Figure 13.4 Block diagram of loosely coupled GNSS/INS integration using an EKF. The dotted lines represent the closed-loop configuration.

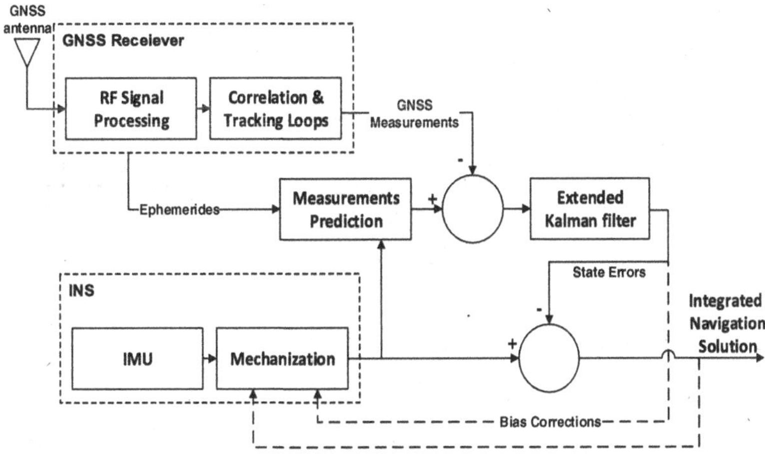

Figure 13.5 Block diagram of tightly coupled GNSS/INS integration using an EKF. The dotted lines represent the closed-loop configuration.

Figure 13.6 shows the two major methods of INS aiding; the first method is called the Doppler-aiding, where the input to the integration filter is the pseudorange, carrier phase, and/or Doppler measurements, and the output is the Doppler estimate. This estimate is fed back to the receiver phase-locked loop (PLL). The PLL detects the deviation from the receiver intermediate frequency, and this deviation consists of the Doppler frequency, clock error, and noise effect. So, by estimating the Doppler frequency as an aiding for the PLL, the bandwidth will be narrowed to handle just the clock error and the remaining Doppler estimate error, and hence, the noise effect is reduced and the performance is enhanced [44,45]. Some researchers consider this type as a tightly coupled integration but with Doppler aiding [42,46].

The second method can be thought of as a deeper integration, where the input to the integration filter is the very raw receiver data, the in-phase (I) and quadrature-phase (Q) correlator outputs, and the filter output is used directly to control the numerically controlled oscillator of the tracking loops. This method is the most used in ultratight integration implementations, and it can be implemented using a centralized filter (one master filter) or a federated filter [42].

The disadvantage of deep integration is the complexity because it requires access to the GNSS receiver tracking loops which is still not possible for the majority of commercial GNSS receivers. Therefore, most of the research on this type of integration is based on receiver simulators [30].

13.3.2.2 General Integration Architectures

The previous section discussed some specific fusion architectures which are practically applied for GNSS/INS integration. When more sensors or navigation systems are incorporated, the fusion architecture gets more complex. The primary goal when selecting an architecture is to maximize the accuracy and robustness of the integrated system, whereas the major challenge is to minimize the system complexity and the computational and financial cost. Furthermore, the choice of the integration approach is constrained by the nature of the information that can be obtained from each sensor. For example, if a low-cost GNSS receiver does not provide its raw measurements, the previously mentioned tightly coupled or ultratightly coupled methods cannot be applied.

There are various integration architectures for multisensor integrated navigation which can be grouped into *epoch-by-epoch* and *filtered* architectures [12]. Figure 13.7 shows these two categories and their subimplementations. Under each category, the implementation might be cascaded or centralized in addition to the federated configuration in the case of filtered architectures.

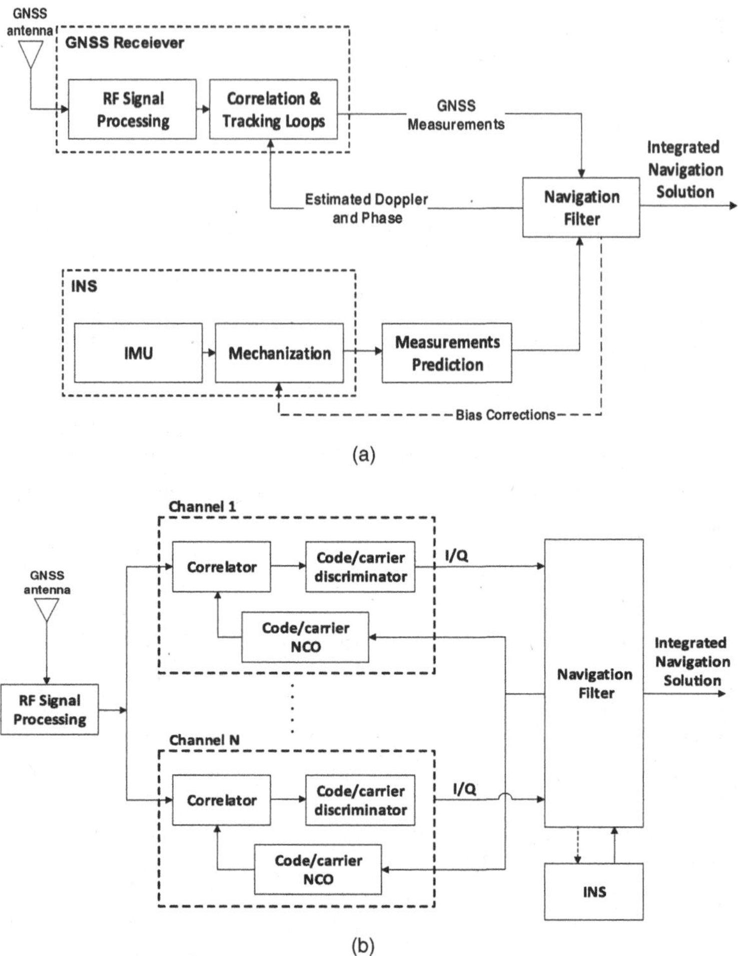

Figure 13.6 The two methods of INS aiding for ultratightly coupled GNSS/INS integration. (a) Doppler-aiding and (b) using I and Q information (centralized filter example).

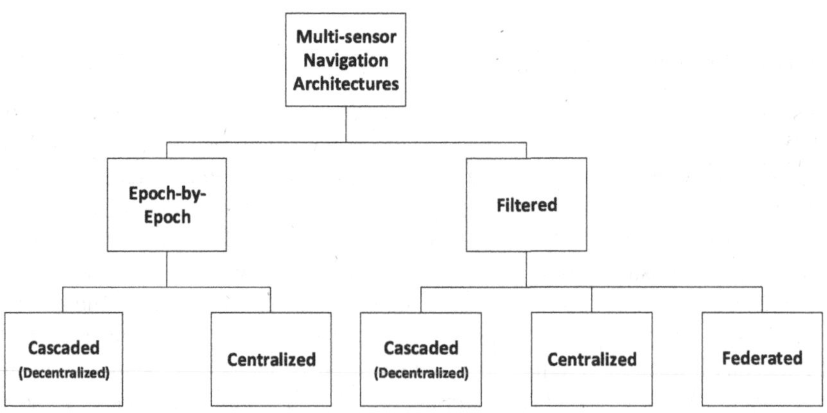

Figure 13.7 General integration architectures for multisensor navigation.

The cascaded configuration represents the case when different navigation systems provide their solution independently before the fusion step, and this is why it is also called *decentralized* architecture. The term *cascaded* is because each navigation solution is typically the output of an estimation process, and then, all the individual solutions are input to the final fusion algorithm. The *centralized* configuration includes one master fusion algorithm, which works on the raw measurements from all the sensors. The *federated* configuration reduces the load on the master fusion filter by adding a stage of local filters as will be discussed later in this section.

In epoch-by-epoch architectures, the information of the current epoch from different navigation systems is combined using weighted least squares. The advantages of epoch-by-epoch processing are the simplicity, low processor load, and fault isolation as an error in one of the epochs will not propagate to the following epochs. Epoch-by-epoch processing can be applied in position-fixing systems such as GNSS, but it cannot work with dead-reckoning systems, especially INS. To be able to calibrate the inertial sensor biases and limit its drift with time, a navigation filter must be employed. A workaround can be to fuse GNSS and INS first through a filter and consider it as one solution and, then, fuse it with other sensors in an epoch-by-epoch approach. However, in applications such as AV/CVs positioning, the filtered architectures are more practical because it is applicable for all sensors and also allows the system model of each sensor to be employed in the estimation process.

In filtered architectures, the fusion is performed through one of the integration filters introduced in Section 13.3.1. In the rest of this section, the cascaded filtered, centralized filtered, and federated filtered architectures are described. Without the loss of generality, the following discussion assumes that the selected integration filter is the EKF, and the system model is based on the error state vector as explained in Section 13.3.1.1. The INS is normally chosen as the reference system to provide the solution prediction, while the other sensors provide the update measurements.

13.3.2.2.1 Cascaded Filtered Architecture

Figure 13.8 shows a block diagram of the cascaded filtered architecture using an EKF. Each navigation system estimates its own solution in a decentralized way before they are input to the fusion

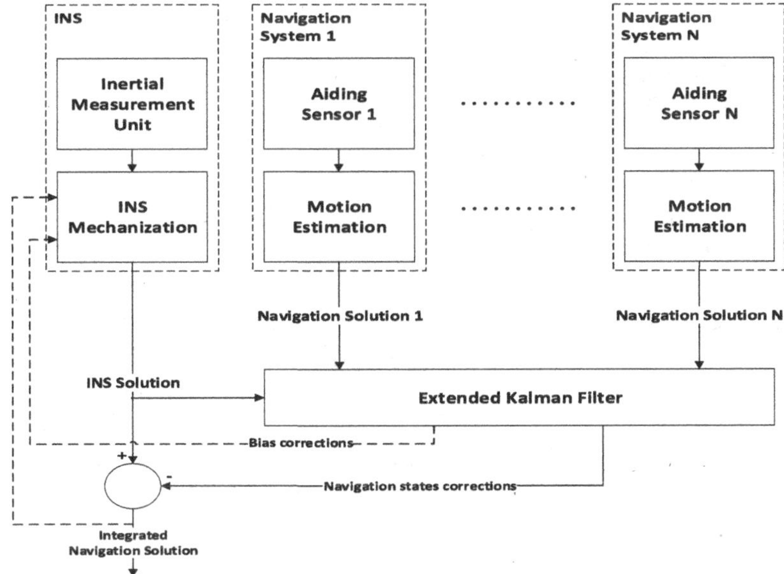

Figure 13.8 Block diagram of cascaded filtered integration architecture using an EKF.

filter. The INS is used as the reference solution to build the system model. The other N navigation systems provide the filter update through their final navigation solutions. A closed-loop implementation can be achieved if the estimated errors are fed back to correct the INS mechanization of the next epoch as shown by the dotted lines in Figure 13.8. If only the GNSS solution is used for aiding, this architecture reduces to the loosely coupled configuration in Section 13.3.2.1. Feedback for aiding systems is also possible if needed; however, this will make the systems interdependent and fault detection harder.

One advantage of the cascaded filter architecture is that the navigation solution can be generated at a faster rate than fusion filter processing which reduces the computational load. Another advantage is the simpler design of the navigation filter as the system states, and the updates are the final navigation parameters, and this also means that it can work with aiding systems that do not provide raw measurements. The limitations of this approach come from the cascaded configuration, which suffers from the possibility of correlated measurements and makes no use of the sensor measurements if no final solution is obtained.

13.3.2.2.2 Centralized Filtered Architecture

In the centralized filtered architecture, the sensor measurements are input directly to the fusion filter as shown in Figure 13.9. This means that all the available measurements contribute to the integrated solution; nevertheless, care must be taken for the time synchronization between different measurements. As there is only one fusion filter, this architecture has no error correlation issues. The disadvantages of this approach are the high computational load, no individual navigation solutions from the aiding systems unless they are generated in parallel, and it does not apply to systems that do not output its raw measurements. If only GNSS measurements are used for aiding, this architecture reduces to the tightly coupled configuration in Section 13.3.2.1.

13.3.2.2.3 Federated Filtered Architecture

The idea of federated architectures is simply to reduce the computational load on the master fusion filter in centralized architectures. The INS solution is first integrated separately with each of the

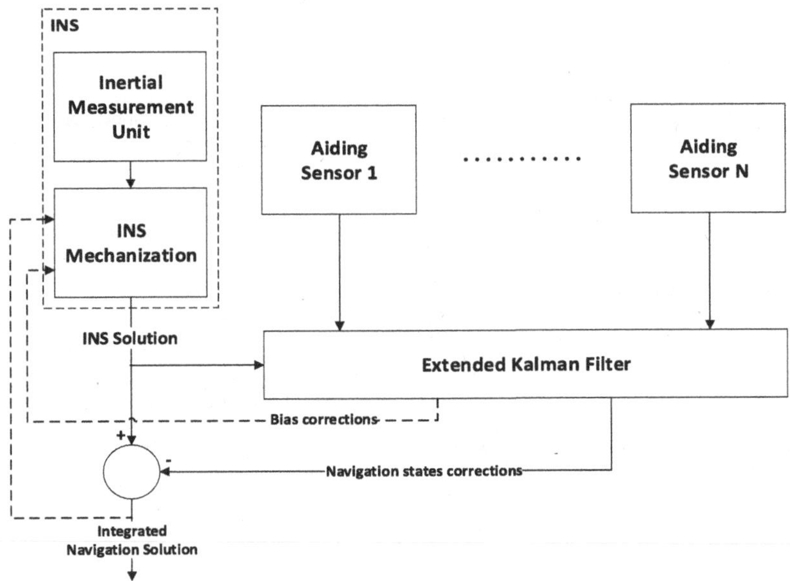

Figure 13.9 Block diagram of centralized filtered integration architecture using an EKF.

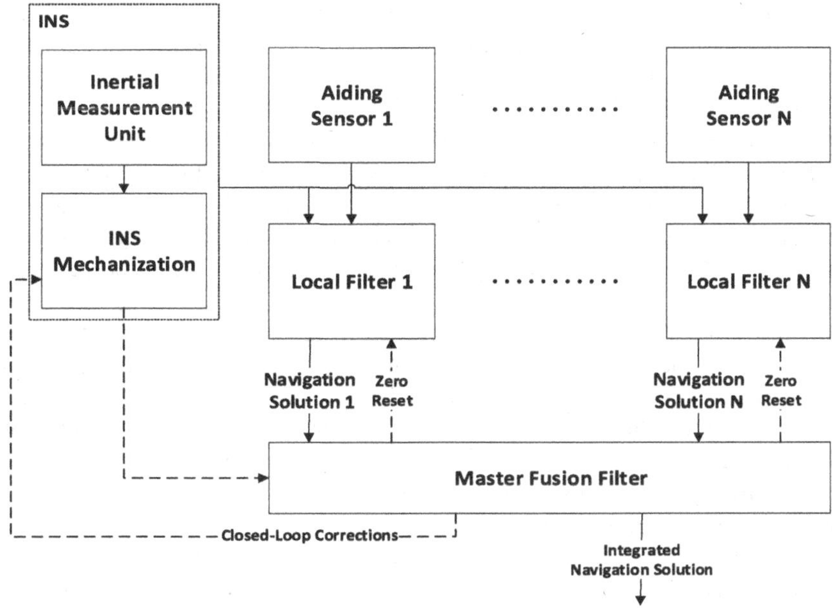

Figure 13.10 Block diagram of federated zero-reset integration architecture using an EKF.

other sensors through a bank of local filters. The solutions from these filters are then passed to another fusion step to generate the final integrated solution.

There are various implementations of the federated architecture. For example, in the first stage, the INS solution can be a common input to all local filters or it can be computed within each local filter to allow different feedback values from each filter. The second stage can be epoch-by-epoch least squares or a fusion filter.

Figure 13.10 shows an example of a federated filtered integration with zero reset. In this architecture, a filter such as EKF is used in the last stage. To avoid the problem of cascaded filters, the states of the local filters, which are passed to the master filter, are reset to zero and to its initial covariance after each epoch processing [12]. The local filters can be designed as cascaded or centralized integration. This federated zero-reset approach can process the input measurements to the local filters with a rate higher than the master filter. An application to this approach is the federated ultratightly coupled GNSS/INS integration [47].

13.4 PPP/INS INTEGRATION: A CASE STUDY FOR AUTOMATED LEVEL 2 DRIVING

The ultimate goal of the multisensor fusion in AV/CVs positioning is to provide continuous, reliable, and ultraprecise positioning information for the safe operation of AV/CVs. Currently, the available land vehicles in the market adopt up to level 2 of automation [48]. This section provides a case study of GNSS PPP/INS integration. This study is an application of the multisensor fusion in AV/CVs for level 2-automated driving on highways with submeter-level accuracy.

GNSS receivers are presently the primary source of outdoor positioning services for automobile navigation systems. A global precise positioning solution can be obtained through GNSS PPP without the need for local reference stations. However, GNSS PPP positioning requires continuous visibility of the GNSS satellites and relatively open-sky conditions, which are not satisfied all the time in the case of land vehicle navigation. For example, the GNSS solution will deteriorate or become unavailable in challenging environments such as urban canyons, tunnels, and under overpasses. On the other hand, INS positioning is continuous and is not affected by external signals or infrastructure, but its performance degrades with time due to the inherent sensor errors.

The integration of GNSS PPP and INS can provide a continuous and reliable navigation solution that comprises the benefits of the two systems and transcends their limitations. The integrated PPP/INS solution is independent of the weather conditions, and it can be applied for level 2-automated driving on highways and suburban areas. For challenging GNSS conditions such as in downtown areas and underground, aiding from perceptive systems, such as cameras and radar, is required. However, this is related to the higher levels of automation which is beyond the scope of this section.

The performance of PPP/INS integration is affected by the accuracy of PPP, the quality of the INS, and the fusion method. The PPP accuracy is influenced by the number of frequencies used, the number of constellations, and the type of corrections, among other factors. The INS quality depends on the grade of the IMU which means it is proportional to the IMU cost. Different integration modes and filters can be used in the integration process as described in Section 13.3. These various factors result in different realizations of the PPP/INS system. The choice of which realization to use depends on the application requirements such as accuracy and cost. The rest of this section includes a brief overview of the GNSS PPP followed by a description of the implemented real-time low-cost PPP/INS for level 2-automated vehicles.

13.4.1 PRECISE POINT POSITIONING

PPP is a technique for GNSS positioning with a high accuracy anywhere on the globe using a single GNSS receiver [9]. Nevertheless, PPP must address the error sources that cancel out in differential GNSS such as atmospheric effect and site displacement effects [5].

PPP is more accurate than the standard point positioning (SPP) because of two reasons; first, it uses the precise GNSS carrier phase measurements in addition to the pseudorange measurements. Second, PPP utilizes precise corrections for the satellite orbit and clock errors. These corrections can be obtained from commercial or open-access sources. The primary source of free PPP corrections is the International GNSS Service (IGS) [49]. IGS has been providing a set of products with different latencies for GPS. The most accurate corrections come from the IGS final products with a latency of 12–18 days, which can be used in postmission and offline applications. Currently, IGS provides corrections for other GNSS constellations through its multi-GNSS experiment project [50,51]. Furthermore, IGS has officially launched its real-time service (IGS-RTS) on April 1, 2013, to support real-time PPP applications [52].

The standard PPP model is based on dual-frequency (DF) GNSS observations, which allows the first-order ionospheric errors to be eliminated. A significant challenge for PPP is the long convergence time (larger than 20 min), which is due to the time needed to resolve the carrier phase ambiguities. The carrier phase ambiguity is the unknown integer number of full carrier cycles between the receiver and the satellite during the initial signal acquisition [5]. The convergence time of the PPP solution depends on several factors such as the number of visible satellites, satellite geometry, quality of the correction products, receiver multipath environment, and atmospheric conditions.

The research is still ongoing to reduce the PPP convergence time to meet the requirements of real-time applications. Recent studies have shown that the convergence time might be shortened by using triple-frequency observations, multiconstellation observations, and using ionospheric corrections rather than ionospheric-free (IF) combinations of the GNSS observations. Furthermore, fixing the ambiguities to its integer value can reduce the convergence time compared to the float solution; however, it requires additional corrections and more complex algorithms, which makes it still not practical for real-time applications.

13.4.1.1 Standard Dual-frequency PPP

The standard PPP model uses the IF combinations of DF-GNSS observations to mitigate ionospheric errors [53]. The basic corrections that are done in SPP also apply to PPP such as relativistic and Sagnac effects [54]. Nevertheless, more error sources have to be addressed in PPP such as phase wind-up [55], antenna phase center, and site displacement effects [10]. To reach centimeter-level accuracy, especially in the height component, only the dry component of the tropospheric delay, which accounts for 90% of the error, is a priori modeled. On the other hand, the wet component is difficult to model, and hence, it is estimated as one of the unknowns of the DF-PPP model. More details about the different PPP error sources can be found in Ref. [56].

After applying the precise satellite orbit and clock corrections and the proper error models, the GNSS IF pseudorange P_{IF} and carrier phase Φ_{IF}observations can be written as

$$P_{IF} = \rho + cdt^r + ISB + m_w Z_w + \varepsilon_{P_{IF}} \tag{13.16}$$

$$\Phi_{IF} = \rho + cdt^r + ISB + m_w Z_w + \lambda_{IF} A_{IF} + \varepsilon_{\Phi_{IF}} \tag{13.17}$$

where ρ is the true geometric range between the satellite and receiver in meters, c is the speed of light, and dt^r is the receiver clock bias in seconds. The term ISB represents the *inter-system bias*, in meters, if other constellations rather than GPS were employed, whereas this term is equal to zero for GPS observations. The zenith wet tropospheric delay (for elevation angles of 90 degrees), in meters, is represented by Z_w, and m_w is an elevation-dependent mapping function. The term A_{IF} denotes the float ambiguity in cycles where each cycle has a wavelength λ_{IF}. The multipath effects and the receiver noise are represented by $\varepsilon_{P_{IF}}$ and $\varepsilon_{\Phi_{IF}}$ for the pseudorange and carrier phase observations, respectively.

The unknowns of the standard PPP model in Eqs. (13.16) and (13.17) are the three position parameters, the receiver clock error, the ISB in case of multi-GNSS, the wet tropospheric delay, and the float ambiguities (one per each satellite). These unknowns can be estimated using the least-squares or the EKF.

13.4.1.2 Single-Frequency PPP

As most of the low-cost GNSS receivers in the mass market provide single-frequency (SF) observations, SF-PPP has become an interesting topic for many researchers [57]. The accuracy of SF-PPP is generally less than DF-PPP because it is limited to the decimeter-to-submeter level in the kinematic mode [58]; nevertheless, this level of accuracy is sufficient for some applications that have low cost and fast convergence as priorities.

Similar to DF-PPP, the precise PPP corrections and error models are applied to the SF pseudorange and carrier phase observations; therefore, they reduce to

$$P_1 = \rho + cdt^r + ISB - B^s_{P_1} + T + I_1 + \varepsilon_{P_1} \tag{13.18}$$

$$\Phi_1 = \rho + cdt^r + ISB + T - I_1 + \lambda_1 A_1 + \varepsilon_{\Phi_1} \tag{13.19}$$

where the subscript 1 denotes the chosen GNSS frequency, B^s is the satellite code biases, T is the total tropospheric delay, and I is the ionospheric delay, all in meters.

Three differences can be noticed between the SF-PPP and DF-PPP models. The first difference is that the ionospheric error is not canceled and, hence, must be addressed. Two approaches can be used to mitigate the ionospheric delays with SF-PPP. The first approach is called Group and Phase Ionospheric Correction (GRAPHIC) [59], which averages the pseudorange and phase measurements of each satellite to obtain a first-order IF observation. However, the resulting observation contains half of the pseudorange noise in addition to the phase integer ambiguities, which results in a relatively large initial convergence time to reach the required accuracy (15 minutes or longer) [10].

Another approach is to use the predicted ionospheric corrections from organizations such as IGS and Jet Propulsion Laboratory (JPL). Utilizing these corrections has been shown to improve the solution accuracy faster than the GRAPHIC approach [60,61].

The second difference between the SF-PPP and DF-PPP models is the satellite code bias term B^s_{P1}. The satellite clock corrections are referred to the IF P1/P2 combination. Therefore, with SF observations, care must be taken to use the proper differential code bias that matches the utilized clock corrections. The last difference between the two models relates to the tropospheric delay. Although the wet component can be estimated as an unknown, similar to the DF-PPP model, the total tropospheric delay can be a priori modeled using models like Saastamoinen's model [62], and the accuracy of the solution can still be sufficient for some applications.

13.4.2 SF-PPP/INS INTEGRATION

This case study describes the implementation and testing results of a developed real-time low-cost PPP/INS integration system. This system was introduced in an article published by the authors of this chapter in the Sensors journal [63]. The developed PPP/INS system aims to provide a low-cost continuous and reliable positioning solution for level 2-automated driving with submeter accuracy. This system is a typical example of the multisensor fusion for AV/CVs positioning.

13.4.2.1 Methodology

Figure 13.11 shows the block diagram of the developed PPP/INS integrated system. The fusion of the INS and SF-PPP solutions is performed in the loosely coupled mode through an EKF. The EKF is implemented using the closed-loop configuration described in Section 13.3.1.1. In the loosely coupled mode, the final PPP solution is integrated with the INS solution, which allows employing low-cost GNSS receivers that do not provide raw measurements. High-end IMUs could be used to maximize the quality of the INS solution, but it comes with a cost overhead. Therefore, a MEMS IMU was used to meet the low-cost requirement. The performance of the developed system relies on utilizing SF-PPP for a high-quality GNSS solution and efficient PPP/INS fusion. In the following, the EKF system and measurement models are described.

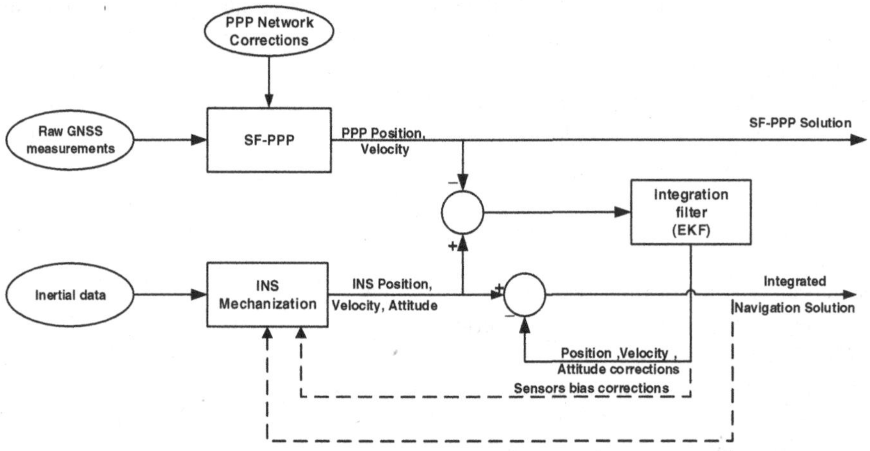

Figure 13.11 Block diagram of the developed SF-PPP/INS integrated system. The integration is performed in the loosely coupled mode where the SF-PPP and INS solutions are integrated through an EKF. The dotted lines represent the closed-loop configuration where the estimated errors are fed back to the INS mechanization module.

13.4.2.1.1 EKF System Model

The system model and the prediction of the solution are obtained from the INS mechanization. A full IMU system, which consists of three accelerometers and three gyroscopes, was employed to measure the accelerations and rotations in all directions in three-dimensional (3D) space. INS mechanization is the process of using the IMU measurements to calculate the position, velocity, and attitude information. Although the IMU measurements are measured in the body (vehicle) frame with respect to the inertial frame, the mechanization was performed in the Local-Level Frame (LLF). The LLF is a practical choice for land vehicle navigation as it provides the position in terms of latitude φ, longitude λ, and altitude h. The LLF shares the same origin with the vehicle frame, and its axes point to either East, North, and Up directions or North, East, and Down directions. In the developed system, the ENU directions of the LLF are adopted. More details about the different navigation reference frames can be found in Ref. [13].

The state error vector for SF-PPP/INS loosely coupled integration comprises fifteen states which are grouped into five 3×1 column vectors

$$\delta \mathbf{x}_{15\times 1} = \begin{bmatrix} \delta \mathbf{r}^{\mathbf{l}} & \delta \mathbf{v}^{\mathbf{l}} & \delta \psi^{\mathbf{l}} & \mathbf{b_a} & \mathbf{b_g} \end{bmatrix}^T \tag{13.20}$$

where the superscript l denotes the LLF; $\mathbf{r}^{\mathbf{l}} = \begin{bmatrix} \varphi & \lambda & h \end{bmatrix}^T$ is the vehicle position vector; $\mathbf{v}^{\mathbf{l}} = \begin{bmatrix} v_e & v_n & v_u \end{bmatrix}^T$ includes the vehicle velocity components in east, north, and up directions; ψ^l is the attitude vector; $\mathbf{b_a}$ and $\mathbf{b_g}$ represent the accelerometer and gyroscope biases, receptively. The vehicle frame is defined such that the y-axis points to the forward direction, the x-axis points to the lateral directions, and the z-axis points to the up direction.

The a priori estimated system states $\hat{\mathbf{x}}^-$ are predicted using the nonlinear INS model given by [13,64]

$$\begin{bmatrix} \dot{\mathbf{r}}^{\mathbf{l}} \\ \dot{\mathbf{v}}^{\mathbf{l}} \\ \dot{C}^l_b \end{bmatrix} = \begin{bmatrix} D^{-1}\mathbf{v}^{\mathbf{l}} \\ C^l_b \mathbf{f}^{\mathbf{b}} - (2\Omega^l_{ie} + \Omega^l_{el})\mathbf{v}^{\mathbf{l}} + \mathbf{g}^{\mathbf{l}} \\ C^l_b(\Omega^b_{ib} - \Omega^b_{il}) \end{bmatrix} \tag{13.21}$$

where C^l_b is the rotation matrix from the body frame to LLF, $\mathbf{f}^{\mathbf{b}}$ is the vector of specific force measurements from accelerometers in the body frame, and $\mathbf{g}^{\mathbf{l}}$ is the gravity vector in the LLF. The notation Ω^p_{mn}, where the subscripts m,n and the superscript p are arbitrary navigation frames, denotes the skew-symmetric matrix form of the angular velocity vector that represents the rotation from n-frame to m-frame measured in p-frame coordinates. The letters i, e, l refer to the inertial frame, Earth-Centered Earth-Fixed frame, and LLF, respectively. D^{-1} is a transformation of the velocity vector $\mathbf{v}^{\mathbf{l}}$ to geodetic coordinates that uses the meridian radius R_M and normal radius R_N of the Earth's ellipsoid and is defined as

$$D^{-1} = \begin{bmatrix} 0 & \frac{1}{R_M+h} & 0 \\ \frac{1}{(R_N+h)\cos\varphi} & 0 & 0 \\ 0 & 0 & 1 \end{bmatrix}. \tag{13.22}$$

The vehicle attitude is represented by the three Euler angles: pitch (p), roll (r), and azimuth (A) [64]. The pitch angle describes the rotation around the x-axis (lateral direction) of the vehicle frame, whereas the roll angle is the rotation around the y-axis (forward direction). The azimuth angle is the rotation around the z-axis (up direction) measured clockwise between the vehicle forward direction and the Earth's north direction. The matrix C^l_b is given by

$$C^l_b = \begin{bmatrix} \cos A\cos r + \sin A\sin p\sin r & \sin A\cos p & \cos A\sin r - \sin A\sin p\cos r \\ -\sin A\cos r + \cos A\sin p\sin r & \cos A\cos p & -\sin A\sin r - \cos A\sin p\cos r \\ -\cos p\sin r & \sin p & \cos p\cos r \end{bmatrix} \tag{13.23}$$

The third equation in Eq. (13.21), $\dot{C}_b^l = C_b^l(\Omega_{ib}^b - \Omega_{il}^b)$, together with Eq. (13.23) is needed to calculate the attitude angles. However, the solution to this problem cannot be obtained in closed form and requires numerical integration methods such as the quaternion approach [64].

The system model in Eq. (13.21) represents the nonlinear state evolution function f in Eq. (13.2). The sensor biases are modeled by a first-order Gauss–Markov stochastic process. Then, Eqs. (13.4) and (13.5) can be used to calculate the state transition matrix ϕ [64].

Because the closed-loop configuration of the EKF is adopted, the state prediction is done by resetting the error states $\delta \mathbf{x}_\mathbf{k}^- = 0$, whereas the covariance prediction is achieved through Eq. (13.7).

13.4.2.1.2 EKF Measurement Model

The primary update for the implemented PPP/INS system comes from the SF-PPP solution. The measurement error vector is calculated as the difference between the predicted INS solution and the update from the SF-PPP solution

$$\delta \mathbf{z}_{\text{PPP}} = \begin{bmatrix} \mathbf{r}_{INS}^l - \mathbf{r}_{\text{PPP}}^l \\ \mathbf{v}_{INS}^l - \mathbf{v}_{\text{PPP}}^l \end{bmatrix}. \tag{13.24}$$

The measurement noise covariance R, which is used to calculate the Kalman gain in Eq. (13.12), can be obtained from the covariance of the SF-PPP solution. In the loosely coupled integration, the relation is linear between the system states and its corresponding update, and hence, the design matrix H is given by

$$H_{\text{PPP}} = \begin{bmatrix} I_{6\times6} & 0_{6\times9} \end{bmatrix}. \tag{13.25}$$

In the case of a GNSS signal blockage, the update from PPP is not available. Using only the INS solution, especially with low-cost sensors, will lead to a large solution drift. For land vehicles, some constraints can aid the INS during GNSS outages. Two examples of these constraints are the zero-velocity update (ZUPT) and the nonholonomic constraints (NHCs) [14].

In ZUPT, when the vehicle is detected to be static, all the velocities should be zero. This fact is used to reset the velocity errors and limit the position error growth. The measurement error vector and design matrix when using ZUPT are

$$\delta \mathbf{z}_{\text{ZUPT}} = \begin{bmatrix} \mathbf{v}_{\text{INS}}^l - \mathbf{0}_{3\times1} \end{bmatrix} \tag{13.26}$$

$$H_{\text{ZUPT}} = \begin{bmatrix} 0_{3\times3} & I_{3\times3} & 0_{3\times9} \end{bmatrix} \tag{13.27}$$

The NHCs in land vehicles are based on the fact that the vehicle does not slip or fly, which means that the vehicle velocity in the lateral and up directions is close to zero. Thus, the measurement vector of the NHC update is represented using the INS velocity in the vehicle frame which is assumed to coincide with the sensor body frame denoted by the superscript b

$$\delta \mathbf{z}_{\text{NHC}} = \begin{bmatrix} \mathbf{v}_{\text{INS,lateral}}^b - 0 \\ \mathbf{v}_{\text{INS,up}}^b - 0 \end{bmatrix} \tag{13.28}$$

In Ref. [65], the error in the velocity in the body frame was related to the velocity error in the LLF and the attitude errors by the formula

$$\delta \mathbf{v}^\mathbf{b} = C_l^b \delta \mathbf{v}^\mathbf{l} - C_l^b (\mathbf{v}^\mathbf{l} \times) \delta \psi^\mathbf{l} \tag{13.29}$$

where $(\mathbf{v}^\mathbf{l}\times)$ is the skew-symmetric form of the velocity error vector in the LLF. Following the ENU directions order and assuming the forward motion is in the y-direction of the body frame, the design matrix of the NHC update can be written as

$$H_{\text{NHC}} = \begin{bmatrix} 0_{1\times3} & C_{11} & C_{21} & C_{31} & -v_u C_{21} + v_n C_{31} & v_u C_{11} - v_e C_{31} & -v_n C_{11} + v_e C_{21} \\ 0_{1\times3} & C_{31} & C_{32} & C_{33} & -v_u C_{32} + v_n C_{33} & v_u C_{31} - v_e C_{33} & -v_n C_{13} + v_e C_{32} \end{bmatrix} \tag{13.30}$$

where C_{ij} is the element at row i and column j of the matrix C_l^b, which is the transpose of the matrix defined in Eq. (13.23).

Finally, the system error states and its posterior covariance matrix are updated using Eqs. (13.12) to (13.14). The final estimated solution at epoch k is calculated as

$$\hat{\mathbf{x}}_{\mathbf{k}}^{+} = \hat{\mathbf{x}}_{\mathbf{k}}^{-} - \delta \mathbf{x}_{k}^{+}. \quad (13.31)$$

13.4.2.2 Results and Discussion

The real-time performance of the developed PPP/INS system was evaluated through two road test trajectories. The first trajectory examines the open-sky and suburban performance, whereas the second trajectory includes more challenging conditions such as high dynamics, overpass bridges, and a complete GNSS outage. First, the experimental setup used in these tests is described, and then, the results obtained from each trajectory are displayed.

13.4.2.2.1 Experimental Setup

The test data were collected using a testing van where the GNSS antennas were put on the roof. The remaining equipment was mounted on a flat platform that is firmly attached to the testing vehicle such that the IMU frame is oriented with the vehicle frame to the maximum possible extent.

The SF-GNSS measurements, for GPS and GLONASS, were obtained from the low-cost u-blox EVK-8MT receiver. Real-time SF-PPP corrections can be freely obtained either from the IGS-RTS or from a satellite-based augmentation system (SBAS) [66]. In the implemented SF-PPP, corrections for the satellite orbit and clock errors, ionospheric delays, and code biases were obtained from the Centre National d'Etudes Spatiales (CNES), one of the IGS-RTS analysis centers. CNES is the only IGS analysis center transmitting ionospheric corrections so far. The CLK91 stream, one of CNES products, transmits orbit, clock, and code biases corrections for GPS and GLONASS every 5 s, and transmits ionospheric corrections every 60 s. The CNES real-time corrections were received through the Internet, while the SBAS messages were also logged using the u-blox receiver to be used in case of an Internet outage.

The SF-PPP/INS system comprises an u-blox EVK-8MT GNSS receiver and an LSM6DSL chip which is a consumer-grade MEMS IMU [67]. The results were compared to u-blox EVK-M8U Untethered Dead Reckoning (UDR) solution, which was a benchmark in the navigation market for low-cost GNSS/INS applications at the time these tests were performed. The u-blox EVK-M8U GNSS receiver was configured to use SBAS corrections to get the best-integrated solution for comparison.

Finally, the reference solution was obtained from a DGNSS/INS integration where the real-time rover data were collected using the NovAtel SPAN on ProPak6 system with IMU-KVH [68] as a tactical-grade IMU. The IGS UCAL station was used as a reference station with a maximum baseline length of 12 km. Furthermore, the reference data were postprocessed using NovAtel's Waypoint Inertial Explorer software.

13.4.2.2.2 Road Test Trajectory 1

The first road test trajectory lasted approximately 35 minutes in Calgary, Alberta, Canada. The trajectory, as shown in Figure 13.12 on a Google map, started from an open-sky condition at the University of Calgary and near the Alberta Children's Hospital, and then, the car moved toward a residential area. The residential area was a typical suburban environment with community houses and trees on both sides of the road. Finally, the car moved back toward the university.

Figure 13.13 shows the vehicle speed during the first trajectory. The car forward speed, measured by a car odometer, was <60 km/h with frequent stops and low-speed periods which is typical for a suburban area. Figure 13.14 shows the number of satellites used in the PPP solution versus time. A

Figure 13.12 The first test trajectory of SF-PPP/INS integration, Calgary, Alberta, Canada.

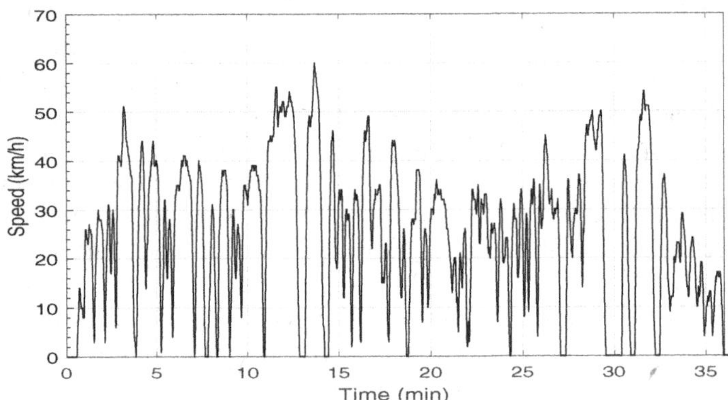

Figure 13.13 Vehicle speed during the first trajectory.

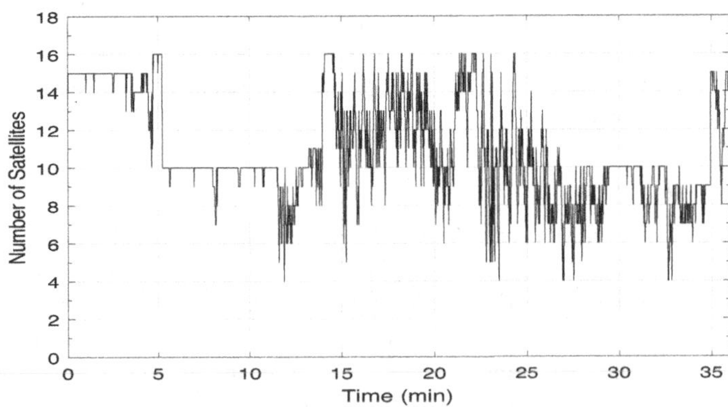

Figure 13.14 Number of satellites used in the PPP solution of the first trajectory.

minimum number of four satellites were seen in this test, whereas the maximum number was 16, thanks to employing both GPS and GLONASS satellites.

Figure 13.15 shows the 3D position errors for the SF-PPP, the integrated SF-PPP/INS, and the u-blox UDR solutions. The integrated SF-PPP/INS solution, in Figure 13.15(b), is dominated by the precision of the SF-PPP solution when there is enough number of visible satellites. On the other hand, the integration with INS smoothed most of the spikes that were in the PPP solution at

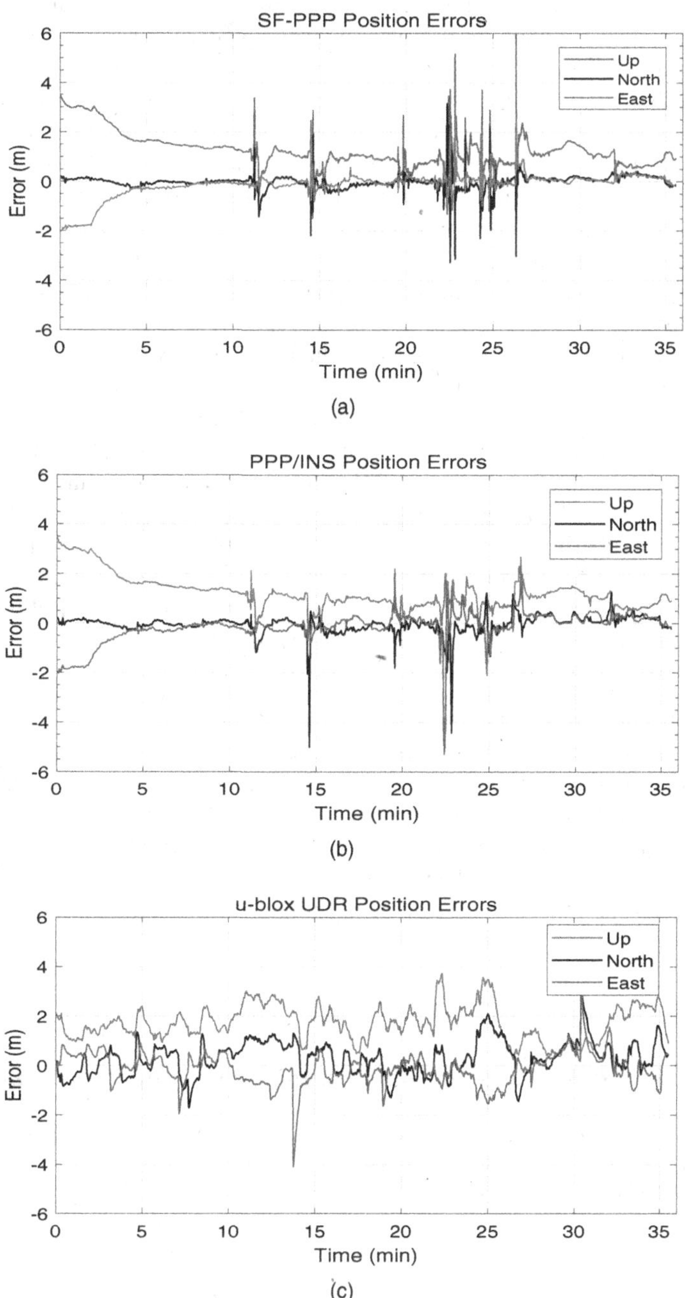

Figure 13.15 The first trajectory position errors versus time of (a) SF-PPP, (b) SF-PPP/INS, and (c) u-blox UDR.

the epochs with a low number of satellites. Nevertheless, the integrated solution still suffers from some error spikes such as the ones before the 15th min and after the 22nd min in Figure 13.15(b). The fusion of the PPP and INS solutions is based on the quality of the two solutions measured by statistics such as the standard deviation. This is why in some cases, an inaccurate standard deviation of the PPP solution can mislead the EKF and cause a drift in the integrated solution especially with varying dynamics such as turning. This can be considered as one of the limitations of the loosely coupled mode of integration.

The u-blox UDR solution in Figure 13.15(c) has fewer error spikes than the developed SF-PPP/INS; nevertheless, the u-blox errors have relatively a wider error range compared to the SF-PPP/INS solution after the first 5 min. Both SF-PPP and SF-PPP/INS solutions seem to have a convergence period during the first 5 minutes of the trajectory that does not exist in the u-blox UDR solution. When this issue was explored further, it was found that on the day of this test, CNES has recently changed the format of its real-time ionospheric corrections such that it became incompatible with the developed code. Since there were no ionospheric corrections from CNES, the code automatically shifted to use the SBAS ionospheric corrections which needed around 5 minutes to be obtained in real time. This indicates that a good practice when using real-time PPP is to have a back-up correction source such as another correction stream or SBAS corrections.

Table 13.3 compares the root mean square (RMS) and maximum errors of both u-blox UDR and the developed SF-PPP/INS solutions after the first 5 minutes to avoid the initial convergence time without ionospheric corrections. The developed SF-PPP/INS system achieved submeter RMS horizontal accuracy, and the results were better than the u-blox UDR solution. Moreover, the RMS error in the vertical direction of the integrated solution is less than the UDR solution.

To further demonstrate the benefit of the PPP/INS integration in maintaining the lane-level accuracy in suburban environments, Figure 13.16 shows two examples, using Google Earth, of how trees and houses in a suburban environment can affect the PPP solution. It can be seen that the PPP-alone solution was affected several times by the partial outages of GNSS satellites, and this drove the solution outside the driving lane. On the other hand, the integrated PPP/INS solution was close to the reference solution and provided a solution within the driving lane.

13.4.2.2.3 Road Test Trajectory 2

The second road test was carried out in Calgary, Alberta, and Canada for approximately 1 hour. Figure 13.17 shows the trajectory where the car has started in a suburban area and moved north to make a few loops in an open-sky environment in the top left part of the trajectory; then, the car moved back south on a highway inside the city with an 80 km/h speed limit passing under several overpasses and experiencing changing dynamics as can be seen from the speed profile in Figure 13.18. The last part of the trajectory included underground parking for 3 minutes, and the

Table 13.3
The First Test Trajectory RMS and Maximum Position Errors after 5 minutes.

	Developed SF-PPP/INS		u-blox UDR	
	RMS Error (m)	MAX Error (m)	RMS Error (m)	MAX Error (m)
Horizontal	0.6	5.4	1.0	4.8
Vertical	1.1	3.4	1.9	3.7

The first 5 minutes were excluded to avoid the effect of the time needed to acquire the ionospheric corrections on the comparison.

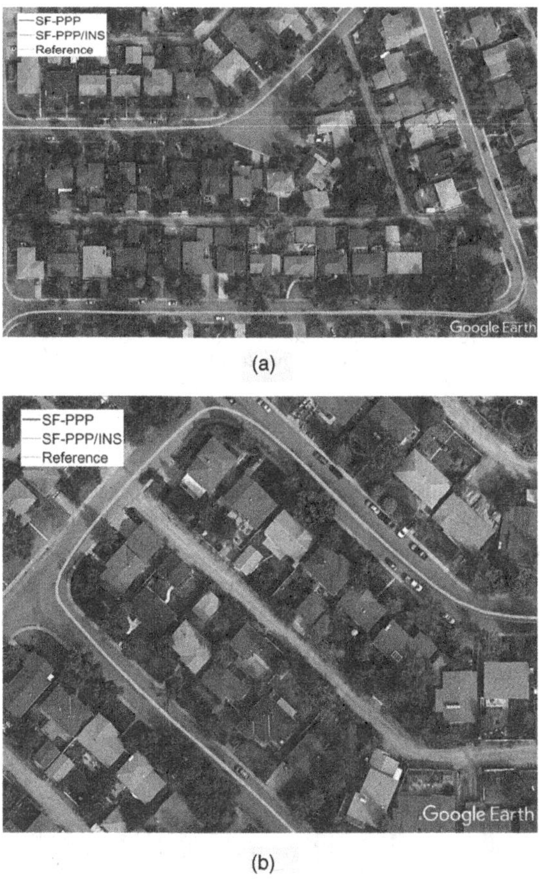

Figure 13.16 Two examples of the developed SF-PPP/INS system performance in suburban areas showing how the system can provide a solution within the driving lane.

Figure 13.17 The second test trajectory of SF-PPP/INS integration in Calgary, Alberta, Canada.

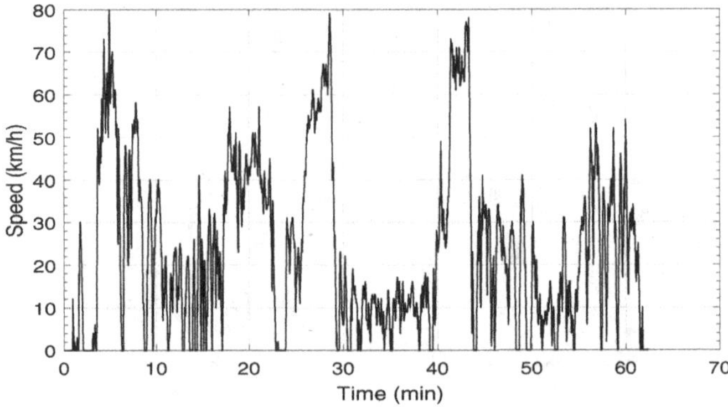

Figure 13.18 Vehicle speed during the second trajectory.

test ended in a suburban area. Figure 13.19 shows the number of satellites used in the PPP solution. The epochs, before the 50th min, at which the number of satellites dropped to five or less correspond mainly to the times when the car moves under overpasses. The long period of zero satellites after the 50th min corresponds to the time when the car went down the underground parking.

Table 13.4 compares the RMS and maximum position errors of the whole trajectory for both u-blox UDR and the developed SF-PPP/INS solutions. The maximum errors mainly occurred in the underground parking where no PPP solution was available. The integrated SF-PPP/INS solution has lower errors compared to the u-blox UDR solution.

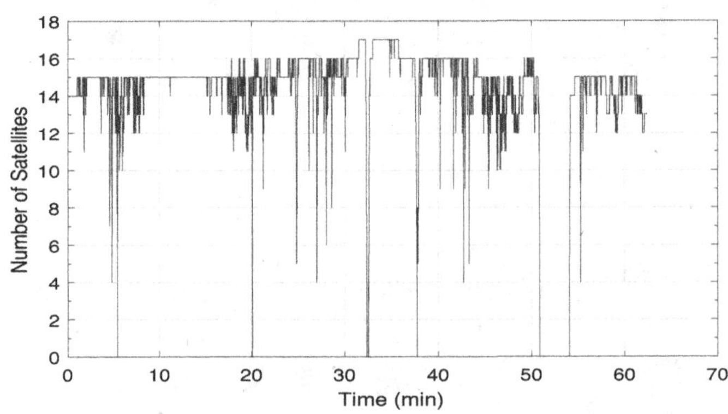

Figure 13.19 Number of satellites used in the PPP solution of the second test trajectory.

Table 13.4
RMS and Maximum Position Errors for the Whole of the Second Test Trajectory

	Developed SF-PPP/INS		u-blox UDR	
	RMS Error (m)	MAX Error (m)	RMS Error (m)	MAX Error (m)
Horizontal	1.5	12.5	8.8	60.8
Vertical	1.5	8.5	3.3	11.9

The developed SF-PPP/INS system errors are less than the UDR errors. The maximum errors occurred when the vehicle entered the underground parking.

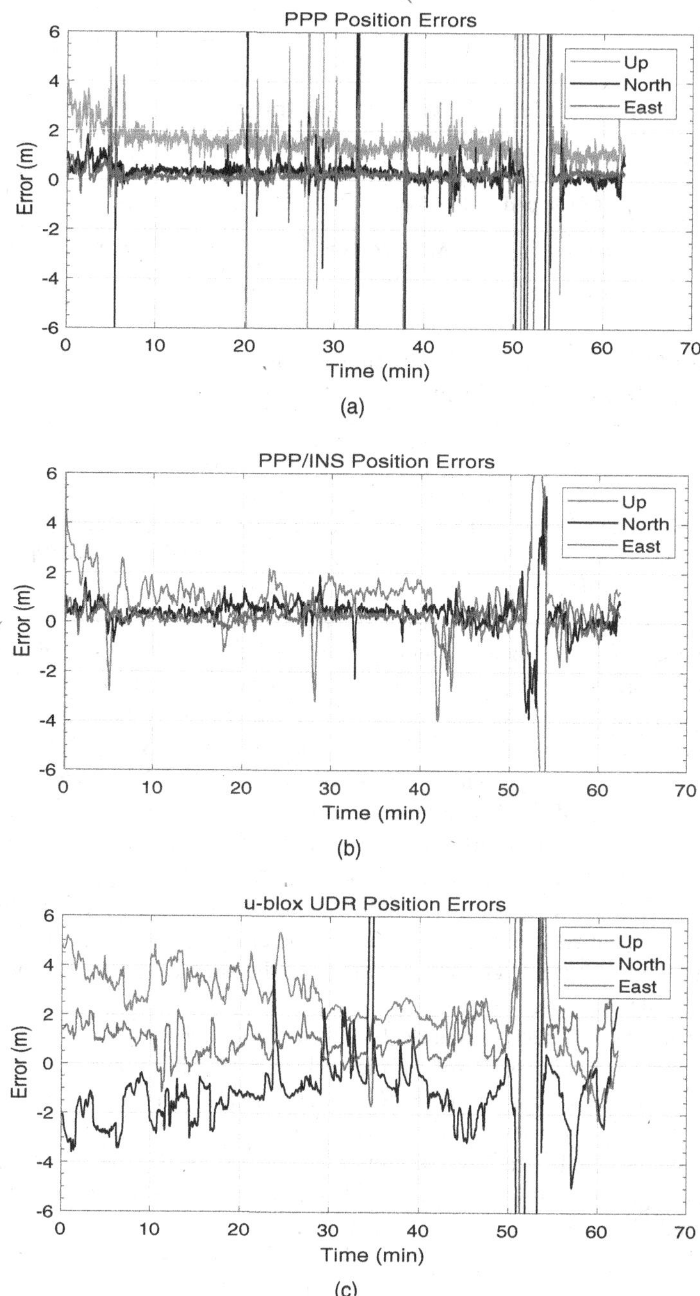

Figure 13.20 Second test trajectory position errors versus time for (a) SF-PPP, (b) SF-PPP/INS, and (c) u-blox UDR.

Figure 13.20 shows the 3D position errors within ±6 m for the SF-PPP, the integrated SF-PPP/INS, and the u-blox UDR solutions. When the number of visible satellites is low, typically five or less but can vary based on the satellite geometry and multipath effects, the PPP solution suffers from large spikes. These spikes are accompanied by high covariance and sometimes with no-fix status which means that there is no SF-PPP solution. The integration with INS smooths these

Table 13.5
RMS and maximum position errors, before entering the underground parking in the second test trajectory.

	Developed SF-PPP/INS		u-blox UDR	
	RMS error (m)	MAX error (m)	RMS error (m)	MAX error (m)
Horizontal	0.7	2.6	2.6	21.1
Vertical	1.4	4.6	2.8	5.3

The developed SF-PPP/INS system maintained submeter horizontal accuracy and meter-level vertical accuracy, and was better than UDR solution.

spikes and assures the continuity of the navigation solution. The fast convergence of the SF-PPP solution after GNSS outages has contributed to a more stable and reliable integrated solution. The u-blox UDR position errors, in Figure 13.20(c), are worse than the developed PPP/INS system in most of the trajectories.

For further analysis, the test results are divided into two parts: the first part ends before entering the parking lot and includes the open-sky and highway driving with several overpass bridges. The second part includes driving through underground parking for three minutes. Table 13.5 shows the position accuracy comparison for the first part of the trajectory. The advantage of using an SF-PPP compared to the SPP, which is the typical solution from low-cost GNSS receivers, is demonstrated by comparing the developed solution with u-blox UDR solution which is an SPP solution augmented with SBAS corrections. The SF-PPP solution has contributed to reducing the horizontal RMS errors of the integrated solution from meter to submeter level of accuracy in suburban environments.

According to the requirements of level 2 of automation, AVs on highways must have a continuous navigation solution with lane-level accuracy. The frequent overpasses impose a challenge on GNSS-based navigation systems including the ones with PPP accuracy. Figure 13.21 shows two examples of how the presented SF-PPP/INS system could maintain the lane-level accuracy even when the car is moving under a wide overpass.

In the second part of the trajectory, Figure 13.22 shows on a Google map how the SF-PPP/INS solution outperforms the u-blox UDR solution in the complete GNSS outage in the underground parking. This is an example of how the PPP/INS system behaves in relatively long GNSS outages. Despite the errors reached the meter level, this performance is still acceptable given the utilized low-cost consumer-grade IMU and the long outage period (3 min). If more accuracy is required with long GNSS outages, such as in the higher level of automation, aiding from perception systems, such as cameras and radars, should be incorporated.

13.5 SUMMARY

This chapter focused on the multisensor fusion in the application of AV/CVs precise positioning. The chapter started with an introduction to the six levels of autonomous driving and how these levels set a guide to the positioning requirements and the sensor technology that should be adopted for each level. The second section of this chapter described the different sensors that are typically used in land vehicle navigation including GNSS, INS, odometers, and perception systems such as cameras, LiDAR, and radar. Each sensor was briefly described showing its advantages and limitations. The complementary characteristics of these sensors were the basis for the multisensor fusion to achieve a continuous, precise, and robust navigation solution. The third section presented the well-known fusion filters such as the EKF and the particle filter. This section also discussed the

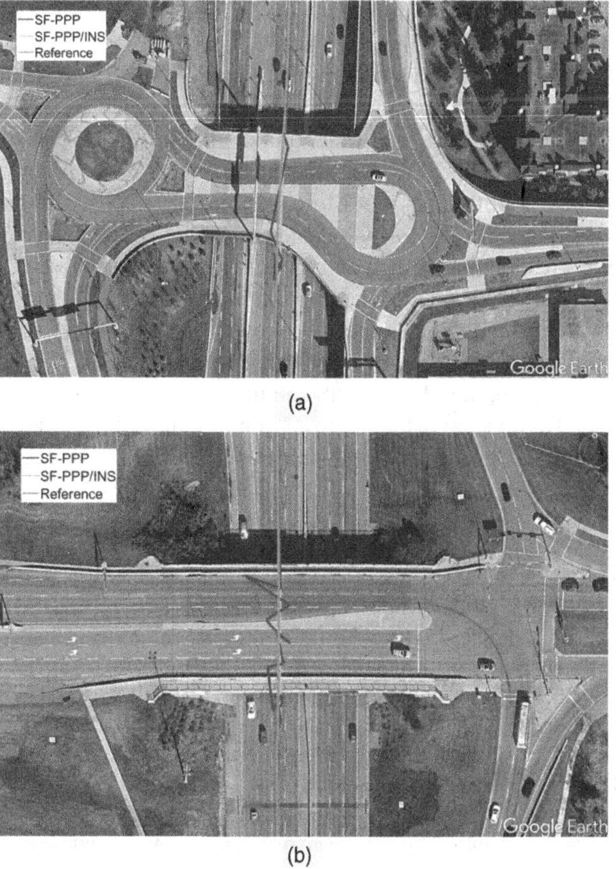

Figure 13.21 Two examples of the developed SF-PPP/INS system performance on highways showing how the system could maintain a solution within the driving lane.

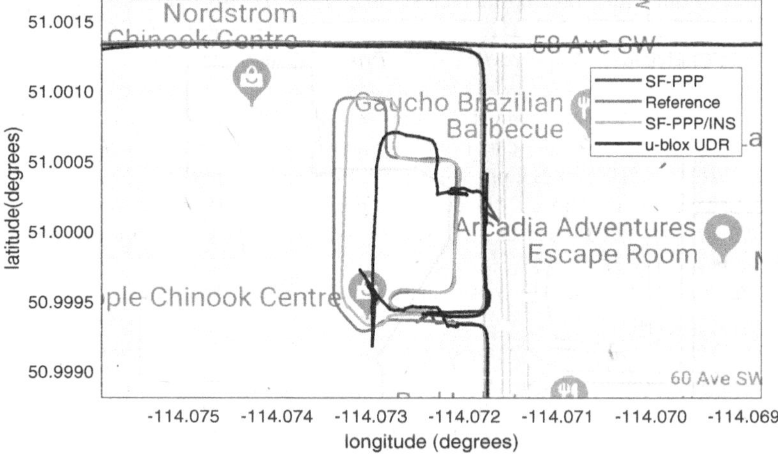

Figure 13.22 Navigation performance comparison between SF-PPP, developed SF-PPP/INS, and u-blox UDR in underground parking for 3 minutes.

different architectures that are used in multisensor navigation starting with GNSS/INS integration and then generalizing for the case with more sensors.

The last section of this chapter provided a case study of GNSS PPP/INS integration targeting submeter-level accuracy for level 2 driving on highways and suburban areas. The presented system objectives were to provide a low-cost, real-time, and continuous precise navigation solution. To fulfill the low-cost requirement, SF-PPP was adopted because it can employ measurements from low-cost SF GNSS receivers in the market. Moreover, low-cost consumer-grade MEMS sensors were utilized for the INS part. The EKF was selected as the integration filter in the developed SF-PPP/INS system. The corresponding system and measurement models were described. The system was tested through two road tests, and the performance was compared with the SPP-SBAS solution provided by u-blox UDR technology. The results showed that the proposed SF-PPP/INS system could provide submeter accuracy over highways with open-sky, suburban, and momentary GNSS outage conditions. Furthermore, the SF-PPP/INS solution was better than the u-blox UDR solution in a long GNSS outage. This system can be utilized in providing submeter accuracy for navigation along highways for automated level 2 driving that is presently desirable by car manufacturers.

REFERENCES

1. T. M. Gasser and D. Westhoff, "BASt-study: Definitions of automation and legal issues in Germany," in *Proceedings of the 2012 Road Vehicle Automation Workshop*, Irvine, CA, June 2012.
2. SAE International, "Taxonomy and definitions for terms related to driving automation systems for on-road motor vehicles", The Society of Automotive Engineers International Std., Rev. J3016-201806, June 2018. [Online] Available: https://www.sae.org/standards/content/j3016_201806/.
3. National Highway Traffic Safety Administration (NHSTA), "Automated driving systems 2.0: A vision for safety," September 2017. [Online] Available: https://www.nhtsa.gov/sites/nhtsa.dot.gov/files/documents/13069a-ads2.0_090617_v9a_tag.pdf.
4. U.S. Department of Defense, "Preparing for the future of transportation: Automated vehicles 3.0," October 2018. [Online] Available: https://www.transportation.gov/av/3/preparing-future-transportation-automated-vehicles-3.
5. P. Misra and P. Enge, *Global Positioning System: Signals, Measurements and Performance*, 2nd ed. Lincoln, MA: Ganga-Jamuna Press, 2006.
6. C. C. Counselman III, I. I. Shapiro, R. L. Greenspan, and D. B. Cox Jr, "Backpack VLBI terminal with subcentimeter capability," *Proceedings of a Conference Radio Interferometry Techniques for Geodesy*, Cambridge, MA, pp. 409–414, 1979.
7. C. C. Counselman and S. A. Gourevitch, "Miniature interferometer terminals for earth surveying: Ambiguity and multipath with global positioning system," *IEEE Transactions on Geoscience and Remote Sensing*, vol. GE-19, no. 4, pp. 244–252, 1981.
8. E. Kaplan and C. Hegarty, *Understanding GPS: Principles and Applications*. Norwood, MA: Artech House, 2006.
9. J. Zumberge, M. Heflin, D. Jefferson, M. Watkins, and F. H. Webb, "Precise point positioning for the efficient and robust analysis of GPS data from large networks," *Journal of Geophysical Research: Solid Earth (1978–2012)*, vol. 102, no. B3, pp. 5005–5017, 1997.
10. J. Kouba, F. Lahaye, and P. Tétreault, "Precise point positioning," in *Springer Handbook of Global Navigation Satellite Systems*, P. J. Teunissen and O. Montenbruck, Eds. Cham: Springer International Publishing, 2017, ch. 25, pp. 723–751.
11. S. Bisnath and Y. Gao, "Precise point positioning: A powerful technique with a promising future," *GPS World*, vol. 40, pp. 43–50, 2009.
12. P. D. Groves, *Principles of GNSS, Inertial, and Multisensor Integrated Navigation Systems*, 2nd ed. Norwood, MA: Artech House, 2013.
13. A. Noureldin, T. B. Karamat, and J. Georgy, *Fundamentals of Inertial Navigation, Satellite-based Positioning and their Integration*. Berlin Heidelberg: Springer-Verlag, 2013.
14. A. Brandt and J. F. Gardner, "Constrained navigation algorithms for strapdown inertial navigation systems with reduced set of sensors," in *Proceedings of the 1998 American Control Conference. ACC (IEEE Cat. No.98CH36207)*, vol. 3, Philadelphia, PA, June 1998, pp. 1848–1852.

15. X. Niu, S. Nasser, C. Goodall, and N. El-Sheimy, "A universal approach for processing any MEMS inertial sensor configuration for land-vehicle navigation," *Journal of Navigation*, vol. 60, no. 2, pp. 233–245, 2007.
16. B. Phuyal, "An experiment for a 2-D and 3-D GPS/INS configuration for land vehicle applications," in *PLANS 2004. Position Location and Navigation Symposium (IEEE Cat. No. 04CH37556)*, Monterey, CA, April 2004, pp. 148–152.
17. D. Sun, M. G. Petovello, and M. E. Cannon, "GPS/reduced IMU with a local terrain predictor in land vehicle navigation," *International Journal of Navigation and Observation*, vol. 2008, p. 15, 2008.
18. U. Iqbal, A. F. Okou, and A. Noureldin, "An integrated reduced inertial sensor system - RISS/GPS for land vehicle," in *2008 IEEE/ION Position, Location and Navigation Symposium*, Monterey, CA, May 2008, pp. 1014–1021.
19. J. Georgy, A. Noureldin, M. J. Korenberg, and M. M. Bayoumi, "Low-cost three-dimensional navigation solution for RISS/GPS integration using mixture particle filter," *IEEE Transactions on Vehicular Technology*, vol. 59, no. 2, pp. 599–615, 2010.
20. C. R. Carlson, J. C. Gerdes, and J. D. Powell, "Error sources when land vehicle dead reckoning with differential wheelspeeds," *Navigation*, vol. 51, no. 1, pp. 13–27, 2004.
21. J. Wilson and M. Slade, "Accelerometer compensated differential wheel pulse based dead reckoning," in *Proceedings of the 22nd International Technical Meeting of The Satellite Division of the Institute of Navigation (ION GNSS 2009)*, Savannah, GA, September 2009, pp. 3087–3095.
22. S. Wang, R. Clark, H. Wen, and N. Trigoni, "DeepVO: Towards end-to-end visual odometry with deep recurrent convolutional neural networks," in *Proceedings of the 2017 IEEE International Conference on Robotics and Automation (ICRA)*, Singapore, 2017, pp. 2043–2050.
23. W. Ci, Y. Huang, and X. Hu, "Stereo visual odometry based on motion decoupling and special feature screening for navigation of autonomous vehicles," *IEEE Sensors Journal*, vol. 19, no. 18, pp. 8047–8056, 2019.
24. M. Aldibaja, N. Suganuma, and K. Yoneda, "Improving localization accuracy for autonomous driving in snow-rain environments," in *2016 IEEE/SICE International Symposium on System Integration (SII)*, Sapporo, Japan, IEEE, December 2016, pp. 212–217.
25. D. Manners, Nissan follows tesla in ruling out lidar for autonomous EVs. [Online] Available: https://www.electronicsweekly.com/news/business/nissan-follows-tesla-ruling-lidar-autonomous-evs-2019-05/.
26. T. Lee, V. Skvortsov, M. Kim, S. Han, and M. Ka, "Application of *w*-band *fmcw* radar for road curvature estimation in poor visibility conditions," *IEEE Sensors Journal*, vol. 18, no. 13, pp. 5300–5312, 2018.
27. M. Rapp, M. Barjenbruch, M. Hahn, J. Dickmann, and K. Dietmayer, "Probabilistic ego-motion estimation using multiple automotive radar sensors," *Robotics and Autonomous Systems*, vol. 89, pp. 136–146, 2017.
28. H. Durrant-Whyte and T. C. Henderson, "Multisensor data fusion," in *Springer Handbook of Robotics*, B. Siciliano and O. Khatib, Eds. Berlin: Springer, 2008, pp. 585–610.
29. R. E. Kalman, "A new approach to linear filtering and prediction problems," *Journal of Basic Engineering*, vol. 82, no. 1, pp. 35–45, 1960.
30. J. Zhou, S. Knedlik, and O. Loffeld, "INS/GPS tightly-coupled integration using adaptive unscented particle filter," *The Journal of Navigation*, vol. 63, no. 03, pp. 491–511, 2010.
31. A. Manzanilla, S. Reyes, M. Garcia, D. Mercado, and R. Lozano, "Autonomous navigation for unmanned underwater vehicles: Real-time experiments using computer vision," *IEEE Robotics and Automation Letters*, vol. 4, no. 2, pp. 1351–1356, 2019.
32. V. Mahboub and D. Mohammadi, "A constrained total extended Kalman filter for integrated navigation," *The Journal of Navigation*, vol. 71, no. 4, pp. 971–988, 2018.
33. N. Ko, W. Youn, I. Choi, G. Song, and T. Kim, "Features of invariant extended Kalman filter applied to unmanned aerial vehicle navigation," *Sensors*, vol. 18, no. 9, p. 2855, 2018.
34. A. Gelb, *Applied Optimal Estimation*. Cambridge, MA: MIT Press, 1974.
35. N. J. Gordon, D. J. Salmond, and A. F. Smith, "Novel approach to nonlinear/non-Gaussian Bayesian state estimation," *IEE Proceedings F-Radar and Signal Processing*, vol. 140, no. 2, pp. 107–113, 1993.
36. F. Gustafsson, F. Gunnarsson, N. Bergman, U. Forssell, J. Jansson, R. Karlsson, and P.-J. Nordlund, "Particle filters for positioning, navigation, and tracking," *IEEE Transactions on Signal Processing*, vol. 50, no. 2, pp. 425–437, 2002.

37. A. Doucet and A. M. Johansen, "A tutorial on particle filtering and smoothing: Fifteen years later," *The Oxford Handbook of Nonlinear Filtering*, vol. 12, no. 656–704, p.3, 2011.
38. M. Nieto, A. Cortés, O. Otaegui, J. Arróspide, and L. Salgado, "Real-time lane tracking using Rao-Blackwellized particle filter," *Journal of Real-Time Image Processing*, vol. 11, no. 1, pp. 179–191, 2016.
39. M. S. Arulampalam, S. Maskell, N. Gordon, and T. Clapp, "A tutorial on particle filters for online nonlinear/non-Gaussian Bayesian tracking," *IEEE Transactions on Signal Processing*, vol. 50, no. 2, pp. 174–188, 2002.
40. M. Speekenbrink, "A tutorial on particle filters," *Journal of Mathematical Psychology*, vol. 73, pp. 140–152, 2016.
41. D. Titterton, J. L. Weston, and J. Weston, *Strapdown Inertial Navigation Technology*, 2nd ed. London: IET, 2004.
42. M. Lashley, D. M. Bevly, and J. Y. Hung, "Analysis of deeply integrated and tightly coupled architectures," in *IEEE/ION Position, Location and Navigation Symposium*, Indian Wells, CA, 2010, pp. 382–396.
43. M. Petovello, C. O'Driscoll, and G. Lachapelle, "Carrier phase tracking of weak signals using different receiver architectures," in *Proceedings of the 2008 National Technical Meeting of The Institute of Navigation*, San Diego, CA, January 2008, pp. 781 – 791.
44. S. Alban, D. M. Akos, S. M. Rock, and D. Gebre-Egziabher, "Performance analysis and architectures for INS-aided GPS tracking loops," in *Proceedings of the 2003 National Technical Meeting of the Institute of Navigation*, Anaheim, CA, January 2003, pp. 611–622.
45. X. Zhang and C. Guo, "Optimal design of loop bandwidth in GPS/INS deeply integration," in *International Workshop on Microwave and Millimeter Wave Circuits and System Technology (MMWCST)*, Chengdu, China. October 2013, pp. 181–183.
46. G. T. Schmidt and R. E. Phillips, "INS/GPS integration architectures," Defense Technical Information Center, Technical Report, 2010.
47. M. Lashley and D. M. Bevly, "Performance comparison of deep integration and tight coupling," *Navigation: Journal of the Institute of Navigation*, vol. 60, no. 3, pp. 159–178, 2013.
48. PPSC Working Group on Automated and Connected Vehicles, *Automated and Connected Vehicles Policy Framework for Canada*. Developed by the Policy and Planning Support Committee (PPSC), January 2019.
49. International GNSS Service. [Online] Available: http://www.igs.org/.
50. O. Montenbruck, P. Steigenberger, R. Khachikyan, G. Weber, R. Langley, L. Mervart, and U. Hugentobler, "IGS-MGEX: Preparing the ground for multi-constellation GNSS science," *Inside GNSS*, vol. 9, no. 1, pp. 42–49, 2014.
51. O. Montenbruck, P. Steigenberger, L. Prange, Z. Deng, Q. Zhao, F. Perosanz, I. Romero, C. Noll, A. Stürze, G. Weber, et al., "The Multi-GNSS Experiment (MGEX) of the International GNSS Service (IGS)–achievements, prospects and challenges," *Advances in Space Research*, vol. 59, no. 7, pp. 1671–1697, 2017.
52. T. Hadas and J. Bosy, "IGS RTS precise orbits and clocks verification and quality degradation over time," *GPS Solutions*, vol. 19, no. 1, pp. 93–105, 2015.
53. J. Kouba and P. Héroux, "Precise point positioning using IGS orbit and clock products," *GPS Solutions*, vol. 5, no. 2, pp. 12–28, 2001.
54. M. Karaim, M. Elsheikh, and A. Noureldin, "GNSS error sources," in *Multifunctional Operation and Application of GPS*, R. B. Rustamov and A. M. Hashimov, Eds. London: IntechOpen, 2018, ch. 4, pp. 69–85.
55. J.-T. Wu, S. C. Wu, G. A. Hajj, W. I. Bertiger, and S. M. Lichten, "Effects of antenna orientation on GPS carrier phase," *Manuscripta Geodaetica*, vol. 18, pp. 91–98, 1993.
56. M. ElSheikh, "Integration of GNSS precise point positioning and inertial technologies for land vehicle navigation," Ph.D. dissertation, Queen's University, 2019.
57. K. Chen and Y. Gao, "Real-time precise point positioning using single frequency data," in *Proceedings of the 18th International Technical Meeting of the Satellite Division of The Institute of Navigation (ION GNSS 2005)*, Long Beach, CA, September 2005, pp. 1514–1523.
58. L. Li, C. Jia, L. Zhao, J. Cheng, J. Liu, and J. Ding, "Real-time single frequency precise point positioning using SBAS corrections," *Sensors*, vol. 16, no. 8, p. 1261, 2016.
59. T. P. Yunck, "Coping with the atmosphere and ionosphere in precise satellite and ground positioning," *Environmental Effects on Spacecraft Positioning and Trajectories*, vol. 21, pp. 1–16, 1993.

60. V. L. Knoop, P. F. de Bakker, C. C. Tiberius, and B. van Arem, "Lane determination with GPS precise point positioning," *IEEE Transactions on Intelligent Transportation Systems*, vol. 18, no. 9, pp. 2503–2513, 2017.
61. S. Choy, K. Zhang, and D. Silcock, "An evaluation of various ionospheric error mitigation methods used in single frequency PPP," *Journal of Global Positioning Systems*, vol. 7, no. 1, pp. 62–71, 2008.
62. J. Saastamoinen, "Atmospheric correction for the troposphere and stratosphere in radio ranging satellites," *The Use of Artificial Satellites for Geodesy*, vol. 15, pp. 247–251, 1972.
63. M. Elsheikh, W. Abdelfatah, A. Nourledin, U. Iqbal, and M. Korenberg, "Low-cost real-time PPP/INS integration for automated land vehicles," *Sensors*, vol. 19, no. 22, p. 4896, 2019.
64. J. Farrell, *Aided Navigation: With High Rate Sensors*. New York: McGraw-Hill, Inc., 2008.
65. S. Eun-Hwan, "Accuracy improvement of low cost INS/GPS for land applications," Master's thesis, University of Calgary, 2001.
66. M. Elsheikh, H. Yang, Z. Nie, F. Liu, and Y. Gao, "Testing and analysis of instant PPP using freely available augmentation corrections," in *Proceedings of the 31st International Technical Meeting of the Satellite Division of the Institute of Navigation (ION GNSS+ 2018)*, Miami, FL, 2018, pp. 1893–1901.
67. ST, "LSM6DSL datasheet." [Online] Available: https://www.st.com/resource/en/datasheet/lsm6dsl.pdf.
68. NovAtel, "IMU-KVH1750 datasheet". [Online] Available: https://www.novatel.com/assets/Documents/Papers/IMUKVH1750D19197v1.pdf.

14 Deploying Wireless Charging Systems for Connected and Autonomous Electric Vehicles

Binod Vaidya and Hussein T. Mouftah
University of Ottawa

CONTENTS

14.1 INTRODUCTION

With internal combustion engine-based vehicles, increasing the demand for mobility in urban areas [1] yields significant greenhouse gas emissions and thus causes environmental problems (i.e., air pollution and climate change).

Automotive/transportation sector is undergoing a phenomenal course of rapid, multifaceted change. The current trend of automotive industry is moving toward electric vehicles (EVs), which can overcome the global environmental crisis and energy challenges. Besides, autonomous vehicle (AV) technologies are also progressing at a very rapid pace. Correspondingly, connected and autonomous electric vehicles (CAEVs) [2] technologies can be considered as highly appealing automotive technologies in the recent years.

EVs and CAEVs could be the potential solutions to urban transportation problems [3]. In the near future, CAEVs could likely be the next transportation revolution, especially in taxi fleets. A wide adoption of CAEVs could reduce environmental degradation through reduced CO_2 emissions in the urban environment while furnishing beneficial economic and social outcomes through improved efficiency, traffic flow, road safety, and greater access.

Many companies are heavily investing in AV technologies and adopting different approaches to bring major changes in the automotive industry. Major players are traditional original equipment manufacturers (OEMs) including Nissan, Ford, Mercedes-Benz that are already in the automotive business and some disruptive tech giants (i.e., Waymo and Uber) that have no significant background in the automotive industry.

CAEV technology is an active area of research and possesses numerous challenging applications. Though CAEV technology shows great promises and has substantial benefits, there are several impediments along the way of its deployment.

It can be observed that relatively short driving range of most EVs and longer recharging times collectively yields a range of anxiety for EV users; this may be one of the impediments preventing the widespread deployment of EVs. To minimize such an anxiety, charging EV in a timely fashion is imperative to guarantee a certain degree of its mobility [4]. However, a lack of public-charging infrastructure could become an obstacle to the electromobility (e-mobility). So the employment of fast-charging and dynamically wireless charging [5] would be crucial to tackle some of these issues.

Wireless power transfer (WPT) [6,7], which is based on inductive power transfer (IPT) technology, is getting increasing attention in the EV domain, as a plug-in or wired charging would be inconceivable while EVs are in the motion. Thus, the WPT would be the only solution for the in-motion charging [8]. This will make EV charging much more convenient as it can be recharged automatically without human intervention [9]. With dynamic wireless charging (DWC) [10], the CAEV can charge itself via charging pads (CPs) embedded into the road surface.

Furthermore, for CAEV charging, the waiting time at the charging stations is still a challenging issue [4]. In order to have a significant growth of CAEVs in the urban areas, an adequate number of charging facilities in urban areas is needed as well as an efficient smart CAEV charging management is required for managing and allocating the charging station resources [4]. In order to mitigate such issues, an effectual charging strategy is required [11].

This chapter presents wireless charging infrastructure for CAEV and depicts implementation of such a system that can provide effective communication between CAEV and EVSE (Electric Vehicle Supply Equipment) as well as furnish automated reservation-based charging strategies, and depicts implementation of a system that can provide effective automated reservation for CAEVs. The significance of this system is to alleviate the shortcomings of the wired charging infrastructure by deploying wireless charging stations (WCSs) and assuring the convenience of automated and coordinated charging strategies. For instance, in automated charging services, CAEV shall have automated reservation and charging functionalities and might be able to charge without a human aid using its built-in software, sensors, actuators, etc. However, a WCS or wireless EV supply equipment (WEVSE) should be able to offer automated charging capabilities by providing a charging service to the CAEV, which includes WEVSE reservation and wireless charging process. The automated reservation allocation scheme is a strategic constituent of the charging strategies. By deploying an automated reservation mechanism, the CAEVs can reserve recharging schedule ahead of time, so such a strategy can not only minimize waiting time but also alleviate congestion at the charging stations (i.e., recharging congestion) [4].

In recent years, there have been many researches and studies undertaken in the transportation domain related to connected and autonomous vehicles (CAVs) [12]. Furthermore, as WPT for EVs is a promising technology, it has also attracted attention of many researchers and automotive industries [6–9].

Similarly, some studies have investigated communication technologies used in the wireless charging systems [11,13] for the EV charging infrastructure. For instance, the paper [12] presents communication networks for the dynamic wireless power transfer using dedicated short-range communication and Wi-Fi technologies. Some works highlight the static wireless charging (SWC) for EVs [5,11], whereas others focus on DWC systems [14–18]. For example, the paper [15] focuses on routing strategy for EVs that require recharging, so that they can wirelessly charge using mobile energy disseminators. Only few studies have investigated wireless charging for CAVs [19,20].

Some works emphasize on charging strategies [21–23] for CAEV including the choice of charging station, the required amount of energy, and optimal scheduling. While some other works depict in-route charging of mobility-on-demand EVs [24] and shared automated EVs [25].

The remainder of this chapter is organized as follows: Section 14.2 describes preliminary that includes CAEVs; Section 14.3 depicts wireless charging systems. Standardization for wireless EV charging and CAEV charging management system are presented in Sections 14.4 and 14.5, respectively. Section 14.6 sums up the summary.

14.2 PRELIMINARY

14.2.1 CONNECTED AND AUTONOMOUS ELECTRIC VEHICLES

Connected and Autonomous Electric Vehicles are basically a combination of connected vehicles (CVs), AV, and EV. A CV is a vehicle with technology that enables it to communicate with nearby vehicles, infrastructure, as well as objects but neither automated nor electrically operated. However, an AV is a vehicle that is, in the broadest sense, capable of driving itself without human intervention. And, EV is a vehicle that powers up and operates with energy stored in the battery.

CAEVs definitely transform the existing mobility paradigm. It can be observed that technological advancements in driving assistants and network connectivity yield further opportunities and services, and meet the sustainable development for cleaner, safer, and smarter mobility [21].

CAEVs offer many potential advantages in terms of sustainable development for environment-friendly urban mobility, which are as follows:

- *Improved Safety*: CAEVs may eliminate many of the accidents caused by human error, estimated at about 90% of all accidents.
- *Greater Mobility*: CAEVs can provide better mobility for those who cannot drive, including elderly, disabled, and youth.
- *Reduced Parking Lots*: CAEVs can drop off passengers at their destinations without needing nearby parking lots.
- *Relaxed Drivers*: Using CAEVs, drivers can rest, work, or entertain themselves during a trip.
- *Increased Car-Sharing*: CAEVs shall reduce the need for individually owned cars, as the deployment of shared CAEV shall increase dramatically.
- *Increased Road Capacity*: CAEVs can be used for fleet platooning such that more predictable traffic flow and reduced congestion can be achieved.
- Fewer CO_2 emissions and pollutants: Using the electric power to operate, CAEVs can reduce greenhouse gas emissions as well as air pollution, minimize environmental impact, and improve the quality of life in urban areas.
- *Less Fuel Costs*: Fossil fuel will not be consumed to run CAEVs, so fuel consumption is significantly reduced.

Typically, CAEV is a vehicle that is not only capable of sensing its environment and navigating with little or no human input and communicating with other entities but also operated by the electric power. The CAEV may collect various sensors data through intravehicle communication system and feed the data to in-vehicle applications. Similarly, the CAEV may communicate with the roadside unit, cloud infrastructure, and other CAEVs to perform some computing tasks collaboratively [22].

Basically, the CAEV is composed of five major components:

- Perception system which is responsible for sensing the environment to understand its surroundings. The CAEV may sense its environment using various sensing devices including Radar, LiDAR (Light Detection and Ranging), image sensors, 3D camera, etc.
- Localization and mapping system that enables the vehicle to know its current location. The most commonly used technique is Global Navigation Satellite System that provides geospatial positioning with global or regional coverage.
- Driving policy refers to the decision-making capability of a CAEV under various situations, such as negotiating at roundabouts, giving way to vehicles and pedestrians, and overtaking vehicles.
- *Communication System*: As CAVs shall be connected to the surrounding environment such as vehicles with vehicle-to-vehicle connectivity, to the infrastructure with vehicle-to-infrastructure, and to anything else such as the Internet, vehicle-to-anything, through wireless communication links.
- *Storage Battery System*: This system includes charger and battery packs in the vehicle. Basically, the state-of-charge (SoC) level determines the amount of charge stored in the battery.

Society of Automotive Engineers (SAE) released SAE International Standard J3016 that sets out taxonomy and standard to define different levels of autonomy. SAE updated its classification in 2016 as SAE J3016-201609. Basically, vehicle automation has been categorized into various levels of AV technology having no automation to full automation. For instance, automated driver-assistance systems such as adaptive cruise control correspond to lower automation levels, while fully automated driverless vehicles correspond to higher automation levels. As per the SAE standard, the defined levels of vehicle automation are as follows:

Level 0 - No automation
Level 1 - Driver assistance
Level 2 - Partial automation
Level 3 - Conditional automation
Level 4 - High automation
Level 5 - Full automation

14.3 WIRELESS CHARGING SYSTEMS

Currently, the wired charging of EVs is used; however, in the near future, the wireless solution will be significantly widespread as a method of charging for EVs and CAEVs.

WPT technology, or as wireless charging technology, is getting attention in recent years for its applicability in charging EVs and CAEVs. Thus, wireless EV charging system is a system for WPT between the ground assembly (GA) and vehicle assembly (VA) including alignment and wireless communications.

In the recent years, OEMs are developing WPT-enabled EVs. This will make EV charging much more convenient as it can be recharged automatically. WPT for EVs and CAEVs has the potential to overcome the shortcomings of wired charging and eliminate some hurdles toward vehicle electrification and sustainable mobility.

Wireless charging has several desirable features in terms of safety and comfort:

- Charging process is simple and automatic; it does not require any human input.
- It is small in size and compact compared to a wired system.

- Compared to a wired system, it requires less space and can be installed underneath the surface.
- As it does not have any contact, there are no exposed electric connections.
- It can avoid electrocution risk typically arising from power cords.
- Newer WPT designs are getting better in efficiency.

The development of wireless EV charging systems is currently at the early stage, and the focus is mainly on unidirectional charging of EVs. The deployment of wireless EV charging systems is dependent on the advancement of WPT technologies.

Based on installation, the wireless market can be segmented into residential and commercial. The residential segment, which is very appealing, has a significant growth, whereas the public and commercial segments are yet to be considered.

Increase in EV adoption and surge in demand for the energy-efficient source as an alternative to fossil fuel may expand the growth of the wireless EV charging market. Nonetheless, the relatively expensive integration of wireless IPT technology, lack of high-power WPT technology, and comparatively slower recharging hinder the market growth. Excessive research in wireless charging technologies, especially focused on high-power WPT technologies as well as technological advancements and adoption of appealing marketing strategy, would be considered as open issues for the advancement of WPT technologies.

With the adoption of CAEVs, wireless charging will become an essential component of the electromobility (e-mobility) in the future. Especially, with the increased demand of shared mobility, CAEV taxi fleets in the urban areas may need to recharge periodically, without leaving their service area. In such a case, wireless charging offers a viable solution that will shape the urban transportation. As zero-emission driving in urban areas and automated driving shall play a major role in the forthcoming automotive development, the wireless power revolution shall be a key player for the smart cities.

14.3.1 TYPES OF WIRELESS EV CHARGING

Wireless EV charging is based on IPT technology, which transfers power between two coupled coils; a primary coil at a wireless charger is connected to the electrical grid, while a secondary coil is located at the EV such that there is a reasonable air gap between them.

In such near-field charging technique, a transmitting coil of the wireless charger produces a magnetic field that transfers energy via induction to a nearby receiving coil of the EV. Some fraction of the magnetic flux generated by the transmitting coil that penetrates the receiving coil contributes to the power transfer. And, the transfer efficiency depends on the coupling between the coils and their quality factor.

Wireless charger is also known as WEVSE. WEVSE is the equipment, which is connected to an electrical power source and having GA that can wirelessly transfer electric energy to an EV.

Mainly, there are two types of IPT for the wireless charging. Static IPT is deployed when the vehicle is spotted in a parking lot, and dynamic or quasi-dynamic IPTs are deployed when the vehicle is either on move or on a brief stop at the traffic red light, respectively.

In the SWC system, CAEVs shall wirelessly communicate with the base controller of the WCS in order to their recharging process. Figure 14.1 shows the overviews of SWC for CAEVs. Basically, the WEVSE has the base controller for communication and the CP for IPT.

It should be noted that as the wired charging would be impossible while the EVs are in the motion, thus, the WPT would be the only solution for the dynamic or quasi-dynamic charging.

In the DWC system, CPs are placed under a portion of the roadbed, and a CAEV's battery is charged wirelessly while the CAEV is being driven over the CPs. Such systems shall enable the moving CAEVs to charge their batteries using the magnetic IPT technology. Thus, it can be referred

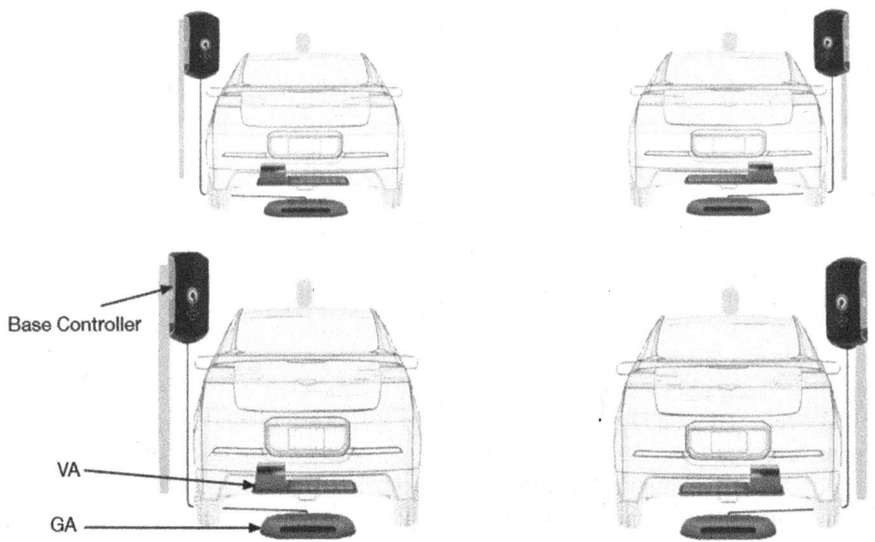

Figure 14.1 SWC system.

as "Roadway IPT". It has been demonstrated that the efficiency of the power transfer in the dynamic charging systems can be more than 80%.

Figure 14.2 depicts the DWC system. In order to achieve such a system, the CAEVs need to communicate with various entities of the system such as a charging service provider, road side units, and CPs. Each charging lane shall have a large number of CPs extended over a long distance (several miles) to allow the EVs to acquire enough amount of power while traveling within this distance.

Some companies such as WiTricity, Qualcomm Halo, and Plugless have demonstrated static wireless EV charging systems for public use [8]. However, some notable pilot projects for DWC are online electric vehicle at KAIST, South Korea; EU (European Union) projects such as FABRIC (Feasibility analysis and development of on-road charging solutions for future EVs) and UNPLUGGED as well as innovative research at Utah State University, US Department of Energy's Oak Ridge National Laboratory (ORNL) [8].

14.4 STANDARDIZATION FOR WIRELESS EV CHARGING

To achieve interoperable, compatible, and vendor-neutral implementation of the wireless charging technology, it is essential to have an establishment of international standards.

14.4.1 ISO/IEC 15118 STANDARD

ISO/IEC 15118: Road vehicles – vehicle-to-grid (V2G) communication interface is an international standard defining a V2G communication interface for bidirectional charging/discharging of EVs [26].

ISO 15118-8:2018, basically, specifies the requirements of the physical and data link layer of wireless high-level communication between EV and the EVSE. It can be noted that the wireless communication technology is used as an alternative to the wired communication technology as defined in the ISO 15118-3 standard.

This applies, in particular, to the communication technologies as communications play a significant role in the wireless charging systems and are used for various purposes including correct

Figure 14.2 DWC system.

positioning, connection establishment, and authentication. With the inception of WPT technologies and the tremendous development of wireless communication, the need for a wireless communication between EV and EVSE infrastructure becomes imperative. Besides, WPT, the information exchange between the vehicle and the charging unit has to take place wirelessly.

The actors involved in the architecture of the ISO 15118-8 standard are classified into two types: primary actors that are directly participated in the charging scenario and the secondary actors that are indirectly involved in it. The primary actors are composed of the EV and the wireless charging point, namely, EVSE, while the central primary actors are the communication controllers at the EV and the EVSE, i.e., Electric Vehicle Communication Controller (EVCC) and Supply Equipment Communication Controller (SECC), respectively. The SECC and the EVCC must exchange information before charging as well as during the charging process.

The ISO 15118-8 standard mainly highlights on the data link layer and the physical layer as these layers are intended to correspond to the lowest layers of ISO model for open systems.

14.4.1.1 Wireless Communication Requirements

ISO 15118-8 specifies requirements for wireless communication module on both EVCC and SECC side. Wireless local area networks (WLANs) as specified in IEEE 802.11-2012 for wireless communication have been considered for both EVCC and SECC. More specifically, they shall implement feature set of high-throughput station or high-throughput access point and operate in ISM 2.4 GHz or 5 GHz bands.

ISO 15118-8 also depicts various use cases in relationship to wireless communication for WPT, considering different range requirements of the communication channel, which are as follows:

- *Discovery*: In this phase, EVCC has entered the communication range of SECC and then associates to appropriate SECC to start a high-level communication for further steps. It takes place when the EV is in typically the 5—30-m range of the EVSE.
- *Fine Positioning*: In this phase, alignment of primary and secondary devices for efficient power transfer in case of WPT. It takes place when the EV is aligning with CP of the wireless charging system.
- *Charging Control*: In this phase, transferring, controlling, and monitoring power requested by EV from EVSE. It takes place when the EV is sufficiently aligned with the CP of the wireless charging system.

In the wireless charging systems for CAEV, requirements for the wireless communication modules on communication controllers, i.e., SECC and EVCC, should be considered for robust communication between the SECC and EVCC. As EVCC and SECC utilize WLAN as specified in IEEE 802.11-2012 for the wireless communication, consideration of WLAN technology requirements for SECC and EVCC is vitally important. Table 14.1 shows the WLAN technology requirements for communication controllers.

The distance between EVCC and SECC for charging control depends on the installation location of wireless communication modules and antennae. As the distance influences reliability of the communication link, the choice of mounting location is extremely critical.

14.4.2 SAE J2954 STANDARD

Wireless charging of electric and plug-in hybrid vehicles

Table 14.1
WLAN Technology Requirements for Communication Controllers

Requirements	Communication Controllers
Use IEEE 802.11 compliant technology	SECC, EVCC
Configure as Access Point according to IEEE 802.11	SECC
Configure as Station according to IEEE 802.11	EVCC
Support feature of High-Throughput Access Point according to IEEE 802.11-2012 on all channels	SECC
Support feature of High-Throughput Station according to IEEE 802.11-2012 on all channels	EVCC
Beacon period shall not exceed T_{beacon} = 105 ms where T_{beacon} is beacon interval	SECC
Use active and/or passive scanning	EVCC

The SAE adopted the standard "SAE J2954 Wireless Power Transfer for Light-Duty Plug-In/Electric Vehicles and Alignment Methodology" to establish WPT between WEVSEs and EVs [27]. This standard defines acceptable criteria for interoperability, electromagnetic compatibility, minimum performance as well as safety and testing for wireless charging of EVs.

SAE J2954 establishes a common frequency band using 85 kHz (81.39–90 kHz) for all light-duty vehicle systems. In addition, four classes of PH/EV of WPT levels are given today.

- 3.7 kW specified in TIR J2954
- 7.7 kW specified in TIR J2954
- 11 kW to be specified in revision of J2954
- 22 kW to be specified in revision of J2954

The current SAE J2954 addresses unidirectional charging; however, bidirectional energy transfer would be addressed in future. Furthermore, this standard only covers the stationary charging of EVs; in the future, it could cover dynamic charging applications, i.e., IPT charging while driving.

14.4.2.1 Overviews of J2954 WPT System and Charging Process

A WPT system is shown in Figure 14.3, which is harmonized with ISO 19363. In such a system, the transmitter system is known as GA and the receiver one is called VA. The WPT technology chosen is IPT, using magnetic resonance. The GA and the VA are coupled through their respective coils, and they use wireless communication.

Wireless charging of EVs, basically, requires to communicate between the VA and the GA over a wireless physical medium. The SAE J2954 relies on various standards such as SAE J2836/6, SAE J2847/6, and SAE J2931/6 to support communications needed for WPT. Wireless charging process flow diagram is shown Figure 14.4.

The details of wireless EV charging process is as follows:

1. *WPT Charging Spot Discovery*: This state allows a driver or a vehicle to locate a WPT location and get information on its capability, compatibility, and availability. Communication supporting this functionality may utilize Internet technologies.
2. *Guidance*: Specified communication technologies are used to facilitate manual or automated positioning of EV for optimal power transfer. Guidance may provide assistance to the vehicle or driver when farther than 1.5 m to the charging location for navigation into the parking bay/slot.
3. *Fine Alignment*: This state provides assistance to the vehicle or driver when closer than 1.5 m from the charging location to facilitate center alignment between the VA coil and the GA coil.

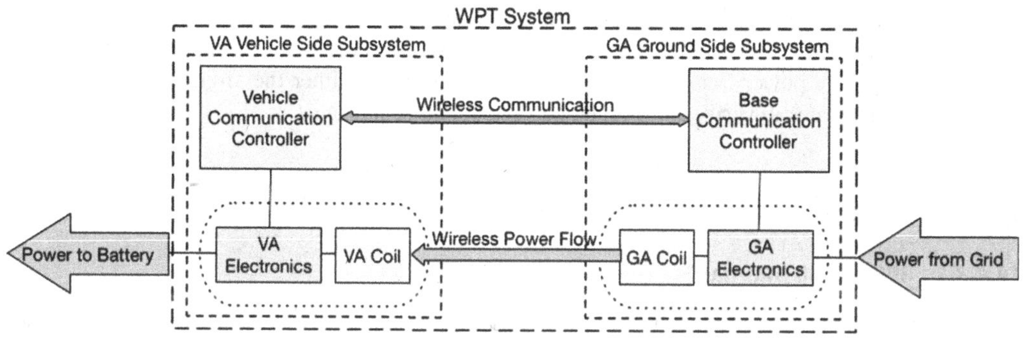

Figure 14.3 SAE J2954 WPT flow diagram.

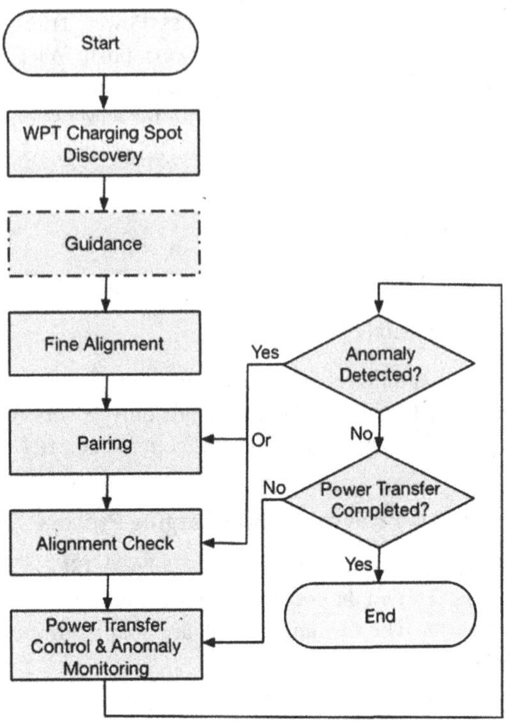

Figure 14.4 Wireless charging process flow diagram.

4. *Pairing*: This state aims to validate that the VA coil is positioned on top of the intended GA coil for power transfer.
5. *Alignment Check*: This state aims to validate that the VA coil is ready to receive power from the GA coil.
6. *Power Transfer Control and Anomaly Monitoring*: After proper alignment, the VA requests power transfer such that the GA provides the required power. A key functionality required for an efficient power transfer is the capability for the EV to control the power transfer process. The messages defined in SAE J2847/6 can provide the following capabilities: verification of compatibility; initiation and termination of a charging cycle; control of the GA current/voltage/energy to match the vehicle's requests; and modification of the power transfer process.

   During power transfer, both GA and VA independently monitor the occurrence of any anomalies such as unexpected changes in power, current, voltage, efficiency, or other measurements within the GA or VA. If the GA or the VA detects any anomalies, the GA shall shut down the power transfer immediately and proceed to either the alignment check or the pairing states to verify appropriate alignment and/or pairing.

14.4.3 OTHER WEVC STANDARDS

14.4.3.1 IEC 61980 Standards

IEC 61980 standards deal with EV WPT systems [28]. Particularly, IEC 61980-2 entitled "Specific requirements for communication between electric road vehicle and infrastructure with respect to wireless power transfer systems" applies to the communication between EV and WPT systems when connected to the electric power. This document includes standards for operational characteristics and functional characteristics of the WPT communication subsystem.

14.4.3.2 ISO/PAS 19363 Standard

The International Organization for Standardization (ISO) has released ISO/PAS 19363:2017 entitled "Electrically propelled road vehicles - Magnetic field wireless power transfer - Safety and interoperability requirements" in January 2017 [29]. This specification is aimed to be compatible with the IEC 61980 standards. In particular, ISO/PAS 19363 considers the magnetic field power transfer for car passengers and light-duty vehicles, in stationary and unidirectional applications.

14.4.3.3 SAE J2847/6 Standard

SAE J2847/6: "Communication between Wireless Charged Vehicles and Wireless EV Chargers" defines abstract messages supporting the wireless transfer of energy between EVs and the Wireless Charger (WEVSE) [30]. The primary purpose of SAE J2847/6 is to provide the communication to achieve wireless charging control irrespective of variations in the wireless charging technology employed.

14.5 CAEV CHARGING MANAGEMENT SYSTEM

In this section, a CAEV charging management system is presented. Our system is based on SWC. It consists of the following key components: SecCharge CAEV System, Wireless Charging Station Operators (WCSO), and Energy Service Providers (ESPs) [31]. Figure 14.5 depicts a high-level diagram for CAEV charging management system.

ESPs are autonomous entities producing and distributing energy to the consumers, whereas WCSOs are fundamentally accountable for operation and maintenance of WCSs. These WCSs are based on SWC. WCSOs can typically manage multiple WCSs at various charging sites/locations through their control centers. As shown in Figure 14.2, a particular EV charging network may have several WCSOs such as WCSO-1, WCSO-2, and WCSO-N.

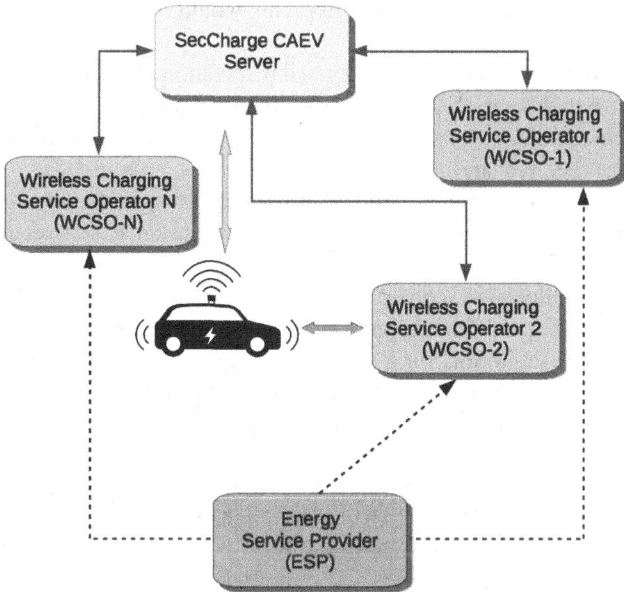

Figure 14.5 High-level diagram for CAEV charging management system.

A SecCharge CAEV System, which includes SecCharge CAEV back-end servers, is a central component of the system architecture. Being a smart management center for CAEV charging, it is designed for facilitating charging services to CAEV by providing effectual coordination with various WCSOs in CAEV charging networks.

In one hand, SecCharge provides charging-related services to CAEVs including finding WCS location, assigning reservation, and managing CAEV information and financial transactions. In another hand, the SecCharge is responsible for interacting with WCSOs for efficient smart charging. SecCharge establishes agreements with different WCSOs in the EV charging network such that CAEVs can charge in the entire charging infrastructure.

SecCharge CAEV system is based on a centralized server that smartly manages charging-related activities. In case of scheduling and reservation services, the system maintains reservation allocation information including occupied and available time slots, so that CAEVs can have automated reservation [32] of a WCS operated by any WCSO.

14.5.1 AUTOMATED RESERVATION MECHANISM

In this section, automated reservation mechanism is proposed for SecCharge CAEV charging management system [31].

A range anxiety is one of the major issues in EV (CAEV) charging. In order to minimize such an anxiety, charging CAEV in a timely fashion is imperative to guarantee a certain degree of its mobility.

Similar to EVs, CAEVs are concerned with charging process as the charging time periods are still significantly long, especially for wireless charging, as it is still in the very early stage. Possibly, long charging times may cause considerable delays, owing to not only the CAEV charging process but also the waiting times due to busy charging stations. Furthermore, unlike EVs, CAEVs', especially level 4 and 5, driver assistance system shall take most of the tasks, so CAEVs would be responsible for timely charging as well.

In the future, the number of CAEVs will increase significantly in the urban areas, then they may experience congestion at several WCSs. Since waiting in queue at the charging stations may not be convenient for CAEVs, the system should be able to provide current waiting time for the WCSs such that the CAEVs would recharge accordingly. This would help to curtail queuing time at the WCSs.

In this paper, we have proposed a novel automated reservation mechanism for charging CAEVs. By deploying an automated reservation mechanism, the CAEVs can reserve recharging schedule ahead of time, so such a strategy can not only minimize waiting time but also alleviate congestion at the charging stations (i.e., recharging congestion).

Primarily, SecCharge CAEV system embraces charging strategies based on slotted reservation where a 24-hour period is divided into time slot intervals, for instance, δ is set at 15 minutes in our case. Table 14.2 shows the notations used in the scheme.

Automated reservation allocation scheme is a strategic constituent of the charging strategies. The automated charging module gets information from the sensing unit to determine the state-of-charge (SoC) of the CAEV. A threshold SoC for a CAEV SoC_{Th}, say 30%, is an initial desired SoC level for recharging a particular CAEV. It should be noted that SoC_{Th} must be greater than SoC_{min}.

Thus, as soon as current SoC reaches SoC_{Th}, it will send an alert and triggers Reservation Request (Reserve_Req).

In automated reservation allocation scheme, the CAEV sends Reserve_Req, which has 3-tuple $\{SoC_{\mathrm{Th}}, \mathrm{Loc}_{\mathrm{Cur}}, \mathrm{Dest}\}$, to the SecCharge CAEV system indicating that the CAEV needs recharging. Figure 14.6 depicts message flows for CAEV Charging.

Upon receiving Reserve_Req, the SecCharge CAEV Server shall deploy an algorithm for determining an appropriate SoC level and reserving time slots with an appropriate charging station. Figure 14.7 depicts an algorithm flow for the automated reservation.

Table 14.2
Notations Used in the Scheme

Symbol	Description
$SoC_{\min}, SoC_{\max}$	Minimum and maximum SoC level
$SoC_{\mathrm{Th}}, SoC_{\mathrm{tar}}$	Threshold and targeted SoC level for a given CAEV
t_{char}	Required charging time
t_{tra}	Travel time to the WCS
$t_{\mathrm{arr}}, t_{\mathrm{wa}}$	Arrival time and waiting time at the WCS
$t_{\mathrm{s}}^{\mathrm{r}}, t_{\mathrm{f}}^{\mathrm{r}}$	Reservation start time and finish time
τ^{r}	Reservation duration
η	Number of time slots
δ	Time slot interval
$\mathrm{Loc}_{\mathrm{Cur}}$	Current location of a given CAEV
Dest	Destination to be reached by a given CAEV

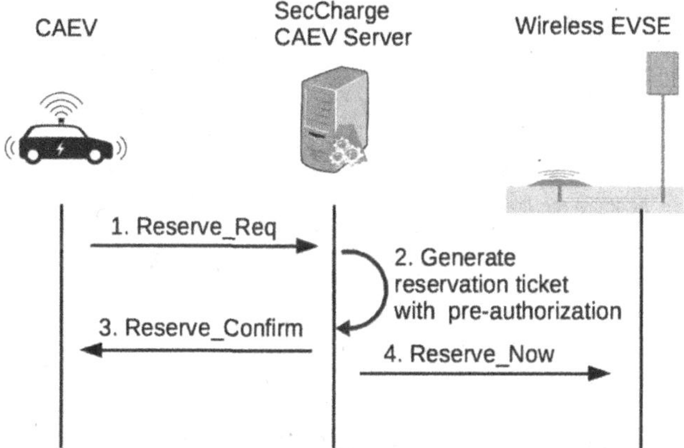

Figure 14.6 Message flows for automated reservation.

Target SoC level for recharging is given as follows:

$$SoC_{\mathrm{tar}} = \alpha\left(SoC_{\max} - SoC_{\mathrm{Th}}\right) \tag{14.1}$$

where α is a coefficient for recharging, such that $\alpha \in \{0,1\}$.

Determination of α depends on various factors such as distance to be traveled, a location of a WSC (i.e., downtown or suburb) and time of day (peak hour or wee hour). As SecCharge CAEV server is responsible for determining α, in turn, SoC_{tar} for each CAEV, a slot allocation in WEVSE will be optimal.

Required charging time t_{char} is computed as follows:

$$t_{\mathrm{char}} = SoC_{\mathrm{tar}} \times \frac{Q}{CR_x \times \mu} \tag{14.2}$$

where Q is CAEV's battery capacity (kWh); CR_x is charge rate; μ is CAEV charging efficiency, typically, $\mu = 0.9$.

While computing t_{char}, CR_x should be considered with a lower value of either the vehicle's acceptance rate or the WSC power output rate.

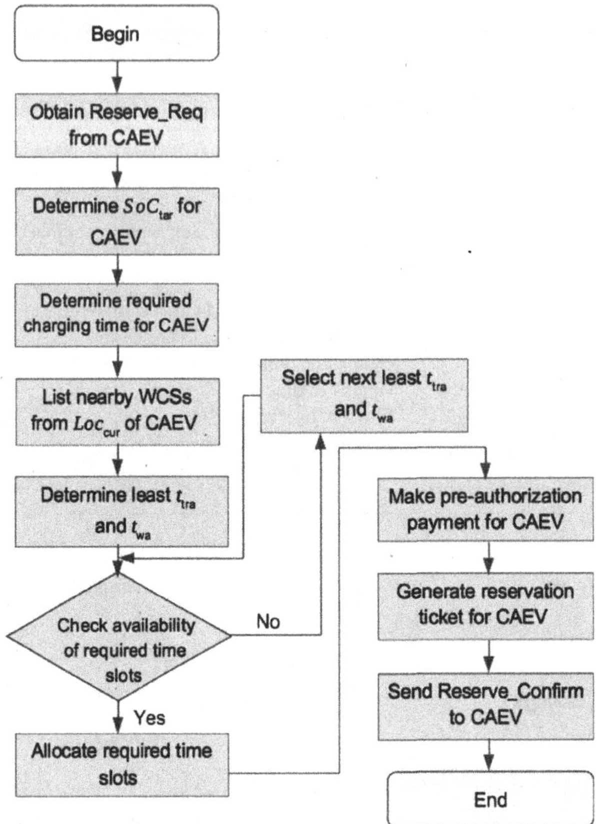

Figure 14.7 Algorithm flow for automated reservation.

With the list of nearby WEVSEs, the SecCharge CAEV Server shall compute travel time (t_{tra}) to each WEVSE as well as waiting time (t_{wa}). The waiting time (t_{wa}) is the time between the arrival time (t_{arr}) of the CAEV to the WEVSE and the time that the CAEV starts to receive the charging service.

After allocating the time slots in the WEVSE, the SecCharge CAEV Server shall generate Reservation ticket, which has the reservation information. At least, the reservation information for the CAEV shall have Reservation start time (t_{s}^{r}), Reservation finish time (t_f^r), and information on WSC location.

Reservation finish time and reservation duration can be computed as follows:

$$t_{\text{f}}^{\text{r}} = t_{\text{s}}^{\text{r}} + \eta \times \delta \tag{14.3}$$

$$\tau^{\text{r}} = t_{\text{f}}^{\text{r}} - t_{\text{s}}^{\text{r}} \tag{14.4}$$

After a successful preauthorization payment with the help of the Payment gateway, the status changes to "payment pre-authorized". Upon successful generation of the reservation ticket, the SecCharge CAEV server shall send a confirmation for automated reservation (Reserve_Confirm) to the CAEV. In the meantime, the SecCharge CAEV server shall also send the automated reservation details (Reserve_Now) with τ^{r} in order to reserve time slot(s) at the given WCS.

Upon receiving Reserve_Confirm, the CAEV can obtain a desired route to reroute via the given WEVSE.

14.5.2 SECCHARGE TEST BED FOR SWC SYSTEM

The basic architecture of the SWC system consists of SecCharge Server, Control center, WCS or EVSE, and CAEV. Figure 14.8 depicts a system architecture for SWC system.

SecCharge software utilizes a combination of client-server, mobile, and web-based architecture. The web-based aspect of SecCharge software is hosted on SecCharge web server. The clients initiate request to the server, and the web server provides appropriate response to the browser. The client-server aspect of the software involves mobile devices initiating request(s) to the server. Hypertext transfer protocol (HTTP) is used for communication between clients and server.

Technical details of the SecCharge System are discussed as follows. Linux Ubuntu is deployed as a basic operating system in the servers. And, the Application Server is based on Apache Tomcat, which provides light-weight web container. However, Oracle database has functioned as an underlying database of the system. Model view controller and Hibernate object relational mapping are utilized as core frameworks. Spring MVC framework, which provides model view controller architecture, facilitates in constructing flexible and loosely coupled web applications. Hibernate provides a framework for mapping object-oriented domain model to relational database.

REST (Representational State Transfer) being lightweight, maintainable, and scalable, RESTful services are implemented as SecCharge web services. And, JSON (JavaScript Object Notation), which is a lightweight data-interchange format, is employed.

For our deployment, the CAEV wireless charging emulator system is used, which is composed of CAEV emulator and EVSE emulator. The CAEV emulator is built on Raspberry Pi 3 having

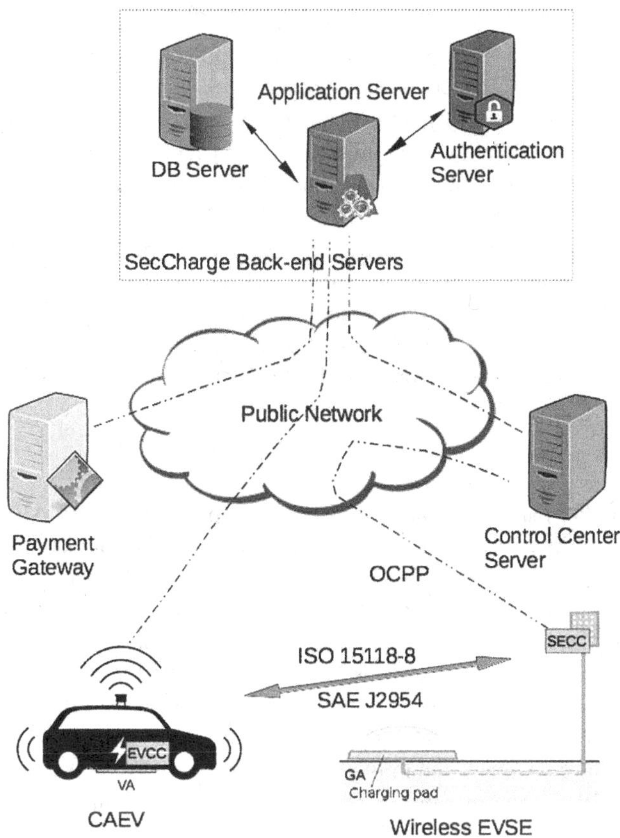

Figure 14.8 System architecture for SWC system.

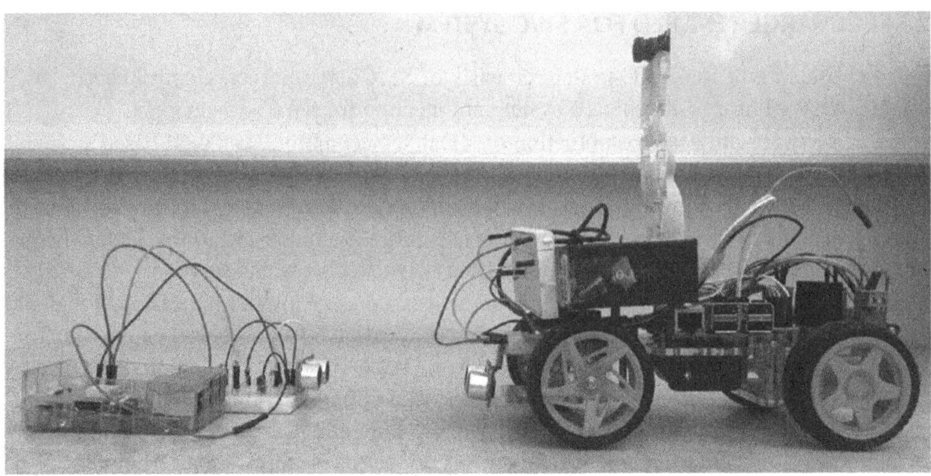

Figure 14.9 CAEV wireless charging emulator system.

in-built WiFi and BLE module, sensors such as wide camera, ultrasonic sensor, and I^2C-bus pulse-width modulation controller, namely PCA9685, whereas the EVSE emulator is built on Raspberry Pi 3 and ultrasonic sensor.

CAEV wireless charging emulator system that is shown in Figure 14.9 supports ISO 15118-8 and SAE J2954, such that data transfer can take place wirelessly from the EVSE emulator to the CAEV emulator and vice versa. And for the communication between the WEVSE and the CAEV, IEEE 802.11n is used as a wireless networking standard.

Open charge point protocol (OCPP), which is an open protocol between charging stations and a managing central system, would allow charging stations and central systems from different vendors to easily communicate with each other. Thus, the OCPP is used as a communication protocol between the EVSE and the control center.

In order to utilize Google Map services, Android app is developed for CAEV such that a desired route can be displayed. Figure 14.10 shows various routes for the particular CAEV, where A is the current location, B is the destination location, and C is the WEVSE location. For instance, Figure 14.10a shows the direct route from A to B before automated reservation process, whereas Figure 14.10b shows a modified route from A to B via C after automated reservation process.

14.6 SUMMARY

Alternative urban mobility paradigms such as CAEVs are appealing since CAEVs can significantly contribute to not only optimize traffic flow and improve road safety but also minimize dependence on fossil fuel and reduce carbon emission in urban areas. For large deployment of CAEVs, the sufficient number of available charging facilities will be required in urban areas. And, an efficient and smart CAEV charging management is required for managing and allocating the charging station resources from different WCSOs. We have designed a system utilizing automated reservation-based charging strategies that include effective reservation management and efficient allocation of time slots of WCSs.

In this chapter, we have designed and implemented a wireless CAEV charging management system that is based on SWC. The proposed system can provide not only effective communication between the CAEV and WSC but also automated reservation with the help of the back-end system. The significance of this system is to alleviate the shortcomings of the wired charging infrastructure and to assure convenience of automated charging strategies. In order to emulate CAEV wireless

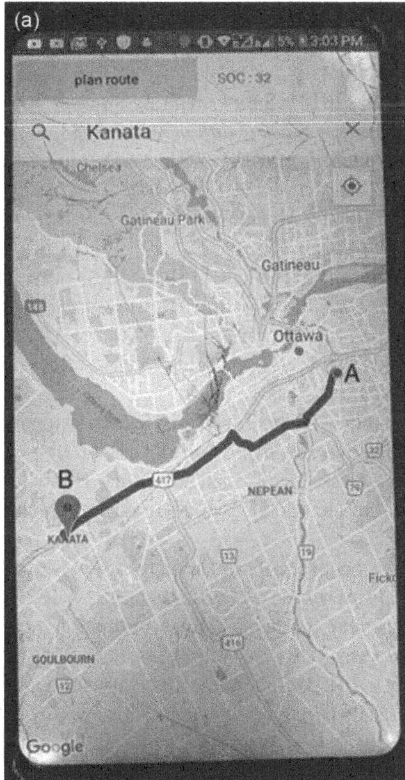

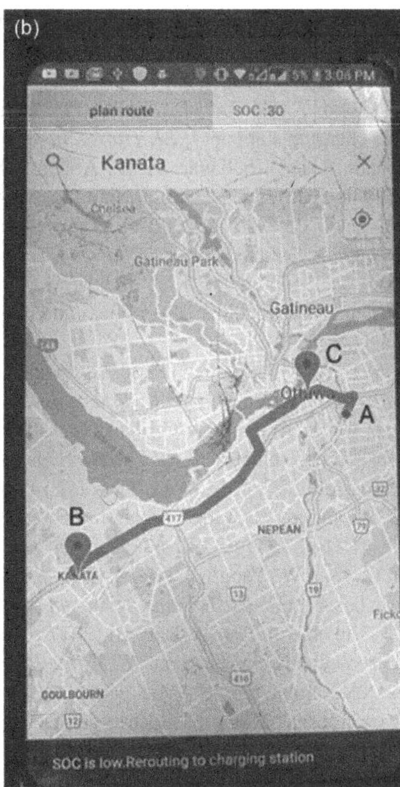

Figure 14.10 Various routes for the CAEV: (a) before and (b) after automated reservation process.

charging, we have considered the CAEV emulator and the EVSE emulator, such that they can communicate through WLAN using WEVC standards such as ISO 15118-8 and SAE J2954. Furthermore, the EVSE can communicate with the back-end server using the OCPP protocol. In this system, charging strategies based on automated reservation are used, which encompass optimum WCS selection and scheduling including reservation slot allocation and preauthorized payment.

As DWC is more challenging than SWC, efficient and smart CAEV charging for the former shall be considered as a future extension. Furthermore, cybersecurity that includes security and privacy is very crucial for wireless charging; hence, our forthcoming direction would be provisioning cybersecurity solutions for wireless charging (i.e., static and dynamic).

ACKNOWLEDGMENT

This research work is supported by Smart Grid Fund (SGF), Ministry of Energy, The Ontario Government and Canada Research Chair (CRC) Fund, Canada.

REFERENCES

1. Cui, Q., Weng, Y., & Tan, C. W. (2019). Electric vehicle charging station placement method for urban areas. *IEEE Transactions on Smart Grid*, 10(6), 6552–6565.
2. Bagloee, S. A., Tavana, M., Asadi, M., & Oliver, T. (2016). Autonomous vehicles: challenges, opportunities, and future implications for transportation policies. *Journal of Modern Transportation*, 24(4), 284–303.

3. Zhang, Y., Zhang, G., Fierro, R., & Yang, Y. (2018). Force-driven traffic simulation for a future connected autonomous vehicle-enabled smart transportation system. *IEEE Transactions on Intelligent Transportation Systems*, 19(7), 2221–2233.
4. Moghaddam, Z., Ahmad, I., Habibi, D., & Phung, Q. V. (2018). Smart charging strategy for electric vehicle charging stations. *IEEE Transactions on Transportation Electrification*, 4(1), 76–88.
5. Manshadi, S. D., Khodayar, M. E., Abdelghany, K., & Uster, H. (2018). Wireless charging of electric vehicles in electricity and transportation networks. *IEEE Transactions on Smart Grid*, 9(5), 4503–4512.
6. Li, S. & Mi, C. C. (2015). Wireless power transfer for electric vehicle applications. *IEEE Journal of Emerging and Selected Topics in Power Electronics*, 3(1), 4–17.
7. Patil, D., McDonough, M. K., Miller, J. M., Fahimi, B., & Balsara, P. T. (2018). Wireless power transfer for vehicular applications: Overview and challenges. *IEEE Transactions on Transportation Electrification*, 4(1), 3–37.
8. Jang, Y. J. (2018). Survey of the operation and system study on wireless charging electric vehicle systems. *Transportation Research Part C: Emerging Technologies*, 95, 844–866.
9. Ahmad, A. M., Alam, S., & Chabaan, R. (2018). A comprehensive review of wireless charging technologies for electric vehicles. *IEEE Transactions on Transportation Electrification*, 4(1), 38–63.
10. Garcia-Vazquez, C. A., Llorens-Iborra, F., Fernandez-Ramirez, L. M., Sanchez-Sainz, H., & Jurado, F. (2017). Comparative study of dynamic wireless charging of electric vehicles in motorway, highway and urban stretches. *Energy*, 137, 42–57.
11. Yang, C., Lou, W., Yao, J., & Xie, S. (2018). On charging scheduling optimization for a wirelessly charged electric bus system. *IEEE Transactions on Intelligent Transportation Systems*, 19(6), 1814–1826.
12. Lin, P., Liu, J., Jin, P. J., & Ran, B. (2017). Autonomous vehicle-intersection co-ordination method in a connected vehicle environment. *IEEE Intelligent Transportation Systems Magazine*, 9(4), 37–47.
13. Echols, A., Mukherjee, S., Mickelsen, M., & Pantic, Z. (2017). Communication infrastructure for dynamic wireless charging of electric vehicles. In *2017 IEEE Wireless Communications and Networking Conference (WCNC)*, San Francisco, CA, 19–20 March 2017.
14. Zaheer, A., Neath, M., Beh, H. Z., & Covic, G. A. (2017). A dynamic EV charging system for slow moving traffic applications. *IEEE Transactions on Transportation Electrification*, 3(2), 354–369.
15. Hwang, I., Jang, Y. J., Ko, Y. D., & Lee, Y. D. (2018). System optimization for dynamic wireless charging electric vehicles operating in a multiple-route environment, *IEEE Transactions on Intelligent Transportation Systems*, 2(6), 1709–1726.
16. Kosmanos, D., Maglaras, L. A., Mavrovuniotis, M., Moschoyiannis, S., Argyriou, A., Maglarass, A., & Janicke, H. (2018). Route optimization of electric vehicles based on dynamic wireless charging. *IEEE Access*, 6, 42551–42565.
17. Chen, H., Su, Z., Hui, Y., & Hui, H. (2018). Dynamic charging optimization for mobile charging stations in internet of things. *IEEE Access*, 6, 53509–53520.
18. Debnath, S., Foote, A., Onar, O. C., & Chinthavali, M. (2018). Grid impact studies from dynamic wireless charging in smart automated highways. In *2018 IEEE Transportation Electrification Conference and Expo (ITEC)*, Long Beach, CA, 13–15 June 2018.
19. Chen, T. D., Kockelman, K. M., & Hanna, J. P. (2016). Operations of a shared, autonomous, electric vehicle fleet: Implications of vehicle & charging infrastructure decisions. *Transportation Research Part A*, 94, 243–254.
20. Doan, V. D., Fujimoto, H., Koseki, T., Yasuda, T. Kishi, H., & Fujita, T. (2018). Allocation of wireless power transfer system from viewpoint of optimal control problem for autonomous driving electric vehicles. *IEEE Transactions on Intelligent Transportation Systems*, 19(10), 3255–3270.
21. Yi, Z. & Shirk, M. (2018). Data-driven optimal charging decision making for connected and automated electric vehicles: A personal usage scenario. *Transportation Research Part C*, 86, 37–58.
22. Lam, A. Y. S., Leung, Y. W., & Chu, X. (2016). Autonomous-vehicle public transportation system: Scheduling and admission control. *IEEE Transactions on Intelligent Transportation Systems*, 17(5), 1210–1226.
23. Yu, J. J. Q., & Lam, A. Y. S. (2018). Autonomous vehicle logistic system: Joint routing and charging strategy. *IEEE Transactions on Intelligent Transportation Systems*, 19(7), 2175–2187.
24. Ammous, M., Belakaria, S., Sorour, S., & Abdel-Rahim, A. (2019). Optimal cloud-based routing with in-route charging of mobility-on-demand electric vehicles. *IEEE Transactions on Intelligent Transportation Systems*, 20(7), 2510–2522.

25. Mohamed, A. A. S., Meintz, A., & Zhu, L. (2019). System design and optimization of in-route wireless charging infrastructure for shared automated electric vehicles. *IEEE Access*, 7, 79968 –79979.
26. ISO. (2018). ISO 15118-8:2018, Road vehicles: V2G communication interface – Part 8: Physical layer and data link layer requirements for wireless communication.
27. SAE International.(2019). Wireless power transfer for light-duty plug-in/electric vehicles and alignment methodology, SAE J2954..
28. International Electrotechnical Commission (IEC). (2019). IEC - TS 61980-2, Electric vehicle wireless power transfer (WPT) systems – Part 2: Specific requirements for communication between electric road vehicle (EV) and infrastructure.
29. ISO. (2017). ISO/PAS 19363:2017, Electrically propelled road vehicles – Magnetic field wireless power transfer – Safety and interoperability requirements.
30. SAE International. (2015). Communication between wireless charged vehicles and wireless EV chargers, SAE J2847-6.
31. Vaidya, B. & Mouftah, H. T. (2018). Wireless charging system for connected and autonomous electric vehicles. In *2018 IEEE Globecom Workshops (GC Wkshps),* Abu Dhabi, UAE, 9–13 December 2018.
32. Vaidya, B. & Mouftah, H. T. (2018). Automated reservation mechanism for charging connected and autonomous EVs in smart cities. In *2018 IEEE 88th Vehicular Technology Conference (VTC2018-Fall) Workshop*, Chicago, IL, 27 August 2018.

15 Dynamic Wireless Charging of Electric Vehicles

Sadegh Vaez-Zadeh, Amir Babaki, and Ali Zakerian
University of Tehran

CONTENTS

15.1 INTRODUCTION

Autonomous vehicles are mainly of the electric type. Therefore, charging infrastructure for electric vehicles (EVs) is an indispensable part of future smart cities. Despite many advantages of EVs, there are deficiencies associated with their batteries. They are heavy and expensive with a limited lifetime in comparison with the EV itself. Moreover, their energy limitation causes range anxiety as an obstacle against their widespread use. Considering these challenges, dynamic wireless power transfer (DWPT) technology for EV charging has become an attractive solution. By using the technology, the battery size and weight reduce and its lifespan increases due to a lower battery depth of discharge. Hence, the battery cost decreases substantially. In addition, the dynamic charging increases the driving range of EVs. Although implementing DWPT systems for EV charging has its own costs, the added expenditure of infrastructure can be compensated by the EV cost reduction. The DWPT technology is well suited to autonomous vehicles of different types. EVs can be charged when they stop at a particular point, e.g., parking lot and bus stations or when they move on urban streets or roads. Traditionally, the power flows only from the grid side to the EVs side (G2V). Nevertheless, in the future smart cities, EVs as potentially distributed sources of energy can share their extra energy with other EVs on the road or with the power grid directly or via the primary track, which is located beneath the road. Thus, the power-sharing includes vehicle to vehicle (V2V) and vehicle to grid (V2G). Accordingly, DWPT technology seems to be an integral part of the future smart cities.

In this chapter, the concept of dynamic wireless power transfer and its advantages compared to the static power transfer are reviewed. Then, the general principles of wireless power transfer and its major technical issues are presented. In addition, the effects of system parameters on the performance of wireless power transfer systems are investigated. Afterward, major parameter estimation and non-parameter estimation methods are presented to track the maximum efficiency in the systems. Finally, the future outlook of dynamic charging of electric vehicles is investigated for adapting the technology to the smart cities including V2V, V2G, and G2V as well as two-side communication between the EVs and the grid.

15.2 LITERATURE REVIEW

Dynamic wireless charging of moving vehicles is becoming an attractive option for future public and private transportation systems [1–3]. In wireless charging of moving EVs, the power request is dynamically met by the primary inverter via the primary winding, which is placed under the road, and a secondary winding underneath the vehicle [4]. A battery is also provided on-board with constant voltage (CV) or constant current (CC) charge states as a supportive source [5]. On both primary and secondary sides, power is transferred to the windings of different shapes [6] via a compensating circuit whose topologies are selected based on design considerations [7].

Depending on the power capability, the dynamic charging increases the driving range and reduces the size and cost of the battery pack. In this way, the added cost of the infrastructure can be compensated by the EV cost reduction [8]. Quantitative cost estimation of dynamic charging is difficult, since the battery cost highly depends on the percentage of the road covered by the power track and its power level. Nevertheless, a dynamic wireless system for a public transportation track of 2% charging coverage with 18 electric buses is compared with a stationary wireless charging system for the same number of electric buses [9]. Using the dynamic charging system, the total cost reduces by about 20.8% in comparison with the stationary system. Such a WPT system requires high power on-road and on-board windings with a potential for substantial power loss. Therefore, maximization of system efficiency is essential in the design stage as well as during the operation [10,11]. The backup battery is usually connected in parallel with the WPT system and the EV motor drive to supply the EV power bus [12], when the on-road WPT supply is not in access. Therefore, it is necessary to regulate the output voltage of the WPT system at the nominal voltage of the battery. Otherwise, the voltage ripple is translated to the current ripple causing excessive heating and battery life reduction. A common voltage ripple of 10% reduces battery life to 82%, whereas its life increases to 95% with the halved voltage ripple [13]. Care must be taken to manage any bus voltage variations due to changing load or magnetic coupling. Besides, dynamic charging with multiple power tracks allows frequent battery charging, which improves the battery life up to ten times compared to the stationary wireless charging [9]. Ideally, the system should be able to maintain high efficiency under varying load and coupling coefficient without severe constraints on the power transfer capability.

In order to achieve the above-mentioned goals, different approaches have been proposed. An optimal output voltage, which results in maximum efficiency is determined in a WPT system with a dc/dc secondary converter [2,10,13,14]. It needs offline information of the coupling coefficient. A cascaded boost-buck converter is used to derive the optimal load [1]. Also, the optimal load and maximum efficiency and extracted with two dc/dc converters on one side or both sides of a WPT system [15,16]. Some other relevant references report the efficiency optimization by adjusting a reflected equivalent load on the primary side via a secondary switching converter [17]. An optimal load is calculated for multiple secondary coils [18]. Impedance matching circuits are also added to WPT systems in order to improve the efficiency [18,19], whereas the control system may become complicated.

In addition, asymmetric switching of the primary inverter is counted to regulate the output voltage [14,20], as well as controlling the inverter duty cycle proposed in [21]. However, it has a start-up problem. Employing two receiver coils with two distinct on-board systems regulates the total received power, too [22]. In a similar study, the output voltage is regulated by a hysteresis controller, where the duty cycle of the primary side inverter is tuned to achieve maximum efficiency [7]. Recently, dynamic retuning controllers have been proposed to enable the primary supply to operate at the fixed frequency and minimize loss by continually switching capacitances within the resonant loop, thereby increasing the complexity of the system and controller design [23]. Bearing in mind the above concise review, some problems need to be emphasized including the system complexity, particularly due to the use of double dc/dc converters, and the required information such as the coupling coefficient, which changes with driving conditions.

15.3 ELECTRIC VEHICLES AND DYNAMIC CHARGING

On-road dynamic charging of electric vehicles by wireless power transfer (WPT) systems has emerged recently as a clean and convenient means for future public and private transportation. Such a WPT system requires high power on-road and on-board coils, together with high switching power converters. The coils usually carry high currents with a potential for substantial power loss. Therefore, maintaining high power transfer efficiency is essential in the system design stage as well as during the operation. Figure 15.1 shows a general schematic view of on-road EVs during charging via the power track beneath the road.

In wireless charging of moving EVs, the power request is dynamically met via the primary inverter and the coils, which are placed beneath the road, and a secondary tuned resonant circuit underneath the vehicle. A battery is also provided on-board with constant voltage or constant current charge states as a supportive source [5]. On the primary side, power is transferred to the coils of different shapes via a compensating circuit whose topology is selected based on design considerations. However, the secondary compensation circuit may be different and optimized at a different resonance frequency. Presently, the international standards are under development for statically

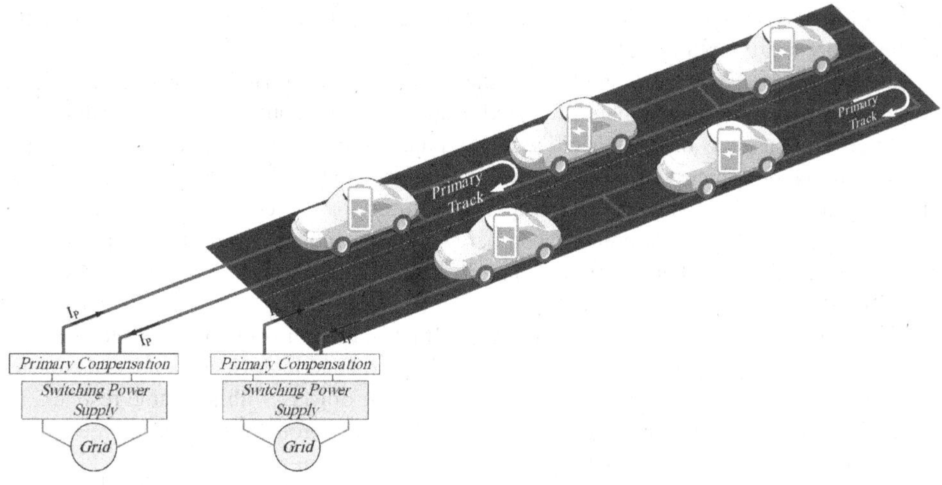

Figure 15.1 Wireless dynamic charging of electric vehicles on the road.

charging such as SAE-j2954 and they define the range of operating frequency, resonant parameters, coil dimension, and some test procedures. In such a high power application, transferring as much power as possible is important, in addition to high efficiency. Ideally, the system should be able to maintain maximum possible efficiency under varying load without severe constraints on the power transfer capability. In addition, the voltage regulation of the EV power bus, supplied by the power transfer system and battery, is needed for a long life of the battery.

15.4 WIRELESS POWER TRANSFER (WPT)

The idea of electrical power transfer without mechanical contact, so-called wireless power transfer, was a human dream at the early stages of the electrical power conversion where Nikola Tesla took the preliminary steps in the late nineteenth and early twentieth centuries [24]. In the course of time, long-distance transfer of communication signals became feasible. It is almost a century that signals are transferred in long distances via electromagnetic waves for applications such as radio, television, and communication systems. However, the methods of wireless power transfer are much less developed due to the fact that the functional limitations and design considerations for power transfer systems are more demanding than those of the signal transfer systems [25]. In recent years, new demands for wireless power transfer have emerged and the research on this kind of power transfer has attracted more attention due to the development of low/high-power electronic equipment. Nowadays, WPT is used widely in low-power applications such as battery chargers for mobile phones, biomedical applications, household apparatuses, and rotating applications [26]. Major opportunities for the static and dynamic charging of EVs are particularly growing due to the technology development in the field of power electronics and high-speed transportation. Wireless power transfer is categorized by three general types in terms of fundamental principles: Electromagnetic Radiation (Microwave), Capacitive, and Inductive.

Electromagnetic radiation-based power transfer in short distances can be used to lighten fluorescent lamps with several watts, for instance. Other products have also been developed by utilizing microwave power transfer methods. The process of converting sunlight into high-frequency microwaves and transmitting them into the earth is also considered [27]. The power of microwaves can also be efficiently transferred using lenses and reflection mirrors. A major problem of this method used to be the conversion of power into usable power by the antennas. This problem has been solved by introducing rectifier antennas, Rectena, in 1963. An important application of wireless power transfer via microwaves was demonstrated through the space solar power program (SPS) in the late 1970s [27]. The wireless power radiation can be used to transfer power to far destinations even up to several tens of kilometers. However, this system is complex and its design and manufacturing are costly with respect to other wireless power transfer methods.

Another wireless power transfer method is the capacitive one in which two separated plates shape a large air gap for power transfer, referred to as capacitive coupled power transfer. Recent research on different aspects of the capacitive method include investigation of the effects of system coupling variations, introducing different analysis methods, using series compensation systems for efficiency improvement, and presenting different structures for device charging [28]. Although, a capacitive system has a simple concept compared with an inductive power transfer system, it requires high electric field intensity higher than 30 kV/cm in the air gap, which results in some practical difficulties.

Inductive power transfer (IPT) is based on the mutual induction principle as well as the resonance phenomenon. An IPT system is composed of electromagnetic devices, control sub-systems, and power electronics circuits. It is known that two resonant objects with the same resonance frequency have the most energy exchange. This fact leads to numerous studies on the performance evaluation of magnetic coupled resonators and their applications in magnetic systems [29]. In this technology, high-frequency AC power is transmitted from a primary coil to a secondary coil, through a time-varying flux linkage. At the secondary side, the receiver coil picks up the AC power and met the load. This system includes electromagnetic and power electronic components as well as compensator circuits and control system. In an IPT system, the secondary side can move with respect to the primary side. Thus, it can be used in transportation applications. Electrical isolation of the secondary side from the primary side, high reliability, and the capability of being used is a wide range of power level (10 W to 100 kW) are among the major advantages of IPT technology [30]. Therefore, it is the most used WPT technology. Accordingly, this chapter is focused on IPT.

There has been a lot of attention to the feasibility of the on-road dynamic charging of Electric Vehicles (EVs). As a result, it has recently appeared as a hot research topic [5,30–32]. EVs widespread adoption has been challenged by factors such as high initial cost, attributed mostly to the battery cost, limited travel distance or so-called "range anxiety", slow charging rates, and lack of charging facilities. Roadway WPT systems where EVs get powered dynamically through WPT can potentially solve most of the aforementioned problems [4,7,33,34]. In particular, the dynamic charging is emphasized as a solution for problems associated with range, cost, and charging rate [24]. Ref. [35] shows a practical project in an England highway, which is equipped with dynamic wireless power transfer (DWPT) charger. The necessity of preparing DWPT infrastructure has been emphasized here against the cost of using batteries with the some disadvantages mentioned before. Ref. [36] proves that on-line EVs with the related under road winding infrastructure will be much more economical than EVs with heavy on-board batteries. This is concluded based on an extensive analysis in [36].

15.4.1 WPT PRINCIPLES

A general schematic view of a WPT system is demonstrated in Figure 15.2. The principles of inductive power transfer can be described in three steps:

1. In the first step, the DC power is converted to high-frequency AC power by using a power electronic inverter.
2. In the second step, the primary (transmitter) coil transmits electromagnetic power to the secondary (receiver) coil via the produced air gap flux linkage.
3. In the third step, the secondary coil picks up the AC power, where it is then converted to DC power by a rectifier and is given to a battery or a load.

High-frequency AC power in the coils is required for a strong magnetic coupling between the two coils to results in effective inductive power transfer. Compensating resonant circuits in the primary and secondary sides are used to enhance the transferred power quality and power transfer efficiency

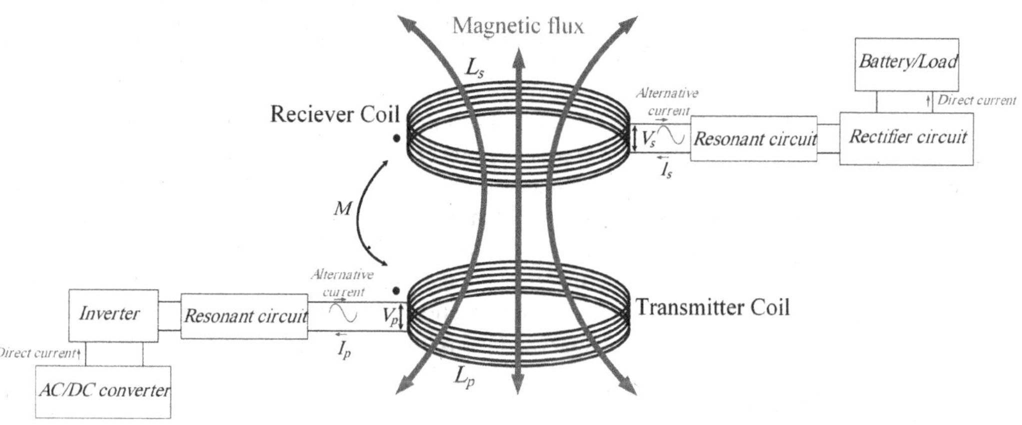

Figure 15.2 Principles of inductive power transfer.

in addition to minimizing the converter switches rating. In Figure 15.2, M is defined as the mutual inductance between the primary and secondary coils and is determined as:

$$M = K\sqrt{L_p L_s}, \tag{15.1}$$

where, K, L_p, and L_s are the coupling coefficients between the primary and secondary coils, primary self-inductance and secondary self-inductance, respectively. I_p and I_s are defined as the primary and secondary currents flowing through the corresponding coils. In WPT systems with low magnetic coupling between the primary and secondary coils, the leakage inductance is higher. In return, the magnetizing inductance is low. So, for the sake of achieving a high efficiency in WPT systems, the operating frequency should be high enough. By this way, the magnetizing impedance increases and the resulted magnetizing current decreases as it is desirable. According to Faraday induction law $\left(V = -\frac{\partial \varphi(t)}{\partial t}\right)$, two coils have some interaction on each other in the form of voltage induction, which can be calculated as:

$$V_p = L_p \frac{\partial I_p}{\partial t} - M \frac{\partial I_s}{\partial t}, \tag{15.2}$$

$$V_s = -L_s \frac{\partial I_s}{\partial t} + M \frac{\partial I_p}{\partial t}, \tag{15.3}$$

where V_p and V_s are the voltages across the primary and secondary coils. Equations (15.2) and (15.3) are achieved by assuming that the self-inductances of both coils are almost the same with different coupling coefficients. The first term in each equation presents the voltage across the coil resulted by its current, while the second term shows the linkage voltage caused by another coil current. Accordingly, the current flowing in the secondary coil I_s is the result of induced voltage in the secondary side $M\frac{dI_p}{dt}$.

15.4.2 WPT SYSTEM

Higher-frequency voltage in the coils provides high efficiency for IPT. However, it has still some drawbacks including low power transfer capability. In other words, it needs a very large primary coil current to transfer the certain active power to the secondary side. To solve this problem, the compensation circuit is designed for both the primary and secondary sides. The secondary compensator increases the system efficiency and power transfer level, while using a primary compensator, the required volt-ampere (VA) rate of inverter for a certain transferred power reduces as a unity power factor (PF) is achieved. Figure 15.3 shows the different compensation circuit types including series or parallel [11]. In addition, mixed series-parallel compensation circuits have been introduced recently.

Figure 15.4 illustrates a typical WPT with a series-series compensation topology. The low-frequency AC power is converted to high-frequency power by means of the passive bridge rectifier and the inverter, which are located at the primary side. A Phase-Shift PWM modulation is used for the inverter switches in order to produce a sufficient three level voltage. Also, a series compensation circuit used in the primary side resonates with the primary coil self-inductance at the inverter operating frequency, which is the same as the inverter voltage frequency. At the secondary side, using a series compensator circuit leads to proper compensation of the secondary reactance. Then the desirable AC power provided by the secondary compensator is rectified again to charge the battery as a load.

The primary and secondary resonance frequencies should be the same as the operating frequency in order to transfer maximum power. For a series-series compensation circuit, the resonance frequencies of both sides are obtained as:

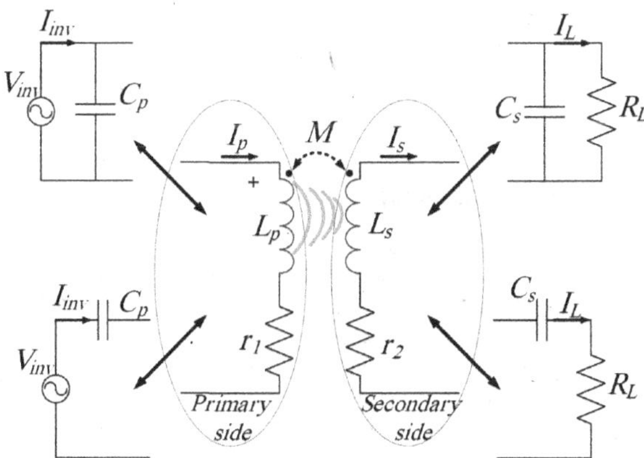

Figure 15.3 Different compensation topologies used in WPT systems [11].

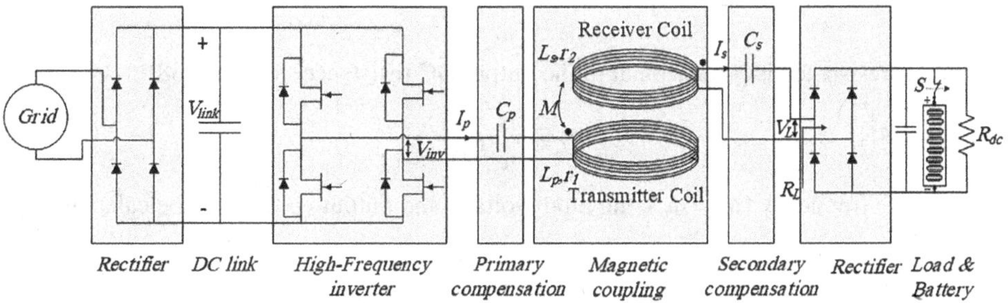

Figure 15.4 A general view of the WPT system showing its components.

$$\omega_p = \frac{1}{\sqrt{L_p C_p}} = \frac{1}{\sqrt{L_s C_s}} = \omega_s, \tag{15.4}$$

where, ω_p and ω_s are the primary and secondary resonance frequencies, respectively. These values also equal the operating frequency, ω_o.

15.4.3 ANALYSIS OF WPT

Considering the fundamental harmonic approximation (FHA) model in a resonant system [37], Figure 15.5 shows the equivalent circuit model of the studied WPT system [36].

The primary inverter voltage and the secondary load voltage are obtained as [36]:

$$V_{inv} = [r_1 + j(\omega L_p - 1/\omega C_p)] I_p - j\omega M I_s, \tag{15.5}$$

$$V_L = -[r_2 + j(\omega L_s - 1/\omega C_s)] I_s + j\omega M I_p. \tag{15.6}$$

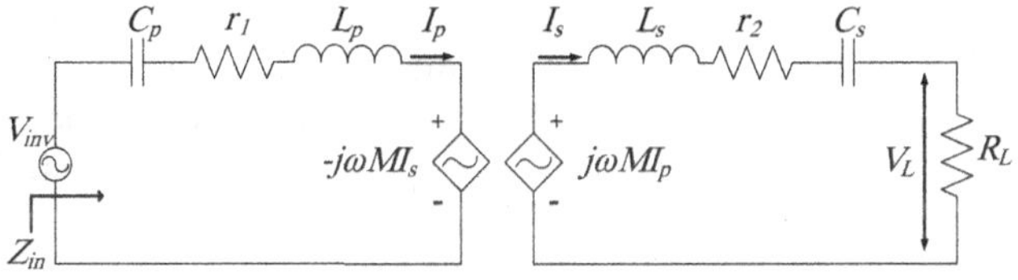

Figure 15.5 Equivalent circuit model of WPT systems [36].

Table 15.1
WPT System Specifications

Symbol	Quantity	Value
V_{link}	DC link voltage	220 V
L_p	Primary inductance	65µH
L_s	Secondary inductance	65µH
C_p	Primary capacitance	61 nF
C_s	Secondary capacitance	61 nF
r_1	Primary resistance	0.33Ω
r_2	Secondary resistance	0.11Ω
f_o	Resonance frequency	80 kHz
f_s	Operating frequency	79 kHz $< f_o <$ 90 kHz
K	Coupling coefficient	0.05–0.4

The output AC resistance is proportional to the output DC resistance, R_{dc}, i.e. [38] :

$$R_L = \frac{8}{\pi^2} R_{dc}. \tag{15.7}$$

At the resonance frequency ($\omega = \omega_o$), the input voltage and output voltage can be calculated as:

$$V_{\text{inv}} = r_1 I_p - j\omega_o M I_s, \tag{15.8}$$

$$V_L = -r_2 I_s + j\omega_o M I_p. \tag{15.9}$$

Considering $V_L = R_L I_s$, the input and output current is determined as [36] :

$$I_s = \frac{\omega_o M I_p}{R_L + r_2}, \tag{15.10}$$

where,

$$I_p = \frac{V_{\text{inv}}}{r_1 + \frac{(\omega_o M)^2}{R_L + r_2}}. \tag{15.11}$$

As a result, the transferred power and the efficiency are calculated as:

$$P_L = R_L \cdot I_s^2, \tag{15.12}$$

$$\eta_{s-s} = \frac{R_L}{r_2 + R_L + \frac{r_1 \cdot \left[(r_2+R_L)^2 + (L_s \cdot \omega_o - 1/C_s \cdot \omega_o)^2\right]}{(\omega_o \cdot K \cdot L_s)^2}}. \tag{15.13}$$

The power transfer and system efficiency depend on the load, mutual inductance, and resonance frequency. The studied WPT system specifications are demonstrated in Table 15.1 [36].

Figure 15.6 represents the efficiency in terms of the frequency for different loads [36]. It is found that for all output loads, the maximum efficiency is obtained in a frequency at or close to the resonance frequency, which is considered f_o = 80 kHz in this chapter [36]. Figures 15.7 and 15.8 show the power transfer and the efficiency in terms of the load and coupling coefficient under the resonance frequency [4]. It is evident that the efficiency increases as the coupling coefficient increases. However, this is not true for power transfer. The output power equation can be written in another form under a resonance frequency as:

$$P_L \cong \frac{(M\omega_o I_p)^2}{R_L} \propto \frac{1}{M^2}. \tag{15.14}$$

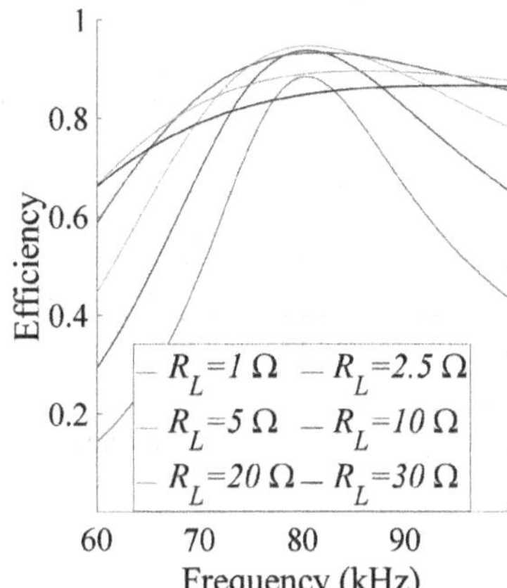

Figure 15.6 WPT system efficiency versus operating frequency for different values of output load resistance [36].

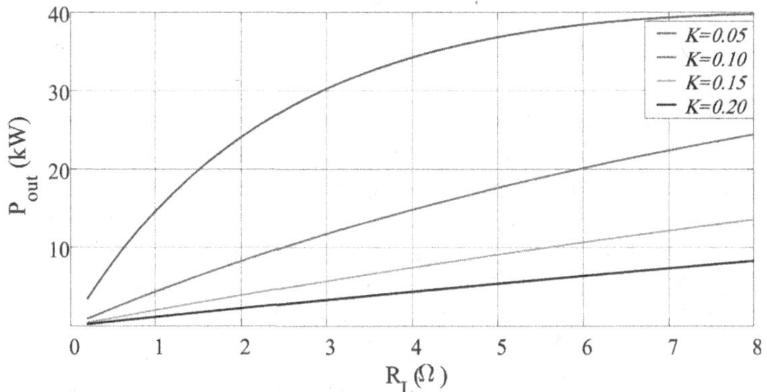

Figure 15.7 The resonant WPT versus the load for different values of coupling coefficient [4].

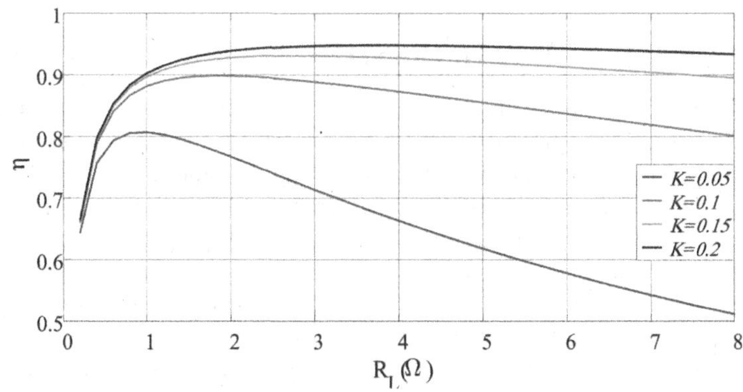

Figure 15.8 The resonant WPT system efficiency versus the load for different values of the coupling coefficient [4].

Therefore, for the series-series compensation circuit, the output power decreases as coupling increases. This feature is desirable for application with low-coupling coefficient like static and dynamic charging of electric vehicle, where the coupling is usually in the range of K = 0.05–0.3 according to the international standard SAE-j2954 for light-duty plug-in/electric vehicles.

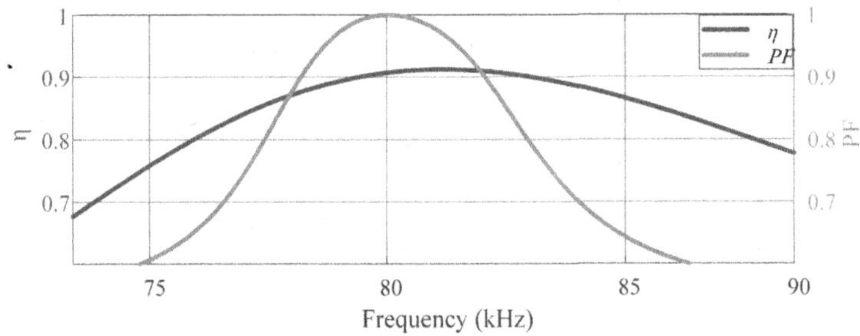

Figure 15.9 WPT system efficiency and power factor of the produced high-frequency AC power in terms of frequency [4].

The resonance frequency of the primary and secondary sides is often considered identical. In this situation, applying the resonance frequency leads to a maximum power transfer efficiency, reduced inverter switching losses, and unity power factor operation. The power factor can be defined for this system as:

$$PF = \frac{P_{in}}{S_{in}} = \cos\left(\tan^{-1}\left(\frac{\text{Im}(z_{in})}{\text{Re}(z_{in})}\right)\right), \tag{15.15}$$

where, $\text{Im}(z_{in})$ and $\text{Re}(z_{in})$ are the imaginary part and the real part of the total reflected impedance, z_{in}, respectively. Also, P_{in} and S_{in} are the active power and apparent power of the primary inverter, respectively. For a certain transferred power, high input power factor causes a reduction of VA rating of the WPT system components including the power electronic switches and the bridge diodes [33] or increases the transferred power for a given rating of the power supply [7]. When the WPT system operates under the resonance frequency, the phase difference between the inverter voltage and current is zero. In this situation, the PF is unity for the high-frequency AC power generated by the primary inverter. This is because the total reflected impedance seen from the primary side z_{in} is a pure resistance at the resonance frequency. Figure 15.9 indicates the efficiency and PF in terms of frequency [4]. It is seen that at the resonance frequency (fo=80 kHz), PF = 1, while the maximum efficiency occurs at a frequency near the resonance one, but not exactly at that frequency.

15.5 DYNAMIC WIRELESS POWER TRANSFER

The demand for EVs and hybrid electric vehicles (HEVs) is increasing due to their higher energy efficiency and lower greenhouse emissions [34]. Charging of these vehicles is of great importance. Plug-in HEVs charged by a connection to the grid through a cable have become more popular during the last two decades due to their superior performances with respect to conventional HEVs. However, this method does not provide ultimate vehicle safety due to an electrical contact between vehicle and grid, which may cause sparks or electrical shocks [11]. In addition, frequent cable charging may be regarded as inconvenient for drivers. Recently, stationary wireless power transfer is presented as an alternative to cable charging, though major problems still remain intact, which are associated with the EV battery itself. Firstly, the vehicle on the road must reach a charging station to charge its battery pack as the main energy storage. Since access to such a station is limited, the vehicle must use batteries with high storage capacity. This increases the total volume, weight, and cost of the vehicle. The dynamic wireless power transfer (DWPT), which is presented in this chapter, fundamentally overcomes the present EV charging issues. Using DWPT systems, EVs and HEVs can be charged with no need to be at a particular stop point, such as a parking lot and a traffic light. Instead, the vehicles can be charged dynamically, i.e., via an on-road power track while moving on the road. In this way, the battery capacity as the back-up energy source reduces up to 80% in comparison with the static WPT system [39].

15.5.1 DWPT SYSTEM

In wireless charging of moving EVs, the power request is dynamically met via the primary coil, sometimes called the primary track, which is placed under the surface of the road, and a secondary coil underneath the vehicle as seen in Figure 15.10 [36]. There has been much attention to the feasibility of on-road dynamic charging of EVs [8]. The recent relevant studies indicate the dynamic charging of in-motion vehicles as a hot research topic [35,40–43] .The corresponding infrastructure can be installed along the main traffic lanes. The EV gets energy from the on-road WPT system and therefore a greater driving range can be achieved. Moreover, a smaller battery needed to be installed in the EV to provide the same driving range. As a result, such a system can be a pathway to overcome the bottlenecks of electric mobility, i.e., the limited driving range and the high cost, which are both related to the technology and the specifications of today's batteries [44]. Nevertheless, the main challenge of DWPT is to provide a trade-off between the back-up battery capacity and the percentage of the road side covered by the power track so that the total system cost is minimized.

15.5.2 DWPT ANALYSIS

In DWPT applications, the on-board backup battery is connected in parallel with the WPT system and the EV motor drive supplies the EV's power bus when the on-road WPT supply is not in access. Therefore, it is necessary to regulate the output voltage of the WPT system at the nominal voltage of the battery. Otherwise, the voltage ripple is translated to the current ripple causing excessive heating and battery life reduction. In such applications, the output load and power demand changes due to the driving conditions, road slope, vehicle speed and acceleration. Also, the coupling coefficient between the primary on-road coil and secondary on-board coil varies according to misalignment or changes in the air gap. So, it is essential to ensure high efficiency and output voltage regulation against the load and coupling coefficient variations, which affect the mentioned criteria. Figure 15.11 shows the efficiency of the DWPT system under the optimum frequency versus output power for different values of coupling coefficient and load in a WPT system without any converter at the secondary side [36]. Each value of the power corresponds to two different load resistances. For each coupling coefficient, the efficiency is maximized at a unique optimum load, which will be explained in the next section.

A 3-D curve of the output voltage in terms of the load and coupling coefficient under an optimum frequency is depicted in Figure 15.12 [36]. Frequent changes in the load and coupling coefficient keep the DWPT system away from operating at the optimum point. Therefore, using a closed-loop control system is evitable in this system with continuous variation of driving conditions. It will be presented after investigating the design considerations.

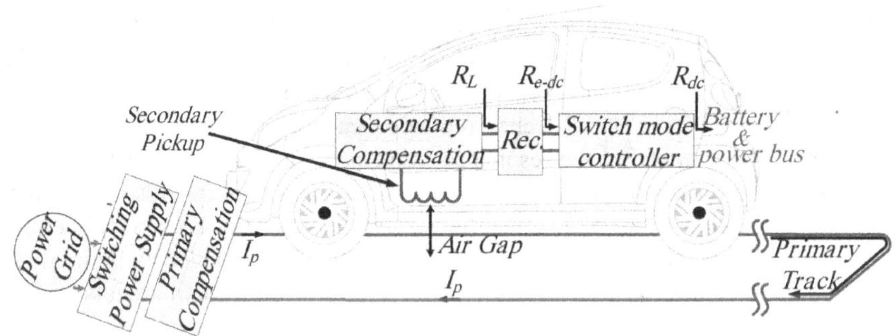

Figure 15.10 Structure of a DWPT system with on-road coil [36].

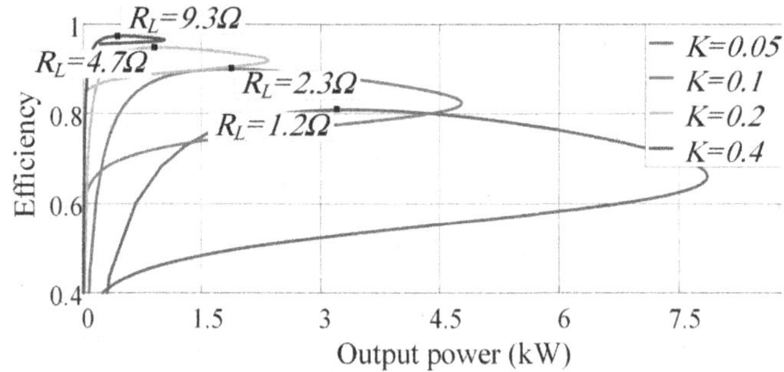

Figure 15.11 The efficiency for different power level at some coupling coefficients within the standard range [36].

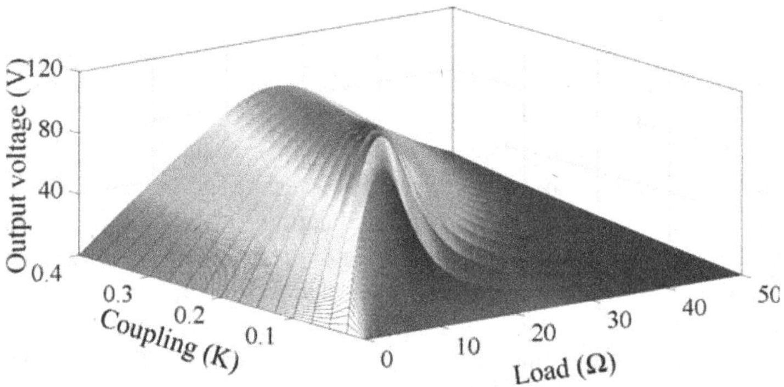

Figure 15.12 Output voltage versus the coupling and terminal load variations under the optimum operating frequency [36].

15.5.3 DESIGN CONSIDERATIONS

In this section, a practical design of wireless power transfer systems with a high efficiency is presented. The main objective is to find the compensating capacitors and the supply frequency. From the electromagnetic perspective, there are some design considerations for the primary transmitter and the secondary receiver coil structures to improve the utilization of the magnetic core and reduce the construction cost and magnetic coil usage [45]. In addition, with two-phase, three-phase and four-phase receivers introduced in [46], the designed fluctuation factor can be reduced from 1.0 to 0.29, 0.13, and even 0.08.

Compensating capacitors of the primary and the secondary side rarely change once the system is designed. However, the primary and secondary inductances as well as the mutual inductance may change due to the air gap variations and load operating conditions. As a result, the efficiency may significantly change. In practice, an efficiency reduction can be prevented by tracing the system behavior and adjusting and applying the resonance frequency. In this section the effects of variations of different parameters on the efficiency are investigated and desirable resonance frequencies together with the corresponding resonant capacitors are determined for both sides' compensating circuits. An algorithm is developed, as presented in Figure 15.13, to achieve an optimal resonance frequency for a high efficiency [47]. At first, the initial values of the system are taken. Then, a set of constraints is considered including the limits of inductances, resonant capacitors, and frequency. Next, using the efficiency equation, the system efficiency is obtained for different resonant parameters. The system frequency concurrently changes with the change in capacitors and inductances.

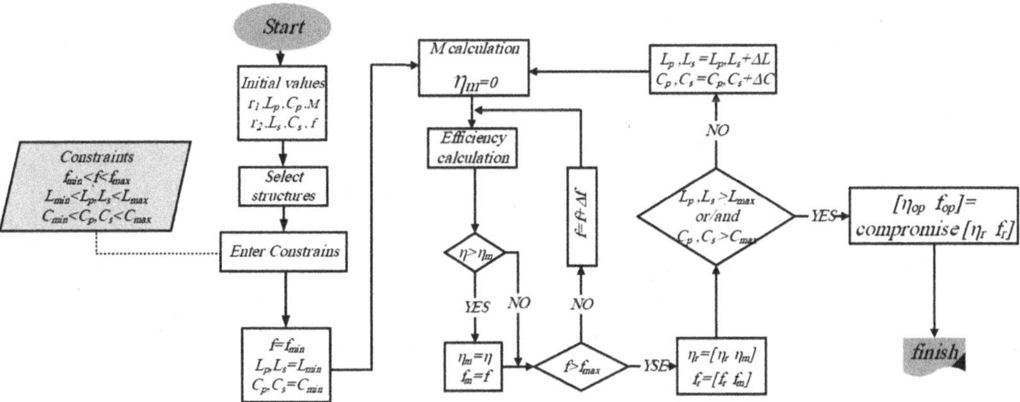

Figure 15.13 Flowchart of maximum efficiency tracking algorithm based on the tuning of resonant component in series-series WPT systems [47].

The resonance frequency and the related efficiency are recorded at each stage of the algorithm. Finally, the efficiency values and resonance frequencies are compared and a proper frequency with its corresponding efficiency is determined.

Recently, an international standard is defined for wireless charging of electric vehicle denoted as "SAE-J2954" [48]. The standard introduces an operating frequency range of 79–90kHz [48]. Also, there are some standard values for the primary and secondary pad dimensions, self-inductance value, and output voltage range depending on the power level and class type, which is related to the air gap between the coils. In addition, equations of Table 15.2 are introduced in order to determine design boundary conditions [49].

In Table 15.2, ω_n and D are per unit inverter frequency and its duty cycle, respectively. Q_s and Q_{smin} denote a load quality factor and its minimum value, respectively. G_{ivsys} and G_{vsys} are the transconductance gain and the voltage gain of the system, respectively. ω_{cvL} and ω_{cvH} denote the per unit

Table 15.2
Equation of Boundry Conditions [49]

No	Symbol	Quantity	Value
1	The required charging current	$G_{\text{ivsys}}(D=1\text{and}\omega_n) > \eta_{iv} \times K_{cc}$	$0 < L_p < \frac{64}{\pi^4 \eta_{iv}{}^2 K_{cc}^2 k^2 \omega_0^2 L_s}$
2	The required charging voltage	$G_{\text{vsys}}(\omega_{\text{cvL}} or \omega_{\text{cvH}} \text{ and } D=1) = \sqrt{L_s/L_p}$	$\frac{L_s}{K_{cv\min}^2} > L_p > \frac{L_s}{K_{cv\max}^2}$
		$K_{cv\max} > \sqrt{L_s/L_p} > K_{cv\min}$	
3	The minimum value of quality factor	$Q_s(\omega_n = 1) > Q_{s\text{min}}$	$L_s > \frac{8R_L Q_{s\text{min}}}{\pi^2 \omega_n \omega_0}$
4	The secondary resonant voltage and current	$I_s < I_{s\max}$	$L_s < \frac{2\sqrt{2}\omega_n V_{cs\max}}{\pi I_s \omega_0}$
		$V_{cs} < V_{cs\max}$	
5	The primary resonant voltage and current	$I_p(\omega_n = \omega_{cvH}) < I_{p\max}$	$L_p > \frac{I_s^2(64R_L^2\omega_n^2 + \pi^4\omega_0^2 L_s^2(\omega_n^2-1)^2)}{8\pi^2 k^2 \omega_0^2 L_s \omega_n^4 I_{p\max}^2}$
		$V_{cp}(\omega_n = \omega_{cvH}) < I_{cp\max}$	$L_p < \frac{8\pi^2 k^2 \omega_0^2 L_s \omega_n^6 V_{cp\max}^2}{I_s^2(64R_L^2\omega_n^2 + \pi^4\omega_0^2 L_s^2(\omega_n^2-1)^2)}$

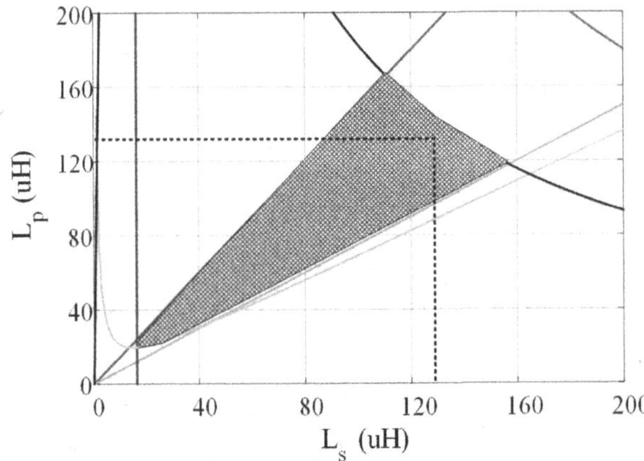

Figure 15.14 The eligible area for primary and secondary self-inductances to ensure Table 15.2 considerations.

frequencies which provide a unity value for G_{ivsys}. K_{cc} and K_{cv} are the required trans-conductance gain and voltage gain, respectively. The factor of the current allowance is shown by η_{iv}. I_{pmax}, V_{c1max}, and V_{c2max} denote the primary maximum resonant current, the primary maximum resonant voltage and the secondary maximum resonant voltage, respectively. Equations of Table 15.2 are plotted in Figure 15.14. The figure shows the acceptable values for the primary and secondary coil inductances, L_p and L_s.

15.6 OPTIMAL PERFORMANCE OF DWPT SYSTEMS

The driving conditions of an EV including the road slope, and vehicle speed and acceleration change during the dynamic charging of the EV. As a result, the power demand varies. In addition, the alignment situation of the vehicle with respect to the charging track affects the coupling coefficient between the primary and secondary coils during the movement. These variations influence the system performance including efficiency and power transfer capability. So, it is essential to always have the system to perform optimally despite the variations of power demand and coupling coefficient. The power demand is modeled by the AC resistance, R_L, in mathematical analysis.

15.6.1 MAXIMUM EFFICIENCY AND OPTIMAL FREQUENCY

In general, the efficiency of a WPT system is calculated by dividing the output DC power, P_{out}, by the input DC link power, P_{in}. Thus, for an IPT system with a series-series compensating circuit, the efficiency is given by [36]:

$$\eta_{s-s} = \frac{R_L}{r_2 + R_L + r_1 \left|\frac{I_P}{I_S}\right|^2}, \tag{15.16}$$

where,

$$\left|\frac{I_P}{I_S}\right| = \frac{\sqrt{(r_2 + R_L)^2 + (\omega_s L_s - \frac{1}{\omega_s C_s})^2}}{\omega_s M}. \tag{15.17}$$

It is evident from Eq. (15.17) that the efficiency becomes a maximum when Eq. (15.17) becomes a minimum. On the other hand, the maximum efficiency occurs at a frequency equals to the secondary side resonance frequency [2,13,14]. Substituting the resonance frequency of $f_o = f_s = \frac{1}{2\pi\sqrt{L_s C_s}}$ in Eq. (15.17) reduces the numerator to a minimum. However, the denominator increases with an increasing frequency beyond the resonance frequency, thus contributing to a maximum efficiency.

Figure 15.15 shows the maximum efficiency for a resonance frequency of f_o = 80 kHz within the standard range. It is seen that the maximum efficiency occurs at a frequency higher than the resonance frequency.

Therefore, a resonance frequency may not be an overall solution for maximizing the efficiency. The argument can further be elaborated by substituting Eq. (15.17) into Eq. (15.16). This yields:

$$\eta_{ss} = \frac{R_L}{r_2 + R_L + \frac{r_1 \cdot \left((r_2+R_L)^2 + (L_s \cdot \omega_s - 1/C_s \cdot \omega_s)^2\right)}{(\omega_s \cdot M)^2}}. \tag{15.18}$$

A solution for the optimum frequency is introduced here to overcome the problems associated with the selection of the resonance frequency. The proposed operating frequency is obtained by solving the derivative of Eq. (15.18) with respect to frequency, i.e.:

$$\frac{d\eta_{s-s}}{d\omega_s} = 0. \tag{15.19}$$

Substituting Eq. (15.18) into Eq. (15.19) yields:

$$(R_L + r_2)^2 + \left(L_s\omega_s - \frac{1}{C_s\omega_s}\right)^2 + \left(\frac{1}{C_s} - L_s\omega^2\right)\left(L_s + \frac{1}{C_s\omega^2}\right) = 0. \tag{15.20}$$

The value of optimal frequency is then determined from Eq. (15.20) as [36]:

$$f_{\text{opt}} = \frac{1}{2\pi\sqrt{L_sC_s - \frac{1}{2}(C_s(r_2 + R_L))^2}}. \tag{15.21}$$

This optimum operating frequency provides higher efficiency in comparison with the resonance frequency as illustrated in Figure 15.16 [36]. The efficiency improvement increases at higher

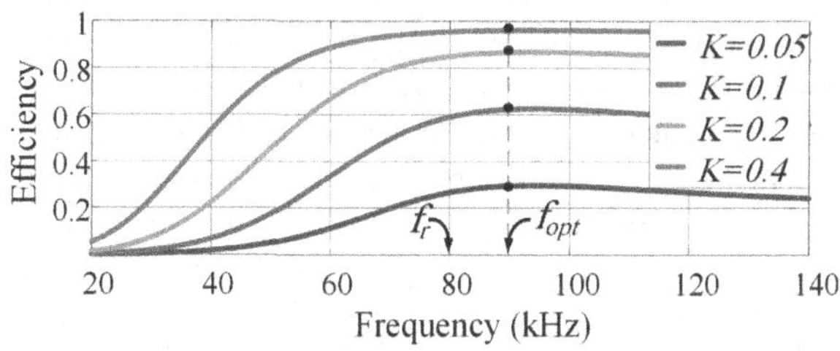

Figure 15.15 WPT system efficiency in terms of operating frequency for various coupling coefficients [36].

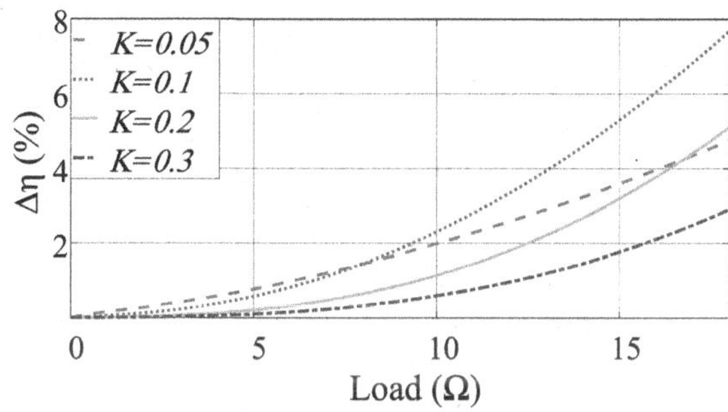

Figure 15.16 Comparison of IPT system efficiency versus load under the resonance frequency and the optimal frequency at different coupling coefficients [36].

load values. According to Eq. (15.21), the optimum frequency is always higher than the resonance frequency. It is stated in Section 15.5 of this chapter that the maximum possible efficiency is achieved under a specific output load for any coupling coefficient. Substituting Eq. (15.21) into Eq. (15.18), the efficiency is obtained in terms of the load resistance as:

$$\eta_{\text{SS-Max}} = \frac{R_L}{r_2 + R_L + \frac{r_1\left(L_sC_s(r_2+R_L)^2 - \frac{1}{4}C_s^2(r_2+R_L)^4\right)}{K^2 L_s{}^2}}. \tag{15.22}$$

A derivative of Eq. (15.22) with respect to R_L can be solved to yield the optimum value of the output load resistance. The derivative equation is presented as:

$$\frac{\partial \eta_{\text{SS-Max}}}{\partial R_L}\Big|_{R_L=R_{\text{opt}}} = 0. \tag{15.23}$$

Now, substituting Eq. (15.22) into Eq. (15.23) yields the optimum load resistance as [36]:

$$R_{\text{opt}} = M\omega_o\sqrt{r_2/r_1} = K\sqrt{(L_s \cdot r_2)/(C_s \cdot r_1)}. \tag{15.24}$$

It is evident that the optimum load depends on the coupling coefficient. Thus the optimum load changes with coupling variation in DWPT systems. So, an optimum load tracking is considered as a means for maintaining maximum efficiency. Figure 15.17 shows the efficiency in terms of load and coupling coefficient [36]. The figure illustrates the optimum load for different coupling coefficients. It also shows that the maximum efficiency increases with an increasing coupling coefficient. This is in agreement with the analytical results presented in Eq. (15.24).

15.6.2 LOAD MATCHING

It is mentioned before that the output load and coupling coefficient may change continually in DWPT applications. On the other hand, the system has maximum possible efficiency under a unique output load for each coupling coefficient. Therefore, it is possible to match the reflected impedance seen from the secondary side to the optimum one to maximize the efficiency at each coupling coefficient. A DC/DC converter located between the load and the rectifier of the secondary side can do the job. The converter duty cycle is adjusted to match every load to the optimum one. Indeed, the turn-on time of the switch converter is tuned for this purpose. Because the output load and the

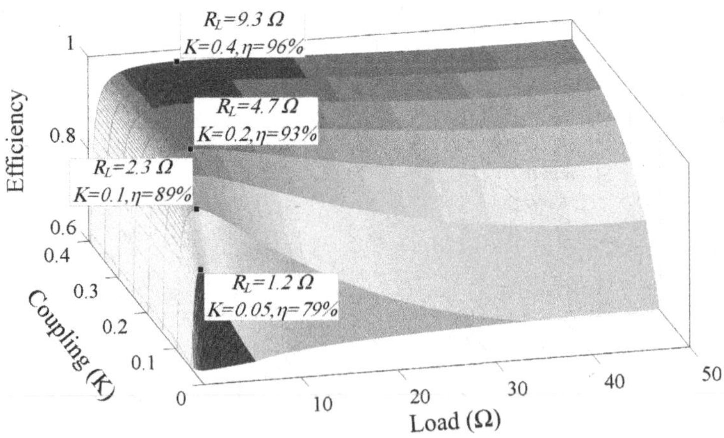

Figure 15.17 A 3-D plot of WPT system efficiency in terms of driving cycle variations [36].

coupling coefficient, and the corresponding optimum load change, the duty cycle must be adjusted continuously during the motion. Figure 15.18 shows the common topologies of such a DC/DC impedance matching converter together with the related conversion factor.

Here, the buck converter is investigated. The reflected impedance by this converter can be obtained as:

$$R_L = \frac{8}{D_r{}^2 \pi^2} R_{dc}. \tag{15.25}$$

On the other hand, the optimum load is determined according to Eq. (15.24). Then, the optimum duty cycle to achieve the optimum equivalent load is calculated as:

$$D_{r-\text{opt}} = \frac{1}{\pi} \cdot \sqrt{8R_{dc}/M\omega_o\gamma}, \tag{15.26}$$

where,

$$\gamma = \sqrt{r_2/r_1} \tag{15.27}$$

It is seen in Eq. (15.26) that the optimum conversion ratio depends on M and thus on the coupling coefficient, which is not fixed in DWPT applications. Indeed, the changes in the road slope or the misalignment between the primary and secondary coils change the coupling coefficient.

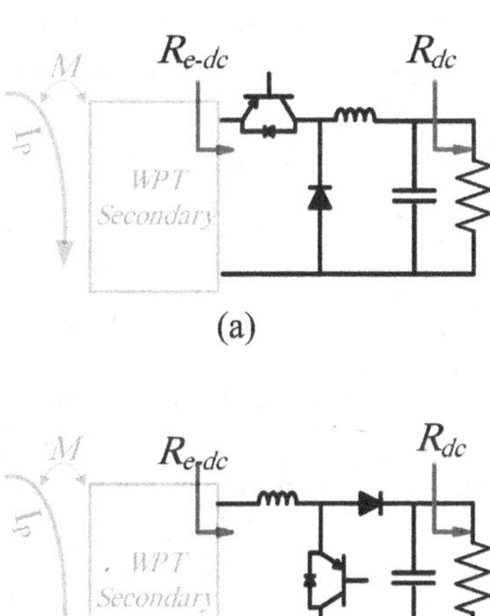

(a)

(b)

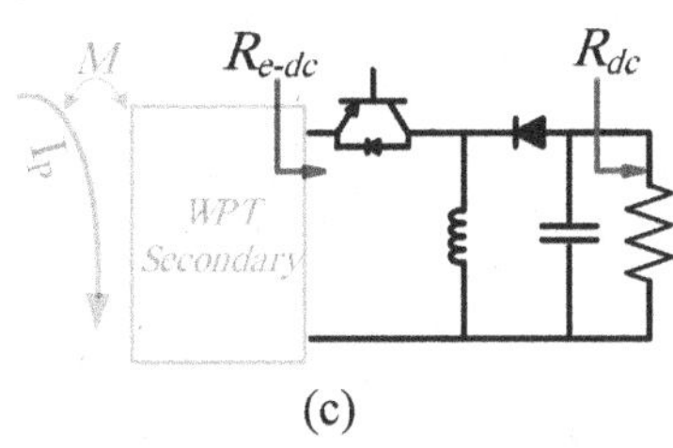

(c)

Figure 15.18 Different types of DC/DC impedance matching converters used in the secondary side of WPT systems; (a) Buck converter $R_{e-dc} = \frac{R_{dc}}{D_r{}^2}$, (b) Boost converter $R_{e-dc} = R_{dc}(1-D_r)^2$, (c) Buck-Boost converter $R_{e-dc} = R_{dc}\left(\frac{1-D_r}{D_r}\right)^2$.

15.7 CONTROL SYSTEM

A closed-loop control for DWPT systems is inevitable to ensure the optimum operating conditions of the system, because there are frequent changes of the power demand and coupling coefficient during the charging state. Achieving a high efficiency of power transfer and maintaining a constant voltage of the power bus at the output of the secondary side are considered as the main criteria for fulfilling an optimum operation. Thus, the closed-loop control must maximize the efficiency and regulate the bus voltage upon the changing power demand (load) and coupling coefficient. The voltage regulation is particularly needed because severe voltage variations could damage the battery [12]. In this section, it is intended to investigate the voltage regulation and the maximum efficiency control separately as much as possible for the sake of clarity. Nevertheless, they are interrelated and the DWPT control system performs these two tasks simultaneously. In fact variations of the system load and coupling coefficient affect both the output voltage and the system efficiency. There are some solutions for voltage regulation and maximum efficiency control. Two tracking control methods are presented here. The first one does not use any parameters of the compensating circuits, whereas the second method includes an estimation scheme for the coupling coefficient.

15.7.1 OUTPUT VOLTAGE REGULATION

The voltage of the power bus at the output of the DWPT system should be regulated to prevent damage to the EV battery, which is installed in parallel with the power bus. Figure 15.4 is referred to investigate the voltage regulation [21]. The fundamental harmonic of the inverter voltage and the output voltage under the optimum frequency are obtained as [36]:

$$V_{\mathrm{inv}(1)} = (4/\pi) \cdot V_{\mathrm{link}} \cdot \sin(\pi D/2), \tag{15.28}$$

$$V_L = R_L \omega K L V_{\mathrm{inv}(1)} / |Z_p| \sqrt{R_L{}^2 + \left(\omega L - \frac{1}{\omega C}\right)^2}, \tag{15.29}$$

where V_{link} is the DC link voltage of the primary side. V_L depends on the system frequency, inverter duty cycle (D), coupling coefficient, and output load. Here, D is defined as a control variable regulating the output voltage at a certain value upon the variations of the load and coupling coefficient. Figure 15.19 indicates the output voltage with and without voltage regulation, while the output load changes at $t = 0.6$ s [50]. The voltage regulation is performed by online adjustment of the duty cycle.

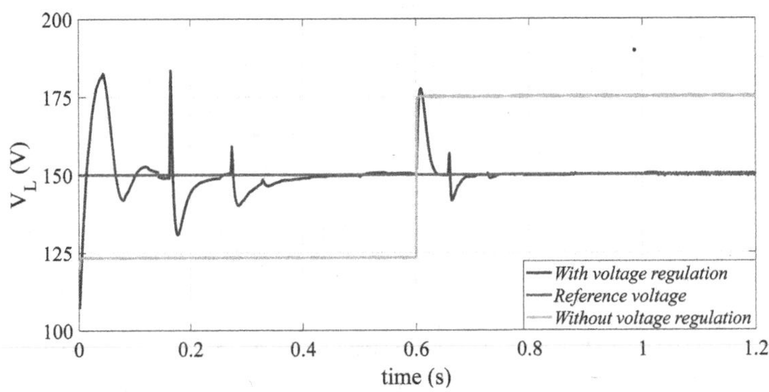

Figure 15.19 Output load voltage during the dynamic EV charging with variable driving conditions [50].

15.7.2 MAXIMUM EFFICIENCY ACHIEVEMENT

In high-power applications of WPT such as dynamic charging of EVs, achieving a high efficiency is more important than transferring a maximum power [10,11]. Ideally, the system should be able to maintain maximum efficiency under varying load, coupling coefficient, and system parameters, without severe constraints on the power transfer capability. Here, maximum efficiency tracking is presented by two methods: independent of the system parameters and with the estimation of the coupling coefficient.

15.7.2.1 Resonant Parameter Independent Method

The control system for this method is shown in Figure 15.20 [36]. It achieves three goals including:

1. *Tracking of Maximum Power Transfer Efficiency by an Optimum Load Matching Algorithm*: The conversion ratio of the secondary DC/DC converter is tuned to so that the resulted efficiency is obtained. The efficiency is monitored after each change in D_r. If the efficiency increases, D_r changes in the same direction. Otherwise, the direction is reversed. In this way, no information of the coupling coefficient is needed as the efficiency is always monitored.
2. *Adjustment of the Switching Frequency of the Inverter to Its Optimum Value to Follow the Maximum Possible Efficiency*: The optimum frequency is related to the equivalent output load. So, after each change in the equivalent load, the new value of the optimum frequency is calculated and applied to the primary side inverter to achieve the maximum possible efficiency.
3. *Regulation of the Output Voltage*: As it is mentioned before, the inverter duty cycle is adjusted to regulate the output voltage. This is done by means of a proportional-Integral (PI) controller. If the output voltage exceeds the reference one (battery nominal voltage), the duty cycle decreases. So, the resulted fundamental harmonic of the inverter voltage decreases according to Eq. (15.28). This causes an output voltage decrement. In this way, the output voltage would be fixed by a closed-loop negative feedback control loop.

The result of the efficiency tracking method is illustrated in Figure 15.21 for two cases, i.e., adjusting the operating frequency to the resonance frequency, and adjusting it to the optimum one [36].

At $t = 1.35$ s, the output load changes. As a result, the efficiency falls down, which makes the control system initiate a new search to find another optimum frequency and a new optimum conversion ratio corresponding to a maximum efficiency. The efficiency is always greater under an optimum frequency compared with the one under the resonance one.

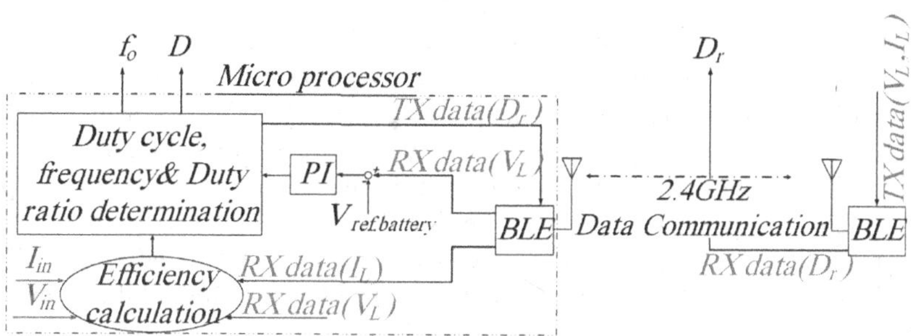

Figure 15.20 The schematic view of the control system to track maximum possible efficiency [36].

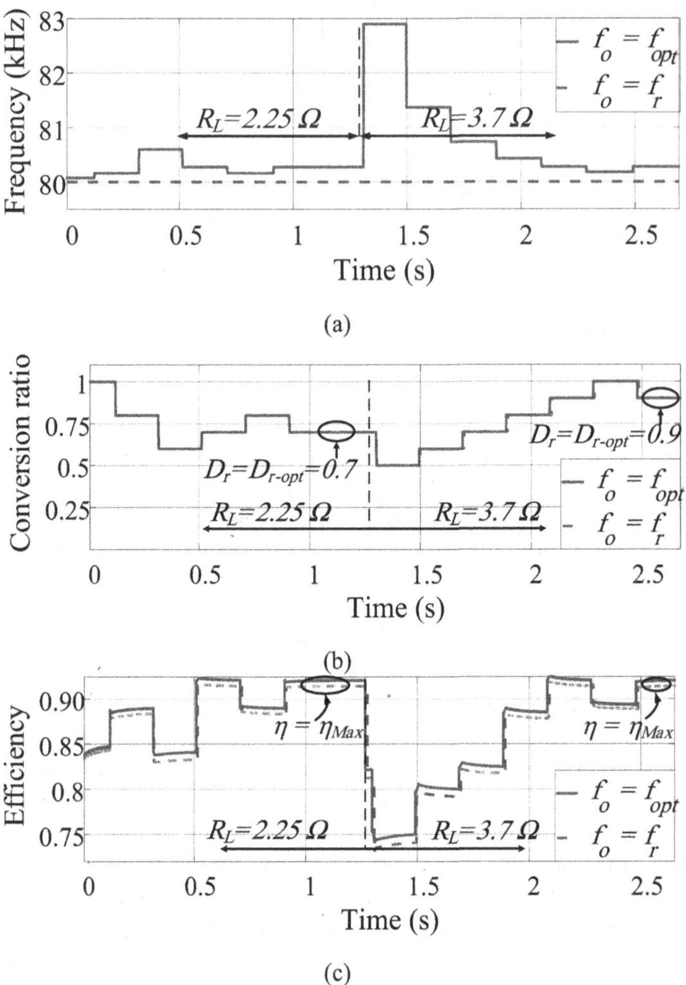

Figure 15.21 Maximum efficiency tracking method without coupling coefficient estimation for two systems, operating under resonance frequency and the optimal one; (a) switching frequency of the inverter, (b) DC/DC conversion ratio tracking, (c) efficiency variation [36].

15.7.2.2 Coupling Coefficient Estimation-Based Method

The mentioned tracking method makes the control system independent from the information of WPT system parameters. Nevertheless, applying the frequency steps in the search of the optimum frequency is time consuming. Furthermore, the proposed algorithm may get stuck in a local maximum efficiency instead of finding the global maximum one. Therefore, parameter estimation methods have been proposed to obtain the optimum operating point of the WPT system more accurately and quickly. As it was shown in Eq. (15.24), the optimum equivalent load resistance is a function of the mutual inductance and the system coupling coefficient. The information of these parameters are critical in dynamic wireless charging since they would vary by EVs movement on the charging track. Some articles have proposed estimation methods to find the mutual inductance [13]. Mutual inductance of the system can be estimated as follows, using KVL equations of the equivalent circuit model as [4]:

$$M \simeq \frac{V_{\text{inv}} - r_1 I_p}{\omega_o I_s}, \tag{15.30}$$

where, V_{inv}, I_p, and I_s are the RMS values of inverter output voltage, primary side current, and secondary side current, respectively. In this method, the system operating frequency is set at the common resonance frequency of the primary and secondary sides. The system efficiency is maximized and the power factor reaches unity under this frequency. The estimation method requires primary and secondary currents and DC link voltage measurements. However, Eq. (15.30) can be simplified as follows by neglecting the primary coil voltage drop.

$$M \simeq \frac{V_{\text{inv}}}{\omega_o I_s}. \tag{15.31}$$

Therefore, the measurement of the primary side current is canceled at the cost of less accuracy. While the mutual inductance is determined, the optimum duty cycle of secondary side Buck converter is calculated and applied to the PWM block to optimize the system efficiency. In addition, the primary-side inverter duty cycle is employed to regulate the load voltage. The control block diagram is depicted in Figure 15.22 [4]. Another method for the mutual inductance estimation needs merely load voltage and current measurements. This method depends on the load voltage regulation. In Section 15.7.1, the voltage regulation method by adjusting the inverter duty cycle was discussed (Figure 15.19). The required inverter duty cycle to maintain the load voltage at its nominal value can be written as [50]:

$$D = \frac{2}{\pi}\sin^{-1}\left(\frac{V_L}{V_{\text{link}}}\frac{r_1(R_L+r_2)+(\omega_o M)^2}{R_L\omega_o M}D_r\right), \tag{15.32}$$

whereas the load voltage regulation subsystem determines the duty cycle, Eq. (15.32) can be stated as:

$$(\omega_o M)^2 - \frac{V_{\text{link}}}{V_L}D_r R_L \sin\left(\frac{\pi}{2}D\right)(\omega_o M) + r_1(r_2+R_L) = 0, \tag{15.33}$$

The polynomial equation with a degree of two can be solved as [50]:

$$\omega_o M = \frac{V_{\text{link}}D_r R_L \sin(\frac{\pi}{2}D) + \sqrt{D_r^2 V_{\text{link}}^2 {R_L}^2 \sin^2(\frac{\pi}{2}D) - 4V_L r_1(r_2+R_L)}}{2V_L}, \tag{15.34}$$

It is notable that another solution to Eq. (15.34) is not acceptable since $\omega_o M$ becomes less than zero. A flowchart of the control method is shown in Figure 15.23 [50].

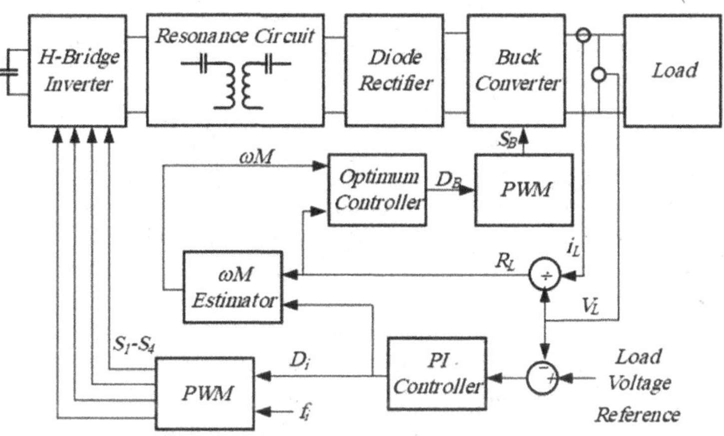

Figure 15.22 Control block diagram [4].

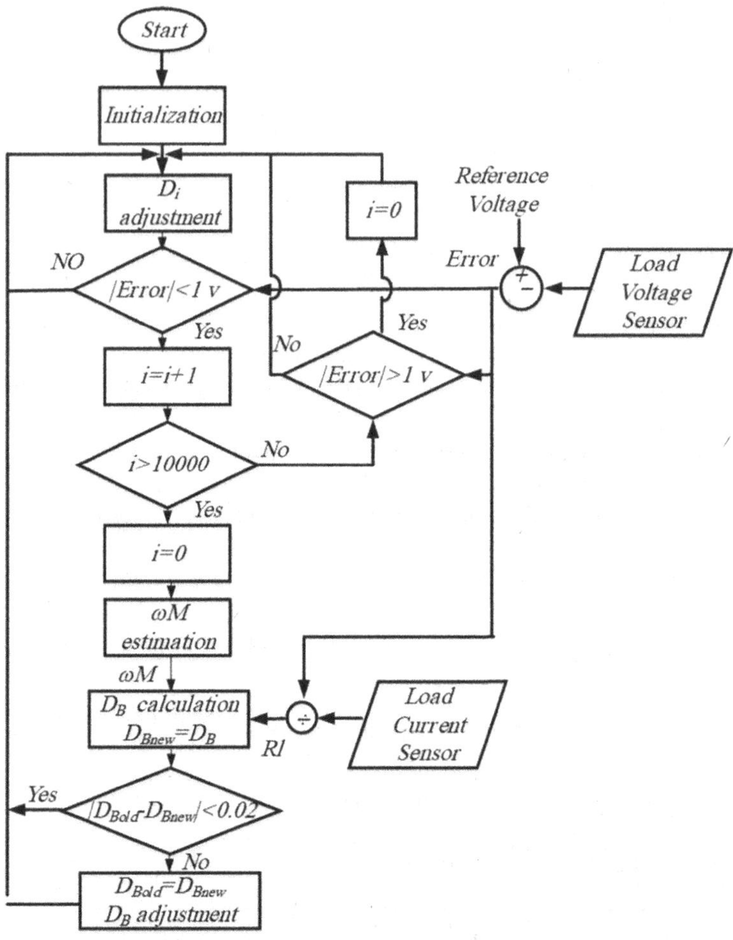

Figure 15.23 The flowchart of the parameter estimation method [50].

As it is shown in Figure 15.23, the load voltage regulation is applied first. The mutual inductance is estimated according to Eq. (15.34), when it is assured that the load voltage is settled at its reference. Therefore, the optimum Buck duty cycle is applied by the PWM subsystem. The changes in the Buck converter duty cycle vary the load voltage. Hence, it is needed to be regulated at its reference again. Therefore the inverter duty cycle is updated by the voltage regulation sub-system. It is recalled that in dynamic wireless charging the mutual inductance may vary frequently. This causes variations in the load voltage and optimum duty cycle of the Buck converter. Thus, the control algorithm monitors load voltage and mutual inductance continuously. The aforementioned methods are limited to work under a unity power factor. From Eq. (15.15), it can be seen that the operation with the common resonance frequency of the primary and secondary sides provides the unity power factor. Nonetheless, the primary and secondary side resonance frequencies may not be exactly the same in practice. In these cases, the unity power factor is realized by the system operation under neither the primary nor the secondary side resonance frequency [50]. Another estimation method has been proposed which can cope with inequality of both sides resonance frequencies [51]. The control method is capable of determining the secondary side circuit parameters, as well as the mutual inductance of the system. Therefore, the primary side circuit parameters are known and the mutual inductance and secondary side circuit parameters are unknown. From the equivalent circuit model, the mutual reactance can be stated as [51]:

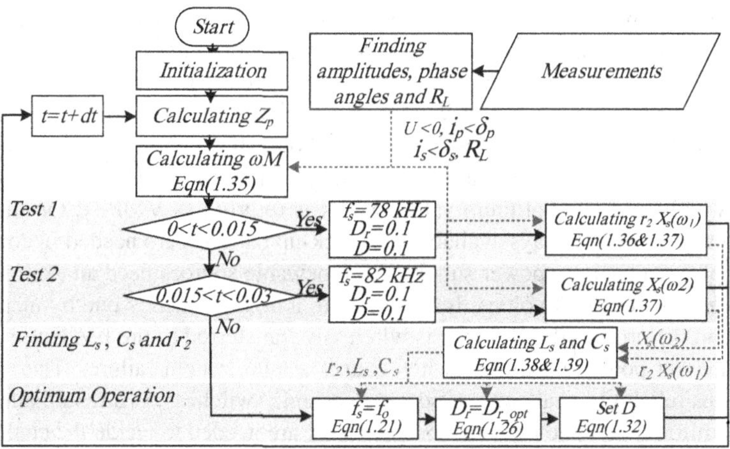

Figure 15.24 The third parameter estimation method [51].

$$X_M = \omega M = \frac{V_{\text{inv}} - i_p |Z_p| \cos(\delta_P - \delta_{ZP})}{i_s \sin(\delta_S)}, \tag{15.35}$$

where, δ_P, δ_S, and $\delta_Z P$ are the phase of the primary coil current, phase of the secondary coil current, and phase of primary side impedance, respectively. On the other hand, the secondary impedance of the system can be decomposed into its real and imaginary parts as:

$$r_2 = -R_L - X_M \left(\frac{i_P}{i_S}\right) \sin(\delta_P - \delta_S), \tag{15.36}$$

$$X_S = \left(L_S\omega - \frac{1}{C_S\omega}\right) = X_M \left(\frac{i_P}{i_S}\right) \cos(\delta_P - \delta_S), \tag{15.37}$$

where X_s is the secondary side reactance in which, L_s and C_s are the two unknown parameters in one equation. Thus, they can be determined by applying two test frequencies denoted as ω_1 and ω_2 at the beginning of charging time. By solving the non-linear equations L_s and C_s can be written as [4]:

$$C_S = \frac{\frac{\omega_1}{\omega_2} - \frac{\omega_2}{\omega_1}}{\omega_1 X_S(\omega_1) - \omega_2 X_S(\omega_2)}, \tag{15.38}$$

$$L_S = \frac{\frac{X_S(\omega_1)}{\omega_2} - \frac{X_S(\omega_2)}{\omega_1}}{\frac{\omega_1}{\omega_2} - \frac{\omega_2}{\omega_1}}. \tag{15.39}$$

A control system flowchart is depicted in Figure 15.24 [51]. In the two short time period two test frequencies are applied to the system at the beginning of the operation. While L_s and C_s are determined, the secondary resonance frequency can be calculated. This is the optimum operating frequency of the system. The optimum duty cycle of the Buck converter is calculated by the obtained r_2 and X_m. Furthermore, the proper duty cycle of the inverter can be obtained directly according to Eq. (15.32) and applied to the inverter PWM block. The control system monitors the value of X_m continuously in order to keep the system in its optimum operating point during the remaining charging period.

15.8 FUTURE OUTLOOK

Dynamic charging of electric vehicles is a new technology with many opportunities; some of them are briefly reviewed in this section. The technology can be used together with renewable power generation to provide a basis for total sustainable systems form the supply to demand. In fact the need for a power source along the whole road provides the opportunity of using the photovoltaic (PV) power sources as input power supplies, especially where the power grid is not readily accessible. Figure 15.25 shows the general view of an EV wireless charging system with a PV source aside the road [52]. The source needs to work at optimal operating conditions to produce maximum power. In PV power generation, the maximum power point tracking (MPPT) fulfills this task by adjusting the PV voltage to its optimum value, which is referred as VMPPT. On the other hand, since the sunlight may not be always available, the back-up batteries are needed in connection with a PV power supply as a back-up power supply. All renewable sources need an exclusive converter to meet the power bus. The link voltage is converted to a high frequency one by means of a high-power inverter and the produced AC power is wirelessly transferred to the EV. But, in this system, a defect in an inverter power electronic switch causes a total system failure. Therefore, the system faces a serious reliability challenge of power electronic switches. As a result, alternative WPT topologies, in addition to reliable inverters and switches, are needed to tackle the challenge. On the other hand, using extra switching converters for the back-up battery source and MPPT in such WPT systems increases the system cost and control complexity. There are new solutions, other than the conventional resonant converters to overcome the problem. The flyback-based inverter is proposed recently, which is compatible with the photovoltaic input source [52]. The photovoltaic standards can be fulfilled by the proposed topology and applied to DWPT systems.

An electric motor drive of DC or AC type is the main component of any EV, which can be fed through a DWPT system. In such an application, the power is transferred to the motor drive through a primary side inverter, the magnetic coupling, and a passive/active rectifier on the secondary side of the system. In this way, a DC-link isolates the WPT system from the motor drive [9,13]. A DC-link electrolyte capacitor with a high degrade factor is thus used. Finally, the DC-link voltage supplies the motor inverter as a drive. With this structure, the whole system requires two inverters, one in the primary side, and the other as a part of the motor drive. Each inverter requires a capacitor filter with its own cost and operating constraints [53]. Therefore, the total system suffers from structure complexity and cost deficiency, in addition to reduced reliability and power transfer efficiency. Thus, eliminating one inverter with its DC electrolyte capacitor and other related elements helps to achieve system simplicity and cost-saving [54]. This solution can be done by using a wireless motor drive system in which an AC motor is derived without needing an inverter on the secondary side.

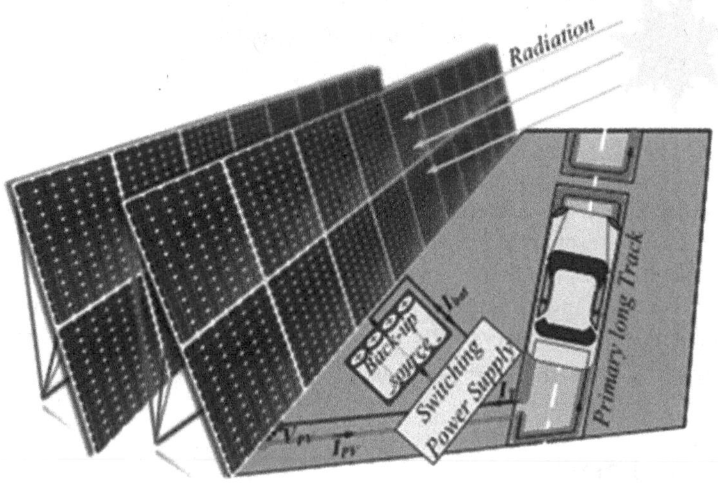

Figure 15.25 On-road primary track supplied by a PV module and a back-up source in an EV dynamic charging application [52].

In other words, the motor is straightly driven by the primary side inverter, wirelessly. Figure 15.26 shows a schematic view of the conventional wireless motor drive system and the one with a single inverter [54].

In the system with one inverter, a simple secondary switching converter replaces the secondary inverter. It can be driven without requiring any communication to the primary side. In this situation, the motor is driven by the primary inverter, which also provides the high-frequency AC power to the coupled magnetic link. This structure will be very interesting from the industry point of view and has potential for further development.

By the advent of distributed generation (DG) of electricity, the power consumer roles can be combined with the one of the power provider, thus introduce power prosumers. In future smart cities, electric vehicles play the roles of prosumers [55,56]. As shown in Figure 15.27, not only they can be fed by the grid (G2V), but also they are able to feed the grid (V2G) or other vehicles (V2V). Thus, appropriate infrastructure is needed to make the roles possible. Dynamic wireless power transfer is a good solution through which electric vehicles can share the power with each other and with the grid via the common charging track which is located beneath the road [57]. To

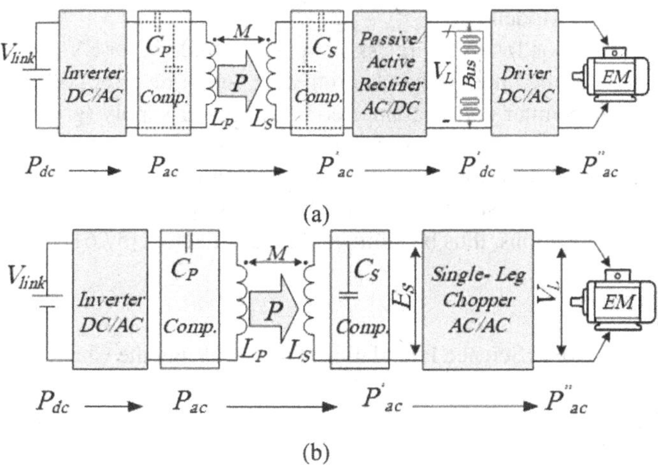

Figure 15.26 Supplying EV motor drives by DWPT systems; (a) conventional structure with two inverters, (b) wireless motor drive with only one inverter [54].

Figure 15.27 The future perspective of the smart cities comes up with WPT and communication technologies for V2G, G2V, and V2V power-sharing.

manage the power-sharing between the EVs and the grid, a secured monitoring service is needed to communicate with both the EVs and the grid.

At present, the biggest drawback of plug-in charging systems is the limited number of charging stations, which negatively affects people's desire to purchase pure battery electric vehicles. In addition, charging stations require regular maintenance and service to ensure the equipment is working properly. In the stationary state, a replacement of the plug-in by the wireless vehicle-to-vehicle (V2V) charging system can help to solve the above issues. This concept is applicable to the dynamic charging of EVs. When the vehicle's battery runs low, the driver can request the help of other EVs on the road to wirelessly charge the vehicle. The process involves several auxiliary technologies, such as vehicle-to-vehicle communication, GPS, and battery state of charge detection, etc. The driver can find suitable vehicles using an application on their smart device that indicates available energy donor vehicle, pricing, distance, and other factors. This concept can reduce the potential EV buyers' anxiety that an EV battery will run flat some distance from the nearest charging station [58,59].

In 2012, some automakers announced their plans to include wireless charging technology on future EVs. In 2014, Plug-less Power began offering WPT kits for the Chevy Volt, Nissan LEAF, and Cadillac ELR. In the years since, Plug-less has expanded its offering to include BMW i3, Mercedes S550e, and Tesla Model S [25].

Currently, the most common form of wireless charging technology for EVs includes a transmitter coil, which is embedded in the floor of the charging area, and a receiver coil that is embedded in the car's chassis. The transmitter coil is connected to the power supply (grid or storage), and the receiver coil is connected to the car's battery [58,59]. Nevertheless, the Smart City concept allocates a significant role in transport issues in the city [60]. According to this concept, EVs can actively reinforce the development of the smart grid if their charging processes are properly coordinated through two-way communications, thus benefiting all types of actors [57,61].

ACKNOWLEDGMENT

The support from Iran National Science Foundation (INSF) through the Chair of Wireless and Contactless Power Transfer is appreciated.

REFERENCES

1. M. Fu, H. Yin, X. Zhu, and C. Ma, "Analysis and tracking of optimal load in wireless power transfer systems," *IEEE Transactions on Power Electronics*, vol. 30, no. 7, pp. 3952–3963, 2014.
2. Y. Narusue, Y. Kawahara, and T. Asami, "Maximum efficiency point tracking by input control for a wireless power transfer system with a switching voltage regulator," in *2015 IEEE Wireless Power Transfer Conference (WPTC)*, Boulder, CO, pp. 1–4, IEEE, 2015.
3. S. Lukic and Z. Pantic, "Cutting the cord: Static and dynamic inductive wireless charging of electric vehicles," *IEEE Electrification Magazine*, vol. 1, no. 1, pp. 57–64, 2013.
4. A. Zakerian, S. Vaez-Zadeh, A. Babaki, and M. F. Moghaddam, "Efficiency optimization of a dynamic wireless EV charging system using coupling coefficient estimation," in *2019 10th International Power Electronics, Drive Systems and Technologies Conference (PEDSTC)*, Iran, pp. 629–634, IEEE, 2019.
5. F. Musavi and W. Eberle, "Overview of wireless power transfer technologies for electric vehicle battery charging," *IET Power Electronics*, vol. 7, no. 1, pp. 60–66, 2014.
6. M. G. S. Pearce, G. A. Covic, and J. T. Boys, "Robust ferrite-less double D topology for roadway IPT applications," *IEEE Transactions on Power Electronics*, vol. 34, no. 7, pp. 6062–6075, 2018.
7. W. Zhang and C. C. Mi, "Compensation topologies of high-power wireless power transfer systems," *IEEE Transactions on Vehicular Technology*, vol. 65, no. 6, pp. 4768–4778, 2015.
8. S. Jeong, Y. J. Jang, and D. Kum, "Economic analysis of the dynamic charging electric vehicle," *IEEE Transactions on Power Electronics*, vol. 30, no. 11, pp. 6368–6377, 2015.

9. S. Hasanzadeh and S. Vaez-Zadeh, "Design of a wireless power transfer system for high power moving applications," *Progress in Electromagnetics Research*, vol. 28, pp. 258–271, 2013.
10. T.-D. Yeo, D. Kwon, S.-T. Khang, and J.-W. Yu, "Design of maximum efficiency tracking control scheme for closed-loop wireless power charging system employing series resonant tank," *IEEE Transactions on Power Electronics*, vol. 32, no. 1, pp. 471–478, 2016.
11. S. Hasanzadeh, S. Vaez-Zadeh, and A. H. Isfahani, "Optimization of a contactless power transfer system for electric vehicles," *IEEE Transactions on Vehicular Technology*, vol. 61, no. 8, pp. 3566–3573, 2012.
12. U. K. Madawala, M. Neath, and D. J. Thrimawithana, "A power–frequency controller for bidirectional inductive power transfer systems," *IEEE Transactions on Industrial Electronics*, vol. 60, no. 1, pp. 310–317, 2011.
13. D. Kobayashi, T. Imura, and Y. Hori, "Real-time coupling coefficient estimation and maximum efficiency control on dynamic wireless power transfer for electric vehicles," in *2015 IEEE PELS Workshop on Emerging Technologies: Wireless Power (2015 WoW)*, Daejeon, pp. 1–6, IEEE, 2015.
14. W. Zhong and S. R. Hui, "Charging time control of wireless power transfer systems without using mutual coupling information and wireless communication system," *IEEE Transactions on Industrial Electronics*, vol. 64, no. 1, pp. 228–235, 2016.
15. H. Li, J. Li, K. Wang, W. Chen, and X. Yang, "A maximum efficiency point tracking control scheme for wireless power transfer systems using magnetic resonant coupling," *IEEE Transactions on Power Electronics*, vol. 30, no. 7, pp. 3998–4008, 2014.
16. D. Ahn and S. Hong, "Wireless power transfer resonance coupling amplification by load-modulation switching controller," *IEEE Transactions on Industrial Electronics*, vol. 62, no. 2, pp. 898–909, 2014.
17. M. Fu, T. Zhang, C. Ma, and X. Zhu, "Efficiency and optimal loads analysis for multiple-receiver wireless power transfer systems," *IEEE Transactions on Microwave Theory and Techniques*, vol. 63, no. 3, pp. 801–812, 2015.
18. T. C. Beh, M. Kato, T. Imura, S. Oh, and Y. Hori, "Automated impedance matching system for robust wireless power transfer via magnetic resonance coupling," *IEEE Transactions on Industrial Electronics*, vol. 60, no. 9, pp. 3689–3698, 2012.
19. O. C. Onar, J. M. Miller, S. L. Campbell, C. Coomer, C. P. White, and L. E. Seiber, "A novel wireless power transfer for in-motion EV/PHEV charging," in *2013 Twenty-Eighth Annual IEEE Applied Power Electronics Conference and Exposition (APEC)*, Long Beach, CA, pp. 3073–3080, IEEE, 2013.
20. H. L. Li, A. P. Hu, G. A. Covic, and C. Tang, "A new primary power regulation method for contactless power transfer," in *2009 IEEE International Conference on Industrial Technology held in Churchill*, pp. 1–5, , Victoria, Australia, IEEE, 2009.
21. R. Tavakoli and Z. Pantic, "Analysis, design, and demonstration of a 25-kW dynamic wireless charging system for roadway electric vehicles," *IEEE Journal of Emerging and Selected Topics in Power Electronics*, vol. 6, no. 3, pp. 1378–1393, 2017.
22. Y. Huang, C. Liu, Y. Zhou, Y. Xiao, and S. Liu, "Power allocation for dynamic dual-pickup wireless charging system of electric vehicle," *IEEE Transactions on Magnetics*, vol. 55, no. 7, pp. 1–6, 2019.
23. A. Kamineni, G. A. Covic, and J. T. Boys, "Self-tuning power supply for inductive charging," *IEEE Transactions on Power Electronics*, vol. 32, no. 5, pp. 3467–3479, 2016.
24. N. Tesla, "System of transmission of electrical energy", US Patent 645,576, March 20 1900.
25. A. P. Hu, Selected resonant converters for IPT power supplies. PhD thesis, ResearchSpace Auckland, 2001.
26. S. Hasanzadeh and S. Vaez-Zadeh, "A review of contactless electrical power transfer: Applications, challenges and future trends," *Automatika*, vol. 56, no. 3, pp. 367–378, 2015.
27. J. O. McSpadden and J. C. Mankins, "Space solar power programs and microwave wireless power transmission technology," *IEEE Microwave Magazine*, vol. 3, no. 4, pp. 46–57, 2002.
28. C. Liu, A. P. Hu, B. Wang, and N.-K. C. Nair, "A capacitively coupled contactless matrix charging platform with soft switched transformer control," *IEEE Transactions on Industrial Electronics*, vol. 60, no. 1, pp. 249–260, 2011.
29. A. Karalis, J. D. Joannopoulos, and M. Soljačić, "Efficient wireless non-radiative mid-range energy transfer," *Annals of Physics*, vol. 323, no. 1, pp. 34–48, 2008.
30. D. Kim, A. Abu-Siada, and A. Sutinjo, "State-of-the-art literature review of WPT: Current limitations and solutions on IPT," *Electric Power Systems Research*, vol. 154, pp. 493–502, 2018.

31. M. F. Moghaddam, S. Vaez-Zadeh, A. Zakerian, and A. Babaki, "Design and analysis of a modified dual phase shift control method for a wireless EV charger considering coupling uncertainty," in *2019 10th International Power Electronics, Drive Systems and Technologies Conference (PEDSTC)*, Iran, pp. 635–640, IEEE, 2019.
32. A. Zakerian, S. Vaez-Zadeh, and A. Babaki, "A dynamic WPT system with high efficiency and high power factor for electric vehicles," *IEEE Transactions on Power Electronics*, vol. 68, no. 11, pp. 10429–10438, 2019.
33. D. Ahn and S. Hong, "A study on magnetic field repeater in wireless power transfer," *IEEE Transactions on Industrial Electronics*, vol. 60, no. 1, pp. 360–371, 2012.
34. S. G. Wirasingha and A. Emadi, "Pihef: Plug-in hybrid electric factor," in *2009 IEEE Vehicle Power and Propulsion Conference*, Dearborn, MI, pp. 661–668, IEEE, 2009.
35. H. England, "Feasibility study: Powering electric vehicles on England's major roads," *Published by: Highways England Company*, vol. 28, no. 07, p. 2015, 2015.
36. A. Babaki, S. Vaez-Zadeh, and A. Zakerian, "Performance optimization of dynamic wireless EV charger under varying driving conditions without resonant information," *IEEE Transactions on Vehicular Technology*, vol. 68, no. 11, pp. 10429–10438, 2019.
37. X. Zhang, T. Cai, S. Duan, H. Feng, H. Hu, J. Niu, and C. Chen, "A control strategy for efficiency optimization and wide ZVS operation range in bidirectional inductive power transfer system," *IEEE Transactions on Industrial Electronics*, vol. 66, no. 8, pp. 5958–5969, 2018.
38. B. K. Kushwaha, G. Rituraj, and P. Kumar, "3-d analytical model for computation of mutual inductance for different misalignments with shielding in wireless power transfer system," *IEEE Transactions on Transportation Electrification*, vol. 3, no. 2, pp. 332–342, 2017.
39. B. J. Limb, Z. D. Asher, T. H. Bradley, E. Sproul, D. A. Trinko, B. Crabb, R. Zane, and J. C. Quinn, "Economic viability and environmental impact of in-motion wireless power transfer," *IEEE Transactions on Transportation Electrification*, vol. 5, no. 1, pp. 135–146, 2018.
40. L. Chen, G. R. Nagendra, J. T. Boys, and G. A. Covic, "Double-coupled systems for IPT roadway applications," *IEEE Journal of Emerging and Selected Topics in Power Electronics*, vol. 3, no. 1, pp. 37–49, 2014.
41. S. Li and C. C. Mi, "Wireless power transfer for electric vehicle applications," *IEEE Journal of Emerging and Selected Topics in Power Electronics*, vol. 3, no. 1, pp. 4–17, 2014.
42. J. M. Miller, P. T. Jones, J.-M. Li, and O. C. Onar, "ORNL experience and challenges facing dynamic wireless power charging of EV's," *IEEE Circuits and Systems Magazine*, vol. 15, no. 2, pp. 40–53, 2015.
43. J. Shin, S. Shin, Y. Kim, S. Ahn, S. Lee, G. Jung, S.-J. Jeon, and D.-H. Cho, "Design and implementation of shaped magnetic-resonance-based wireless power transfer system for roadway-powered moving electric vehicles," *IEEE Transactions on Industrial Electronics*, vol. 61, no. 3, pp. 1179–1192, 2013.
44. T.-E. Stamati and P. Bauer, "On-road charging of electric vehicles," in *2013 IEEE Transportation Electrification Conference and Expo (ITEC)*, Dearborn, MI, pp. 1–8, IEEE, 2013.
45. Z. Wang, S. Cui, S. Han, K. Song, C. Zhu, M. I. Matveevich, and O. S. Yurievich, "A novel magnetic coupling mechanism for dynamic wireless charging system for electric vehicles," *IEEE Transactions on Vehicular Technology*, vol. 67, no. 1, pp. 124–133, 2017.
46. S. Cui, Z. Wang, S. Han, and C. Zhu, "Analysis and design of multiphase receiver with reduction of output fluctuation for ev dynamic wireless charging system," *IEEE Transactions on Power Electronics*, vol. 34, no. 5, pp. 4112–4124, 2018.
47. S. Hasanzadeh and S. Vaez-Zadeh, "Efficiency analysis of contactless electrical power transmission systems," *Energy Conversion and Management*, vol. 65, pp. 487–496, 2013.
48. J. Schneider, "SAE J2954 overview and path forward," SAE International, Warrendale, PA, 2013.
49. Y. Jiang, L. Wang, Y. Wang, J. Liu, M. Wu, and G. Ning, "Analysis, design, and implementation of WPT system for EV's battery charging based on optimal operation frequency range," *IEEE Transactions on Power Electronics*, vol. 34, no. 7, pp. 6890–6905, 2018.
50. A. Zakerian, S. Vaez-Zadeh, and A. Babaki, "Efficiency maximization control and voltage regulation for dynamic wireless EV charging systems with mutual induction estimation," in *IECON 2019-45th Annual Conference of the IEEE Industrial Electronics Society*, Lisbon, vol. 1, pp. 4298–4303, IEEE, 2019.
51. A. Zakerian, S. Vaez-Zadeh, and A. Babaki, "Maximum efficiency control of a wireless EV charger with on-line parameter calculation," in *2019 IEEE PELS Workshop on Emerging Technologies: Wireless Power Transfer (WoW), held in London, UK*, IEEE, 2019.

52. A. Babaki, S. Vaez-Zadeh, M. F. Moghaddam, and A. Zakerian, "A novel multi-objective topology for in-motion WPT systems with an input DG source," in *2019 10th International Power Electronics, Drive Systems and Technologies Conference (PEDSTC)*, Iran, Shiraz, pp. 787–792, IEEE, 2019.
53. R. Mecke, C. Rathge, W. Fischer, and B. Andonovski, "Analysis of inductive energy transmission systems with large air gap at high frequencies," in *European Conference on Power Electronics and Applications*, Toulouse, 2003.
54. A. Babaki, S. Vaez-Zadeh, and A. Zakerian, "Wireless motor drives with a single inverter in primary side of power transfer systems," in *Presented in 2019 IEEE PELS Workshop on Emerging Technologies: Wireless Power Transfer (WoW)*, London, UK, IEEE, 2019.
55. T. Shiramagond and W.-J. Lee, "Integration of renewable energy into electric vehicle charging infrastructure," in *2018 IEEE International Smart Cities Conference (ISC2)*, Kansas City, MI, pp. 1–7, IEEE, 2018.
56. Y. Nie, X. Wang, and K.-W. E. Cheng, "Multi-area self-adaptive pricing control in smart city with EV user participation," *IEEE Transactions on Intelligent Transportation Systems*, vol. 19, no. 7, pp. 2156–2164, 2017.
57. L. A. Maglaras, F. V. Topalis, and A. L. Maglaras, "Cooperative approaches for dymanic wireless charging of electric vehicles in a smart city," in *2014 IEEE International Energy Conference (ENERGYCON)*, Dubrovnik, Croatia, pp. 1365–1369, IEEE, 2014.
58. X. Mou, R. Zhao, and D. T. Gladwin, "Vehicle to vehicle charging (V2V) bases on wireless power transfer technology," in *IECON 2018-44th Annual Conference of the IEEE Industrial Electronics Society*, Washington, DC, pp. 4862–4867, IEEE, 2018.
59. X. Mou, R. Zhao, and D. T. Gladwin, "Vehicle-to-vehicle charging system fundamental and design comparison," in *2019 IEEE International Conference on Industrial Technology (ICIT), held in Melbourne, Australia*, pp. 1628–1633, IEEE, 2019.
60. L. Grackova, I. Oleinikova, and G. Klavs, "Electric vehicles in the concept of smart cities," in *2015 IEEE 5th International Conference on Power Engineering, Energy and Electrical Drives (POWERENG), held in Riga, Latvia*, pp. 543–547, IEEE, 2015.
61. W. Shuai, P. Maillé, and A. Pelov, "Charging electric vehicles in the smart city: A survey of economy-driven approaches," *IEEE Transactions on Intelligent Transportation Systems*, vol. 17, no. 8, pp. 2089–2106, 2016.

16 Wirelessly Powered Unmanned Aerial Vehicles (UAVs) in Smart City

Malek Souilem
Instituto de Telecomunicações, Aveiro, Portugal
École Nationale d'Ingénieurs de Sousse, Université de Sousse, Sousse, Tunisia
Laboratory of Elec, and Microelec. Université de Monastir, Tunisia

Wael Dghais
Instituto de Telecomunicações, Aveiro, Portugal
Laboratory of Elec, and Microelec. Université de Monastir, Tunisia
Institut Supérieur des Sciences Appliquées et de Technologie de Sousse, Tunisia

Ayman Radwan
Instituto de Telecomunicações and Universidade de Aveiro, Aveiro, Portugal

CONTENTS

16.1 INTRODUCTION

Smart city concept is vastly being adopted by various cities, all around the globe. The smart city concept mainly depends on information and communication technology (ICT), to provide the required information, such as measurements from various sensors, in addition to the required connectivity, which would enable the overall concept of smart city. A new player, which is foreseen to have a significant impact on the smart city, is an unmanned aerial vehicle (UAV). UAVs are expected to play vital roles in different directions, within the smart city's overall ecosystem. They are forecasted to be used in various domains, such as environmental hazards monitoring, traffic management, and pollution monitoring, all of which contribute greatly to the development of the smart city. However, the major issue, facing the wide adoption of UAVs in various solutions of smart city, is the limited flying time, due to the dependability on chargeable batteries, which usually drain fast, due to the high power consumption of UAVs. Addressing such drawback of UAVs, Wireless Power Transfer (WPT) has been proposed as a solution, to enable the charging of UAVs, during their flight, to reduce operation interruption for re-charging. Recent advances in WPT techniques would assist in the vast adoption of UAVs in the smart city scenario, in addition to the assistance in 5G network connectivity and performance. Multiple efforts have been dedicated toward enhancing WPT for UAVs, in addition to the use of Power Line Communication (PLC) in the assisting of charging of UAVs.

This chapter reviews the state-of-the-art and recent advances and applications of wireless charging applied to UAVs, especially as an assisted 5G network. Thanks to their 3D mobility, UAVs are foreseen among the key players/components that bring an additional degree of freedom for researchers to explore the space mobility for an improved 5G network performance, such as throughput, adaptability, maintenance, and self-organization network. Wireless charging scenario enables multiple UAVs charging, which could increase the battery autonomy of a fleet of UAVs, whenever a critical situation arises that may require high-rate transmissions or in search rescue agency and assisting civil security.

The chapter provides a brief background about UAVs, and their usage in smart cities (Section 16.2). This is followed by different efforts reported in the literature, targeting the energy efficiency of UAVs (Section 16.3). The chapter then presents different techniques for WPT, covering 2D and 3D, design and performance issues, in addition to simultaneous power and data transmission (Sections 16.4 and 16.5).

16.2 UAVs IN SMART CITY

UAV is an aircraft that does not require a human pilot, such as drones, small aircrafts, balloons, and airships [1]. The UAV design varies from a few inches in size to the size of a full aircraft. Generally, UAVs are differentiated based on the stringent constraint imposed by the SWAP (Size, Weight, and Power), since the SWAP constraint directly impacts the maximum operational altitude, communication, coverage, computation, and endurance capabilities of each UAV [2]. UAV architecture consists of multiple main components consisting of the control system, the monitoring system, the data processing system, and the landing system [3]. The internal system provides a wide range of functions ranging from navigation to providing data transfer to the ground. Consequently, they have been used in smart cities, as 5G-aided connectivity and supporting intelligent transportation system (ITS) and civil/healthcare rescue operations. The European Platform for Intelligent Cities and the European Network of Living Labs defined smart cities as "the use of discrete new technology applications such as radio-frequency identification (RFID) and the Internet of things through more holistic conception of intelligent, integrated working that is closely linked to the concept of living and user generated services" [4]. Therefore, ITS is a key component in the smart city, since it relies on the effective integration of connected and autonomous vehicles, such as the UAV due to their mobility, autonomous operation, and communication/processing capability [5].

Generally, the smart city seeks to achieve the objectives of a future city by utilizing ICT solutions and trends. There is increased interest to focus on utilizing ICT services and smart solutions in long-term smart city development. The design of such a smart city requires huge and full integration of ICT and its trends. UAVs contribute to these goals. That is why UAVs are involved in a wide range of applications and functions in smart cities [4]. A representation of Smart City and the usage of UAVs within is shown in Figure 16.1.

The main goals of the smart city are to optimize the consumed resources, to reduce cost, and to improve interactivity and the life quality of its citizens, by incorporating ICT to manage urban life, agriculture, air pollution, traffic, emergencies, energy source and consumption, government, healthcare, water, and others. Toward this trend, UAVs can provide a number of useful tools to monitor and operate a smart city in order to achieve its main goals; provide an efficient infrastructure

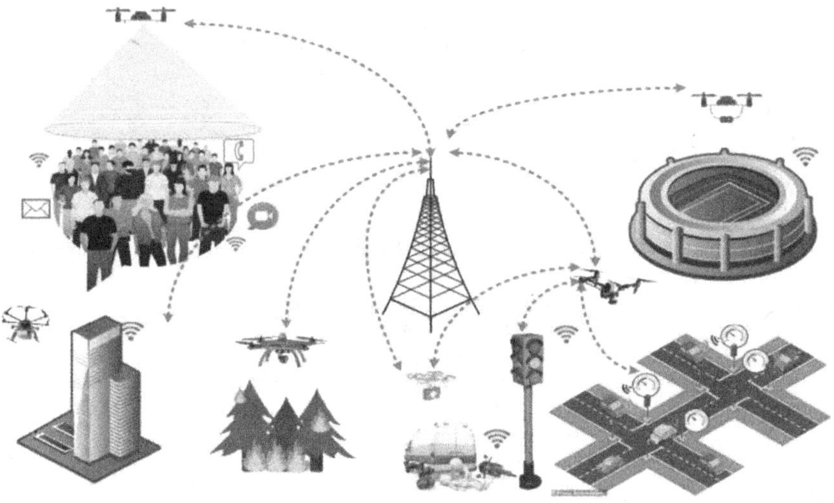

Figure 16.1 Smart city and UAV application.

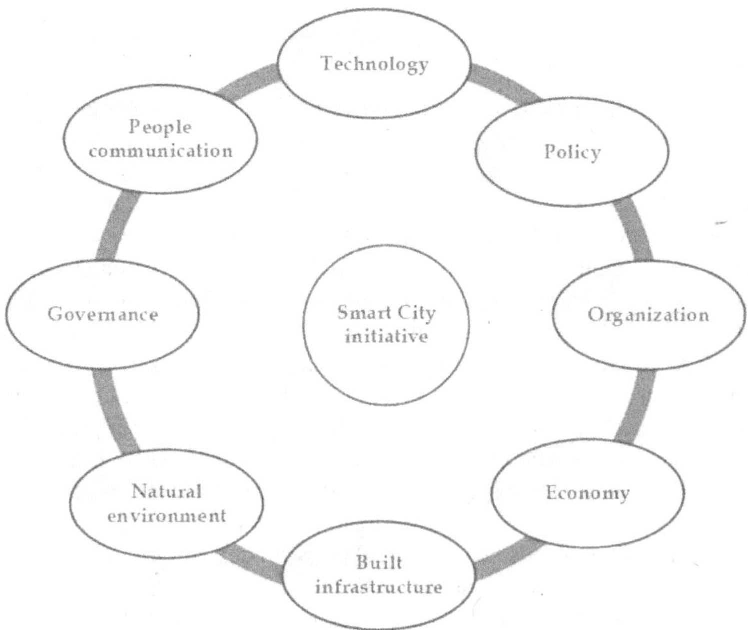

Figure 16.2 Key factors of a smart city.

and services with reduced cost [1]. Some efforts have defined eight core factors, that influence smart city initiatives [2,6]. Those factors are presented in Figure 16.2, while Table 16.1 further elaborates the factors and their significance.

Initially, UAVs were mainly used in military environments and security fields. Recently, UAVs have been integrated in a wide range of applications in civil society, within the ecosystem of a smart city. UAVs have been involved in various domains, including, to name a few, environmental hazards monitoring, traffic management, and pollution monitoring, all of which contribute greatly to the development of the smart city.

Table 16.1
Key Factors Impacting the Smart City Development

Factors	Description
Management and organization	The effectiveness and efficiency of the smart city
Technology	- Smart computing technologies applied to critical infrastructure components and services. - Help to make an intelligent decision.
Governance	Exchange information according to rules and standards in order to achieve goals and objectives
Policy-context	Understand the use of information systems in appropriate ways
People and communities	Active members in the governance and management of the city
Economy	Business creation, job creation, workforce development, and improvement in productivity
ICT infrastructure	Fundamental element in a smart city's development and related to its availability and performance factors
Natural environment	- Increase sustainability and to enhance natural resource management. - Protect the natural resources and related infrastructure.

As UAVs have a wide range of applications and models, some researchers have classified them into three categories, which are safety control, scientific research, and commercial applications. According to the authors of [2], the different beneficial opportunities for UAV can be summarized into different main categories, which are further elaborated as follows. First, geospatial surveying is considered as one of the newest field that incorporated UAVs. The design of UAV facilitates the installation of wireless sensors on-board that enable UAVs to be used in the geospatial fields, such as land surveying, Geographic Information System (GIS), and environmental analysis. Second, thanks to the advanced research in data mining, UAVs have the opportunity to be effectively used in civil security activities, like security surveillance in the smart city. Urban security increases the attractiveness of cities and enables cities to host big public events. For instance, the ease of deployment of UAVs would allow city management to rapidly and easily set up a security surveillance for upcoming public events, even with a huge number of attendees. UAVs can provide wide security coverage. Traffic and crowd management is another important field, where UAVs were applied in smart policing activities, which have lately been supported by the US congress and top-level federal agencies, such as the Bureau of Justice Assistance, and the US Department of Justice. Using UAVs in disaster situations like fires, floods, and earthquakes, will help authorities in the control of such situations more efficiently. Generally speaking, UAVs are a helpful tool in natural disaster control and monitoring activities. In agriculture and environment management, UAVs can be used to fertilize crops through controlled fertilizer dropping, to monitor the growth of crops, and to monitor the environment. UAVs are also useful for the big data processing of smart infrastructure in the context of smart cities. Finally, UAVs can act as a third party technology to coordinate information from various systems. The ground system, controlling the UAV, can send commands to UAV to direct the information to another ground system or another UAV.

The latest dimension for UAVs usage is assistance in networking. Recently, UAVs are seen as another dimension of enhancing networking; hence, they are considered as a complementary element in the 5G communication ecosystem. UAVs can be used as BSs (more specifically on-demand small cells) that potentially facilitate wireless broadcast and support high-rate transmissions, since it is characterized by outstanding features, such as easy, flexible, and cheap deployment [1–3]. Furthermore, the design of UAVs has a degree of freedom with the controlled mobility, compared to the communication with a fixed infrastructure. Current telecommunication systems are facing a capacity demand challenge, very expensive operation cost, and high capital investments. Therefore, the concept of a heterogeneous network (HetNet), which is defined as a hybrid network composed of macro BSs and various sized BS, has appeared as an attractive solution to meet the high data communication demands, in addition to the foreseen increased number of connected devices [7,8]. The HetNet topology brings the network closer to mobile users, which further helps in achieving the future demands of 5G, such as low latency, and high data rates [8]. However, in big public events, such as a festival or concert, or during unexpected or emergency situations, such as disaster relief and service recovery, the deployment of terrestrial infrastructures is economically infeasible and challenging due to high operational expenditure, as well as sophisticated and volatile environments, not to mention, the long deployment period.

Currently, it is widely believed that individually existing networks cannot meet the requirement to process enormous volumes of data and execute substantial applications, such as IoT, cloud computing, and big data. Therefore, there is an agreement among scientific communities and main players in the networking area, to develop an integrated network architecture from the space-based network, air-based network, and ground-based network. An intuitive solution to offload the cellular traffic is to deploy small cells (e.g., pico and femto cells). However, in unexpected or temporary events, the deployment of terrestrial infrastructures is challenging since the mobile environments are sophisticated, volatile, and heterogeneous. A newly adopted potential solution resorts to the usability of drone-cells [2,3], which has been proven to be instrumental in supporting ground cellular networks in areas of erratic demand. The idea is to bring the drone-cells close to the ground users, in order to improve their QoS, due to the short-range LoS connections from the sky. Figure 16.3 shows a typical UAV-assisted intelligent HetNet architecture, with one macro-BS (MBS) and multiple drone-cells [2,9]. Such deployment has shown various advantages, such that using UAVs as small BSs is considered a novelty that will be the solution of such situations and will facilitate the use of wireless small BS. For instance, a UAV can be the solution in situations such as network service recovery in a disaster-stricken region, or enhancement of public safety. Particularly, UAV-aided eMBB will soon be, not a complementary, but an essential component of 5G network ecosystem [2,10].

In summary, the aerial base stations (AeBS) represent the promising solution that enhances the capacity and coverage of terrestrial cellular system infrastructure, specifically in some exceptional

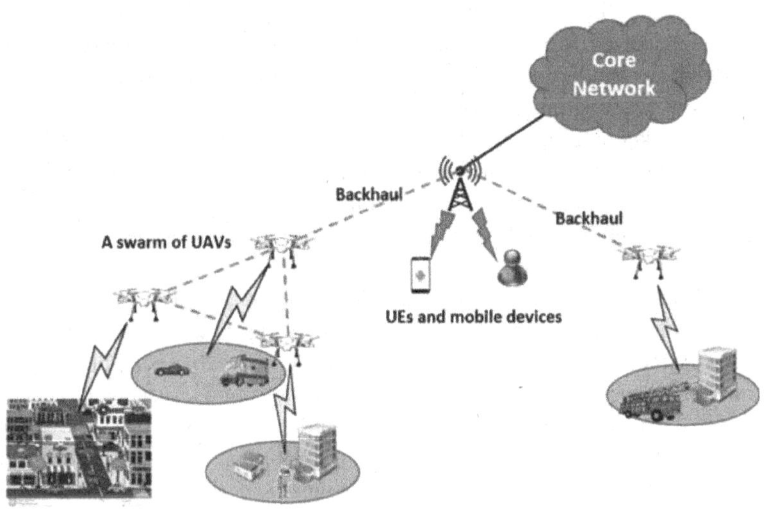

Figure 16.3 An intelligent HetNet incorporating drone-cells.

situations when the performance, coverage, and capacity of the terrestrial infrastructure are limited [11]. A major advantage of the AeBSs over static terrestrial BSs is their ability to change their positions to serve as a dynamic network for users, optimally. An AeBS can be efficiently integrated into terrestrial cellular wireless networks to either serve the ground users directly or relay traffic to another part of the terrestrial network [12]. The envisioned HetNet can take advantage of the high performance of different types of BSs, to optimize network performance. For example, dense deployment of terrestrial networks, in urban areas, supports high data rate access, satellite communication systems can provide wide coverage and seamless connectivity to remote and sparsely populated areas, while UAV communications can assist the existing cellular communications for the rapid service recovery and offer the traffic offloading of the extremely crowded areas, in a cost-effective fashion [13]. As illustrated in Figure 16.3, the next-generation networks should intelligently and seamlessly integrate multiple nodes to form a multitier hierarchical architecture, including a drone-cell tier for large radio coverage areas, a ground small cell tier for small radio coverage areas, a user device tier with D2D communications, and so forth [2,9].

With the forthcoming era of 5G, densely populated users are thirsty for broadband wireless communications and network operators are expected to support diverse services, with high wireless data demands, such as multimedia streaming and video downloads. The unrelenting increment in mobile traffic volumes imposes an unacceptable burden on the operators in terms of increased capital expenditure (CAPEX) and operating costs (OPEX). All these factors derive operators to adopt the heterogeneous concept, including terrestrial, satellite, and aerial tiers.

16.3 ENERGY EFFICIENCY AND CONNECTIVITY OF UAVs

In the previous section, we have shown the importance of UAVs in the ecosystem of smart cities, in addition to becoming a vital component of future wireless communications, i.e., 5G and B5G. UAVs have proven to be a great asset for smart city deployment. Additionally, UAVs are seen to fill multiple gaps in the provision of ubiquitous, reliable, low-cost networking, in the overall ecosystem of 5G. However, all those advantages do not come easy. UAVs face multiple challenges, which can be expressed by SWAP (i.e., Size, Weight, and Power). Size and weight are big challenges but are controlled by the requirements of the application/service that the UAV is targeting. For instance, if the UAV needs to carry a certain object, its size will be dependent on such element; hence, optimizing size and weight is a big challenge, but are more controlled by the target application. On the other hand, power consumption is another challenge that limits the flight time of UAVs; therefore, can limit their application and bound the range of their applications. Toward addressing such a challenge, researchers are heading into multiple directions. First, research into better battery technologies are going and will continue; however, from the previous decades, it is clear that progress in such direction is not keeping up with the increase in power consumption requirements of mobile devices. Another direction is optimizing the energy consumption of UAVs. This direction has shown potential advantages and is further discussed in this section.

Energy efficiency is a fundamental issue, generally in the design of any wireless network, and more specifically in a futuristic network system that incorporates UAVs, as essential system elements to provide connectivity within the 5G integrated ecosystem. The flight act of UAV normally requires a large amount of power; therefore, energy efficiency is a key design challenge. Additionally, UAVs are characterized by large range availability, since they are able to roam the whole network environment. However, UAVs are unlike traditional ground transceivers, deprived of external power suppliers. UAVs are only powered by a limited capacity battery as the energy resource. Moreover, UAVs face the challenge of limited energy availability, mainly due to performing multiple energy-hungry activities, such as flight control, sensing, the transmission of data, and/or running some applications. Generally, UAVs are required to return to their base, for battery charging, frequently. Thus, energy efficiency is critical, though challenging, to guarantee stable and sustainable communication services, and to avoid being the performance bottleneck [1,2]. Energy is a scarce resource for AeBSs; hence, energy efficiency of such networks is essential, given that the entire network infrastructure, including battery-operated ground terminals, which means they need to operate under power-constrained situations. In conclusion, wise management of energy is quite beneficial for the lifetime of the network [12].

It is thus challenging to control a group of UAVs, to maintain certain communication coverage in the long run; hence, minimizing energy consumption is critical in such type of network (i.e., UAV-assisted communications network). Energy efficiency optimization would result in enhancing the battery lifetime of the UAV-assisted networks. Recently, multiple research studies have focused on energy-aware deployment and improving the operation charging mechanisms of UAVs. Here, we present a review of various studies, trying to optimize energy efficiency or UAVs' trajectory. **Mozaffari et al.** studied the energy-efficient deployment of multiple UAV-BSs, for minimizing the total required transmission power of UAVs while satisfying the data rate requirements of the ground users [14]. **Islam et al.** studied how to reduce the total energy consumption, by designing and implementing an energy-efficient AeBSs [12]. Implementing the sleep mode in BSs was proposed, for improving the energy efficiency; moreover, a novel strategy is proposed, for further improving energy efficiency, by considering ternary state transceivers for AeBSs. **Lui et al.** investigated the problem of controlling a group of UAVs, to achieve certain communication coverage in the long run, while preserving their connectivity and minimizing their energy consumption [15]. The emerging Deep Reinforcement Learning (DRL) is leveraged to control UAVs, to achieve high communication coverage and fairness, at the low energy consumption level. **Ruan et al.** investigated the UAV coverage problem and proposed a multi-UAV coverage model, based on energy-efficient communication [16]. The proposed model is composed of two steps: coverage maximization and power control. **Zeng and Zhang** studied the energy-efficient UAV communications, via optimizing the UAV's trajectory with a fixed attitude [17]. The work proposes a new design paradigm that jointly considers both the throughput and the UAV's energy consumption of fixed-wing. **Hua et al.** addressed the problem of maximizing the system energy efficiency of the UAV while satisfying the fairness among sensor nodes, by jointly optimizing the UAV trajectory and UAV time allocation [18]. In Ref. [19], **Mozaffari et al.** explored the efficient deployment and mobility of multiple UAVs, used as AeBSs to collect data from ground Internet of Things (IoT) devices. A new framework was proposed, for jointly optimizing the 3D placement and the mobility of the UAVs, device-UAV association, and uplink power control, in order to enable reliable uplink communications for IoT devices, with a minimum total transmit power. **Zeng et al.** [20] studied UAV-enabled wireless communications, where a rotary-wing UAV is dispatched to communicate with multiple ground nodes (GNs). The authors aimed to minimize the total UAV energy consumption, including both propulsion energy and communication-related energy, while satisfying the communication throughput requirement of each GN. To this end, they first derived a closed-form propulsion power consumption model, for rotary-wing UAVs. They, then, formulated the energy minimization problem, by jointly optimizing the UAV trajectory and communication time allocation among GNs, as well as the total mission completion time. **Kuru et al.** analyzed several delivery schemes within a platform, such as delivery with and without using air highways, and delivery using a hybrid scheme, along with several delivery methods (i.e., optimal, premium, and first-in first-out) to explore the use of UAV swarms as part of the logistics operations [21].

Meanwhile, there are various research efforts that examine the UAV energy efficiency issue, from the operating UAVs battery researching mechanism. Generally, the energy consumption of the battery-powered UAV is usually split into two components: the energy consumed by the communication unit and the energy used for the hardware and mobility of UAVs. Hence, energy harvesting UAV is crucial to prolong its flight duration, without adding significant mass or size of the fuel system. In recent applications, it was shown to be very advantageous, to harvest energy from ambient sources, for recharging UAVs' batteries, which is referred to as wireless powered UAV networks. A lot of works, existing in the literature, have proceeded to improve the endurance of electrically powered UAVs, through enhancing their operation and investigating energy harvesting. **Sun et al.** investigated resource allocation design, for multicarrier systems, employing a solar-powered UAV for providing communication services to multiple downlink users [22]. Their work presented a joint design of the three-dimensional positioning of the UAV and the power and subcarrier allocation for system throughput maximization. The same team, also, proposed resource allocation algorithm design for multicarrier solar-powered UAV communication systems [23]. In particular, the UAV is powered by solar energy, enabling sustainable communication services to multiple ground users. The authors study the joint design of the 3D aerial trajectory and the wireless resource allocation for maximizing the system overall throughput, over a given period of time. **Sowah et al. [24]** presented a rotational energy harvester, based on a brushless DC generator to harvest ambient energy, for prolonging the flight time of a quadcopter, while a prototype of the rotational energy harvesting system was also implemented, which comprised a quadcopter, power management system, and a battery charging system. **Yang et al.** analyzed the outage performance of UAV harvesting energy from the ground BS, where both shadowed-Rician fading and shadowed-Rayleigh fading were, respectively, considered [25].

Moreover, recent studies have focused on WPT charging systems of the UAVs, which aim to create more flexibility and improve the device agility. The next two sections will discuss Wireless Power Transfer (WPT) in more details; however, this section provides some efforts proposed for WPT specifically for energy efficiency of UAVs. **Xu et al.** studied a new UAV-enabled WPT system, where a UAV-mounted energy transmitter broadcasts wireless energy, to charge distributed energy receivers (ERs) on the ground [26]. In particular, the authors considered a basic two-user scenario and investigated how the UAV can optimally exploit its mobility, via trajectory design, to maximize the amount of energy transferred to the two energy receivers, during a finite charging period. **Long et al.** proposed the architecture of energy neutral IoUAVs [27], where recharging stations were used to energize UAVs, via WPT with RF signals, which significantly enabled the continuous operation lifetime. The work by **Vincent et al.** introduced a wireless charging method that employs a capacitive wireless charging scenario for a UAV, via a master-slave arrangement [28]. This work increases the flight autonomy and range of UAVs; thus, making them more usable. A 150×170 mm plate pair, for wirelessly charging a drone battery, is studied with different dielectrics. Finally, a comparison of different configurations of the capacitive plate interface was outlined, based on simulation results. **Lu et al.** focused on studying wireless techniques, available for drone mission duration improvement [4]. They also discussed and practically examined the most feasible and reliable technique to charge UAV, using power lines.

Finally, **Galkin et al.** outlined three battery charging options to be considered by a network operator [29]. The authors used simulations to demonstrate the performance impact of incorporating those options into a cellular network, where UAV infrastructure provides wireless service.

To conclude the survey of research efforts toward enhancing energy efficiency of UAVs, all those efforts have been summarized, using one sentence, in Table 16.2.

16.4 WIRELESS POWER TRANSFER FOR UAVs

Up till now in this section, we have illustrated the potential of UAVs in the ecosystem of future smart cities, and we have demonstrated the challenge facing UAVs, namely issues with power consumption, limiting flying time. In the previous section, we have listed different research efforts, in the literature, which address the energy efficiency of UAVs, including the usage of WPT. In this section, we concentrate on WPT.

In recent years, researchers have achieved great advances in the field of Wireless Power Transfer (WPT). WPT can be one solution to overcome the issue of a short flight time of UAVs. WPT can be used for charging UAVs' batteries, during their flight; thus, enhancing their autonomy and mobility. The concept of WPT in collaboration with UAVs is further discussed in this section. Different WPT techniques are elaborated in Section 16.5, with emphasis on design issues and performance analysis.

Table 16.2
Related Work of the UAV's Energy Efficiency

	Publication	One Sentence Summary
SCs deployment/ technology	Mozaffari et al. [14]	Energy-efficient deployment of multiple UAV-BSs
	Islam et al. [12]	Implementing the sleep mode in AeBSs
Algorithmic/ modeling	Lui et al. [15]	Emerging Deep Reinforcement Learning for UAVs used to control UAVs
	Ruan et al. [16]	Multi-UAV coverage model based on energy-efficient communication
Optimization	Zeng et al. [17]	Optimizing the UAV's trajectory with fixed attitude
	Hua et al. [18]	Optimizing the UAV trajectory and UAV time allocation
	Mozaffari et al. [19]	Optimizing the 3D placement and the mobility of the UAVs, device-UAV association, and uplink power control
	Zeng et al. [20]	Optimizing the UAV trajectory and communication time allocation among GNs, as well as the total mission completion time
	Kuru et al. [21]	Analyze several delivery methods (i.e., optimal, premium, and first-in first-out)
Harvest energy	Sun et al. [22]	A resource allocation design for multicarrier systems, employing a solar-powered UAV
	Sun et al. [23]	A resource allocation algorithm design for multicarrier solar-powered UAV communication systems
	Sowah et al. [24]	A rotational energy harvester for prolonging the indoor flight time of quadcopter
	Yang et al. [25]	Analysis of the outage performance of UAV harvesting energy from the ground BS
WPT	Xu et al. [26]	Enabled WPT system, where a UAV-mounted ET broadcasts wireless energy to charge distributed ERs on the ground
	Long et al. [27]	Energy neural IoUAVs architecture: recharging stations were used to energize the UAVs via WPT with RF signals
	Vincent et al. [28]	A wireless charging method that employs a capacitive wireless charging scenario
	Lu et al. [4]	Wireless techniques to charge UAV using power lines
	Galkin et al. [29]	Simulation of three options of battery charging, where UAV provides wireless services

Additionally, some efforts toward energy efficiency of UAVs, using WPT, have been listed in previous Section 16.3.

The significant rise of renewable energy generators and the wide adoption of the smart grid concept have created the need for constant information exchange between different components of the grid. Furthermore, the increasing adoption of UAVs and drones is an additional boost to this situation [1–4]. Power line communication (PLC) is revived, by establishing itself as one of the possible ways to ensure data communications in power grids. The recent development of Home-Plug Green PHY shows the future trend, by reaching communication speeds of hundreds of Mbits/s [3,30]. Moreover, Broadband over Power Line (BPL) physical layer uses multi-carrier orthogonal frequency division multiplex modulation carriers spaced at 24.414 kHz, with carriers from 2 to 30 MHz, to enable high throughput in a noisy environment. However, the IEEE 1901 and Home-Plug AV2 standards extend this range to 50 and 86 MHz, respectively, achieving hundreds of Mbps

[4,5,31]. Moreover, BPL has a number of advantages making it an appealing complement to other RF wireless technologies, since it does not require any new wiring installations, which significantly reduces deployment costs, enabling a new generation of high-speed indoor communications. Another advantage of BPL is enabling communications with hard-to-reach nodes, where the RF signal suffers from high attenuation, as in underground buildings and hospitals.

On the other hand, recent advances in WPT techniques for charging would assist the 5G network connectivity and performance using UAVs. With the issue of UAVs reduced flying time due to fast battery drain, WPT via a resonant inductive channel (RIC) is an appealing solution that can boost longer flying time of UAVs. These WPT and BPL technologies can be integrated to share the RIC for transmitting power and convening bidirectional data signals. The concept of a coupled WPT–BPL system was recently demonstrated and analyzed [1,3,5]. Design guidelines were presented to guarantee an efficient power transfer and BPL communications in a planar (i.e., 2-dimensional (2D)) coils. The inductive channel is formed by the transmitting (T_x) and receiving (R_x) coils, which are separated by a distance (d) and misalignment (θ). The communication and control unit (CCU) is used to activate, monitor, and control the devices used in the power and data links of the WPT-BPL system to ensure high power-efficient transfer and reliable BPL communications. It is important to note here that this is one possible solution; however, other methods for providing both communications and WPT are available and should be exploited. Additionally, the hardware optimization of the planar wireless power and data links are separately investigated [5,32–36].

This idea enables UAV-assisted 5G communications and backhauling, which can be potentially applied to cellular offloading, and data collection from consumer electronic appliances and healthcare devices, fast service recovery after natural disasters, as shown in Figure 16.4.

The planar (2D) and three dimensional (3D) are the two main two types of WPT. They can be classified according to the dimension of the magnetic field, generated by the transmitting coils, and will be received and rectified by the receiving antenna at the consumer electronic devices, drones or UAVs. For instance, the 2D WPT case imposes the placement of the receiving coils in the $x-y$ plane; therefore, restricting the flexibility and the mobility of the receiving devices in the vertical axes (i.e., z-axis), as depicted in Figure 16.5a and b. In fact, the receiver should be touching in parallel or in proximity (i.e., max. few centimeters) of the wireless power bank. The planar resonant magnetic inductive channel can improve the vertical charging distance and enable multiple receivers charging. However, angle misalignment limits the efficiency of the overall system, composed of multiple wireless power transmitting coils and multiple receiving coils, as shown in Figure 16.6.

It is important to note here that these studies [32–36] overlook the omnidirectional antenna design challenges and opportunities, and disregard the study of the misalignment and angle deviation,

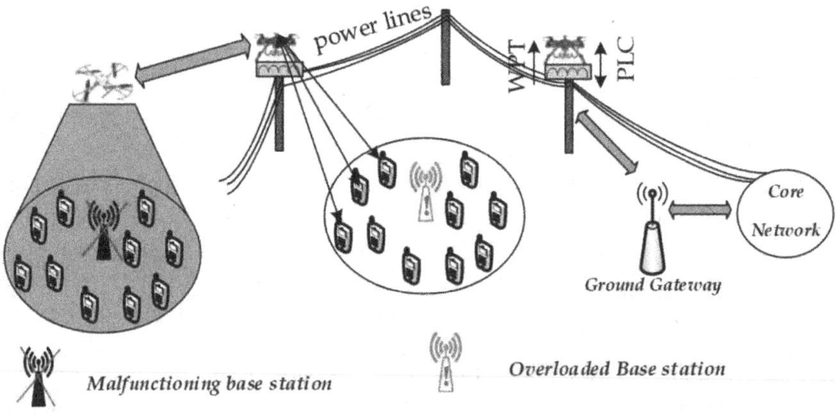

Figure 16.4 Integrated BPL and WPT for UAV-assisted 5G communications.

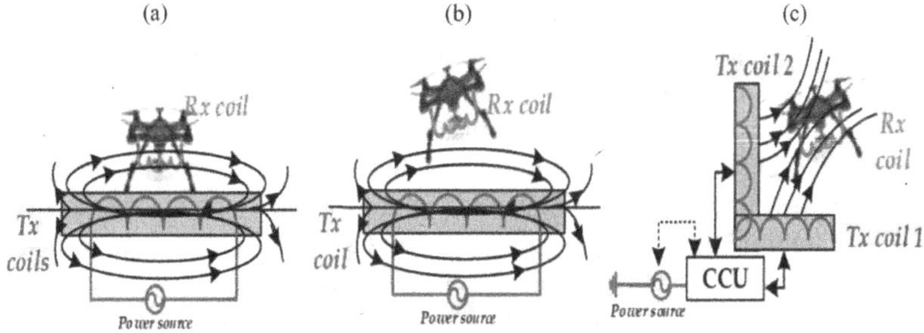

Figure 16.5 Illustration of the planar 2D WPT: (a) chargeable fixed device; (b) non-rechargeable device (in flight); (c) controlled magnetic field for wireless charging.

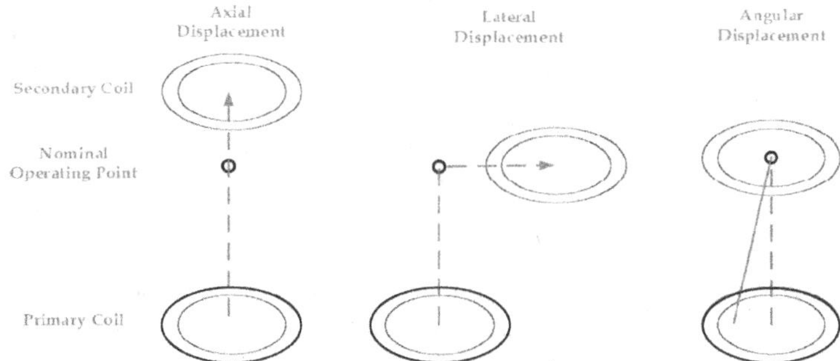

Figure 16.6 Misalignment cases that affect the performance of WPT and data transmission between UAVs and charging modules.

which is critical in charging UAVs. Moreover, only planar two and four coils of the power transmitting structure have been analyzed, while investigating different data access point to achieve the BPL. Recently, two-phase and three-phase 3D omnidirectional transmitting power channels have been investigated, along with control mechanisms to only achieve a high cover space of the magnetic field distribution and effective wireless inductive power link [33,34]. Therefore, the induced voltage in the receiving coil could be retrieved in different positions in a 3D planar (i.e., x, y, z). This scenario enables mobility of the charging device and could be applied for in-flight charging of UAVs. Consequently, designing a 3D near-field inductive power transmitter is really challenging, for ensuring reliable in-flight UAV charging. Such a challenging research topic has been attracting the interest of the scientific committee, from both academia and industry. For instance, an efficient WPT is achieved by controlling the direction of the multiple transmitting magnetic fields, by means of a communication and controlling unit (CCU), which enables wireless charging for different receiver's position and orientation as illustrated in Figure 16.5c.

Several approaches were proposed to enable the design of 3D effective wireless charging, which can be classified as follows:

a. Algorithmically control the orientation and position of the R_x coils array in a planar WPT;
b. Develop new 3D WPT transmitters to generate a magnetic field that can cover the space of the interest for device charging;
c. Study of different coil geometries (i.e., circular, square, pentagon, hexagon) and how they are organized and controlled to generate a uniform magnetic field distribution and high coupling coefficient for multiple device charging;
d. Review of low-frequency and high-frequency power line application for WPT and data communication, simultaneously, through the near-field resonant inductive channel.

Furthermore, the three-phase power cloud transmitter system, presented by the global energy transmission system (GET7) that consists of a hexagonal frame of wires, strung between poles, that represent the transmitting coils as shown in Figure 16.7. The sprung wires create an inductive charging field that can charge the batteries of drones in mid-air. The transmitting coil is driven by a three-phase inverter that creates an alternating electromagnetic field. A second coil in the drone then takes in power from the electromagnetic field and converts it back into electric current, to charge the battery. In fact, multiple drones can be charged simultaneously. Moreover, the system is quite portable, to be moved where required, enabling drones to fly continuously over large areas. As shown in Figure 16.7, the transmitting hexagonal coil is formed by several lines. The hotspots, in red, indicate the magnetic field density.

It is important to elaborate that this solution still presents technical challenges, with regards to its feasibility and deployment for multiple UAVs in-flight wireless charging.

Moreover, data communications, through the RIC formed by the power lines and UAVs coils, can be technically realized and experimentally tested. However, the power signal operates in the KHz-MHz ISM frequency bandwidth, which is in proximity with the BPL data signal; this issue may raise EMC issues, due to the power levels of the harmonics, generated by means of the WPT transceiver's nonlinearity. This nonlinearity has to be controlled in order to avoid the interference with the frequency band of data signals. Therefore, there is a requirement for the design of a 3D inductive channel, with the required filters and processing algorithm for the high-power transfer and low-power data signal communication. The feasibility of this idea will further be extrapolated, among the enabling technologies that explore the power line channel for creating a UAV energy network, which is crucial for continuously assisting the 5G communications networking.

16.5 WPT-BPL FOR UAVs

Having introduced UAVs in smart city ecosystem and emphasized the importance of energy efficiency of UAV operation, it is clear to emphasize that the challenge is to ensure UAVs simultaneous continuous efficient WPT and broadband data communication, via magnetically coupled coils. This section introduces the main variables affecting WPT, while highlighting solutions that can be implemented to overcome WPT inefficiency. We present the basic electromagnetics theories to understand the antenna coils, a procedure for coil design, calculation and measurement of inductance, an antenna-tuning method, and the relationship between read range vs. size of the antenna coil. Finally, the evaluation performance of the power and data links are presented.

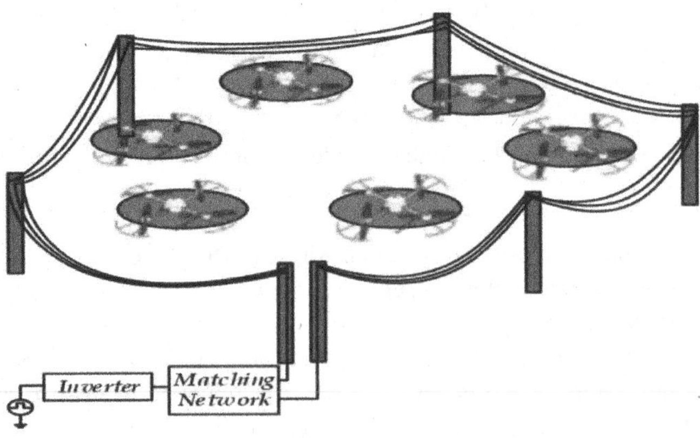

Figure 16.7 High-frequency and power wireless power station.

16.5.1 INDUCTION OF MAGNETICALLY COUPLED CIRCUIT

The time-varying magnetic field, $B(t)$, can be generated by several wiring geometry of coils, such as a finite conducting wire or a loop wire, driven by a time-varying current i(t), as shown in Figure 16.8 [37,38].

The magnetic field produced by the current on a round conductor (wire), with a finite length is given by [37]:

$$B(t) = \frac{\mu_0 i(t)}{4\pi r}(\cos(\delta_1) - \cos(\delta_2)) \tag{16.1}$$

where r is the distance between the wireless power transmitter and receiver. μ_0 is the permeability of free space and given as $\mu_0 = 4\pi 10^{-7}$ Hm^{-1}. In a special case with an infinitely long wire, where $\delta_1 = -180°$ and $\delta_1 = 0°$, Eq. ((16.1)) can be rewritten as [37]

$$B(t) = \mu_0 i(t)/4\pi r \tag{16.2}$$

Figure 16.8b presents a wireless power transmitting circular loop coil (e.g., antenna), the produced $B(t)$ with N-turns can be written as:

$$B(t) = \frac{\mu_0 N i(t)}{2(r^2 + a^2)^{1.5}} a^2 \tag{16.3}$$

where a is the radius of the coil. It is worth noting that the magnetic field, $B(t)$, produced by a loop antenna decays with $1/r^3$ as $r \gg a$. This near-field decaying behavior of the magnetic field is the main limiting factor in the wireless power transfer. The field strength is maximum in the plane of the loop and directly proportional to the current (i), the number of turns (N), and the surface area of the loop.

Let L be a curve in space enclosing a surface S, and let $\Phi(t)$ be the magnetic flux that measures the number of magnetic field lines passing through a surface, which is expressed as

$$\Phi(t) = \iint_S \vec{B}(\vec{r},t) \cdot d\vec{S} \tag{16.4}$$

If the magnetic field is constant (e.g., uniform) in the area S, then the magnetic flux can be re-written as

$$\Phi(t) = B(t) \cdot S \cdot \cos(\theta) \tag{16.5}$$

To maximize the flux with an area vector of constant magnitude and a field of constant magnitude, we need to maximize the factor of $\cos(\theta)$ that reaches its maximum magnitude of 1, when $\cos(\theta)$), the angle between the area vector and the magnetic field, is either 0° or 180°. In other words, the area vector must be parallel to the magnetic field, which is the case when the plane of the receiver is perpendicular to the magnetic field.

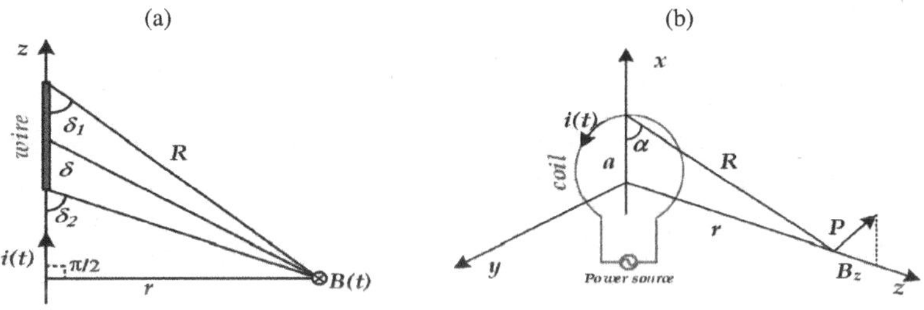

Figure 16.8 Magnetic field, $B(t)$, at location p due to current $i(t)$ (a) on a straight conducting wire; (b) on the loop.

If the curve L represents an electric circuit, the Faraday's law of induction states that the electromotive force (EMF), $U(t)$, generated in the circuit and induced by a change in magnetic flux is

$$U(t) = -\frac{\partial \Phi(t)}{\partial t} = -N\frac{\partial B(r,t)}{\partial t} \cdot S \cdot \cos\theta \cong N \cdot B(r) \cdot S \cdot \cos\theta \frac{di(t)}{dt} \tag{16.6}$$

A closed electric loop will generate a magnetic field. Since the loop encloses a surface, it will experience a flux and EMF due to the field, it has generated. If I is the current through the circuit, the EMF generated by self-induction is:

$$U(t) = -L\frac{di}{dt} \tag{16.7}$$

where L is the self-inductance, which characterizes the conductor loops (coils). The inductance of a conductor loop (coil) depends totally on the material properties (permeability) of the space that the flux flows through and the geometry of the layout. If the diameter of the wire, d, is very small compared to the diameter D of the conductor coil ($d/D \ll 1$), a simple approximation can be used to deduce

$$L = \frac{\Phi}{i} = \frac{N^2 \mu_0 D}{2} Ln\left(\frac{D}{d}\right) \tag{16.8}$$

Assuming that the magnetic flux is generated by a sinusoidal current source, then it has the expression of $B(t) = B_{\max}\cos(2\pi f_0 t)$. Therefore, $U(t)$ can be written by the following expression:

$$U(t) = 2\pi f_0 B_{\max} S\cos(\theta) \cdot \sin(2\pi f_0 t) \tag{16.9}$$

This equation summarizes the main variables affecting the induced voltage, at the WPT receiver, which are:

- $B_{\max}$: which depends on the electrical current amplitude, at the WPT $I_{\max}$, the antenna/coil geometry (e.g., radius or shape) and the distance between the WPT transmitting and receiving coils;
- f_0: is the frequency of the alternating transmitting signal, at which the receiving part should be accorded;
- S: the geometry of the receiving coil that collects the magnetic field;
- $\cos(\theta)$: is the orientation angle between the magnetic field vector and the surface normal.

We, now, move to describe the details of the inductively coupled coils. Considering a second coil (area A_2) is near the first coil (area A_1), through which a current is flowing, then this will be subject to a proportion of the total magnetic flux Φ flowing through A_1. The two circuits are connected together, by this partial flux or coupling flux. The magnitude of the coupling flux, Φ_{21}, depends on the geometric dimensions of both conductor loops, the position of the conductor loops in relation to one another, and the magnetic properties of the medium (e.g., permeability) in the layout. Likewise the self-inductance definition L, the mutual inductance M_{21}, between the conductor's loop 2 and loop 1, is defined as the ratio of the partial flux Φ_{21}, enclosed by conductor loop 2, to the current i_1 in conductor loop 1.

$$M_{21} = \frac{\Phi_{21}(i_1)}{i_1} = \oint_{A_2} \frac{B_2(i_1)}{i_1} \cdot dA_2 \tag{16.10}$$

Moreover, current i_2 flows through loop 2, thereby generating the coupling flux Φ_{12} in loop 1; therefore, a mutual inductance, $M_{12} = M_{21}$, also exists. The graph of mutual inductance shows a strong similarity to the graph of magnetic field strength B along the x-axis. Assuming a homogeneous magnetic field, the mutual inductance M_{12}, between the two coils, can be calculated using (16.10) and substituting πr^2 for A, thus obtaining:

$$M_{21} = \frac{B_2(i_1) \cdot N_2 \cdot A_2}{i_1} = \frac{\mu_0 N_1 r_1^2 N_2 r_2^2 \pi}{2(r_1^2 + a^2)^{1.5}} \tag{16.11}$$

16.5.2 DESIGN ISSUES AND PERFORMANCE ANALYSIS

Designing UAVs and charging base stations that combine WPT and data communication, through power lines and integrates the smart grid facilities, will boost the convergence of these technologies to help implementing the 5G networks and beyond. However, the design of a maximum power transfer and high data rate communication between the power and data transmitters and receivers, through the wireless inductive channel can face a bandwidth challenging. In fact, high-Q and narrow bandwidth wireless channel of the power transmitters and receivers are required to ensure efficient WPT, which conceptually limits the high data rate communication, because it requires wider bandwidth communication channel.

16.5.2.1 Output Power Analysis

The maximum power that can be drawn from the secondary circuit is in most application not enough. To increase the maximum power output, a capacitor is added in the pickup circuit and the system is operated in resonance with the primary circuit. It has been shown that in resonance the maximum output power is [37–39]:

$$P_{\text{out}} = \omega I_1^2 \frac{M^2}{L_2} Q_2 = \omega I_1^2 k^2 L_1 Q_2 \tag{16.12}$$

where ω is the operating frequency and I_1 is the current in the primary circuit. $M = k\sqrt{L_1 L_2}$ is the mutual inductance [39], which is a function of the relative distance between the two coils and the coupling coefficient $k \in [0, 1]$. L_2 is the inductance in the pickup circuit and Q_2 is a tuned quality factor. The equation captures the basic challenges with inductive WPT and controllability. In fact, k is equal to one for a zero-flux leakage and perfect coupling, which is more applicable for transformers than for WPT, where distance and/or orientation between the coils result in a change in the magnetic flux. Equation (16.11) gave the relationship between output power and different system variables. It is clear that the output power depends on the coupling coefficient and mutual inductance. For instance, if the new coupling coefficient $k_1 = k/\sqrt{2}$, the frequency at the power transmitter or the quality factor of the secondary circuit should be doubled in order to compensate the performance drop.

Furthermore, we see that the coupling coefficient (squared) has more influence on the mutual inductance than the coil selection. The coupling coefficient is strongly connected to the relative distance and orientation between the two coils. The exact relationship between the coupling coefficient and the displacement is a complex function, which requires an analysis of the magnetic field strength and shape.

16.5.2.2 Design for Broadband Communication

The performance of an inductive channel, in high data rate communications, is evaluated by estimating the channel capacity, (C), based on the Shannon–Harltey approach [40].

$$C = \int_0^{BW} \log_2 \left(1 + \left(\frac{S(f)}{N(f)}\right)\right) df \tag{16.13}$$

where C is the channel capacity in bits per second (bps), BW is the bandwidth of the channel. $S(f)$ and $N(f)$ are the signal and the noise power spectrum, respectively. The noise is considered as additive white Gaussian noise at the receiver. Therefore, the signal to noise ratio (SNR) is then defined as SNR$= S_I/N$. The received power spectrum can be expressed as a function of the injected power spectrum and the transfer function

$$S(f) = |H(f)|^2 S_I \tag{16.14}$$

16.5.3 2D AND 3D POWER AND DATA CHANNEL

Due to the manufacturing variability and the imprecisely landing position that may occur, several hardware solutions are recently proposed to mitigate the mutual inductance variability, due to the space variation. These solutions were separately acting on the transmitting and/or receiving coils of the inductive channel. For instance, position adjustment and the mechanical control of the orientation of the receiver coil can be used at the WPT receiver side. Besides, some actions were recently undertaken at the WPT transmitter side, in order to generate an omnidirectional magnetic field, by proposing several 3D architectures based on a special arrangement of several coils at the wireless transmitter. Surely, this complex power transmitter facilitates the UAVs wireless charging, even during their flight.

Most wireless systems transfer power in a directional manner. For instance, a novel method has been presented to tune Class E inverters that are used as the primary coils driver in WPT systems based on the resonant inductive coupling. By varying the inductance of the DC-feed inductor via a DC current source and the switching frequency, the Class E inverter can operate in an optimized switching condition achieving ZVS and ZVDS, as the coupled coils of the inductive link are radially displaced [41]. It has to also be noted that the proposed tuning method can be efficiently implemented electronically and may be applied to other WPT applications, such as inductive charging for electric vehicles and mobile devices, where displacements and coils' misalignments are likely to occur. Future work should include further investigation into other possible tuning methods, such as adjusting the duty cycle of the MOSFET gate driving signal, while maintaining the frequency at a constant value.

Moreover, coil misalignment leads the primary coil driver to operate in an un-tuned state, which causes non-optimum switching operation and results in an increase in switching losses [41]. Several research efforts, along with experimental validation, present various methods to electronically tune a Class E inverter, used as a primary coil driver in an inductive WPT system to minimize the detrimental effects of misalignment between the inductively coupled coils, which may occur during operation. The tuning method uses current-controlled inductors and a variable switching frequency to achieve optimum switching conditions regardless of the misalignment.

Recently, 3D WPT has been addressed by several research groups. They describe a new discrete magnetic field control technique that can ensure omnidirectional wireless power transfer in an efficient manner [42–45]. The proposed control has been successfully implemented. Both theoretical and experimental results are included to confirm the validity of the control method. For instance, a discrete magnetic field vector control method is proposed. It has been successfully implemented in a 3D omnidirectional wireless power system. It enables the load positions to be determined so that magnetic flux can be focused to the targeted loads to maximize the energy efficiency. In this work, simulated and measured results for a single load are included, and the experiment results matched the theoretical model very well. The model used in this paper would also be applied to multiple loads conditions, however, the power distribution requires more advanced control methods.

The works in Refs [42,43] suggested the use of 3D orthogonal to form an omnidirectional wireless transmitter and the same 3D coil structure was adopted in the receiver. Although 3D orthogonal coil is impractical to use for wireless charging of mobile phones, and RFID tags, due to space arrangement at the receiver side, the potential application can be foreseen for drones and UAV's applications. Omnidirectional WPT transmitter with planar 2D receiver architecture has been proposed in Refs. [44–47]. While the current control method in [44] does not offer the genuine omnidirectional features, such limitation has recently been dealt with by the authors in [45].

Furthermore, the position control in inductive power transfer, and how it compares to state-of-the-art methods for inductive power transfer control were investigated [41]. The use of position control to keep the mutual inductance constant and thus optimize the efficiency of an IPT system has been investigated. With operations in the nearfield region, the mutual inductance is highly volatile, both for axial and lateral movement. Maintaining a constant reference position in this region imposes demanding requirements, on both the controlling unit and the mechanical apparatus responsible for the physical movement.

While generating a magnetic field in an omnidirectional manner offers great flexibility in free-positioning of the loads in a 3-dimensional manner, such approach has been known to be energy inefficient, because most of the magnet flux flows into regions without the loads. For example, the magnetic field is controlled to rotate in Refs. [42,43], covering the entire sphere, to ensure the planar receiving coil always picks up power from the transmitter, regardless of which angle the coil is placed.

16.6 CONCLUSION

This chapter has presented the concept of UAVs within the ecosystem of smart city. It is clear that UAVs have multiple potential uses in the vastly adopted concept of a smart city. UAVs can help in different areas, such eHealth, Intelligent Transport System (ITS), monitoring of smart grid, and emergency situations. More importantly, UAVs can provide another dimension for enhancing the upcoming 5G networking concept, by extending networking connectivity into areas, hard to reach by wired networking, or for occasional events, such as concerts or sports events. Although UAVs have great potential within a smart city, they suffer from short flight periods, due to the limited capacity of their rechargeable batteries. To take full advantage of UAVs, solutions to overcome their short lifetime of missions have to be found. This chapter has provided some solutions for the energy efficiency of UAVs, which mainly focus on the periodic sleeping of certain modules, the optimization of the paths of UAVs, in addition to decreasing and enhancing the data transmission.

Another solution, which shows high potential, is Wireless Power Transfer (WPT). WPT is foreseen to enable longer lifetime of UAVs' flight time/missions, through wireless charging of their batteries, while in flight mode. Moreover, the development of WPT technology brings a new source of vitality for the areas of electrified transportation, robotics, medically implanted devices, service industries, etc. Meanwhile, these areas pose new challenges to the WPT system, where an urgent challenge is to provide highly efficient power transfer to large regions of space, thus enabling device charging in an unencumbered and seamless fashion.

This chapter provided a brief overview of WPT, its different techniques, design issues, in addition to some performance issues. Within such an overview, different WPT techniques, such as the two- and three-phase topologies, along with the planar and three-dimensional power transmitter architectures, have been presented. Moreover, simultaneous WPT and data communication through the fixed low-frequency power lines infrastructure or the mobile high-frequency power lines infrastructure for in-flight charging for multiple UAVs, are also discussed.

To conclude, UAVs provide another dimension of freedom, for advancing the smart city concept, in addition to enhancing the performance of current and future wireless networking of those cities. Additionally, UAVs and WPT present high potential of benefits and foreseen services, whether combined or used separately.

However, based on the review given in this chapter, multiple challenges are still to be addressed, for those two technologies to achieve their highest foreseen potential. For instance, problems still exist with wireless charging, such as misalignment of transmitting and receiving coils in certain techniques, the short range of power transfer, and the difficulty in achieving both power and data transfer simultaneously. UAVs limited power has to be addressed in a more inclusive way, including how to optimize their wireless charging, when to charge, and where to locate the charging stations.

ACKNOWLEDGMENT

This work is funded by FCT/MCTES through national funds and when applicable co-funded EU funds under the project UIDB/50008/2020-UIDP/50008/2020; and co-funded by FEDER – PT2020 partnership agreement under the project, 5G-AHEAD IF/FCT- IF/01393/2015/CP1310/CT0002;

This work is also supported by the European Regional Development Fund (FEDER), through the Competitiveness and Internationalization Operational Programme (COMPETE 2020) of the Portugal 2020 framework [Project MUSCLES with Nr. 017787 (POCI-01-0145-FEDER-017787)];

REFERENCES

1. Y. Zeng, R. Zhang, & T. J. Lim, (2016). Wireless communications with unmanned aerial vehicles: Opportunities and challenges. *IEEE Communications Magazine*, 54(5): 36–42.
2. B. Li, Z. Fei, & Y. Zhang, (2019). UAV communications for 5G and beyond: Recent advances and future trends. *IEEE Internet of Things Journal*, 6(2): 2241–2263.
3. M. Mozaffari, W. Saad, M. Bennis, Y. N. Nam, & M. Debbah, (2019). A tutorial on UAVs for wireless networks: Applications challenges and open problems. *IEEE Commununications Surveys and Tutorials*, 21(3): 2334–2360.
4. M. Lu, M. Bagheri, A. P. James, & T. Phung, (2018). Wireless charging techniques for UAVs: A review, reconceptualization, and extension. *IEEE Access*, 6: 29865–29884.
5. M. Lu, A. James, & M. Bagheri, (2017). Unmanned aerial vehicle (UAV) charging from powerlines. *IEEE PES Asia-Pacific Power and Energy Engineering Conference*, Bangalore, India, pp. 1–6.
6. F. Mohammed, A. Idries, N. Mohamed, J. Al-Jaroodi, & I. Jawhar, (2014). UAVs for smart cities: Opportunities and challenges. *International Conference on Unmanned Aircraft Systems (ICUAS)*, pp. 267–273.
7. A. Radwan, K. M. S. Huq, S. Mumtaz, K. Tsang, & J. Rodrigeuz, (2016). Low-cost on-demand C-RAN based mobile small-cells. *IEEE Access*, 4: 2331–2339.
8. A. Radwan, M. F. Domingues, & J. Rodriguez, (2017). Mobile caching-enabled small-cells for delay-tolerant e-health apps. *IEEE Conference on Communications (ICC)*, Kansas City, MO, pp. 103–108.
9. H. Ullah, N. G. Nair, A. Moore, C. Nugent, P. Muschamp, & M. Cuevas, (2019). 5G communication: An overview of vehicle-to-everything drones and healthcare use-cases. *IEEE Access*, 7: 37251–37268.
10. Y. Huo, X. Dong, T. Lu, W. Xu, & M. Yuen, (2019). Distributed and multilayer UAV networks for next-generation wireless communication and power transfer: A feasibility study. *IEEE Internet of Things Journal*, 6(4): 7103–7115.
11. B. Galkin, J. Kibilda, & L. A. DaSilva, (2019). A stochastic model for UAV networks positioned above demand hotspots in urban environments. *IEEE Transactions on Vehicular Technology*, 68(7): 6985–6996.
12. N. Islam, K. Sithamparanathan, K. Chavez, J. Scott & H. Eltom, (2019). Energy efficient and delay aware ternary-state transceivers for aerial base stations. *Digital Communication and Network*, 5(1): 40–50.
13. X. Liu, M. Qiu, X. Wang, W. Liu, & K. Cai, (2017). Energy efficiency optimization for communication of air-based information network with guaranteed timing constraints. *Journal of Signal Processing Systems*, 86(2–3): 299–312.
14. M. Mozaffari, W. Saad, M. Bennis, & M. Debbah, (2016). Optimal transport theory for power-efficient deployment of unmanned aerial vehicles. *IEEE Conference on Communications (ICC)*, Kuala Lumpur, Malaysia, pp. 1–6.
15. C. H. Liu, Z. Chen, J. Tang, J. Xu, & C. Piao, (2018). Energy-efficient UAV control for effective and fair communication coverage: A deep reinforcement learning approach. *IEEE Journal on Selected Areas in Communications (JSAC)*, 36(9): 2059–2070.
16. L. Ruan, J. Wang, J. Chen, Y. Xu, Y. Yang, H. Jiang, Y. Zhang, & Y. Xu, (2018). Energy-efficient multi-UAV coverage deployment in UAV networks: A game-theoretic framework. *China Communications*, 15: 194–209.
17. Y. Zeng & R. Zhang, (2017). Energy-efficient UAV communication with trajectory optimization. *IEEE Transactions on Wireless Communications*, 16(6): 3747–3760.
18. M. Hua, Y. Wang, Z. Zhang, C. Li, Y. Huang, & L. Yang, (2019). Energy-efficient optimization for UAV-aided wireless sensor networks. *IET Communications*, 13(8): 972–980.
19. M. Mozaffari, W. Saad, M. Bennis, & M. Debbah, (2017). Mobile Unmanned Aerial Vehicles (UAVs) for energy-efficient internet of things communications. *IEEE Transactions on Wireless Communications*, 16(11): 7574–7589.
20. Y. Zeng, J. Xu, & R. Zhang, (2019). Energy minimization for wireless communication with rotary-wing UAV. *IEEE Transactions on Wireless Communications*, 18(4): 2329–2345.

21. K. Kuru, D. Ansell, W. Khan, & H. Yetgin, (2019). Analysis and optimization of unmanned aerial vehicle Swarms in logistics: An intelligent delivery platform. *IEEE Access*, 7: 15804–15831.
22. Y. Sun, D. W. K. Ng, D. Xu, L. Dai, & R. Schober, (2018). Resource allocation for solar powered UAV communication systems. in *Proceedings of IEEE Workshop on Signal Processing Advances in Wireless Communications (SPAWC)*, Kalamata, Greece, June 2018.
23. Y. Sun, D. Xu, D. W. K. Ng, L. Dai, & R. Schober, (2019). Optimal 3D-trajectory design and resource allocation for solar-powered UAV communication systems. *IEEE Transactions on Commununications*, 67(6): 4281–4298.
24. R. A. Sowah, M. A. Acquah, A. R. Ofoli, G. A. Mills, & K. M. Koumadi, (2017). Rotational energy harvesting to prolong fight furation of quadcopters. *IEEE Transactions on Industry Applications*, 53(5): 4965–4972.
25. L. Yang, J. Chen, M. O. Hasna, & H.-C. Yang, (2018). Outage performance of UAV-assisted relaying systems with RF energy harvesting. *IEEE Communications Letters*, 22(12): 2471–2474.
26. J. Xu, Y. Zeng, & R. Zhang, (2018). UAV-enabled wireless power transfer: Trajectory design and energy optimization. *IEEE Transactions on Wireless Communications*, 17(8): 5092–5106.
27. T. Long, M. Ozger, O. Cetinkaya, & O. B. Akan, (2018). Energy neutral internet of drones. *IEEE Communications Magazine*, 56(1): 22–28.
28. D. Vincent, P. S. Huynh, L. Patnaik, & S. S. Williamson, (2018). Prospects of Capacitive Wireless Power Transfer (CWPT) for unmanned aerial vehicles. *IEEE PELS Workshop on Emerging Technologies: Wireless Power Transfer (Wow)*, Montréal, QC, Canada.
29. B. Galkin, J. Kibilda, & L. A. DaSilva, (2019). UAVs as mobile infrastructure: Addressing battery lifetime. *IEEE Communications Magazine*, 57(6): 132–137.
30. J. Zyren, (2011). The HomePlug green PHY specification & the in-home smart grid. *IEEE International Conference on Consumer Electronics (ICCE)*, Las Vegas, NV, USA.
31. X. Liu, D. He, & H. Ding, (2019). Throughput maximization in UAV-enabled mobile relaying with multiple source nodes. *ScienceDirect Physical Communication*, 33: 26–34.
32. Z. Liang, et al. (2019). A compact spatial free-positioning wireless charging system for consumer electronics using a three-dimensional transmitting coil. *Energies*, 12(8): 1409–1418.
33. M. Kim, H. Kim, D. Kim, Y. Jeong, H. H. Park, & S. Ahn, (2015). A three-phase wireless-power-transfer system for online electric vehicles With reduction of leakage magnetic fields. *IEEE Transactions on Microwave Theory and Techniques*, 63(11): 3806–3813.
34. H. H. Lee, S. H. Kang, & C. W. Jung, (2017). 3D-spatial efficiency optimisation of MR-WPT using a reconfigurable resonator-array for laptop applications. *IET Microwave, Antennas & Propagation*, 11(11): 1594–1602.
35. W. Zhang, T. Zhang, Q. Guo, L. Shao, N. Zhang, X. Jin, & J. Yang., (2018). High-efficiency wireless power transfer system for 3D, unstationary free-positioning and multi-object charging. *IET Electric Power Applications*, 12(5): 658–665.
36. M. Su, Z. Liu, Q. Zhu, & A. P. Hu, (2018). Study of maximum power delivery to movable device in omnidirectional wireless power transfer system. *IEEE Access*, 6: 76153–76164.
37. Y. Lee, (2003). Antenna circuit design for RFID applications. Microship Application notes AN710. http://ww1.microchip.com/downloads/en/appnotes/00710c.pdf, available online.
38. R. Lourens, (2008). Low-frequency magnetic transmitter design. Microship Application notes AN2320. http://ww1.microchip.com/downloads/en/AppNotes/00232B.pdf, available online.
39. X. Mou & H. Sun, (2015). Wireless power transfer: Survey and roadmap. in *Proceedingds of IEEE Vehicular Technology Conference, (VTC Spring)*, Glasgow, UK, May 2015.
40. S. Zouaoui, M. Souilem, W. Dghais, A. Radwan, S. Barmada, & M. Tucci, (2019). Wireless power transfer and data communication through two-coil inductive channel. *IEEE Global Communications Conference*, Waikoloa, HI, USA, December 2019.
41. S. Aldhaher, P. C. Luk, & J. F. Whidborne, (2014). Electronic tuning of misaligned coils in wireless power transfer systems. *IEEE Transactions on Power Electronics*, 29(11): 5975–5982.
42. K. O'Brien, (2007). *I*nductively Coupled Radio Frequency Power Transmission System for Wireless Systems and Devices. Shaker Verlag GmbH, Germany.
43. O. Jonah, S. V. Georgakopoulos, & M. M. Tentzeris, (2013). Orientation insensitive power transfer by magnetic resonance for mobile devices. *IEEE Wireless Power Transfer Conference*, Perugia, Italy, pp. 5–8, May 2013.

44. D. Wang, Y. Zhu, Z. Zhu, T. T. Mo, & Q. Huang, (2012). Enabling multi-angle wireless power transmission via magnetic resonant coupling. *International Conference on Computing and Convergence Technology (ICCCT)*, Seoul, Korea, pp. 1395–1400, 2012.
45. W. Ng, C. Zhang, D. Lin, & S. Y. R. Hui, (2014). Two-and three-dimensional omni-directional wireless power transfer. *IEEE Transactions on Power Electronics*, 29(9): 4470–4474.
46. S. Barmada, M. Tucci, N. Fontana, W. Dghais & M. Raugi, (2019). Design and realization of a multiple access wireless power transfer system for optimal power line communication data transfer. *Energies*, 12(6): 988.
47. W. Dghais & M. Alam, (2016). Wireless power transfer and in-vehicle networking integration for energy-efficient electric vehicles. *Springer Mobile Networks and Applications*, 23: 1151–1164.

17 Cyber Security Considerations for Automated Electro-Mobility Services in Smart Cities

Binod Vaidya and Hussein T. Mouftah
University of Ottawa

CONTENTS

17.1 INTRODUCTION

Connected and Autonomous Electric Vehicles (CAEVs) are at the forefront of a technological evolution within the automotive and transportation sectors. As CAEVs can significantly contribute to optimize traffic flow, improve road safety, minimize dependence on fossil fuel, and reduce carbon emission in urban areas, attraction to these potential mobility paradigms in the smart cities will increase significantly in coming years. In a smart automated electro-mobility (AEM) system, a number of services can be offered to support shared mobility [1] with the use of an electric energy system. Even though the advent of AEM services in the urban areas would bring unprecedented opportunities for safety, innovation, and urban growth, these appealing possibilities could yield new challenges in cyber security [2]. CAEVs potentially become more vulnerable to cyber attacks, in turn, cyber threats would be a serious issue, thus robust protection mechanisms against such attacks are urgently needed to provide safety and security on AEM-related services [3]. This book chapter shall provide overviews of cyber security considerations for AEM services and discuss cyber security challenges in peer-to-peer (P2P) car-sharing [4] schemes in the urban areas. We also present security solution for such a car-sharing scheme based on the current state of the art such as Conjugated Authentication and Authorization [5].

The remainder of this book chapter is organized as follows: Section 17.2 describes Connected and Autonomous Electric Vehicles; Section 17.3 depicts Automated Electro-Mobility Services including Peer-to-peer Car-sharing services. Cyber Security for AEM services and Proposed Security Solution for P2P car-sharing systems are presented in Sections 17.4 and 17.5, respectively. Finally, Section 17.6 outlines a summary.

17.2 CONNECTED AND AUTONOMOUS ELECTRIC VEHICLES

Connected and Autonomous Electric Vehicles (CAEVs) are basically a combination of connected vehicles (CV), autonomous vehicles (AV), and electric vehicles (EV). A connected vehicle (CV) is a vehicle with technology that enables it to communicate with nearby vehicles, infrastructure, as well as objects; but may not be automated nor electrically operated. While an autonomous vehicle (AV) is a vehicle that is, in the broadest sense, capable of driving itself without human intervention. An electric vehicle (EV) is a vehicle that powers up and operates with energy stored in the battery.

Typically, CAEV is a vehicle that is not only capable of sensing its environment and navigating with little or no human input and communicating with other entities but also is operated by electric power. The CAEV may collect various sensors data through intra-vehicle communication system and feed the data to in-vehicle applications. Similarly, the CAEV may communicate with roadside unit (RSU), cloud infrastructure, and other CAEVs, to perform some computing tasks collaboratively [1].

Basically, the CAEV is composed of five major components: perception system; localization and mapping system; driving policy; communication system; and storage battery system.

CAEVs offer many potential advantages in terms of sustainable development for environment-friendly urban mobility, which are improved safety, greater mobility, reduced parking lots, relaxed drivers, increased car-sharing, increased road capacity, fewer CO_2 emissions, and less fuel costs.

CAEVs definitely transform the existing mobility paradigm. It can be observed that the technological advancements in driving assistants and network connectivity yield further opportunities and services and meet the sustainable development for cleaner, safer, and smarter shared mobility [1,2].

17.3 AUTOMATED ELECTRO-MOBILITY SERVICES

With the advent of modern transportation, shared mobility services [1] such as free-floating car-sharing, peer-to-peer (P2P) car-sharing [2], and ride-sharing or ride-hailing, yield viable alternatives to traditional public transport or taxis and personal cars. Mobility as a Service (MAAS) is the buzzword for modern transportation services in the smart cities since shared mobility services have exalted probable environmental, social, and economic benefits of reduced road traffic and parking congestion, reduced vehicle ownership, increased public transit ridership, reduced vehicle meters traveled (VMT), and minimized greenhouse gas (GHG) emissions. Shared mobility shall support multi-modal transport services that include first/last mile connections between outlying communities and public transit services [1].

Automated Electro-Mobility (AEM) services include various CAEV-related mobility services. For instance, CAEV will be capable to recharge at the wireless EVSE (EV supply equipment) without human intervention or the wired EVSE with the help of robotic arms.

17.3.1 PEER-TO-PEER CAR-SHARING SERVICES

In this section, we discuss Peer-to-peer (P2P) car-sharing services [2].

A car-sharing is one aspect of shared mobility, which is getting attention in the urban areas. Due to its potential benefits, the car-sharing has gained popularity and is becoming a major solution to more efficient mobility. Mainly, there are three forms of car-sharing: one-way free-float car-sharing; round-trip car-sharing; and peer-to-peer car-sharing.

Peer-to-Peer (P2P) car-sharing is a form of person-to-person lending as part of the sharing economy that enables individuals to share a car. Such a car-sharing is an innovative approach to vehicle sharing in which vehicle owners temporarily rent their personal vehicles to other users in their geographical areas. They are typically coordinated through private networks or car-sharing service providers. As with person-to-person lending, the adoption of location-based services and wide-spread smartphone technology have significantly contributed to the advancement of P2P car-sharing.

Basically, P2P car-sharing mechanism operates similarly to round-trip car-sharing in trip and payment type. However, the vehicles themselves are primarily privately owned and operated by third-party service providers. And vehicle owners receive rental costs to rent out their vehicles.

In modern days, more and more people are relying on such mobility services that have several benefits. For instance, P2P car-sharing alleviates upfront vehicle costs and scales more economically than traditional car-sharing. It is believed that P2P car-sharing can provide greater potential for vehicle accessibility than traditional car-sharing. Furthermore, P2P car-sharing has the potential to reduce the number of vehicles on the road and lower carbon emissions.

Figure 17.1 shows the conceived CAEV P2P car-sharing platform. It is based on the interaction between different groups of players through the digital platform: mobility claimants (e.g., CAEV users/riders), vehicle CEAV owners (can be personal or collective ownership), and P2P car-sharing service provider (can be public or private).

Typically, the conceived CAEV P2P car-sharing platform is subscription-based. That means, prior to the P2P car-sharing process, CAEV owners should sign-up to the P2P Car-sharing platform and provide their CAEVs' information. Similarly CAEV users/riders should also register to the same system.

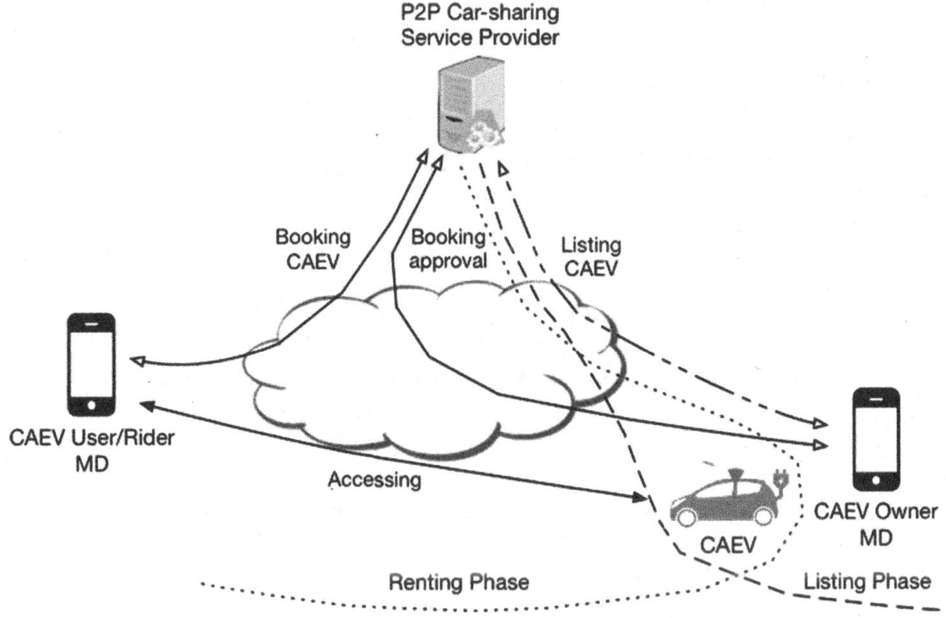

Figure 17.1 CAEV P2P car-sharing platform.

CAEV P2P car-sharing scheme may have two phases: the listing phase and the renting phase. In the listing phase, the CAEV owner shall register their vehicles for P2P car-sharing services. Whereas, the renting phase, CAEV users can book and access their desired vehicle by using their smartphones.

Once the CAEV is rented through P2P car-sharing platform, CAEV user/ rider shall be able to perform IoT-enabled CAEV services including automated recharging and automated valet parking.

17.4 CYBER SECURITY FOR AEM SERVICES

17.4.1 SECURITY GOALS

The main security goals for AEM services are authentication, authorization, integrity, confidentiality, non-repudiation, and availability.

- *Authentication*: The process of confirming and insuring the identity of objects. In AEM services, each entity should have the ability to identify and authenticate all other entities in the system.
- *Authorization*: The process of giving permission to an entity to do or have something.
- *Integrity*: The process of ensuring that information in the AEM ecosystem is trustworthy and accurate. In such an ecosystem, the alteration of basic information or even the infusion of invalid information could prompt major issues, e.g., identity theft, financial theft, accidents. A loss of integrity yields the unauthorized modification or destruction of information.
- *Confidentiality*: The process of ensuring that the information is only accessed by authorized people. A loss of confidentiality yields the unauthorized disclosure of information.
- *Non-repudiation*: The way toward guaranteeing the ability to demonstrate that a task or event has occurred (and by whom), with the goal that this cannot be denied later.
- *Availability*: The process of ensuring timely and reliable access to and use of information for the intended users. A loss of availability yields the disruption of access to or use of information or an information system.

17.4.2 POTENTIAL THREATS

Though AEM services have lots of potential in smart cities, they also bring new risks and challenges [6]. AEM services are vulnerable to various kinds of cyberattacks [7]. Typically, potential attacks may include as follows.

- *Eavesdropping*: It is a passive attack in which the malicious entity secretly listens to the communications of others without their consent.
- *Spoofing Attack*: It is an attack where a malicious actor successfully identifies as another by falsifying data, to gain an illegitimate advantage.
- *Man in the Middle (MiTM) Attack*: It is an attack where the malicious actor secretly relays and possibly alters the communications between two communicating parties who believe they are directly communicating with each other. Such an attack is possible when little or no authentication exists for communication between parties.
- *Replay Attacks*: It is an attack in which a valid data transmission is maliciously or fraudulently repeated or delayed. It is one of the effective MiTM attacks.
- *Denial of Service (DoS) Attack*: Due to the inherent real-time communication requirements in AEM services, the entities are vulnerable to DoS attacks. A DoS attack is when a network is flooded with excessive messages that restrict or block the transmission of legitimate traffic.

- *Ransomware Attack*: In such an attack, malicious software shall be employed to lock a system and/or its files and prevent access by legitimate users until ransom (i.e., fee) is paid. For example, WannaCry ransomware attack struck several organizations around the world in 2017 [2]. With the advent of CAEVs, ransomware is also a growing concern and it is believed that such vehicles become targets of specialized ransomware such as WannaDrive [8], which would disable a vehicle until the ransom has been paid.

17.4.3 SECURITY REQUIREMENTS

This section contains some of the technical requirements for providing security to AEM services [9,10].

- *Cryptographic Algorithms and Protocols*: The requirements aim for choosing cryptographic tools, key lengths, a pseudo random generator to be used for AEM services [11].
  - *Cryptographic Algorithms and Key Lengths*: Basically, cryptographic protocols are implemented using cryptographic algorithms such as symmetric and asymmetric and hash functions. They depend on certain parameters such as key size. For instance, if key size is too small, then the algorithm may be vulnerable to various attacks including brute-force attacks.
  - *Cryptographic Random Number Generator (RNG)*: Dedicated RNG should be devised to generate random numbers that are used for digital signatures or authentication protocols.
  - *Key Management*: The system should support establishing a fresh session key for each communication session. Such a key should be time-limited.
- *Communication Security*: The requirements address required security mechanisms for implementing end-to-end security for AEM services [12].
  - *Confidentiality*: Appropriate technique should be used to protect the confidentiality of the communication in the AEM system.
  - *Message Integrity*: As message integrity is usually verified using a message authentication code (MAC) or a block cipher in authenticated encryption mode, a suitable method should be deployed to verify the integrity of application layer messages in the AEM system.
  - *Message Freshness*: In order to mitigate replay attacks, the messages in the AEM system should be secured by deploying a counter or a nonce.
  - *Message Authentication*: The entity in the AEM system should be capable of determining the authenticity of a source of a message.
  - *Non-Repudiation*: Implementing some technique such as digital signatures, the AEM system would be able to support non-repudiation such that a sender would not be able to deny that he sent it.
- *Access Control*: The requirements intend to define a proper authorization method for the AEM system [13].
  - *Access Control*: The AEM system should allow to restrict access to certain individual.
  - *User Authentication*: An entity in the AEM system should be able to authenticate corresponding communication parties using a proper method such as a challenge-response protocol.

17.5 PROPOSED SECURITY SOLUTION FOR P2P CAR-SHARING SYSTEM

In this section, the proposed security solution for P2P car-sharing system based on Conjugated authentication and authorization (CAA) method [5] is discussed.

17.5.1 ISSUES AND CHALLENGES

Due to the continually evolving nature of security risks, it may be one of the problematic elements in cyber security domain. With advancements in the emerging technologies in the automotive domain, new security threats shall also arise. Keeping up with these continual changes and advances in security attacks to protect against them may be challenging [14].

Particularly, in the P2P car-sharing platform, the CEAV possesses multiple attack surfaces that might be exposed to possible cyber attacks. For instance, an adversary may breach vehicle remote applications (i.e., remote unlocking and locking) [15,16] on smartphones that interact with vehicular systems. Connected vehicle services such as over-the-air firmware updates provide potential threats. Infotainment systems could offer easy access for hackers to have direct access to onboard systems. Moreover, telematics systems may provide vulnerabilities to those vehicles as well as autonomous vehicle systems such as driving and parking assistance systems may become a potential foothold for a cyber attack [17,18].

The P2P car-sharing system may have a challenging task for establishing trust among end entities. Furthermore, CAEVs can be used by individuals who do not own them, so liability issues may be critical for P2P car-sharing services. Cyber attacks due to fraudulent user accounts may pose a growing risk to the P2P car-sharing systems.

17.5.2 SECURITY CONSIDERATIONS

Following security considerations are taken into account to tackle potential threats or vulnerabilities in the CAEV P2P car-sharing system :

- The platform should not necessitate imperishable access to the centralized backend system;
- The platform should have capabilities of distributed authentication and authorization;
- There should have security association between communication participating entities;
- Unique authentication and authorization should be provided for end entities;
- Integrity and authenticity of communications should be ensured;
- There should be a Credentials service authority (CSA) that acts as a root of trust;
- The CSA should issue enrollment and security credentials;
- The CSA should provide mechanisms to renew security credentials;
- Security credentials shall contain immutable attributes reflecting access rights and privileges of an entity;

17.5.3 CONJUGATED AUTHENTICATION AND AUTHORIZATION

Conjugated authentication and authorization (CAA) method comprised of the entity authentication of the participating entities and authenticated prior binding (APB). The CAA method yields comprehensive multi-level authentication, thus it provides more fine-grained access control for accessing the particular CAEV by the specified CAEV user. It should be noted that the CAA method is substantially different than multi-factor authentication.

In recharging use cases, only with a valid CAA, the permission is granted for the charging process; only then shall the CAEV user be able to perform charging activities for the respective CAEV in any EV charging network. That means, the CAA method allows the CAEV user to access for charging activities for the respective CAEV only when the authenticity and integrity of the participating entities are validated at two different levels. Figure 17.2 depicts the conjugated authentication and authorization (CAA) method that can be used for the recharging use cases.

Thus by employing the CAA method, adversaries would not have access for charging operation without successfully conducting two-level authentication and authorization processes, namely entity authentication and authenticated prior binding (APB).

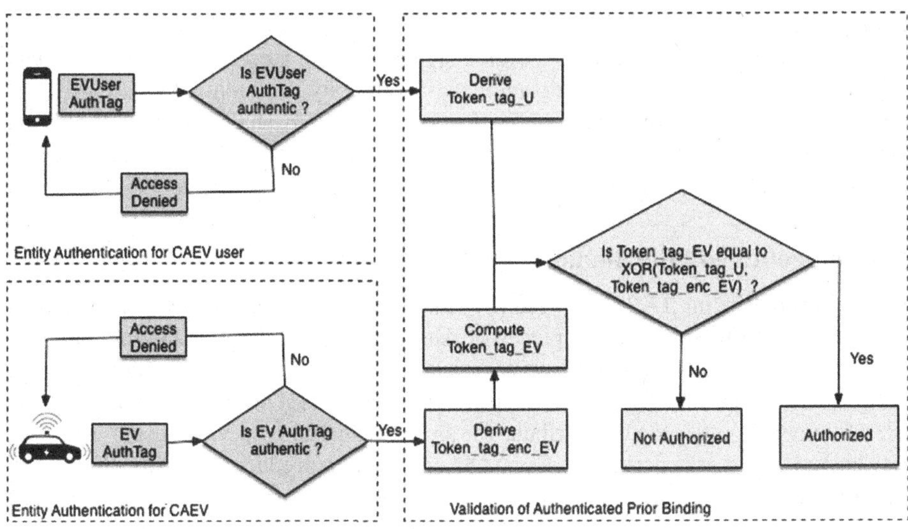

Figure 17.2 Conjugated authentication and authorization (CAA) method.

The CAA method is devised to furnish an explicit non-repudiation of the participating entities (i.e., CAEV user and CAEV) for the access control to the charging activities. So the authorized CAEV user would not be able to deny the charging activities later on.

Table 17.1 shows notations used in the CAA mechanism.

17.5.3.1 Entity Authentication

Entity authentication [5] is a mechanism, in which a party (i.e., validation authority) can gain assurance that the identity of another party (i.e., claimant) is same as stated and is actually

Table 17.1
Notations Used in the CAA Mechanism

Symbol	Description
ID_U, ID_R, ID_V	Identity for CAEV owner, CAEV user, and CAEV, respectively
CS_U, CS_R	Claim sets for CAEV owner and CAEV user, respectively
CS_V, CS_{V1}	Claim sets for CAEV wrt CAEV owner and CAEV user respectively
Tok_U, Tok_R, Tok_V	Tokens for CAEV owner, and CAEV user, respectively
Tok_V, Tok_{V1}	Tokens for CAEV wrt CAEV owner and CAEV user, respectively
$TTag_U, TTag_R$	Token tags for CAEV owner, and CAEV user, respectively
$TTag_V, TTag_{V1}$	Token tags for CAEV wrt CAEV owner and CAEV user, respectively
$TTagEnc_V, TTagEnc_{V1}$	Encrypted token tag for CAEV wrt CAEV owner and CAEV user, respectively
$LS_U(t_{exp} - t_{iss})$	Time duration of security token
GSK	Group shared key
TK_V	Transit key for CAEV
SK_{XX}, SK_{XX1}	Session keys for wrt CAEV owner and CAEV user, respectively
R_x	Nonce
AT_X	Authentication Tag

participating in the authentication process. Entity authentication may be achieved with an authentication tag, which shall be constructed as a digital signature using a private key of the participating entity. The authentication tag can provide authenticity, integrity, and non-repudiation for the individual entity. That means, with a valid authentication tag, the VA shall confirm not only that the request was sent by a particular claimant and was not altered in transit but also that the claimant cannot deny having sent the request.

17.5.3.2 Authenticated Prior Binding

Authenticated Prior Binding (APB) scheme [5] is a technique to cryptographically bind the participating entities (and their respective pieces of data) together so that this binding can be easily verified by a third party. The APB scheme is basically constructed by combining a cryptographic hash function (i.e., keyed MAC) and an encryption scheme (i.e., XOR operation) in such a manner that each participating entity shall contribute for an individual hashed value using a keyed MAC then these hashed values are encrypted using XOR operation to produce cipher-text based on both.

In the APB scheme, $TTag$ is formed using a keyed MAC algorithm. Credentials service authority (CSA) is responsible for creating $TTag_U$ and $TTagEnc_V$ for EV user and EV, respectively. Thus during generating individual specialized tokens for participating entities (i.e., EV user and EV), the CSA shall create an EV user $TTag_U$. As a $TTagEnc$ is an outcome of the APB scheme, the CSA shall construct $TTagEnc_V$ by XORing $TTag_V$ & $TTag_U$.

Authenticated Prior Binding (APB) is based on underlying "MAC-then-Encrypt concept, in which the user needs to perform two different operations (i.e., hash function and encryption) on a piece of data to provide authenticity, integrity, and confidentiality. Specifically, APB is meant for pre-binding of participating entities as well as for providing authenticity and integrity along with confidentiality.

17.5.4 TOKEN-BASED AUTHENTICATION AND AUTHORIZATION

A specialized token-based security technique [5] shall be encompassed in order to provide superior security, scalability, and flexibility as well as user convenience. In the token-based authentication and authorization, the system shall securely provide an authorization token to the participating entity (i.e., CAEV and CAEV user), which can provide authenticity, integrity, non-repudiation, and achieve convincingly faster re-authentication. One of the significant benefits of such token-based authentication and authorization is that it shall eliminate time-consuming lookup into the database every time when authenticating the participating entities in the P2P car-sharing systems.

A token should be self-contained, and claim-based, that means, it shall contain all the required information within itself, and has a set of claims that define about the participating entity, thus, eliminating the need to query the database during charging sessions.

The token should have a header, a payload, and a signature; similar to a JSON Web Token (JWT). The header shall at least encompass a type, an algorithm, and may contain certificate information.

In general, the payload shall comprised of at least a claim set (CS), which should include the following claims: (TokenID, Issuer, Subject, Pair, Nonce, Issued date, Expiration date).

In case of the EV owner, the payload shall also have $TTag$, whereas in the case of the EV, instead of $TTag$ field, it shall have $TTagEnc$. The signature is constructed by signing the header and the payload with the help of either a secret (with HMAC) or a private key (with RSA or ECC). The information in the token can be validated and trusted, since it is digitally signed. And the token must be renewed in a periodic manner to thwart replay attacks.

And for an enrolling entity, a cross-related token having tuples {Subject, Pair} is constructed. For instance, the owner ID (ID_U) as a subject and the electric vehicle ID (ID_V) as a pair shall be designated in the token of the CAEV owner (Tok_U), whereas ID_V as a subject and $ID_U)$ as a pair shall be designated in the token of the EV (Tok_V).

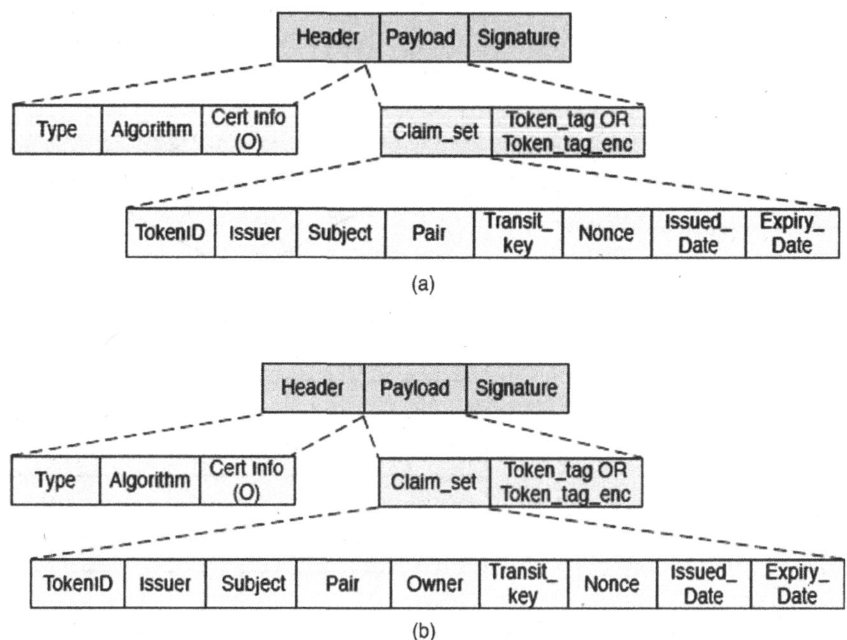

Figure 17.3 Token formats: (a) for wrt CAEV owner and (b) for wrt CAEV user.

Figure 17.3a and b show token formats for with respect to CAEV owner and with respect to CAEV user, respectively. It should be noted that $TTagEnc, TK$ shall be used for CAEV only.

17.5.4.1 Token Generation Phases

Credentials service authority (CSA) shall be responsible for generating security tokens to the CAEV owner and the CAEV user/ rider. APB token generation shall be divided into two phases. The first phase is for CAEV owner, which ensues during the registration of CAEV owner. Whereas, the second one is for CAEV user/ rider that occurs during the renting phase of the P2P car-sharing process. It should be noted that the CSA shall share a group shared key GSK to all the participating entities during the entity registration.

17.5.4.1.1 Token Generation for CAEV Owner

During this phase, security tokens for CAEV owner and corresponding CAEV shall be generated.

Upon validating CAEV owner credentials, the CSA shall set CS_U and also compute SK_{UD} and $TTag_U$ to generate a EVOwner token. The EVOwner token generation procedure is shown as follows.

$Step$ 1. Set $CS_U \in \{TokID_i, Iss, ID_U, ID_V, R_i, LS_U(t_{exp} - t_{iss})\}$
$Step$ 2. Compute $SK_{UD} := KDF(R_i, x_D.X_U)$
$Step$ 3. Compute $TTag_U := HMAC(SK_{UD}, CS_U)$
$Step$ 4. Compute $Sign$
$Step$ 5. Construct $Tok_U \in \{Type, Algo, Cert\_Info, CS_U, TTag_U, Sign\}$
$Step$ 6. Hold $TTag_U$ until Tok_V is constructed
$Step$ 7. Send Tok_U to CAEV owner

Figure 17.4 depicts a flow diagram for the generation of EVOwner token that shall be used by the CAEV owner.

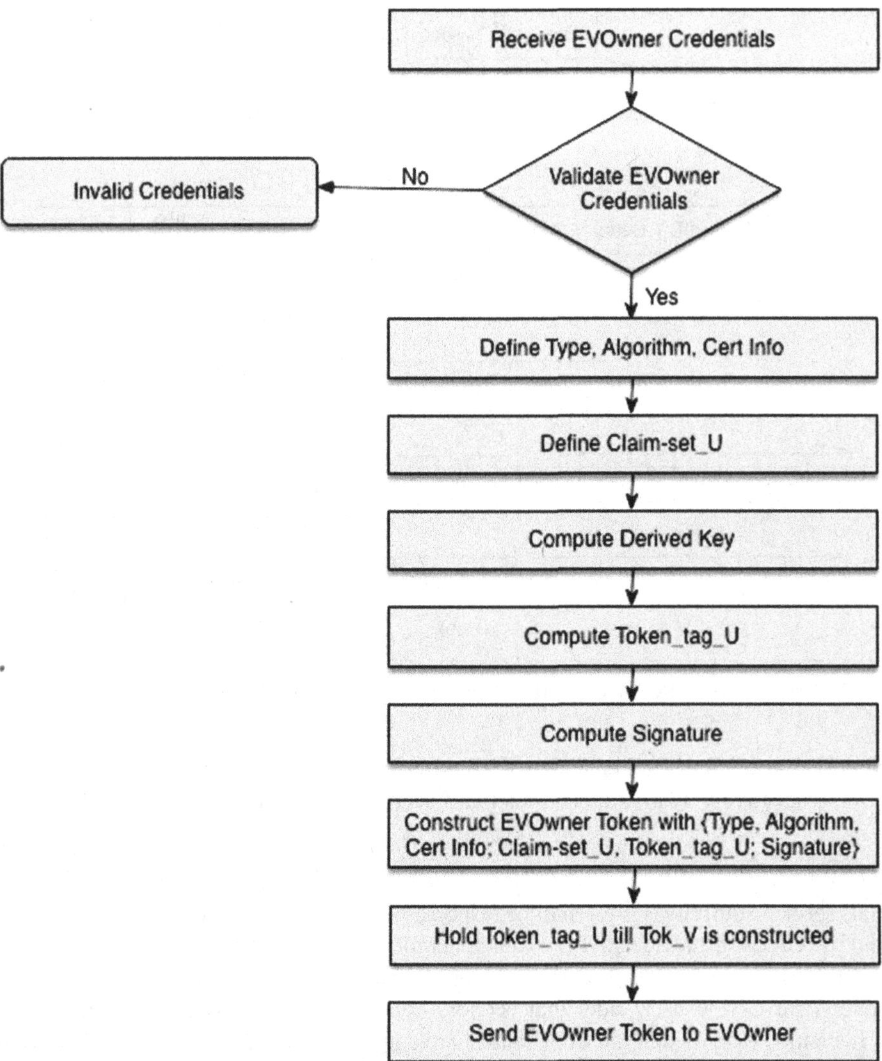

Figure 17.4 Flow diagram for the generation of EV owner token.

Similarly, upon validating CAEV credentials, firstly, CSA shall compute TK_V and then set CS_V. It shall also compute SK_{VD}, $TTag_V$, and $TTagEnc_V$. The EV token generation process is shown as follows.

Step 1. Compute $TK_V := XOR(x_D.X_V, GSK)$
Step 2. Set $CS_V \in \{TokID_j, Iss, ID_V, ID_U, TK_V, R_j, LS_V(t_{exp} - t_{iss})\}$
Step 3. Compute $SK_{VD} := KDF(R_j, x_D.X_V)$
Step 4. Compute $TTag_V := HMAC(SK_{VD}, CS_V)$
Step 5. Compute $TTagEnc_V := XOR(TTag_V, TTag_U)$
Step 6. Construct $Tok_V \in \{Type, Algo, Cert\_Info, CS_V, TTagEnc_V, Sign\}$
Step 7. Send Tok_V to CAEV

Figure 17.5 depicts a flow diagram for the generation of EV token for the corresponding EVOwner that shall be used by the CAEV.

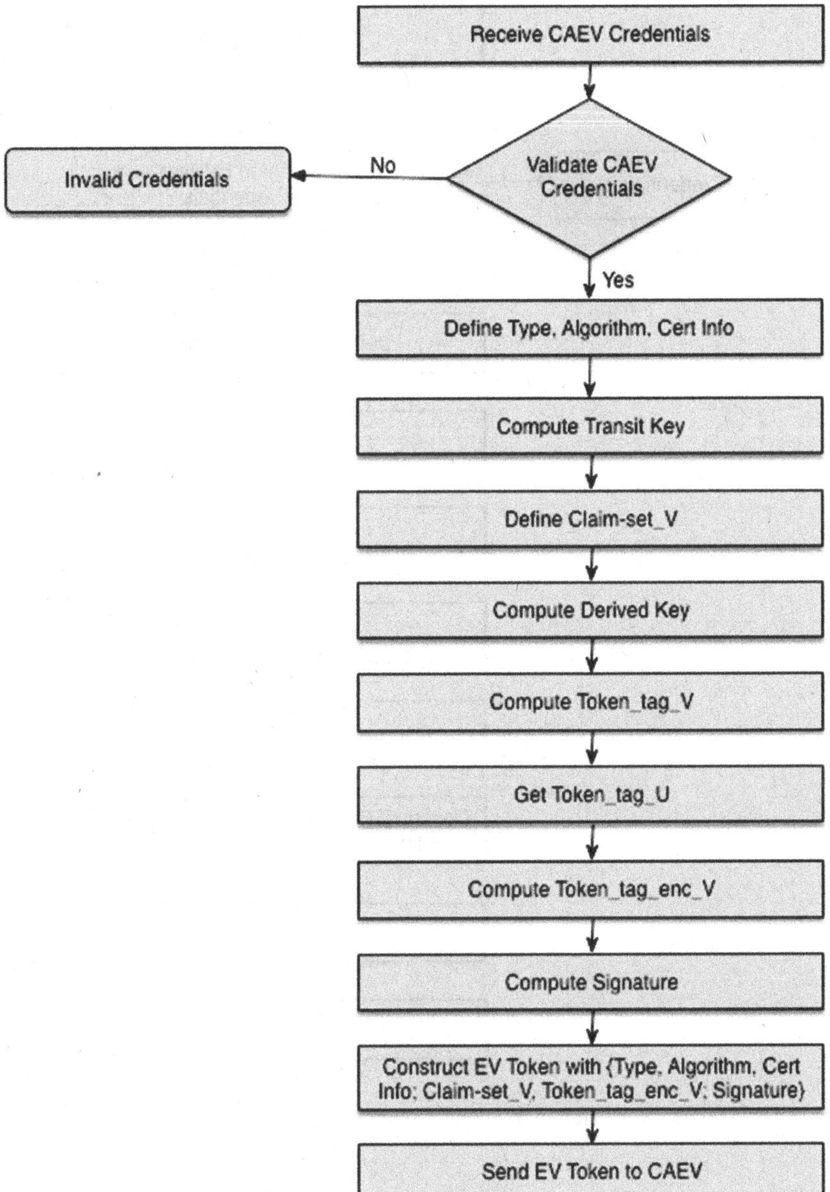

Figure 17.5 Flow diagram for the generation of EV token for the corresponding EV owner.

17.5.4.1.2 Token for CAEV User/ Rider

During this phase, security tokens for CAEV user/rider and corresponding CAEV shall be generated.

Once validation of CAEV user credentials, the CSA shall set CS_R and then compute $SK_{DR}, TTag_R$. The EVUser token generation procedure is depicted as follows.

$Step$ 1. Set $CS_R \in \{TokID_k, Iss, ID_R, ID_V, ID_U, R_k, LS_R(t_{exp} - t_{iss})\}$
$Step$ 2. Compute $SK_{DR} := KDF(R_k, x_D.X_R)$
$Step$ 3. Compute $TTag_R := HMAC(SK_{DR}, CS_R)$

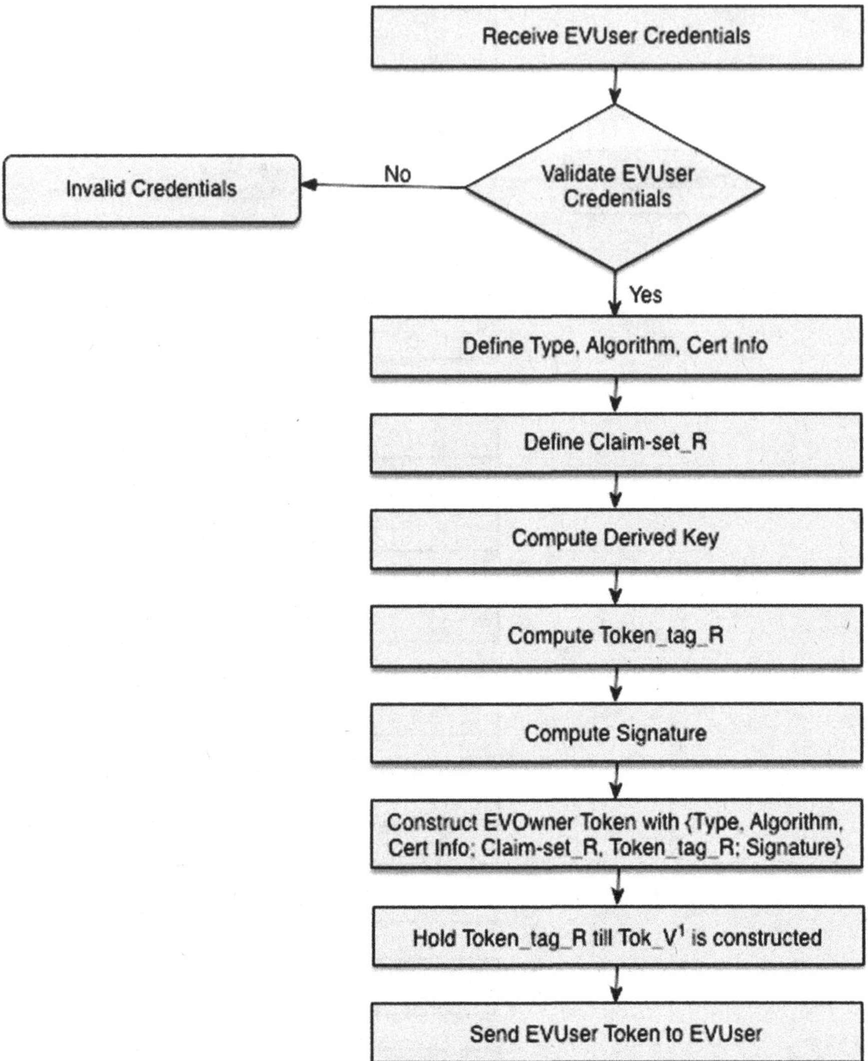

Figure 17.6 Flow diagram for generation of EV user token.

Step 4. Compute *Sign*
Step 5. Construct $Tok_U \in \{Type, Algo, Cert\_Info, CS_R, TTag_R, Sign\}$
Step 6. Hold $TTag_R$ until Tok_{V^1} is constructed
Step 7. Send Tok_R to CAEV user

Figure 17.6 depicts a flow diagram for the generation of EV User token that shall be used by the CAEV user/rider.

Likewise, upon validating CAEV credentials, the CSA shall compute TK_V and then set CS_{V^1}. It shall also compute $SK_{V_D^1}$, $TTag_{V^1}$, and $TTagEnc_{V^1}$. To generate such an EV token, the process is shown as follows.

Step 1. Compute $TK_V := XOR(x_D.X_V, GSK)$
Step 2. Set $CS_{V^1} \in \{TokID_l, Iss, ID_V, ID_R, ID_U, TK_V, R_l, LS1_V(t_{exp} - t_{iss})\}$
Step 3. Compute $SK_{V_D^1} := KDF(R_l, x_D.X_V)$

Step 4. Compute $TTag_{V^1} := HMAC(SK_{VD^1}, CS_{V^1})$
Step 5. Compute $TTagEnc_{V^1} := XOR(TTag_{V^1}, TTag_R)$
Step 6. Construct $Tok_{V^1} \in \{Type, Algo, Cert\_Info, CS_{V^1}, TTagEnc_{V^1}, Sign\}$
Step 7. Send Tok_{V^1} to CAEV

Figure 17.7 depicts flow diagram for the generation of EV token for corresponding EVUser that shall be used by the CAEV.

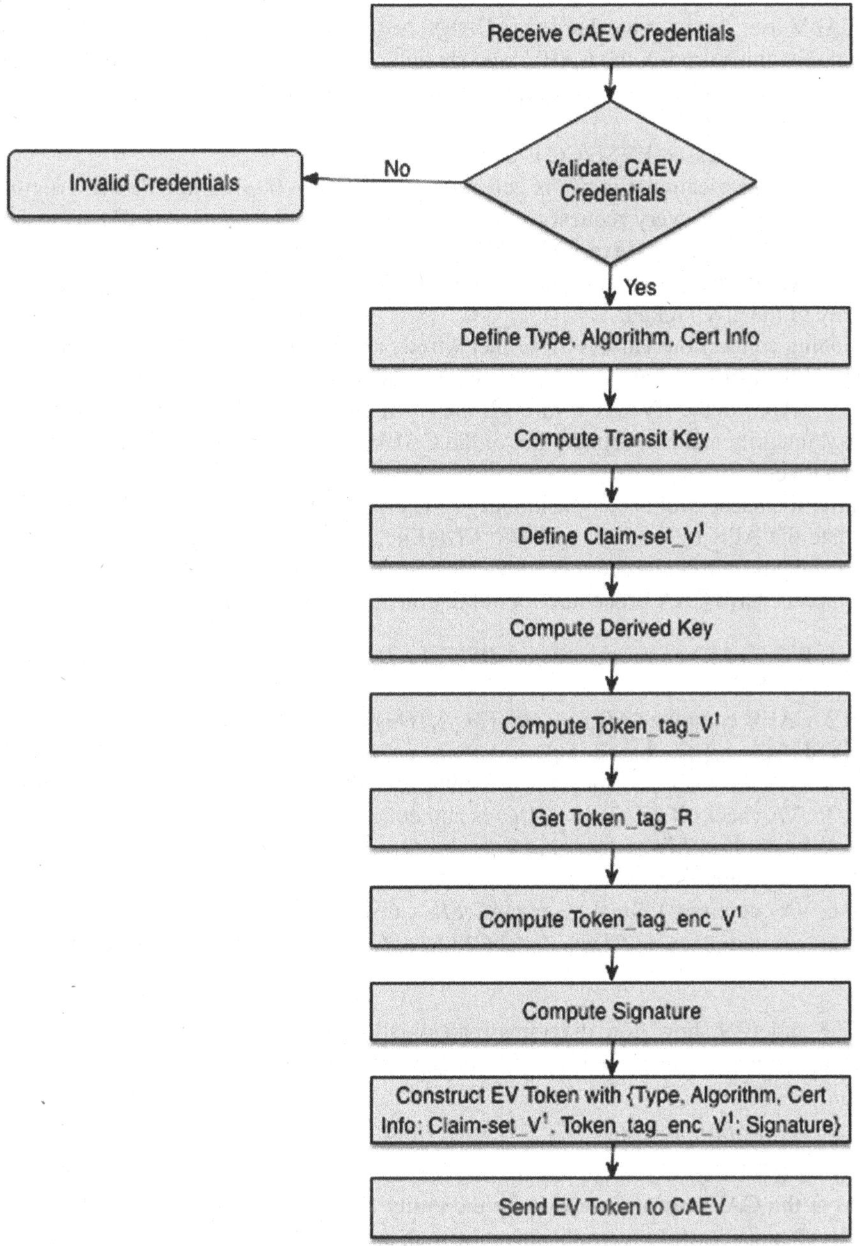

Figure 17.7 Flow diagram for generation of EV token for corresponding EV user.

17.5.4.2 Validation Phase

User identification and authentication can be performed by various entities. A validation authority (VA) shall be reside at various entities such as EVSE Control Center (EVSE CC) and CAEV.

For instance, one of the use cases is the automated recharging. CAEV may need to recharge at the EVSE, which can be wired or wireless. In the case of wired EVSE, some robotic arms shall be used to handle the charge process. Whereas, the wireless EVSE shall have a ground assembly or charging pad such that the CAEV needs to align with the charging pad.

17.5.4.2.1 Verification at EVSE CC

When a CAEV user desires to recharge his CAEV, he can initiate the recharging process using the mobile app. For this purpose, the CAEV user shall choose proper settings (i.e., mode of payment) in advance. CAEV can perform recharging automatically only after the valid operation of the CAA method.

The validation process at EVSE CC is discussed below.

Initially, an authentication tag AT_R is generated as $DS(Tok_R, ID_R)$ (where DS is a digital signature). Subsequently, for every request to the EVSE CC, the mobile device (MD) shall submit the $\{Tok_R\}$ along with $\{AT_R\}$. Upon receiving $\{ID_R, Tok_R, AT_R\}$, the VA at the EVSE CC shall validate the authenticity, integrity, and non-repudiation of the CAEV user as well as verify the authenticity and integrity of the given token.

Provisioning connection with EVSE (either wired or wireless), the EV shall generate an authentication tag (AT_V) using its private key such that it is the digital signature $DS(ID_V, Tok_{V1})$ and send $\{ID_V, Tok_{V1}, AT_V\}$ to the EVSE CC through the EVSE. The VA at the EVSE CC shall validate the authenticity, integrity and non-repudiation of the CAEV as well as verify authenticity and integrity of the given token.

Only after the successful accomplishment of the entity authentication, the VA at the EVSE CC shall validate the APB by comparing $XOR(TTagEnc_{V1}, TTag_R)$ with computed $TTag_{V1}$. If the result is true, the recharging process occurs.

The comprehensive CAA procedure for validation at the EVSE CC is given as follows.

Step 1. EVuser MD computes $AT_R := DS(Tok_R, ID_R)$ and sends $\{ID_R, Tok_R, AT_R\}$ to EVSE CC

Step 2. CAEV computes $AT_{V1} := DS(Tok_{V1}, ID_V)$ and sends $\{ID_V, Tok_{V1}, AT_{V1}\}$ to EVSE CC

Step 3*a*. VA checks if $DS(Tok_R, ID_R)$ is authentic, then proceed furtherotherwise reject

Step 3*b*. VA checks if $DS(Tok_{V1}, ID_V)$ is authentic, then proceed furtherotherwise reject

Step 4. VA derives $TTag_R$ from Tok_R

Step 5. VA computes $SK_{V1} := KDF(R_l, XOR(TK_V, GSK))$

Step 6. VA computes $TTag_{V1} := HMAC(SK_{V1}, CS_V)$

Step 7. VA compares if $TTag_{V1} = XOR(TTagEnc_{V1}, TTag_R)$, then proceed further, otherwise reject

Figures 17.8 and 17.9 show flow diagrams for validation of entity authentication of participating end-entity and validation of Authenticated Prior Binding, respectively.

17.5.4.2.2 Verification at CAEV

For deploying IoT-based CAEV applications such as locking, unlocking, the CAA method can be utilized at the CAEV. In this case, only the entity authentication for the CAEV user shall be performed. After successful accomplishment of such an entity authentication, the VA at the CAEV shall validate the APB by comparing $XOR(TTagEnc_{V1}, TTag_R)$ with computed $TTag_{V1}$. If the result is true, it shall proceed further.

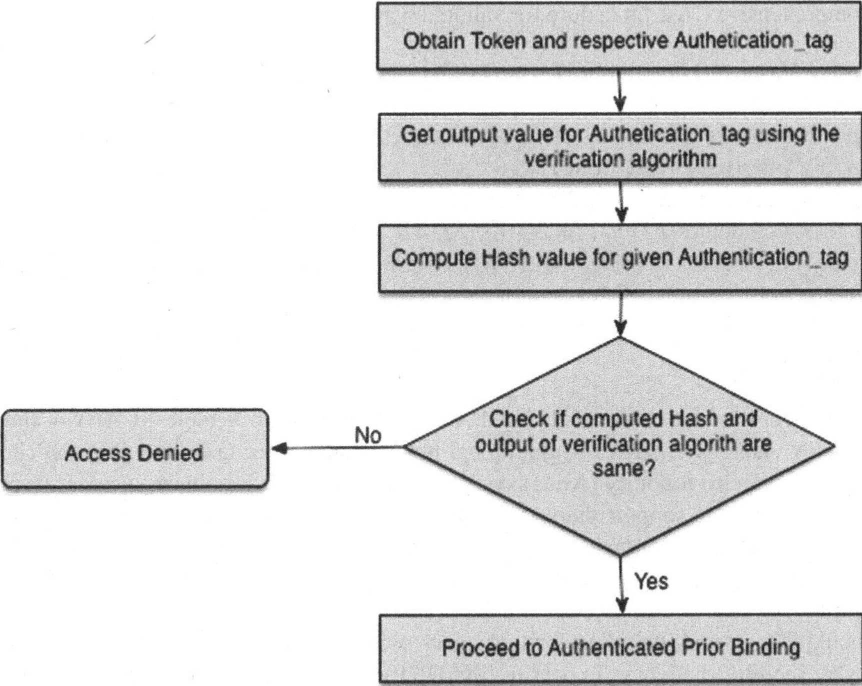

Figure 17.8 Flow diagram for the validation of entity authentication.

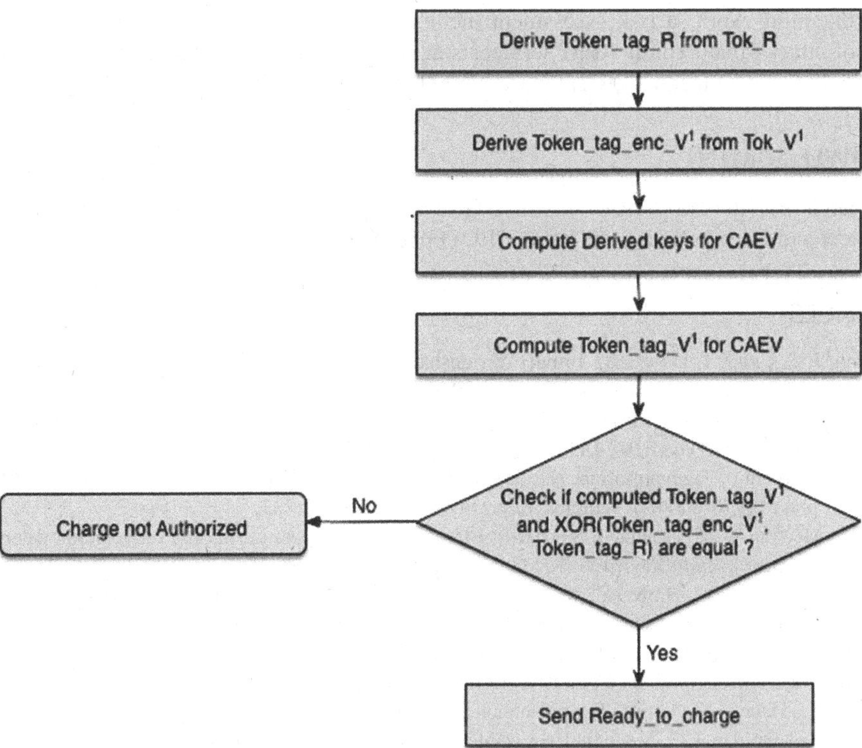

Figure 17.9 Flow diagram for the validation of authenticated prior binding.

The comprehensive CAA procedure for validation at the CAEV is given as follows.

Step 1. EVuser MD computes $AT_R := DS(Tok_R, ID_R)$ and sends $\{ID_R, Tok_R, AT_R\}$ to CAEV
Step 2. VA checks if $DS(Tok_R, ID_R)$ is authentic, then proceed further otherwise reject
Step 3. VA derives $TTag_R$ from Tok_R
Step 4. VA computes $SK_{V1} := KDF(R_I, x_D.X_V)$
Step 5. VA computes $TTag_{V1} := HMAC(SK_{V1}, CS_V)$
Step 6. VA compares if $TTag_{V1} = XOR(TTagEnc_{V1}, TTag_R)$, then proceed further, otherwise reject

17.6 SUMMARY

With the wide deployment of Connected and Autonomous Electric Vehicles (CAEVs), the shared mobility yields viable alternatives to traditional public transport or taxis and personal cars. In a smart automated electro-mobility (AEM) system, a number of services including peer-to-peer car-sharing can be offered to support shared mobility with the use of the electric energy system. Even though the advent of AEM services in the urban areas would bring unprecedented opportunities for safety, innovation, and urban growths, these appealing possibilities could yield new challenges in cyber security.

This book chapter provides overviews of cyber security considerations for AEM services and discuss cyber security challenges in peer-to-peer (P2P) car-sharing schemes in the urban areas. We also present a security solution for such a car-sharing scheme based on Conjugated Authentication and Authorization method such that entities such as CAEV and EVSE CC can have VA to validate the authenticity of the participating entities.

In forthcoming work, a risk assessment methodology including a threat model to enable an analysis of attacks that exploit AEM services shall be considered. Furthermore, distributed trust adopted by blockchain technology can be exploited in future enhancement.

ACKNOWLEDGMENT

This research work is supported by Smart Grid Fund (SGF), Ministry of Energy, The Ontario Government and Canada Research Chair (CRC) Fund, Canada.

REFERENCES

1. Farhan, J. & Chen, T. D. (2018). Impact of ridesharing on operational efficiency of shared autonomous electric vehicle fleet. *Transportation Research Part C* , 93, 310–321.
2. Takahashi, J. (2018). An overview of cyber security for connected vehicles. *IEICE Transactions on Information and Systems*, E101-D(11), 2561–2575.
3. U.S. Department of Transportation, National Highway Traffic Safety Administration. (2016). Cybersecurity best practices for modern vehicles. Report No. DOT HS 812 333.
4. Dill, J., McNeil, N., & Howland, S. (2019). Effects of peer-to-peer carsharing on vehicle owners' travel behavior. *Transportation Research Part C: Emerging Technologies*, 101, 70–78.
5. Mouftah, H.T., Vaidya, B., & Adams, C. (2018). Methods and systems for conjugated authentication and authorization. US Patent Application 20180336551.
6. Thai, J., Yuan, C., & Bayen, A.M. (2018). Resiliency of mobility-as-a-service systems to denial-of-service attacks. *IEEE Transactions on Control of Network Systems*, 5(1), 370–382.
7. Hodge, C., Hauck, K., Gupta, S., & Bennett, J. (2019). *Vehicle Cybersecurity Threats and Mitigation Approaches*. Golden, CO: National Renewable Energy Laboratory. NREL/TP-5400-74247.

8. Wolf, M., Lambert, R., Enderle, T., & Schmidt, A.D. (2017). WANNADRIVE? feasible attack paths and effective protection against ransomware in modern vehicles. In *15th Embedded Security in Cars (ESCAR) Conference Europe,* Berlin, Germany, November 7–8, 2017.
9. Li, X., Yu, Y., Sun, G., & Chen, K. (2018). Connected vehicles' security from the perspective of the in-vehicle network. *IEEE Network*, 32(3), 58–63.
10. SAE International. (2017). Security for plug-in electric vehicle communications J2931/7_201710.
11. Chen, Q.A., Yin, Y., Feng, Y. Mao, Z.M., & Liu, H.X. (2018). Exposing congestion attack on emerging connected vehicle based traffic signal control. In *Network and Distributed Systems Security (NDSS) Symposium 2018,* San Diego, CA, February 18–21, 2018.
12. Burzio, G., Cordella, G.F., Colajanni, M., Marchetti, M., & Stabili, D. (2018). Cybersecurity of connected autonomous vehicles : A ranking based approach. In *2018 International Conference of Electrical and Electronic Technologies for Automotive,* Milan, Italy, July 9–11, 2018.
13. Parkinson, S., Ward, P., Wilson, K., & Miller, J. (2017). Cyber threats facing autonomous and connected vehicles: Future challenges. *IEEE Transactions on Intelligent Transportation Systems*, 18(11), 2898–2915.
14. Lim, H.S.M. & Taeihagh, A. (2018). Autonomous vehicles for smart and sustainable cities: An in-depth exploration of privacy and cybersecurity implications. *Energies*, 11(5:1062), 1–23.
15. Bird, J. (2019). Car hacking threatens vision of connected mobility. Special report: The future of the car, *The Financial Times*. URL: https://www.ft.com/content/163f08c6-6ce3-11e9-9ff9-8c855179f1c4.
16. Hunt, T. (2016). Controlling vehicle features of Nissan LEAFs across the globe via vulnerable APIs. URL: https://www.troyhunt.com/controlling-vehicle-features-of-nissan/.
17. Lee, C.W. & Madnick, S. (2018). A system theoretic approach to cybersecurity risk analysis and mitigation for autonomous passenger vehicles. Working Paper CISL# 2018-09, MIT SLOAN. URL: http://web.mit.edu/smadnick/www/wp/2018-09.pdf.
18. Nash, L., Boehmer, G., Wireman, M., & Hillaker, A. (2017). *Securing the Future of Mobility: Addressing Cyber Risk in Self-Driving Cars and Beyond.* Deloitte University Press. URL: https://www2.deloitte.com/us/en/insights/focus/future-of-mobility/cybersecurity-challenges-connected-car-security.html.

18 Incentivized and Secure Blockchain-based Firmware Update and Dissemination for Autonomous Vehicles

Mohamed Baza
Sam Houston State University

Joe Baxter
Western Kentucky University

Noureddine Lasla
Hamad bin Khalifa University

Mohamed Mahmoud
Tennessee Tech University

Mohamed Abdallah
Hamad bin Khalifa University

Mohamed Younis
University of Maryland

CONTENTS

18.1 INTRODUCTION

Over the last few years, the automobile industry has achieved a notable leap toward the realization of practical Autonomous Vehicles (AVs). AVs are equipped with sophisticated systems and subsystems to provide vehicles with advanced communication capabilities, computer vision, autonomous decision-making capability, etc., to enable them to autonomously drive without any human intervention [1]. AVs have the potential to revolutionize our current transportation system by reducing congestion and travel time, increasing fuel efficiency, and improving road safety [2,3].

AVs are composed of many subsystems running specific firmware programs that enable performing all control, monitoring, and data manipulation operations. However, by controlling the functionality of the subsystems through the installation of infected versions of the corresponding firmware, an attacker can successfully hack AVs and fully/partially control them, e.g., to involve the vehicle in accidents deliberately, which may lead to damages and kill people. As an example of this attack, Chrysler company announced a recall for 1.4 million vehicles after hackers have managed to turn off the engine remotely while the vehicles were on motion by exploiting a hackable software vulnerability via the internet-connected entertainment system [1,11]. Therefore, ensuring the integrity and authenticity of AVs' firmware update is primordial and must be carefully investigated. In addition, it may happen that multiple AVs with their various subsystems need to be updated urgently and simultaneously, e.g., to fix newly discovered bugs, thus a high availability of the updates is required.

Most of the existing solutions for firmware update depend on the client-server model in which a manufacturer delegates the process of firmware distribution to trusted cloud providers, such as Microsoft Azure and IBM Cloud [4,5]. However, this central client-server architecture has the single point of failure problem. In case the server is not available, the clients (AVs) cannot access the resources (updates) no matter how powerful the server is. For AVs, there are several factors that make the availability and security of the firmware updates a challenging task. To elaborate, the number of autonomous vehicles on roads is expected to reach 20.8 million in the U.S. alone [4]. Also, each AV has many subsystems that run different programs designed to accomplish specific functions. This creates a tremendous load on the server-side and can broaden the sources of cyberattacks. Moreover, Vehicles may last for many years (more than 20 years); thus the integrity and authenticity of the AVs' firmware should be guaranteed for a very long time.

Recently, blockchain with its capabilities to provide a verified, transparent and distributed ledger without a need for a trusted third party, has drawn the attention of both academia and industry across a wide range of domains, including finance, health-care, and energy [6]. It has evolved, afterward, beyond that to support the deployment of more general-purpose distributed applications. This concept has been introduced by Vitalik Buterin and refers to it as smart-contracts or decentralized autonomous organizations [7]. Smart-contract can be described as an autonomous computer program running on a blockchain network. This program acts as a contract where the terms of the contract can be pre-programmed with the ability to self-execute and self-enforce itself without the need for trusted authorities [8,9].

18.1.1 CONTRIBUTIONS

In this chapter, we propose an incentivized blockchain-based firmware update scheme tailored for AVs. We use blockchain and smart-contract technology to guarantee the authenticity and integrity of the updates. We also exploit the AVs' inter-communication capability and incentivize AVs to participate in the distribution of new firmware updates from one to another, therefore, ensuring high availability and fast delivery of the updates. The main contributions of the proposed chapter are outlined as follows:

- A consortium blockchain created by multiple manufacturers of the AVs and its subsystems is proposed. Each consortium member has the permission to write a smart contract that handles the logic ensuring the authenticity and integrity of its firmware updates without the need for a trusted third party. In addition, in our scheme, the blockchain ledger only stores the hash of the firmware updates to reduce the storage overhead on the blockchain leger.
- A high availability and reliability are ensured by incentivizing AVs to participate in the distribution of the firmware updates. Distributor AVs are rewarded for their honest participation and the smart contract is used to manages the reward system and keeps track of the reputation credit of each AV.
- Attribute-based encryption (ABE) technique is used to allow manufacturers to set a policy about who has the right to download and use an update. The access policy is defined on the smart contract that enforces its execution without an intermediary, so only authorized AVs can request and receive the update.
- Since AVs do not trust each other, a Zero-Knowledge Proof protocol is employed. A distributor can exchange an encrypted version of the update in return for proof of distribution from a receiver AV. The delivery of the decryption key is ensured by the smart contract which reveals the key once the proofs are collected. The smart contract also increments the distributor's reputation based on the received proofs.
- For efficient on-chain computation overhead, we used aggregate signature technique, where the distributor AVs can aggregate multiple proofs of distribution into a single short signature, and then send it to the smart contract.
- A detailed proof-of-concept implementation using real devices to emulate both the blockchain nodes (i.e., validators/miners) and the AVs. Our scheme has been implemented in smart contracts. The results indicate that on-chain overheads are practical and do not affect the operation of AVs network.

The rest of this chapter is organized as follows. Section 18.2 discusses the related work. In Section 18.3, we discuss some preliminaries. Then, our proposed scheme is presented in detail in Section 18.4. Detailed performance evaluations including our proof-of-concept implementations are provided in Section 18.5. Security analysis is given in Section 18.6 respectively. Finally, we give concluding remarks in Section 18.7 followed by Acknowledgement.

18.2 LITERATURE REVIEW

As the number of autonomous vehicles increases, the security and reliability of AVs require complex server infrastructure, thus making it very expensive for the manufacturers. In the literature, the security of autonomous vehicles has been studied in different works [10,12–14]. More specifically, the firmware update has been discussed in several contexts,including wireless sensor network [15,16], IoT [17,18], vehicular network [19], etc. The existing works can be classified either as centralized (client-server model) or decentralized. In the following we review some of the existing solutions in both classes.

In Ref. [19], Nilsson et al. proposed a firmware update protocol for modern intelligent vehicles over the internet. The authors suggested a client-server method using a web portal that delivers the firmware update in fragments. The fragments are protected using a hash chain and ensuring, therefore, the integrity of updates. However, the system is vulnerable to a DoS attack as it relies on a central server. Also, the central solution does not ensure the availability of the update when several vehicles on the road request the firmware updates at the same time.

Lee [20] proposed an incentive-based framework on top of a blockchain ledger for IoT devices update distribution. The proposed scheme solved the fair exchange problem between the distributors and receivers by issuing a single receipt for each update distribution to ensure payment is made only to the distributor who made the transfer. To do so, a key registration phase is introduced where the distributors register the encrypting keys to the manufacturers; then it encrypts each update with a different registered key. However, the proposed scheme has some limitations. First, the unique package property requires that the manufacturer should distribute different patches proportional to the number of IoT devices multiplied by the number of distributors. Therefore, the proposed scheme requires the same data transfer volumes (bandwidth) by the manufacturers, compared to a client-server solution, and thus does not make use of distributed software delivery network. Second, creating a contract for each update distribution leads to more transaction fees. Third, the scheme depends on the ability of IoT devices to pay the cost of the new update by using digital wallets. However, this does not happen in practice, where manufacturers who afford the expenses of the update process.

In sensor networks, several schemes such as [15,16] have been proposed to improve the reliability of delivering new updates/security patches by ensuring their integrity. However, these schemes depend on a single entity to manage the distribution of firmware updates and not scale for large networks.

In Ref. [17], the authors proposed a decentralized solution based on a permission-less blockchain to ensure the integrity of updates by having multiple verification nodes instead of depending on a private centralized vendor network. For the distribution of updates, a peer-to-peer file-sharing network such as BitTorrent is proposed to ensure integrity and versions tractability of updates. Boudguiga et al. [21] suggested enhancing the previous work by adding a trusted checking node to verify the update before deploying it to IoT devices [22–26]. However, the scheme does not provide any incentive for devices to participate and distribute firmware updates to others.

In Ref. [18], the authors proposed a software update framework for the Internet of Things (IoT) devices. The framework allows other parties to deliver the updates in return for digital currency paid by the vendor. However, the scheme incurs high financial cost since it depends on Etherum blockchain [27] which applies fees for each transaction sent to the network.

18.3 PRELIMINARIES

In this section, we present the necessary background on blockchain, smart contracts and some cryptographic primitives that are used in our scheme. The notation details used in the remaining chapter are listed in Table 18.1

18.3.1 BLOCKCHAIN AND SMART CONTRACTS

Blockchain was first introduced in 2008 as the underline technology behind the cryptocurrency known as Bitcoin to help make the peer-to-peer exchange of value without a centralized third party. A blockchain is a distributed, immutable, and append-only data structure formed by a sequence of blocks that are chronologically and cryptographically linked together [8]. Fundamentally, a network composed of a set of nodes called miners or validators is responsible for keeping a trustworthy record of all transactions through a consensus algorithm in a trust-less environment. To exchange some coins from one account to another, for instance, a new transaction is generated and broadcast to the network. Each user is identified by a pseudonym address, usually generated from its public key, and the transaction is authenticated through a digital signature computed using the user's private key.

Table 18.1
System Notations

Symbol	Description
$\mathcal{M}_\theta$	A manufacturer company for AVs
PK_θ/SK_θ	Public/private key pair for manufacturer $\mathcal{M}_\theta$
PK_{V_j}/SK_{V_j}	Public/private key pair for vehicle V_j
$\mathcal{PK}_{U_i}/\mathcal{VK}_{U_i}$	zk-SNARK proving/verifying key pair for update (U_i)
U_i	ith firmware update version
P_i	Access policy defined by the manufacturer for U_i
AC_i	Authentication code of update U_i and policy (P_i)
V_j	A responder vehicle j that receives an update U_i
k_j	Encryption key for U_i of vehicle V_j
h_i	Hash of k_j
$\hat{U}_i$	The firmware update (U_i) encrypted with k_j
C_{V_j}	A concatenation of AC_i and h_j
σ_j	A signature of receiving update U_i from vehicle V_j

One of the exciting applications of blockchains is smart contracts, which are defined as computer codes running on top of a blockchain and is correctly executed without fraud or any interference from a third party [28]. Each contract has a unique address on the blockchain to identify itself and to allow users or other contracts to interact with it. The most popular smart contracts platform is Ethereum [29–31], and the de-facto language for creating contracts in Ethereum is Solidity.[1]

18.3.2 CRYPTOGRAPHIC TOOLS

18.3.2.1 Attribute-Based Encryption

Attribute-based encryption (ABE) is an encryption scheme that allows access control over encrypted data. In ABE, each user is assigned a set of secret keys corresponding to his/her set of attributes. Then, a message is encrypted under an access policy formed from the system's set of attributes. The message can only be decrypted by the users who have the attributes that can satisfy the policy. In our scheme, we use the attribute-based encryption scheme proposed in Ref. [32] to enable the distributor AV to identify the neighboring AVs who have the required features (attributes) to download a firmware update. This scheme is a ciphertext-policy attribute-based encryption (CP-ABE), where the access policy is embedded in the ciphertext.

18.3.2.2 Zero-Knowledge Succinct Non-Interactive Argument of Knowledge (zk-SNARK):

zk-SNARK is a proof construction in which one, called the prover, can prove possession of a specific information, called a witness (w), e.g., a secret key, to someone else, called the verifier, without revealing that information. zk-SNARK does not require any interaction between the prover and verifier. Moreover, these schemes are efficient in the sense that the zero-knowledge proof can be verified quickly.

We adopt the zk-SNARK scheme in Ref. [33]. Formally speaking, let L be an NP language with C as its decision circuit. Two keys play an essential role, namely, the proving key ($\mathcal{PK}$) and the verifying key ($\mathcal{VK}$). The proving key allows any prover to compute a proof π for a statement $y \in L$

[1] https://solidity.readthedocs.io/en/develop/.

with a witness w. Typically, a zk-SNARK scheme consists of the following three polynomial-time algorithms:

1. $\text{Gen}(1^{\lambda}, C) \rightarrow (\mathcal{PK}, \mathcal{VK})$: Given a security parameter λ and C as a decision circuit, the Gen algorithm generates two public keys, including $\mathcal{PK}$ and $\mathcal{VK}$, that are used to prove/verify the membership in L.
2. $\text{Prove}(\mathcal{PK}, y, w) \rightarrow \pi$: Given $\mathcal{PK}$, instance y, and witness for an NP statement w, the Prove algorithm generates a proof π for the statement $x \in L_c$.
3. $\text{Verify}(\mathcal{VK}, y, \pi) \rightarrow \{0, 1\}$. Given $\mathcal{VK}$, instance y, and the proof π, the Verify algorithm outputs 1 if $y \in L_c$, allowing the verifier to verify the instance y.

18.3.2.3 Aggregate Signatures

Given n signatures $(\sigma_1, \ldots, \sigma_n)$ on n distinct messages from n users, the aggregate signature scheme can be used to aggregate all these signatures into a single short signature (σ_{agg}). Then, given σ_{agg} and the n messages, a verifier can efficiently ascertain that the n users indeed signed the messages. In our scheme, we use the aggregate signature scheme proposed in Ref. [34] to reduce computations overhead on the blockchain. The idea is that instead of sending a transaction to the blockchain each time a distributor AV distributes a firmware update and gets a proof from other AV, it can aggregate several proofs to create one short aggregated proof to reduce the number of transactions sent to the blockchain.

18.4 PROPOSED SYSTEM

In this section, we present our scheme that aims to ensure secure and scalable delivery of firmware updates from automobile manufacturers to AVs. We first present a general architecture for the system, followed by system initialization, the smart contract creation, firmware update dissemination, and rewarding.

18.4.1 SYSTEM ARCHITECTURE

Figure 18.1 shows the system architecture, which is comprised of two manufacturers with their AVs, two smart contracts for firmware updates of each manufacturer, and a consortium blockchain. A sketch of the possible interactions between the different system entities is shown in Figure 18.2. The role of each entity in the system is discussed in the following paragraphs.

Manufacturer: The manufacturer is responsible for keeping its manufactured AVs updated with the latest versions of the different firmware of the subsystems that control the AVs. During the manufacturing process of AVs, the manufacturer uploads each AV with a set of cryptographic keys and public parameters that will be used to ensure the secure distribution of firmware updates. Also, each time a new update is released, a corresponding smart contract is deployed by the manufacturer to allow AVs to check the integrity and authenticity of the update. In addition, to attract AVs to participate in the distribution of an update, the manufacturer compensates the participants through a rewarding mechanism, e.g., momentary rewards and free or reduced-price maintenance services.

Autonomous Vehicles: We distinguish between two types of AVs, *distributors* and *responders*. The distributor AV disseminates a new firmware update to other AVs (responders) in its vicinity. Each responder AV that receives an update can also act as a distributor of that update. In this way, we can ensure the large-scale dissemination of the update quickly. Initial distributors are selected by the manufacturer based on their reputations which are recorded in a smart-contract.

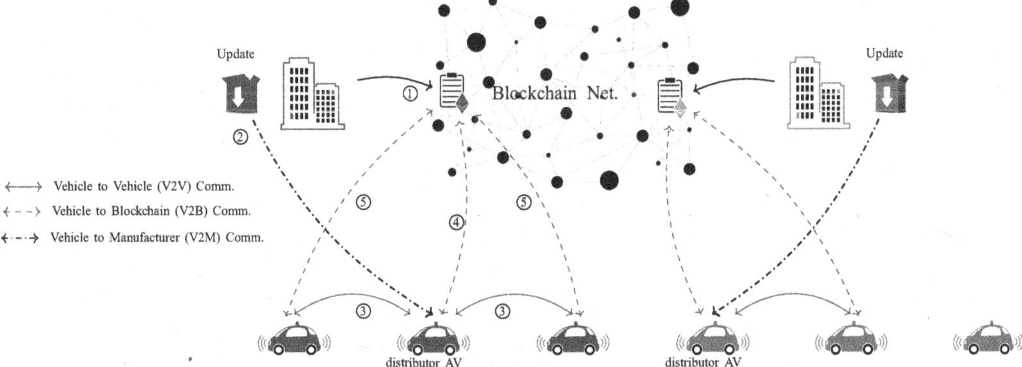

Figure 18.1 System architecture: (1) The manufacturer creates a smart-contract for a new firmware update by including its hash code for authenticity checking by AVs. (2) The manufacturer sends the new update to top-reputation AVs (distributors). (3) A distributor exchanges an encrypted version of the update in return for proof of reception of the update by a responder AV. (4) A redeem transaction, containing multiple proofs, is sent to the smart contract to update the distributor's reputation. (5) The responder AV receives the decryption key of the firmware update from the smart contract.

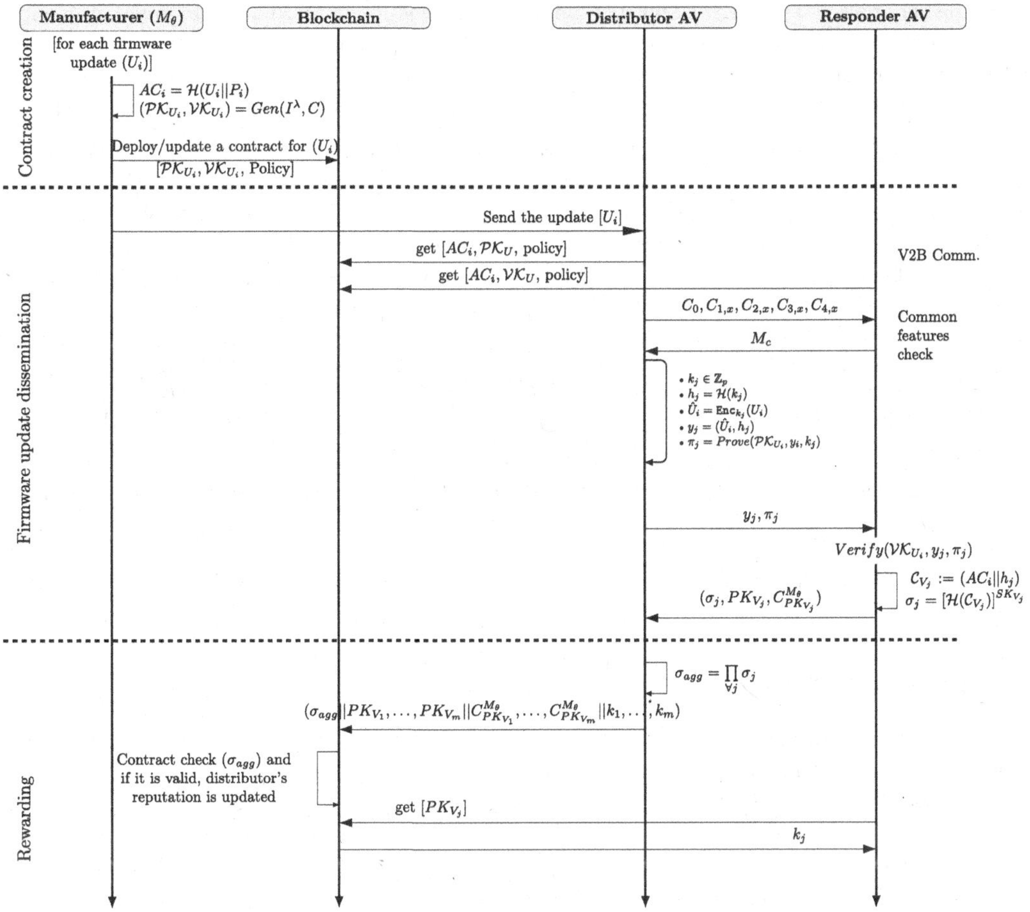

Manufacturer (M_θ) | Blockchain | Distributor AV | Responder AV

Contract creation

[for each firmware update (U_i)]

$AC_i = \mathcal{H}(U_i||P_i)$

$(\mathcal{PK}_{U_i}, \mathcal{VK}_{U_i}) = Gen(I^\lambda, C)$

Deploy/update a contract for (U_i) [$\mathcal{PK}_{U_i}, \mathcal{VK}_{U_i}$, Policy]

Firmware update dissemination

Send the update [U_i]

get [$AC_i, \mathcal{PK}_U$, policy]

get [$AC_i, \mathcal{VK}_U$, policy]

V2B Comm.

$C_0, C_{1,x}, C_{2,x}, C_{3,x}, C_{4,x}$

M_c

Common features check

- $k_j \in \mathbb{Z}_p$
- $h_j = \mathcal{H}(k_j)$
- $\hat{U}_i = \mathrm{Enc}_{k_j}(U_i)$
- $y_j = (\hat{U}_i, h_j)$
- $\pi_j = Prove(\mathcal{PK}_{U_i}, y_i, k_j)$

y_j, π_j

$Verify(\mathcal{VK}_{U_i}, y_j, \pi_j)$

$\mathcal{C}_{V_j} := (AC_i||h_j)$

$\sigma_j = [\mathcal{H}(\mathcal{C}_{V_j})]^{SK_{V_j}}$

$(\sigma_j, PK_{V_j}, C^{M_\theta}_{PK_{V_j}})$

Rewarding

$\sigma_{agg} = \prod_{\forall j} \sigma_j$

$(\sigma_{agg}||PK_{V_1}, \ldots, PK_{V_m}||C^{M_\theta}_{PK_{V_1}}, \ldots, C^{M_\theta}_{PK_{V_m}}||k_1, \ldots, k_m)$

Contract check (σ_{agg}) and if it is valid, distributor's reputation is updated

get [PK_{V_j}]

k_j

Figure 18.2 Firmware update scheme sketch.

Smart Contract: For each new firmware update, a smart contract is created. The contract contains the necessary credentials allowing any receiver of the update to authenticate it and verify its integrity. In addition, the contract implements the reputation logic that evaluates and keeps track of the AVs' activities in the distribution of the firmware update. More specifically, the contract increases the reputation score of a distributor AV after receiving proofs of the firmware distributions from responder AVs.
Blockchain Network: Blockchain network is at the center of our system and it is responsible for the execution of the smart contracts in a distributed manner without relying on a single trusted central party. This is mandatory to ensure a scalable and secure firmware update dissemination. Moreover, to improve the efficiency of the system, we opt for a *consortium blockchain*, where the validators, i.e., nodes with write permission on the shared ledger, are known and trusted entities. In our case, the validators can be the manufacturers of the different automobile brands.

18.4.2 SYSTEM INITIALIZATION

A multi-authority attribute-based encryption scheme is used, where each manufacturer is considered as an authority that decides a set of attributes (or features) for its AVs. $\mathbb{M}$ is the set of all available manufacturers and a manufacturer $\mathcal{M}_\theta \in \mathbb{M}$. Let $\mathbb{A}$ be the set of all attributes (or features) in the system, an access policy (A, δ) on $\mathbb{A}$ with $A \in \mathbb{Z}_p^{l \times n}$, called the share generating matrix in the field $\mathbb{Z}_p$ of prime order p with l rows and n columns, and a function δ that labels the rows of A with attributes from $\mathbb{A}$, i.e., $\delta : [l] \to \mathbb{A}$. In addition, let ρ be a function that maps attributes in rows to its manufacturers, where $\rho : [l] \to \mathcal{M}_\theta$. Consider $e : \mathbb{G}_1 \times \mathbb{G}_2 \to G_T$ a cryptographic bilinear map with generators $g_1 \in \mathbb{G}_1$ and $g_2 \in \mathbb{G}_2$, where $\mathbb{G}_1$ and $\mathbb{G}_2$ are the multiplicative groups. Each manufacturer $\mathcal{M}_\theta \in \mathbb{M}$ should select two random elements $(\alpha_\theta, y_\theta) \xleftarrow{R} \mathbb{Z}_p^*$ as its secret keys, and then, $\mathcal{M}_\theta$ can compute its public key as $PK_\theta = \{e(g_1, g_2)^{\alpha_\theta}, g_1^{y_\theta}\}$. Besides, a public hash function $\mathcal{H} : \{0,1\}^* \to G_1$ is used to map an AV global identifier GID to a point in G_1, a public hash function $F : \{0,1\}^* \to G_1$ that maps an attribute $a \in \mathbb{A}$ to G_1, and a function T that maps an attribute $a \in \mathbb{A}$ to the manufacturer $\mathcal{M}_\theta$, hence, the function $\rho(\cdot)$ can be redefined as $\rho(\cdot)$: $\mathrm{T}(\delta(\cdot))$. The global parameters are then defined as $GP = \{\mathbb{G}_1, \mathbb{G}_2, G_T, \mathbb{Z}_p, \mathcal{H}, F, T, \mathbb{A}, \mathbb{M}\}$.

Besides, during the production, $\mathcal{M}_\theta$ should assign each AV a key for each assigned attribute $a \in \mathbb{A}$ using the AV global identity GID as follows: $\mathcal{M}_\theta$ chooses a random $t \xleftarrow{R} , \mathbb{Z}_p^*$ and outputs to the AV attributes secret keys as $SK_{GID,a} = \{K_{GID,a} = g_2^{\alpha_\theta} H(GID)^{y_\theta} F(a)^t, K'_{GID,a} = g_1^t\}$. Finally, for each AV, $\mathcal{M}_\theta$ generates a public/private key pair as follows: a random number $x_a \xleftarrow{R} \mathbb{Z}_P$ is selected as the private key and the corresponding public key is $PK_{V_j} = g_2^{x_a}$. The manufacturers should obtain a certificate ($C_{PK_{V_j}}^{M_\theta}$) for its public key from the certificate authority. Then, the public/private key pair (PK_{V_j}, x_a) and $C_{PK_{V_j}}^{M_\theta}$ should be added to the AV's tamper-proof device along with the manufacturer's public key PK_θ.

18.4.3 SMART CONTRACT CREATION

Upon releasing a new update by a subsystem's manufacturer, denoted by U_i, the AV manufacturer that uses the subsystem should first test the update. Note that a subsystem's manufacturer may be different from the AV manufacturer. If the AV manufacturer decides to use it on its AVs, it starts the firmware update as follows. The manufacturer creates a smart contract and initializes it by two attributes: (1) A proving/verifying key pair: $(\mathcal{PK}_{U_i}, VK_{U_i}) = \mathrm{Gen}(1^\lambda, C)$, required for the execution of the zk-SNARK protocol; (2) An authentication code for the new firmware update: $AC_i = \mathcal{H}(U_i || P_i)$, where P_i is the access policy defined by the manufacturer to deliver the update to only AVs that have the features defined in the policy.

Algorithm 18.3: Pseudocode for the *Firmware Update* Contract

```
contract FirmwareUpdate
    mapping(address => int) Reputation // Mapping for distributors
        reputation
    mapping(address => int) UpdatedAVs // Mapping for AVs with the No.of
        obtained updates
    function FirmwareUpdate(_PK, _VK, _AC_i, _P_i, X)
        PK ← _PK // Proving Key
        VK ← _VK // Verifying key
        AC_i ← _AC_i // authentication code
        P_i ← _P_i // ABE Policy
        MaxUpdate ← X // Max. No. of download per Update
    function RecieveProof(σ_agg, PK[ ], C[ ] keys[ ])
        address [] RecievedAVs // Received AV list
        for s ← 0 to PK.lengh do
            if verifySig(pk_M, PK[s], C[s])
                return
            end
            if UpdatedAVs[PK[s]] > MaxUpdate
                return
            end
            h_s ← H(keys[s])
            C_{V_s} ← H(AC_i, h_s) RecievedAVs.push(Pairing(PK[s], C_{V_s}))
        end
        if Pairing(g_1, σ_agg)=Prod(RecievedAVs)
            UpdateReputation(msg.sender, PK.length)
            for i ← 0 to PK.lengh do
                emitEvent("KeyRevealed", PK_i, keys[i])
                UpdatedAVs[PK_i]← UpdatedAVs[PK_i]+1
            end
        end
    function UpdateReputation(Dist, N)
        // increase reputation distributors
        Reputation[Dist]← Reputation[Dist]+=N
```

The manufacturer deploys a smart-contract by broadcasting a transaction to the blockchain network. The deployed smart-contract is described in Algorithm 18.1: and includes the following main functions:

- *Authenticity and Integrity of a Firmware Update:* Since the update's authentication code and verification key are stored in the contract, AVs, by consulting the blockchain, can check whether a received firmware update is the same one that was originally approved by the AV manufacturer.
- *AVs' Reputation*: When an AV participates in the distribution of a new update, the proof of distribution is sent to the smart contract which in turn increases its reputation. The manufacturer rewards the highly-reputed AVs, i.e., the active AVs in distributing the firmware. The reward can be momentary, free or reduced-price maintenance service, etc.
- *Firmware Access Control*: Each firmware update has an access policy set by the manufacturer, and the AVs that have enough features to satisfy the policy can receive the firmware.

This can restrict the distribution of the firmware to only certain AVs. The access policy of an update is included in the update's smart-contract.

18.4.4 FIRMWARE UPDATE DISSEMINATION

In this stage, the manufacturer starts the dissemination process of a new update by first selecting the most active AVs in distributing updates (based on their reputations) to act as the initial distributors. As discussed before, rewards are used to incentivize the AVs to act as distributors and actively distribute the new firmware, but the rewarding system should be secure to ensure that only honest distributors which distribute the firmware are rewarded. In practice, each AV manufacturer will use many subsystems made by different companies, and therefore, it is very frequent that different AV manufacturers may use the same subsystems produced by the same company in their vehicles. Thus, it is very important for each manufacturer to ensure that its distributors will deliver a particular update to only certain models of its AVs.

Thanks to ABE, as presented in Section 18.3, each manufacturer can define an access policy for each update on the associated smart contract, where only AVs that belong to the same manufacturer and have enough features, such as model, year of manufacturer, etc, can decrypt and use the firmware update they got from a distributor.

For a distributor to find other AVs which can satisfy the access policy of an update and deliver it, the following steps should be taken.

1. A distributor AV first queries the blockchain for the AC_i, proving key $\mathcal{PK}_{U_i}$, and the manufacture's access policy (A, δ).
2. Then, distributor AV should broadcast an encrypted challenge message (M_c) using the ABE to the nearby AVs. This message is encrypted using the manufacture's public key PK_θ under the access policy (A, δ) set by the manufacturer. Hence, only the AVs which owns the set of attributes that satisfy the policy are able to decrypt the ciphertext CT. To encrypt M_c, distributor AV first creates two random vectors $v = (z, v_2, \ldots v_n)^T$ and $w = (0, w_2, \ldots w_n)^T$, where $\{z, v_2, \ldots v_n, w_2, \ldots w_n\}$ are elements randomly selected from Z_p^*. We denote λ_x as the share of the random secret z corresponding to row x, i.e., $\lambda_x = (A_x \cdot v)$ and w_x denotes the share of zero, i.e., $w_x = (A_x \cdot w)$, where A_x is the x-th row of access matrix A. The distributor AV chooses a random element $t_x \xleftarrow{R} \mathbb{Z}_p^*$ for each row in the policy matrix A and computes the CT as:

$$
\begin{aligned}
&C_0 = M_c \cdot e(g,g)^z;\\
&\{C_{1,x} = e(g_1,g_2)^{\lambda_x} e(g_1,g_2)^{\alpha_{\rho(x)} t_x};\\
&C_{2,x} = g_1^{-t_x};\\
&C_{3,x} = g_1^{y_{\rho(x)} t_x + w_x};\\
&C_{4,x} = F(\delta(x))^{t_x}\}_{x \in [l]}
\end{aligned}
$$

3. After a responder AV receives CT, it first queries the smart contract for the access policy (A, δ), $\mathcal{VK}_{U_i}$ and AC_i. Then, to decrypt M_c, the AV should use the policy (A, δ) from the blockchain and its secret keys $(K_{GID,a}, K'_{GID,a})$ for the subset of rows A_x of satisfied attributes and for each row x to compute

$$
\begin{aligned}
C_{1,x} \cdot e(K_{GID,\delta(x)}, C_{2,x}) \cdot e(H(GID), C_{3,x}) \cdot e(K'_{GID,\delta(x)}, C_{4,x})\\
= e(g_1,g_2)^{\lambda_x} \cdot e(H(GID), g_2)^{w_x}
\end{aligned}
$$

Then, AV calculates the constants $c_x \in \mathbb{Z}_p^*$ such that $\Sigma_x c_x A_x = (1, 0, \ldots 0)$ and computes:

$$\Pi_x (e(g,g)^{\lambda_x}.e(H(GID),g)^{w_x})^{c_x} = e(g,g)^z$$

This is true because $\lambda_x = (A_x \cdot v)$ and $w_x = (A_x \cdot w)$, where $\langle (1,0,\cdots,0) \cdot v \rangle = z$ and $\langle (1,0,\cdots,0), w \rangle = 0$. Hence, the challenge message can be decrypted as $M_c = C_0/e(g,g)^z$.

4. Finally, once a responder AV manages to get M_c, it then replies to the distributor AV with the correct M_c. Henceforth, both the distributor and responder AV can proceed with the firmware update transfer.

A distributor AV sends the firmware update to a responder AV in return for a signature, i.e., a proof for disseminating the firmware. This exchange of firmware update and proof can be made in a trust-less way using zk-SNARK protocol as follows.

1. The distributor AV generates a secret key $k_j \in \mathbb{Z}_\mathbb{P}$ and calculates $h_j = \mathcal{H}(k_j)$.
2. Then, it computes $\hat{U}_i = \mathtt{Enc}_{k_j}(U_i)$, where $\mathtt{Enc}$ is a symmetric-key encryption algorithm.
3. For zk-SNARK protocol, the secret witness is the instance $y_j = (\hat{U}_i, h_j)$ and k_j. The NP statement is as follows:

$$\exists k_j : \mathcal{H}(k_j) = h_j \wedge \mathcal{H}(Dec_{k_j}(\hat{U}_i), P_i) = AC_i \tag{18.1}$$

Which attests that the distributor AV has a key, k_j, such that its hash is h_j, and if k_j is used to decrypt $\hat{U}_i$, it will match the update authentication code AC_i. After that, the distributor computes a zero-knowledge proof $\pi_j = Prove(\mathcal{PK}_{U_i}, y_j, h_j)$ and sends $y_j||\pi_j$ to the responder.

4. Upon receiving $(y_j||\pi_j)$, the responder first verifies that $Verify(\mathcal{VK}_{U_i}, y_j, \pi_j) = 1$, and then computes a signature $\sigma_j = [\mathcal{H}(\mathcal{C}_{V_j})]^{SK_{V_j}}$, where SK_{V_j} is its private key and $\mathcal{C}_{V_k} = (AC_i, h_j)$.
5. Finally, it sends $\left(\sigma_j||PK_{V_j}||C_{PK_{V_j}}^{M_\theta}\right)$ to the distributor.

18.4.5 REWARDING

In this phase, the distributor AV sends a redeem transaction, containing multiple proofs, to the smart contract to update its reputation proportionally to the number of AVs which received the update. This rewarding process is done as follows.

1. To reduce the number of transactions sent to the blockchain, instead of making a transaction each time a firmware is transferred, one transaction can be sent for several firmware transfers efficiently, as follows. Once the distributor AV gets the proofs of transferring an update (σ_j) from other vehicles, it aggregates multiple signatures into a single signature (σ_{agg}) as follows: $\sigma_{\text{agg}} = \prod_{\forall j} \sigma_j$. Note that, a signature σ_j should be different from other received signatures. In other words, the distributor should generate a distinct k_j for each time he sends the new update to other vehicles.
2. The distributor sends a transaction to the blockchain containing

$$(\sigma_{\text{agg}}||PK_{V_1},\ldots,PK_{V_m}||C_{PK_{V_1}}^{M_\theta},\ldots,C_{PK_{V_m}}^{M_\theta}||k_1,\ldots,k_m)$$

where m is the number of vehicles that received the update.

3. The smart contract method `RecieveProof` first verifies that the received public key is one of the certified keys by the manufacturer (see `verifysign` in Algorithm 18.1) as well as many number of times that AV gets the update. Then, it computes $h_j = \mathcal{H}(k_j)$ and $\mathcal{C}_{V_j} = \mathcal{H}(AC_i, h_j)$ for all j. Thereafter, it verifies the aggregated signature by checking if $e(g_1, \sigma_{\text{agg}}) = \prod_{\forall j \in m} e(PK_{V_j}, \mathcal{C}_{V_j})$ or not (see `pairing` in Algorithm 18.1 which can be executed by a pre-compiled contract for elliptic curve pairing operations available at [35]).

4. Finally, the distributor is rewarded by increasing its reputation index proportionally to the number of vehicles that received the update (see the method `UpdateReputation` in Algorithm 18.1).
5. A responder AV queries the contract for a relevant event associated with its public key for the decryption key k_j it needs to decrypt $\hat{U}_i$ to get U_i (see event `''KeyRevealed''` in Algorithm 18.1).

18.5 PERFORMANCE EVALUATIONS

In this section, we evaluate our proposed scheme. We first present a proof-of-concept implementation of our proposed scheme and demonstrate its feasibility. Finally, we evaluate the cost of the proposed implementation in terms of computation and storage overhead.

18.5.1 ON-CHAIN COST

18.5.1.1 Methodology/Experiment Setup

Figure 18.3 shows an overview of the proof-of-concept implementation of our proposed scheme. More specifically, we implemented a private blockchain that is run using instances of Geth inside of Docker containers, built on a modification of the code seen in[2] Geth is an implementation of Ethereum blockchain in Go, and allows us to set up nodes (including validators/miners) with our blockchain. We have set-up several Docker containers running on a workstation machine acting as blockchain nodes (i.e., miners/validators). The blockchain is set up using a Proof-of-Authority (PoA) consensus algorithm. To emulate AV hardware, we used Raspberry Pi devices. Table 18.2 gives the system specifications for the workstation and the Raspberry Pi's used in our implementations

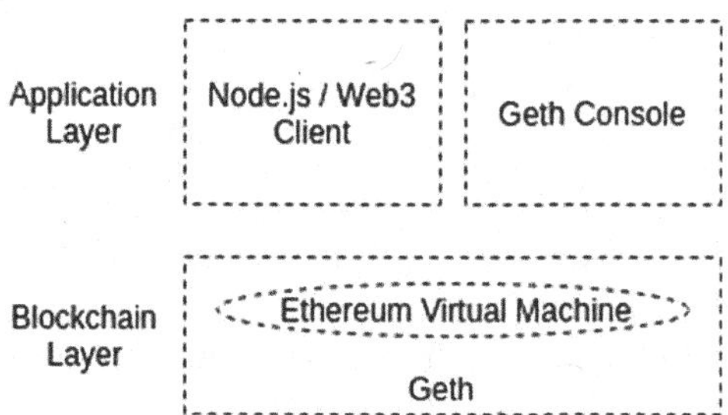

Figure 18.3 Implementation overview.

Table 18.2
Hardware Used in Our Test-Bed Implementation

	PC	Raspberry Pi
CPU	2.00 GHz 8-core Intel i7-4765T	1.2 GHz 4-core ARM Cortex-A53
Memory	7.7 GB	1 GB
Operating System	Ubuntu 18.04.2	Raspbian 9.9

[2]https://github.com/a1brz/eth-private-network.

18.5.1.2 Performance Metrics

We define the following key performance metrics for evaluating the on-chain performance in our proposed scheme.

- *BLS/BGLS Verification Time:* The time elapsed between sending a transaction by an AV and returning a transaction hash by the blockchain indicating that the transaction is successful.

  The specific transaction, in this case, is a call to a function which takes up to three signatures (aggregated or otherwise), messages, and elliptic curve point pairs before sending the information to be verified using the appropriate verification algorithm (BLS or BGLS).
- *Total Time for the Firmware Update:* This time includes the overall time for updating an AV with the new update. Specifically, it includes the off-chain computation overhead for firmware distribution and the on-chain computation time for the BLS verification time, reputation update, and key retrieval.

  We measured the on-chain time by measuring the total time elapsed between sending a series of transactions from an AV and the blockchain network returning a transaction hash indicating that the final transaction was successful. The specific transactions, in this case, are a call to a function that takes up to three signatures (aggregated or otherwise), messages, and elliptic curve point pairs before sending the information to be verified using the appropriate verification algorithm (BLS or BGLS). If the verification is successful, the sender's reputation is updated appropriately, and the decryption key is revealed.
- *Storage cost overhead.* Is the cost of storing the data in a blockchain. According to the Ethereum Yellow Paper [29], storage cost is calculated by measuring the gas cost of transactions as follows

$$\text{Storage cost} = \frac{\text{gasCost} - 21,000}{68} \tag{18.2}$$

Specifically, there is a flat gas fee $G_{\text{transaction}}$ of 21,000 gwei paid for every transaction, and there is a fixed gas fee $G_{\text{txdatanonzero}}$ of 68 gwei paid for every non-zero byte of data or code for a transaction. Therefore, subtracting $G_{\text{transaction}}$ from the transaction gas cost and dividing the result by $G_{\text{txdatanonzero}}$ gives the number of bytes used for the transaction.

18.5.1.3 Results and Discussion

Using the above key metrics, we evaluated the performance of our scheme. The experiments were executed using a program written in Node.js.[3]

The first two experiments are related to the BLS/BGLS verification time with two and six validators, and their results are shown in Figure 18.4. In these experiments, 1300 transactions were sent to the blockchain (simulating 1300 distributor AVs), with each being timed individually. Figure 18.4 shows the execution time of validating proofs sent by distributors. It can be clearly seen that in case of using BLS signatures: as the number of distributors increases, the time taken by the validators increases to reach consensus. It can also be seen that in the case of using BGLS, the execution time is lower than that of BLS. Moreover, as the number of distributors increases, there is no significant increase in the execution time. This is because in BGLS, the validators only verify a batch of proofs in one time, while in BLS, they should verify every proof individually.

In another part of the experiment, we increased the number of validators to assess the impact on execution time. The results indicate that two validators are quicker compared to having six validators case. This is due to two reasons:

---

[3] https://nodejs.org/en/.

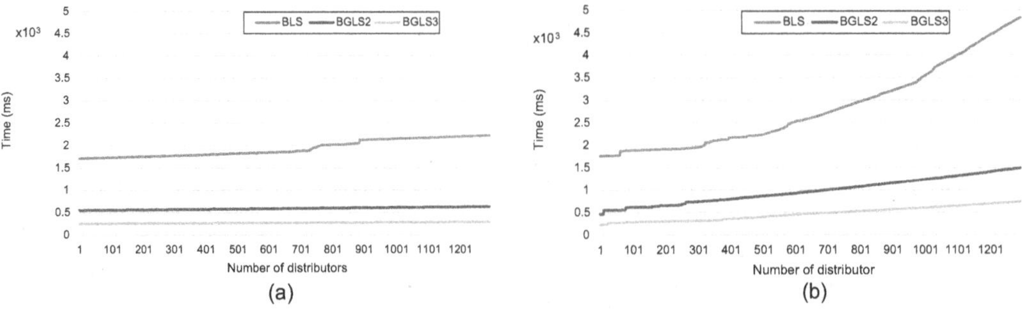

Figure 18.4 Execution time for BLS/BGLS verification with two validators (a) and six validators (b).

- Having more validators increases the amount of time for the validators to reach a consensus before a block is sealed and added to the blockchain.
- Since the validators are run inside of docker containers on a single machine, the available resources are greater per validator when only two are in use. Comparatively, the same resources are divided among six validators, resulting in less overall computation power per validator.

We also run experiments to measure the total execution time for signature verification, reputation update, and key retrieval. Similar to the previous experiments, 1300 transactions were sent to blockchain simulating 1300 distributors with each being timed individually. Figure 18.5 shows the results of these experiments. It can be seen that the total execution time increases as more distributors interact with the contract. Also, the execution time is greater for six validators compared to two validators. In can be concluded the use of BGLS is efficient since the time needed for the updating the AVs is very short comparing to the BLS.

The last experiment was conducted on our blockchain relates to the overhead cost for the storage of transactions. Transactions on Ethereum blockchains have a fixed gas cost measured in gwei (1 ether = 10^{18} gwei), which means that we can accurately predict the gas cost for large amounts of transactions. The following gas costs correspond to each verification type:

- *BLS:* 40,237 gwei
- *BGLS with two signers:* 61,016 gwei
- *BGLS with three signers:* 67,866 gwei

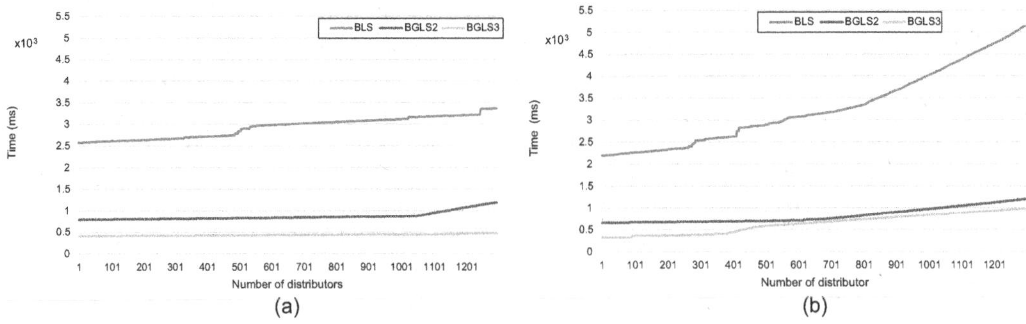

Figure 18.5 Execution time for verification, reputation increase, and key reveal with two validators (a) and six validators (b).

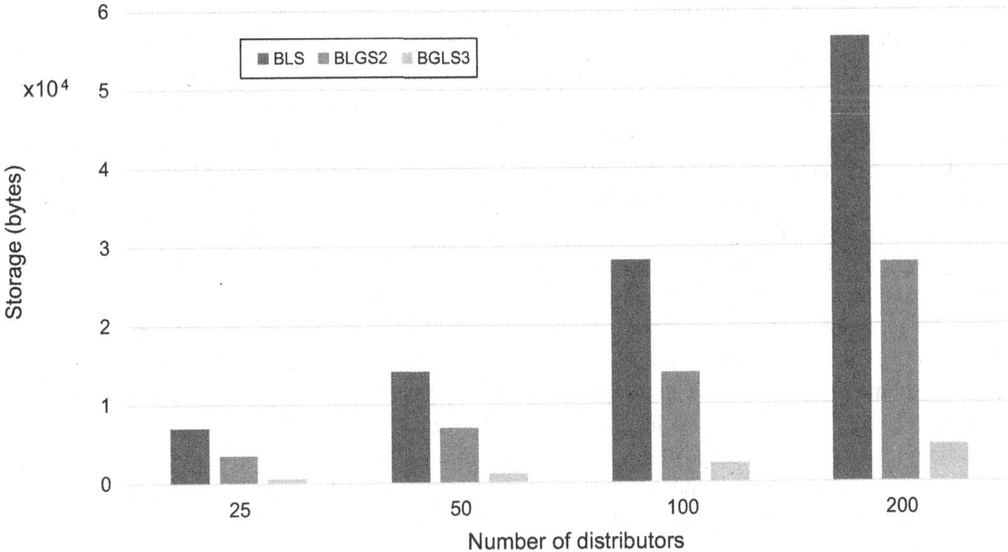

Figure 18.6 Storage overhead comparison.

Eq.(18.2) is used to convert the gas cost into total storage cost used. Figure 18.6 shows the number of bytes used for BLS and BGLS verification for 25, 50, 100, and 200 distributors. BLS consistently requires more storage space, as a new function call must be made for each verification, whereas BGLS can perform multiple verifications with a single function call.

18.5.2 OFF-CHAIN COST

In this section, we evaluate the computation overhead for the cryptography operations used in our scheme.

The computation times of ABE are measured using Intel Core i7- 4765T 2.00 GHz and 8GB RAM machine and Python charm cryptographic library in Ref. [1]. The off-chain computation overhead on each entity is summarized in Table 18.3. In our scheme, a distributor AV needs to broadcast a challenge packet (M_c) encrypted by a number of attributes (γ) specified in the smart contract. According to [1], the required time for encryption is ($10.9\times\gamma + 1.35$) ms. In addition, a responder AV needs to decrypt (M_c) with total decryption time that can be formulated as ($4.03\times\gamma + 0.01$) ms. After running the ABE scheme, zk-SNARK protocol should be run. In this protocol,

Table 18.3
Off-Chain Computation Overhead α Is the Number of Attributes of the Underlying ABE

Entity	Operation	Computation Overhead
Distributor	ABE encryption	$(10.9 \times \alpha + 1.35)$ ms
	Proof generation for ZKSNARK	6 s
Responder	Decrypting the challenge packet	$(4.03 \times \alpha + 0.01)$ ms
	Verifying ZKSNARK	5 ms
	Composing a response packet	40 ms

a proof is generated by the distributor and then verified by the responder. We implemented the NP statement in Eq. (18.1) using Zokrates[4] toolbox. MIMC is used for encryption/decryption due to its efficiency with zk-SNARK proofs, and SHA256 is used for hashing. The time to generate the proof is 6 seconds, whereas, the verification is 5 milliseconds. It should be noted that the distributor AV can generate multiple proofs offline before starting the communication session with the responder AVs. Hence, the total computation time needed to run ABE and zk-SNARK verification of our scheme is low, which is suitable for our application because the AVs are in motion and their communication time is short.

In our scheme, blockchain is required to ensure the authenticity and integrity of the new update. To reduce the cost needed to execute our scheme on the blockchain, most of the computations to secure the scheme are done outside the blockchain. Using a consortium blockchain will remove any constraint on the amount of data that should be sent and stored on the blockchain. Additionally, our scheme reduces the on-chain operations by reducing the number of transactions sent to the blockchain by aggregating several firmware transfers in one transaction using the aggregate signature scheme. Also, as discussed before, the computation cost to run our scheme is low. For the actual time of exchanging the update, according to [2], the mean throughput for delivering data to and from moving vehicles that use IEEE 802.11 protocol is equal to 760 kbit/s. If we assume that the size of a firmware update equals to 1 MByte, then the time required to transfer the update is 1.3 seconds. Therefore, given this transfer time and the short time needed for the cryptographic computations, our scheme can be executed during the contact time of two moving AVs.

18.6 SECURITY ANALYSIS

In this section, we discuss the possible security threats and how our scheme thwarts them.

- *Firmware Integrity:* Since the authentication code of each new firmware update is recorded in the smart contract, our scheme resists any attempt by an adversary to distribute malicious updates. This is because the hash of the new update is stored on the blockchain which is an immutable ledger.
- *Firmware Distribution and Access Control:* To prevent unauthorized AVs from accessing a new firmware update, our scheme allows each AV manufacturer to determine the AVs that can receive the new firmware update. Through the ABE access policy, which is registered in the blockchain, a distributor can prescribe the AVs that have the right to receive the update and prevent unauthorized AVs from receiving it. In addition, because the access policy is embedded in the update authentication code (*AC*), any responder AVs (receiver) can decide if it is targeted by a particular update or not. If a compromised/malicious distributor AV changes the access policy used to encrypt the challenge message, the responder AV can detect this change during the zk-SNARK verification. Hence, the distributor (attacker) is not able to get a proof of distribution from the responder.
- *DoS attack resistance.* The proposed scheme resists Denial-of-service (DoS) attacks [36] that aim to disable the system and the rewarding mechanism. This attack is not possible since there is no central unit that distributes the firmware or runs the scheme. For this attack to succeed, the attackers need to control the majority of the validators (manufacturers) of the blockchain network, which is presumably impossible.
- *Update audibility.* In our scheme, the AVs that have distributed or received an update are recorded in the blockchain. This gives the manufacturer an accurate insight about the percentage of AVs that have received a particular update.

[4]https://github.com/Zokrates/ZoKrates.

18.7 CONCLUSION

In this chapter, a firmware update dissemination scheme leveraging blockchain and smart contracts has been introduced for autonomous vehicles. A smart contract is used to ensure the authenticity and integrity of firmware updates, and more importantly to manage the reputation scores of AVs that transfer the new updates to other AVs. The use of ABE allows AV manufacturer to target a specific set of AVs that have certain features defined by the manufacturer to download the firmware. A zero-knowledge proof protocol is used to enable the AVs to exchange an update for proof of distribution in a trust-less way. To improve the efficiency, an aggregate signature scheme is used to allow a distributor to combine multiple proofs to make only one transaction on the blockchain when it redeems the rewards. Our proposed scheme has been evaluated using real implementations. The experimental results indicate that the on-chain and off-chain computation overheads are in milliseconds, thus is practical to the moving AVs to distribute the updates successfully.

ACKNOWLEDGMENT

This research work was financially supported in part by NSF grants 1619250 and 1852126. In addition, parts of this paper, specifically Sections I, III, and IV were made possible by NPRP grants NPRP10-1223-160045 from the Qatar National Research Fund (a member of Qatar Foundation). The statements made herein are solely the responsibility of the authors

REFERENCES

1. Mohamed Baza, Mahmoud Nabil, Noureddine Lasla, Kemal Fidan, Mohamed Mahmoud, and Mohamed Abdallah. Blockchain-based firmware update scheme tailored for autonomous vehicles. *Proceedings of the IEEE Wireless Communications and Networking Conference (WCNC)*, Marrakech, Morocco, April 2019.
2. Jakob Eriksson, Hari Balakrishnan, and Samuel Madden. Cabernet: Vehicular content delivery using wifi. In *Proceedings of the 14th ACM International Conference on Mobile Computing and Networking*, San Francisco, CA, pp. 199–210, ACM, 2008.
3. Mohamed Baza, Mahmoud Nabil, Niclas Bewermeier, Kemal Fidan, Mohamed Mahmoud, and Mohamed Abdallah. Detecting sybil attacks using proofs of work and location in vanets. arXiv preprint arXiv:1904.05845, 2019.
4. Geng Lin, David Fu, Jinzy Zhu, and Glenn Dasmalchi. Cloud computing: It as a service. *IT Professional*, 2:10–13, 2009.
5. Wesam Al Amiri, Mohamed Baza, Mohamed Mahmoud, Waleed Alasmary, and Kemal Akkaya. Towards secure smart parking system using blockchain technology. *Proceedings of 17th IEEE Annual Consumer Communications & Networking Conference (CCNC)*, Las Vegas, NV, 2020.
6. Mohamed Baza, Mahmoud Nabil, Muhammad Ismail, Mohamed Mahmoud, Erchin Serpedin, and Mohammad Rahman. Blockchain-based charging coordination mechanism for smart grid energy storage units. *Proceedings Of IEEE International Conference on Blockchain*, Atlanta, GA, 2019.
7. Mohamed Baza, Noureddine Lasla, Mohamed Mahmoud, Gautam Srivastava, and Mohamed Abdallah. B-ride: Ride sharing with privacy-preservation, trust and fair payment atop public blockchain. *IEEE Transactions on Network Science and Engineering*, 99:1–1, 2019.
8. Konstantinos Christidis and Michael Devetsikiotis. Blockchains and smart contracts for the internet of things. *IEEE Access*, 4:2292–2303, 2016.
9. Mohamed Baza, Mahmoud Nabil, Muhammad Ismail, Mohamed Mahmoud, Erchin Serpedin, and Mohammad Rahman. Blockchain-based privacy-preserving charging coordination mechanism for energy storage units. arXiv preprint arXiv:1811.02001, 2019.
10. Mohamed I. Baza, Mostafa M. Fouda, Adly S. Tag Eldien, and Hala A.K. Mansour. An efficient distributed approach for key management in microgrids. *Proceedings of the Computer Engineering (ICENCO)*, Cairo, Egypt, pp. 19–24, 2015.

11. Mohammed, Hawzhin and Tonyali, Samet and Rabieh, Khaled and Mahmoud, Mohamed and Akkaya, Kemal Efficient privacy-preserving data collection scheme for smart grid AMI networks. *2016 IEEE Global Communications Conference (GLOBECOM)*, Washington, DC. USA, pp.1–6, 2016
12. Mohamed Baza, Marbin Pazos-Revilla, Mahmoud Nabil, Ahmed Sherif, Mohamed Mahmoud, and Waleed Alasmary. Privacy-preserving and collusion-resistant charging co-ordination schemes for smart grid. arXiv preprint arXiv:1905.04666, 2019.
13. Ahmed Shafee, Mohamed Baza, Douglas A. Talbert, Mostafa M. Fouda, Mahmoud Nabil, and Mohamed Mahmoud. Mimic learning to generate a shareable network intrusion detection model. arXiv preprint arXiv:1905.00919, 2019.
14. M. Baza, Mostfa Fouda, Mahmoud Nabil, Adly S. Tag, Hala Mansour, and Mohamed Mahmoud. Blockchain-based distributed key management approach tailored for smart grid. In *Combating Security Challenges in the Age of Big Data*. Springer, pp. 237–263, 2019.
15. Daehee Kim, Heungwoo Nam, and Dongwan Kim. Adaptive code dissemination based on link quality in wireless sensor networks. *IEEE Internet of Things Journal*, 4(3):685–695, 2017.
16. Prabal K. Dutta, Jonathan W. Hui, David C. Chu, and David E. Culler. Securing the deluge network programming system. In *Proceedings of the 5th International Conference on Information Processing in Sensor Networks*, Nashville, TN, pp. 326–333, ACM, 2006.
17. Boohyung Lee and Jong-Hyouk Lee. Blockchain-based secure firmware update for embedded devices in an internet of things environment. *The Journal of Supercomputing*, 73(3):1152–1167, 2017.
18. Oded Leiba, Yechiav Yitzchak, Ron Bitton, Asaf Nadler, and Asaf Shabtai. Incentivized delivery network of IoT software updates based on trustless proof-of-distribution. arXiv preprint arXiv:1805.04282, 2018.
19. Dennis K. Nilsson. Secure firmware updates over the air in intelligent vehicles. In *2008 IEEE International Conference on Communications Workshops (ICC Workshops' 08)*, Beijing, China, pp. 380–384, 2008.
20. JongHyup Lee. Patch transporter: Incentivized, decentralized software patch system for WSN and IoT environments. *Sensors*, 18(2):574, 2018.
21. Aymen Boudguiga, Nabil Bouzerna, Louis Granboulan, Alexis Olivereau, Flavien Quesnel, Anthony Roger, and Renaud Sirdey. Towards better availability and accountability for IoT updates by means of a blockchain. In *2017 IEEE European Symposium on Security and Privacy Workshops (EuroS&PW)*, pp. 50–58, Paris, France, 2017.
22. Tolulope A. Odetola, Oderhohwo Ogheneuriri, and Syed Hasan. A scalable multilabel classification to deploy deep learning architectures for edge devices. arXiv preprint arXiv:1911.02098, 2019.
23. Tolulope A. Odetola, Hawzhin Raoof Mohammed, and Syed Hasan. A stealthy hardware trojan exploiting the architectural vulnerability of deep learning architectures: Input interception attack (IIA). arXiv preprint arXiv:1911.00783, 2019.
24. Tolulope A. Odetola, Katie M. Groves, and Syed Hasan. 2l-3w: 2-level 3-way hardware-software co-verification for the mapping of deep learning architecture (DLA) onto FPGA boards. arXiv preprint arXiv:1911.05944, 2019.
25. Ibrahim Yilmaz and Rahat Masum. Expansion of cyber attack data from unbalanced datasets using generative techniques. arXiv preprint arXiv:1912.04549, 2019.
26. Ibrahim Yilmaz. Practical fast gradient sign attack against mammographic image classifier. arXiv preprint arXiv:2001.09610, 2020.
27. Wesam Al Amiri, Mohamed Baza, Karim Banawan, Mohamed Mahmoud, Waleed Alasmary, and Kemal Akkaya. Privacy-preserving smart parking system using blockchain and private information retrieval. arXiv preprint arXiv:1905.04666, 2019.
28. Nick Szabo. The idea of smart contracts. *Nick Szabo's Papers and Concise Tutorials*, 6, 1997.
29. Gavin Wood. Ethereum: A secure decentralised generalised transaction ledger. *Ethereum Project Yellow Paper*, 151:1–32, 2014.
30. Mohamed Baza, Mohamed Mahmoud, Gautam Srivastava, Waleed Alasmary, and Mohamed Younis. A light blockchain-powered privacy-preserving organization scheme for ride sharing services. *Proceedings of the IEEE 91th Vehicular Technology Conference (VTC-Spring)*, Antwerp, Belgium, May 2020.
31. Mohamed Baza, Andrew Salazar, Mohamed Mahmoud, Mohamed Abdallah, and Kemal Akkaya. On sharing models instead of the data for smart health applications. *Proceedings of IEEE International Conference on Informatics, IoT, and Enabling Technologies (ICIoT'20)*, Doha, Qatar, 2020.

32. Yannis Rouselakis and Brent Waters. Efficient statically-secure large-universe multi-authority attribute-based encryption. In *Proceedings of the International Conference on Financial Cryptography and Data Security*, San Juan, Puerto Rico, pp. 315–332, Springer, 2015.
33. Eli Ben-Sasson, Alessandro Chiesa, Eran Tromer, and Madars Virza. Succinct non-interactive zero knowledge for a von neumann architecture. In *USENIX Security Symposium*, San diego, California, pp. 781–796, 2014.
34. Dan Boneh, Craig Gentry, Ben Lynn, and Hovav Shacham. Aggregate and verifiably encrypted signatures from bilinear maps. In *International Conference on the Theory and Applications of Cryptographic Techniques*, Warsaw, Poland, pp. 416–432, Springer, 2003.
35. Precomiled contract for BGLS signature. https://github.com/project-arda/bgls-on-evm.
36. Danny Dolev and Andrew Yao. On the security of public key protocols. *IEEE Transactions on Information Theory*, 29(2):198–208, 1983.

Index

Printed in the United States
By Bookmasters